Herbal Pharmacopeia

This book comprehensively explores the intersection between traditional herbal medicine and cutting-edge nanotechnology. The chapters introduce modern techniques used in herbal extraction and analysis. The principles of drug discovery from plants are discussed, with a focus on the identification and development of bioactive compounds that have therapeutic potential. It discusses the pharmacological properties, biotechnological approaches in drug development, and challenges in the formulation and standardization of herbal medicines. Emerging trends and applications of nanotechnology in herbal pharmacotherapy, such as nanoparticle synthesis, enhanced bioavailability using nanocarriers, safety assessments, novel and targeted delivery systems, and regulatory considerations, are thoroughly discussed. Additionally, it includes a comparative analysis of traditional and nano-formulated approaches and their implementation in clinical settings. Towards the end, the book reviews the regulatory considerations for herbal products and future perspectives in herbal pharmacopeia. This book is intended for researchers, clinicians, and professionals in herbal medicine, pharmacology, and nanotechnology.

Dr. Arshad Farid currently serves as an Assistant Professor in the Gomal Center of Biochemistry and Biotechnology, Gomal University, D.I. Khan, Pakistan. He has approximately 17 years of teaching and research experience and has supervised 13 MPhil. research students while mentoring numerous PhD candidates. Farid has more than 100+ publications of international and national repute, the total cumulative impact factor of his published papers being 450+. His scientific activities are highly interdisciplinary, ranging from medicinal plants, natural product chemistry, herbal drug development, and ethnopharmacology to nanotechnology, drug delivery, and microbiology. He has served as the editor for six books accepted for publication and has written more than 50 book chapters for various publishers. With a notable count of 27 international patents (Germany, UK, Japan, and South Africa etc.), he has published more than 15 abstracts in national and international conference proceedings and has contributed significantly to the scientific community by peer-reviewing for more than 30 esteemed journals.

In 2023, Dr. Farid was a fellow of the British Council-Charles Wallace Fellowship program at the University of Aberdeen, Scotland, UK. Prior to that, in 2007, he received a fellowship for "Training in the Environmental, Agricultural, and Analytical Chemistry Laboratory Techniques" from the Department of Chemistry at the University of Glasgow, Scotland, UK. He is a Research Fellow of INTI International University.

Herbal Pharmacopeia

Nanotechnology and Advancing Drug Discovery

Edited by
Arshad Farid

CRC Press
Taylor & Francis Group
Boca Raton New York London

CRC Press is an imprint of the
Taylor & Francis Group, an **informa** business

First edition published 2025
by CRC Press
2385 NW Executive Center Drive, Suite 320, Boca Raton FL 33431

and by CRC Press
4 Park Square, Milton Park, Abingdon, Oxon, OX14 4RN

CRC Press is an imprint of Taylor & Francis Group, LLC

ISBN: 978-1-032-89288-7 (hbk)
ISBN: 978-1-032-89289-4 (pbk)
ISBN: 978-1-003-54441-8 (ebk)

DOI: 10.1201/9781003544418

Typeset in Times
by SPi Technologies India Pvt Ltd (Straive)

Contents

1 Introduction to Herbal Pharmacopeia

Kushagra Sharma
Department of Horticulture, Jawaharlal Nehru Krishi Vishwa Vidyalaya, Jabalpur, India

Kondle Ravi, Ab Waheed Wani, and Sanjeev Kumar
Department of Horticulture, School of Agriculture, Lovely Professional University, Phagwara, India

Shuja Qureshi
KVK Ganderbal, SKUAST-Kashmir, Kashmir, India

1.1 INTRODUCTION

The integration of herbal remedies into contemporary medicine underscores a profound and enduring relationship between human health and plant-based therapies. Modern pharmacopoeias, meticulously crafted to standardize medicinal products, encompass both synthetic and natural substances [1]. Among the notable herbal pharmacopoeias are the Indian Herbal Pharmacopoeia (IHP), the Korean Herbal Pharmacopoeia (KHP), the American Herbal Pharmacopoeia (AHP) and the British Herbal Pharmacopoeia (BHP). Each of these resources offers detailed and rigorously reviewed monographs on a diverse range of botanicals, including Ayurvedic, Chinese, and Western herbs. For instance, the AHP provides an extensive repository of information on herbal medicines, serving as an indispensable reference for academics, healthcare professionals, manufacturers, and regulators [2]. Herbal medicines are a vital component of primary healthcare worldwide, with around 80% of the global population depending on them for various health-related needs, as reported by the World Health Organization (WHO) [3]. This reliance reflects the swift expansion of the phytopharmaceutical industry, particularly within the dietary supplement market. Despite this growth, many supplements suffer from inadequate research and variable quality.

The relationship between traditional herbal practices and modern therapeutic applications is significant. Of the 120 active compounds extracted from plants and employed in current medical treatments, 80% show a strong alignment with their traditional uses [4]. Usage statistics reveal that herbal medicines are widely embraced across different regions: over 90% of individuals in Africa, 70% in India, and significant percentages in Germany, Australia, Canada, the United States, Belgium, and France use herbal remedies to meet their healthcare needs [5]. Economically, herbal medicines present a compelling alternative to synthetic drugs. Developing a new synthetic drug generally takes about 12 years and can cost roughly $230 million. In contrast, the research and development of herbal medicines tend to be faster and more cost-effective. The global market for herbal medicines is experiencing rapid growth, with an annual increase of over 15%, and it is projected that the trade in herbal products in the United States will reach $70 billion by 2050 [6].

Countries in the developing world, which are rich in plant species with medicinal properties, play a crucial role in this field. It is estimated that around 35,000 plant species have medicinal uses, with over 7,000 compounds in modern pharmacopoeia originating from these plants. The WHO defines

DOI: 10.1201/9781003544418-1

herbs as crude plant materials used in various forms and recognizes herbal medicines as preparations and mixtures derived from these plant materials [7].

The emergence of nanomedicine has brought revolutionary advancements to the field, enabling the manipulation of nanoparticles for drug development. Nanotechnology, involving particles smaller than 100 nanometers, offers new opportunities for addressing challenges in medicine and pharmacy. This technology enables the exact control of atoms and molecules, resulting in the creation of innovative materials and systems with distinctive properties. Recent advancements in herbal technology showcase a refined combination of traditional knowledge and modern scientific methods [8]. Precision farming and controlled environments optimize the cultivation of medicinal herbs, while nanocarriers improve the targeted delivery of herbal extracts. The incorporation of artificial intelligence further speeds up drug discovery from herbal sources. This merging of ancient wisdom with advanced science marks a new era in the development of potent herbal products and medicines, ready to tackle current health challenges with enhanced effectiveness and understanding [3].

1.2 THE EVOLUTION OF HERBAL MEDICINE: A HISTORICAL PERSPECTIVE

Herbal medicine, as articulated in the ancient Chandogya Upanishad—"The essence of all beings is Earth. The essence of the Earth is Water. The essence of Water is Plants. The essence of Plants is the Human Being"—has historically unfolded along two primary trajectories. The first trajectory represents a tradition of simplicity where community-based practitioners, including folk healers and familial custodians of herbal knowledge, employed medicinal plants to provide care [9]. These practitioners, who often held a profound understanding of local flora, have played a crucial role in maintaining health and wellness within their communities through generations. The second pathway includes the professional medical practitioners of ancient times, who were often among the most educated scholars of their era. This path is well recorded in foundational texts such as Dioscorides' *De Materia Medica* and the *Hippocratic Corpus* from Greece, India's *Charaka Samhita*, China's *Shen Nong Ben Cao*, and Avicenna's *Canon of Medicine* from Persia. During the Renaissance, herbal medicine was further advanced by prominent figures such as Fuchs, Gerard, Mattioli, Parkinson, and Salmon [10]. These established medical traditions were closely linked with sophisticated concepts of anatomy, physiology, and pathology. On the other hand, folk healing practices were rooted in hands-on experience and collective community knowledge. While early Western medicine, shaped by the humoral theories of Hippocrates, Galen, and Avicenna, has progressed and moved away from these old frameworks, the core principles of Ayurvedic and Chinese medicine have remained steadfast. Both systems maintain that humans are an integral part of nature and continue to evolve while adhering to their historical principles, reflecting a harmonious alignment with natural processes for healing.

1.3 DIVERSITY OF HERBAL PHARMACOPOEIAS ACROSS THE GLOBE

Traditional medicines have been used worldwide for millennia, providing critical therapeutic resources to nearly 85% of the global population and contributing substantially to contemporary healthcare. These ancient remedies remain influential in drug discovery, with around 46% of newly approved drugs between 1981 and 2019 either sourced from, or inspired by, natural products [11]. During the COVID-19 pandemic, traditional Chinese medicines played a significant role in alleviating symptoms, slowing disease progression, and decreasing both mortality and recurrence rates. Currently, there are over 60 recognized pharmacopoeias globally, with 42 of them (70%) including documentation of herbs or natural products [12].

1.3.1 The Indian Pharmacopoeia (IP)

The Indian Pharmacopoeia (IP) is the official benchmark for drug quality in India. Created by the Indian Pharmacopoeia Commission (IPC), which operates under the Ministry of Health and Family Welfare of the Government of India, the IP sets forth criteria for the identity, purity, and potency of pharmaceuticals in order to guarantee their quality. The IP is periodically updated with addenda to incorporate urgent revisions and new entries, and these updates are considered as authoritative as the main version of the IP.

The roots of the Intellectual Property (IP) can be traced to the 1844 publication of the *Bengal Pharmacopoeia and General Conspectus of Medicinal Plants*. This work concentrated on native drugs but also featured certain European remedies. India's inaugural official pharmacopoeia was issued in 1868, blending components from the 1867 British Pharmacopoeia with indigenous medicines [13]. After India gained independence, the Indian Pharmacopoeia Committee was established in 1948, and it published the first edition of the Indian Pharmacopoeia in 1955.

1.3.2 The European Pharmacopoeia (Ph. Eur.)

The European Pharmacopoeia (Ph. Eur.) characterizes a herbal drug as an entire, fragmented, or cut plant or its parts, encompassing algae, fungi, or lichens, in its natural state, which is usually dried but can occasionally be fresh [11]. It also covers specific exudates that have not been subject to particular processing. Founded in 1964 through the Convention on the Elaboration of a European Pharmacopoeia under the Council of Europe, the Ph. Eur. sets quality standards for the production and regulation of medicines across Europe and has an impact on other nations, including Moldova [9]. The 11th edition, effective from January 1, 2023, is the latest, although the 10th edition was used in this review due to availability constraints [14]. The Ph. Eur. has expanded its reach, with Moldova joining the Convention in 2017 and countries such as India, Japan, and Uzbekistan becoming observer states, demonstrating its ongoing relevance and dynamism.

In the European Pharmacopoeia (Ph. Eur.), herbal medicines are identified using a binomial nomenclature system that includes botanical scientific names [11]. The initial term specifies the genus, species, or variety, while the second term describes the type of plant part, such as underground parts (e.g., radix, rhizome, tubera, bulbus), bark (cortex), or aerial parts (e.g., herba, folium, flos, fructus, pseudofructus, pericarpium, semen, seminis tegumentum, gemmae). The 10th edition of the Ph. Eur. features more than 2,500 monographs and around 370 general texts [13]. This includes six general monographs (e.g., herbal drugs, herbal drug extracts, essential oils), 27 general texts (e.g., pesticides, heavy metals in herbal drugs, High performance thin layer chromatography (HPTLC) of herbal drugs and preparations), and 315 individual monographs [15].

1.3.3 United States Pharmacopoeia (USP)

The United States Pharmacopoeia (USP) is a self-governing, scientific, nonprofit entity committed to advancing public health by setting public standards. Its goal is to enhance global health by guaranteeing the quality, safety, and effectiveness of medicines and food products [12]. Within the Dietary Supplements section (Volume 3) of the USP, there are detailed monographs for herbal drugs [7]. These monographs adhere to established nomenclature and quality control protocols, listing herbal drug names in English and their scientific names in Latin, and specifying the minimum content of quantified constituents when relevant [16].

1.3.4 The Russian Federation's State Pharmacopoeia (SPRF)

The 14th edition of the State Pharmacopoeia of the Russian Federation (SPRF), introduced in 2018, includes 107 monographs detailing medicinal plants [15]. This edition adds 25 new monographs and eliminates one from the 11th edition. It also covers plants which are not indigenous to Russia, which do not have a traditional usage history, or which are commonly used in Western medicine [17]. Each SPRF monograph presents the name of the herbal drug in both Russian and Latin. The Russian name is given first, followed by the Latin name, which is in the plural form.

1.3.5 Romanian Pharmacopoeia (RPh)

The 10th edition of the Romanian Pharmacopoeia (RPh), which has been in use since 1993, consists of 1,315 pages. It is complemented by three supplements: Supplement I (154 pages, 2000), Supplement II (313 pages, 2004), and Supplement III (370 pages, 2006) [11]. This edition marks a notable decrease in the number of herbal drug monographs, now listing only 48—34 of which are indigenous and 14 imported—compared to the 180 monographs found in the first edition [18]. Each monograph for herbal drugs includes the title in Latin, starting with the genus or species name followed by the organ used in the nominative singular form, as well as Romanian names, synonyms, the full scientific name of the plant, and, where relevant, the minimum content of quantified constituents [4].

1.3.6 Hausa Herbal Pharmacopoeia

The Hausa herbal pharmacopoeia, deeply rooted in the traditional medicinal practices of the Hausa community in northern Nigeria, serves as a valuable repository of indigenous knowledge. This compendium offers insights into the herbal medicine traditions of this rural community, highlighting their rich heritage in utilizing plant-based remedies for various health conditions. One of the focal points of the Hausa pharmacopoeia is the use of plants to treat oral diseases. This aspect of their traditional medicine underscores the community's reliance on natural resources to address specific health concerns [19]. By examining the potential efficacy of these treatments through the lens of pharmacological literature, we can gain a better understanding of how these age-old practices align with modern scientific knowledge. The conservative approach taken by the Hausa herbalists mirrors the broader practices of folk medicine, where the effectiveness of a remedy is often judged by its ability to alleviate specific, well-defined symptoms. The Hausa herbal pharmacopoeia represents a critical link between traditional knowledge and contemporary scientific inquiry [20]. By documenting and analyzing these practices, we can appreciate the enduring relevance of indigenous medicine in modern healthcare, especially in the context of complex diseases like malaria.

1.4 COMBINING HERBAL MEDICINE, ENZYME ENGINEERING, AND NANOTECHNOLOGY

Nanoparticles (NPs) have unique attributes, including a large surface area and beneficial physicochemical properties, that allow them to significantly modify the pharmacokinetics and pharmacodynamics of bioactive compounds they encapsulate. When flavonoids are encapsulated in NPs, their stability, solubility, and permeability are notably improved [21]. This encapsulation extends the drug's circulation time in the bloodstream, improves targeted delivery, and increases tumor penetration. These attributes collectively contribute to superior therapeutic outcomes, such as enhanced anti-osteonecrosis and anti-tumor efficacy, compared to free drugs. The solubility and permeability of drugs are significantly improved when encapsulated within NPs, resulting in increased bioavailability [3]. Moreover, nanoparticles (NPs) can be designed to target specific tissues or cells through either passive or active mechanisms. This customization helps prolong their circulation time in the bloodstream and improves targeted delivery [12]. Delivery systems utilizing nanotechnology for active herbal components offer numerous advantages, such as enhanced solubility, increased bioavailability, better pharmacological effectiveness, and improved stability of the active compounds. They also provide protection against chemical and physical breakdown, enabling the use of lower dosages [22].

Although herbal medicine holds significant potential as an alternative therapy, incorporating nanotechnology into these treatments comes with a number of challenges. Key concerns include the high costs of production, challenges associated with scaling up the process, and the limited information available regarding the safety and toxicity of herbal formulations that use nanotechnology [17].

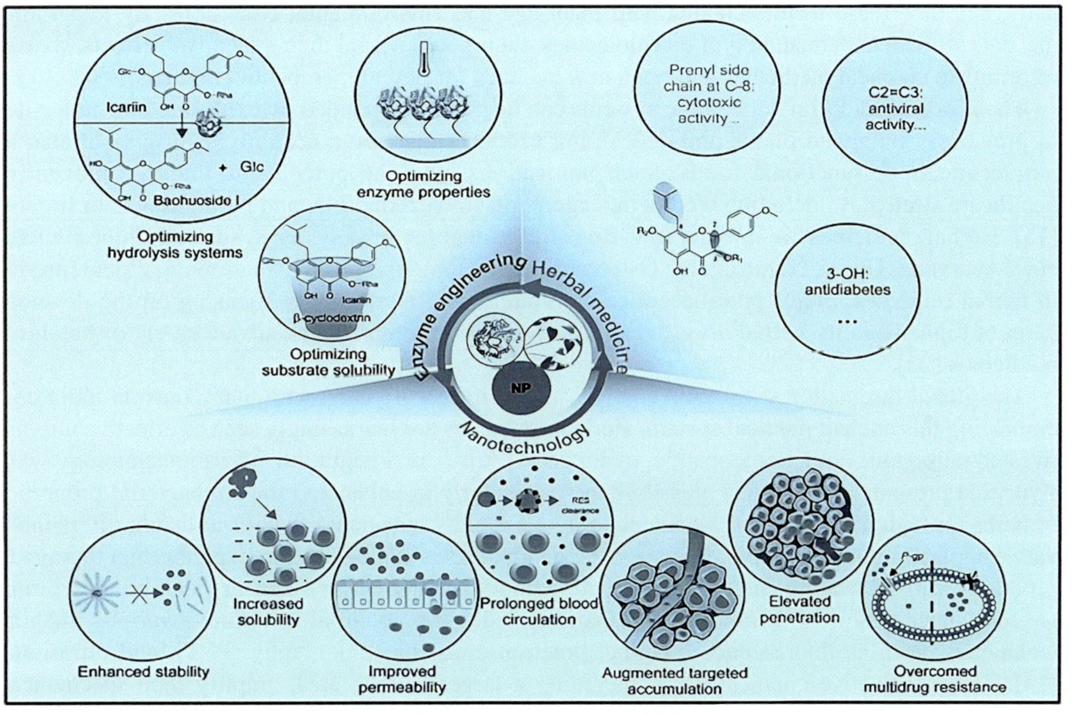

FIGURE 1.1 Schematic representation of the integration of herbal medicine, enzyme engineering, and nanotechnology aimed at enhancing productivity, activity, and therapeutic effects of epimedium flavonoids. This approach also seeks to improve in vivo delivery efficiency. Key elements include nanoparticles (NP), glucose (Glc), rhamnose (Rha), the reticuloendothelial system (RES), and P-glycoprotein (P-gp).

The stability of nanoparticles is another significant concern, as some may exhibit a tendency to release drugs upon contact with blood components. The high production costs associated with nanosystems for phytochemicals result in elevated market prices. In European countries, where drug selection and public funding are based on rational criteria, these high costs pose a substantial barrier. Consequently, the likelihood of high-priced nanomedicines reaching the market and patients in these regions is diminished [23]. Nonetheless, the development of herbal medicines utilizing nanotechnology-based delivery systems remains a promising strategy for enhancing their pharmacological activity (Figure 1.1).

1.5 AYURVEDA AND THE INTEGRATION OF NANOTECHNOLOGY

Ayurveda, a time-honored system of medicine rooted in the five-element principle of Pancha Mahabhuta—Aakash (Space), Vayu (Air), Tej (Fire), Jal (Water), and Prithvi (Earth)—provides a comprehensive framework for understanding the effects of food and medicine. This system uses physical attributes, manifesting as one or more of the six tastes, combined with principles such as temperature (hot or cold), post-digestive effects, and a range of stimulating, transformative (inflammatory), or unifying activities to create a holistic therapeutic effect [24]. India's prominence in herbal medicine is significantly attributed to its traditional systems, including Ayurveda and Unani, both of which rely heavily on herbal substances. Siddha, another traditional system, also incorporates herbs, albeit to a lesser extent. A rich history of empirical observation, experimentation, and refinement has shaped the evolution of herbal formulations in India. However, modernizing these traditional systems to meet international standards for both raw and processed herbal products is crucial [25]. This involves standardizing detailed information on the chemical profiles of plants, their various parts, and how these factors change with plant age and environmental conditions. By integrating this data with an understanding of bio-molecules, their potency, and their synergistic effects, we can reformulate classical medicines and open new avenues for herbal therapeutic products.

The scope of traditional medicine systems can be greatly expanded by exploring bio-molecules in previously untapped plants and developing products that serve not only as drugs but also as nutraceuticals or functional foods. Such innovations have the potential to enhance preventive healthcare strategies, including weight management, stress reduction, and immune system support [15]. Herbal medicines are integral to various traditional medical systems, such as Chinese medicine, Ayurveda, Unani, Naturopathy, Osteopathy, and Homeopathy. With a growing global interest in herbal remedies, major pharmaceutical companies are increasingly focusing on the development of higher-quality herbal drugs, creating ripe opportunities for the advancement of botanical medicines [22].

The global fascination with Ayurveda has surged in recent years, with many outside India now embracing this ancient medical system. Herbal medicines are increasingly seen as effective alternatives to allopathic and homeopathic treatments [26]. The integration of nanotechnology with Ayurveda presents exciting new possibilities, particularly in enhancing the antibacterial properties of herbs through nanoparticles. Nanotechnology, a rapidly advancing field of research, offers innovative applications in medicine. For example, nanoparticles can be used in hyperthermia therapy to produce localized heating and tissue destruction under an alternating magnetic field, holding promise for cancer treatment. Moreover, nanoparticles have the potential to improve various imaging techniques, such as fluorescence imaging, positron emission tomography (PET), and ultrasound [24]. Their distinctive characteristics, including a larger surface area, amplify their mechanical, magnetic, optical, and catalytic properties, thereby expanding their use in medicine.

Utilizing Ayurvedic nanomedicines for cancer therapy has the potential to transform the field by allowing for more precise drug delivery, thereby enhancing treatment effectiveness and minimizing adverse effects. Traditional Ayurvedic bhasmas, which have been used historically to treat a range of conditions, inherently incorporate principles of nanotechnology [27]. These bhasmas possess Rasayana properties, known for immune modulation and anti-aging benefits, and Yogavahi characteristics, which facilitate targeted drug delivery, making them highly valuable in contemporary medical treatments.

1.6 ENHANCING HERBAL MEDICINES THROUGH NANOTECHNOLOGY

The development of herbal medicines presents considerable challenges for pharmaceutical companies, primarily due to the factors influencing the biological efficacy and therapeutic consistency of plant-based treatments [28]. Conditions such as asthma, pain, and fever demand rapid therapeutic action, whereas chronic ailments like hypertension, cancer, and diabetes require sustained effects over time. The inherent physical and chemical properties of herbal medicines often restrict their effectiveness at both ends of this spectrum.

Nanotechnology presents a promising approach by utilizing nanomaterials—such as polymer nanoparticles, solid lipid nanoparticles, lipid crystal systems, liposomes, and nanoemulsions—as delivery systems to protect herbal medicines from external degradation and improve their bioavailability [29]. Research indicates that nanotechnology can significantly amplify the effectiveness of herbal medicines by improving drug potency, facilitating the sustained release of active ingredients, reducing dosage requirements, and enhancing overall biological activity.

Over the past decade, the study of polymer nanoparticles has seen substantial progress, owing to their ability to target specific sites and respond to external stimuli. When designing polymer nanoparticles for herbal formulations, careful consideration of biotoxicity and stability is crucial. Utilizing biodegradable and biocompatible polymers, such as polylactic acid (PLA), polylactic-glycolic acid (PLGA), and chitosan, can greatly optimize the delivery mechanisms for herbal compounds [30] (Table 1.1).

TABLE 1.1
Overview of Selected Herbs: Scientific Names, Uses, and Historical Applications

Herb	Scientific Name	Common Uses and Properties	Historical Uses	References
Swertia chirata	*Swertia chirata*	Antifungal, antibacterial, useful for controlling blood sugar levels.	Used to treat diabetes, malaria, and liver diseases.	[2]
Cinnamomum verum	*Cinnamomum verum*	Eases menstrual discomfort, anti-diabetic, antibacterial, antioxidant, anti-inflammatory, anticancer effects.	Women are advised to drink warm cinnamon water to reduce menstrual pain.	[4]
Thymol crystal	*Thymol (2-isopropyl-5-methylphenol)*	Strong antiseptic property, pleasant aromatic odor, strong flavor.	Known as "Ajwain ke phool."	[11]
Tribulus terrestris	*Tribulus terrestris Linn.*	Annual shrub, grows in Mediterranean, subtropical, and desert climates.	Used in Ayurveda for various medicinal purposes.	[4]
Curcuma longa	*Curcuma longa*	Utilized in Ayurveda, Unani, and Siddha medicine for sinusitis, rheumatoid arthritis and loss of appetite.	Commonly known as "Haldi Ka Phool."	[23]
Tinospora cordifolia	*Tinospora cordifolia*	Scavenges free radicals generated during aflatoxicosis.	Used in traditional medicine for various ailments.	[3]

1.7 APPROACHES OF NANOTECHNOLOGY IN HERBAL MEDICINE

Recently, there has been a growing interest in applying nanotechnology to herbal medicines. Different techniques are used to create nanoformulations, such as nanoparticles, nanoemulsions, and liposomes [31]. These colloidal systems have particle sizes ranging from 10 nm to 1000 nm. Some nanoparticle systems, however, have been observed with average particle sizes greater than 100 nm. For example, nanonized curcuminoids, paclitaxel, and praziquantel have average particle sizes of 450 nm, 147.7 nm, and over 200 nm, respectively [23].

1.7.1 Solid Lipid Nanoparticles (SLN)

Solid lipid nanoparticles (SLNs) offer a unique method for drug delivery that sets them apart from other colloidal systems like emulsions, liposomes, and polymeric nanoparticles. SLNs combine the benefits of these systems while addressing their limitations, making them a promising alternative [28]. Their production involves simple processes suitable for large-scale manufacturing, resulting in improved physicochemical stability and protection for sensitive drugs. SLNs are made from lipids that are solid at room temperature. They are created by melting the lipid and incorporating the drug, with stabilization provided by a surfactant [32]. This robust lipid structure offers effective protection for drug molecules against chemical breakdown. SLNs are created by integrating a liquid lipid (oil) into an oil/water emulsion that contains solid lipids or a blend of solid lipids. SLNs, with particle dimensions between 50 and 1,000 nm and demonstrating biocompatibility, are appropriate for a range of pharmaceutical delivery methods, such as oral, injectable, and transdermal administration [30]. Unlike liposomes and nanoemulsions, polymeric nanoparticles offer specific advantages, such as enhanced drug and protein stability and controlled release properties. Typically composed of physiological lipids, SLNs are preferred for their low risk of acute and chronic toxicity and their suitability for large-scale production. They are employed in delivering diverse phytoconstituents and in treating chronic conditions like lymphatic infections, cancer, and neurodegenerative diseases [33]. Surface modification of SLNs can address the challenge of burst release upon oral administration, enabling more targeted drug delivery.

1.7.2 Nanoemulsions

Nanoemulsions are stable, isotropic systems that remain thermodynamically balanced. These systems are clear dispersions of oil and water, held together by a layer of surfactant molecules at the interface. Typically, the droplets in nanoemulsions range from 10 to 100 nanometers in size and can be classified as either oil-in-water (o/w) or water-in-oil (w/o). Their clarity is attributed to the droplets being smaller than one-quarter of the wavelength of visible light, which is around 150 nanometers [34]. Unlike basic micellar solutions, nanoemulsions demonstrate greater solubilization capacity and enhanced stability compared to more unstable systems like emulsions and suspensions. They can be generated with minimal energy input, either by heat or agitation, and possess an extended shelf life. The tiny droplet size amplifies the interfacial area, thereby improving drug transport characteristics [35]. Nanoemulsions also exhibit enhanced consistency in plasma concentration profiles and drug bioavailability. Their tiny droplet size provides stability by preventing sedimentation and creaming, with Ostwald ripening being the main mechanism of degradation. One significant use of nanoemulsions is in creating nanoparticles. These uniform, kinetically stable, and isotropic systems, which include oil, water, and a mix of surfactants and cosurfactants, feature droplet sizes between 20 and 200 nanometers [36]. Nanoemulsions enhance the permeability, stability, and solubility of both hydrophilic and lipophilic phytoconstituents, thereby improving bioavailability and therapeutic efficacy.

1.7.3 Liposomes

Liposomes are tiny vesicles made of materials similar to cell membranes, which makes them particularly efficient for drug delivery, especially in targeting cancer and various diseases [25]. These spherical, enclosed structures have a lipid bilayer that can range from nanometers to several micrometers in size and display distinct amphiphilic characteristics. For example, liposomal vesicles loaded with silybin, created through the ethanol injection method, showed improved liver-protective and stomach-protective effects in mice suffering from liver damage induced by carbon tetrachloride [37]. The research found that silymarin liposomes demonstrated a 55% increase in hepatoprotective efficacy compared to silymarin that was not encapsulated in liposomes. Additionally, silymarin proliposomes were created to evaluate their pharmacokinetic properties and bioavailability in rats and beagle dogs after oral ingestion, revealing enhanced bioavailability in comparison to unformulated silymarin [7]. Additionally, a liposomal buccal formulation of silymarin significantly improved its hepatoprotective effect in albino rats. Liposomes, with their bilayer lipid membranes and hydrophilic core, benefit from their amphiphilic nature, enhancing the efficacy and safety of drugs due to their biocompatibility and biodegradability [11]. They can function as carriers for drugs that are either hydrophilic or hydrophobic. These carriers are categorized into three types based on their size and number of layers: small unilamellar vesicles, large unilamellar vesicles, and multilamellar vesicles [17].

1.7.4 Ethosomes, Transferosomes, and Transethosomes

Ethosomes are a distinct class of ethanolic liposomes that include phospholipids, cholesterol, stabilizers, and elevated levels of ethanol. They are specifically engineered for topical and transdermal drug delivery [31]. The substantial ethanol concentration greatly improves the stability and permeability of these vesicles, facilitating efficient delivery of drugs into deeper layers of the skin. Conversely, transferosomes are flexible and adaptable vesicles composed of phospholipids, stabilizers, edge activators, alcohol, and high amounts of hydrophilic modulators such as organic ions [38]. They are categorized into first, second, and third generations based on their composition and structural characteristics. These vesicles can penetrate the skin's deeper layers without compromising their integrity. Transethosomes are an advanced variant of both ethosomes and transferosomes, incorporating additional edge activators such as Tween 80, Span 80, and sodium cholate. This advanced formulation enhances drug permeation into deeper skin layers, facilitating both topical and systemic drug delivery with controlled and sustained release of molecules of varying molecular weights [26].

1.7.5 Niosomes and Phytosomes

Niosomes are vesicular structures formed from non-ionic surfactants like Tween, Span, and Brij. They notably improve the solubility and stability of phytoconstituents. The process involves blending these surfactants with cholesterol, which provides mechanical stability and minimizes leakage, thereby enhancing the efficiency of entrapment [39]. Niosomes are generally categorized into small unilamellar vesicles, multilamellar vesicles, and large unilamellar vesicles. Phytosomes, a technology pioneered by the Italian pharmaceutical and nutraceutical firm Indena, are molecular assemblies made up of phospholipids combined with plant extracts that are high in hydrophilic phytoconstituents [9]. These complexes, usually formed in ratios of 1:1 or 1:2, enhance stability and bioavailability through hydrogen bonding. Known also as herbosomes, phytosomes can be incorporated into various formulations including creams, solutions, lotions, emulsions, and gels [40].

1.7.6 Micelles, Dendrimers, and Nanostructured Lipid Carriers (NLCs)

Micelles are colloidal systems composed of surfactants or amphiphilic substances like pluronics, polyethylene glycol (PEG), and polycaprolactone (PCL). These structures generally range from 5 to 100 nm in size [36]. They form stable aggregates when the concentration reaches a critical micellar concentration and at specific temperatures known as the critical micellar temperature. Micelles are highly effective at solubilizing hydrophobic drugs and improving their ability to penetrate interstitial spaces [41]. Dendrimers are complex, highly branched macromolecules with customizable dimensions and forms. They feature a central core surrounded by both hydrophobic and hydrophilic groups. These characteristics make dendrimers suitable for use as drug delivery systems and therapeutic agents [38]. Some dendrimers are commercially produced and incorporated into available products. Nanostructured lipid carriers (NLCs) are an advanced evolution of solid lipid nanoparticles (SLNs). By blending solid and liquid lipids, NLCs form a matrix that enhances the capacity for drug encapsulation. The presence of liquid lipids creates structural imperfections in the matrix, which facilitates increased drug loading [42]. NLCs are engineered to stay solid at room temperature by meticulously regulating the amount of liquid lipid present. They are categorized into three types: imperfect matrix, amorphous matrix, and multiple matrix types [11].

1.7.7 Nanoparticles, Nanocapsules, and Nanogels

Nanoparticles, with sizes between 10 and 100 nanometers, are highly effective carriers for both hydrophilic and hydrophobic drugs. They safeguard the encapsulated drugs from chemical and enzymatic breakdown [39]. For controlled and targeted drug delivery, biodegradable polymers are becoming more preferred compared to non-biodegradable options. Nanocapsules, made from biodegradable polymers like poly(lactide) (PLA), poly(e-caprolactone) (PCL), and poly(lactide-co-glycolide) (PLGA), are often referred to as layer-by-layer nanoparticles [27]. These nanocapsules consist of a drug core encased in polymers or polyelectrolytes with opposing charges, facilitating targeted delivery to specific locations and enhancing the pharmacokinetics, stability, and solubility of phytoconstituents that are poorly soluble in water [31]. Nanogels, on the other hand, are three-dimensional, cross-linked polymer networks that can be hydrophilic, hydrophobic, or amphiphilic. They are divided into hydrogels and organogels, with dimensions varying from 1 to 1000 nanometers [43].

1.8 TYPES OF NOVEL DRUG DELIVERY SYSTEMS (NDDS)

This section describes the classification of novel drug delivery systems (NDDS) specifically utilized with herbal drugs and phytochemicals (Table 1.2).

TABLE 1.2
Comparison of Various Drug Delivery Systems

Type	Composition	Method of Preparation	Advantages	Disadvantages	Application	References
Liposomal Drug Delivery System	– Phospholipids – Water	– *Passive Loading Systems*: – Mechanical dispersal – Solvent dispersal – Detergent expulsion	– Reduces tissue exposure to harmful drugs – Increases stability – Adaptable for targeting – Reduces drug toxicity	– Short half-life – Low solubility – Leakage – Phospholipid degradation	Used for various drug deliveries, e.g., actinomycin-D	[21]
Ethosomal Drug Delivery System	– Water – Phospholipids – Ethanol	– *Cold Method*: – Mix drug, ethanol, and phospholipid; heat at 30°C – *Hot Method*: – Combine heated aqueous and organic phases and add drug	– Enhances delivery of large molecules – Promotes skin permeability – High patient compliance	– Product loss during transfer – Potential skin irritation – Suited for slow-release drugs	Topical and transdermal delivery, enhancing permeability and delivery of peptides and proteins	[43]
Transferosomal Drug Delivery System	– Phospholipids – Surfactants – Water	– Self-assembly with unique malleable film	– Versatile with high entanglement efficiency – Suitable for both topical and systemic delivery – Sustained release	– Chemically unstable – Not cost-effective – Depends on phospholipid purity	Effective DDS for both topical and systemic delivery	[44]
Phytosomal Vesicular Drug Delivery System	– Bioactive herb extract – Phospholipid	– Formation of strong bond between bioactive herb and phospholipid	– Overcomes traditional DDS limitations – Stable in gastrointestinal environment – Enhances penetration and bioavailability	– None specified	CNS stimulant, antioxidant, immune booster (e.g., Grape Seed), hepatic illnesses, skin infections	[43]

1.9 NANOTECHNOLOGY AND ITS APPLICATIONS

Nanotechnology involves the manipulation of materials at the nanoscale, offering transformative potential in various fields. In the realm of herbal pharmacopeia, researchers are increasingly exploring ways to integrate phytoconstituents—bioactive compounds derived from plants—into nano-sized carriers [45]. This approach aims to achieve precise drug delivery, enhance the absorption of these compounds, and improve overall treatment efficacy. Encapsulating phytoconstituents at the nanoscale helps to overcome several challenges. It reduces their susceptibility to degradation, enhances their solubility in aqueous solutions, and facilitates controlled, gradual release. These advancements not only bolster the effectiveness of phytoconstituents in treating a range of diseases but also expand their potential applications across medicine, agriculture, energy, and environmental remediation.

The fusion of nanotechnology with phytoconstituents has ushered in innovative opportunities, paving the way for novel therapeutic strategies and applications [44]. This synergy highlights the evolving landscape of herbal pharmacology and underscores the significant impact that nanotechnology can have on advancing drug discovery and delivery systems (Table 1.3).

TABLE 1.3
Potential Applications of Nanotechnology in Herbal Pharmacopeia

Sr. No.	Futuristic Approaches	Significant Potential	References
1	**Nanoparticle Delivery System**	Encapsulation of phytoconstituents within nanoparticles enhances targeted delivery to specific cells or tissues. This improves bioavailability and therapeutic efficacy, reduces side effects, and increases impact in drug delivery and personalized medicine.	[45]
2	**Cancer Therapeutics**	Incorporating phytoconstituents into nano-sized formulations can enhance cancer treatment by precisely targeting cancerous cells, reducing harm to normal tissues, and boosting the efficiency of therapies like chemotherapy.	[18]
3	**Agriculture and Crop Enhancement**	Nanoencapsulation of phytoconstituents can develop smart agrochemicals for targeted delivery of pesticides, herbicides, and growth-promoting compounds. This minimizes environmental impact and maximizes crop yields.	[34]
4	**Nanobiosensor**	Integration of phytoconstituents into nanosensor platforms for detecting specific molecules or pathogens. Applications include environmental monitoring, disease diagnosis, and food safety.	[44]
5	**Antimicrobial Nanomaterial**	Creation of novel antimicrobial materials incorporating phytoconstituents for use in wound dressings, surface coatings, or water treatment systems to combat bacterial and fungal infections.	[45]
6	**Nanogel and Drug Release**	Nanogels loaded with phytoconstituents provide controlled and sustained drug release, useful for conditions requiring long-term treatment such as chronic pain management.	[45]
7	**Energy Storage and Conversion**	Nanostructured materials with phytoconstituents could be used in energy storage (e.g., batteries and supercapacitors) and energy conversion (e.g., solar cells and fuel cells).	[9]
8	**Environmental Remediation**	Phytoconstituent-loaded nanoparticles designed for efficient pollutant removal from soil and water, aiding in cleaning up contaminated environments and improving ecosystem health.	[14]
9	**Personal Care and Cosmetics**	Nanoencapsulation of phytoconstituents in skincare and cosmetic products can enhance stability, controlled release, and efficacy of the products.	[9]
10	**Anti-inflammatory and Immunomodulator Applications**	Phytoconstituents with anti-inflammatory or immunomodulatory properties can be engineered into nanoparticles for more effective treatment of inflammatory diseases and autoimmune disorders.	[34]

1.10 EFFICACY AND SAFETY OF HERBAL MEDICINE

The efficacy and safety of traditional herbal medicines have been established through their extensive use over millennia. Nonetheless, to address public concerns about quality control and clinical validation, more rigorous research is essential. As herbal medicines become increasingly popular, especially within the sphere of complementary and alternative medicine (CAM), there is a growing demand for robust clinical evidence. In this context, reverse pharmacology presents a valuable approach [41]. This approach utilizes well-established experiential knowledge to identify potential research opportunities, which are subsequently explored through experimental and clinical studies to develop drug candidates. Reverse pharmacology is especially beneficial in countries such as India, where varied healthcare systems support the comprehensive documentation of new pharmacodynamic effects through interdisciplinary teamwork [46]. Often, observational therapeutics precede reverse pharmacology in the development of new natural drugs.

Herbal medicines are often regarded as safe with minimal risk of severe side effects, though there are notable exceptions. Some herbs can lead to adverse reactions or interact with other medications, either enhancing or diminishing their effects. This issue is highlighted in the three-volume series *Adverse Effects of Herbal Drugs*, published by Springer in the 1990s. Furthermore, herbal products might interact with foods, conventional medications, or other drugs [38]. Nonetheless, there is still a lack of extensive data on herb–drug interactions, and many reports are not supported by thorough laboratory research (Table 1.4).

1.11 CONCEPT OF BHASMA AND NANOTECHNOLOGY

The preparation of Bhasma, a traditional Ayurvedic formulation, involves two primary processes: the extraction of metals from their mineral forms (known as Satpavna) and the conversion of purified metals or alloys into non-toxic Bhasma. In the Satpavna process, raw mineral materials are subjected to a series of treatments that induce physiochemical transformations and incorporate additional constituents [21]. This meticulous process is crucial for ensuring the efficacy and safety of the final product. A pivotal step in Bhasma preparation is Bhasmikaran, which involves the transformation of metallic compounds combined with organic substances into an ash-like residue [48]. This technique contrasts with modern methods used to engineer nanoparticles. Throughout Bhasma preparation, metals are specifically converted into desired chemical forms, effectively eliminating their inherent toxicity. This transformation involves converting metals from their zero-valent state to higher oxidation states, a critical process for reducing toxicity [49]. Consequently, metal particles are downsized, the toxic effects of metal oxides are neutralized, and the medicinal properties are effectively enhanced.

1.11.1 Nanoparticle Nature of Bhasma

Nanoparticles are at the forefront of both biomedical and commercial innovation due to their unique properties and applications. Their nanoscale dimensions enable them to penetrate cellular structures, engage in cellular metabolism, interact with DNA and proteins, and potentially modify gene expression [13]. Metal nanoparticles, characterized by their extremely small size and distinctive chemical properties, play a crucial role in the development of diverse therapeutic applications [36]. In Ayurvedic medicine, Bhasma bears a resemblance to nanocrystalline materials. These materials consist of crystallites with at least one dimension smaller than 100 nm [41]. Ayurvedic metallic nanocrystallites, referred to as Bhasma, exhibit remarkable physicochemical characteristics, such as biocompatibility and ease of surface functionalization. The production of Bhasma utilizes traditional methods like Mardana (trituration) and Bhavana (levigation), which play a crucial role in reducing particle size and forming these metallic nanocrystallites [47].

Recent advancements in nanotechnology have significantly enhanced the production of Bhasma. Nanotechnology focuses on materials at dimensions ranging from 1 to 100 nm. When materials are reduced to this scale, their mechanical, thermal, optical, magnetic, and other properties can change

TABLE 1.4
Summary of Nanoparticles for Various Therapeutic Applications, Including Synthesis Methods and Functionalization

Sr. No.	Nanoparticle Name	Functionalization	Uses	Method of Synthesis	References
1	**Curcumin**	Anticancer	Potent anticancer and antitumor	Wet-milling technique	[32]
2	**Paclitaxel**	Anticancer	Several tumors, ovarian and breast tumor	Nano precipitation	[46]
3	**Berberin**	Antineoplastic	Inflammation and several cancer	Emulsion and ionic gelatin	[46]
4	**Camptothecin**	Anticancer	Potent anticancer	Encapsulated with hydrophobically modified glycol	[47]
5	**Ginkgo biloba**	Alzheimer's dementia	Loss of memory, thinking, language, behaviour	Combination of dry and wet process	[8]
6	**Triptolide**	Anti-arthritis	Rheumatoid arthritis, inflammatory and autoimmune diseases	Nano encapsulation	[15]
7	**Salvia miltiorrhiza**	Anti-hyperlipidemia	Cerebrovascular diseases	Phospholipid complex loaded	[11]
8	**Quercetin**	Antioxidant	Potent anticancer	Gelatin and chitosan loaded	[46]
9	**Breviscapine**	Anticardiovascular	Cerebrovascular and cardiovascular diseases	Lipid encapsulation	[48]
10	**Naringenin**	Antioxidant, anti-inflammatory	Several tumors and hepatoprotective	Nano precipitation	[9]
11	**Dodder**	Antioxidant	Carcinogenesis and hepatoprotective	Nano precipitation	[22]
12	**Silymarins**	Hepatoprotectives	Hepatoprotectives, several liver diseases and breast cancer	Cold homogenization	[19]
13	**Genistein**	Antioxidant	Cardiovascular, breast and uterine cancer also in osteoporosis	Nano emulsion and chitosan microspheres	[17]
14	**Annual magwort**	Antimalarial	Antimalarial, also used for asthma	Hydrophilic encapsulation	[43]

dramatically. For example, as a sphere's size decreases from 1 micrometer to 1 nanometer, its surface area-to-volume ratio increases by a factor of 10^9, which can greatly affect its catalytic properties [50]. The particle size of Bhasma can be precisely controlled by adjusting the number of Putapaka (heat treatment) steps during its preparation. For therapeutic uses, particle sizes in the range of 10–100 nm are generally employed, while for aphrodisiac applications, smaller sizes of 10–50 nm are preferred. Both Mardana and Bhavana processes contribute to achieving nanoscale particle sizes (less than 100 nm in any dimension) [45].

Advancements in nanotechnology have been greatly aided by sophisticated analytical methods, including transmission electron microscopy (TEM), scanning tunneling microscopy (STM), and atomic force microscopy (AFM). These techniques enable precise confirmation of the nanometric properties of Bhasma [38]. Nanoparticles can have either positive or negative charges on their surfaces, depending on their preparation method. Their applications include drug delivery, diagnostic purposes, and targeted therapeutic interventions [51]. In modern medicine, metals such as silver, gold, zinc, copper, and calcium are employed as Bhasmas, often demonstrating efficacy without adverse effects.

1.12 SUPERMOLECULES AND NANOTECHNOLOGY

Supermolecules, formed through the binding of two or more molecules via intermolecular interactions, create intricate and organized aggregates with unique microstructures and macroscopic properties. The concept of the supermolecule emerged in the mid-1930s, with supramolecular chemistry being formally introduced by French scientist Jean-Marie Lehn in 1987. This field emphasizes molecular aggregates held together by non-covalent interactions, highlighting the pivotal role of these interactions.

At present, supermolecules are primarily classified into two categories: those that arise from the self-assembly of small molecules and those that are made up of small molecules combined with inorganic complexes. It is worth noting that numerous supermolecules resulting from the self-assembly of small molecules have been discovered in Traditional Chinese Medicines (TCMs) [52]. Examples include combinations such as berberine with rhein, berberine with baicalin, and berberine with cinnamic acid. These combinations have shown enhanced bioactivity and reduced toxicity compared to their individual components.

The superior bioactivity and effective toxicity reduction of these supermolecules contribute to more stable drug delivery systems. However, the stability of supermolecules remains a concern due to their dependence on non-covalent forces, which can lead to variations in activity, toxicity, and stability. Consequently, they hold potential as quality markers (Q-markers) for ensuring the safety and efficacy of herbal medicinal products (HMPs). Despite their potential, several challenges persist. The complexity of TCM compositions and current limitations in isolating their active ingredients make it difficult to identify and extract active supermolecules. Furthermore, while supermolecules demonstrate promising in vitro and in vivo activity, it is uncertain whether they maintain their supermolecular form within the body. Additionally, detecting supermolecules poses challenges, as chromatographic separation techniques can disrupt their structures. Advances in mass spectrometry imaging may provide new methods for detecting and localizing supermolecules in vivo [53]. Given their pharmacological potential, supermolecules are anticipated to receive increased research focus and become integral to future HMP quality control practices. As small molecules dominated the past and macromolecules are the current focus, supermolecules are expected to be the next focal point, underscoring their significance alongside small molecules and macromolecules in quality control practices.

1.13 PHARMACOVIGILANCE IN HERBAL MEDICINE: CHALLENGES AND GLOBAL PERSPECTIVES

The relentless quest to combat life-threatening diseases has spurred the development of diverse treatment approaches, with cancer therapy standing out as a key area of focus. Despite the progress in conventional cancer treatments, these therapies face several significant limitations. Issues such as restricted drug accessibility to various tissues, the need for high dosages, undesirable cytotoxicity, the emergence of drug resistance, and non-specific targeting highlight the urgent need for novel technologies to overcome the deficiencies of traditional cancer treatments [54].

Nanotechnology has emerged as a transformative force, facilitating the creation of advanced nanoparticulate drug delivery systems. Modern innovations, including polymeric nanocapsules, liposomes, nanoemulsions, nanoparticles, and solid lipid nanoparticles, offer substantial improvements in drug delivery. These systems enhance drug solubility and bioavailability, increase pharmacological activity, optimize tissue distribution, and mitigate toxicity from prolonged exposure. Additionally, they provide protection against physical and chemical degradation, addressing the limitations of conventional drugs, which often suffer from poor release profiles and absorption issues.

In Europe, the use of herbal medicines is steeped in a rich tapestry of traditions from Chinese, Indian, North and South American, African, and European systems. This diversity poses significant challenges for herbal pharmacovigilance, particularly in selecting appropriate naming systems for herbs—whether botanical, common, pharmaceutical, or herbal drug names—and in validating the botanical identity of herbal ingredients. Unlike synthetic medicines, these challenges are not easily addressed by existing pharmacovigilance systems or electronic data systems originally designed for pharmaceuticals (e.g., DynPage, UMC) [55]. Problems such as naming inconsistencies or adulterations do not integrate seamlessly into current pharmacovigilance frameworks. While adjustments to these systems may be necessary, establishing entirely separate systems for herbal medicines could introduce additional complexities and confusion, potentially reducing reporting rates.

The global rise in herbal medicine use underscores a significant gap in safety information across diverse patient populations. Variations in critical factors such as pharmacogenomics, metabolization profiles, and gut microflora composition can profoundly impact both the bioactivity and safety of these medicines. To develop effective and safe guidelines for herbal medicine use, robust pharmacovigilance is crucial for collecting reliable safety data. Although herbal medicine is gaining increasing global acceptance, the cultivation of medicinal plants encounters significant obstacles. The infrastructure for supporting herbal medicine industries remains underdeveloped, with shortcomings in quality control and monitoring of medicinal toxicity [21]. There is limited oversight of side effects and a lack of international standards governing the methods and dosages of herbal medicines [56]. Additionally, there is a lack of public awareness about the risks associated with self-medication and the possible interactions between herbal and pharmaceutical drugs. The significance of using herbal remedies correctly and at the right time is frequently underestimated, which could greatly affect their effectiveness in healthcare [17].

1.14 FUTURE PROSPECTS OF NANOMEDICINES

The escalating global interest in herbal remedies and natural products has spurred significant advancements in the field of nanomedicine, particularly within drug delivery systems. Numerous institutions are dedicated to both fundamental research and clinical trials aimed at enhancing these systems. The focus is on delivering precise dosages to targeted sites within the body, which not only mitigates side effects such as toxicity and hypersensitivity but also bolsters overall patient health and vitality [57].

Looking ahead, incorporating herbal nanoparticles into cancer drug delivery is a highly promising research direction. This advancement could yield groundbreaking outcomes. With the advancement of science and technology, nanoparticles are becoming more common in various aspects of daily life, often bringing about transformative effects that may not be immediately noticeable [49]. The future of nanotechnology in herbal drug delivery appears exceptionally promising, with interdisciplinary collaborations set to elevate nanoscience as a leading treatment modality in the near future.

The use of herbal nanoparticles in the treatment of serious conditions like cancer, diabetes, and anemia is expected to garner significant research interest. As natural products and herbal treatments are increasingly combined with nanocarriers, there is anticipated to be a notable enhancement in the effectiveness of drug delivery systems [58]. This progress will further solidify nanotechnology's pivotal role in modern medicine.

1.15 CONCLUSION

In the dynamic realm of drug discovery and therapeutic advancement, the fusion of herbal pharmacopeia with modern scientific innovations, particularly nanotechnology, presents an exciting frontier. This chapter has delved into the historical importance and enduring relevance of herbal medicine, highlighting its potential to tackle contemporary health issues. By integrating ancient wisdom with cutting-edge technologies like nanoscale drug delivery systems, we can unlock new levels of efficacy, safety, and precision in herbal treatments. The convergence of traditional herbal knowledge with modern nanotechnology not only improves the bioavailability and targeted delivery of herbal compounds but also paves the way for the development of novel therapeutic agents. This integration represents a balanced approach that merges time-honored natural remedies with rigorous scientific research, paving the way for more personalized and effective healthcare solutions. As we further explore the extensive repository of herbal knowledge, it is vital to uphold principles of sustainability, ethical sourcing, and respect for indigenous practices. The future of herbal pharmacopeia hinges on its ability to innovate while preserving the rich cultural heritage that underpins it. By doing so, we ensure that herbal medicine remains a vital component of global healthcare, contributing to the well-being of future generations. This chapter introduces the vast possibilities at the intersection of herbal pharmacopeia and nanotechnology, setting the stage for subsequent exploration and discovery. The journey of blending ancient traditions with modern science is not merely about harnessing the therapeutic potential of herbs but about redefining the future of medicine itself.

REFERENCES

1. Bhowmick, S., Singh, T., & Chauhan, P. S. (2024). Harnessing the power of aromatic and medicinal plants for natural product innovation. In *Medicinal and Aromatic Plants: Current Research Status, Value-Addition to Their Waste, and Agro-Industrial Potential* (Vol I, pp. 211–222). Springer Nature Switzerland.
2. Devangan, P., Bajad, G., Loharkar, S., Wadate, N., Gollapalli, S., Dutta, T., & Madan, J. (2024). Quality-by-design strategy for developing novel herbal products. In *Introduction to Quality by Design (QbD) From Theory to Practice* (pp. 263–295). Springer Nature Singapore.
3. Shokeen, A., Dhawan, B., Sarwat, M., & Gupta, S. (2024). Advancement in herbal drugs for the treatment of Parkinson's disease. In *Targeting Angiogenesis, Inflammation, and Oxidative Stress in Chronic Diseases* (pp. 251–276). Academic Press.
4. Verma, R., Bhardwaj, S., & Gupta, J. (2018). Broad view of nano herbal medicine. *International Journal of Advanced Research in Science and Technology*, *5*(2), 1–5.
5. Rh, R. U. (2015). Traditional herbal medicine, pharmacognosy, and pharmacopoeial standards: A discussion at the crossroads. In *Evidence-based validation of herbal medicine* (pp. 45–85). Elsevier.
6. Majidzadeh, H., Araj-Khodaei, M., Ghaffari, M., Torbati, M., Dolatabadi, J. E. N., & Hamblin, M. R. (2020). Nano-based delivery systems for berberine: A modern anti-cancer herbal medicine. *Colloids and Surfaces B: Biointerfaces*, *194*, 111188.
7. Zahoor, I., Mir, T. A., Ganaie, T. A., Allai, F. M., Ayoub, W. S., & Farooq, S. (2024). Antidiabetic potential from selected Himalayan underutilized herbs: A review. *Food and Humanity*, 100297. https://doi.org/10.1016/j.foohum.2024.100297
8. Hu, D., Gao, J., Yang, X., & Liang, Y. (2021). Chinese pharmacopoeia revisited: A review of anti-depression herbal sources. *Natural Product Communications*, *16*(12), 1934578X211059312.
9. Jurel, P., Bahadur, S., & Bajpai, M. (2024). Emerging trends in pharmacological and therapeutic potential of glycyrrhizic acids: Traditional and Nanotechnological approach. *Pharmacological Research - Modern Chinese Medicine*, 100461. https://doi.org/10.1016/j.prmcm.2024.100461
10. Khan, E., Kesari, K. K., & Arif, J. M. (2018). Importance, challenges and future of traditional herbal medicines for cancer care. *Biochemical and Cellular Archives*, *18*(1), 1–9.
11. Sen, S., & Chakraborty, R. (2015). Toward the integration and advancement of herbal medicine: A focus on traditional Indian medicine. *Botanics: Targets and Therapy*, *2005*, 33–44.
12. Kumar, K. S. (2015). Herbal Pharmacopoeias—An overview of international and Indian representation. *Journal of Ayurvedic and Herbal Medicine*, *1*, 59–60.
13. Cheng, W., Li, S., Han, J., Su, J., & Cai, W. (2022). Supermolecules as a quality markers of herbal medicinal products. *Heliyon*, *8*(12), e12497.
14. Elena, D. L. (2012). Pharmacognostic methods for analysis of herbal drugs, According to European Pharmacopoeia. *Promising Pharmaceuticals*, *2012*, 38–62.
15. Kitanov, G., Karcheva, D., & Lukova, P. (2015). Comparative analysis of monographs on herbal drugs and herbal drug preparations included in the European Pharmacopoeia (Ph. Eur. 8). *Annales UMCS, Pharmacia*, *62*, 20–27.
16. Kumari, R., & Kotecha, M. (2016). A review on the standardization of herbal medicines. *International Journal of Pharmaceutical Sciences and Research*, *7*(2), 97–106.

17. Shikov, A. N., Tsitsilin, A. N., Pozharitskaya, O. N., Makarov, V. G., & Heinrich, M. (2017). Traditional and current food use of wild plants listed in the Russian Pharmacopoeia. *Frontiers in Pharmacology*, *8*, 841.
18. Benedec, D., Oniga, I., & Hanganu, D. (2017). Medicinal herbs in Romanian Pharmacopoeias-155 years of history. *Hop and Medicinal Plants*, *25*, 69–78.
19. Etkin, N. L. (1981). A Hausa herbal pharmacopoeia: Biomedical evaluation of commonly used plant medicines. *Journal of Ethnopharmacology*, *4*(1), 75–98.
20. Ge, W., Gao, Y., He, L., Jiang, Z., Zeng, Y., Yu, Y., & Zhou, F. (2024). Developing Chinese herbal-based functional biomaterials for tissue engineering. *Heliyon*. *10*, e27451.
21. Lu, Y., Luo, Q., Jia, X., Tam, J. P., Yang, H., Shen, Y., & Li, X. (2023). Multidisciplinary strategies to enhance therapeutic effects of flavonoids from *Epimedii Folium*: Integration of herbal medicine, enzyme engineering, and nanotechnology. *Journal of Pharmaceutical Analysis*, *13*(3), 239–254.
22. Saxena, S. (2020). Medicinal biome: From tradition to modernized pharmacy. *Natural Products and Their Utilization Pattern* (pp. 43–70), Nova Science Publishers.
23. Krishnaswamy, S. (2024). Phytopharmaceutical biotechnology: Integration of botany, pharmacology and plant biotechnology to deliver the best therapeutic potential of herbs. In *Concepts in Pharmaceutical Biotechnology and Drug Development* (pp. 437–464). Springer Nature Singapore.
24. Balkrishna, A., Solleti, S. K., Singh, H., Tomer, M., Sharma, N., & Varshney, A. (2020). Calcio-herbal formulation, Divya-Swasari-Ras, alleviates chronic inflammation and suppresses airway remodelling in mouse model of allergic asthma by modulating pro-inflammatory cytokine response. *Biomedicine & Pharmacotherapy*, *126*, 110063.
25. Farooq, S., Mehmood, Z., Qais, F. A., Khan, M. S., & Ahmad, I. (2019). Nanoparticles in ayurvedic medicine: Potential and prospects. In *New Look to Phytomedicine* (pp. 581–596). Academic Press.
26. Usman, M. R. M., Jain, B. V., Patil, K. G., & Sonawane, Y. P. (2023). Concomitant Use of Local Herbal Cornucopia in Providing Relief from Respiratory Disorders. *Journal of Survey in Fisheries Sciences*, *10*(2S), 1054–1061.
27. Kunle, O. F., Egharevba, H. O., & Ahmadu, P. O. (2012). Standardization of herbal medicines-A review. *International Journal of Biodiversity and Conservation*, *4*(3), 101–112.
28. Bilia, A. R. (2015). Herbal medicinal products versus botanical-food supplements in the European market: State of art and perspectives. *Natural Product Communications*, *10*(1), 1934578X1501000130.
29. Patil, R. Y., Patil, S. A., Chivate, N. D., & Patil, Y. N. (2018). Herbal drug nanoparticles: Advancements in herbal treatment. *Research Journal of Pharmacy and Technology*, *11*(1), 421–426.
30. Marella, S., & Tollamadugu, N. V. K. V. P. (2018). Nanotechnological approaches for the development of herbal drugs in treatment of diabetes mellitus—A critical review. *IET Nanobiotechnology*, *12*(5), 549–556.
31. Ali, H., Alvi, A., Nawab, A., Salman, S., Zafar, F., & Naveed, S. (2019). Novel drug delivery approaches in formulation development; stability considerations and quality features of herbal products. *RADS Journal of Pharmacy and Pharmaceutical Sciences*, *7*(3), 155–164.
32. Pavithra, K., & Manimaran, V. (2024). A Review of Safety, Quality, Regulation, and Delivery Approaches for Phytopharmaceuticals. *Jordan Journal of Pharmaceutical Sciences*, *17*(2), 316–332.
33. Qadir, A., Khan, N., Singh, S. P., Akhtar, J., & Arif, M. (2015). Nanotechnological approaches to herbal drugs used in cancer therapy. *International Journal of Pharmaceutical Sciences and Research*, *6*(10), 4137.
34. Saidakbarovich, R. K., Kizi, S. I. B., & Kizi, R. S. U. (2023). Development of technology of production and quality assessment of drug substance" dry extract of scutellaria iscanderi l. with silver nanoparticles". *Frontline Medical Sciences and Pharmaceutical Journal*, *3*(03), 1–16.
35. Choudhary, N., & Sekhon, B. S. (2011). An overview of advances in the standardization of herbal drugs. *Journal of Pharmaceutical Education and Research*, *2*(2), 55.
36. Bavya, S., Nagalakshmi, R., & Sundari, U. T. (2022). Anatomical and preliminary phytochemical evaluation of the medicinal plant: Enicostemma littorale. *Medicina*, *2022*, 31.
37. Eloff, J. N., Gurib-Fakim, A., & Phillips, L. D. (2010). *African Herbal Pharmacopoeia* (Vol. 53). T. Brendler (Ed.). Association for African Medicinal Plants Standards.
38. Pallavi, M. C., & Aarti, V. P. (2020). Herbal nanogel formulation: A novel approach. *Journal of Science & Technology (JST)*, *5*(5), 149–153.
39. Fursenco, C., Drăgălin, E. A., Calalb, T., & Uncu, L. (2023). Comparative assessment of pharmacopoeia requirements regarding the standardization of herbal drugs. In *Direcţii de reformare a sistemului farmaceutic din perspectiva cursului european al Republicii Moldova* (pp. 104–116).

40. De, A., De, S., Saha, N., Das, B., Naskar, S., & Samanta, A. (2024). Pharmacopoeias, national formulary and extra pharmacopoeia. In *Dosage Forms, Formulation Developments and Regulations* (pp. 83–98). Academic Press.
41. Bik-San Lau, C., Yue, G. G. L., Lau, K. M., Chan, Y. Y., Shaw, P. C., Kwok, H. F., & Wong, L. S. (2019). Method establishment for upgrading chemical markers in pharmacopoeia to bioactive markers for biological standardization of traditional Chinese medicine. *Journal of Traditional and Complementary Medicine, 9*(3), 179–183.
42. Kumar, M., Ranjan, R., Dandapat, S., Srivastava, R., Raipat, B. S., & Sinha, M. P. (2023). Green nanotechnology: Synthesis of silver nanoparticles using aqueous leaf extracts of Swertia chirayita and Punica Granatum. *The Bioscan, 18*(3), 167–176.
43. Prakash, J., Srivastava, S., Ray, R. S., Singh, N., Rajpali, R., & Singh, G. N. (2017). Current status of herbal drug standards in the Indian pharmacopeia. *Phytotherapy Research, 31*(12), 1817–1823.
44. Izah, S. C., Ogidi, O. I., Ogwu, M. C., Salimon, S. S., Yusuf, Z. M., Akram, M., … & Iyingiala, A. A. (2024). Historical Perspectives and Overview of the Value of Herbal Medicine. In *Herbal Medicine Phytochemistry: Applications and Trends* (pp. 3–35). Springer International Publishing.
45. Ahlawat, J., Verma, N., & Sehrawat, A. R. (2014). Globalisation of herbal drugs: A bliss and concern. *International Journal of Science and Research, 3*(11), 466–474.
46. Pathak, A., Gupta, A. P., & Pandey, P. (2024). Herbal medicine and sustainable development challenges and opportunities. *Herbal Medicine Phytochemistry: Applications and Trends*, 1–26.
47. Sonu Ambwani, S. A., Roopali Tandon, R. T., Ambwani, T. K., & Malik, Y. S. (2018). *Current knowledge of nano delivery systems and their beneficial applications in enhancing the efficacy of herbal drugs.*
48. Shinde, V. M., Dhalwal, K., Potdar, M., & Mahadik, K. R. (2009). Application of quality control principles to herbal drugs. *International Journal of Phytomedicine, 1*(1), 4–8.
49. Sindhu, R. K., Gupta, R., Wadhera, G., & Kumar, P. (2022). Modern herbal nanogels: Formulation, delivery methods, and applications. *Gels, 8*(2), 97.
50. Govindaraghavan, S., & Sucher, N. J. (2015). Quality assessment of medicinal herbs and their extracts: Criteria and prerequisites for consistent safety and efficacy of herbal medicines. *Epilepsy & Behavior, 52*, 363–371.
51. Gaud, S., Jain, N. K., & Kannojia, P. (2018). Standardization of an Ayurvedic Bhasma preparation and its evaluation as a potential haematinic agent. *Journal of Drug Delivery and Therapeutics, 8*(6-A), 19–24.
52. Kumar, R. (2023). Nanotechnology in herbal medicine: Challenges and future perspectives. In *Nanotechnology in Herbal Medicine* (pp. 515–548). Woodhead Publishing.
53. Mirzaeian, R., Sadoughi, F., Tahmasebian, S., & Mojahedi, M. (2021). The role of herbal medicines in health care quality and the related challenges. *Journal of Herbmed Pharmacology, 10*(2), 156–165.
54. Parveen, B., Parveen, A., Parveen, R., Ahmad, S., Ahmad, M., & Iqbal, M. (2020). Challenges and opportunities for traditional herbal medicine today, with special reference to its status in India. *Annals of Phytomedicine, 9*(2), 97–112.
55. Peng, B., Xie, Y., Lai, Q., Liu, W., Ye, X., Yin, L., & Chen, H. (2024). Pesticide residue detection technology for herbal medicine: Current status, challenges, and prospects. *Analytical Sciences, 40*(4), 581–597.
56. Kamel, R. (2021). Contribution of novel delivery systems in the development of phytotherapeutics. Herbal medicine: Back to the future: Volume 4, *Infectious Diseases, 4*, 148.
57. Atmakuri, L. R., & Dathi, S. (2010). Current trends in herbal medicines. *Journal of Pharmacy Research, 3*(1), 109–113.
58. Chiu, P. (2024). Modernity, Pharmacy Law, and The Chinese Pharmacopoeia. In *A History of Western Pharmacy in China* (pp. 167–192). Singapore: Springer Nature Singapore.

2 Historical Development of Herbal Medicine

Roomah Javed, Maria Faraz, and Arshad Farid
Gomal Center of Biochemistry and Biotechnology, Gomal University, D.I. Khan, Pakistan

Samy Selim
Department of Clinical Laboratory Sciences, College of Applied Medical Sciences, Jouf University, Sakaka, Saudi Arabia

2.1 INTRODUCTION

Herbal medicine (herbalism) refers to plant-based substances that possess nutritional, therapeutic, or preventive properties (Sharma, Sabharwal et al. 2021). It is a fascinating field that connects herbal medicine and Ayurveda, covering a wide range of areas such as botany, medicinal plant research, pharmacognosy, phytochemistry, phytotherapy, botanical medicines, Ayurveda, natural chemistry, agriculture science, Unani medicine, biotechnology, and biochemistry (Hoffmann 2003). The plant parts used in herbal therapy contain seeds, berries, roots, leaves, fruits, bark, flowers, or even whole plants (Pan, Litscher et al. 2014). Herbal medicine dates back to prehistoric times (Qadir and Raja 2021). Early humans used plants for healing, and archaeological evidence supports their acquaintance of herbal remedies. In the 19th century, the germ theory of disease transformed medicine. Evidence-based practices emerged, leading to pharmaceutical drugs replacing many herbal treatments. Despite modern advancements, traditional and alternative medicine systems still incorporate herbs. People pursue natural remedies alongside evidence-based treatments. Historically, the use of herbs and spices in cooking also had medicinal benefits. Some compounds in culinary herbs suggest antimicrobial properties (Martínez-Graciá, González-Bermúdez et al. 2015). In recent times, herbal medications have expanded widespread interest due to their various benefits. These formulations, resulted from natural herbs, are increasingly documented as effective remedies for various health conditions. Surprisingly, despite their unconventional nature, over 80% of the global population depends on herbal products and medicines to maintain well-being (Khan and Ahmad 2019). However, this flow in herbal usage has also led to product manipulations and adulterations, disappointing both consumers and manufacturers and, in some cases, resulting in serious costs. Scientists grapple with the challenge of developing consistent analytical techniques to quantitatively assess marker/bioactive chemicals and other essential ingredients, confirming consistent profiling of the phytochemical composition. Standardization is a critical step in establishing quality control (Verma and Affairs 2016).

In recent times, phytomedicine, also known as herbal medicine, has increased mainstream recognition due to developments in analysis, quality control, and clinical research (Barkat, Goyal et al. 2021). According to the World Health Organization (WHO), phytomedicine incorporates the knowledge, skills, and practices rooted in various cultures (Nigam 2021). These practices, whether scientifically explainable or not, contribute to sustaining health and addressing physical and mental illnesses (Organization 2009). Plant-derived substances contain historical herbal medicines that do not require

DOI: 10.1201/9781003544418-2

industrial processing. These remedies have been practiced for centuries in local and regional healing traditions. Long before the advent of allopathic drugs, herbal medicines served humanity, with preparations like tinctures, teas, poultices, and powders originating from crude plant materials (Li and Weng 2017; Nasim, Sandeep et al. 2022). The historical use of plants for healing predates registered human history and has had a significant influence on the development of modern medicine (Chaachouay, Zidane et al. 2024). Early medications, such as aspirin, digitoxin, morphine, quinine, and pilocarpine, were primarily derived from plant-based sources (Bhardwaj, Verma et al. 2018).

In many developing countries, a significant portion of the population rely on traditional healers and their stock of medicinal plants to address healthcare needs. Despite the coexistence of modern medicine, herbal remedies endure popular due to historical and cultural factors (Jamshidi-Kia et al. 2017). These products are increasingly accessible commercially, particularly in developed nations. However, some ingredients are now advertised for purposes not originally planned within traditional healing systems. For instance, ephedra is sponsored for weight loss or athletic performance enhancement. While certain countries impose stringent manufacturing standards for herbal medicines, this consistency is not universal. In Germany, herbal products sold as 'phytomedicines' adhere to the same efficacy, safety, and quality criteria as other drug products (Liu and Salmon 2010). In contrast, in the USA, most herbal products in the market fall under the type of dietary supplements, which do not need pre-approval based on these criteria (Kumar and Kumar 2009).

2.2 PREHISTORY

The prehistoric herbal medicine refers to the period before humans could read and write. During the prehistoric period, human populations were distributed across the world (Balick and Cox 2020). These populations formed isolated communities and cultures that interrelated periodically. Archaeologists have studied the history of herbalism into different periods by considering different factors (Giannenas, Sidiropoulou et al. 2020).

Prehistoric medicine is highly circumstantial, varying based on location and the people involved (Winkelman and Theory 2022). As a result, it represents different levels of societal development according to the period being studied. In archaeology, the study of medicine focused on various techniques such as observing illnesses in the remains of human beings, analyzing plant remnants, and conducting archaeological sites to expose ancient healthcare applications. During this ancient time, people likely combined natural and supernatural methods to treat conditions and diseases (Brown, McIlwraith et al. 2020). While there was no formal research, trial and error played a role. However, due to the lack of antiseptics, proper facilities, and an ignorance of the role of germs, even minor injuries like cuts, bruises, and broken bones could become serious if infected. Additionally, evidence recommends that conditions like osteoarthritis, rickets, and bone deformities existed, likely influenced by factors such as heavy lifting and malnutrition. Life expectancy was low, with women frequently dying in childbirth. Evidence from prehistoric burial practices suggests some knowledge of bone structure, but public health concepts were likely absent. Interestingly, Neanderthal tombs in Iraq dating back around 60,000 years contain pollen and flower fragments from numerous medicinal plants, highlighting the ancient origins of herbal medicine (Izah, Ogidi et al. 2024). For example, the Neanderthal burial site "Shanidar IV" in northern Iraq revealed the pollen from seven plant species which are still used as herbal remedies (Chazan 2021). Researchers believe that the rodent, *Meriones tersicus*, played a significant role in transporting pollen to its current location. This finding underscores the Neanderthals' knowledge of herbal medicine, as they utilized plants like Ephedra, Centaurea, Senecio, Althea, and Achillea for health purposes (Hunt, Pomeroy et al. 2023). This implies that herbal remedies were widely used during that era.

Similarly, over 5,000 years ago there is that the 'Iceman' known as called Otzi had also used medicinal herbs, likely to treat parasites in his intestines. The study of prehistoric medicine provides valuable awareness into the early applications of healing and the role of plants in promoting health and well-being (Qadir and Raja 2021).

2.2.1 Ancient Civilization

Ancient history of herbalism spans from approximately 3000 BCE to 500 CE (Oberhelman 2020). While the periodization may suggest a uniform history, it's crucial to recognize that sociocultural norms and technological advancements varied significantly over this period, both locally and globally (Cruse 2021).

Early civilizations across the world harnessed the healing properties of plants, laying the groundwork for modern herbal medicine (Elendu 2024). Ancient herbal medicine, which also covers a similar timeframe, featured various methods about healing that come from different parts of the world (Sharma et al. 2021). These theories often intertwined nature, religion, and human health, emphasizing concepts of fluids and energy mixing. While well-known scholars and important texts offered detailed medical perceptions, which come across several difficulties. Early civilizations used a variety of herbs for medicinal purposes (Giannenas, Sidiropoulou et al. 2020). Chamomile, which was used by the ancient Romans and Egyptians, was known for its calming effects and treated digestive issues and insomnia (Sarkar and Chawla 2022). Garlic, which was widely used in ancient Egypt, Greece, and Rome, was valued for its antibacterial properties and used in the treatment of infections and respiratory problems (Sayed 2023). Similarly, in ancient China and India, ginger was used to treat nausea, digestive issues, and inflammation (Komiljonova 2024). The Romans utilized lavender for its antiseptic and anti-inflammatory properties, as well as its calming scent to reduce stress and anxiety (Grecu et al. 2021). Mint was employed by the ancient Greeks and Romans to aid digestion and relieve headaches (Stapley 2023). Turmeric, a key component in Ayurvedic medicine in ancient India, was known for its anti-inflammatory and antioxidant properties (Akaberi et al. 2021). Rosemary, used by the ancient Greeks and Romans, was believed to enhance memory and act as an antiseptic (Stapley 2023). The ancient Egyptians called aloe vera the "plant of immortality" and used it for treating wounds and skin conditions. These herbs were essential in ancient medicinal practices and continue to be used in modern herbal medicine (Albahri et al. 2023) (Table 2.1).

TABLE 2.1
Potential Uses of Some Common Herbs

Herbs	Scientific Names	Potential Uses
Aloe	Aloe vera/Aloe barbadense	Antibacterial & Moisturizer
Cranberry	*Vaccinium macrocarpon*	Urinary Health, Heart Health & Diabetes
Chameleon Plant	*Houttuynia cordata*	Chronic Sinusitis
Chamomile	*Matricaria recutita*	Digestive Issues & Insomnia
Echinacea	*Coneflower*	Boils, Fever & Herpes
Fenugreek	*Trigonella foenum-graecum*	Diabetes & Satiety
Garlic	*Allium sativum*	Heart & Dental Health, Sickle Cell Anemia & Diabetes
Ginger	*Zingiber officinale*	Osteoarthritis, Nausea, Indigestion & Painful Menstruation
Ginkgo	*Ginkgo biloba*	Cognition, Schizophrenia, Glaucoma, Blood Vessels, Migraines & Vertigo
Green Coffee Bean	*Coffea*	Mild Hypertension
Lavender	*Lavandula angustifolia*	Stress & Anxiety
Mint	*Mentha*	Digestion & Relieve Headaches
Rosemary		Improve Memory & Served an Antiseptic
Turmeric	*Curcuma longa*	Osteoarthritis & Joint Pain
Wheatgrass	*Triticum aestivum*	Chemotherapy induced Bone Marrow Suppression & Inflammatory Bowel Disease

2.2.1.1 Mesopotamia

Mesopotamia is a region which is situated between the Tigris and Euphrates rivers in the Middle East (Al Bomola 2011). It served as the cradle for some of the world's earliest civilizations, including the Sumerians and the Akkadians (a broad classification, which included both the Assyrians and the Babylonians). In early Mesopotamian healing practices, there was an intriguing blend of naturalistic and supernatural beliefs, intertwining medicine, science, magic, and religion (Pouyan 2016).

The earliest recorded evidence of using medicinal plants for drug preparation comes from a Sumerian clay slab discovered in Nagpur, and which dates from the 3rd millennium BCE. These tablets contained detailed information about drug prescriptions, surgical procedures, and even exorcisms. This ancient artifact, dating back approximately 5,000 years, contains 12 recipes for drug formulations, referencing over 250 different plants. Among these plants are alkaloid-containing species like poppy, henbane, and mandrake. Additionally, the Indian Vedas, sacred texts, also mention the use of plants for healing, reflecting their abundant presence in the country (Singh 2016). Highly specialized professionals, including seers, exorcists, and physician-priests, administered and performed these practices. One early example of a prescription-like medication can be found in Sumerian texts from the Third Dynasty of Ur (around 2112 BCE to 2004 BCE).

Following the Akkadian Empire's conquest of the Sumerian civilization, the Babylonian civilization gained in prominence. A notable contribution to their medicinal literature was the Diagnostic Handbook, authored by Esagil-kin-apli during the reign of Babylonian king Adadapla-iddina. It provided detailed observations and logical rules for connecting symptoms to diagnoses and prognoses (Žuškin et al. 2008). Most artifacts from ancient Mesopotamian civilizations originate from the neo-Assyrian and neo-Babylonian periods, offering insights into medical practices despite gaps due to document damage.

In the Mesopotamian civilizations, various medical innovations emerged (Pouyan 2016). These included practices related to prophylaxis (preventing the spread of disease), documentation of strokes, and an awareness of mental illnesses.

2.2.1.2 Ancient Egypt

Ancient Egypt, a remarkable civilization, thrived along the River Nile, which flourished for nearly 30 centuries. However, it faced many challenges: Persian conquests weakened it in 525 BCE, and Alexander marked its ultimate downfall in 332 BCE. Ancient Egyptian texts hold a fascinating allure, especially because of the language and translation debates that surround them (Schiødt 2023). Many translations rely on calculations between ancient and modern concepts, leaving room for uncertainty about conditions and circumstances. Even though physical documents from that historical period are very limited, the Papyrus Ebers provide valuable information about ancient herbal practices (Gonzalez 2021).

The Papyrus Ebers contains detailed lists of various health issues and their medications in relations to matters such as skin and limb issues (Von Klein 1905). It comprises the study of various herbal plants such as cannabis, garlic, juniper, aloe, castor bean, and devil's apple. Actions primarily focused on alleviating prevalent symptoms, as these symptoms were often considered the basic problem. Unfortunately, information about the assortment and preparation of these medications remains mostly unknown. Several translated information conclude that physicians previously have knowledge of treatment techniques, so they don't restate them explicitly. While our modern understanding of Egyptian herbals relies on translations of ancient information that spread this tradition to various areas worldwide, manipulating and shaping medical applications sin different cultures (Heinrich et al. 2012; David and Forshaw 2023). This glimpse into ancient Egyptian medicine is invaluable.

Egyptian healers primarily relied on locally sourced herbs, although they did occasionally import some herbs from regions like Lebanon (Abdel-Azim et al. 2011).

2.2.1.3 India, China, Greece, & Rome

2.2.1.3.1 India

Ayurveda, India's traditional healthcare system, dates back over 5,000 years. Ancient sages developed a holistic approach to health and well-being during the Vedic period in India by confining the body, mind, and soul. More than 114 Ayurvedic treatments for various diseases were covered in the ancient book 'Atharvaveda'.

Ayurveda is an ancient medical system of India, which blends two main strands from different traditions. Ayurveda combines traditional herbal practices with new theories and treatments that started around 600 BCE (Kizhakkeveettil, Parla et al. 2024). The Sushruta Samhita records 700 herbal plants (Jaiswal, Liang et al. 2016) with different treatments about some diseases.

The good health condition and diseases are not pre-planned and can be regulated by the human force, stated that in *Charaka Samhita*. The information collected from the mentioned books is concerned with the examination, diagnosis, treatment, and prognosis of various diseases. Sushruta is famous for his surgical actions, including rhinoplasty, ear lobe repair, lithotomy, and cataract surgery, earning him the title "father of plastic surgery" (Champaneria, Workman et al. 2014). It also described over 125 surgical instruments.

Ayurvedic typically contains different categories of medicines such as surgery, psychiatric medicine, ENT diseases, toxicology, rejuvenation, and fertility (Chaudhry 2019). Teaching was integrated with clinical subjects, such as anatomy with surgery and embryology with pediatrics.

2.2.1.3.2 China

Different plant seeds were used as herbal medicine for several diseases in the Bronze Age Shang dynasty. Shennong Ben Cao Jing, the first Chinese pharmacopoeia, records 365 herbal plants and (Sheng, Shang et al. 2019) their applications, including Ephedra, Cannabis and Hydnocarpus which is used for the treatment of leprosy.

The Chinese empires significantly regulate the methods of healing by using traditional Chinese medicine (TCM). This system began developing around the Zhou dynasty and is evident in early texts like the Classic of Changes (Yi Jing) and Classic of Poetry (Shi Jing). Traditional Chinese medicine, rooted in Taoist physicians' empirical observations, reflects the belief that human experiences mirror universal principles (Zhan and values 2014). The foundational text, Huangdi Neijing (Yellow Emperor's Inner Canon), was written between the 5th and 3rd centuries BCE. In the late 2nd century CE, Zhang Zhong Jing's Treatise on Cold Damage referenced the Neijing Suwen. Huangfu Mi, a Jin dynasty practitioner, also cited the Yellow Emperor in his Jiayijing around 265 CE. The Tang dynasty saw the expansion and revision of the Suwen, solidifying its role in TCM. For thousands of years, TCM has been used for different health treatments through various ways such as surgery and other therapies (Dong and Medicine 2013).

2.2.1.3.3 Greece and Rome

The early Greece exposed the herbal remedies and their recipes about several health problems in the Hippocratic Corpus. In contrast to the religious healing practices of the time, the Corpus notably lacks rites, prayers, or chants, highlighting the Hippocratic emphasis on logic and reason in medicine.

The Corpus mentioned a variety of herbs, some of them imported from distant places such as Arabia (McCabe 2009). Although many imported ingredients were likely too costly for everyday use, some recommended herbs, such as elderberries and St. John's Wort, were more affordable and accessible (Laird, Laird et al. 2019).

A Greek physician, Galen of Pergamon, was extremely creative in detailing his medical knowledge in Rome about herbalism. Among his many writings on herbs and their properties, his most notable work is the "Works of Therapeutics." Galen discusses the integration of various medical disciplines to rehabilitate and treat health-related problems. Diocles of Carystus was a well-respected and prolific writer. He supervised certain herbal medicines and their actions which play a significant role in the area of health. Although his original texts are lost, many medical scholars have quoted him extensively, allowing us to learn about his work through these fragments. Galen, Celsus, and Soranus were the prominent sources of information about the teaching of Diocles with relation to herbalism (Cartwright and Armstrong 2020).

Ancient Rome explored the complete history documenting over 900 herbs and their medicinal uses. Their work is a crucial resource for understanding the early herbal medicines and remedies (Jarmusch 2015). They saw illnesses as natural phenomena and believed that using drugs could help maintain balance and harmony in nature.

Pedanius Dioscorides authored his *De Materia Medica*, in which he detailed over 1,000 medicines from herbs, minerals, and animals (Duffin 2018) and provided a classified study of the plants which has proved essential for later herbalists and botanists.

2.3 MIDDLE AGES AND BEYOND

During the Middle Ages, the study and practice on herbal medicines was crucial for survival in the absence of modern prescription drugs and this period laid the foundation for the natural therapies that are still in use today. As detailed, the use of herbal medicines had taken place since the Stone Age, which significantly influenced the technological advancement of these medicines throughout history (Pan, Litscher et al. 2014). The facts and data collected about various herbals based on the processes developed over the centuries. Most of the botanical texts in use at this time were in Greek or Latin (Lord 2021).

The widespread study of herbals began in the Middle East and Asia, where new research work about herbals and herbal medicines was conducted by Dioscorides (Lord 2021). He made extensive discoveries of herbs which could be employed to generate new ideas for producing new medicines which have strong healing properties (Staub et al. 2016). As a result of his innovative techniques, herbalists collectively practiced these herbs for several problems (Kay 2022), but due to lack of evidence, it is very complicated to learn about these herbs and their properties, while some herbs were found near the sea, adding to their variety. Dioscorides' volumes provided valuable information on useful properties and their physical distribution in the environment. Some herbs naturally grow naturally in specific regions with high healing properties and were exported due to social and atmospheric limitations (Némethy et al. 2020). Some extrinsic herbs, such as ginger, citrus and echinacea, were unfamiliar in the West, where herbalists studied and increased their knowledge about these plants for health practices due to their flavors and colors. In the Middle Ages, some essential herbs were used widely due to their particular properties and were in common use for the treatment of minor injuries and common diseases such as colds (Sams 2015). Furthermost, some herbs exploited during that period were harvested directly from nature and used in their pure form. By contrast, other wild herbs were processed in order to study their beneficial properties and were readily available to the home-grown population (Voeks 2004). Herbalists utilized both natural and wild-grown herbs for treating slight infections and wounds, and occasionally even in major surgical procedures. For example, cannabis was used significantly for nervousness (Jahromi et al. 2021). There is a substantial increase in the use of herbs for various health issues in the late Middle Ages and many of these, as used for example in essential oils and ointments, remain in use today. These new procedures were used for curing diseases through providing raw materials for treatment. During this period, people practiced and prescribed herbal remedies for skin and other disorders (Kalu 2022).

During this period, a researcher from the Arabian School introduced the experimental study of physiology to better understand how the human body works. These studies were particularly focused on finding treatments for sexually transmitted diseases. This was a significant advancement in medical science at the time, as it helped develop new ways to diagnose and treat these ailments. He established new medicine to identify and control specific infections. It has been estimated that around 760 herbal medicinal plants and medicines were described during this period (Tobyn et al. 2010).

2.3.1 Translation of Herbals

As stated, during the Middle Ages, herbalists made intensive studies of different plants and discovered their traditional and medicinal uses by focusing on different parameters of such plant parts. These herbalists characterized the structure and environmental uses through the collection, growth, and storage of these plants (Wachtel-Galor and Benzie 2012).

In this period, both botany and gardening emphasized the efficacy of plants with common herbs essential for medical purposes. Translation of herbals and their usage started from Baghdad and expanded outward to the regions of Europe (Rakow and Lee 2015). This collaborative process involved many people translating and adding to texts.

The translation of herbs, their facts, and images during the Middle Ages resulted in several versions of manuscripts from various different sources. This vital process of translation significantly advanced the scientific knowledge about the use of specific herbs in different regions around the world at that time. It involved continuous revisions and additions, reflecting its lively nature. Benedictine monasteries, which were renowned for their extensive knowledge of herbals, cultivated gardens with herbs deemed useful for treating various human ailments (Beltrán Peralta et al. 2022). These efforts laid the groundwork for modern medical education, which has been heavily influenced by monastic practices. Conventual institutes were established where Greek manuscripts were translated into Latin, for the preservation and dissemination of knowledge.

Medieval botanical knowledge was intricately linked to medicine, mainly for the treatment of various diseases. Some essential factors of plants played a significant role in the preparation of specific herbals and sometimes included the preparation and usage instructions (Pengelly 2020). To certify current medical use, a reference book was formed which is practically designed for famous and essential herbals.

A Greek philosopher described the different features of 500 herbal plants in different papers, which classified them according to their structure and morphology. He developed *The Causes of Plants*, which was later translated into Latin and has continued to be used for centuries. Known as the "grandfather of botany," Theophrastus' contributions were foundational. Another writer, Crateuas, produced the first pharmacological book on medicinal plants, in work that had an influence for many later generations. A Greek physician illustrated around 600 plants and their medicinal uses, with his illustrations remaining influential in pharmacology and medicine until the Renaissance (Stapley 2023).

At this time monasteries served as health services, where monks passed on herbal knowledge among themselves and to their patients. These herbals, with their complex illustrations, were intended for those who were already in possession of prior knowledge. Their usefulness has been questioned due to unrealistic depictions and multiple plants demanding to treat the same disorder. However, practiced healers effectively employed these herbal plants without needing detailed instructions (Upadhyay, Roy et al. 2007). Monks collected and organized texts to make them useful in their monasteries, adapting classical remedies to their own and local needs. This adaptation may explain why existing collections of remedies differ from volume to volume.

Oral transmission was another method by which medical knowledge was passed down the centuries. It is a misconception that early medieval medicine can be fully understood by identifying texts alone, as prior knowledge is also essential. Many factors influenced the translation of herbals; writing or explaining was just a small part. These remedies evolved from numerous previous translations, incorporating diverse influences (Anderson 1997).

2.3.2 Early Modern Era

Herbalism plays a vital role in bridging the gap between traditional knowledge with emerging scientific approaches. The incorporation of American plants and the advancement of modern medicine shaped the landscape of healing practices during that time. During this period, there was a flourishing in the publication of herbals (books about medicinal plants). Many herbals became available in languages other than Latin or Greek, thanks to efforts such as those of the monks in the Middle Ages. These herbals contained detailed information about various plants, their properties, and their uses in healing. People relied on these texts for practical knowledge about herbal medicine (Crellin, Philpott et al. 1990).

During the 16th and 17th centuries, herbalists played a significant role in English literature. *The Grete Herball* (1526) was the first illustrated herbalist in English, followed by well-known works like those of John Gerard (1597) and Nicholas Culpeper (1653). Despite ridicule from contemporary physicians, both Gerard and Culpeper achieved high degrees of fame. In addition, the new medicinal plants presented valuable insights into herbal knowledge from Mexico to Europe (Bye et al. 2016).

Later, the several dynamic chemical drugs were introduced and employed in medicine.

During the 18th and 19th centuries, there was a further expansion in herbal knowledge. European explorers and settlers encountered new plants in the Americas. These explorations led to the incorporation of American plants into herbal medicine. The exchange of botanical knowledge between continents also enriched herbal practices at the time. Simultaneously, modern medicine was evolving. Scientific discoveries, such as the germ theory of disease, transformed the state of medical understanding (Weindling and - 2004). Simultaneously, traditional herbal practices in use drew significantly on scientific research techniques. In the Americas, medical knowledge was primarily derived from herbal books due to the scarcity of physicians (Pan, Litscher et al. 2014), while European settlers brought their own knowledge of native plants and their remedies with patriots. A botanist played a key role in studying and recording Native American plant remedies, often including this information in printed almanacs.

In the 19th century, pharmacology became more formalized, leading to a better understanding of how drugs affect the body. Samuel Thomson, an uneducated yet respected herbalist, had a great influence on medical professionals, who began calling themselves Thomsonians to distinguish from traditional doctors who used methods like calomel and bloodletting (Janik 2015). In this period, there is great stimulation of experimental methods in herbal medicine.

2.4 MODERN TIMES

In 1910, Abraham Flexner, an education reformer, published a comprehensive report on American medical schools. His work exposed inadequacies in medical education, including uneven instruction and pseudoscientific fields like electrotherapy, homeopathy, chiropractic, and naturopathy. Consequently, many substandard medical schools were closed, reorienting medical education towards a science-based approach. Following the Flexner Report, traditional herbalism was considered an alternative medicine. Eclectic medical schools that focused on botanical medicine faced closure. Herbalism was marginalized in favor of more conventional approaches (Russo and Dougherty 2013).

In 1949, Mao Zedong reintroduced traditional Chinese medicine (TCM) into China's healthcare system. TCM relies heavily on herbalism, acupuncture, and other holistic practices (Chan 2016). Thousands of practitioners, including Americans, received training in TCM for hospital use. TCM practitioners focus on balancing the states of yin and yang within the body. They use 12 meridians (energy pathways) to regulate the flow of Qi (vital energy) throughout the body. TCM is practiced worldwide for health promotion, disease prevention, and treatment. It integrates various approaches, with herbal medicine being a central component. TCM considers the whole person, rather than just isolated symptoms. Diagnosis and treatment are based on the balance of yin (earth, cold, femininity) and yang (sky, heat, masculinity). Yin and yang affect interactions within the universe's five elements: metal, wood, water, fire, and earth.

In the 1930s, Britain grappled with challenges related to herbalism practices. Meanwhile, in the US, regulations and restrictions were imposed on herbal medicine. Licensing and oversight became more stringent (Wahlberg 2006).

The WHO estimates that 80% of people worldwide use herbal medicines for some part of their primary healthcare. Herbal remedies have a long history and are still widely used across cultures (Halberstein 2005). In Germany, there are about 600 to 700 plant-based medicines available, and around 70% of German doctors prescribe them. This reflects the integration of herbal medicine into the country's healthcare system. In the US, prescribing treatments requires a legal medical license, which is regulated by each state. There is no specific licensing for herbalists, so anyone can employ and supply herbs. Conventional herbal medicine involves a multidisciplinary approach. It often combines ancient wisdom with modern scientific knowledge (Heinrich et al. 2023). Many modern alternative doctors use herbalism because plants have various benefits and few side effects. Herbal remedies can complement conventional treatments and provide alternative options for patients.

2.5 CURRENT STATUS

Herbal medicines (phytomedicines) are rapidly expanding globally. People increasingly turn to herbal products for health challenges. Interest in natural therapies has surged in both developed and developing countries. Africans and Indians mainly depend on conventional physicians for initial healthcare by using the herbal remedies. China integrates conventional herbal medicine into overall healthcare delivery (it accounts for around 40% of all treatments). Many Chinese hospitals used traditional herbal medicines for various diseases to treat the patients. Herbs are used to treat chronic and acute diseases, including cardiovascular issues, prostate problems, depression, and inflammation (Tachjian et al. 2010). Herbal medicine has been used to treat HIV-related wasting symptoms in Africa. Approximately 90% of new drug molecules are believed to originate from nature. Effective agents include anticancer drugs (e.g., paclitaxel), antimalarials (e.g., artemisinin), and antidiabetic compounds (e.g., metformin). India used 25,000 operative plant-based preparations traditionally. Over 1.5 million practitioners follow the traditional medicinal system. India's 7,800 medicinal drug manufacturing units consume about 2,000 tons of herbs annually. Exported herbal products are worth about 1 billion rupees (Surve, Shyamsundar et al. 2024) (Figure 2.1).

The WHO predicts that the global herbal market will reach $5 trillion by 2050 (currently $62 billion). India and China contribute over 70% of global herbal diversity.

Herbal medicine is often preferred because it's more affordable, it aligns with patients' beliefs, and it is perceived as being safer than synthetic medicines (Judith, Ijeoma et al. 2016). It is principally used for promoting health and treating chronic conditions. People turn to traditional remedies when conventional treatments fail, especially in severe diseases such as advanced cancer. However, the belief that traditional medicines are always safe is not necessarily true, particularly when they are being combined with other medications.

Herbs are used for both chronic and acute conditions. They address various health challenges, including cardiovascular issues, prostate problems, depression, and inflammation (O'Hara et al. 1998). Traditional herbal medicines used significantly during the 2003 SARS outbreak in China. Herbs were used as essential oils, tablets, and ointments to treat different skin and health issues (Alamgir et al. 2017). Extracts vary based on solvent, temperature, and extraction time (e.g., tinctures, decoctions, macerates). However, there's no standardization, and components can vary significantly between batches and producers.

Plants contain a diverse array of compounds, including secondary metabolites such as aromatic substances (often phenols or their derivatives, such as tannins). Many of these compounds have antioxidant properties. Ethnobotanicals play a crucial role in pharmacological research and drug development. They serve not only as direct therapeutic agents but also as starting materials for drug synthesis or models for active compounds. This breakthrough demonstrated that drugs from plants could be purified and given in precise dosages, regardless of their source or age. Subsequent discoveries, like penicillin, further advanced this approach. Today, products inspired by plants and natural sources (including fungi and marine microorganisms) contribute significantly to commercial drug

FIGURE 2.1 Common Healing Herbs in Asia.

preparations (Pagare, Bhatia et al. 2015). These drugs include antibiotics, antimalarials, and lipid-lowering agents, which were obtained from fungi.

Natural products also play a significant role in cancer therapeutics. Over 60% of the cancer treatments either on the market or in testing are based on natural sources. Among 177 approved cancer drugs, more than 70% originate from natural products or mimetics. Additionally, about 25% of globally prescribed drugs come from plants, with 121 active compounds in use. Between 2005 and 2007, some 13 drugs derived from natural products were approved in the United States. Over 100 natural product-based drugs are in clinical studies, and 11% plant-derived drugs were comprised in WHO's essential medicine (Aware et al. 2022).

2.6 CHALLENGES ASSOCIATED

The use of herbal medicines has grown significantly over the past few decades, with up to 80% of people worldwide relying on them for primary healthcare. Herbal medicines often enter the market without mandatory safety or toxicological evaluations, and many countries lack effective regulatory mechanisms for manufacturing practices and quality standards (Bandaranayake 2006). This means that their effects on health are not thoroughly assessed before they become available to consumers. Many countries lack effective machinery to regulate manufacturing practices and quality standards for herbal medicines. As a result, these products are continuously made available to consumers without proper oversight. Poor quality control can lead to variations in the composition and potency of herbal products. Inadequate quality standards can lead to reduced efficacy or unexpected adverse effects. Monitoring the safety of herbal medicines is difficult due to poor reporting systems and limited knowledge of traditional and complementary alternatives, causing adverse reactions to go unnoticed or unreported (Ekor 2014). Establishing scientific evidence for health claims of herbal medicines is challenging, as they are not subject to the rigorous clinical trials required for pharmaceutical drugs.

2.6.1 Regulation and Safety of Herbal Medications

Regulating herbal medicines on a global basis is difficult due to diverse cultural practices, varying quality standards, limited scientific evidence on efficacy and safety, and the misconception that herbal products are inherently safe. Despite increased cooperation among regulatory agencies, harmonization remains lacking due to specific national laws and various categories of herbal products. Regulatory authorities often face challenges with insufficient staffing and funding, making adequate oversight and enforcement essential (Zhou, Li et al. 2019). Developing suitable testing procedures and high-quality standards for complex mixtures of herbal medicines is challenging due to the presence of multiple active compounds.

Registering or listing herbal medicines can be burdensome, particularly for manufacturers with limited education. The streamlining of these processes is essential to ensure safety and quality. There is often insufficient research data on the efficacy of herbal medicines, making it challenging to establish evidence-based efficacy. Additionally, mechanisms to control and regulate advertising, clinical claims, and health benefits of herbal products are inadequate (Zhou, Li et al. 2019). Clear guidelines are necessary to avoid misleading claims.

To ensure the safe and effective use of herbal medicines in public health services, it is crucial for regulators, healthcare professionals, manufacturers, and patients to work together.

2.6.2 Quality Control of Herbal Medicine

Ensuring the quality of herbal medicines involves several major hurdles. A key problem is the adulteration and contamination of these products. Herbal medicines frequently consist of a blend of different plant materials, which can be mixed with other substances either deliberately or accidentally. This can lead to inconsistencies in the levels of active ingredients, potentially diminishing their effectiveness or causing adverse side effects (Muyumba, Mutombo et al. 2021). Additionally, there is a lack of standardized methods for growing, harvesting, and processing medicinal plants. Genetic differences, environmental conditions, and harvesting methods can greatly affect the quality and consistency of herbal products. Without standardized methods, it is difficult to ensure that each batch of herbal medicine meets the required quality standards.

The quality of raw materials is very important for herbal medicines, inclined by genetic and environmental factors. These factors stimulate the quality control of herbal medicines (Zhang, Wider et al. 2012). To control the quality of finished herbal products, the main challenges to be located particularly those that are in mixture form. Therefore, the general standards and measures for confirming the quality of finished herbal products are significantly more complicated compared to other types of pharmaceuticals.

Analyzing herbal medicines is challenging due to the many phytochemicals they contain, making it difficult to identify and measure active compounds accurately. Advanced techniques are needed, which may not be widely available or affordable. Additionally, the lack of strict regulatory oversight for herbal medicines leads to inconsistent quality control and gaps in safety and efficacy information (Thakkar et al. 2020). Establishing international standards and strong regulatory frameworks is crucial to address these issues.

Finally, the overharvesting of wild medicinal plants raises sustainability issues, potentially depleting natural resources and impacting the quality of herbal medicines. To preserve these valuable resources and maintain their quality, it's essential to adopt sustainable farming and conservation practices.

Tackling these issues demands a comprehensive strategy that includes better standardization, advanced analytical methods, strict regulatory frameworks, and sustainable practices to guarantee the safety, effectiveness, and quality of herbal medicines. WHO supports the quality and control procedures which are important for herbal materials, production, safety and efficiency of herbal medicines (NAUMOVIĆ, ARSIĆ et al. 2014).

2.6.3 Safety Monitoring of Herbal Medicines

Ensuring the safe and effective use of herbal medicines requires diligent safety monitoring. A key part of this is pharmacovigilance, which systematically gathers, analyzes, and interprets data on adverse effects linked to herbal medicines (Ekor 2014). This information comes from healthcare professionals, consumers, and manufacturers. By reviewing these reports, authorities can spot potential safety issues and take steps to reduce risks.

Another key element is evaluating case reports. Each reported adverse reaction is thoroughly examined in order to determine its cause and severity. This involves a detailed review of the patient's medical history, the specific herbal product used, and the context of the adverse event. These assessments help in understanding the safety profile of herbal medicines and identifying any patterns or trends that might suggest a wider safety concern (Ekor 2014).

Regulatory frameworks are crucial for safety monitoring. Many countries have set guidelines to ensure the quality and safety of herbal medicines, requiring thorough testing and safety evidence from manufacturers before products can be marketed (Al-Worafi 2020). Additionally, post-marketing surveillance monitors the safety of these medicines once they are available to the public, helping to identify any new or rare adverse effects not detected during pre-market testing.

Risk communication is essential for informing healthcare professionals and the public about the potential risks of herbal medicines (Al-Worafi 2020). Effective communication certifies users are aware of probable side effects and know how to use these products safely. This can involve labeling requirements, public health advisories, and educational campaigns.

Ensuring the safety of herbal medicines is a complex process that involves cooperation among regulatory authorities, healthcare professionals, manufacturers, and consumers. By establishing strong pharmacovigilance systems, performing detailed assessments, enforcing regulatory standards, and maintaining effective risk communication, the safety and effectiveness of herbal medicines can be improved for everyone (Barnes 2003).

2.6.4 Bioavailability of Herbal Medicines

Bioavailability refers to the amount of a drug or a substance which actually reaches the bloodstream and becomes active in the body. For herbal medicines, this concept is vital because it influences how effective and beneficial the treatment will be.

One key challenge with the bioavailability of herbal medicines is their complex chemical composition (Mukherjee et al. 2015). They contain various active compounds with different solubility and absorption properties, leading to inconsistent absorption and varying therapeutic effects. For example, some compounds may not dissolve well in water, making them harder for the body to absorb efficiently.

Poor solubility and stability are major issues for herbal compounds. Many have high molecular weights and don't dissolve well in water, limiting absorption in the gut. Additionally, some degrade before absorption, reducing bioavailability (Zhao, Yang et al. 2019). Advanced delivery systems like nanoparticles and liposomes are being explored to enhance their solubility and stability.

First-pass metabolism significantly affects the bioavailability of herbal medicines. When ingested, herbal compounds pass through the liver and can be metabolized before entering the bloodstream, reducing their active amount. Strategies like sublingual or transdermal delivery are being explored to bypass this process (Gupta, Chang et al. 2017). Additionally, interactions with foods and medications can influence absorption, making it essential to understand these interactions for optimal therapeutic use.

In summary, the bioavailability of herbal medicines is exaggerated by their chemical complexity, solubility, stability, first-pass metabolism, and interactions with other substances. Enhancing bioavailability is crucial for maximizing their therapeutic benefits. Ongoing research aims to change advanced delivery systems and strategies to overcome these challenges (Mukherjee, Harwansh et al. 2015).

2.6.5 Clinical Trials

Randomized controlled clinical studies are crucial for herbal medicines to be accepted in conventional medicine. These studies verify the effectiveness and safety of medicinal plants. Since it's sometimes hard to measure the benefits of herbal treatments before clinical trials, these studies are especially important. Detailed case reports can also offer valuable insights and help generate new research ideas.

In clinical trials, herbal formulas are often indicated for specific patterns rather than diseases. Therefore, differentiating a disease into several patterns is a practical approach (Firenzuoli, Gori et al. 2007). The endpoints measured should include modern parameters, associated indices, and overall improvement. Despite the challenges, it is possible to align traditional diagnostic and therapeutic systems with modern research methodologies. Quantitative standardization of pattern diagnosis helps re-evaluate the efficacy of herbal formulas. The methods and guidelines for validating modern medicines should also apply to herbal products, even though they take a holistic approach to treatment (Firenzuoli, Gori et al. 2007). However, conventional clinical research designs can be challenging to apply to traditional medicine due to the individualized nature of herbal therapies. Large patient numbers are required for clinical trials to ensure adequate and statistically appropriate study designs, control groups, and sufficient evidence for efficacy. This increases the time and expenses involved in conducting these trials.

Conducting clinical trials with herbal medicines presents unique challenges. For instance, using an inactive drug can be difficult when the herbal preparation has a distinct smell or taste, as seen with certain essential oils. Additionally, patients who have previously used the herbal medicine being studied, especially if it has noticeable sensory properties, cannot be easily randomized into control groups. While clinical trials for herbal drugs are possible, a review of the literature shows that few well-controlled, double-blind trials have been conducted (Guo, Pittler et al. 2007). Meta-analyses from reputable medical journals like *The Annals of Internal Medicine*, JAMA (*Journal of the American Medical Association*), *The British Medical Journal*, *The Lancet*, and *The British Journal of Clinical Pharmacology* support this observation. Several factors contribute to these discrepancies, including the lack of standardization and quality control of herbal drugs, varying dosages, inadequate randomization, insufficient patient numbers for statistical significance, and difficulties in creating appropriate inactive drugs due to the distinct tastes and aromas of herbal medicines (Guo et al. 2007). Additionally, there are wide variations in the duration of treatments using these medicines.

2.7 FUTURE PERSPECTIVES

The future encompasses the cumulative demand of herbal medicines and herbal healthcare products all over the world. One significant perspective is the integration of herbal medicines into mainstream medical practices. As research confirms the efficacy and safety of herbal treatments, their acceptance in the medical community is likely to grow (Bhardwaj et al. 2018). This trend is driven by the rise in chronic diseases requiring long-term care, which has increased interest in alternative medicine.

Advances in scientific research and technology are promising. Modern techniques such as genomics, proteomics, and bioinformatics help us understand how herbal compounds work (Bhardwaj et al. 2018). These advancements lead to consistent, high-quality herbal products and help discover new, effective herbal drugs.

Herbal medicines are increasingly recognized for their role in combating antimicrobial resistance. As antibiotic-resistant bacteria pose a growing threat, the diverse bioactive compounds in herbal medicines offer a promising alternative (Álvarez-Martínez et al. 2020). Research is investigating how combining herbal extracts with conventional antibiotics can enhance effectiveness and reduce resistance. This strategy not only introduces new treatment options but also promotes sustainable healthcare practices.

The future of herbal medicines looks promising, with ongoing research and technological advancements facilitating their broader acceptance and integration into modern healthcare systems.

2.8 CONCLUSION

The historical development of herbal medicine highlights its enduring significance and contributions to human health. From ancient civilizations to the present, herbal medicine has evolved, reflecting its deep-rooted connection with human culture and well-being. Civilizations like the Babylonians, the Egyptians, the Chinese, and the Greeks extensively used plants for medicinal purposes, laying the foundations of traditional medicine systems. Over time, these practices were refined and meticulously documented, creating a vast repository of knowledge. Despite the rise of modern pharmaceuticals, herbal medicine remains crucial, with a significant portion of the global population relying on it for primary healthcare. Herbal medicines are used and rapidly increasing worldwide. Herbal products are characteristically safe because they are obtained from natural herbs but they also have some bad reactions. This underscores the need for standardized and strengthened regulatory policies worldwide. Regulatory bodies must use clinical trials to ensure the precise and important flow of herbal products for their increasing growth. This historical journey underscores the resilience and adaptability of herbal medicine, securing its role in modern healthcare and highlighting its potential for future advancements.

REFERENCES

Abdel-Azim, N. S., et al. (2011). "*Egyptian herbal drug industry: challenges and future prospects.*" 5(2): 136–144.

Akaberi, M., et al. (2021). "*Turmeric and curcumin: from traditional to modern medicine.*" 15–39.

Al-Worafi, Y. M. (2020). Herbal medicines safety issues. *Drug safety in developing countries*, Elsevier: 163–178.

Al Bomola, A. J. T. (2011). "*Temporal and spatial changes in water quality of the Euphrates river-Iraq.*"

Alamgir, A., et al. (2017). "*Herbal drugs: their collection, preservation, and preparation; evaluation, quality control, and standardization of herbal drugs.*" 453–495.

Albahri, G., et al. (2023). "*The therapeutic wound healing bioactivities of various medicinal plants.*" 13(2): 317.

Álvarez-Martínez, F. J., et al. (2020). "*Tackling antibiotic resistance with compounds of natural origin: A comprehensive review.*" 8(10): 405.

Anderson, F. J. (1997). *An illustrated history of the herbals*, iUniverse.

Aware, C. B., et al. (2022). "*Natural bioactive products as promising therapeutics: A review of natural product-based drug development.*" 151: 512–528.

Balick, M. J. and P. A. Cox (2020). *Plants, people, and culture: the science of ethnobotany*, Garland Science.

Bandaranayake, W. M. J. M. p. t. m. p. i. d. (2006). "*Quality control, screening, toxicity, and regulation of herbal drugs.*" 25–57.

Barkat, M. A., et al. (2021). "*Herbal medicine: Clinical perspective and regulatory status.*" 24(10): 1573–1582.

Barnes, J. J. D. S. (2003). "*Pharmacovigilance of herbal medicines: a UK perspective.*" 26: 829–851.

Beltrán Peralta, N., et al. (2022). "*Wine and monasteries: Benedictine monasteries in Europe.*" 25(6): 652–683.

Bhardwaj, S., et al. (2018). "*Challenges and future prospects of herbal medicine.*" 1(1): 12–15.

Brown, N., et al. (2020). "*Health and Medicine.*"

Bye, R., et al. (2016). "*Ethnobotany and ethnohistorical sources of Mesoamerica.*" 41–65.

Cartwright, A. C. and N. A. Armstrong (2020). *A history of the medicines we take: from ancient times to present day, Pen and Sword History.*

Chaachouay, N., et al. (2024). "*Plant-derived natural products: a source for drug discovery and development.*" 3(1): 184–207.

Champaneria, M. C., et al. (2014). "*Sushruta: father of plastic surgery.*" 73(1): 2–7.

Chan, K. J. W. J. o. T. C. M. (2016). "*The evolutional development of Traditional Chinese Medicine (TCM) outside the Chinese Mainland: Challenges, training, practice, research, and future development.*" 2(4): 6–28.

Chaudhry, B. (2019). *A handbook of common medicinal plants used in Ayurveda*, Kojo Press.

Chazan, M. (2021). *World prehistory and archaeology: pathways through time*, Routledge.
Crellin, J. K., et al. (1990). *Herbal medicine past and present*, Duke University Press.
Cruse, M. (2021). *A History of Science: From Agriculture to Artificial Intelligence*, Arcturus Publishing.
David, R. and R. Forshaw (2023). *Medicine and Healing Practices in Ancient Egypt*, Liverpool University Press.
Dong, J. and A. Medicine J. E. B. C. (2013). "*The relationship between traditional Chinese medicine and modern medicine*." 2013(1): 153148.
Duffin, C. J. (2018). *The historical roles of mineral materials in folk medicine and the development of the materia medica*, Kingston University.
Ekor, M. J. F. I. P. (2014). "*The growing use of herbal medicines: issues relating to adverse reactions and challenges in monitoring safety*." 4: 177.
Elendu, C. J. M. (2024). "*The evolution of ancient healing practices: From shamanism to Hippocratic medicine: A review*." 103(28): e39005.
Firenzuoli, F., et al. (2007). "*Herbal medicine today: clinical and research issues*." 4: 37–40.
Giannenas, I., et al. (2020). *The history of herbs, medicinal and aromatic plants, and their extracts: Past, current situation and future perspectives*. Feed additives, Elsevier: 1–18.
Gonzalez, S. R. (2021). "*Examining health inequity in ancient Egypt*."
Grecu, M., et al. (2021). "*Benefits and uses of lavender essential oil as a complementary and alternative therapy-a short review*." 64(3).
Guo, R., et al. (2007). "*Herbal medicines for the treatment of allergic rhinitis: a systematic review*." 99(6): 483–495.
Gupta, R. C., et al. (2017). "*Interactions between antidiabetic drugs and herbs: an overview of mechanisms of action and clinical implications*." 9: 1–12.
Halberstein, R. A. J. A. O. E. (2005). "*Medicinal plants: historical and cross-cultural usage patterns*." 15(9): 686–699.
Heinrich, M., et al. (2012). *Plants as medicines*. The Cultural History of Plants, Routledge: 208–241.
Heinrich, M., et al. (2023). "*Herbal medicine use in the UK and Germany and pharmacy practice-A commentary*." 19(3): 535–540.
Hoffmann, D. (2003). *Medical herbalism: the science and practice of herbal medicine*, Simon & Schuster.
Hunt, C. O., et al. (2023). "*Shanidar et ses fleurs? Reflections on the palynology of the Neanderthal 'Flower Burial' hypothesis*." 105822.
Izah, S. C., et al. (2024). *Historical Perspectives and Overview of the Value of Herbal Medicine*. Herbal Medicine Phytochemistry: Applications and Trends, Springer: 3–35.
Jahromi, B., et al. (2021). "*Herbal medicine for pain management: efficacy and drug interactions*." 13(2): 251.
Jaiswal, Y., et al. (2016). "*Botanical drugs in Ayurveda and traditional Chinese medicine*." 194: 245–259.
Jamshidi-Kia, F., et al. (2017). "*Medicinal plants: Past history and future perspective*." 7(1): 1–7.
Janik, E. (2015). *Marketplace of the marvelous: the strange origins of modern medicine*, Beacon Press.
Jarmusch, S. A. (2015). *Ancient Pharmacology: Theophrastus' Historia Plantarum and Pliny the Elder's Historia Naturalis*, University of Liverpool.
Judith, O., et al. (2016). "*Popularity and customer preferences for herbal medicines in Nigeria: a questionnaire based survey*." 4(3): 69–76.
Kalu, U. N. (2022). *Intergenerational knowledge transfer in traditional herbal medicine (THM) practices among the Igbo tribe in Nigeria: A qualitative study*, The University of Regina (Canada).
Kay, E. (2022). "*A history of herbalism: cure, cook and conjure*."
Khan, M. S. A. and I. Ahmad (2019). *Herbal medicine: current trends and future prospects*. New look to phytomedicine, Elsevier: 3–13.
Kizhakkeveettil, A., et al. (2024). *History, Present and Prospect of Ayurveda. History, Present and Prospect of World Traditional Medicine, World Scientific*: 1–72.
Komiljonova, O. J. Ц. ж. о. и. и. (2024). "*The Use of Ginger for Medicinal Diseases Based On Traditional Medicine*." 3(1): 203–211.
Kumar, V. and V. J. Kumar I. J. P. S. (2009). "*An overview of herbal medicine*." 1(1): 1–20.
Laird, S. A., et al. (2019). The botanical medicine industry. *The commercial use of biodiversity*, Routledge: 78–116.
Li, F.-S. and J.-K. Weng J. N. P. (2017). "*Demystifying Traditional Herbal Medicine with Modern Approach*." 3(8): 1–7.
Liu, F. X. and J. W. Salmon J. I. M. (2010). "*Herbal Medicine Regulation in China, Germany, and the United States*." 9(6): 123–140.
Lord, R. (2021). *Clinical Herbalism-E-Book: Clinical Herbalism-E-Book, Elsevier Health Sciences*.

Martínez-Graciá, C., et al. (2015). "*Use of herbs and spices for food preservation: Advantages and limitations.*" 6: 38–43.

McCabe, A. (2009). "*19. Imported materia medica, 4th—12th centuries, and Byzantine pharmacology.*"

Mukherjee, P. K., et al. (2015). Bioavailability of herbal products: approach toward improved pharmacokinetics. *Evidence-based validation of herbal medicine*, Elsevier: 217–245.

Muyumba, N., et al. (2021). "*Quality control of herbal drugs and preparations: The methods of analysis, their relevance and applications.*" 4: 100070.

Nasim, N., et al. (2022). "*Plant-derived natural products for drug discovery: current approaches and prospects.*" 65(3): 399–411.

Naumović, S., et al. (2014). "*From plant materials to quality herbal products.*" 121.

Némethy, S., et al. (2020). "*Collection, cultivation and processing of medical plants, herbs and spices in the Balaton Ecomuseum–herbal medicine as intangible cultural heritage.*" 6(1): 52–87.

Nigam, M. (2021). Phytomedicine: Scope and current highlights. *Preparation of Phytopharmaceuticals for the Management of Disorders*, Elsevier: 39–54.

O'Hara, M., et al. (1998). "*A review of 12 commonly used medicinal herbs.*" 7(6): 523.

Oberhelman, S. M. (2020). *Healing Manuals from Ottoman and Modern Greece: The Medical Recipes of Gymnasios Lauriōtis in Context*. Walter de Gruyter GmbH & Co KG.

Organization, W. H. (2009). *The use of herbal medicines in primary health care, WHO Regional Office for South-East Asia.*

Pagare, S., et al. (2015). "*Secondary metabolites of plants and their role: Overview.*" 9(3): 293–304.

Pan, S.-Y., et al. (2014). "*Historical perspective of traditional indigenous medical practices: the current renaissance and conservation of herbal resources.*" 2014(1): 525340.

Pengelly, A. (2020). *The constituents of medicinal plants: an introduction to the chemistry and therapeutics of herbal medicine*, Routledge.

Pouyan, N. W. J. O. P. R. (2016). "*Mesopotamia, the cradle of civilization and medicine.*" 5(4): 192–225.

Qadir, S. U. and V. Raja (2021). Herbal medicine: Old practice and modern perspectives. *Phytomedicine*, Elsevier: 149–180.

Rakow, D. A. and S. A. Lee H. R. V. (2015). "*Western botanical gardens: history and evolution.*" 269–310.

Russo, E. B. and A. Dougherty (2013). *Herbal voices: American herbalism through the words of American herbalists*, Routledge.

Sams, T. (2015). *Healing Herbs: A Beginner's Guide to Identifying, Foraging, and Using Medicinal Plants/ More Than 100 Remedies from 20 of the Most Healing Plants*, Fair Winds Press (MA).

Sarkar, M. and P. Chawla (2022). Curative Properties of Chamomile in Gastrointestinal Disorders. *Herbs, Spices, and Medicinal Plants for Human Gastrointestinal Disorders*, Apple Academic Press: 125–140.

Sayed, S. F. (2023). Herbal drugs as antibiotics. *Antibiotics-Therapeutic Spectrum and Limitations*, Elsevier: 479–532.

Schiødt, S. Z. F. Ä. S. U. A. (2023). "*"Its Leaves Are Like the Daughter of the Willow": The Herbal Genre in Ancient Egypt.*" 150(2): 263–280.

Sharma, A., et al. (2021). Herbal medicine—An introduction to Its history. *Herbal Medicine in Andrology*, Elsevier: 1–8.

Sheng, P., et al. (2019). "*Archaeobotanical evidence for early utilization of cockleburs (Xanthium strumarium L., Asteraceae) in the Xinjiang Uyghur Autonomous Region of China.*" 11: 2027–2038.

Singh, A. K. P. G. R. (2016). "*Exotic ancient plant introductions: part of Indian 'Ayurveda' medicinal system.*" 14(4): 356–369.

Stapley, C. (2023). *A History of Plant Medicine: Western Herbal Medicine from the Ancient Greeks to the Modern Day, Aeon Books.*

Staub, P. O., et al. (2016). "*Back to the roots: A quantitative survey of herbal drugs in Dioscorides' De Materia Medica (ex Matthioli, 1568).*" 23(10): 1043–1052.

Surve, M. V., et al. (2024). "*A review on current scenarios of pharmaceutical and herbal medicine and future prospects.*" 27(2): 049–059.

Tachjian, A., et al. (2010). "*Use of herbal products and potential interactions in patients with cardiovascular diseases.*" 55(6): 515–525.

Thakkar, S., et al. (2020). "*Regulatory landscape of dietary supplements and herbal medicines from a global perspective.*" 114: 104647.

Tobyn, G., et al. (2010). *The Western Herbal Tradition E-Book: The Western Herbal Tradition E-Book*, Elsevier Health Sciences.

Upadhyay, B., et al. (2007). "*Traditional uses of medicinal plants among the rural communities of Churu district in the Thar Desert, India.*" 113(3): 387–399.

Verma, N. J. C. R. and R. Affairs (2016). "*Current regulatory challenges and approaches in the registration of herbal drugs in Europe*." 33(1): 9–24.
Voeks, R. A. A. O. T. A. O. A. G. (2004). "*Disturbance pharmacopoeias: medicine and myth from the humid tropics*." 94(4): 868–888.
Von Klein, C. H. J. O. T. A. M. A. (1905). "*The Medical Features of the Papyrus Ebers*." 45(26): 1928–1935.
Wachtel-Galor, S. and I. F. Benzie (2012). "*Herbal medicine: an introduction to its history, usage, regulation, current trends, and research needs*."
Wahlberg, A. (2006). *Modernisation and its side effects: an inquiry into the revival and renaissance of herbal medicine in Vietnam and Britain*, London School of Economics and Political Science.
Weindling, P. J. M. T. H. (2004). "*From germ theory to social medicine*." 239.
Winkelman, M. J. J. O. A. M. (2022). "*An ethnological analogy and biogenetic model for interpretation of religion and ritual in the past*." 29(2): 335–389.
Zhan, M. J. S. T. H. V (2014). "*The empirical as conceptual: Transdisciplinary engagements with an "experiential medicine"*." 39(2): 236–263.
Zhang, J., et al. (2012). "*Quality of herbal medicines: challenges and solutions*." 20(1–2): 100–106.
Zhao, J., et al. (2019). "*Improvement strategies for the oral bioavailability of poorly water-soluble flavonoids: An overview*." 570: 118642.
Zhou, X., et al. (2019). "*Current status and major challenges to the safety and efficacy presented by Chinese herbal medicine*." 6(1): 14.
Žuškin, E., et al. (2008). "*Ancient medicine-a review*." *Acta Dermatovenerologica Croatica*, 16(3): 0–0.

3 Modern Techniques in Herbal Extraction and Analysis

Sadaf Javaria
Institute of Food science and Nutrition, Gomal University, D.I. Khan, Pakistan

Muhammad Zareef
School of Food and Biological Engineering, Jiangsu University, Zhenjiang, China

Muhammad Nadeem
Institute of food science and Nutrition, University of Sargodha, Sargodha, Pakistan

3.1 INTRODUCTION

Since ancient communities, phytochemicals from plant sources have benefited many communities around the world. Healthcare systems use metabolites as drugs in the management of various diseases. Additionally, phytochemicals are utilized as lead compounds in the synthesis of medicines. The chapter gives an overview of updated methodologies for the extraction, isolation, and analysis of bioactive compounds from herbal sources in relation to the keen interest in therapeutic value concerning naturally occurring phytochemicals of plant origin (Abubakar & Haque 2020). Products of herbal origin in medical applications have been used since time immemorial; however, modern technologies drove this very ancient practice into a new era of precision (Dastangoo et al., 2020). It calls, therefore, for the adoption of more sensitive and advanced extraction methods in the identification and exploitation of this bank of phytochemicals in herbs that can be used as medicine (Abhari & Khaneghah 2020). The chapter thus acts as a single tool for researchers, practitioners, and aspirants in the industry who mean to direct state-of-the-art methodologies in herbal extraction and analysis toward drug discovery, quality control, and the development of evidence-based herb formulations. It has done so by pulling together theoretical underpinning, helpful methodologies, and incipient trends.

3.2 HERBAL EXTRACTION

Extraction in herbalism is one of the most important processes in releasing the potent core of botanical blessings. To harness the potential of herbs, one needs to know and work under the rules of extraction. Here, it is defined what the basic conception of herbal extraction, solvent, variables of the extraction efficiency, and principles are.

3.2.1 Principles of Extraction: Unlocking Nature's Bounty

Essentially, extraction is a procedure of separating the active ingredient from the matrix of the plant. The technique can be proven to arise out of the basic concepts of physics and chemistry. The basic premise of extraction is the concept of solubility, which can be described as the capability of

 DOI: 10.1201/9781003544418-3

a compound to dissolve well in a solvent. Hence the nature of what is extracted, and the amount of its extraction from the matrix into the solvent, is based on the extractability of the solute in the solvent. For example, alkaloids and flavonoids are pungent compounds that find solace in dissolving in polar solvents such as ethanol or water. On the other hand, more non-polar substances from the plants, which are essential oils, are effectively extracted by a non-polar solvent like hexane or ether. Another subsection of the principal part is diffusion, which is the process by which the molecules go from an area of high concentration to one of lower concentration (Brunner, 2005). It is as if they are spreading out to achieve a uniform concentration. This helps to ensure that the diffusion process extracts all of the compounds from the plant material. Among the basic scientific concepts that arise from this focus on the areas of chemistry and physics are solubility, diffusion, and mass transfer. These are essentially the principles employed to extract the active compound from herbs. Let us delve a little further into each of these topics.

1. **Solubility**: Solubility refers to the ability of a solute to dissolve in a solvent to form a solution. Each solute and solvent pair has a specific saturation point, beyond which no further solute can dissolve. During plant material extraction with a solvent, the release of plant material involves the extraction of various compounds. The quantities of these extracted compounds depend on their solubility in the chosen solvent. The degree of solubility depends on their composition, with different substances having different solubility. Meanwhile, nonpolar compounds, such as essential oils, are well extracted in nonpolar solvents such as hexane or ether, whereas polar compounds, such as alkaloids, flavonoids, and sugars, prefer better solubilization in polar solvents like water or ethanol (Li et al., 2012). Selecting the solvent for best extraction requires knowledge of the nature of solubility of the target compound.
2. **Diffusion**: This is the process by which the solute or, in some cases, even the solvent molecules are pushed into the solution from a high concentration of the solute to a lower concentration of the conditioning solvent. Importantly, this process works toward equilibrating the solutes and conditioning solvents, hence pushing the solutes in the solvent into it (Adegbola et al., 2017). This allows complete extraction of plant material in compounds in herbal extraction. During the extraction process, targeted compounds are dissolved into the solvent phase from the plant matrix. The rate of diffusion depends on numerous factors, such as particle size, the characteristics of the solvent, temperature, and agitation (Afroz et al., 2015). To control and optimize these mentioned factors improves the diffusion and hence an increase in the extraction efficiency.
3. **Mass Transfer**: This process refers to the transfer of solutes among phases, such as from solid plant material to a liquid solvent. The mechanism of mass transfer between solid plant material and the solvent phase, thus controls the dissolution and diffusivity mechanism of the target chemicals in terms of their extraction from the plant matrix and During desorption, the chemical compounds are released from the plant surface into the solvent For instance, dissolution is the process that relates to the solubilization of molecules in a solvent; and diffusion is what allows the molecules to move around freely in the solvent (Chen et al., 2011). In the same article, Chen and his colleagues proceeded to describe desorption process as chemicals migrating from the plant's surface and moving into a liquid.

The net effect of this means, therefore, is that efficient mass transfer will proceed when the principles are followed by high extraction yields because a value addition aspect will make extractions of the targeted species very complete (Dababi, et al., 2020). It may also repay even those herbalists who want to extract more than they need of some components without contaminating their extracts with unwanted substances. Indeed, this optimization of the extraction conditions, based on parameters such as solubility by diffusivity and mass transport rate, may enable the attainment of bioactive chemical-rich extracts that have been shown in the past to produce desirable medicinal effects.

3.2.2 Choice: Solvent Selection of a Suitable Medium

Menstruum is a solvent that is used for the extraction of compounds in the medicinal plant. Selecting an appropriate solvent is one of the best solutions to proper extraction efficiency. The choice of menstruum depends very much on the type of the herb, on the part of the herb from which extraction is to be done, and the nature of the biologically active compounds. Generally, polar solvents such as water, methyl alcohol, or methanol and ethyl alcohol or ethanol are used for the extraction of polar compounds; for nonpolar compounds, hexane, and dichloromethane are used. In liquid-liquid extraction, two miscible solvents are taken, for example, water and dichloromethane; water-hexane; water-ether, etc. In all mixtures, however, water is an indispensable component because of its high polar power and its miscibility with organic solvents (de Menezes Rodrigues et al., 2017).

Properties of the Solvent

Water: Water is considered a solvent that is universal. In the extraction of polar compounds, water is considered to be the most popular polar solvent. It is cheap, dissolves a broad range of solvents, is highly polar, and is non-flammable and non-toxic (Zahari et al., 2020). However, it also has some disadvantages: it promotes microbial growth, it may cause hydrolysis and a large amount of heat is required in case of concentrating the extract of water.

Alcohol: Alcohol is polar in nature like water, also. The miscibility in water is excellent and has a property of extracting polar secondary metabolites. When its concentration exceeds 20%, then it acts as a natural self-preservative, but it is nontoxic in lower concentrations. Unlike water, a little heat is called for in extracting the concentrate but none of the fats, waxes, and gums are soluble in it. Alcohols are volatile and flammable.

Chloroform: This is non-polar in nature. It is extensively used in the extraction of fats, oils, terpenoids, and flavonoids, and so on. Chloroform is soluble in alcohols, has a sweet smell, and is colorless. However, used widely and in large quantities, it has a mild sedative effect. It has recently been reported to have carcinogenic properties (Zhang et al., 2018).

Ether: The solvent is non-polar and has been used in extracting alkaloids, terpenoids coumarins and fatty acids. It is water-soluble and tasteless, and has a low boiling point. It is also a very stable and not reactive compound given that it doesn't react with acids, bases, and metals. However, it is very volatile and flammable (Díaz-Batalla et al., 2006).

Extraction efficiency: factors affecting maximum yield and quality Solvent is an important factor influencing efficiency, selectivity, and safety in the extraction process, and thus critically affects overall herbal extraction. In pharmacopeia, the alcohols used as solvents are specified, but various parameters have to be considered in selecting a solvent appropriate for extracting with herbs (Diep, et al., 2020). We shall consider them in detail.

Herb extraction is a complex process in which many levels of interactions blend to finally determine how well one can extract targets out of one's material. High efficiency in extraction is essentially required to maximize yield, keep optimum product quality, and, above all, to minimize the use of resources.

The following factors are responsible for this phenomenon:

1. **Particle Size**: This is the particle size for which the herbal material is composed that dominates the efficiency of extraction.

   Crushing to a fine powder exposes a large surface area for the permeation of the solvent into the particle and also increases the contact of the solvent with the target compounds or solutes. Since divalent ions and carboxylates cannot interdiffuse, a higher surface needs to be achieved. The finer particle sizes present an increased rate and higher yield of extraction compared to a larger surface area. For this reason, the herbal materials must be ground or milled accordingly so as to maximize the efficiency of their extraction.

2. **Temperature**: The temperature is known to significantly influence kinetics involved in extraction by changing solvent solubility and diffusivity, apart from chemical reactions.

   Generally, high extraction temperatures enhance the rate of mass transfer by increasing the solvent's capacity to dissolve solutes, thereby accelerating the kinetics of the process. However, if the temperature goes beyond a certain limit, the destruction of the thermolabile stuff occurs, causing the synthesis of unwanted byproducts (Dymek et al., 2021). Optimal temperatures have to be selected carefully between target compound stability and its rate of extraction within a reasonable period.
3. **The Extraction Time**: This clearly determines the yield and composition of the final extract. The extraction times clearly determine the concentration of extracted compounds. Unwanted compounds may be extracted or degradation of heat-labile components takes place within an extended extraction time (Fomo et al., 2020). Crude time, such as a long extraction time in favor of a longer duration, allows for the diffusion and equilibrium of solutes. Balance of these two assures best possible yield with good quality extracts.
4. **Solvent Volume-to-Raw Material Ratio**: The ration of the volume of the solvent to the material plays a factor in terms of how well the compound will be extracted, as this will determine the velocity and penetration of the solvent into the particles and hence will simultaneously affect the rate of dissolution of the compound.

   There is a general rule of thumb that dictates that the greater this ratio of solvent to material used, the more efficient the extraction will actually be, because the available quantity of solvent is increased as is the mass transfer rate (Geng et al., 2015). However, this tends to escalate processing costs and environmental concerns in case too many solvents are used for that purpose. The optimization of the ratio between the solutes and the liquid medium is required in order to get the maximum extraction efficiencies but also the minimum solvents for the extraction process.
5. **pH**: The pH of the extraction medium mostly affects the ionization state and the solubility of some of the compounds amounting to their extraction efficiency.

   At specific pH levels, some substances, in particular, exhibit high solubilities whereas others undergo chemical decompositions or precipitation. Modification of the pH of the extracting solvent to be in line with this range can hence be used to enhance extraction selectivity as well as efficiency. Alkaloids, phenolics, and organic acids are mostly picked in herbal substances (Zygler et al., 2012).
6. **Agitation**: In mass transfer, we need agitation to achieve adequate permeation of the solvent through the plant matrix, thereby enabling the effective exposure of the desired compounds during extraction.

   This will, however, be greatly enhanced by mechanical stirring, shaking, or agitation through the use of ultrasound, among others, which increases the rate at which extraction occurs because of increased surface area available for contact with the solvent and, therefore, the diffusing mass transfer resistance (Omeroglu et al. 2019). The proper observation of such agitation techniques allows the proper mixture circulation, thereby achieving an even mixture and maximizing yields from the extractions process.
7. **Plant Variability**: Different types of plants, variation within plant species, how they are grown, and how they are harvested all lead to a difference in the composition and concentration of beneficial compounds in the herbal materials.

   This makes the extraction process complicated; hence, it is just as important to decide on plants under standardized methods of extraction so that there is always consistency in quality and effective results. These criteria include the age of the plants, its habitat, and the way it is to be handled after harvesting (Omeroglu et al. 2019).
8. **The Method of Extraction**: The extraction method to be used for the extraction of compounds from your herbal material is of great concern with regard to efficiency in extraction.

There are some commonly available methods, which include maceration, percolation, Soxhlet extraction, supercritical fluid extraction, and ultrasound-assisted extraction. Each method, therefore, achieves a balance between efficiency, selectivity, and processing time (Tsao and Deng Z., 2004). The selection of the appropriate extraction method in relation to the nature of the herbal material and the desired product characteristics are thus very essential and play a crucial role in gaining good results from an extraction process.

1. Modern Extraction Techniques
    - **Supercritical Fluid Extraction (SFE)**
    - **Microwave-Assisted Extraction (MAE)**
    - **Ultrasound-Assisted Extraction (UAE)**
    - **Pressurized Liquid Extraction (PLE)**
    - **Subcritical Water Extraction (SWE)**

3.3 SUPERCRITICAL FLUID EXTRACTION (SFE)

Supercritical Fluid Extraction (SFE) is a technique which is used for the extraction of specific components from the source or material, in which supercritical fluids act as the extracting solvent. In other words, SFE is a process of separating extract from the matrix with the help of supercritical fluids which act as an extracting solvent.

A supercritical fluid is an enormously compressed fluid that signifies the properties of both gases and liquids. They are created by raising temperature and pressure beyond the critical point of a substance. These have high density and are non-compressible (Shinde et al., 2009).

Supercritical fluids (SF) have exclusive characteristics whereby they can diffuse through solids like a gas and simultaneously be able to dissolve materials like a liquid. The most common supercritical fluids in use within modern industry are carbon dioxide and water. Carbon dioxide is the SF which is most commonly used for extraction purposes, because it has relatively low critical temperature and pressure. It is also non-toxic and readily available.

3.3.1 Working Principle of SFE

The working principle of Supercritical Fluid Extraction relies on the unique characteristics of supercritical fluids. When the substance is heated above its critical temperature and critical pressure, it changes into a supercritical state where it shows the characteristics of both liquids and gases. At this stage, the fluid can penetrate the solid material and can dissolve the desired components effectively.

3.3.2 Parts of the SFE System

1. **Pump**: A pump is used to pressurize the carbon dioxide and convey it to the extraction vessel.
2. **Extractor**: The extraction vessel in which the raw materials (the herbs) are kept. Here the supercritical carbon dioxide interacts with the material for the extraction of the desired compounds.
3. **Heater**: The heater maintains the desired temperature at which the carbon dioxide remains in its supercritical state.
4. **Separator**: Once the extraction process is completed, the carbon dioxide and the extracted compounds are transferred to the separator where the pressure is decreased, and the carbon dioxide is reverted to the gas, leaving the extract to separate out.
5. **Condenser**: This apparatus condenses the carbon dioxide gas back into its liquid form for reuse in the system.
6. **Flow Meter**: This monitors and controls the rate of carbon dioxide.

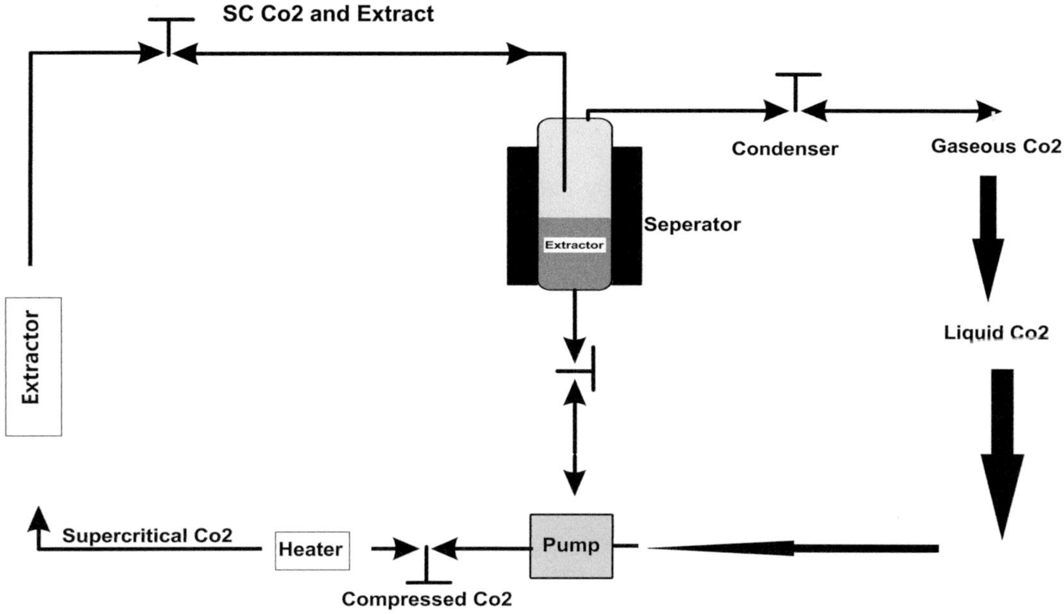

FIGURE 3.1 Method of extraction from herbs using SFE.

3.3.3 Process of Extraction

1. **Selection of Herbs**: For extraction, the herbs which pose essential oils or active compounds are selected.
2. **Preparation**: In the SEF of herbs fresh plant material is used commonly. When a fresh sample is extracted from herbs, it contains a high degree of moisture. This could result in mechanical difficulties, for example, the restrictor may become clogged because of ice formation. To avoid this problem, the solution is mixed with anhydrous Na_2SO_4 (Figure 3.1).
3. **Grinding**: The raw material (herbs) is ground into a fine powder to enhance surface area for extraction.
4. **Loading**: The powdered herbs are loaded into the extractor vessel through a pump which exerts high pressure.
5. **Pressurization**: CO_2 is pumped into the extractor vessel and heated to reach its supercritical state.
6. **Extraction**: The supercritical CO_2 penetrates the herb matrix and dissolves the target compounds.
7. **Separation**: The mixture of CO_2 and dissolved compounds is transferred to the separator, where the pressure is lowered. The CO_2 becomes gaseous and separates from the extracted compounds.
8. **Collection**: The extract is collected from the separator.
9. **Recycling**: The gaseous CO_2 is condensed back to a liquid state and recycled back into the system.

3.3.4 Applications

1. It is used for the extraction of compounds used in flavours, fragrances, and essential oils.
2. It is used for the extraction of bioactive compounds from herbs, which are used in nutraceuticals and pharmaceutical products.
3. It is also used for the purification of natural compounds and products

3.4 MICROWAVE-ASSISTED EXTRACTION (MAE)

Microwave-Assisted Extraction (MAE) is a method which is used for the extraction of bioactive compounds from raw materials such as herbs and plants through the use of microwave energy. Microwave energy has the strong ability to warm up solvents and plant materials in a very short time for the extraction of required compounds. Heat is generated following ionic conduction and dipole rotation mechanism through microwaves by interacting with polar compounds for example water and organic components in the plant medium. In MAF, heat and mass are transferred in the same path which creates a synergistic effect, this quickens the extraction process, and extraction yield is also improved. It is also regarded as a "green technology" because in the MAE process the usage of organic solvent is decreased. There are two types of Microwave-Assisted Extraction method:

1. Solvent-free extraction: usually used for volatile compounds.
2. Solvent extraction: usually used for non-volatile compounds.

3.4.1 Working Principle

The MAE works on the principle that with microwave radiations polarizable materials and dipoles of polar solvent are changed. Microwaves are a type of electromagnetic radiations which have frequencies ranging from 300 MHz to 300 GHz. The polar molecules and ions in the solvent and plant material are oscillated when microwaves are applied to the plant material and heat is produced through dipole rotation and ionic conduction. The mass transfer rate, breaking the cell walls of plant material and releasing the required compounds into the solvent, are enhanced due to this localized heating .

3.4.2 Components of a Microwave-Assisted Extraction System

1. **Microwave Generator**: This part of microwave-assisted extraction system generates microwaves of a specific frequency. A frequency of 2450 MHz is generally used for MAE.
2. **Microwave Cavity**: It is extraction chamber. The microwave cavity is made to contain and homogeneous reflection of microwaves.
3. **Sample Holder**: This is a container which holds the plant/herb material and extraction solvent during the process of extraction.
4. **Temperature and Pressure Sensors**: The temperature and pressure inside the extraction chamber can be monitored and controlled to provide and ensure the optimum extraction conditions.
5. **Cooling System**: This keeps the temperature within safe limits to avoid the decomposition or degradation of required compounds.
6. **Magnetic Stirrer**: This ensures the homogenous mixing of the solvent and sample to increase extraction efficiency.
7. **Control Unit**: This is the unit that manages temperature, pressure, microwave power and extraction time settings.

3.4.3 Method of Extraction from Herbs by MAE

1. **Preparation of Plant Material**: Herbs are washed, dried, and ground up to make a fine powder to increase the surface area for extraction. Dried plant material still contains traces of moisture, which acts as a target for microscopic heating.
2. **Selection of Solvent**: A proper and well-suited solvent which must be transparent is chosen on the basis of polarity of required compounds. The solvent used must be transparent to microwave radiation to ensure efficient absorption by the sample. Commonly used solvents include water, ethanol, methanol, or their mixtures.

Flow chart of MAE process

Application of microwave radiations
↓
Heating and evaporation of moisture
↓
High pressure application on the cell wall of plant material
↓
Swelling and rupturing of cell wall of plant material
↓
Oozing out of required components.

FIGURE 3.2 Flow chart diagram of Ultrasound-Assisted Extraction (UAE).

3. **Loading the Sample**: In the sample holder, a prepared sample of herb is placed along with the chosen solvent.
4. **Microwave Treatment**: The cavity sample holder is placed in the microwave. The microwave generator is turned on and the microwave increases the temperature of the sample quickly by penetrating in it. This results in the dehydration of the cellulose of the plant material as the temperature increases. This reduces its mechanical strength.
5. **Extraction Process**: The cell walls of the herb's material are ruptured due to localized heating. It helps to release the bioactive compounds into the extraction solvent. The yield and quality of the extracted compounds could be enhanced by the adjustment of extraction time and microwave power (Figure 3.2)
6. **Cooling and Filtration**: The mixture is cooled after extraction is completed. Then it is filtered to separate the plant residue from the solvent containing the extracted compounds.
7. After the extraction process, the mixture is allowed to cool. It is then filtered to separate the solvent containing the extracted compounds from the plant residue. After the extraction process, the mixture is allowed to cool. It is then filtered to separate the solvent containing the extracted compounds from the plant residue.
8. **Concentration and Purification**: The obtained extract could be concentrated by evaporating the solvent, while the isolation of desired compounds will require additional purification steps.

3.5 ULTRASOUND-ASSISTED EXTRACTION (UAE)

Ultrasound waves: the waves which have exceeded the audible frequency range > 20 kHz are designated as ultrasound waves. When these waves pass and spread through the medium, the compression and rarefaction of particles of the medium results. UAE technique could be combined either with temperature called "thermosonication" or pressure "manosonication."

Ultrasound-assisted extraction is the technique in which the application of ultrasonic waves improves the extraction of bioactive compounds from a wide range of material herbs. This method improves efficiency and extraction yield by using sound waves of high frequency which create cavitation bubbles in the solvent used for extraction. It then bursts on the surface of plant matrix, a shock damage to plant call wall increases the mass transfer of required compounds from the call membrane into the solution.

3.5.1 Working Principle

This phenomenon of UAE works on the principle of cavitation. This phenomenon is created when ultrasonic waves pass through a liquid; cycles of high and low pressure are developed. In the cycle of low pressure, small vacuum bubbles are formed in the liquid. These bubbles grow over successive cycles until they reach a critical size. At that time, during a high-pressure cycle, the bubbles burst viciously. And this produces strong energy. This leads to:

3.5.1.1 Cell Disruption

The collapse of cavitation bubbles disrupts the plant cell walls, thereby increasing the contact area between the applied solvent and the plant material.

3.5.1.2 Increased Mass Transport

Due to the burst of cavitation bubble, the micro-jets and shock waves are created which increases the penetration of applied solvent into the plant matrix and bioactive compounds ooze out.

3.5.1.3 Enhanced Solvent Effectiveness

The localized exerted pressure and temperature can increase the solubility of required compounds in the solvent.

3.5.2 Parts of the Ultrasound-Assisted Extraction System

1. **Ultrasonic Generator**: It converts the electrical energy into high-frequency ultrasonic waves.
2. **Transducer**: The electrical ultrasonic waves produced by the generator is then converted by the transducer into mechanical vibration
3. **Sonotrode (Probe)**: This probe is immersed in the extraction solvent. It transmits the generated ultrasonic energy into the liquid.
4. **Extraction Vessel**: This vessel holds the solvent and plant material. It could be a simple vessel or more complex setup with temperature and pressure controls.
5. **Cooling System**: This is needed in keeping the temperature low in the extraction system to prevent the thermal degradation of temperature-sensitive compounds.
6. **Power Supply**: This provides energy to the ultrasonic generator and transducer.

3.5.3 Method of Extraction from Herbs

1. **Preparation of Plant Material**: Drying and grinding of the herbs is carried out. The plant material is reduced to particle size to provide a greater surface area for extraction.
2. **Selection of Solvent**: Suitable solvent is chosen depending on its polarity and solubility of target compounds. This might include, for example, methanol, water, ethanol etc.
3. **Loading of the Extraction Vessel**: The powdered herb material and the selected solvent are loaded into the extraction vessel.
4. **Ultrasonication**: The ultrasonic probe is submerged into the solvent. The ultrasonic generator is switched on to generate the ultrasonic waves, which produce cavitation in the solvent.
5. **Extraction**: Extraction time, temperature required and herb sample are optimized on the basis of the specifications of the extraction process. This process may take either a few minutes or several hours.
6. **Filtration and Separation**: The mixture obtained after extraction is filtered to remove all solid remains; the filtrate contains the required bioactive compounds.
7. **Concentration and Purification**: Concentration or purification can be achieved by the use of various techniques such as evaporation, chromatography, or distillation (Figure 3.3).

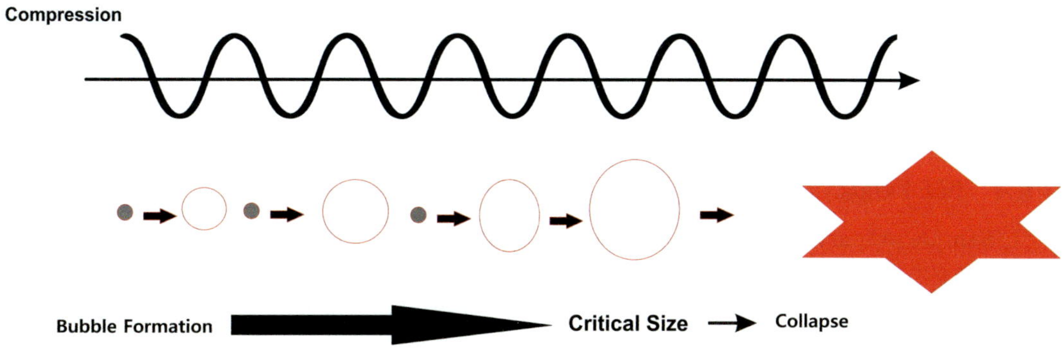

FIGURE 3.3 How the plant cell bursts by ultrasonic waves.

3.6 PRESSURIZED LIQUID EXTRACTION (PLE)

3.6.1 Definition

Pressurized Liquid Extraction (PLE), also known as "Accelerated Solvent Extraction," is the process of extracting required bioactive compounds from solid and/or semi-solid samples by using solvents at high temperature and/or high pressure. The efficiency and speed of the extraction process are greatly boosted by the technique, which is driven by enhanced solubility and the mass transfer of target compounds (Goettel et al., 2013).

3.6.2 Working Principle

PLE operates in conditions in which the solvent is maintained above its boiling point but below its critical point. The high temperature enhances the capacity of the solvent to dissolve the required compounds. While, the high pressure prevents the boiling off of the solvent and keep it in the liquid state. This combination of temperature and pressure improves extraction efficiency in:

1. **Improving Solubility**: The solubility of the required phyto-compounds in the solvent is increased by higher temperatures.
2. **Improvement in Mass Transfer**: Increased temperature and pressure reduces the surface tension and viscosity of the extraction solvent. It improves the diffusion of the extraction solvent in the plant matrix and the release of required compound.
3. **Reducing Extraction Time**: The improved mass transfer and solubility result in a faster extraction rate, which reduces the overall extraction time.

3.6.3 Parts of the PLE System

1. **Solvent Reservoir**: It holds the extraction solvent used for extraction.
2. **Pump**: This is also known as delivery system. It adds the solvent to the SPE cartridge at a controlled flow rate and pressure.
3. **Extraction Cell**: This is used to hold the plant sample and the solvent. It is designed to withstand very high temperatures and pressures.
4. **Heater**: This heats the extraction solvent to the chosen temperature before it reaches the extraction cell.
5. **Pressure Regulator**: This is used to maintain the required pressure within the extraction system.

6. **Collection Vials**: These are vessels which collect the extract after it has passed through the extraction cell.
7. **Control Unit**: This is used to monitor and control all of the changes which take place together with the control of pressure, temperature, and flow rate during the extraction process.

3.6.4 PLE Extraction Method

1. **Sample preparation**: The plant sample is dried, ground up, and then kept into the extraction cell. The cell is then tightly sealed.
2. **Solvent selection**: Keeping in view the polarity and solubility of required compound, the solvent is selected.
3. **Selection of solvent**: A proper selection of the solvent based on the polarity and solubility of the analytes of interest is made.
4. **System setup**: It involves setting up the system with the extraction cell loaded with samples. A heater, collection vials, and filling of the solvent reservoir are the parts of this system.

Heating and Pressurization: The required temperature for the heating of extraction solvent is maintained. It is then pumped at high pressure into the extraction cell.

Extraction: In the extraction cell, the extraction solvent streams through the sample and dissolves the required compounds. The increased temperature and pressure boost the extraction efficiency.

Collection: The solvent with the extracted required compounds leaves the extraction cell and is collected in collection vials.

Repeat Cycles: This process is repeated a number of times with a fresh extraction solvent to confirm the complete extraction of required compounds.

Post-Extraction Processing: The collected extract of required compounds may then be concentrated or further purified by using any techniques such as evaporation, chromatography, or distillation (Figure 3.4).

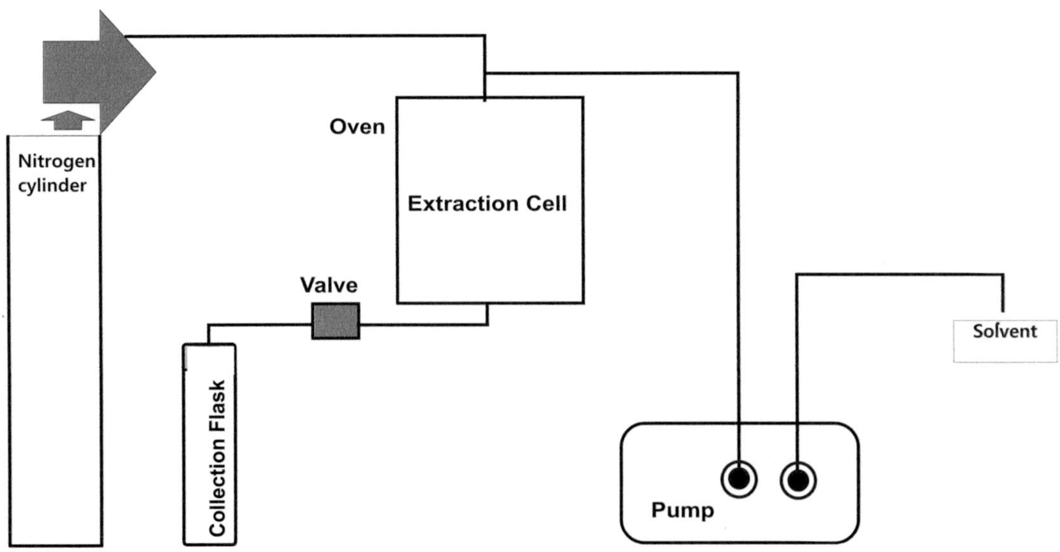

FIGURE 3.4 Process diagram of Subcritical Water Extraction (SWE).

3.7 SUBCRITICAL WATER EXTRACTION (SWE)

3.7.1 Supercritical fluids

The supercritical state is a state of any matter that occurs when any substance or matter is brought to the zone of its "critical point," at which it experiences high temperature and pressure. At this stage the substance shows intermediate characteristics of both liquid and gas and acts both like a "reactant" and like a "catalyst." This state of substance is described as a supercritical fluid.

3.7.2 Supercritical Fluid Extraction (SFE)

This is a sample extraction method which is used to extract the required compound from the sample by using the special characteristics of supercritical fluids. Supercritical fluids become diffused quickly like gases and can be dissolved as easily as liquids.

Subcritical Water Extraction (SWE): This is one of the most advanced techniques, and is used for the extraction of required compounds from herbs. In this technique, water is used as a supercritical fluid (at between 100°C and 374°C and high pressure) to extract the required compounds from herbs. This method takes advantage of the distinctive solvent properties of supercritical water.

3.7.3 Working Principle of Subcritical Water Extraction (SWE)

The working principle of SWE is based on the fact that the behavior of water in the subcritical phase is unique. When temperature of water is increased to between 100°C and 374°C and pressure is increased (to keep the water in liquid state) its "dielectric constant" is decreased, resulting in a decrease in polarity. This subcritical water dissolves more compounds than is the case when cold water is used.

3.7.4 Parts of the Subcritical Water Extraction System

1. **Extraction Vessel**: Herb material is placed in the extraction vessel. This vessel should be made in a manner to withstand high temperature and pressure.
2. **Heater**: The heater is used to heat the water to acquire the subcritical temperature.
3. **Pump**: The pump maintains the higher pressure to ensure that the water doesn't boil and that it remains in the supercritical phase during extraction.
4. **Cooling System**: After extraction, this is used to separate the extracted compounds from water. The solution is cooled within this system.
5. **Separator**: The "separator" separates the extracted compounds from the water.
6. **Control System**: This system controls the temperature, pressure, and flow rate and maintain them to ensure have the best conditions of extraction (Figure 3.5).

3.7.5 Process of Subcritical Water Extraction

1. **Preparation of Plant Material**: The sorting, cleaning, and drying of herb material is carried out. This plant material is then ground into fine powder so that the surface area could be increased for extraction.
2. **Loading the Extraction Vessel**: The sample is then loaded into the extraction vessel and water is added into it.
3. **Heating**: The water is heated to its supercritical stage; it then acts as a supercritical fluid. The system heats the water to the subcritical range. The water acts as a solvent in its subcritical state, effectively dissolving the bioactive compounds.

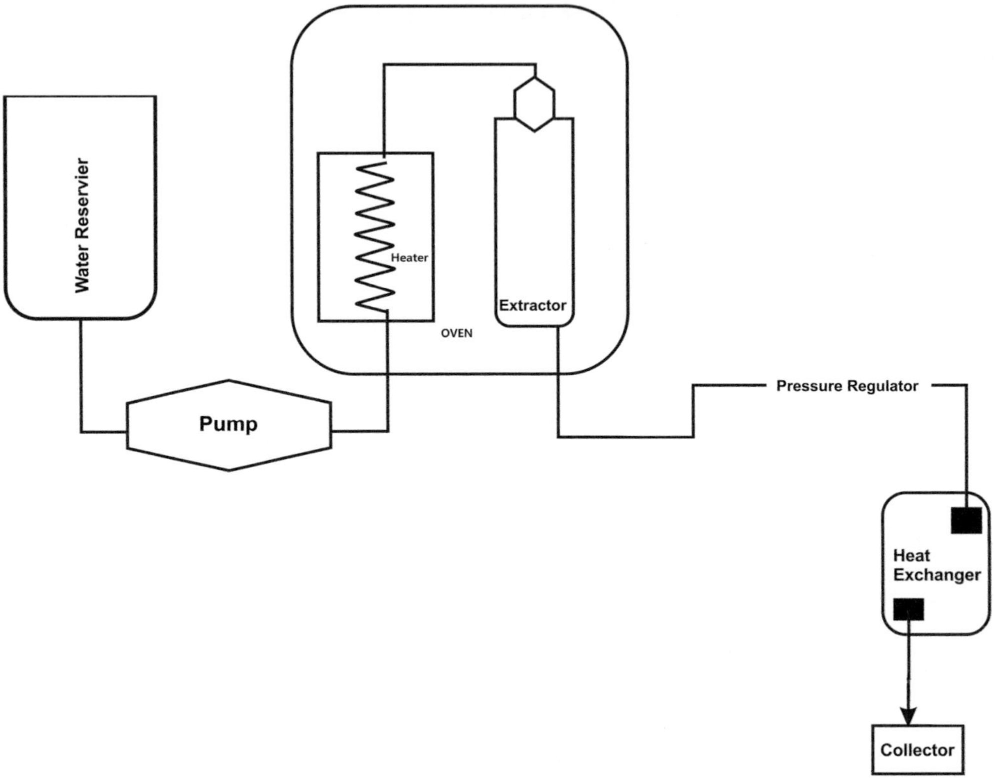

FIGURE 3.5 Flow chart diagram of Pressurized Liquid Extraction (PLE).

4. **Extraction**: The water, which is hot and pressurized, is passed through herbal material. The heated, pressurized water passes through the plant material, extracting the desired compounds.
5. **Separation and Cooling**: The extracted material is cooled down, and bioactive compounds are precipitated or separated out from the water.
6. **Collection**: The extract is cooled down; this process precipitates or separates the bioactive compounds from the water.

Analytical Techniques for Herbal Analysis

- High-Performance Liquid Chromatography (HPLC)
- Gas Chromatography-Mass Spectrometry (GC-MS)
- Liquid Chromatography-Mass Spectrometry (LC-MS)
- Fourier Transform Infrared Spectroscopy (FTIR)
- Nuclear Magnetic Resonance (NMR) Spectroscopy

3.8 HIGH-PERFORMANCE LIQUID CHROMATOGRAPHY (HPLC)

The use of herbs in medicine is as ancient as humankind. Medicinal herbs have vast usage, not only used as just medicine, but also in the cooking and perfumery purposes. Herbs generally are medicinal, culinary, and aromatic plants rich in bioactive compounds and phytochemicals that form the development row of therapeutic effects. The identification and valuation of these compounds set the core of health benefits as well as quality control for herbal products. High-performance liquid chromatography (HPLC) is the very efficient analytical technique used in the separation, identification, and quantitation of bioactive compounds in complex herbal matrices. The principles of HPLC and updates in its application to the separation of bioactive compounds in herbs are outlined in this paper.

3.8.1 Principles of HPLC

Basically, HPLC is based on liquid chromatography, where the mobile phase—liquid—is carried in a column that has been filled with a stationary phase to carry out a sample. It is a separation-based technique in which the interaction of the sample with the stationary phase allows the separation of compounds based on different affinities. Key components making up an HPLC system include a solvent reservoir, a pump, an injector, a column, a detector, and a data acquisition system. Selection of the stationary phase is silica-based for the most part, and the mobile phase would be a mixture of solvents, relying upon the nature of compounds to be analyzed. RP-HPLC is a mode most commonly used to analyze the bioactive compounds in herbs: nonpolar stationary phases and polar mobile phases are utilized. Detection varies, but often includes UV-Vis detectors, photodiode array detectors, and mass spectrometry due to their high sensitivity and specificity.

Sample preparation is an essential part of HPLC analysis to achieve accurate and reproducible results. Herbals generally need to be dried and powdered into a fine powder prior to extraction. The preparation of solutions begins with the selection of the extraction solvent, taking into account the polarity of the target compounds. Generally utilized solvents fall in the class of water, methanol, ethanol, and mixtures of the above. Available methodologies for extraction are maceration, reflux, ultrasound-assisted extraction, microwave-assisted extraction for best results .

These extracted samples, however, block require further purification to remove interferences from the substances. Techniques such as SPE and LLE can be used to concentrate and purify the target analytes. The purified extract that remains is filtered, and is injected into the HPLC system for analysis.

3.8.2 Bioactive Compounds Analysis

3.8.2.1 Phenolic Compounds

Among the wide variety of compounds, great attention has been paid to phenolic compounds such as flavonoids, phenolic acids, and tannins in herbs since these are bioactive compounds. HPLC has been applied to a wide range of phenolic compound analyses. In general, RP-HPLC separation employs a C18 column with a mobile phase consisting of water and acetonitrile or methanol, often modified with an acid like formic acid or acetic acid to improve peak shape and resolution.

3.8.2.2 Alkaloids

Alkaloids are nitrogen-containing compounds that have very important pharmacological activities. When it comes to the analysis of alkaloids, generally HPLC methods use mobile phases with ion-pairing reagents so that such basic compounds are strongly retained and separated. Detecting these compounds will basically use UV-Vis and MS detectors, hence providing contributions to qualitative and quantitative analysis.

3.8.2.3 Terpenoids

Terpenoids make up a class of bioactive compounds with vast diversity in biological activities, and they include monoterpenes, sesquiterpenes, diterpenes, and others. Terpenoids are analyzed by HPLC on practically nonpolar phases with gradient elution for good separation. Additionally, since a good number of them have volatility, it is very advantageous to analyze them in HPLC coupled to MS.

Being one of the most common constituents in herbs, glycosides carry another non-sugar moiety with a sugar moiety. For the most part, glycosides exhibit very varied therapeutic effects. The usual methodology used to decipher the glycosides is RP-HPLC with gradient elution. The enzymatic hydrolysis is performed before the analysis to raise the sensitivity of the detection based on the release of an aglycone.

3.8.3 Recent Advances in HPLC Techniques

3.8.3.1 Ultra-High-Performance Liquid Chromatography

Ultra-high-performance liquid chromatography (UHPLC) provides a higher resolution compared with traditional HPLC. Besides, it reduces analysis time, as it is also quite sensitive. It is comprised mainly of a column with smaller particles, about sub-2 μm in diameter, used at higher running pressures, thereby providing better separation efficiency. Nowadays, this technique has gained popularity for the analysis of complex herbal extracts.

3.8.3.2 HPLC-MS

Thus, the coupling of HPLC with MS could help not only as a multiple separation tool but also as an atomic level structural elucidator. It would make it possible to identify those compounds not only with their mass-to-charge ratio (m/z) but also by the fragmentation patterns. In essence, HPLC-MS is a very powerful tool for the analysis of a group of compounds which are devoid of chromophores and therefore cannot be detected with UV-Vis detectors.

The major advantage of PDA detectors is that they record multiple wavelengths simultaneously, give additional spectral detail on the identity of the analyte. This is highly useful for the identification of co-eluting compounds and to calculate the purity of peaks. HPLC-PDA is utilized for both qualitative and quantitative analysis of phenolic compounds and other phytochemicals in herbs.

3.8.3.3 Chiral HPLC

Sometimes enantiomers are compounds related to chiral experimentations in HPLC. This kind of analysis, in analyzing bioactive compounds with a chiral center, is quite significant, because mostly the enantiomers can have very divergent biological activities. They use chiral stationary phases or chiral additives in the mobile phase to gain enantiore solution.

3.8.4 Applications of Herbal Medicine

3.8.4.1 Quality Control

The relation of the drug from the pharmacokinetic data, or the relation of the data, is often difficult to interpret to the pharmacologist; hence, HPLC plays an important role in the quality assurance of herbal products. It fills the gap in ensuring that consistency in herbal preparations is maintained as over-the-counter drugs, thus confirming the amount of bioactive compound present and their exact concentration for safety. It is also a very common practice in setting a benchmark in the standardization of herbal extracts toward common reference compounds to confirm quality assurance of the herbal products.

3.8.4.2 Pharmacokinetic

The pharmacokinetic study describes the absorption, distribution, metabolism, and secretion of bioactive compounds from herbal remedies into blood circulation. HPLC will also characterize their interaction with biological systems. Metabolomics will, thus, provide a holistic description of the bioavailability and therapeutic efficacy of herbal drugs in human health.

Metabolomics is a definition of the analytical activity of a biological system within a biological system. In metabolomic investigations, this is one of the principal tools combined with HPLC and either MS or NMR. It allows the profiling of phytochemicals in herbs and also finds biomarkers that are related to therapeutic action/toxicity.

HPLC, therefore, is very necessary in the process of the isolation and identification of potential candidates for new drugs. The high-throughput methods of HPLC are feasible for the quick screening of a large number of samples so that the speed of discovering new compounds can be greatly improved.

3.8.4.3 Challenges and Prospects for Further Study

Though HPLC is a powerful technique in terms of the analysis of bioactive compounds from herbs, it does suffer from some challenges. This might cause complications in herbal matrices in terms of the separation and identification of compounds. In addition, matrix effects are also well known to affect the quantification in terms of accuracy. In addition, high volumes of organic solvents are used during sample preparation, which brings an environmental impact into play and raises questions about the sustainability of the technique.

Future trends in research on herbs with HPLC embrace the design of more selective and sensitive detectors, the application of various green chemistry principles to bring reduction in the overall environmental footprint, and the incorporation of advanced chemometrics for the interpretation of complex data sets. HPLC, coupled with other techniques such as supercritical fluid chromatography and capillary electrophoresis, offers advanced resolution and speed for the analysis. HPLC analysis of bioactive compounds and phytochemicals from herbs provides one of the cornerstones in this context, given the techniques' versatility, precision, and complexity of its matrices. Therefore, this is a very important tool for research and quality control purposes in herbal medicine. Further improvements in the technology of HPLC, combined with a large number of new developed analytical techniques in which it is implicated, make it an increasingly powerful tool by which to bring out the therapeutic potential of the herbal compounds. This means that HPLC has been so important for the safety, efficacy, and consistency in herbal products with increased demands for the products.

3.9 GAS CHROMATOGRAPHY-MASS SPECTROMETRY (GC-MS)

Herbs have been used since ancient times not only as plants for diet, but also to improve human health and nutrition. They are an immense source of bioactive compounds and phytochemicals responsible for the therapeutic activity of herbs, and they possess a wide array of bioactive constituents, which include alkaloids, flavonoids, terpenoids, and phenolic compounds coupled with essential oils. The analysis and characterization of these compounds are important to understand their role in the promotion of health and the prevention of diseases. Amongst the various analytical techniques, GC-MS has been reported to be a promising tool for the qualitative and quantitative analysis of bioactive compounds in herbs. This chapter focuses on describing the principles, methodology, and applications of GC-MS analysis of phytochemicals from herbs.

3.9.1 Principles of GC-MS

GC-MS is an analytical technique which combines the features of Gas Chromatography (GC) with those of Mass Spectrometry (MS). A sample is vaporized and passed through a column that is coated with a stationary phase. As components will interact with the stationary phase differently, they will be separated based on their varying volatility and affinity.

The chromatographically separated compounds enter the mass spectrometer, in which they become ionized, usually by electron impact ionization. The ions are then detected after being separated based on their mass-to-charge ratio (m/z). The mass spectrometer produces a spectrum that shows the relative abundance of ions at each m/z value. It then yields a molecular fingerprint that can be used in identifying the compound .

3.9.2 Sample Preparation

Proper sample preparation is essential for accurate GC-MS analysis. Most of the herbs contain a matrix of compounds: both volatile and non-volatile. These should be isolated and concentrated before instrument injection. Common sample preparation techniques include:

1. **Solvent Extraction**: This process uses solvents like methanol, ethanol, or hexane to take out bioactive components from the herb matrix. The choice of the solvent used depends on the polarity of the target compounds.
2. **Steam Distillation**: The process that distills the herb with water to separate the volatile compounds is used to extract essential oils.
3. **Solid-Phase Micro Extraction**: Solid-phase micro extraction (SPME) is a solvent-free technique where volatile compounds from the herb matrix are adsorbed onto a fiber coated with a specific extracting phase.
4. **Supercritical Fluid Extraction**: SFE employs supercritical CO_2 to efficaciously lift out bioactive compounds. It's also employed widely in thermolabile substances.
5. **Instrumental Setups**: The general configuration of GC-MS comprises an injection port, a GC column, a mass spectrometer, and a data processing system. The injection port used is heated to a very high temperature for the purpose of the vaporization of the sample. The GC column is, in most cases, a fused silica capillary column and it has to be selected depending on the nature of the compounds to be analyzed. Such a column is kept in a programmable temperature oven that facilitates the separation of compounds.

The GC-MS system has an ion source, a mass analyzer, and a detector. The ion source ionizes molecules whereas the mass analyzer resolves such ions according to their m/z ratio, which the detector records in terms of ion abundance. Most modern instruments of GC-MS are coupled with large libraries of mass spectra that help in the identification of compounds through spectral matching.

3.9.3 Applications of GC-MS in Herbal Analysis

GC-MS has been excellently applied in the analysis of several bioactive compounds in herbs. Some of the prime applications are discussed below:

3.9.4 Endowed Oil Analysis

These oils, as well as their therapeutic functions, are essentially composed of volatile aromatic compounds that can be classified into four classes: terpene, alcohol, ester, and phenolic. The sensitivity and resolution needed for the analysis of essential oils usually make GC-MS a technique of choice. For instance, analysis of the essential oil from *Lavandula angustifolia*, or lavender, shows that important constituents responsible for the sedative activity of lavender oil include linalool, linalyl acetate, and camphor.

3.9.5 Alkaloids and Phenolic Compounds

Alkaloids represent nitrogen-containing compounds of important pharmacological activity, for example, morphine from the opium poppy (*Papaver somniferum*) and caffeine from the coffee plant (*Coffea arabica*). GC-MS will be able to quantify and identify alkaloids. Similarly, phenolic compounds such as flavonoids and tannins having strong free-radical scavenging activity will lend them very well to analysis by the GC-MS method.

3.9.6 Terpenoids

Terpenoids are large and diverse classes of naturally occurring organic chemicals formed by five-carbon isoprene units. They give the aroma to many herbs and flavorings. The GC-MS analysis of terpenoids from herbs like rosemary (*Rosmarinus officinalis*) and thyme (*Thymus vulgaris*) is aimed at furnishing information regarding the chemistry of their composition and associated health benefits.

3.9.7 Quantitative Analysis

Quantitative analysis using GC-MS is performed by determining the concentration levels of some compounds in an herb sample. The method therefore includes the plotting of calibration curves using known concentrations of standard compounds. Because the peak area in the chromatogram is directly proportional to the concentration of the compound, the quantitation is quite accurate.

3.9.8 Data Analysis and Interpretation

GC-MS data are analyzed by matching the recorded mass spectra with the reference spectra in the libraries such as NIST and Wiley. The retention time, along with the mass spectrum, helps in the proper identification of the compounds. The software tools of different levels do facilitate easier automatic matching and peak selection, thus enabling faster data analysis.

3.10 LIQUID CHROMATOGRAPHY-MASS SPECTROMETRY (LC-MS)

Herbal bioactive compounds and phytochemicals occupy a great deal of interest in research with regard to their therapeutic and nutritional applications, as the popularity of natural products is increasing. Within this context, LC-MS has proven to be very powerful; the major advantages of this technique over the others considered in this chapter are very high sensitivity and specificity, and the fact that complex mixture analysis can be performed. In essence, this chapter outlines the identification and quantification of bioactive compounds and phytochemicals from herbs using LC-MS.

In LC-MS, liquid chromatography provides physical separation, while mass spectrometry supplies the mass analysis, thereby allowing a combination to both identify and quantify mixtures of compounds based on a mass-to-charge ratio. The scope covers a broad array of bioactive compounds and phytochemicals, such as alkaloids, flavonoids, terpenoids, and phenolic acids.

3.10.1 Principles of Liquid Chromatography-Mass Spectrometry

Liquid chromatography (LC) is a general technique, mainly based on the principle that different compounds have differential interactions between a stationary and a mobile phase. In most cases, the stationary phase refers only to a column packed with some solid particles, while the mobile phase refers to a liquid solvent flowing through the column. These will allow other compounds contained in a sample the possibility to interact to varying degrees with the stationary and mobile phase, permitting them to elute at different times—retention times. Separation is the real essence of analyzing complex mixtures, such as herbal extracts.

Mass spectrometry is the technique used to quantify the mass-to-charge ratio of ions. In support of LC-MS, the compounds separated by LC are ionized and introduced into the mass spectrometer. There is a different way of carrying out the process of ionization, including electrospray ionization and atmospheric-pressure chemical ionization. Once these compounds are ionized, their m/z ratios are detected to obtain both qualitative and quantitative information.

3.10.2 Methods for LC-MS Detection Analysis

Critical steps in the analysis by LC-MS of bioactive compounds and phytochemicals are of sample preparation, chromatographic separation, mass spectrometric detection, and data analysis.

Sample Preparation: The preparation of herbal samples should be carried out carefully to ensure the exactness and reproducibility of bioactive compound analyses. The major steps in herbal sample preparation are extraction, filtration, and concentration. The most common extraction is believed to be solvent extraction, and, therefore, methanol, ethanol, and water are very common because they are successful in extracting a diverse compound spectrum.

Column type, mobile phase composition, and the gradient elution have very strong effects on the chromatographic separation. Typically, reverse-phase liquid chromatography is in use in the separation of bioactive compounds using a non-polar stationary phase and a polar mobile phase. Gradient elution usually is in use in enhancing the efficiency of separation through alterations to the composition of the mobile phase over some time during analysis.

Mass spectrometric detection: The technique of ionization properties of on and the mass analyzers should be chosen accordingly since the technique involves different the targeted compounds. By and large, ESI is used for the ionization of polar and semi-polar compounds, and APCI is used for less polar compounds. Once more, it will depend on the mass analyzer, which could be a quadrupole, time-of-flight, or your ion trap, and that also defines the resolution and accuracy of the mass measurement.

Data Analysis: Data obtained by analysis using the LC-MS method require meticulous interpretation for the identification and quantification of the analyzed bioactive compounds. Introduction of software programs, which allow for the processing and interpretation of chromatographic and mass spectrometric data through peak integration and m/z identification, followed by calibration curve-based quantification, has heralded a new era in scientific progress. Unknown compounds may also be identified using databases and metabolomic libraries based on the comparison of obtained spectra with reference spectra.

3.10.2.1 Applications of LC-MS in Herbal Analysis

A large number of applications have been documented for LC-MS in herbal bioactive compounds and phytochemicals. The major applications are:

Alkaloids are a class of nitrogen-containing compounds that exert some pharmacological activity. In this technique, alkaloids such as *Ephedra sinica*, *Catharanthus roseus*, and *Papaver somniferum* have been detected and quantified in some medicinal herbs. The mentioned technique could discriminate between alkaloids that are structurally similar and measure the concentration of each in herbal extracts.

Flavonoids are widespread in plants, with polyphenolic compounds that are antioxidants, as well as being anti-inflammatory and anticancer in nature. The LC-MS analysis of flavonoids from the herbs *Ginkgo biloba*, *Camellia sinensis* (green tea), and *Hypericum perforatum*, St. John's Wort, identified a collection of flavonoid glycosides and aglycones which elucidate their associated health benefits and therapeutic possibilities.

Terpenoids contribute to the aroma and other medicinal values of herbs. LC-MS is reported to investigate terpenoids in herbs such as *Cannabis sativa*, *Salvia officinalis*, and *Zingiber officinale*. This method easily resolves the monoterpene, sesquiterpene, and diterpene classes and also expounds their possible roles in the physiology of the plant and health of humans.

Phenolic Acids are phenolic compounds that exhibit antioxidant and anti-inflammatory activities. LC-MS analysis of phenolic acids of *Rosmarinus officinalis*, *Origanum vulgare*, and *Curcuma longa* herbs has shown a large variety of hydroxycinnamic acids and hydroxybenzoic acids. This means the technique has been applied to give relevant information about the concentration of these metabolites and the distribution of these compounds in various plant parts. Although its huge advantage has been elucidated in the determination of herbs for bioactive compounds and phytochemicals using LC-MS, it is not devoid of a number of challenges associated with that. These include herbal matrices, the existence of isomeric compounds, variability in compound concentration, and other complex interactions. Further, sample preparation is a painstaking and labor-intensive activity. It involves proper care concerning ion suppression or enhancement during ionization.

High-resolution MS and MS/MS are advantageous in overcoming these difficulties. HRMS enables better mass accuracy and resolution and will enable compounds to be identified with an enhanced level of confidence. MS/MS allows for the acquisition of structural information through the fragmentation of ions and analysis of the fragments formed; hence, it helps to identify unknown compounds.

Bioinformatics and chemometrics will also further enrich complex LC-MS data analysis. Principal component analysis and partial least squares discrimination analysis are the two multivariate statistical methods among several other classification and discrimination techniques for herbal samples according to their phytochemical profiles. To gain full information about herbs, it would be important to couple LC-MS with other analytical techniques such as nuclear magnetic resonance and infrared spectroscopy.

The report concluded that probably the most potent and flexible tool in the analysis of bioactive compounds and phytochemicals of herbs was liquid chromatography-mass spectrometry. Thus, the capability of the separation and identification of complex mixtures makes LC-MS one of the indispensable tools for natural product research, in which the detailed information is provided on composition and concentration of bioactive compounds—such analysis for our understanding regarding therapeutic and nutritional potential of herbs. This is sure to be a really great tool, given the constant development in the technology of LC-MS and techniques associated with data handling/analysis, setting the stage for new discoveries in the herbals of science and medicine.

Herbs have been associated with human culture since time immemorial and are known to have therapeutic potential properties related to benefits in health. These benefits, to a great extent, emanate from bioactive compounds and phytochemicals such as flavonoids, alkaloids, terpenes, and phenolic acids. Proper identification and quantification of such compounds are therefore important to assure the medicinal property potential and also for the purpose of standardization of herbs of medicine formulations. Among different analytical techniques, FTIR has proved to be a very strong nondestructive tool in the analysis of bioactive compounds and phytochemicals. The section on the principles of FTIR focuses on its applications in the analysis of herbs, methods of sample preparation, data interpretation, and advantages and limitations.

3.10.3 Principles of FTIR

FTIR spectroscopy is based on the interference of matter interaction with infrared radiation. Infrared light, propagating through a sample, will be absorbed at specific wavelengths corresponding to vibrational frequency of chemical bonds in molecules. Such an absorbed pattern is popularly addressed as an infrared spectrum, since it is unique for every compound.

The basic part of an FTIR spectrometer generally is a Michelson interferometer that divides the infrared beam into two paths. The interfered beams recombine to form what is measured as an interferogram. The interferogram then is processed mathematically with a Fourier transform to convert the interferogram into an infrared spectrum that plots intensity vs. wavenumber in cm^{-1}.

3.10.4 Application of FTIR in Herb Analysis

Thus FTIR has been a very useful tool in providing information at a molecular level regarding the bioactive compounds and phytochemicals of herbs, with minimal preparation of the sample. Here are key applications:

3.10.5 Phytochemical Identification

Each phytochemical exhibits a characteristic infrared spectrum. Thus, the identification of particular compounds present in herbs by comparing their spectrum with that of the reference spectrum of specific compounds present in an unknown sample. For example, a strong leading peak in the region 1600–1500 cm^{-1} is demonstrated by flavonoids due to vibrations of C=C stretching; another important example is the presence of strong absorption bands at around 1700 cm^{-1} in the spectrum attributed to phenolic acids arising from C=O stretching.

3.10.6 Quantitation of Bioactive Compounds

It also can be applied to the quantitative analysis of bioactive compounds. Calibration curves constructed with standard solutions of known concentrations can be used to find the concentration of a component from a herbal extract by its absorbance value. This quantitative strength of FTIR is important for the standardization issue of herbal products to produce consistent therapeutic efficacy.

3.10.7 Structural Elucidation

FTIR provides valuable information on the functional groups present in bioactive compounds and further aids in the elucidation of their structures. For example, hydroxyl groups (-OH) correspond to a broad absorption band placed at about 3400 cm^{-1}, while the presence of carbonyl groups (C=O) is confirmed by the existence of sharp peaks around 1700 cm^{-1}. The structural information can be used in combination with other machine spectroscopy techniques, such as NMR and mass spectroscopy.

In cases where bioactive compounds are extracted from herbs, FTIR can be used to monitor their extraction process. Samples at each step of the extraction process are analyzed to optimize extraction parameters for maximum yield and purity. This online monitoring ability greatly increases the efficiency and reproducibility of herbal extraction.

3.10.8 Sample Preparation for FTIR Analysis

The accuracy of FTIR analysis is greatly influenced by proper sample preparation. The following are some of the most common methods of sample preparation applied in herbal analysis:

3.10.9 Direct Analysis

In some instances, samples can be analyzed directly without the need to prepare them. Solid herbs can be powdered and placed directly on an ATR crystal for direct analysis without any hassle. Given this fact, this is a fast method while reducing the level of handling or changes to the sample and thereby maintaining its integrity.

3.10.10 Extraction

Often, herbs are extracted to obtain a bioactive principle. Extracts are prepared in routinely used solvents and concentrated either using evaporation, distillation, or sublimation techniques. These extracts are subsequently dried to eliminate the residual solvents during FTIR. The final solvent and the technique adopted for extraction influence the extract's yield and its quality, eventually reflecting on the FTIR results.

3.10.11 Pellet Preparation

Sometimes herbs are mixed with a non-absorbing matrix such as potassium bromide (KBr) and pressed to obtain pellets. This approach is appropriate for obtaining high-quality spectra from solid samples, but it requires highly skilled handling since there is a need for uniform mixing and pellet formation.

3.10.12 Thin Films

There are several methods to prepare thin films of liquid samples, such as the preparation of a drop of sample on an ATR crystal or another substrate, followed by drying and so on. This method is very simple, and so used to analyze the liquid extracts and the essential oils from herbs

3.10.13 Data Analysis and Interpretation

The FTIR spectra are interpreted with a good knowledge of the characteristic absorption bands due to different functional groups. Some common bands and their corresponding functional groups are given below:

3500–3200 cm^{-1}: O-H stretching of hydroxyl groups, alcohols, phenols
3000–2850 cm^{-1}: C-H stretching of alkanes
1750–1700 cm^{-1}: C=O stretching of carbonyl groups, ketones, and aldehydes
1600–1500 cm^{-1}: C=C stretching of aromatic rings, alkenes
1300–1000 cm^{-1}: C-O stretch (alcohols, ethers, esters)

Analyzing these bands allows the researcher to understand what functional groups are in the sample and what type of bioactive compound and phytochemical are present. Advanced software tools and databases enhance the accuracy of spectral interpretation through the availability of reference spectra and automatic assignment of peaks.

3.10.14 Advantages of FTIR on Herb Analysis

The following are some of the advantages that make FTIR one of the greatest instruments in analyzing herbs:

3.10.15 Non-Destructive

FTIR is a non-destructive technique that leaves the sample intact for further analysis. This fact becomes important in the case of herbal material that is expensive or scarce.

3.10.16 Fast and Easy

FTIR gives quick results with minimum sample preparation, and is therefore particularly apt for high-throughput screening and quality control in herbal products.

3.10.17 Rich Information

FTIR spectra detail molecular structure and functional group, which can be used to conduct an in-depth analysis of complex herbal mixtures.

3.10.18 Versatility

FTIR is able to analyze all states of matter—solids, liquids, and gases—and is therefore particularly versatile with regard to various forms that herbal samples could exist in, such as raw herbs, extracts, and essential oils.

3.10.19 Cost-Effective

Compared to techniques like NMR and mass spectrometry, FTIR is relatively cheap; hence, it can easily access routine analyses in research and industry.

Despite the benefits, FTIR has associating limitations that must come into consideration:

3.10.20 FTIR Limitations and Low Sensitivity

FTIR may not make it possible to detect compounds that are in a very low concentration. In this case, either more sensitive techniques can be used for trace analysis such as mass spectrometry, or the compound of interest concentrated.

3.10.21 Overlapping Bands

In the case of complex mixtures, the absorption bands of the various compounds may overlay, thereby making spectral interpretation difficult. However, this problem can be reduced by using advanced data analysis methods coupled with complementary techniques.

3.10.22 Preparation of the Sample

Some of the sample preparation methods, such as pellet formation, are tedious and may introduce some artifacts if not handled properly. It is, therefore, imperative that standardized protocols be used to ensure reproducible results.

3.10.23 Conclusion

FTIR is a very powerful tool in the analysis of bioactive compounds among the herbs and phytochemicals, which consequently attests to its non-destructive nature, speed of analysis, and especially its capability to yield detailed information on molecular structure. It is therefore important to remember that reliable and valid results depend on detailed sample preparation and proper data interpretation. Although FTIR has some limitations, the advantages prove to be an asset in the analytical toolkit within the rich and diverse world of herbal bioactive compounds and phytochemicals. With technological advancement, FTIR is sure to go on playing a significant role in unlocking herbal therapeutic potential through the help of safe and effective development of herbal products.

3.11 NUCLEAR MAGNETIC RESONANCE SPECTROSCOPY (NMR)

Analysis of bioactive compounds and phytochemicals has, of late, received much attention due to the potential health benefits associated with antioxidant, anti-inflammatory, and anticancer qualities. Of the several techniques of analyses available, nuclear magnetic resonance (NMR) spectroscopy has emerged outstanding in being non-destructive and highly reproducible, and especially where comprehensive structures need to be elucidated. The bioactivity effects of the compounds, molecular structure, the dynamics of molecular relations, and interactions all emerge from NMR spectroscopy. In this light, NMR spectroscopy establishes details regarding the molecular structure, dynamics, and intermolecular interactions of the bioactive compounds and hence remained the tool to analyze phytochemicals.

One of the principles underlying NMR spectroscopy is that atomic nuclei with magnetic moments associated with spin angular momentum will interact with an externally applied magnetic field. Such nuclei, when subjected to a magnetic field, absorb electromagnetic radiation at characteristic frequencies and re-emit the same. That is the aspect from which information regarding the chemical environment surrounding these nuclei can be derived, hence determination and quantification of various molecular structures.

Perhaps the most common nuclei used for NMR spectroscopy are 1H and 13C, although 15N and 31P also have widespread applications. The most important quantity in an NMR spectrum is the chemical shift, measured in parts per million (ppm). It provides a measure of the influence of the electron environment on the nuclei being studied. Other parameters in the determination of the molecular structure include coupling constants, signal multiplicities, and peak area integration.

3.11.1 Sample Preparation and Instrumentation

Before actual NMR analysis can commence, herb samples must be prepared for bioactive compound extraction and purification. The drying and grinding of herb materials and their extraction with solvents such as methanol, ethanol, or water are the most common extraction methods needed, depending on polarity. The prepared extract must then be reconcentrated and further purified by methods which include liquid–liquid extraction, solid-phase extraction, or chromatography.

The purified extract is dissolved in a suitable deuterated solvent such as deuterated chloroform $CDCl_3$ or deuterated dimethyl sulfoxide DMSO-d_6. Deuterated water (D_2O) is also used to exclude the interference of the signals of the solvents. Finally, the sample gets transferred into one of the NMR tubes and is placed in the NMR spectrometer.

Modern NMR spectrometers are ultrahigh-field magnets that reach frequencies of up to 400–800 MHz for 1H NMR or higher, equipped with cryogenic probes and high-tech software for data acquisition and processing. With the exception of these ultra-sensitive features, they also have excellent resolution and with one-dimensional and two-dimensional NMR spectroscopy.

3.11.2 One-Dimensional NMR Spectroscopy

One-dimensional NMR spectroscopy, in particular 1H NMR and 13C NMR, is the technique most commonly used for the preliminary structural elucidation of bioactive compounds. 1H NMR provides information on the number and kinds of hydrogen atoms in a molecule and their chemical environment, as well as their interaction with adjacent hydrogen atoms. For example, the presence of aromatic protons, aliphatic protons, and hydroxyl protons may easily be identified from their characteristic chemical shift and splitting pattern.

^13C NMR spectroscopy gives information concerning the carbon skeleton of a molecule and hence supplements ^1H NMR. The natural abundance of this isotope is lower and so is its gyromagnetic ratio, which makes it less sensitive compared to ^1H. Nevertheless, a very good-quality ^13C NMR spectrum can be measured on most modern NMR machines by high sensitivity and extended acquisition times. Due to a different local environment, chemical shifts for the carbon atoms of the carbonyl, carboxyl, and alkyl groups are quite different, making them useful in the identification of bioactive compounds.

3.11.3 Two-Dimensional NMR Spectroscopy

This structural tool has become quite effective, especially when enhanced, because it provides better structural information. It does this through signal correlation derived from two different nuclei at two different locations within the same nucleus. Possible 2D NMR experiments include COSY, HSQC, HMBC, and NOESY.

COSY provides information about proton–proton coupling and, therefore, defines connectivity around hydrogen atoms in a molecule. Both HSQC and HMBC are experiments that correlate proton and carbon signals providing information on carbon–hydrogen connectivity and carbon–carbon framework, respectively. NOESY provides information of the spatial proximity of the nuclei and hence is useful for the assignment of three-dimensional structure of bioactive compounds.

3.11.4 Phytochemical Applications

NMR spectroscopy is a widely used tool in the analysis of bioactive compounds and phytochemicals from most herbs, plants used as traditional medicines, culinary herbs, and dietary supplements. NMR is a versatile and effective tool for phytochemical analysis:

Some of these groups of bioactive compounds are alkaloids, which have significant pharmacological activity. Studies to identify and characterize alkaloids from *Rauvolfia serpentina*, *Catharanthus roseus*, and *Erythroxylum coca* included applications of NMR spectroscopy. The analysis of the whole data from the 1 H and 13 C NMR spectra and 2D NMR experiments helps to elicit a detailed structural characteristic, which led to the discovery of some novel alkaloids and their respective biosynthetic pathways.

Characterization of Polyphenols in Culinary Herbs: Polyphenols are powerful antioxidants and represent the main group of phytochemicals in many widely consumed dietary culinary herbs, such as rosemary, thyme, and oregano. Numerous recent studies have reported the use of NMR spectroscopic methods for the direct profiling and quantification of polyphenolic compounds, namely, the flavonoids, phenolic acids, and tannins. From NMR spectra, in theory, the chemical shift patterns and the coupling constants can provide structural information for polyphenols and the substitution patterns, thus making the evaluation of their antioxidant activity much easier.

Terpenoids represent a big class of biologically active compounds that can be found in essential oils, which can be isolated from herbs like lavender, peppermint, or eucalyptus. The terpenoids, as they were identified—monoterpenes, sesquiterpenes, and diterpenes compounds—were quantified using NMR spectroscopy. According to the characteristic chemical shifts of the terpenoids, The multiplicity of protons and carbons can aid in the identification of compounds and the estimation of their bioactivity. Structural

Elucidation of Saponins in Medicinal Herbs: Saponins are known to be functional glycosides of triterpenoids or steroids that have been reported to contain a varied number of biological activities, including antimicrobial and anticancer properties. Among this, NMR spectroscopy has been largely put to use for the study of saponins of important medicinal herbs like *Panax ginseng*, *Glycyrrhiza glabra*, and *Quillaja saponaria*. Details of the structure of the saponin aglycones and sugar moieties could be further elaborated on the basis of 2D NMR experiments combined with information from ^1H NMR and ^13C NMR experiments.

Metabolomics and chemotaxonomy: NMR metabolomics is an exhaustive profiling of any herb extract metabolites taken for the study of their chemical composition and variation. It has found use in the chemotaxonomic classification, identification of bioactive compounds, and quality control of herbal products. For example, NMR metabolomics-based experiments have been performed to classify various Salvia species by their chemotaxonomic characteristics and detect biomarkers related to their medicinal properties.

Despite its great benefits, there are many drawbacks tied to NMR spectroscopy in the analysis of bioactive compounds and phytochemicals. The fact that NMR sensitivity is fairly low in comparison to techniques like mass spectrometry brings a major problem with regard to how this method may turn out to be time-consuming or may fail in detection for compounds present in trace amounts. Along with this, sample preparation and solvent used are optimal variables that can highly influence the quality of NMR spectra and hence always demand close optimization.

Some of these limitations of NMR are now beginning to be overcome as a result of developments in technology, which includes high-field magnets, cryogenic probes, and hyperpolarization techniques for higher sensitivity and resolution (Dymek et al., 2021). Furthermore, the integration with other analytical techniques, particularly mass spectrometry and chromatography, further augments the ability of NMR in phytochemical analysis.

In this respect, future research in NMR spectroscopy may be expected to focus on the development of new pulse sequences and data analysis algorithms that are able to go even further in their application to complex mixtures and low-abundance compounds. In this respect, further NMR metabolomic applications are expected to clarify systems biology issues that are relevant to questions of personalized medicine and the multi-way complex interactions of bioactive compounds with human health.

3.11.5 Quality Control and Standardization of Bioactive Compounds from Herbs: Extraction and Analysis

Herbal products are becoming more popular every day due to their therapeutic benefits associated with the promotion of health and wellness. However, bioactive compounds in herbs may be quite variable; therefore, very strict quality control and standardization measures are necessary to ensure the efficacy, safety, and consistency of herbal products. This can be achieved only by developing standardized extraction methodologies, bioactive constituent analysis, and adherence to the regulations in force.

The quality control of the herbal drugs includes various parameters to ensure safety, efficacy, and consistency. These include:

Identity Confirmation: The correct identification of the species of herbs is against adulteration and assurance of the correct therapeutic properties. This has to be even further aided by botanical authentication and DNA barcoding.

Purity and Contaminant Testing: Herbal products should be free from heavy metals, pesticides, microbial pathogens, and adulterants. Inductively coupled plasma mass spectrometry is one of the common techniques used in determining heavy metals, whereas chromatography is used to detect traces of pesticides.

Quantification of Bioactive Compounds: Major bioactive compounds should be concentrated in uniformity within limits that arise to have a good effect of the herbal product.

Physicochemical Parameters: Moisture content, ash value, extractive values, and solubility will be among the parameters that will influence the stability and efficacy of the remedy.

Organoleptic Properties: Sensory properties used for analysis include color, odor, taste, and texture. In order to determine acceptability, they are accessed.

3.11.6 Techniques of Standardization

Standardization involves defining specific quality parameters to ensure that each batch of the herbal product consistently meets these established specifications. Techniques of standardization include:

Chemical Fingerprinting of Herbs: This can be said to be the recording of their unique profile of bioactive compounds, which characterizes them. In these techniques, high-performance liquid chromatography, gas chromatography, and mass spectrometry are used. This will form a reference profile to which different batches of a herbal product are compared.

Marker Compound Analysis: The specific bioactive compounds, known markers for their bioactive therapeutic potential, are quantified for concentration by HPLC, GC-MS, or LC-MS. This ensures a consistent specified quantity in every batch of the mentioned marker compounds.

Spectroscopic Methods: These include nuclear magnetic resonance spectroscopy and infrared spectroscopy by Fourier transform. All those provide information on molecular structure and the composition of bioactive compounds which constitute an important step in standardization (Fomo et al., 2020).

Biological Assays: These are conducted to determine the biological activity of the herbal product, including antioxidant properties, anti-inflammatory, or antimicrobial activity. Another parameter for standardization is the consistency in the batches with respect to their biological activity.

3.11.7 Extraction and Analysis of Bioactive Compounds

Extraction methods greatly affect the yield as well as the composition of bioactive compounds. Standardized extraction lends a hand in terms of better reproducibility and a consistent regime. There are a number of techniques of extraction:

Solvent Extraction: Extraction of bioactive compounds is done using water, ethanol, methanol, or a mixture of these. The choice of a solvent depends on the polarity of the targeted compound.

SFE: Supercritical fluids like CO_2 are used for extraction of bioactive compounds. This process is very efficient and may lead to solvent-free extracts.

Another methodology includes MAE and UAE, which in turn increases the efficiency of extraction by disrupting the plant cell walls through the pulsation of microwave or ultrasonic energy.

After extraction, the bioactive compounds are analyzed by using advanced techniques as follows:

- **HPLC and ultra HPLC**: These are the most used techniques for separation and identification of bioactive compounds quantified in an herbal extract.

GCMS is the abbreviation of gas chromatography with mass spectrometry. This technique is carried out on volatile and semi-volatile compounds. GC-MS elucidates the detail of the composition of essential oils and other volatile compounds of herbs.

NMR Spectroscopy detects bioactive compounds and offers a full description of their structure and quantification.

The regulatory bodies that have given guidelines for quality control and standardization of herbal products, respectively, are FDA, EMA, and WHO. The herbal products have abided by these regulations to offer safe and effective high-quality herbal products. The principal regulatory requirements would be:

Good Manufacturing Practices: Compliance with GMP ensures uniform manufacturing and control of herbal products according to the defined quality standards.

Labeling Requirements: Proper labeling is required to provide information on botanical name, part used, extraction method, and concentration of bioactive compounds.

Clinical Trials and Safety Assessments: Herbal products proposed to be used for treatment shall be passed through clinical trials for their safety and efficient conduct. The adverse effects and drug interaction reports are to be assessed and documented.

3.11.8 Conclusion

It is in light of this that quality control and standardization of herbal bioactive compounds, therefore, gain further importance to ensure their effectiveness, safety, and consistency in therapeutic purposes. Different extraction processes, advanced analytical techniques in conformity with the regulatory requirements, give herbal products a chance to keep their quality and feel that consumers and health professionals are certain. In a nutshell, since there is a rising demand for herbal medicinal products, the continued development of quality control and standardization of the products will be pivotal toward the development of safe and effective herbal therapeutics.

REFERENCES

Abhari, K., & Khaneghah, A. M. (2020). Alternative extraction techniques to obtain, isolate and purify proteins and bioactive from aquaculture and by-products. In J. M. Lorenzo & F. J. Barba (Eds.), *Advances in Food and Nutrition Research* (Vol. 92, pp. 35–52). Academic Press.

Abubakar, A. R., & Haque, M. (2020). Preparation of medicinal plants: basic extraction and fractionation procedures for experimental purposes. *Journal of Pharmacy & Bioallied Sciences*, 12(1), 1–10.

Adegbola, P., Aderibigbe, I., Hammed, W., & Omotayo, T. (2017). Antioxidant and anti-inflammatory medicinal plants have potential role in the treatment of cardiovascular disease: a review. *American Journal of Cardiovascular Disease*, 7(2), 19.

Afroz, R.D., Nurunnabi, A.S.M., Khan, M.I., & Jahan, T. (2015). Effect of wheatgrass(Triticum aestivum) juice on high density lipoprotein (HDL) level in experimentally induced dyslipidaemic male Long Evans rat. *Delta Medical College Journal*, 3(1), 18–24.

Brunner, G. (2005). Supercritical fluids: Technology and application to food processing. *Journal of Food Engineering*, 67(1–2), 21–33.

Chen, S., Xing, X. H., Huang, J. J., & Xu, M. S. (2011). Enzyme-assisted extraction of flavonoids from Ginkgo biloba leaves: Improvement effect of flavonol transglycosylation catalyzed by Penicillium decum-bens cellulase. *Enzyme and Microbial Technology*, 48(1), 100–105.

Dababi, I., Gimello, O., Elaloui, E., & Brosse, N. (2020). Water Extraction of Tannins from Aleppo Pine Bark and Sumac Root for the Production of Green Wood Adhesives. *Molecules (Basel, Switzerland)*, 25(21), 5041. doi:10.3390/molecules25215041 PMID:33143002

Dastangoo, S., Hamed Mosavian, M. T., & Yeganehzad, S. (2020). Optimization of pulsed electric field conditions for sugar extraction from carrots. *Food Science & Nutrition*, 8(4), 2025–2034.

De Menezes Rodrigues, G., Cardozo-Filho, L., & Da Silva, C. (2017). Pressurized liquid extraction of oil from soybean seeds. *Canadian Journal of Chemical Engineering*, 95(12), 2383–2389.

Díaz-Batalla, L., Widholm, J. M., Fahey, G. C., Castaño-Tostado, E., & Paredes-López, O. (2006). Chemical components with health implications in wild and cultivated Mexican /common bean seeds (Phaseolus vulgaris L.). *Journal of Agricultural and Food Chemistry*, 54(6), 2045–2052.

Diep, T., Pook, C., & Yoo, M. (2020). Phenolic and anthocyanin compounds and antioxidant activity of tamarillo (Solanum betaceum Cav.). *Antioxidants*, 9(2), 169.

Dymek, A., Widelski, J., Wojtanowski, K. K., Płoszaj, P., Zhuravchak, R., & Mroczek, T. (2021). Optimization of Pressurized Liquid Extraction of Lycopodiaceae Alkaloids Obtained from Two Lycopodium Species. *Molecules (Basel, Switzerland)*, 26(6), 1626.

Fomo, G., Madzimbamuto, T. N., & Ojumu, T. V. (2020). Applications of nonconventional green extraction technologies in process industries: Challenges, limitations and perspectives. *Sustainability*, 12(13), 5244.

Geng, S., Liu, Y., Ma, H., & Chen, C. (2015). Extraction and antioxidant activity of phenolic compounds from okra flowers. *Tropical Journal of Pharmaceutical Research*, 14(5), 807–814.

Li, H., Deng, Z., Wu, T., Liu, R., Loewen, S., & Tsao, R. Microwave-assisted extraction of phenolics with maximal antioxidant activities in tomatoes. *Food Chemistry* 2012, 130, 928–936.

Omeroglu, P.Y., Acoglu, B., Özdal, T., Tamer, C. E., & Çopur, Ö. U. (2019). Extraction techniques for plant-based bio-active compounds. In *Natural Bio-active Compounds* (pp. 465–492). Springer.

Shinde, V.M., Dhalwal, K., Potdar, M., & Mahadik, K. (2009). Application of quality control principles to herbal drugs. *International Journal of Phytomedicine*, 9(1), 4–8.

Tsao, R., & Deng, Z. (2004). Separation procedures for naturally occurring antioxidant phytochemicals. *Journal of Chromatography B*, 812, 85–99.

Zahari, N. A. A. R., Chong, G. H., Abdullah, L. C., & Chua, B. L. (2020). Ultrasonic-Assisted Extraction (UAE) Process on Thymol Concentration from Plectranthus amboinicus Leaves: Kinetic Modeling and Optimization. *Processes (Basel, Switzerland)*, 8(3), 322.

Zhang, Q. W., Lin, L. G., & Ye, W. C. (2018). Techniques for extraction and isolation of natural products: A comprehensive review. *Chinese Medicine*, 13(1), 1–26.

Zygler, A., Słomińska, M., & Namieśnik, J. (2012). Soxhlet extraction and new developments such as soxtec. In J. Pawliszyn (Ed.), *Comprehensive Sampling and Sample Preparation* (pp. 65–82). Academic Press.

4 Principles of Drug Discovery from Plants

Sayyeda Sabahat Babar, Zarghoona Jawad, Fatima Ayub, and Muhammad Imran Khan
Department of Biomedical Sciences, Pak Austria Fachhochschule: Institute of Applied Sciences and Technology, Haripur, Pakistan

4.1 INTRODUCTION

Plants have played a role in providing substances for many years, contributing greatly to the advancement of modern medicine [1]. The standing connection between humans and plants has resulted in the identification of healing compounds that continue to be essential in today's medical field [2]. This section aims to investigate the principals involved in discovering, extracting, and utilizing plant-based compounds for purposes. By combining knowledge of plants with cutting-edge techniques in photochemistry and biotechnology, we can better tap into the healing properties of plants [3]. Additionally, this section will explore methods for isolating compounds through testing the impact of metabolomics and genomics on drug development and how nanotechnology can improve the effectiveness and delivery of plant-derived substances. It will also showcase examples of medications derived from plants to illustrate both the promise and the obstacles associated with plant-based drug discovery within today's pharmaceutical industry.

4.2 HISTORICAL CONTEXT OF PLANT-BASED MEDICINES

4.2.1 Early Use and Cultural Significance

Throughout history civilizations have relied on plants for medicinal purposes, with local healers harnessing the power of flora to treat a variety of health issues. Ancient records from Egypt, China, India, and Greece highlight the practice of utilizing plant-based remedies [4]. For example, the Ebers Papyrus, a manuscript dating back to around 1550 BCE, documents more than 700 plant-derived medicines [5]. Similarly, traditional healing systems like Ayurveda in India and traditional Chinese medicine (TCM) have traditions of employing plants for such purposes. Texts such as the *Charaka Samhita* and the *Compendium of Materia Medica* offer insights into plants and their therapeutic properties [6].

4.2.2 Development of Pharmacognosy

The scientific study of medicinal plants, known as pharmacognosy, emerged as a formal discipline in the 19th century. Pioneering work by scientists such as Friedrich Sertürner, who extracted morphine from opium in 1805, marked the beginning of the isolation and characterization of active compounds from plants [7]. This period saw significant advancements in extracting and isolating compounds, laying the groundwork for modern pharmacology.

4.2.3 Impact of Plant-Based Medicines on Modern Pharmacology

Many of the medicines we use today have their origins in plants, showing how plant-based remedies have greatly influenced medicine. For example, the active ingredient of aspirin was sourced from

DOI: 10.1201/9781003544418-4

the bark of willow trees, and that of quinine, which was used to treat malaria, comes from the bark of cinchona trees. These discoveries led to the exploration of plant-derived molecules, which led in turn to the development of drugs and treatments for various conditions. According to the World Health Organization (WHO), more than a quarter of medications stem from plants, with many others being versions of natural compounds [2].

4.3 DIVERSITY AND SIGNIFICANCE OF BIOACTIVE COMPOUNDS IN MODERN PHARMACOLOGY

Plants produce metabolites, such as terpenoids, alkaloids, flavonoids, and phenolics. These compounds have a range of effects and can be used for different purposes in nature, such as defending against infections and herbivores. For example, the alkaloid vincristine, which is taken from the Madagascar periwinkle, is used to treat cancers while a range of fruits and vegetables contain quercetin known for its anti-inflammatory and antioxidant properties [3].

4.3.1 Integration of Ethnobotanical Knowledge

Traditional knowledge about plants, known as ethnobotanical knowledge, is crucial in helping researchers find medicinal compounds. By studying how indigenous peoples use plants for healing purposes scientists can pinpoint plants that show potential for use and give them priority for study [8]. This approach has led to the discovery of medications such, as paclitaxel, an anti-cancer drug originally derived from the bark of the Pacific yew tree (*Taxus brevifolia*) and used in traditional medicine by Native American tribes [9].

4.3.2 Advanced Phytochemical Techniques

Advancements in phytochemistry technology have significantly enhanced the ability to isolate and detect compounds from plants. Modern analytical methods such as nuclear magnetic resonance (NMR) spectroscopy, mass spectrometry (MS), and high-performance liquid chromatography (HPLC) can accurately analyze complex plant extracts. These techniques simplify the process of discovering compounds and comprehending their compositions, paving the way for developing new medications [10].

4.3.3 Bioassay-Guided Fractionation

Bioassay-guided fractionation plays a role in uncovering substances derived from plants. This process entails methodically dividing a plant extract into its parts, examining each part for its biological effects. Once active components are identified, they undergo purification and examination to isolate the compounds responsible for the desired results [11]. This ongoing process integrates assessments, with scrutiny to pinpoint and understand the most effective compounds.

4.3.4 Role of Metabolomics and Genomics

The study of metabolomics and genomics presents perspectives on how plants produce metabolites through metabolic pathways and genetic factors. Metabolomics involves identifying compounds and studying their creation and regulation within a system while genomics focuses on understanding the genetic components responsible, in order to generate secondary metabolites. Integrating these approaches allows researchers to improve the production of substances and gain insights into the molecular mechanisms that drive plant metabolism [9].

4.3.5 Integration of Nanotechnology

Nanotechnology presents solutions to enhance the effectiveness and strength of plant-based compounds. By designing nanoparticles to encase these compounds derived from plants, their solubility, stability, and ability to be absorbed by the body can be enhanced. This method can boost the effectiveness of these compounds. It can facilitate drug delivery, thereby reducing side effects and improving treatment outcomes [12]. For instance curcumin, a substance found in turmeric (*Curcuma longa*), usually has bioavailability; however, when enclosed in nanoparticles its therapeutic benefits are greatly enhanced [13].

4.4 ETHNOBOTANICAL APPROACHES

For generations, the practice of ethnobotany has played a role in uncovering the benefits of plant-based substances. Drawing on wisdom and customs, these methods help pinpoint plants that may offer value. Through documentation and the study of plant usage across societies ethnobotanical studies lay a solid groundwork, for today's pharmaceutical research endeavors [14].

4.4.1 Traditional Knowledge and Indigenous Applications

Ethnobotany delves into how individuals within a culture utilize plants, forming crucial groundwork for uncovering medicinal compounds. The wealth of wisdom amassed over generations provides insights into the healing properties of plants. Passed down through lines, indigenous practices involve leveraging plants to address a range of health issues [15]. This accumulated knowledge isn't just anecdotal; it often stems from observations and real experiences.

One notable illustration of the significance of knowledge lies, as mentioned in passing above, in the utilization of the Cinchona tree (*Cinchona* spp.) by indigenous communities to combat malaria [16]. This accumulated knowledge isn't just anecdotal; it often stems from observations and real experiences.

A notable illustration of the significance of knowledge lies in the utilization of the Cinchona tree (*Cinchona* spp.) by indigenous communities to combat malaria. The extraction of quinine its component, proved pivotal in the development of an antimalarial medication. Likewise the rosy periwinkle (*Catharanthus roseus*), traditionally used in Madagascar's folk medicine, contributed to the discovery of vincristine and vinblastine drugs in cancer treatment [17].

The wealth of information possessed by groups regarding plant-based medicine serves as an asset for contemporary drug exploration. These communities often hold knowledge about biodiversity and the therapeutic applications of diverse plant species. This expertise is frequently embedded within traditions, rituals, and traditional healthcare systems—underscoring a bond between community members and their natural surroundings [14].

Studies show that indigenous healers utilize plants not only for their immediate benefits but also taking into account the broader context, such as spiritual and cultural elements of healing. For example, the use by Native American tribes of Echinacea for treating infections and wounds integrates both the spiritual aspects and the healing traditions. This comprehensive perspective can assist researchers in exploring the roles of plant compounds beyond their direct medicinal impacts [18].

4.4.2 Ethnopharmacological Surveys and Their Relevance

Ethnopharmacological surveys are organized studies that record the wisdom and utilization of healing plants, among cultural communities. These surveys entail thorough on-site research, including discussions with healers and community members well as the gathering of plant samples for further examination. They play a role in identifying plants that may have properties and in grasping the cultural significance of their usage. One illustrative case demonstrating the impact of surveys is the unearthing of the inflammatory attributes of turmeric (*Curcuma longa*) in traditional Indian medicine [19]. Research into ethnopharmacology revealed that turmeric was widely used in Ayurveda due to its inflammatory and antioxidant qualities [20]. Subsequent scientific investigations validated these properties resulting in the application of curcumin, the active component of turmeric, in various therapeutic contexts [13].

Ethnopharmacological surveys act as a link between knowledge and contemporary scientific exploration. By documenting the application of plants, these surveys aid in conserving wisdom and serve as a foundation for validating scientifically and developing medications. They frequently lead to the discovery of compounds and therapeutic substances that might otherwise remain undiscovered. For instance, studying the herb *Artemisia annua* through ethnopharmacology. A herb long used in medicine to combat fevers led to the discovery of artemisinin. This particular substance has become an element in combating malaria, thereby showcasing the promise of studying traditional medicinal practices [21].

4.5 PHYTOCHEMICAL TECHNIQUES

4.5.1 Methods of Plant Extraction and Isolation

The extraction and retrieval of bioactive compounds from plants are fundamental steps in phytochemical research, providing the initial materials for further analysis and development. Various methods have evolved to efficiently isolate these compounds, each with its own advantages and limitations (Figure 4.1).

4.5.1.1 Solvent Extraction

One of the methods employed for creating plant extracts is through solvent extraction. This process involves dissolving the phytochemicals using solvents such as ethanol, methanol, hexane, or chloroform [23]. The choice of solvent is influenced by the polarity of the target molecules. For instance polar solvents such as ethanol and methanol are typically used to extract chemicals like flavonoids and phenolics while non-polar solvents such as hexane are preferable for extracting compounds such as terpenoids and essential oils. Percolation, maceration and soxhlet extraction are among the techniques employed in this type of extraction process [24].

4.5.1.2 Supercritical Fluid Extraction (SFE)

Supercritical fluid extraction (SFE) is a technique that utilizes fluids, with carbon dioxide (CO_2) being a prominent choice as the extracting agent. When subjected to pressures and temperatures beyond its critical point (31.1°C and 73.8 bar), CO_2 transforms into a state displaying unique properties of both liquid and gas that set it apart from traditional liquids and gases.

This approach is known for its efficiency, eco friendliness and effectiveness in extracting compounds without causing degradation. The safe nature of CO_2 and its ability to target compounds through temperature and pressure adjustments have established SFE as a preferred method for extracting phytochemicals [25].

4.5.1.3 Microwave-Assisted Extraction (MAE)

By using microwave energy to heat up the solvent and plant material, microwave-assisted extraction (MAE) enhances the efficiency of extracting bioactive components. This method offers advantages, including extraction times, reduced solvent usage, and increased yields and quality of extracts [26]. The heating and uniform energy distribution from microwaves facilitates the breakdown of plant cell walls allowing for the penetration of solvents and the release of intracellular chemicals. The application of MAE has shown success in extracting stable compounds such as polyphenols, alkaloids, and essential oils [27].

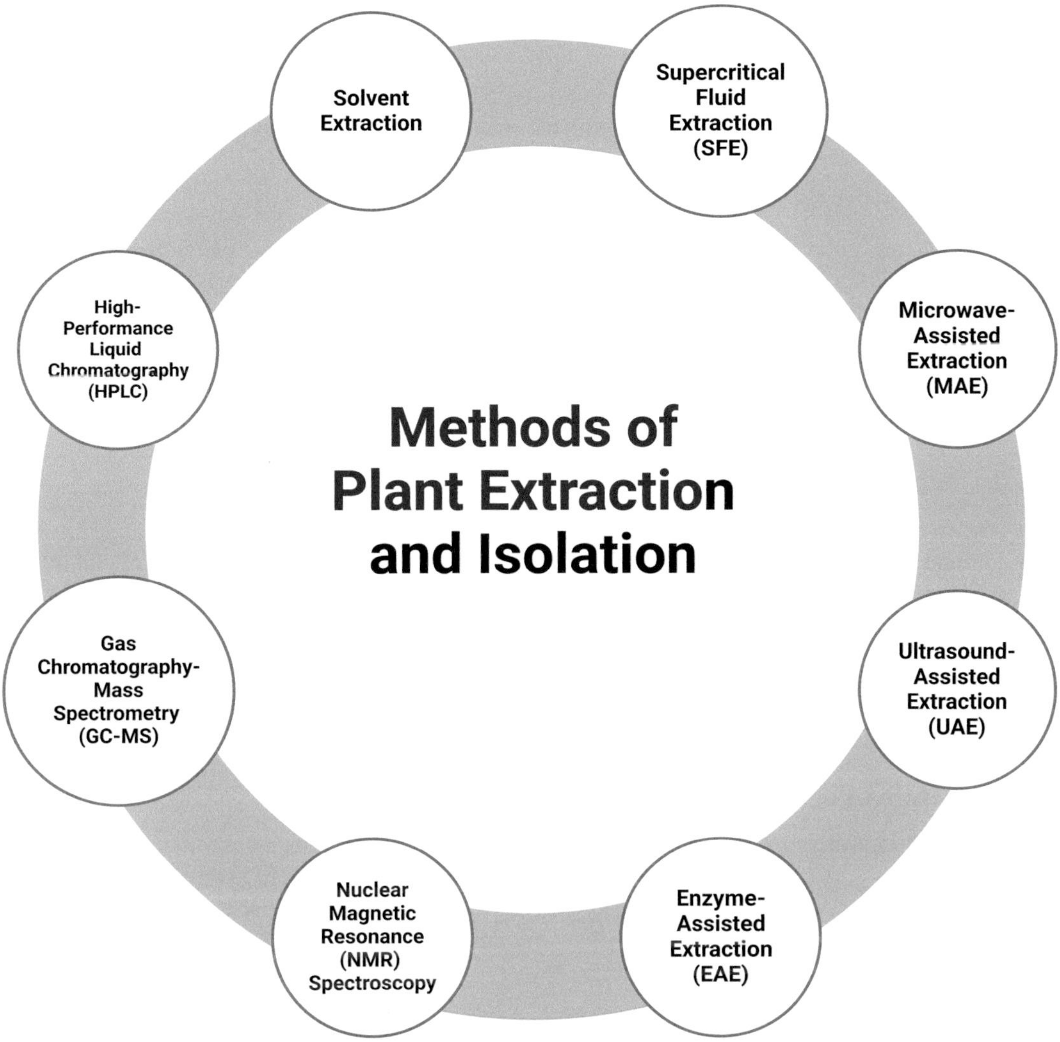

FIGURE 4.1 Methods of plant extraction and isolation.

4.5.1.4 Ultrasound-Assisted Extraction (UAE)

Ultrasound-assisted extraction (UAE) involves using waves to create cavitation in the solvent, which helps break down plant cell walls and facilitates the transfer of phytochemicals into the solvent. This method is recognized for its simplicity, cost-effectiveness, and ability to function at room temperature, making it ideal for extracting heat compounds [28]. UAE has proven effective in extracting phytochemicals such as flavonoids, terpenoids, and glycosides, resulting in extraction efficiency and higher yields compared to traditional methods [28].

4.5.1.5 Enzyme-Assisted Extraction (EAE)

Enzyme-assisted extraction (EAE) uses specific enzymes to hydrolyze plant cell walls, thereby facilitating the delivery of bioactive compounds. Enzymes such as cellulase, pectinase, and hemicellulase break down the structural polysaccharides in plant cell walls, increasing the permeability and solubility of the target compounds [29]. EAE is considered to be a 'green' extraction method due to its mild operating conditions and the reduced use of harsh solvents. This technique is particularly effective for extracting polysaccharides, saponins, and other complex macromolecules [30].

4.6 BIOASSAY-GUIDED FRACTIONATION

One systematic approach to isolating and characterizing compounds from plant extracts involves the use of bioassay-guided fractionation. This method entails dividing the extract into fractions, evaluating each fraction for its effects and further refining the fractions to obtain pure bioactive substances [31].

4.6.1 Fractionation Techniques

Plant extracts are initially separated using techniques such as liquid–liquid extraction, solid phase extraction, or column chromatography. These methods divide the extract into parts based on variations in polarity, solubility, or molecular size [32]. Liquid–liquid extraction involves dividing the extract between two solvents; by contrast, solid phase extraction uses an adsorbent to selectively retain specific compounds. Column chromatography, which includes methods like silica gel chromatography and Sephadex LH 20, sorts compounds according to their interactions with the mobile phases [33].

4.6.2 Biological Assays

The effectiveness of each portion is evaluated through a variety of tests, which can range from lab tests, animal studies to the use of tissues or organs. Lab tests, like cell cultures or enzyme inhibition, are commonly preferred for their simplicity and cost-effectiveness. Animal studies are conducted to assess the effects and toxicity of the fractions [34]. Ex vivo tests involve the use of tissues or organs to study the compounds' biological activity in a natural setting. The choice of test depends on the intended purpose and the specific type of effect being studied, such as antimicrobial, anticancer, or anti-inflammatory properties [35].

4.6.3 Iterative Purification

After determining fractions through bioassay testing they are purified using chromatographic methods, such as high-performance liquid chromatography (HPLC) or preparative thin layer chromatography (TLC). HPLC stands out for its clarity, sensitivity, and capacity to manage mixtures. The goal of each fractionation and bioassay stage is to extract the substances in their most refined state while considering their bioactivity retention [36].

4.7 CHROMATOGRAPHIC AND SPECTROSCOPIC METHODS FOR COMPOUND IDENTIFICATION

Once bioactive compounds are isolated, their identification and structural elucidation are critical steps in phytochemical research. Chromatographic and spectroscopic methods are essential tools for this purpose [32].

4.7.1 High-Performance Liquid Chromatography (HPLC)

High-performance liquid chromatography (HPLC) is a method used for analyzing and distinguishing molecules in a mixture [37]. In this process, a column filled with a phase filters the plant extract while a liquid mobile phase carries the sample through the column. The separation of compounds occurs due to factors such as size and polarity influenced by how the stationary phase interacts with the chemicals in the sample [38].

4.7.2 Gas Chromatography-Mass Spectrometry (GC-MS)

The method of gas chromatography-mass spectrometry (GC-MS) combines the detection and identification capabilities of mass spectrometry, with the separation abilities of gas chromatography.

This technique is particularly effective for analyzing semi-volatile substances [39]. By utilizing a column and an inert gas to transport the vaporized sample, GC-MS separates components based on their volatility and interactions with the columns phase. After the compounds are separated, the mass spectrometer breaks them down to generate a mass spectrum containing structural and molecular weight information. GC-MS is commonly used for analyzing phytochemicals, such as terpenoids and essential oils [40].

4.7.3 Nuclear Magnetic Resonance (NMR) Spectroscopy

Nuclear magnetic resonance (NMR) spectroscopy, an invasive analytical technique, offers detailed insights into the dynamics, environment, and structure of molecules [41]. By utilizing radiofrequency radiation and a powerful magnetic field to interact with nuclei, NMR spectroscopy generates resonance signals for deciphering molecule structures. The used ^1H NMR and ^13C NMR techniques reveal information about hydrogen and carbon atoms. This method plays a role in determining the structure of complex phytochemicals, including stereochemistry and functional group identification.

4.7.4 Fourier Transform Infrared (FTIR) Spectroscopy

Fourier transform infrared (FTIR) spectroscopy is utilized to detect groups and describe structures by analyzing their vibrational changes. When a molecule absorbs infrared light it experiences excitement, leading to an IR spectrum that showcases its molecular characteristics [42]. FTIR spectroscopy is commonly employed for recognizing groups, like hydroxyl, carbonyl, and amine groups, in plant compounds. This method is frequently combined with techniques to verify particular functional groups and offer additional structural insights.

4.7.5 Metabolomics and Genomics in Plant Drug Discovery

The emergence of metabolomics and genomics has brought about a transformation in the realm of plant-based drug exploration. These innovative techniques offer in-depth understandings into the metabolic and genetic structures of plants, facilitating the detection and improvement of substances. In the next section we delve into the functions of metabolomics and genomics in pinpointing compounds the genetic strategies employed to uncover and amplify phytochemicals as well as real-world examples showcasing the utilization of genetic technologies, in drug discovery [43].

4.8 ROLE OF METABOLOMICS IN IDENTIFYING BIOACTIVE COMPOUNDS

Metabolomics involves examining the metabolites in a biological system, thereby providing an overview of the organism's metabolic condition. In the realm of plant-based drug exploration, metabolomics plays a role in pinpointing substances uncovering how they are produced and grasping their significance in plant function and environmental interactions (Figure 4.2).

4.8.1 Identification of Bioactive Compounds

At first, we extract substances from plants, and separate and analyze them using methods like gas chromatography mass spectrometry (GC-MS), liquid chromatography mass spectrometry (LC-MS), and nuclear magnetic resonance (NMR) spectroscopy. These techniques help us recognize and quantify substances, both additional ones that could potentially offer medicinal benefits [44].

When scientists analyze the compositions of plants grown in environments or with different genetic makeups they can pinpoint certain substances linked to their health benefits. For example, research has revealed that studying the metabolites of healing plants such as *Catharanthus roseus*

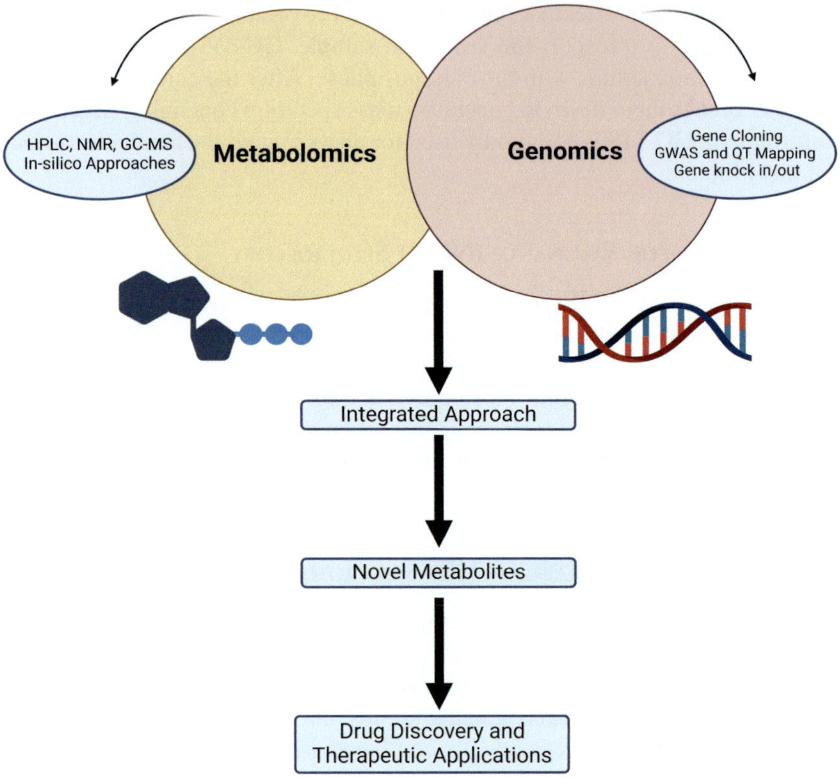

FIGURE 4.2 Role of metabolomics and genomics in plant drug discovery.

has uncovered vinca alkaloids, which play a role in their ability to combat cancer. Furthermore, delving into metabolomics can shed light on how these substances work, thereby offering information on their medical uses [45].

4.8.2 Explanation of Biosynthetic Pathways

Studying how bioactive compounds are produced and improved requires an understanding of their pathways. Metabolomics, conjugating with omics techniques, helps in mapping out these pathways. For instance the process of creating taxol, a substance with anticancer properties found in the Pacific yew tree (*Taxus brevifolia*) has been extensively researched using metabolomics. This method has unveiled the web of enzymes and substances that play a role in taxol production, making it easier to develop approaches to boost its output [46]. The identification and explanation of plant genomes through advancements in technologies have uncovered the genes responsible for producing bioactive substances [47].

4.8.3 Discovery of Biosynthetic Genes

Detecting the biosynthetic gene clusters (BGCs) plays a role in the creation of metabolites through genome-based approaches. These clusters contain genes that produce enzymes for guiding the stages of biosynthesis. Through methods like association studies (GWAS) and Mapping Quantitative Trait Loci (QTL) Mapping scientists can associate particular genes with the generation of bioactive substances [48]. For example, determining the gene cluster in *Artemisia annua* has been crucial for genetically modifying the plant to produce the chemical [49].

4.8.4 Enhancement of Phytochemical Production

Genetic tools also play a role in boosting the production of plant chemicals through methods such as metabolic engineering and biology. Researchers can enhance the output of compounds by amplifying biosynthetic genes or introducing new pathways [50]. For example, boosting the activity of the gene for diphosphate synthase in tomato plants has resulted in higher levels of carotenoids, which are known for their antioxidant properties [51]. Furthermore, scientists have utilized synthetic biology techniques to recreate pathways in microbial hosts, allowing for the efficient production of plant-based compounds, on a larger scale [2].

4.9 CASE STUDIES OF GENOMIC APPLICATIONS IN DRUG DISCOVERY

4.9.1 Case Study 1: Artemisinin Production in *Artemisia annua*

Artemisinin, a compound used to combat malaria which is extracted from the *Artemisia annua* plant, showcases how genomics has played a role in discovering drugs. By studying the composition of *Artemisia annua*, scientists have been able to modify the plants genes to increase production. This was achieved by boosting the activity of enzymes like amorpha 4,11 diene synthase in the pathway.

4.9.2 Case Study 2: Taxol Biosynthesis in *Taxus* spp.

Taxol, a cancer medication, is naturally found in the Pacific yew tree (*Taxus brevifolia*). Through research, scientists have identified the gene clusters responsible for Taxol production and the intricate enzymatic processes involved. Through genetic manipulation techniques, these genes have been transferred into plants and microorganisms to enhance Taxol production. For instance, by introducing taxadiene synthase into *Escherichia coli* bacteria researchers have successfully generated precursors of Taxol without relying on its extraction from yew trees [52].

4.9.3 Case Study 3: Resveratrol Production in Vitis vinifera

Resveratrol, a compound found in grapes, has been linked to health benefits such as protecting the heart and reducing inflammation. Scientists can now boost resveratrol production in grapevine cell cultures through modifications following the discovery of the genes, thereby achieving resveratrol synthesis. By enhancing the expression of the gene that produces synthase researchers have successfully raised the resveratrol levels in grapevine tissues, showcasing how genomics can enhance the yield of plant compounds [54].

4.10 BIOTECHNOLOGICAL ADVANCES

The use of methods has completely transformed the realm of discovering plant-based medicines allowing for the production and increased effectiveness of natural compounds. This section delves into the progress in biotechnology that has shaped this field, highlighting advancements in tissue culture, the modification of sustainable production practices through biotechnology, and the impact of synthetic biology on developing plant-based medications.

4.10.1 Tissue Culture and the Genetic Modification of Medicinal Plants

Tissue culture and genetic modification also play a role as tools in the arsenal of biotechnology for improving the cultivation of plants. Tissue culture, also known as in vitro culture, involves growing plant cells, tissues, or organs on a medium under sterile conditions. This technique eliminates the need for farming methods by facilitating plant growth and phytochemical production [55].

Micropropagation: Micropropagation stands out as an application of tissue culture that allows for the mass production of identical plants. This process ensures uniformity and consistency among plants, which is crucial for synthesizing bioactive compounds. For example, micropropagation techniques have successfully generated Madagascar periwinkle plants that yield quantities of anticancer substances such as vincristine and vinblastine [56].

Cell Suspension Cultures: Another important aspect is the development of cell suspension cultures, where plant cells are grown in liquid media. It is possible to produce secondary metabolites on a large scale by scaling up these cultures in bioreactors. It has been shown that shikonin, a substance which has anti-inflammatory and wound-healing qualities, may be produced from *Lithospermum erythrorhizon* using cell suspension cultures [57].

Genetic Modification: Advancements in engineering have improved the manufacturing of substances by introducing new genes or altering existing ones within plant genomes. This method enables the amplification of enzymes involved in the production pathways of desired compounds. For instance, modifying *Artemisia annua* to boost the expression of amorpha 4,11 diene synthase enzyme has resulted in yields of artemisinin, an antimalarial medication [58].

4.10.2 Sustainable Production of Phytochemicals through Biotechnology

Biotechnological approaches offer sustainable solutions to produce phytochemicals, addressing issues such as the overharvesting of wild plants and the environmental impact of traditional agricultural practices (Figure 4.3).

Bioreactor Systems: Using bioreactor systems to grow plant cells and tissues creates a controlled setting for the scalable production of compounds [59]. These systems can be fine-tuned for factors such as supply, oxygen levels, and pH to enhance the output of desired substances. Bioreactors have proven effective in producing paclitaxel (Taxol), an anticancer agent, from *Taxus* species [60].

Elicitation: Stimulating the production of metabolites in plant cultures involves triggering responses through stimuli. Examples of elicitors that can boost production include yeast extract, salicylic acid, and methyl jasmonate. For instance, research has shown that adding methyl jasmonate to cell suspensions of *Hypericum perforatum* (St. John's Wort) enhances the yield of hypericin, a substance with antidepressant properties [61].

Metabolic Engineering: The process of altering a plant's metabolic pathways to increase the synthesis of chemicals is referred to as metabolic engineering. This can be achieved by introducing genes or blocking competing pathways. One illustration is to incorporate genes from the grape stilbene biosynthesis pathway into *Escherichia coli* to produce resveratrol, a compound known for its anti-inflammatory properties [62].

4.10.3 Role of Synthetic Biology in Plant-Based Drug Development

The interdisciplinary area of synthetic biology, which blends engineering and biology, has evolved the field of plant-based medication discovery. It entails the creation of novel biological components, tools, and systems in addition to the functional redesign of already-existing natural biological systems

Pathway Reconstruction: Synthetic biology makes it possible to rebuild pathways in microbial hosts, as happens, for example, with bacteria and yeast which enables the production of plant-based compounds within these systems. This method overcomes the challenges linked to growing plants and extracting substances. For example, scientists have effectively inserted the pathway for the antimalarial medication artemisinin into *Saccharomyces cerevisiae*, allowing for the creation of artemisinic acid, a key component of artemisinin [63].

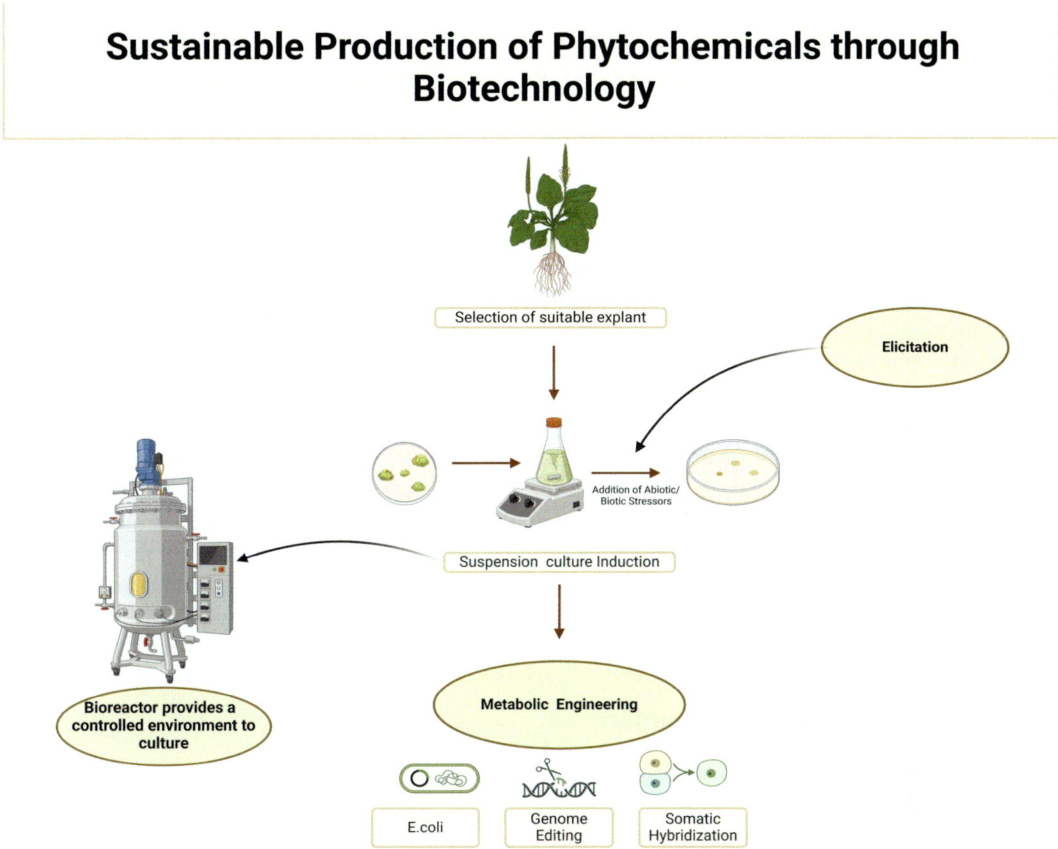

FIGURE 4.3 Biotechnology use in the sustainable production of biochemicals.

Standardized Parts and Modular Systems: The incorporation of components like promoters, ribosome attachment sites, and coding sequences simplifies the piecemeal construction of biosynthetic routes. This modularity enables experimentation and the enhancement of pathways for the generation of substances. Synthetic biology techniques have been leveraged to create pathways for production in yeast, offering a regulated substitute for conventional opium poppy farming [64].

CRISPR/Cas9 Genome Editing: The field of plant genome editing has seen a transformation, due to the application of the CRISPR/Cas9 method. This innovative technology allows for the enhancement of the production of chemicals, through the control of gene activity, the introduction of genes, or the removal of unwanted ones. For example, by removing genes linked to competing pathways, CRISPR/Cas9 has been used to boost the *Camptotheca acuminatas* production of camptothecin, a cancer compound [65].

4.11 NANOTECHNOLOGY IN PHYTOCHEMICAL DELIVERY

4.11.1 Enhancing the Bioavailability of Plant-Derived Drugs with Nanocarriers

The pharmaceutical industry has undergone changes due to the integration of nanotechnology into drug delivery systems. This advancement has opened up possibilities for enhancing the effectiveness of plant-derived medications. Despite their properties, phytochemicals often face challenges related to low bioavailability, which stem from their inherent physical and chemical characteristics, such as instability, limited water solubility, and difficulty in crossing biological barriers. Liposomes, nanoparticles, and nanoemulsions are among the nanocarriers that have emerged as solutions to overcome these limitations [66].

4.11.1.1 Nanoparticles

Nanoparticles can be used to protect phytochemicals, which are usually between 1 and 100 nanometers in size. This helps prevent the degradation of phytochemicals and improves their ability to dissolve. For instance, by encapsulating curcumin—a hydrophobic compound extracted from turmeric (*Curcuma longa*)—within polymeric nanoparticles, there is evidence of absorption. Studies suggest that incorporating curcumin into nanoparticles significantly enhances the stability and absorption of the compound, leading to treatment outcomes across various health conditions [13] (Figure 4.4).

4.11.1.2 Liposomes

A protective layer made of phospholipids surrounds liposomes, which are spherical containers capable of holding both water-soluble and fat-soluble medicines. These vesicles offer advantages such as toxicity, compatibility with the body, and the ability to contain plant-based compounds. Studies have shown that quercetin, a compound known for its inflammatory and antioxidant properties, can be effectively enclosed in liposomes, leading to enhanced absorption and effectiveness in treatment. Research suggests that quercetin within liposomes demonstrates drug distribution in the body and increased medicinal effects compared to its form [67].

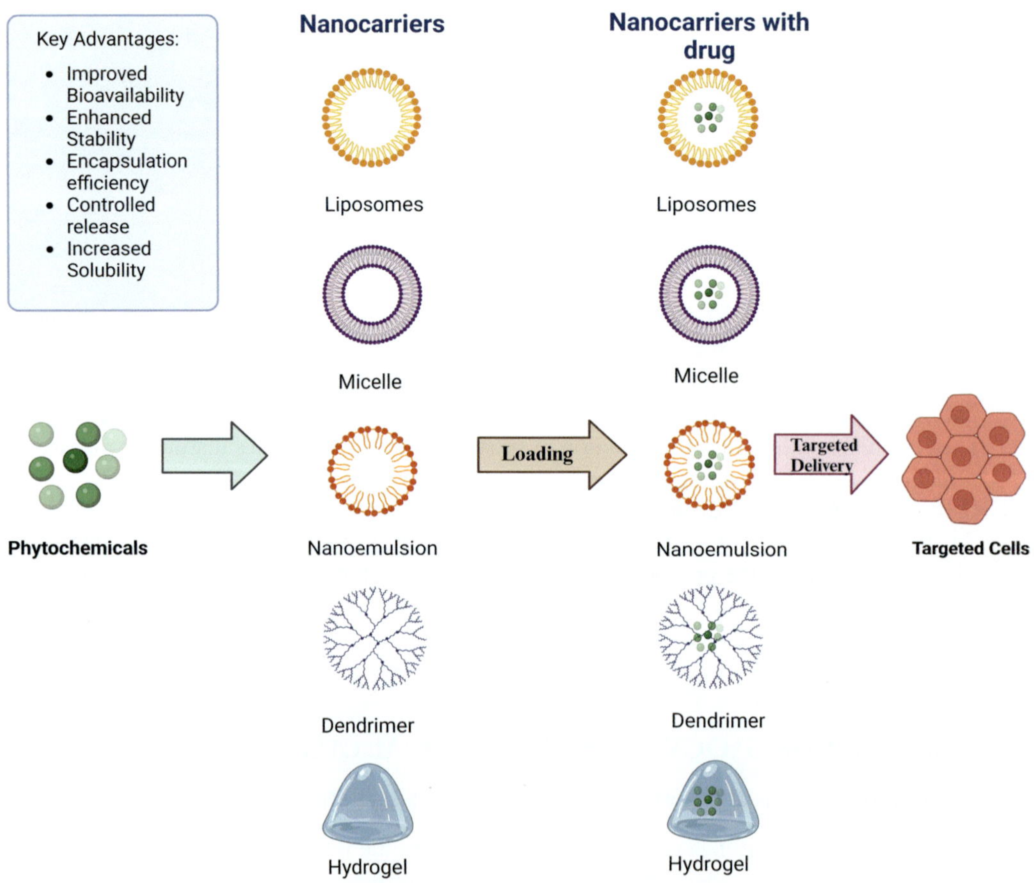

FIGURE 4.4 Nanotechnology in phytochemical delivery.

4.11.1.3 Nanoemulsions

Nanoemulsions are emulsions that help boost the solubility and stability of hydrophobic plant compounds. These special formulations hold the potential to enhance how poorly water-soluble substances are absorbed by the body. For instance, nanoemulsion versions of oils, like peppermint and rosemary, have shown improved absorption and effectiveness, making them more useful, for purposes [68].

4.11.2 Targeted Delivery Systems Using Nanotechnology

Targeted drug delivery systems using nanotechnology aim to minimize side effects and improve treatment effectiveness by delivering chemicals to affected tissues or cells. Nanocarriers can be tailored to identify and attach to biomarkers on target cells for the delivery of phytochemicals to the intended site of action [69].

4.11.2.1 Active Targeting

Active targeting involves attaching molecules, like aptamers, peptides, or antibodies, to nanocarriers to enable them to bind to receptors on target cells. One strategy for delivering curcumin to cancer cells that have an abundance of receptors is through the use of nanoparticles conjugated with folate. Studies in animal models have shown that this targeted delivery approach enhances the uptake of curcumin by cancer cells leading to heightened cell death and reduced tumor growth [70].

4.11.2.2 Passive Targeting

In tumor tissues, nanoparticles tend to build up due to the characteristics of blood vessels and inefficient lymphatic drainage. Passive targeting exploits the Enhanced Permeability and Retention (EPR) effect in these tissues. Paclitaxel, a plant-based cancer drug, has been formulated into nanoparticles to leverage the EPR effect, thereby allowing for higher drug concentrations in tumor tissues and improving treatment outcomes [71] (Figure 4.5).

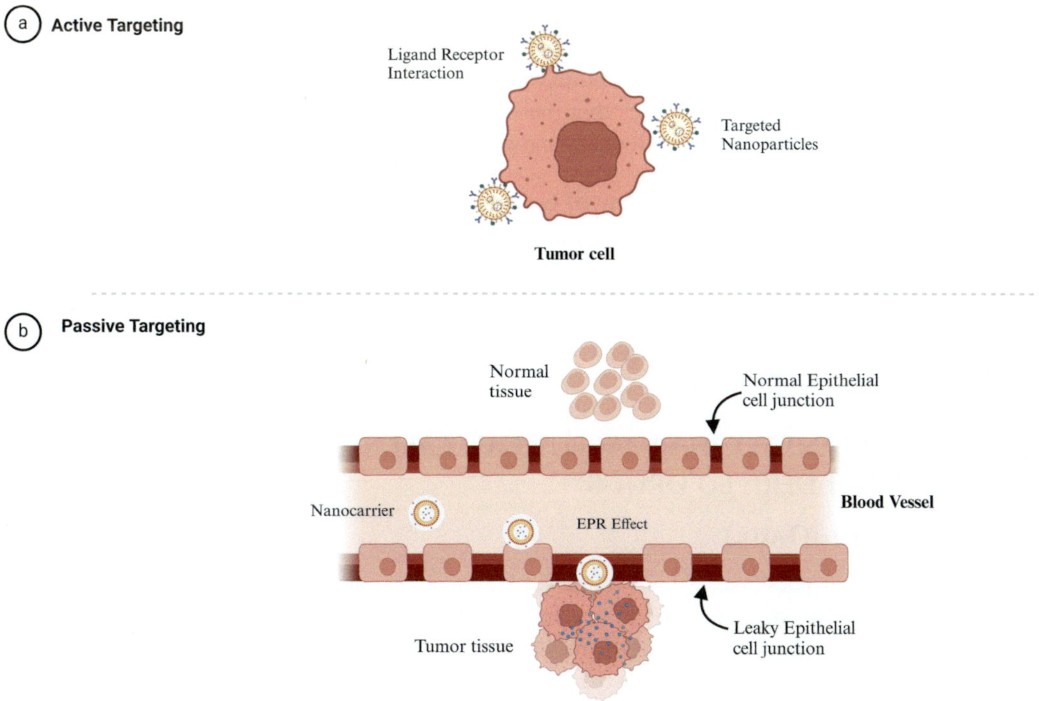

FIGURE 4.5 Targeted delivery systems using nanocarriers: a) active targeting and b) passive targeting.

4.11.2.3 Multifunctional Nanocarriers

Nanoparticles with functions are created to serve purposes, including delivering drugs conducting imaging tests and monitoring therapies. These innovative setups incorporate targeting molecules, imaging substances and therapeutic ingredients in one package, making it possible to track drug delivery and treatment progress in time. For instance, gold nanoparticles modified with targeting molecules and imaging agents have been applied to deliver resveratrol, a compound known for its cancer properties, thereby enabling the simultaneous visualization and treatment of cancer cells [72].

4.11.3 Case Studies of Nano-Formulated Phytochemicals

4.11.3.1 Curcumin-Loaded Nanoparticles

Curcumin, an anti-inflammatory and anticancer substance, faces challenges in being absorbed effectively by the body because of its limited solubility and quick breakdown. When curcumin is enclosed in nanoparticles, there are enhancements in how it moves through the body. Research has revealed that polymeric nanoparticles carrying curcumin stayed in the bloodstream longer and that they were taken up readily by cells and showed effectiveness against cancer, in breast cancer models when compared to plain curcumin [73] .

4.11.3.2 Quercetin-Loaded Liposomes

Quercetin is recognized for its ability to combat inflammation and act as an antioxidant. Its use in settings is restricted by its low solubility in water and quick elimination from the body. By encapsulating quercetin in liposomes, these obstacles have been overcome, leading to improved absorption and effectiveness. Studies show that loading quercetin into liposomes boosts its stability, raises its availability in the body, and heightens its inflammatory properties when tested on animals with inflammatory conditions [74].

4.11.3.3 Resveratrol-Functionalized Gold Nanoparticles

Resveratrol, an occurring polyphenol known for its anticancer and heart properties, encounters obstacles due to its limited absorption in the body and quick breakdown. Researchers have found success in enhancing the stability, absorption, and effectiveness of resveratrol by attaching it to nanoparticles. Recent research indicates that these modified gold nanoparticles boost resveratrol's ability to fight cancer by triggering cell death and halting tumor development in cancer types [75].

4.11.3.4 Nanoemulsion Formulations of Essential Oils

Peppermint and rosemary essential oils offer healing benefits. Their effectiveness is hindered, however, by low solubility and stability. Nanoemulsion formulations have enhanced the absorption and therapeutic qualities of these oils. Studies indicate that nanoemulsions containing peppermint and rosemary oils show absorption, stability, and biological activity, resulting in antimicrobial and anti-inflammatory effects [76].

4.12 FROM PLANT TO PHARMACEUTICAL: CASE STUDIES OF PLANT-DERIVED DRUGS

4.12.1 Paclitaxel (Taxol)

Taxol, also known as paclitaxel, stands out as a recognized example of a medication derived from plants. In the 1960s, during an investigation carried out by the National Cancer Institute (NCI) paclitaxel was first extracted from the bark of the Pacific yew tree (*Taxus brevifolia*). This compound, called diterpenoid, showed effectiveness in combating tumors in cases of ovarian and breast cancers. The journey to its discovery involved in-depth exploration and collaboration among chemists, pharmacologists, and medical practitioners. Paclitaxel functions by stabilizing microtubules, which hinders cell division and encourages programmed cell death in cells. By establishing a means of producing paclitaxel through synthetic methods using the needles of the European yew tree (*Taxus baccata*), its importance in oncology was further cemented [77].

4.12.2 Artemisinin

Artemisia annua, a plant that produces artemisinin, serves as an example of plant-based medicine. In the 1970s the Chinese chemist Tu Youyou discovered this while searching for treatments. Artemisinin, and its variations such as artesunate and artemether, are now crucial in treating malaria, especially when combined to prevent resistance issues [78]. This compound effectively fights the malaria parasite at stages of its life cycle, due to its peroxide bridge. The global impact of malaria has significantly reduced thanks to the effectiveness of artemisinin-based combination therapies (ACTs) [79].

4.12.3 Morphine

Morphine, a compound derived from the opium poppy plant, has played a role in pain relief for more than two hundred years. When Friedrich Sertürner discovered and isolated morphine in the 1800s, it marked an advancement in the field of alkaloid chemistry and led to the development of modern painkillers. By binding to receptors in the brain and altering how pain is perceived, morphine effectively reduces discomfort [80]. Despite its risks of dependency and misuse, morphine continues to be a treatment option for managing pain in medical settings, especially following surgeries, injuries, and cancer diagnoses [81].

4.12.4 Quinine

Quinine, obtained from the bark of the cinchona tree (*Cinchona officinalis*), was among the substances derived from plants to gain widespread use in Western medicine. Its effectiveness against malaria was initially acknowledged by communities in South America. It was later embraced by Europeans during the 17th century. The introduction of quinine marked an advancement in combating malaria. Played a role in enabling European expansion into regions plagued by the disease. This compound functions by disrupting the parasite's ability to metabolize hemoglobin, ultimately causing its demise. While newer antimalarial medications have largely supplanted quinine, its historical significance as a plant-based discovery remains noteworthy [82].

4.12.5 Challenges and Limitations in Plant-Based Drug Development

4.12.5.1 Complexity of Plant Extracts

Creating plant-based medicines faces an obstacle due to the nature of plant extracts. Plants generate a range of compounds, some of which might contribute to the claimed healing properties. Uncovering and separating the elements demands resources and time. Moreover the interaction of chemicals in a plant extract can lead to combined effects that complicate the identification of active components, requiring advanced analytical and bioassay methods [83].

4.12.5.2 Variability in Chemical Composition

The composition of plant materials can differ greatly depending on a number of factors, as with where they grow, the conditions they are grown in, and how they are harvested. These variations can impact the reliability and uniformity of medications derived from plants. It is crucial to standardize plant extracts and uphold quality control measures in order to create pharmaceuticals. Modern methods, like metabolomics and chemometrics, are now being employed often to tackle these challenges by offering chemical analyses of plant extracts [84].

4.12.5.3 Sustainable Sourcing and Conservation

The responsible gathering of plants is an issue, especially for species which grow slowly or which are found in limited areas. Excessive harvesting of plants can exhaust populations and put biodiversity at risk [85]. The responsible gathering of plants is an issue, especially for species that grow slowly or are found in limited areas. Excessive harvesting of plants can exhaust populations and put biodiversity at risk. It is important to use eco-cultivation methods, such as controlled farming and the use of plant cell cultures to ensure a source of raw materials. Additionally, global regulations and conservation efforts such as the Convention on Biological Diversity play a role in protecting endangered plant species [86] (Figure 4.6).

4.12.5.4 Regulatory and Approval Processes

The journey to receiving approval for plant-based medications from bodies such as the US Food and Drug Administration (FDA) and the European Medicines Agency (EMA) can be quite demanding. This is mainly because of the nature of plant extracts and the strict requirements for testing their safety and effectiveness. In order to meet these standards, detailed information on how these drugs interact with the body, their toxicity levels, and how well they work in settings is crucial. The presence

FIGURE 4.6 Challenges in plant-based drug development.

of components in plant extracts adds another layer of complexity to this process, calling for testing methods and strong clinical trial frameworks [87].

4.12.6 Intellectual Property and Benefit Sharing

In the realm of developing plant-based medicines, it is crucial to address issues surrounding intellectual property rights and agreements and on benefit sharing, especially when ancestral knowledge comes into play. It is vital to uphold fairness by providing compensation and acknowledgment to communities and nations of origin, thereby fostering ethical and just practices in drug exploration. While global frameworks like the Nagoya Protocol offer direction on genetic resource access and the fair sharing of benefits, putting these guidelines into action can be an intricate process and may provoke disagreements [88].

4.13 FUTURE PERSPECTIVES

4.13.1 Emerging Trends in Plant-Based Drug Discovery

On occasion, there have been advancements in the realm of discovering medicinal compounds from plants offering promising prospects for pharmaceutical research. One key trend is the integration of high-throughput screening (HTS) techniques, which significantly accelerates the discovery phase by allowing the testing of plant extracts and chemicals against various biological targets. Through the use of bioinformatics tools, scientists can promptly understand how potential drug candidates work [37].

Metabolomics has recently been utilized in drug discovery, providing a new perspective in drug discovery. By delving into the metabolites in a biological system, metabolomics aims to uncover the chemical reactions occurring in plants. This field of research holds potential in discovering compounds and revealing the intricacies of their production processes. Integrating metabolomics with other omics disciplines, like proteomics and genomics, enhances our grasp of plant biochemistry, facilitating the discovery of medicinal treatments [100].

The realm of exploring plant-based medicine is increasingly being impacted by advancements in nanotechnology. Thanks to nanoparticle-based delivery techniques, plant-derived compounds now exhibit solubility, stability, and bioavailability. These innovative methods enable the delivery of plant chemicals to tissues or cells, thereby enhancing their therapeutic effectiveness. For instance, research has shown that curcumin, known for its bioavailability, can be efficiently delivered using nanoparticles, thereby bolstering its potential as a substance [73].

Furthermore, the use of machine learning (ML) and artificial intelligence (AI) in drug discovery is gaining momentum. AI and ML algorithms have the ability to identify patterns and predict activities by analyzing datasets generated through plant research. This approach can streamline the drug development process, thereby reducing both the time and costs associated with methods. By combining AI and ML with established knowledge, scientists can focus on plants and substances with the potential for medical benefits [101].

4.13.2 Integrating Traditional Knowledge with Modern Science

The collaboration between wisdom and modern scientific approaches shows great promise in the development of plant-based medicines [102]. Drawing from the wealth of knowledge accumulated over centuries by civilizations, valuable insights into the healing properties of plants can be gained. This knowledge serves as a guide for researchers in identifying plants with potential, thereby expediting the drug discovery process. One key method to merge wisdom with science involves engaging in collaborative research with indigenous communities [19]. The mechanisms behind conventional plant-based medicines can be better understood and validated through the use of contemporary scientific methods such as computer modelling and molecular docking. These techniques give researchers insight into the potential therapeutic benefits of plant compounds by predicting how they may interact with particular biological targets. For example, computational studies have been used to explore the binding affinity of phytochemicals to enzymes involved in disease pathways, offering a scientific basis for their traditional use [103].

When merging wisdom with science it is crucial to establish standardized procedures for extracting and studying plant compounds. By ensuring consistency and reliability in research outcomes, standardization facilitates comparisons across studies. Techniques such as mass spectrometry and high-performance liquid chromatography are used to analyze the chemical profiles of plant extracts forming the basis for quality control and standardization. Moreover, advancements in biotechnology, such as biology and genetic engineering, can enhance the production of substances obtained from plants. For instance, one way in which to boost the supply of compounds in plants involves altering their metabolic pathways using metabolic engineering techniques. This method has proven successful in creating artemisinin, an anti-malarial compound, within genetically modified yeast cells, thereby showcasing the potential of biotechnology in discovering pharmaceuticals from plants [37].

4.13.3 Potential of Plant Genomics and Biotechnology

The field of creating plant-based medicines is experiencing advancements thanks to progress in plant genetics and biotechnology. By delving into a plant's composition, known as 'plant genomics,' researchers are uncovering the genes and processes responsible for producing beneficial compounds. Through cutting-edge next generation sequencing (NGS) technology scientists can swiftly and accurately decode the genomes of plants [47]. The availability of plant genome sequences has simplified the identification of genes for producing substances. By comprehending these pathways, researchers can make modifications to enhance the production of desired chemicals. For instance, pinpointing the genes for generating the anticancer compound paclitaxel in yew trees Species has paved the way for creating genetically modified plants that yield higher levels of paclitaxel [53].

The use of engineering and synthetic biology is a method for generating plant-based substances in different systems. By moving the pathways of plants into microorganisms like bacteria or yeast, scientists can create significant amounts of beneficial compounds in a regulated and environmentally friendly way. This technique has proven effective in manufacturing substances like artemisinin and taxol, thereby illustrating the potential for the production of plant-based medications [9]. Furthermore, the advancement of CRISPR/Cas9 genome editing technology has unlocked possibilities for enhancing plant genomes. This technique empowers scientists to pinpoint genes and make modifications facilitating the study of gene functions and the development of plants with enhanced medicinal properties. For instance, CRISPR/Cas9 has been applied to enhance the production of metabolites in plants, thereby elevating their potential as therapeutic agents. Alongside manipulation, utilizing biotechnology in plant tissue culture offers an scalable approach to producing bioactive compounds [104]. In controlled conditions, the process of growing plant cells or tissues in a laboratory setting is referred to as 'plant tissue culture.' This method allows for the production of compounds without needing to harvest the whole plant. Researchers have successfully used this technique to produce substances such as shikonin and ginsenosides, thereby demonstrating its viability for large-scale manufacturing [105].

4.14 CONCLUSION

The exploration of substances derived from plants plays a role in modern pharmacology, combining traditional knowledge with state-of-the-art scientific approaches. While older practices have set the foundation, new techniques such as analyzing plant chemicals guided fractionation through bioassays Similarly, the use of nanotechnology has greatly improved the identification and application of plant-based compounds. Despite facing challenges such as the complexity of plant extracts and sustainability concerns incorporating these methods shows promise for success in creating effective, safe, and accessible treatments. By honoring both the traditions of healing practices and the advancements in science we ensure that the discovery of plant-based medications will continue to drive progress in medical treatments.

REFERENCES

1. Rathor L. Medicinal Plants: A Rich Source of Bioactive Molecules Used in Drug Development. In: Mandal SC, Chakraborty R, Sen S, editors. *Evidence Based Validation of Traditional Medicines: A comprehensive Approach [Internet]*. Singapore: Springer; 2021 [cited July 30, 2024]. pp. 195–209. Available from: https://doi.org/10.1007/978-981-15-8127-4_10
2. Making Plants Modern: Medicinal Plants in Twentieth-Century British Pharmacy—ProQuest [Internet]. [cited July 30, 2024]. Available from: https://www.proquest.com/openview/192b568a4c16245b0633599064b10c64/1?pq-origsite=gscholar&cbl=2026366&diss=y
3. Altemimi A, Lakhssassi N, Baharlouei A, Watson D, Lightfoot D. Phytochemicals: Extraction, Isolation, and Identification of Bioactive Compounds from Plant Extracts. *Plants*. 2017 Sep 22;6(4):42.
4. Maldonado Miranda JJ. Medicinal plants and their traditional uses in different locations. In: Bhat RA, Hakeem KR, Dervash MA, editors. *Phytomedicine [Internet]*. Academic Press; 2021 [cited July 30, 2024]. pp. 207–23. Available from: https://www.sciencedirect.com/science/article/pii/B9780128241097000145
5. History and Present of European Traditional Herbal Medicine (Phytotherapy) | History, Present and Prospect of World Traditional Medicine [Internet]. [cited Jul 30, 2024]. Available from: https://www.worldscientific.com/doi/abs/10.1142/9789811282171_0003
6. Ayurveda and Traditional Chinese Medicine: A Comparative Overview—PMC [Internet]. [cited Jul 30, 2024]. Available from: https://www.ncbi.nlm.nih.gov/pmc/articles/PMC1297513/
7. History of Research on Medicinal Plants in India | SpringerLink [Internet]. [cited Jul 30, 2024]. Available from: https://link.springer.com/chapter/10.1007/978-3-030-98701-5_2
8. Sustainability | Free Full-Text | Role of Traditional Ethnobotanical Knowledge and Indigenous Communities in Achieving Sustainable Development Goals [Internet]. [cited Jul 30, 2024]. Available from: https://www.mdpi.com/2071-1050/13/6/3062
9. Molecules | Free Full-Text | Exploiting the Biosynthetic Potency of Taxol from Fungal Endophytes of Conifers Plants; Genome Mining and Metabolic Manipulation [Internet]. [cited Jul 30, 2024]. Available from: https://www.mdpi.com/1420-3049/25/13/3000
10. Review of the recent developments in metabolomics-based phytochemical research: Critical Reviews in Food Science and Nutrition: Vol 63, No 19 [*Internet*]. [cited Jul 18, 2024]. Available from: https://www.tandfonline.com/doi/abs/10.1080/10408398.2021.1993127
11. Sasidharan S, Chen Y, Saravanan D, Sundram KM, Yoga Latha L. Extraction, Isolation and Characterization of Bioactive Compounds from Plants' Extracts. *Afr J Tradit Complement Altern Med*. 2010 Oct 2;8(1):1–10.
12. Zhang Z, Li X, Sang S, McClements DJ, Chen L, Long J, et al. Polyphenols as Plant-Based Nutraceuticals: Health Effects, Encapsulation, Nano-Delivery, and Application. *Foods*. 2022 Jan;11(15):2189.
13. Curcumin, the active substance of turmeric: its effects on health and ways to improve its bioavailability—Abd El-Hack—2021—Journal of the Science of Food and Agriculture—Wiley Online Library [Internet]. [cited July 30, 2024]. Available from: https://scijournals.onlinelibrary.wiley.com/doi/full/10.1002/jsfa.11372?casa_token=aIM5XSOCQmwAAAAA%3AUHUfRFNnCwlrq5RpBNwOSgXyM-OK6rJ0sZzmkZy2zs4efO1G3INS3qtBXVu2TRvqblqKCcTrRRgOOGyx4w
14. Well grounded: Indigenous Peoples' knowledge, ethnobiology and sustainability - Turner - 2022 - People and Nature - Wiley Online Library [Internet]. [cited Jul 30, 2024]. Available from: https://besjournals.onlinelibrary.wiley.com/doi/full/10.1002/pan3.10321
15. Plants, People, and Culture | The Science of Ethnobotany | Michael J B [Internet]. [cited Jul 30, 2024]. Available from: https://www.taylorfrancis.com/books/mono/10.1201/9781003049074/plants-people-culture-michael-balick-paul-alan-cox
16. Milliken W, Walker BE, Howes MJR, Forest F, Nic Lughadha E. Plants used traditionally as antimalarials in Latin America: Mining the tree of life for potential new medicines. *J Ethnopharmacol*. 2021 Oct 28;279:114221.

17. Goswami S, Ali A, Prasad ME, Singh P. Pharmacological significance of *Catharanthus roseus* in cancer management: A review. *Pharmacol Res—Mod Chin Med*. 2024 Jun 1;11:100444.
18. Gupta A, Yadav A, Rajan N, Kulshrestha V, Singh H, Priya, et al. Unforgettable Impressions: A Captivating Review Of Echinacea (Purple Coneflower). *Eur Chem Bull*. 2023 Aug 11;12:2408–28.
19. Medicinal Plant Knowledge of the Temuan Tribe in Kampung Orang Asli Hulu Kemensah, Ampang, Selangor and the Evaluation of Anti Inflammatory Properties of Three Selected Plants—ProQuest [Internet]. [cited Jul 30, 2024]. Available from: https://www.proquest.com/openview/639a91b89782e4accec66cec2ea9d950/1?pq-origsite=gscholar&cbl=2026366&diss=y
20. Joshi P, Yadaw GS, Joshi S, Semwal RB, Semwal DK. Antioxidant and anti-inflammatory activities of selected medicinal herbs and their polyherbal formulation. *South Afr J Bot*. 2020 May 1;130:440–7.
21. Ekiert H, Świątkowska J, Klin P, Rzepiela A, Szopa A. Artemisia annua—Importance in Traditional Medicine and Current State of Knowledge on the Chemistry, Biological Activity and Possible Applications. *Planta Med*. 2021 Jan 22;87:584–99.
22. Exploring Indigenous and Local Knowledge and Practices (ILKPs) in Traditional Jhum Cultivation for Localizing Sustainable Development Goals (SDGs): A Case Study from Zunheboto District of Nagaland, India | Environmental Management [Internet]. [cited Jul 30, 2024]. Available from: https://link.springer.com/article/10.1007/s00267-021-01514-6
23. Abubakar AR, Haque M. Preparation of Medicinal Plants: Basic Extraction and Fractionation Procedures for Experimental Purposes. *J Pharm Bioallied Sci*. 2020 Mar;12(1):1.
24. Molecules | Free Full-Text | Major Phytochemicals: Recent Advances in Health Benefits and Extraction Method [Internet]. [cited 2024 Jul 30]. Available from: https://www.mdpi.com/1420-3049/28/2/887
25. Ahangari H, King JW, Ehsani A, Yousefi M. Supercritical fluid extraction of seed oils – A short review of current trends. *Trends Food Sci Technol*. 2021 May 1;111:249–60.
26. Microwave Heating: Electromagnetic Fields Causing Thermal and Non-Thermal … - Google Books [Internet]. [cited 2024 Jul 30]. Available from: https://books.google.com.pk/books?hl=en&lr=&id=c34_EAAAQBAJ&oi=fnd&pg=PA3&dq=Microwave-assisted+extraction+utilizes+microwave+energy+to+heat+the+solvent+and+plant+matrix,+thereby+enhancing+the+extraction+efficiency+of+bioactive+compounds&ots=eaJrlBskgh&sig=QNoxRxq7IqIBuScpuJDE4bbLm8Q&redir_esc=y#v=onepage&q=Microwave-assisted%20extraction%20utilizes%20microwave%20energy%20to%20heat%20the%20solvent%20and%20plant%20matrix%2C%20thereby%20enhancing%20the%20extraction%20efficiency%20of%20bioactive%20compounds&f=false
27. Recent Advances in Microwave Assisted Extraction of Bioactive Compounds from Complex Herbal Samples: A Review: Critical Reviews in Analytical Chemistry: Vol 51, No 2 [Internet]. [cited Jul 30, 2024]. Available from: https://www.tandfonline.com/doi/abs/10.1080/10408347.2019.1686966
28. Ultrasound Assisted Extraction - an overview | ScienceDirect Topics [Internet]. [cited Jul 30, 2024]. Available from: https://www.sciencedirect.com/topics/biochemistry-genetics-and-molecular-biology/ultrasound-assisted-extraction
29. Rosario GG, Miranda-Lopez R. Cellulases, hemicellulases and ligninolytic enzymes: mechanism of action, optimal processing conditions and obtaining value-added compounds in plant matrices. *MOJ Food Process Technol*. 2022 Dec 23;10:30–7.
30. CIMB | Free Full-Text | Current Status of Mining, Modification, and Application of Cellulases in Bioactive Substance Extraction [Internet]. [cited Jul 30, 2024]. Available from: https://www.mdpi.com/1467-3045/43/2/50
31. Plants | Free Full-Text | Bioassay Guided Fractionation Protocol for Determining Novel Active Compounds in Selected Australian Flora [Internet]. [cited Jul 30, 2024]. Available from: https://www.mdpi.com/2223-7747/11/21/2886
32. Plants | Free Full-Text | Extraction Techniques and Analytical Methods for Isolation and Characterization of Lignans [Internet]. [cited Jul 30, 2024]. Available from: https://www.mdpi.com/2223-7747/11/17/2323
33. Assis RC, Mageste AB, de Lemos LR, Orlando RM, Rodrigues GD. Application of aqueous two-phase system for selective extraction and clean-up of emerging contaminants from aqueous matrices. *Talanta*. 2021 Feb 1;223:121697.
34. Kifle ZD, Yesuf JS, Atnafie SA. Evaluation of in vitro and in vivo Anti-Diabetic, Anti-Hyperlipidemic and Anti-Oxidant Activity of Flower Crude Extract and Solvent Fractions of Hagenia Abyssinica (Rosaceae). *J Exp Pharmacol*. 2020 Jun 9;12:151–67.

35. Stevens LJ, Donkers JM, Dubbeld J, Vaes WHJ, Knibbe CAJ, Alwayn IPJ, et al. Towards human ex vivo organ perfusion models to elucidate drug pharmacokinetics in health and disease. *Drug Metab Rev*. 2020 Jul 2;52(3):438–54.
36. Wells M, Fossépré M, Hambye S, Surin M, Blankert B. Uncovering the antimalarial potential of toad venoms through a bioassay-guided fractionation process. *Int J Parasitol Drugs Drug Resist*. 2022 Dec 1;20:97–107.
37. High-Performance Liquid Chromatography (HPLC): A review [Internet]. [cited Jul 30, 2024]. Available from: https://www.advancechemjournal.com/articles/aac-aid1026.php
38. Reverse Phase High Performance Liquid Chromatography—an overview | ScienceDirect Topics [Internet]. [cited Jul 30, 2024]. Available from: https://www.sciencedirect.com/topics/materials-science/reverse-phase-high-performance-liquid-chromatography
39. Gruber B, David F, Sandra P. Capillary gas chromatography-mass spectrometry: Current trends and perspectives. *TrAC Trends Anal Chem*. 2020 Mar 1;124:115475.
40. GCMS Spectrum—an overview | ScienceDirect Topics [Internet]. [cited Jul 30, 2024]. Available from: https://www.sciencedirect.com/topics/chemistry/gcms-spectrum
41. Dar AH, Makroo HA, Shah S, Khan S. Nuclear Magnetic Resonance (NMR) Spectroscopy for Quality Determination of Fruits and Vegetables. In: *Sensor-Based Quality Assessment Systems for Fruits and Vegetables*. Apple Academic Press; 2020.
42. Moro MK, Neto ÁC, Lacerda V, Romão W, Chinelatto LS, Castro EVR, et al. FTIR, 1H and 13C NMR data fusion to predict crude oils properties. *Fuel*. 2020 Mar 1;263:116721.
43. Nascimento TA, Lopes TIB, Nazario CED, Oliveira SL, Alcantara GB. Vegetable oils: Are they true? A point of view from ATR-FTIR, 1H NMR, and regiospecific analysis by 13C NMR. *Food Res Int*. 2021 Jun 1;144:110362.
44. Seukep AJ, Kuete V, Nahar L, Sarker SD, Guo M. Plant-derived secondary metabolites as the main source of efflux pump inhibitors and methods for identification. *J Pharm Anal*. 2020 Aug 1;10(4):277–90.
45. Banyal A, Tiwari S, Sharma A, Chanana I, Patel SKS, Kulshrestha S, et al. Vinca alkaloids as a potential cancer therapeutics: recent update and future challenges. *3 Biotech*. 2023 May 24;13(6):211.
46. Tong Y, Luo YF, Gao W. Biosynthesis of paclitaxel using synthetic biology. *Phytochem Rev*. 2022 Jun 1;21(3):863–77.
47. Bielecka M, Pencakowski B, Nicoletti R. Using Next-Generation Sequencing Technology to Explore Genetic Pathways in Endophytic Fungi in the Syntheses of Plant Bioactive Metabolites. *Agriculture*. 2022 Feb;12(2):187.
48. Foods | Free Full-Text | QTL Mapping and GWAS Reveal the Genetic Mechanism Controlling Soluble Solids Content in Brassica napus Shoots [Internet]. [cited Jul 31, 2024]. Available from: https://www.mdpi.com/2304-8158/10/10/2400
49. Wani KI, Choudhary S, Zehra A, Naeem M, Weathers P, Aftab T. Enhancing artemisinin content in and delivery from Artemisia annua: a review of alternative, classical, and transgenic approaches. *Planta*. 2021 Jul 15;254(2):29.
50. IJMS | Free Full-Text | Recent Advanced Metabolic and Genetic Engineering of Phenylpropanoid Biosynthetic Pathways [Internet]. [cited Jul 31, 2024]. Available from: https://www.mdpi.com/1422-0067/22/17/9544
51. Dong C, Wang Z, Qin L, Zhang C, Cao L, Li H, et al. Overexpression of geranyl diphosphate synthase 1 (NnGGPPS1) from Nelumbo nucifera enhances carotenoid and chlorophyll content and biomass. *Gene*. 2023 Sep 25;881:147645.
52. Molecules | Free Full-Text | Biotechnological Approaches for Production of Artemisinin, an Anti-Malarial Drug from Artemisia annua L. [*Internet*]. [cited Jul 31, 2024]. Available from: https://www.mdpi.com/1420-3049/27/9/3040
54. Molecules | Free Full-Text | Potential Associations among Bioactive Molecules, Antioxidant Activity and Resveratrol Production in Vitis vinifera Fruits of North America [Internet]. [cited Jul 31, 2024]. Available from: https://www.mdpi.com/1420-3049/27/2/336
55. Babich O, Sukhikh S, Pungin A, Ivanova S, Asyakina L, Prosekov A. Modern Trends in the In Vitro Production and Use of Callus, Suspension Cells and Root Cultures of Medicinal Plants. *Molecules*. 2020 Jan;25(24):5805.
56. Biotechnological advancements in Catharanthus roseus (L.) G. Don | Applied Microbiology and Biotechnology [Internet]. [cited Jul 31, 2024]. Available from: https://link.springer.com/article/10.1007/s00253-020-10592-1

57. Cell Suspension Culture-Mediated Secondary Metabolites Production from Medicinal Plants with Antiallergy Properties | SpringerLink [Internet]. [cited Jul 31, 2024]. Available from: https://link.springer.com/chapter/10.1007/978-981-97-1467-4_13
58. Amorpha-4,11-diene synthase: a key enzyme in artemisinin biosynthesis and engineering | aBIOTECH [Internet]. [cited Jul 31, 2024]. Available from: https://link.springer.com/article/10.1007/s42994-021-00058-x
59. Plants | Free Full-Text | Engineering Considerations to Produce Bioactive Compounds from Plant Cell Suspension Culture in Bioreactors [Internet]. [cited Jul 31, 2024]. Available from: https://www.mdpi.com/2223-7747/10/12/2762
60. Yang YH, Mao JW, Tan XL. Research progress on the source, production, and anti-cancer mechanisms of paclitaxel. *Chin J Nat Med.* 2020 Dec 1;18(12):890–7.
61. Methyl jasmonate and salicylic acid as powerful elicitors for enhancing the production of secondary metabolites in medicinal plants: an updated review | Plant Cell, Tissue and Organ Culture (PCTOC) [Internet]. [cited Jul 31, 2024]. Available from: https://link.springer.com/article/10.1007/s11240-023-02485-8
62. Sharifi-Rad J, Quispe C, Durazzo A, Lucarini M, Souto EB, Santini A, et al. Resveratrol' biotechnological applications: Enlightening its antimicrobial and antioxidant properties. *J Herb Med.* 2022 Mar 1;32:100550.
63. Yeast Synthetic Biology for Production of Artemisinin as an Antimalarial Drug | SpringerLink [Internet]. [cited Jul 31, 2024]. Available from: https://link.springer.com/chapter/10.1007/978-3-030-89680-5_6
64. Frontiers | Promising approaches for simultaneous enhancement of medicinally significant benzylisoquinoline alkaloids in opium poppy [Internet]. [cited Jul 31, 2024]. Available from: https://www.frontiersin.org/journals/plant-science/articles/10.3389/fpls.2024.1377318/full
65. Shi M, Gong H, Cui L, Wang Q, Wang C, Wang Y, et al. Targeted metabolic engineering of committed steps improves anti-cancer drug camptothecin production in *Ophiorrhiza pumila* hairy roots. *Ind Crops Prod.* 2020 Jun 1;148:112277.
66. Gorain B, Karmakar V, Sarkar B, Dwivedi M, Leong JTL, Toh JH, et al. Biomacromolecule-based nanocarrier strategies to deliver plant-derived bioactive components for cancer treatment: A recent review. *Int J Biol Macromol.* 2023 Dec 31;253:126623.
67. Halevas EG, Avgoulas DI, Katsipis G, Pantazaki AA. Flavonoid-liposomes formulations: Physicochemical characteristics, biological activities and therapeutic applications. *Eur J Med Chem Rep.* 2022 Aug 1;5:100059.
68. Anti-osteoarthritis potential of peppermint and rosemary essential oils in a nanoemulsion form: behavioral, biochemical, and histopathological evidence | BMC Complementary Medicine and Therapies [Internet]. [cited Jul 31, 2024]. Available from: https://link.springer.com/article/10.1186/s12906-021-03236-y
69. Solanki R, Jodha B, Prabina KE, Aggarwal N, Patel S. Recent advances in phytochemical based nano-drug delivery systems to combat breast cancer: A review. *J Drug Deliv Sci Technol.* 2022 Nov 1;77:103832.
70. Full article: A Novel Folic Acid Receptor-Targeted Drug Delivery System Based on Curcumin-Loaded β-Cyclodextrin Nanoparticles for Cancer Treatment [Internet]. [cited Jul 31, 2024]. Available from: https://www.tandfonline.com/doi/full/10.2147/DDDT.S320119
71. Ying N, Liu S, Zhang M, Cheng J, Luo L, Jiang J, et al. Nano delivery system for paclitaxel: Recent advances in cancer theranostics. *Colloids Surf B Biointerfaces.* 2023 Aug 1;228:113419.
72. Jurczyk M, Kasperczyk J, Wrześniok D, Beberok A, Jelonek K. Nanoparticles Loaded with Docetaxel and Resveratrol as an Advanced Tool for Cancer Therapy. *Biomedicines.* 2022 May;10(5):1187.
73. Development, Characterization and Pharmacokinetic Profile of Chitosan-Sodium Tripolyphosphate Nanoparticles Based Drug Delivery Systems for Curcumin - PMC [Internet]. [cited Jul 31, 2024]. Available from: https://www.ncbi.nlm.nih.gov/pmc/articles/PMC7961214/
74. Tatipamula VB, Kukavica B. Phenolic compounds as antidiabetic, anti-inflammatory, and anticancer agents and improvement of their bioavailability by liposomes. *Cell Biochem Funct.* 2021;39(8):926–44.
75. Wang X, Parvathaneni V, Shukla SK, Kulkarni NS, Muth A, Kunda NK, et al. Inhalable resveratrol-cyclodextrin complex loaded biodegradable nanoparticles for enhanced efficacy against non-small cell lung cancer. *Int J Biol Macromol.* 2020 Dec 1;164:638–50.
76. Nanoemulsions of essential oils to improve solubility, stability and permeability: a review | Environmental Chemistry Letters [Internet]. [cited Jul 31, 2024]. Available from: https://link.springer.com/article/10.1007/s10311-020-01142-2
77. Liu Y, Zhao F, Wang Q, Zhao Q, Hou G, Meng Q. Current Perspectives on Paclitaxel: Focus on Its Production, Delivery and Combination Therapy. *Mini Rev Med Chem.* 2023 Oct 1;23(18):1780–96.

78. Pluijm RW Van Der, AC, Dhorda M, Dondorp AM. Triple Artemisinin-Based Combination Therapies for Malaria—A New Paradigm? *Trends Parasitol.* 2021 Jan 1;37(1):15–24.
79. Kokori E, Olatunji G, Akinboade A, Akinoso A, Egbunu E, Aremu SA, et al. Triple artemisinin-based combination therapy (TACT): advancing malaria control and eradication efforts. *Malar J.* 2024 Jan 18;23(1):25.
80. Frontiers | Opioid Receptors in Immune and Glial Cells—Implications for Pain Control [Internet]. [cited Jul 31, 2024]. Available from: https://www.frontiersin.org/journals/immunology/articles/10.3389/fimmu.2020.00300/full
81. JCM | Free Full-Text | The Pain Management of Trauma Patients in the Emergency Department [Internet]. [cited Jul 31, 2024]. Available from: https://www.mdpi.com/2077-0383/12/9/3289
82. Dagen M. Chapter 1—History of malaria and its treatment. In: Patrick GL, editor. *Antimalarial Agents [Internet].* Elsevier; 2020 [cited Jul 31, 2024]. p. 1–48. Available from: https://www.sciencedirect.com/science/article/pii/B9780081012109000019
83. Antibiotics | Free Full-Text | Interactions between Medical Plant-Derived Bioactive Compounds: Focus on Antimicrobial Combination Effects [Internet]. [cited Jul 31, 2024]. Available from: https://www.mdpi.com/2079-6382/11/8/1014
84. Review On Quality Control Parameters For Standardisation Of Herbal Drug | Journal of Advanced Scientific Research [Internet]. [cited Jul 31, 2024]. Available from: https://www.sciensage.info/index.php/JASR/article/view/687
85. Medicinal Plant Resources: Threat to Its Biodiversity and Conservation Strategies | SpringerLink [Internet]. [cited Jul 31, 2024]. Available from: https://link.springer.com/chapter/10.1007/978-3-030-58975-2_28
86. Sustainability | Free Full-Text | Regulatory Mechanisms for the Conservation of Endangered Plant Species, Chlorophytum tuberosum—Potential Medicinal Plant Species [Internet]. [cited 2024 Jul 31]. Available from: https://www.mdpi.com/2071-1050/15/8/6406
87. Regulatory Guidance: ICH, EMA, FDA | SpringerLink [Internet]. [cited Jul 31, 2024]. Available from: https://link.springer.com/referenceworkentry/10.1007/978-3-319-68864-0_58
88. Access and Benefit Sharing to Genetic Resources and International Policies and Convention 2 Part 2 Indian Journal of Integrated Research in Law 2022 [*Internet*]. [cited Jul 31, 2024]. Available from: https://heinonline.org/HOL/LandingPage?handle=hein.journals/injloitd3&div=99&id=&page=
89. Access and Benefit Sharing in Indigenous Knowledge Stewardship and Sustainable Development | SpringerLink [Internet]. [cited Jul 31, 2024]. Available from: https://link.springer.com/chapter/10.1007/978-3-031-16186-5_8
90. Radha S, Kosuri N, Kumar B. Ethnobotany and Intellectual Property Rights: Balancing Access, Benefit Sharing, and Traditional Knowledge Protection. *Int J Food Nutr Sci.* 2023 Nov 14;11:2022.
91. Dietary Supplements | FDA [Internet]. [cited 2024 Jul 31]. Available from: https://www.fda.gov/food/dietary-supplements
92. Krishna PD, Gowrav MP, Bhaskaran M, Kruthika M.R. Current Regulations of Herbal Medicines in the US and EU. *Curr Tradit Med.* 2024 Dec 1;10(6):141–51.
93. Thakkar S, Anklam E, Xu A, Ulberth F, Li J, Li B, et al. Regulatory landscape of dietary supplements and herbal medicines from a global perspective. *Regul Toxicol Pharmacol.* 2020 Jul 1;114:104647.
94. Traditional ecological knowledge in restoration ecology: a call to listen deeply, to engage with, and respect Indigenous voices - Robinson - 2021 - Restoration Ecology - Wiley Online Library [Internet]. [cited Jul 31, 2024]. Available from: https://onlinelibrary.wiley.com/doi/full/10.1111/rec.13381
95. Ombella JS. Regulation of Natural Resources Located in Indigenous Communities Territory under the Principles of Consultation and Free, Prior-Informed Consent: Perspectives in Selected Countries. *Afr J Int Comp Law.* 2021 Nov 1;29(4):499–520.
96. CEEOL—Article Detail [Internet]. [cited Jul 31, 2024]. Available from: https://www.ceeol.com/search/article-detail?id=1012495
97. Khachigian LM. Pharmaceutical patents: reconciling the human right to health with the incentive to invent. *Drug Discov Today.* 2020 Jul 1;25(7):1135–41.
98. Frontiers | Access and Benefit Sharing Under the Nagoya Protocol—Quo Vadis? Six Latin American Case Studies Assessing Opportunities and Risk [Internet]. [cited Jul 31, 2024]. Available from: https://www.frontiersin.org/journals/pharmacology/articles/10.3389/fphar.2020.00765/full
99. San Reach Landmark IPR Benefit-Sharing Accord for Diet Pill | Cultural Survival [Internet]. [cited Jul 31, 2024]. Available from: https://www.culturalsurvival.org/es/publications/cultural-survival-quarterly/san-reach-landmark-ipr-benefit-sharing-accord-diet-pill

100. Waris M, Koçak E, Gonulalan EM, Demirezer LO, Kır S, Nemutlu E. Metabolomics analysis insight into medicinal plant science. *TrAC Trends Anal Chem*. 2022 Dec 1;157:116795.
101. Applications of Artificial Intelligence and Machine Learning Algorithms to Crystallization | Chemical Reviews [Internet]. [cited Jul 31, 2024]. Available from: https://pubs.acs.org/doi/abs/10.1021/acs.chemrev.2c00141
102. Making small molecules in plants: A chassis for synthetic biology-based production of plant natural products - Liu - 2023 - Journal of Integrative Plant Biology - Wiley Online Library [Internet]. [cited Jul 31, 2024]. Available from: https://onlinelibrary.wiley.com/doi/full/10.1111/jipb.13330
103. Singh H, Bharadvaja N. Treasuring the computational approach in medicinal plant research. *Prog Biophys Mol Biol*. 2021 Sep 1;164:19–32.
104. Rinschen MM, Ivanisevic J, Giera M, Siuzdak G. Identification of bioactive metabolites using activity metabolomics. *Nat Rev Mol Cell Biol*. 2019 Jun;20(6):353–67.
105. Marchev AS, Stoykova ID, Georgiev MI. Large-Scale Production of Specialized Metabolites In Vitro Cultures. In: Loyola-Vargas V, Ochoa-Alejo N, editors. *Plant Cell Culture Protocols [Internet]*. New York, NY: Springer US; 2024 [cited Jul 31, 2024]. p. 303–22. Available from: https://doi.org/10.1007/978-1-0716-3954-2_21

5 Bioactive Compounds in Herbal Remedies

Raheela Sarwar
Center of Biotechnology and Microbiology, University of Peshawar, Peshawar, Pakistan

Muhammad Maisam
Institute of Molecular Biology and Biotechnology, The University of Lahore, Lahore, Pakistan

Muhammad Waseem Khan
Institute of Pharmaceutical Sciences, Khyber Medical University, Peshawar, Pakistan

Youngbo Xue
School of Pharmaceutical Sciences (Shenzhen), Sun Yat-Sen University, Shenzhen, China

Said Hassan
Institute of Biotechnology and Microbiology, Bacha Khan University Charsadda, Charsadda, Pakistan

5.1 INTRODUCTION

The World Health Organization (WHO) estimates that nearly 50,000 people worldwide lose their lives to infectious diseases every day, accounting for almost one-third of all deaths [1]. Antibiotics are an important and significant therapeutic discovery for the management of infections; yet their arbitrary usage has resulted in virulent types of strains becoming more resistant to them [2]. The prevalence of multidrug resistance is heightened by the pathogen resistance caused by antibiotic efflux mechanisms [3]. Despite the availability of multiple pathogen-specific and broad-spectrum medications to control microorganisms, the constant emergence of new pathogen-resistant strains necessitates the exploration of novel antimicrobial agents [4].

Multidrug resistance, which frequently extends to related antimicrobial compounds and makes ordinary microorganisms become opportunistic pathogens, is a major challenge for the contemporary world. The indiscriminate use of antibiotics, which can bioaccumulate and harm host immunity, is alarming in both developing and developed countries [5]. Consequently, there is a pressing need for alternative antimicrobial strategies, with a growing focus on plant sources due to their perceived safety compared to conventional chemotherapeutics [6]. Ancient phytotherapy is gaining interest for its enhanced therapeutic potential and reduced side effects, evolving from traditional folk medicine to modern allopathic treatments as endorsed by the WHO [7]. Historically, herbs have been known for their antimicrobial properties, containing secondary metabolites (SMs) such as essential oils (EOs), alkaloids, terpenoids, flavonoids, and their glycosides, etc. Modern research continues to

DOI: 10.1201/9781003544418-5

document and explore the effectiveness, protection, and potential, of various herbs, many of which possess antimicrobial and immunomodulatory activities [8–10].

Plant-based antimicrobial chemicals are more advantageous than allopathy medications in many ways, including fewer side effects, economic, improved patient tolerance, eco-friendliness, acceptable for long-term use, higher utilization efficiency, renewable nature, and prolonged bioavailability [11]. Furthermore, new research indicates that they can effectively manage non-communicable diseases by preventing the growth of pathogens through the use of herbal extracts in combination with protein inhibitors. This has led to global interest in developing novel antimicrobial agents as alternatives to traditional chemotherapy [12]. Research indicates that these medications may lessen the harmful effects of synthetic pharmaceuticals on people's health when treating and managing cancer. An additional innovative strategy makes use of the synergistic effects of bioactive plant extracts and antibiotics to create more potent complexes that either kill target microorganisms or prevent the formation of their cell walls [13].

The world's population is expected to surpass 7.5 billion people in the next 10 to 15 years, making the management of multidrug-resistant pathogens even more imperative. Nowadays, approximately 70–80% of people use herbal medicines for basic healthcare since they are culturally appropriate and have minimal side effects [14]. Between 60 and 90% of people in developing nations obtain their family healthcare from medicinal plants. Herbal medicines were practiced historically by Egyptian, Middle Eastern, Indian, and Chinese civilizations as early as 3000 BC [15]. Medicinal herbs like *Cimicifuga* species are being used today in Korea, Japan, and China to treat inflammation, pain, and fever. Native Americans treated irregular menstruation, malaria, renal issues, menopause, and sore throats with herbal remedies [16]. In contrast to synthetic pharmaceuticals that contribute to environmental issues through industrial waste, medicinal plants are essential to pharmacological research and drug development since they can be used as models for novel drugs as well as direct therapies [13].

Plants-derived secondary metabolites, including EOs, alkaloids, flavonoids, steroids, and their glycosides, are vital for metabolism but do not directly influence the host's metabolic functions [17]. Historically, botanicals have also been employed as anthelmintic, stomachic, anti-inflammatory, diuretic, and anti-malarial treatments. Additionally, they are employed in conventional and allopathic therapy to prevent neurological, autoimmune, and cardiovascular diseases [18, 19].

The assessment of herbal remedies' clinical potential requires the discovery of new compounds and an understanding of their mechanisms of action. People turned to ethnopharmacognosy, discovering thousands of plant-based phytochemicals as safer and more effective substitutes for chemically manufactured medications due to side effects and microbial resistance [20]. Novel therapeutic approaches are desperately needed since misfiring NF-κB stimulation is associated with a number of illnesses. Despite the high cost of innovative therapies, plant-based products offer effective alternatives. Clinical trials are required to confirm the safety, efficacy, pharmacokinetics, bioavailability, and medication interactions of these bioactive compounds, even though traditional claims suggest advantages [21]. The central theme of this chapter is to explore the potential health benefits of bioactive compounds and herbal medicines for humans.

5.2 TRADITIONAL PHYTOMEDICINE

Traditional medicine relies on the indigenous knowledge, practices, and skills of various cultures to prevent, diagnose, and treat physical and mental illnesses, guided by their unique theories and beliefs [22]. Traditional medicine has long been the most economical and readily available resource for primary healthcare. People initially utilized plants for their nutritional values; however, with the discovery of their therapeutic qualities, these natural resources became vital for the treatment of many illnesses and the enhancement of general health in many human societies [23].

Currently, several branches such as Ayurveda, homeopathy, and Unani Siddha are all aspects of the Indian medical system. Approximately 8,000 plant species have been codified as herbal medicines through folk knowledge and traditions [24]. According to pollen research, 3,500 years ago, Ancient Egyptians had used various herbs for medicinal purposes, specifically for heart, circulatory, and pain ailments, with this practice dating back 4,000 years. Thousands of instructions found in

Egyptian papyrus demonstrated the therapeutic, preservation, and cosmetic benefits of castor oil and coriander [25]. Approximately 500 rare items, processed by methods such as shade drying, grinding, and treating with wine or vinegar, are part of the over 12,000 commonly used by folk therapists [26].

Greek and Roman scholars, including Hippocrates, Celsus, Theophrastus, and Dioscorides, documented a wide array of medicinal applications for herbal plants [27]. Elsewhere in Europe, Romanians also have a rich history of employing medicinal herbs; for instance, in the 5th century BC, Herodotus noted the use of *Leonurus cardiaca* (Motherwort) by individuals north of the Danube River. Romanian pharmacopoeia started incorporating herbal products in the 19th century, and in 1904, the city of Cluj established its inaugural institute specializing in medicinal herbs [22, 28].

Traditional Chinese medicine (TCM), which now has a history of over 3,000 years, exemplifies the application of ancient, accumulated knowledge in holistic healthcare today. China's rural areas are the main regions, where approximately 5,000 traditional herbal treatments are commonly utilized. The Divine Farmer's Classic of Herbalism, originating in China about two millennia ago, is recognized as the eldest herbal manuscript in existence [29]. Numerous herbal pharmacopoeias and monographs on specific herbs have been constructed from the knowledge found in this work. Medicinal herbs are used in therapeutic procedures by African tribes, native American societies, and other traditions developed by TCM, Ayurveda, and Siddha [15]. In countries such as India (Ayurveda or Siddha), Japan (Kampo medicines), China (TCM), and Southern Asia (Unani remedies), these traditional methods continue to be widely used despite the emergence of well-documented synthetic pharmaceuticals in the West [24].

For example, India continues to practice traditional systems such as homeopathy, Ayurveda, naturopathy, Siddha, yoga, and Unani. Ayurveda, meaning "the Science of Life," encompasses a comprehensive and holistic strategy for the maintenance of good health. The origins of traditional Indian medicine are found in Vedic Sanskrit texts dating from 1600 to 3500 BC [26]. The roots of ancient medical knowledge stem from Indian physicians who empirically recorded the therapeutic and medicinal properties of plants within their medical framework. These insights are indeed a valuable resource, providing the basis for the advancement of natural drug research and development. Historical, cultural, and allopathic factors contribute to the reliance of many people in developing nations on folk specialists who use medicinal plants for their healthcare [30].

Ancient practices involving herbal plants depict the history of bioactive compounds [31]. Though bioactive chemicals were unfamiliar to most in the past, their varied applications merited exploration. The generation of plant bioactive substances usually occurs as secondary metabolites (SMs), which are discussed below. The field of natural products has recently drawn significant attention because of several factors, including the discovery of distinct chemical structures and their purposes in naturally occurring SMs, unexplored remedial applications, advancements in analytical techniques, the effectiveness of newly discovered bioactive substances in biochemical research, and the isolation, purification, and characterization of natural compounds [32]. Herbal therapy is extensively used throughout the world since it is safe, non-toxic in nature, has few adverse effects, and is affordable [22].

5.3 MODERN PHYTOMEDICINE

The development of innovative medication formulations hinges significantly on medicinal plants. Throughout history, plant-derived medications have been integral to modern medicine, whether in their full form or through the use of specific plant components, as evidenced by traditional systems [20]. The worldwide emergence of multidrug-resistant pathogens threatens the clinical efficiency of existing antibiotics, making the study of infectious diseases a global concern and a leading cause of death. Throughout the annals of history, herbal medicines have treated a wide array of infectious diseases, utilizing the unparalleled chemical diversity found in plant extracts, whether as standardized natural products or as pure compounds, promising limitless avenues for new pharmaceuticals [33].

A diverse set of methods, including synthetic chemistry, combinatorial chemistry, bioinformatics, and the extraction and purification of bioactive compounds from medicinal herbs and other natural sources, has been used to produce compounds for drug development [34]. Before the 19th century, plant medicines were used mainly in crude forms as tinctures, infusions, syrups, and decoctions, and externally as ointments and washes. Researchers in developing countries conduct extensive studies of native plants, which are sold in local marketplaces. The WHO advises incorporating traditional herbal medicines with Western medicine for effective healthcare in developing countries and achieving primary healthcare goals [35]. Patients in developed countries with chronic diseases are increasingly using herbal treatments as alternatives to synthetic drugs. Approximately one hundred novel medications derived from higher plants are available in the US drug market. Among these are Moderil, Velban, Canescine, Raudixin, and Leurocristine, which were approved between 1950 and 1970 [33, 34].

Testicular, lung, and lymphoma cancers were successfully treated with the phytophyllotoxin that was extracted from *Podophyllum emodi* [36]. Serpentine, a 1953 herbal medication made from *Rauwolfia serpentina* root, lowers blood pressure and hypertension [37]. Treatment recommendations for childhood lymphocytic leukemia, lung, uterine, and breast cancer include vincristine isolated from *Catharanthus roseus*. Similarly, several plant-based medications, known for their anticancer, antidiabetic, antibacterial, and antimicrobial properties, are commonly used to treat conditions like jaundice, hypertension, diabetes mellitus (DM), mental health issues, tuberculosis, and skin disorders [38]. Factors that contribute to the increasing interest in herbal medicines by the members of developed countries include [39]:

1. **Efficiency**: Herbal drugs are supposed to be efficient and gentle and usually act on precise body organs or systems. They are considered to cure particular diseases against which conventional drugs do not work.
2. **No Side Effects**: Herbal medications, having been utilized for an extended period, are generally believed to be free from the potential risks and side effects associated with synthetic treatments.
3. **Economical**: Herbal medicines are generally cheaper than synthetic drugs and contribute key constituents for new drug synthesis. Researchers worldwide are investigating medicinal plants to find more phytochemicals and lead compounds for treating diseases.

5.4 SYNTHESIS AND PURPOSE OF BIOACTIVE COMPOUNDS

All living beings, from the simplest bacteria to the most complex plants, process various chemical compounds essential for their sustenance and survival. Compounds in biological systems are classified into primary metabolites, which support growth and development (including amino acids, carbohydrates, lipids, and proteins), and secondary metabolites (SMs), which enhancing a plant's survival and environmental interactions [40]. SMs are compounds produced post-growth, contributing to survival rather than growth, characterized by unique chemical structures, and frequently forming as mixtures of related compounds. The production of SMs varies among species due to evolutionary selection and unique ecological necessities. Flowering plants, for example, release fragrances to attract pollinators, while diseases and herbivores produce toxins to inhibit neighboring plant growth [41]. Within the realm of SMs, some substances interact with biological systems and are termed bioactive. Thus, a simple way to define bioactive compounds in plants is SMs that cause toxicological or pharmacological effects in animals and humans [42].

Bioactive compounds generally fall into a few distinct families, each with unique structural features derived from their biosynthetic assembly. There are four key pathways for the synthesis of bioactive compounds or SMs [43]: (a) mevalonic acid, (b) shikimic acid, (c) non-mevalonate (MEP), and (d) the malonic acid pathway. The malonic acid and shikimic acid pathways are responsible for the synthesis of phenolic compounds. Alkaloids are derived from aliphatic amino acids (via the tricarboxylic acid cycle) and aromatic amino acids (via the shikimic acid pathway). The synthesis of terpenes occurs through the MEP and mevalonic acid pathways [42, 44]. Moreover, hybrid biosynthesis is valuable in agriculture for their significant gains in biomass and seed yield, a phenomenon known as hybrid vigor or heterosis [45]. In crops like Chinese cabbage, maize, and Arabidopsis, hybrids have larger leaves and taller plants due to increased leaf cell number and size. Hybrid mimics overcome the F1/F2 hurdle, allowing for high-yielding crops from kept-seed planting [46].

5.5 BIOACTIVE COMPOUNDS AND ITS CLASSIFICATION IN MEDICINAL HERBS

Bioactive compounds are crucial to the therapeutic potential of plants, offering a rich and varied assortment of chemical constituents that contribute to their medicinal properties. The categorization of bioactive compounds is still inconsistent, often relying on their intended applications rather than multiple standard categories. For example, classifications based on biosynthesis, designed for simplicity in describing biosynthetic pathways, do not correspond to the extent of pharmacological classification. Based on their chemical structures, plant bioactive compounds are divided into the following main classes [47]:

5.5.1 Phenolic Compounds (PCs)

Among plant SMs, phenolic compounds (PCs) represent one of the most significant categories. These compounds are identified by the presence of one or more phenol groups, a hydroxyl (-OH) function group linkages to aromatic rings, such as benzene or other complicated aromatic ring structures [48]. This chemically heterogeneous class of plant phenolics includes compounds that are water-soluble, organic solvent-soluble, and insoluble polymers. UV radiation activates several simple PCs. The presence of abundant PCs in plants profoundly influences their color, flavor, and taste [49]. Pharmaceutical properties including antioxidant, anti-inflammatory, and antihepatotoxic actions are exhibited by PCs biosynthesized from the shikimate pathway. PCs derivatives encompass benzoic acid derivatives, simple phenylpropanoids, isoflavones, anthocyanins, lignin, tannins, and flavonoid compounds that begin with phenylalanine [50].

Flavonoids, the most prominent class of naturally occurring phenols, occur primarily in higher plants and are essential for floral pigmentation, defense, and nitrogen fixation [51]. In the stage of symbiotic infection, it is secreted by the roots with the assistance of *Rhizobia*. Flavonoids can be categorized into several classes based on the oxidation level of the central ring [52]. Scientists have isolated more than 5,000 flavonoids from various plants, classifying them into flavanonols, anthocyanidins, flavanones, flavans, and isoflavonoids according to their chemical structures (Figure 5.1) [53]. Other flavonoid groups with minor components include dihydrochalcones, dihydroflavonols, coumarins, flavan-3, 4-diols, aurones, and chalcones. Syringic acid, a phenolic substance produced

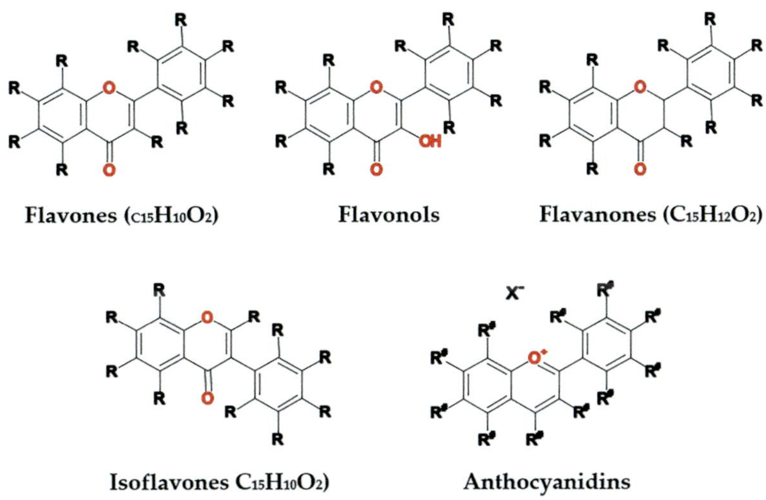

FIGURE 5.1 Common herbal flavonoids and their chemical structures.

through the shikimic acid pathway and frequently found in vegetables and fruits, demonstrates various therapeutic uses for DM prevention [49].

Flavones, which are found in buckwheat, red wine, fruit skins, red pepper, and tomato skin, exhibit anti-inflammatory antiviral, and antimutagenic effects. Similarly, flavonols, found in plants like olive oil, onions, berries, red wine, and grapefruit, show antimutagenic activity and antiviral effects by disrupting the binding between the ACE2 receptor S and protein, inhibiting virus entry, and halting viral replication enzymes [53]. Flavanones, such as naringenin and naringin, have potent antioxidant properties and contribute to the protection of the central nervous system (CNS). Flavanones, present in citrus fruits like oranges, grapefruits, and lemons, demonstrate antibacterial and antimutagenic activities [50, 51].

Isoflavones, largely derived from beans such as soybeans, demonstrate antibacterial and antimutagenic activity and are considered dietary supplements, although research on their health benefits is limited [51]. Anthocyanidins, which are mainly found in raspberries, cherries, and strawberries, are prominent plant pigments acknowledged for their health benefits against diverse oxidative agents (Table 5.1). The three main phenylpropanoid alcohols involved in lignin formation are sinapyl, coniferyl, and coumaryl [52]. Tannins, originally used in tanning hides, are generally toxic to herbivores, reducing their growth and survival when consumed. Found in fruits like blackberries, apples, red wine, tea, and, tannins are common in woody plants, especially in heartwood. Gallic acid is a notable derivative of tannins [49, 53].

5.5.2 Terpenes

The largest class of SMs, referred to as terpenoids, comprises a structurally diverse group with more than 35,000 identified terpenes. Terpenes, generally flammable and insoluble in water, originate from the combination of five-carbon atoms forming an isopentane skeleton. Terpenes are modifiable through cyclization reactions into isoprene units, the basic structural elements that make them easily recognizable. In several significant plant species, including polypodium, lemon, peppermint, basil, cotton, sage, corn, and wild tobacco, terpenoids play a crucial defensive role. They are categorized into various groups based on the quantity of isoprene units [47].

TABLE 5.1
Biological Activities and Herbal Source of Some Common Flavonoids

S#	Class	Example	Activity	Plant Source	Reference
01	Flavanones	Naringenin, Hesperetin, Homoeriodictyol, Eriodictyol.	Antibacterial and antimutagenic	Lemons and oranges, Grapefruits, and Citrus fruits.	[47, 54, 55]
02	Flavonols	Galangin, Quercetin, Myricetin, Kaempferol, Rhamnazin, Isorhamnetin.	Antiviral and antimutagenic	Olive oil, Onion, berries, red wine, and grapefruit.	[56–58]
03	Isoflavones	Glycitein, Genistein, Daidzein.	Antibacterial and antimutagenic	Soybean	[59, 60]
04	Flavones	Luteolin, Tangeritin, Apigenin.	Anti-inflammatory, antiviral, and antimutagenic.	red pepper, red wine, fruit skins, tomato skin, and buckwheat.	[61, 62]
05	Anthocyanidins	Peonidin, Cyanidin, Pelargonidin, Delphinidin.	Essential plant pigments play a crucial role in health by shielding the body from oxidative agents.	Strawberry, Cherry, and raspberry.	[63, 64]

(1): Monoterpenes, composed of two (2) isoprene units are crucial in essential oils (EOs) from plant families like Pinaceae, Apiaceae, Lamiaceae, and Rutaceae. Its molecular formula is $C_{10}H_{16}$ and they encompass alcohols (e.g., linalool), ketones (e.g., carvone), unsaturated hydrocarbons (e.g., limonene), aldehydes (e.g., citronellal), and alcohol esters (e.g., linalyl acetate) [65]. Monoterpenoids, found in the fragrant oils of many plants and marine organisms, are known for their halogenated forms in marine life. They are key components of EOs, serving as pollinator attractants and defense compounds. Additionally, monoterpenoids have potential antioxidant properties and are used in medicines for their antimicrobial, antiseptic, disinfectant, and wound-healing properties [66]. Common herbal EOs compounds and their chemical structures are represented in Figure 5.2. **(2): Sesquiterpenes** are represented by the chemical formula $C_{15}H_{24}$ and structured with three isoprene units. These are categorized into three groups: (a) acyclic (e.g., farnesol); (b) monocyclic (e.g., bisabolol); and (c) bicyclic (e.g., caryophyllene). They all participate in multiple developmental processes, such as tuberization using antagonism. **(3): Diterpenes**, composed of four isoprene units (C20H32), are classified as acyclic or macrocyclic compounds. They include vital bioactive compounds like vitamin K1 and A, which are found in medicinal plants [67]. **(4): Sesterterpenes** are structured with five isoprene units ($C_{25}H_{40}$). **(5): Triterpenes** are structured with six isoprene units (C30H48). **(6): Sesquarterpenes** are made up of seven isoprene units ($C_{35}H_{56}$). **(7): Tetraterpenes** consist of eight isoprene units with the molecular formula $C_{40}H_{64}$. **(8): Polyterpenes** are characterized by long chains of multiple isoprene units. **(9): Norisoprenoids** are distinguished by chain shortening through methyl side chain substitution with hydrogen atoms or methylene group removal [68].

Bornyl acetate exhibits strong antibacterial properties, while camphene is a minor component in some aromatic EOs and a major one in others like oregano and thymus, known for anticancer, antibacterial, and pulmonary disease treatment benefits [69]. Azadirachtin, from neem tree seeds, is a potent insecticide found in pesticides like azaSol, treeAzin, terramera Cirkil, and azaMax. Pinene, derived from coniferous trees, is antioxidant, antimicrobial, anti-carcinogenic, anti-inflammatory, and insecticidal. Limonene, from citrus oils, offers antifungal, antibacterial, and food preservation uses [70, 71]. Saponins, found in *asparagus*, *Allium* species, spinach, oats, tea, sweet potato, and sugar beet possess hypolipidemic properties and can lower lipoprotein, cholesterol levels, and exhibit anticytotoxic antidiabetic and effects [72, 73]. 3-Carene, occurring naturally in aromatic plants like citrus, rosemary, cedar, cannabis, basil, and pine, serves as a flavoring agent and food additive in the food industry [74]. They act as an antimicrobial against food-borne pathogens (both Gram-positive and Gram-negative) [75]. Some common herbal terpenoids and their structures are represented in Figure 5.3, and the biological activities of various herbal terpenoids are expressed in Table 5.2.

Steroids are a subclass of terpenoids, biosynthesized from terpene precursors. These organic compounds have four rings in a specific configuration and are produced by plants, animals, and microorganisms. Due to their high biological activity, steroids are widely studied in chemistry and biochemistry and are in high demand for medical use. This drives the synthesis of various steroids in research and development labs. Steroids are effective in preventing and treating a wide range of diseases, from common infections to severe cancers [76]. Plant steroids are diverse secondary metabolites with important physiological and pharmacological properties. Phytosterols, key components of plant cell membranes, control permeability and fluidity, and exhibit hypocholesterolemic and anticancer activities. They differ from cholesterol by having an alkyl substituent at C-24, while anolides have C-22 and C-26 oxidized to form γ-lactone [77]. Brassinosteroids are plant hormones crucial for development, classified based on the number of carbon atoms in their side chain. Steroid alkaloids, polar compounds effective against insects and pathogens, have a modified steroid skeleton with a nitrogen atom [76, 78]. Some common steroids and their structures are represented in Figure 5.4.

5.5.3 Nitrogen-Containing Compounds

As a structural component, nitrogen is found in a variety of SMs, including cyanogenic glucosides, alkaloids, and glucosinolates. **Alkaloids**, a large family of over 15,000 nitrogen-containing

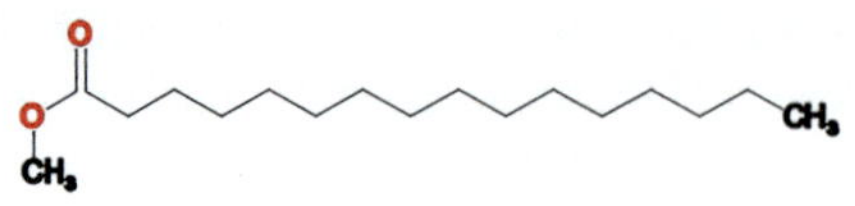

Methyl palmitate

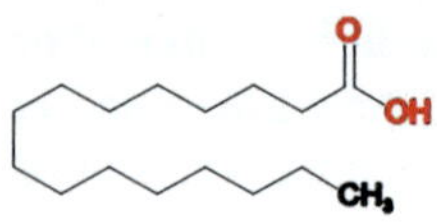

Hexadecanoic acid

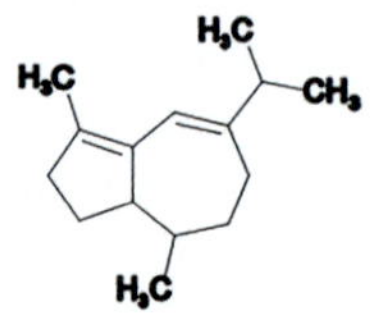

Guaia-4, 6-diene

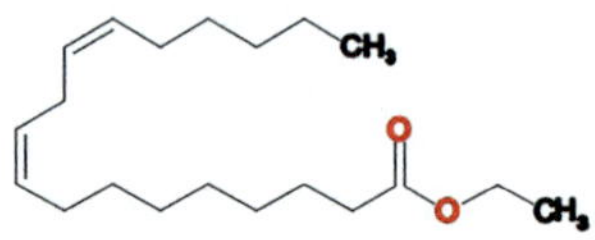

Ethyl linoleate

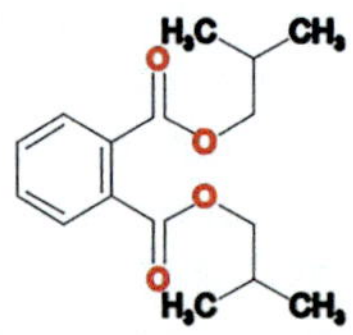

Diisooctyl phthalate

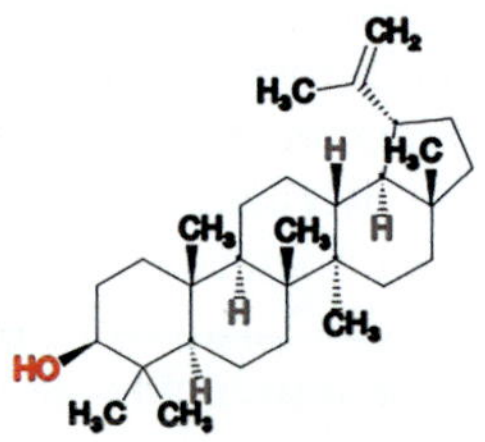

Lupeol

FIGURE 5.2 Common herbal Essential oil (EOs) compounds and their chemical structures.

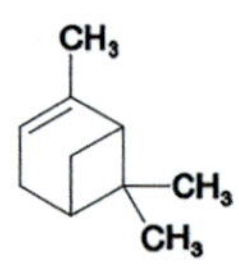

α-Pinene ($C_{10}H_{16}$)

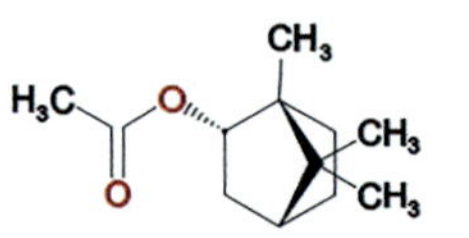

Bornyl acetate ($C_{12}H_{20}O_2$)

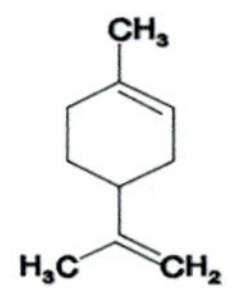

Limonene ($C_{10}H_{16}$)

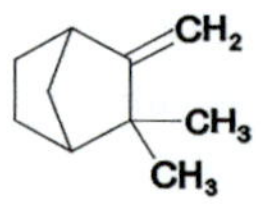

Camphene ($C_{10}H_{16}$)

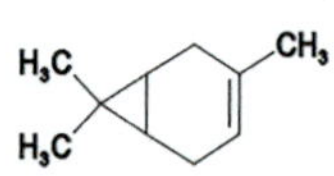

3-Carene

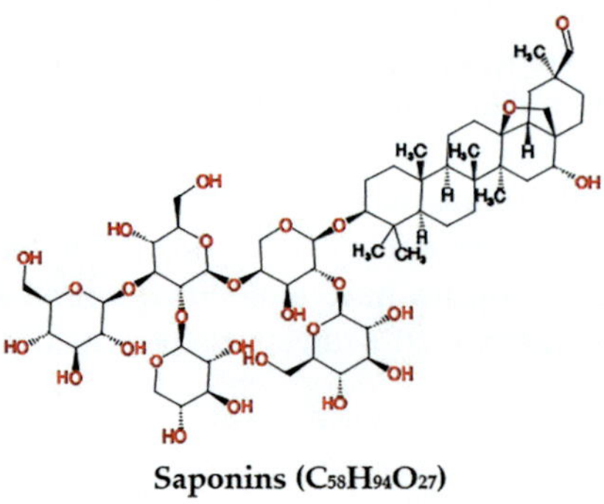

Saponins ($C_{58}H_{94}O_{27}$)

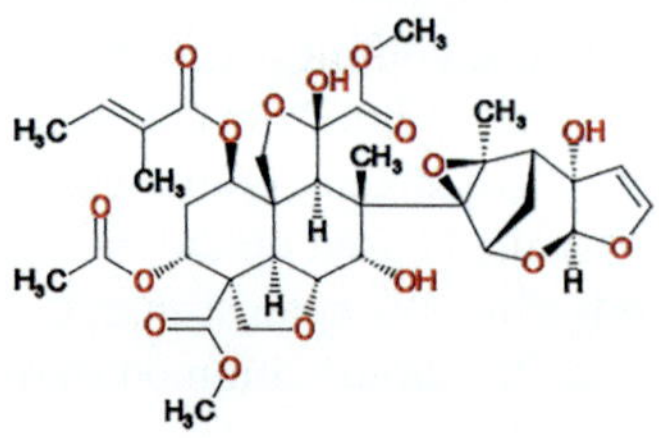

Azadirachtin ($C_{35}H_{44}O_{16}$)

FIGURE 5.3 Common herbal terpenoids and their chemical structures.

TABLE 5.2
Biological Activities and Herbal Source of Some Common Terpenoids

S#	Compound	Nature	Activity	Plant Source	Reference
01	Bornyl acetate	Ester	Antibacterial	Many plant EOs such as pines, cedars, spruces, and hemlocks.	[69, 79]
02	α-Pinene	monoterpene (Unsaturated and bicyclic)	anti-inflammatory, anti-carcinogenic, antioxidant, and antimicrobial.	Many coniferous trees	[72, 80]
03	Azadirachtin	Heterotetracyclic	Active component in pesticides like AzaMax, Cirkil, AzaSol, TreeAzin, Terramera, and insecticidal.	*Azadirachta indica* (Seeds of the neem tree)	[70]
04	Camphene	Bicyclic monoterpene	Used in pulmonary disease, anticancer, and antibacterial.	a significant component of oregano, thyme, and sage EOs, and a minor component of numerous aromatic Eos, including ginger, camphor, and citronella.	[75, 81]
05	Saponins	Triterpene glycosides	Hypolipidemic,Reduce level of lipoprotein and cholesterol, anti-carcinogenic, and antidiabetic.	*Allium* species (asparagus, onion, garlic), Spinach, tea, oats, sweet potato, and sugar beet.	[73, 82]
06	Limonene	Cyclic momoterpene	Antifungal, antibacterial, and food preservation.	Citrus fruit peel oil's primary ingredient	[71]
07	3-Carene	Bicyclic monoterpene	Used as a food additive with antimicrobial properties against food-borne pathogens.	Citrus, rosemary, cedar, pine, basil, and cannabis	[74]

SMs found in about 20% of vascular plants, are generally alkaline [47]. Alkaloids can be divided into heterocyclic and non-heterocyclic types and are common in orders like Gentianales, Rosales, Caryophyllales, and Magnoliales. They exhibit significant biological properties, including cytotoxic [83], pharmacological, antiviral, and antimicrobial effects [84]. In plants, alkaloids act as storage for nitrogen, protection from predators, and regulators of growth. Many alkaloids, such as ephedrine, quinine, and homoharringtonine, are important medicinal agents for treating diseases like DM, malaria, cardiac dysfunction, and cancer [85].

Morphine, derived from the poppy plant, is a vital alkaloid known for its analgesic, anesthetic, and anxiety-reducing properties [86]. Ephedrine, derived from Ephedra species, possesses anti-inflammatory properties and prevents low blood pressure during anesthesia [85]. Quinine, sourced from *Remijia* species, is a vital antimalarial medication first isolated in 1820 and recognized on the WHO's List of Essential Medicines [84]. Homoharringtonine, extracted from *Cephalotaxus fortunei*, is an important anticancer agent, approved by the FDA for chronic myeloid leukemia and used in China for over 50 years to treat myeloid leukemias [87]. *Chelidonium majus* produces chelerythrine, which possesses antibacterial and anticancer properties, offering the potential for the development of new anticancer therapies [88]. Piperine, extracted from *P. officinarum* and *Piper longum* fruits, is utilized as an antihyperglycemic agent [89]. *Vinca minor* produces vincamine, a vasodilator utilized as a dietary supplement in the USA and for treating dementia in Europe [90]. Cyanogenic glucosides, found in over 2500 plant species, defend against herbivores by providing a bitter taste and releasing toxic hydrogen cyanide when tissue is disrupted. They notably originate from apricots, seeds of bitter almonds, peaches, almonds, apples, and wine, varying with alcohol concentration, fruit levels, and temperature [91]. Some common herbal alkaloids and their structures are represented in Figure 5.5. Furthermore, Table 5.3 displays the biological activity of some herbal alkaloids.

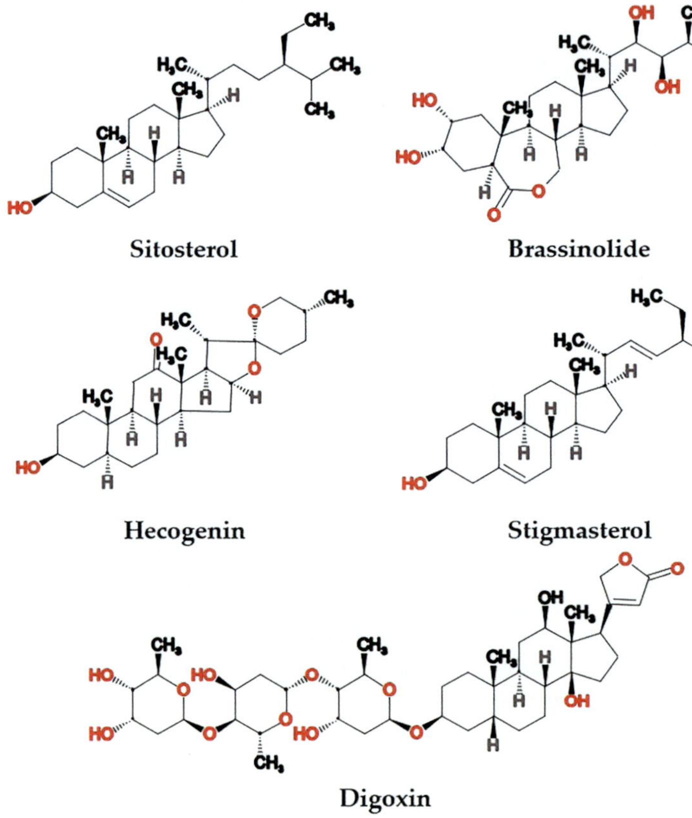

FIGURE 5.4 Common herbal steroids and their chemical structures.

5.6 EXTRACTION OF BIOACTIVE COMPOUNDS

Taking into account the extensive diversity of bioactive compounds and the wide variety of plant species, establishing a standardized and comprehensive strategy to screen out these compounds with potential health benefits is crucial. Extraction is the essential primary phase in studying medicinal plants, as it is essential to isolate the target chemical components for subsequent separation and characterization [96]. To extract metabolites, several methods have been developed, including novel, high-tech methods. The essential steps encompass pre-washing, drying (or freeze-drying), crushing plant materials for a consistent sample, and improving extraction kinetics by maximizing the contact between the sample's surface and the solvent. The extract should follow the traditional healer's procedures if it is based on traditional applications. It is imperative to maintain the active ingredients throughout the preparation process [97].

The selection of a solvent system is based on the characteristics of the target bioactive molecule. Polar solvents like ethanol, methanol, or ethyl-acetate extract hydrophilic compounds, while dichloromethane or a methanol/dichloromethane mixture (1:1) is used for lipophilic compounds [98]. The extraction method's suitability depends on the range of non-polar to polar and thermally labile target compounds. Identifying and characterizing compounds in plant extracts, which frequently consist of a mixture of bioactive compounds with different polarities, presents a considerable challenge. Various methods like reflux heating, sonification, soxhlet extraction, percolation, maceration, and modern extraction techniques such as supercritical-fluid, pressurized-liquid, solid-phase, microwave-assisted, and surfactant-mediated, are used for plant sample extraction [99]. Pure compounds are frequently isolated to determine their biological activity and structures using methods like Thin Layer Chromatography (TLC), flash chromatography, column chromatography, High-Performance Liquid Chromatography (HPLC), and Sephadex chromatography. Non-chromatographic methods

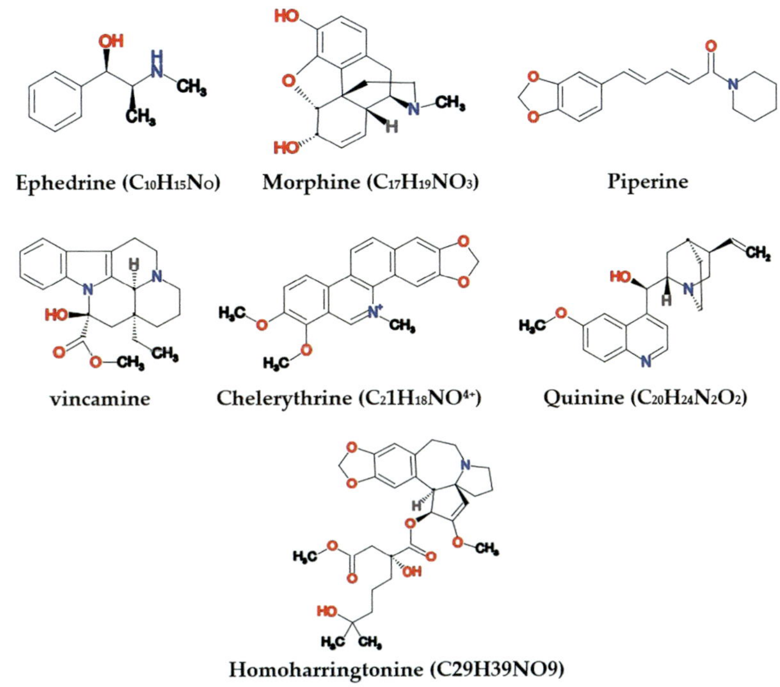

FIGURE 5.5 Common herbal alkaloids and their chemical structures.

TABLE 5.3
Biological Activities and Herbal Source of Some Common Alkaloids

S#	Compounds	Activity	Plant Source	Reference
01	Vincamine	Vasodilatory	*Vinca minor*	[90, 92]
02	Morphine	Analgesic	*Papaver somniferum*	[86, 93]
03	Ephedrine	Antiasthma (Anti-inflammatory)	*Ephedra sp.*	[85]
04	Quinine	Antimalarial (Antimicrobial)	*Remijia sp.*	[84, 94]
05	Piperine	Antihyperglycemic	Fruits of *Piper longum* and *Piper officinarum*	[89]
06	Homoharringtonine	Anticancer activity	*Cephalotaxus fortunei*	[87, 95]
07	Chelerythrine	Antimicrobial	*Chelidonium majus*	[88]

like immunoassay with monoclonal antibodies, Fourier-transform infrared spectroscopy (FTIR), and phytochemical screening assays also assist in identifying bioactive compounds [100]. For the extraction of plant material, these techniques improve extraction efficiency, selectivity, and automation ease by reducing solvent consumption, the degradation of the sample, and the elimination of additional clean-up steps [101].

5.7 ROLE OF HERBS IN DRUG DISCOVERY

Since ancient times, medicinal plants have been utilized in all cultures. Originally employed to fulfill nutritional needs, they have become essential for health improvement and disease treatment. Various plant species are still used in parts of South America, Asia, and Africa for remedies [102]. Many biologically active plant species remain undiscovered, even though the WHO reports that traditional medicine, especially herbal medications, is the primary healthcare for 60% of the world's population and is essential for the development of modern medicine [103]. Traditional medicines based on generations of experience, including herbal products, organic matter, and minerals, are widely accepted for their compatibility with the human body, fewer side effects, and cultural acceptance. Primary healthcare is provided by traditional medicine made from more than 35,000 plant species approximately 80% of the world's population [104].

Ethnomedicinal studies are essential for finding new therapeutics from medicinal plants. The chemical diversity, specific activities, and unique mechanisms of green pharmaceuticals are making them increasingly attractive. Unlike synthetic drugs, which often have unforeseen side effects, herbal medicines are considered safer and offer additional benefits from their medicinal constituents, minerals, and vitamins. This has shifted scientific focus towards ethnomedicines, increasing the demand for herbal medicines and natural products worldwide [105]. Ethnobotany, the study of traditional knowledge about medicinal plants, is largely responsible for developing modern medications. Approximately 25% of modern pharmacopeia drugs and many synthetic analogs are plant-derived. Promising treatments for infectious diseases can be obtained from approximately 78% of newly developed chemical ingredients, whether or not they are derived from natural materials. Plants have been essential in medicine for thousands of years, with significant discoveries like morphine, quinine, codeine, cocaine, and digitoxin still in use [84, 86]. Advances in genome sequencing continue to reveal new drug targets from medicinal plants [32].

Several medicinal plant extracts, such as those containing antifungal proteins like glucanase and chitinase, effectively combat microbial and parasitic infections, protecting developing embryos [106]. Pakistan's rich botanical diversity includes an estimated 400–600 medicinal plant species out of 5700, yet only a small fraction has been subjected to biochemical study. Historically, 84% of Pakistan's population relied on traditional medicines for basic healthcare; this is now largely limited mainly to remote areas due to modernization. Despite efforts to document Indigenous medicinal plants, comprehensive information remains incomplete [107].

5.8 GLOBAL TRADE OF HERBAL MEDICINES

In the pharmaceutical sector, medicinal herbs are now pivotal exports, available in diverse forms such as fruits, bark, tubers, flowers, roots, leaves, seeds, and extracts, contributing significantly to international trade for both developed and developing nations [108]. For medical care, over 75% of people worldwide use herbal remedies, particularly in developing countries where they are accepted culturally, they have few adverse effects, and are physiologically compatible with the human body. The FDA's adjustments to regulations on herbal supplements have driven substantial expansion in the herbal products market. Consumers increasingly prefer herbal medicines because of their fewer adverse effects when compared with modern remedies, and also their trusted reputation for safety. These medications offer excellent treatment for specific disorders and are a good substitute for preventive healthcare [109].

The global herbal products market is valued at around $83 billion annually, with demand growing at 5 to 15% per year according to the World Bank [110]. In the European Union, the herbal market surpasses $20 billion, with notable sales in Germany ($3 billion), Japan ($1.5 billion), France ($1.6 billion), and Italy ($0.6 billion) [111]. Herbal essences are recognized as prescribed medications by national health insurance programs in Germany and France. Between 2004 and 2012, the worldwide pharmaceutical market expanded from $550 billion to $900 billion. About $10 billion is spent on herbal products in India each year, with $1.1 billion of those exports being made up mostly of traditional Ayurvedic products [112]. China exports $3.6 billion of its annual herbal commerce, worth around $48 billion. The European herbal products market was estimated at $7 billion in 1997, with Germany leading the way at $3.5 billion. In 1995, India shipped 32,600 tons of raw pharmaceuticals worth $46 million, while China exported 120,000 tons worth $264.5 million. The global herbal market is currently valued at over $250 billion, with the value of TCM products surpassing $400 billion in 2010. The herbal remedy is also a necessary component of the bulk of modern formulations intended to lower mental disease and hypertension [113, 114].

5.9 EFFECTS OF HERBS AND THEIR BIOACTIVE COMPOUNDS IN HUMAN HEALTHCARE

Bioactive compounds in herbs are naturally occurring chemicals that offer health benefits and play a significant role in alternative and traditional medicine. Recently, herbal bioactive compounds have garnered interest for their potential to improve health and manage conditions like DM, cancer, and cardiovascular disorders (Table 5.4) [18]. Derived from the Greek "bios" (life) and the Latin "activus" (active), these bioactive compounds, both essential and nonessential, can influence human health. Plant bioactive substances are SMs that can affect an animal's or human's physiology or

TABLE 5.4
General Bioactivities of Various Herbal Medicinal Plants

S#	Plant Name	Mechanism/Bioactivity	References
01	*Azadirachta indica*	Constipation, leprosy, respiratory disorders, helminthiasis, and skin infections.	[118]
02	*Angelica keiskei*	Hypertension and coronary heart disease.	[119]
03	*Bauhinia variegata*	Dysentery, malaria, ulcers, skin ailments, leprosy, tuberculosis, and snakebite.	[120]
04	*Bacopa monnieri*	Damaged nerve cell repair, antistress, action of kinase enzyme, transmission of neural impulse, improve synaptic impulse transmission, and increase memory.	[121]
05	*Curcumin*	Immunomodulatory, anti-inflammatory, anti-angiogenic, anti-mutagenic, neuroprotective, and wound healing.	[122]
06	*Euphorbia hirta*	Worm infestation, asthma, diuretic, inflammation, scavenging activity, respiratory tract infection, coughs, purgative, sore, and wound healing.	[123]
07	*Fraxinus hookeri*	pneumonia and typhoid fever.	[124, 125]
08	*Garcinia xanthochymus*	Prostaglandin E2 and cyclooxygenase inhibition.	[126]
09	*Hoslundia opposita*	Sore throats, venereal diseases, epilepsy, fever, anti-bacterial, antiseptic, purgative, diuretic, and febrifuge.	[127]
10	*Lycium barbarum*	Hepatitis, hyperlipidemia, male infertility, cancer, thrombosis, hypo-immunity, and anti-aging.	[128]
11	*Murraya koenigii*	Swollen hemorrhoids, amoebiasis, intestinal inflammation, hepatitis, fresh cuts, pruritus, DM, nausea, burses, oedema, and snakebite.	[129]
12	*Nerium oleander*	Diaphoretic, warts, cancerous, abdomen pain, hyperglycemic, and cardiotonic.	[130]
13	*Ocimum sanctum*	Scavenging, anti-asthmatic, hepatoprotective, hypolipidemic, antidiabetic, expectorant, anticancer, and anti-fertility.	[131]
14	*Polygonum bistorta*	Demulcent, laxative, cholera, bowel syndrome, ulcer colitis, astringent, peptic ulcers, styptic, bleeding, diuretic, and menstruation.	[132]
15	*Randia dumetorum*	Piles, demulcent, diarrhea, gonorrhea, diuretic, asthma, antidysenteric, and emetic.	[133]
16	*Selaginella lepidophylla*	Immunomodulating, antiallergic, anti-viral, antiphlogostic, anti-hepatotoxic, cytotoxic antitumor, and anti-fungal.	[134]
17	*Terminalia chebula*	Renoprotective, antioxidant, anti-bacterial, adaptogenic, anticancer, antiviral, antidiabetic, and radioprotective.	[135]
18	*Urtica diocia*	Urticaria, eczema, prostatic hypertrophy, alopecia, allergic rhinitis, and gout.	[136]
19	*Withania somnifera*	Lung inflammations, immunomodulatory, arthritis, anticancer, radiosensitizing, thyro-regulatory, anti-aging, anti-inflammatory, anxiolytic, dropsy, and cardiotonic.	[137]
20	*Xanthium strumarium*	Leucorrhoea, menorrhagia, herpes, scrofula bladder infections, diaphoretic, renal complaints, CNS depressant, and inflammatory swellings.	[138]
21	*Ziziphora tenuior*	uterus infection, vomiting, dysentery, gastrointestinal, fever, and diarrhea.	[139]

toxicity [115]. Present in various plant parts like roots, flowers, and leaves, they often exist in trace amounts and may require extraction to isolate the desired compounds. Compounds like chromones, flavanones, and steroids from *Viscum coloratum* inhibit superoxide anion production by neutrophils [116]. Plant phenolic compounds are popular for their health advantages and bioactive characteristics, with increased consumption recommended for preventive health. Significant antioxidant properties are attributed to chlorophyll, and parthenolide and other metabolites have been shown to decrease human blood platelet activity. Traditional medicine uses a variety of plant components to treat a wide range of illnesses, including fevers, diarrhea, ulcers, wounds, cholera, and mental disorders. However, challenges such as plant identification, mechanism of action, chemical composition, toxicity, dosage, and cost limit the use and widespread adoption of herbal preparations [117].

5.9.1 Herbal Compounds for the Human Immune System

Diverse solvent compounds/extracts from herbal components increase the generation of antibodies by stimulating human T-lymphocytes, neutrophils, peripheral blood mononuclear cells (PBMCs), and Jurkat (JKT) cells (Table 5.5) [140]. For optimal health, both innate and acquired immunities play a vital role as the immune system targets infectious pathogens or foreign antigens. Even minor deficiencies in either type of immunity can cause substantial problems in infection prevention. First-line infection protection is provided by non-specific/innate immunity, but the adaptive immune response, which involves antibodies, lymphocytes, and other molecules, possesses two essential features: memory and specificity [141]. Innate immunity contains various components that resist specific pathogens through cellular and molecular recognition. This system eliminates some pathogens, and evolutionary innovations have led to adaptive immune responses. Enhancing immune function is vital to controlling epidemic diseases, as immunological dysfunction is a major factor in the emergence of diseases like cancer, respiratory issues, and allergies [142].

Splenocytes and lymphocytes play an integral role in acquired immunity because they are either directly or indirectly engaged in the manufacture of antibodies. They have acid phosphatases, which help kill bacteria and their antigens. Lymphocytes, derived from bone marrow via hematopoiesis, circulate through the blood and lymphatic systems, utilizing surface receptors to confer diversity, immune specificity, memory, and the ability to distinguish self from non-self. T and B lymphocytes are the two primary types of lymphocytes. T cells secrete lymphokines that protect against cancer, microbes, and type IV hypersensitivity reactions, hence mediating cell-mediated protection [143]. IgG and IgM, two key antibody molecules required for humoral immune responses, are produced by

TABLE 5.5
Different Herbal Plants Promote the Enhancement of Human Immune Cells and Antibody Production

S#	Plant Name	Mechanism/Bioactivity	References
01	*Andrographis paniculata*	Increased antibody production.	[150]
02	*Baliospermum montanum*	Chemotactic, phagocytosis, and neutrophils killing intracellular potency.	[151]
03	*Calendula arvensis*	Immunomodulation of T-lymphocytes.	[152]
04	*Glebionis coronaria*	Cytotoxicity.	[153]
05	*Inula crithmoides*	Immunomodulation of T-lymphocytes.	[154]
06	*Rhinacanthus nasutus*	Increased antibody.	[155]
07	*Tripterygium wilfordii*	Immunosuppression of T-cell proliferation	[156]
08	*Bauhinia variegata*	Enhance neutrophils, and antibody.	[120]
09	*Citrus aurantifolia*	Activation of mononuclear cells.	[157]
10	*Gymnema sylvestre*	Chemotaxis, neutrophils, and phagocytosis.	[158]

B-lymphocytes and plasma cells. These immunoglobulins regulate complement activation, the opsonization of microorganisms, and the neutralization of toxins. B-lymphocytes recognize antigens via surface receptors, leading to antigen processing and presentation to TH_2 cells via MHC class II. This initiates clonal expansion, resulting in memory cell formation and differentiation into antibody-secreting plasma cells. Secondary immune responses typically show a shift from initial IgM secretion to a predominant IgG production against recurring antigens [144].

Efficient phagocyte activity in innate immunity, led by macrophages and neutrophils, eliminates microorganisms and dead cells, acting as the initial defense against infections and regulating immune functions. Neutrophils and macrophages are pivotal in host defense, utilizing phagocytosis to combat infections and regulate immune responses, while their oxidative metabolism generates ROS crucial for intracellular killing, confirming their role as primary defenders against foreign invaders. In addition to neutrophils, eosinophils, and natural killer cells, macrophages are essential components of both specific and nonspecific defensive systems against toxins and microorganisms [145]. Their activation states affect the course of disease by intimately interacting with T- and B-lymphocytes to initiate adaptive immune responses. They also maintain tissue balance by recognizing, engulfing, and neutralizing pathogens. During injury, macrophages capture pathogens and collaborate with other immune cells to produce antibodies and orchestrate cell-mediated immunity, while also releasing ROS, cytokines, NO, and lipid mediators crucial for effective inflammatory responses [146].

An immunomodulator alters immune system components to induce either immunostimulation or immunosuppression effects, enhancing complement proteins, granulocytes, NK cells, activated lymphocytes, and macrophages to produce effector molecules. Unlike drugs which target single receptors, immunomodulators can activate both pathways simultaneously, aiming to enhance immune responses (immunostimulatory therapy) by boosting macrophages, granulocytes, T-lymphocytes, complement proteins, and effector substances. Conversely, immunosuppression reduces immune resistance. Both aspects are crucial for normal immune function and disease management. Ayurveda's "Rasayana" emphasizes immunomodulatory activities, driving the search for effective immunotherapeutic agents [147, 148].

Evidence suggests that plant-based substances possess immunomodulatory properties that improve innate immune responses. Immunomodulatory agents, encompassing both immunostimulatory and immunosuppressive effects, are vital for regulating normal immune function and adapting to pathological conditions. Natural compounds provide an alternative approach to conventional chemotherapy for the treatment of many diseases. Medicinal plants are highly rich in compounds that modulate the immune system, especially by activating NK cells, granulocytes, and macrophages, enhancing complement functions and immunity [149].

5.9.2 Bioactive Compounds in Herbs For Cancer Treatment

Worldwide, people are impacted by a range of malignant cancers. Despite the wide range of treatments available, such as immunotherapy, chemotherapy, radiotherapy, and surgery their harmful side effects frequently cause harm to patients and prevent efforts to control cancer. Plant metabolites, long utilized in medical treatments for millennia, are now emerging as new leads for anticancer drug development. Plants have long been employed in medicine; they are the source of over 60% of treatments for cancer and roughly 25% of all modern drugs. This highlights the need for new, more effective cancer treatments that are less toxic, and it suggests that plant-derived metabolites could be a source of these drugs [159].

Numerous plant extracts have demonstrated potential bioactivities, some of which are represented in Table 5.6. *A. pilosa* root aqueous extract protects against cytotoxicity to hepatic cells. The hydroalcoholic extract of *Ebenus boissieri* has the ability to suppress cell proliferation, enhance caspase activity, and elevate IFN-γ and TNF-α levels in MDA-MB231 cells, indicating that it may have potential anti-tumor effects on breast cancer cells. In vitro, the rhizome of *A. officinarum* exhibits anti-inflammatory qualities, whereas the growth of cancers (lung, breast, and colon) is inhibited

TABLE 5.6
Herbal Bioactive Compounds and Their Anticancer Mechanisms

S#	Name of Compounds	Bioactivity/Functions/Mechanisms/Properties	Cell Line	References
01	Alisol B acetate	Induces apoptosis and Bax translocation via Bcl-2 pathways.	Prostate (PC-3)	[164]
02	6-gingerol, 6-paradol	Suppressed cellular proliferation through apoptotic mechanisms.	Promyelocytic (HL-60)	[165]
03	5-Fluorouracil	Bak, Bad, Bcl-2, Bax, and p53 protein	Colon	[166]
04	10-paradol, 6-dehydroparadol, 6-paradol	Caused proteolytic cleavage of pro-caspase-3, induced apoptosis.	Oral squamous (KB)	[167]
05	Allyl isothiocyanate (AITC)	Cyclin B1, control G2/M, reduce Bcl-2 expression, cell division cycle (Cdc-25B, and Cdc-25C).	Prostate (PC-3, LNCaP)	[168]
06	Alkaloid	Inhibit the growth.	Leukemia (Jurkat J6)	[169]
07	Aloe-emodin	DNA fragmentation, induced apoptosis, Bag-1, Bcl-X(L), translocation of Bak, Bak expression, activated caspase-3, -8, -9, Bax.	Lung (CH27)	[170]
08	Anthocyanin	Inhibit cancer cell growth	Lung, colon, gastric, and breast.	[171]
09	Angelicin	Increased cellular cytotoxicity, DNA fragmentation, up-regulation of caspase-9, -3 activity, down-regulate Mcl-1, Bcl-2, Bcl-xL, induced MAP kinases/PI3K/ AKT/GSK-3β, and caspase-8 activity.	Neuroblastoma (SH-SY5Y)	[172]
10	Apigenin	Increased p53 accumulation, G2/M phase arrest, induce apoptosis, induction of p21 expression, enhanced WAF1/p21 level through the p53-dependent pathway.	Hepatoma (Hep G2, Hep 3B, and PLC/ PRF/5).	[173]
11	Baicalein, a flavonoid	DNA fragmentation, induction apoptosis	HL-60	[174]
12	Berberine	ROS, induces apoptosis, inhibition cell growth in G0/G1-phase, cellular DNA, Ca2+ production.	Oral (HSC-3)	[175]
13	Chrysophanol	Induces necrosis by generating ROS	Liver (J5)	[176]
14	Curcumin	Activation caspase-9, -3. up-regulate Bax, decrease cyclin A, B, Cdk1, G2/M phase arrest, down-regulation of Bcl-2.	Nasopharyngeal (NPC-TW 076)	[177]
15	Gypenosides (Gyp)	Inhibits NAT mRNA expression.	Cervix (Ca Ski)	[178]
16	Diallyl trisulfide (DATS)	Induce apoptosis	Gastric (BGC823)	[179]
17	Diosgenin and hecogenin	DNA fragmentation, induce apoptosis.	Cervical (CaSki)	[115]
18	Emodin	DNA disintegration, arrest cell cycle, enhance caspase-3, p53/21, Fas/APO-1.	Hepatoma(PRF/5/ PLC, HepC3A/G2, and SK-HEP-1)	[180]
19	Flavonoid glycoside	DNA fragmentation, anti-proliferative and cytotoxicity effect.	Breast (A375, and HL60)	[181]
20	Fucoidans	Anti-tumor activity.	Melanoma, Colon.	[182]
21	Fucoxanthin (FX)	Induction of Gi arrest, induced GADD45A.	Hepato (HepG2, DU145)	[183]
22	Gammalinolenic acid	Production of IL-1β	Monocytes	[184]
23	Ganoderic acids (GAs)	Anti-angiogenesis and Cytotoxic mechanisms	Cancer cell lines	[185]

(*Continued*)

TABLE 5.6 (CONTINUED)
Herbal Bioactive Compounds and Their Anticancer Mechanisms

S#	Name of Compounds	Bioactivity/Functions/Mechanisms/Properties	Cell Line	References
24	Gallic acid	Induced apoptosis, DNA fragmentation	Stomach colon (Colo 205), and (KATO III).	[186]
25	Ginseng saponin K	Induced apoptotic morphology, caspase-3, interfered Bcl-2.	Prostate (LNCaP)	[187]
26	Ginsenoside-Rb1	Estrogen-responsive luciferase reporter gene activation.	Breast (MCF-7)	[188]
27	Gossypol	Cytotoxic activity	Breast (HBL-100), Malignant (MCF-7/adr, MCF-7).	[189]
28	HMJ-30/quinazoline	Enhance apoptosis, inhibit cell growth, caspase-8, -9, -3 pathways, DNA break.	Osteogenic (U-2 OS, HOS, and 143B).	[190]
29	Limonin	Enhanced doxorubicin cytotoxicity, decreased P-gp activity.	Leukemia (CEM/ADR5000)	[191]
30	Peridinin	Activating caspase-8, -9, decrease cell viability.	Colorectal (DLD-1)	[192]
31	Phenolics	Protection of oxidative induced stress, reduced cell damage.	Lung fibroblasts (CCD-25LU)	[193]
32	Polysaccharides (PS)	Involve defense mechanisms via phagocytosis	Brain tumors	[194]
33	Quercetin	Inhibit p53 protein expression, arrest G2-M phase.	Breast (MDA-MB468)	[195]
34	Rhein	Induced apoptosis, chemotaxis, and nitrate production.	Colonic (CaCo-2)	[196]
35	Tetrandrine, fangchinoline	Reduced P-gp expression	MDR Caco-2, and ADR5000/CEM.	[197]
36	Triptolide	Enhance p17 cleaved, and caspase 3 activity.	Proximal tubular (HK-2)	[198]

by *Withania somnifera*. Additionally, *Actaea dahurica* and *Adina rubella* extracts demonstrate anti-tumor and anticancer activities, respectively [160].

Quercetin stops the growth of breast and liver cancer cells, and saponins have anticancer and hypocholesterolemic properties. Triterpenes derived from *Ganoderma lucidum* exhibit substantial pharmacological potential in cancer treatment since Ganodermanontriol inhibits the proliferation of HCT116 and HT-29 (colon cancer cells) [161]. While these triterpenes do not affect normal human liver cells, they can cause apoptosis in many cancer cell lines, including hepatocellular carcinoma (HuH-7). Flavonoids and phenolics have anti-tumor and cytotoxic effects. Alkaloids that target cancer include paclitaxel, vinca alkaloids, and camptothecin [162]. In cancer therapy, alkaloids from *Chelidonium* extracts inhibit ABC transporters in cancerous cells, whereas camptothecin inhibits DNA topoisomerase. Many different plant compounds have to exhibit anticancer characteristics in a variety of human cell lines [163].

5.9.3 Bioactive Compounds for Neurodegenerative Diseases

Natural plants are a crucial source of bioactive compounds historically used in medicine. They produce SMs for defense and signaling, which often have significant biological and pharmacological activities [199]. These SMs are isolated for therapeutic uses, including oncology, due to their lower toxicity and better patient acceptance. For example, paclitaxel and docetaxel, used in breast cancer treatment, are derived from the Pacific yew tree (*Taxus brevifolia*) and the European yew tree (*Taxus baccata*),

TABLE 5.7
Some Natural Bioactive Compounds for Neurodegenerative Diseases

S#	Bioactive Compounds	Source	Effects	Reference
01	Dihydromyricetin	*Ampelopsis grossedentata*	Inhibition and neuroprotective activity of α-synuclein fibril formation	[206]
02	Flavonoids 1-9	*Oxalis corniculate L.*	Inhibition of carbonic anhydrases II, butyrylcholine esterase (BChE), and acetylcholine esterase (AChE)	[207]
03	(-)-narcissidine, (+)-9-O-demethyl-2-ahydroxyhomolycorine, (-)-9-Omethylpseudolycorine, (-)-pancratinine-C	*Narcissus tazetta L.*	Inhibition of BChE, and AChE	[208]
04	Desmethoxyangonin	*Renealmia Alpinia*	Inhibition of monoamine oxidases (MAOs)	[209]
05	Emodin, Physcion, Helminthosporin, chryso-phanol	*Rumex abyssinicus*	Inhibition of BChE and AChE	[210]
06	(-)-maackian and others	*S. flavescens*	Inhibition of MAOs	[211]

respectively. Additionally, the vinblastine and vinca alkaloids vincristine from the periwinkle plant *Catharanthus roseus*, are used in anticancer treatments [200, 201].

Polyphenols are a class of plant-derived secondary metabolites (SM) with protective effects against neurodegenerative disorders, diabetes, cardiovascular diseases, and cancer. They include coumarins, flavonoids, lignins, stilbenes, phenolic acids and tannins [202]. Coumarins are found in plants like *Melilotus sp., Galium odoratum, Dichanthelium clandestinum, Hierochloe odorata, Anthoxanthum odoratum, Dipteryx odorata*, and *Verbascum* spp. [203]. Resveratrol, a stilbenoid in fruits and plants like *Vitis vinifera* and *Polygonum cuspidatum*, has various biological properties, including neuroprotective effects [201].

The treatment and prevention of neurodegenerative diseases like Alzheimer's and Parkinson's are crucial due to their rising prevalence in the aging population [204]. Their multifactorial nature complicates diagnosis and treatment, with few available drugs. Lifestyle factors, including diet, influence their development, highlighting the role of plant SMs in maintaining nervous system health [205]. Table 5.7 summarizes research on plant SMs with potential activity against neurodegenerative diseases. Various foods and drinks have shown significant antioxidant properties, and recent studies have explored different and novel sources [201].

5.9.4 Bioactive Compounds for Viral Diseases

Human immunodeficiency virus (HIV) causes acquired immune deficiency syndrome (AIDS), with heterosexual sex being the predominant global transmission route, accounting for approximately 87% of cases, as noted by the National AIDS Control Organization (NACO) [212]. HIV can be transmitted from mother to child and through blood products. Once inside the body, it uses CD_4 receptors and either $CXCR_4$ or CCR_5 to multiply in macrophages or $CD4^+$ cells. HIV-1 gradually increases in lymphoid tissues, which causes an increasing decline in immunity and ultimately results in AIDS [213]. The risk of opportunistic infections and neoplasia rises with a decline in $CD4^+$ helper cell counts below 200 cells/mm^3 and an increase in plasma HIV-RNA levels [212]. These processes are caused by activation-induced cell death (AICD) and apoptosis, which also impact non-infected cells. Highly active antiretroviral therapy (HAART) employing protease inhibitors and reverse transcriptase analogs, either nucleoside or non-nucleoside, has been the latest treatment for HIV. HAART successfully inhibits HIV-1 replication; however, additional inhibitors are required for drug-resistant patients. Although these medications aid in immune response modulation, their toxicity limits their long-term efficacy [214].

In HIV/AIDS patients, several medicinal plants prevent the spread of the virus and strengthen the immune system against opportunistic infections. Given the genetically diverse RNA genome of HIV, herbal plants' active compounds present an economical and efficient substitute that may possess anti-retroviral characteristics. Herbal therapy, though historically limited, is regarded as a complementary medicine for HIV and several viral infection patients in Europe [215]. Herbalists use these plants to provide alternative or supplemental treatments in addition to HIV/AIDS medications (Table 5.8). Active compounds from plants, such as glycyrrhizin from *Glycyrrhiza uralensis* and *Moringa oleifera*, have shown efficacy in enhancing immune function in HIV patients [216]. Plant-based bioactive compounds like flavones, tannins, polysaccharides, alkaloids, coumarins, lignans, and terpenes exhibit antiviral properties, making them promising candidates for developing new herpes simplex virus (HSV), Epstein–Barr Virus [217], anti-HIV, hepatitis C virus (HCV), and respiratory syncytial virus (RSV) therapies [213].

Hepatitis is a severe and potentially fatal disease that causes liver inflammation, posing significant health risks. It is commonly known as viral hepatitis and can cause an estimated mortality of 1–4 million annually worldwide [218]. Various viruses, like Herpes simplex, Epstein-Barr, and Cytomegalovirus, can cause liver inflammation, but hepatitis viruses A, B, C, D, and E are the primary culprits. Types B, C, and E can cause chronic hepatitis, which can progress to potentially fatal disorders, including liver cirrhosis or hepatocellular carcinoma [219]. Natural materials derived from plants, herbs, and animals have been utilized recently to create novel antiviral medications to treat viral hepatitis [220]. Medicinal plants provide an economical and low-side-effect treatment option, and, due to the higher toxicity of chemical drugs, the use of more efficient herbal products has increased over the last decade. These natural compounds show significant antiviral effects by interfering with various stages of lifecycle of the hepatitis virus, including replication, viral release, and host-specific interactions [218].

TABLE 5.8
The Effectiveness of Some Medicinal Herbal Plants in HIV and Hepatitis Inhibition and Control in Humans

S#	Plant Name	Mechanism/Bioactivity	References
01	*Ancistrocladus kor*	Inhibits reverse transcriptase	[221]
02	*Banksia micrantha*	RNAse H Pascal activity and RDDP inhibition of HIV-1 RT	[222]
03	*Curcuma longa*	Inhibition of HIV integrase	[223]
04	*Dryopteris crassirhizoma*	Ant-HIV-1 protease activity	[224]
05	*Epimedium grandiflorum*	Inhibitory activity against HIV	[225]
06	*Flammulina velutipes*	HIV-1 RT Inhibition	[226]
07	*Myrothamnus flabellifolius*	Anti-HIV RT action of free radicals is blocked by polyphenols on cell membranes	[227]
08	*Blueberry Proanthocyanidins*	Interruption of binding of HAV and its entry into the cell	[228]
09	*W. chamaedaphne*	Exhibited strong anti-HBV activity	[218]
10	*M. peregrinum*	Inhibited HCV infection and efficient against all major HCV genotypes	[229]

5.9.5 Anti-Inflammatory Bioactive Compounds in Herbs

Inflammation is a multifaceted biological defense mechanism that involves molecular mediators, blood vessels, and immune cells in the body's tissues in reaction to tissue damage, microbial infection, or irritants [230]. Natural products have recently been discovered to have anti-inflammatory properties, providing comprehensive explanations in addition to molecular docking techniques for substances that occur naturally. Research has documented the anti-inflammatory properties of certain herbs, including *Zingiber officinale*, *Borago officinalis*, *Rosmarinus officinalis*, and *Curcuma longa* [47]. These plants hold potential therapeutic use in various clinical contexts. Recently, the development of anti-inflammatory compounds derived from plant SMs has shown notable effectiveness [231]. These compounds include fatty acids, terpenes, polyphenols, and many other bioactive components. Specifically, Aswad et al. reported that numerous plant SMs that are derived from chili peppers, *Erythrina velutina*, and *Zanthoxylum beecheyanum* [47], respectively, such as moupinamide, capsaicin, and hypaphorine, can be employed as novel, potential anti-inflammatory medicines [232].

5.9.6 Antidiabetic Bioactive Compounds in Herbs

Diabetes mellitus (DM) is a metabolic disease defined by persistently high levels of blood sugar and abnormalities in the metabolism of fats, proteins, and carbohydrates as a result of deficiencies in the secretion or action of insulin. Multiple organ failure, malfunction, and long-term damage are caused by diabetic mellitus. There are three main types of DM [233]: Insulin-dependent type 1 diabetes is an autoimmune disease in which the pancreas's insulin-producing cells are destroyed, resulting in little or no insulin production. DM (type 1) typically affects children and young adults and requires daily insulin doses. Type 2 diabetes (insulin-independent): Accounts for over 90% of adult DM cases [234]. The body's inability to effectively use the ample insulin produced by the pancreas is known as insulin resistance. Gestational DM is a common metabolic condition during pregnancy that is characterized by glucose intolerance that is initially identified during the second or third trimester of pregnancy and is caused by either a shortage of insulin or pregnancy hormones [232].

Hyperglycemia causes damage to blood vessels, kidneys, eyes, heart, and nerves. Recently, many medicinal plants have shown antidiabetic and antihyperlipidemic properties. There are around 400 known plant species that have hypoglycemia activity, and because natural plants include safe and useful phytoconstituents such as phenolics, alkaloids, terpenoids, flavonoids, and carotenoids, there is continued interest in developing new antidiabetic medications from these sources; a few of these are listed in Table 5.9 [232]. Because of their greater body compatibility, lower side effects, and cultural acceptability, these medications are frequently used for primary healthcare [235].

5.9.7 Antibiotics

Antibiotic actions are demonstrated by numerous plant SMs against diverse pathogenic microorganisms, targeting key cellular processes including protein assimilation, DNA/RNA replication, and

TABLE 5.9
The Antidiabetic Effects of Common Bioactive Compounds

S#	Compounds	Antidiabetic Properties
01	Momordin	PPAR δ activation
02	Polypeptide-p	Decrease blood glucose level, function as a protein similar to insulin.
03	Saponins	Reduce blood sugar and increase insulin production
04	Conjugated linolenic acid (9c, 11t, 13t)	Activation of PPAR δ
05	Momordicosides	Enhance the uptake of glucose

cell wall synthesis. For centuries, the food industry, agriculture, and pharmacology have used plant-derived natural products to control phytopathogens, weeds, and insects, and to preserve food and develop medicines [236]. Many bioactive compounds from medicinal plants, such as alkaloids like sanguinarine from the *Papaveraceae* family and berberine from Berberis spp., and flavonoids like quercetin and kaempferol from *Camellia sinensis* and *Allium cepa*, respectively, have significant antibacterial activity by inhibiting enzymes and disrupting cell membranes. From natural sources, notable recent antibiotics include retapamulin (2007), daptomycin (2003), and fidaxomicin (2010) [47]. Twelve antibiotics received medical approval between 1935 and 1968; however, between 2003 and 2015, the number of approvals rose to 20, with 16 of those being derived from natural sources. In 2018, Cragg and Newman reported changes to aminoglycosides that produced eravacycline, omadacycline, lefamulin, and sarecycline. They also highlighted plant SMs having antibacterial action, such as sisomicin and plazomicin [237]. Medicinal plant bioactive compounds possess antifungal properties. Green tea's polyphenols, such as epigallocatechin gallate (EGCG), inhibit the growth of many fungal species (e.g., *Aspergillus* and *Candida*), while the EOs of tea tree and oregano demonstrate potent antifungal effects against a range of pathogenic fungi [47, 238].

5.10 SUMMARY

This chapter explores the potential of herbal drugs and bioactive compounds in human healthcare, gathering information on herbal compounds and related products that demonstrate their ability to halt or slow the progression of fatal diseases. Natural products are nature-derived compounds that generally possess pharmacological or biological properties, making them essential in pharmaceutical drug design and discovery. Phytochemicals are classified into primary metabolites, such as amino acids, carbohydrates, chlorophylls, and proteins, and Secondary metabolites, such as terpenoids, tannins, alkaloids, saponins, flavonoids, steroids and their glycosides, etc. In animal models, herbal compounds have demonstrated therapeutic results in the treatment of sarcomas, leukemia, and skin cancer. These compounds are used in conjunction with chemotherapy to minimize adverse effects in cancer patients. Research has shown that herbs can help control antimicrobial and viral disease, nonetheless, there is still insufficient knowledge regarding the clinical efficacy and toxicity of many herbal medications, and, despite being the world's leading cause of death, little is known about their potential anticancer properties. Therefore, identifying new compounds in herbs and understanding their mechanisms are crucial for evaluating their potential clinical applications.

LIST OF ABBREVIATIONS

ACE2	Angiotensin-Converting Enzyme 2
AICD	Activation-Induced Cell Death
AIDS	Acquired Immune Deficiency Syndrome
CD4	Cluster of Differentiation 4
CNS	Central Nervous System
DM	Diabetes Mellitus
EGCG	Epigallocatechin gallate
EOs	Essential Oils
FDA	Food and Drug Administration
FTIR	Fourier-Transform Infrared Spectroscopy
HAART	Highly Active Antiretroviral Therapy
HCV	Hepatitis C Virus
HIV	Human Immunodeficiency Virus
HPLC	High-Performance Liquid Chromatography
HSV	Herpes Simplex Virus
IFN-γ	Interferon-γ

IgG	Immunoglobulin G
IgM	Immunoglobulin M
MEP	Methylerythritol 4-Phosphate
MHC	Major Histocompatibility Complex
NACO	National AIDS Control Organization
NF-κ	Nuclear Factor kappa
NK	Natural Killer Cell
NO	Nitric Oxide
PBMCs	Peripheral Blood Mononuclear Cells
PCs	Phenolic Compounds
RNA	Ribonucleic Acid
ROS	Reactive Oxygen Species
RSV	Respiratory Syncytial Virus
SMs	Secondary Metabolites
TCM	Traditional Chinese medicine
TH2	T Helper cell type 2
TLC	Thin Layer Chromatography
TNF-α	Tumor Necrosis Factor alpha
USA	United States of America
WHO	World Health Organization

REFERENCES

1. Biselli, R.; Nisini, R.; Lista, F.; Autore, A.; Lastilla, M.; De Lorenzo, G.; Peragallo, M. S.; Stroffolini, T.; D'Amelio, R. A historical review of military medical strategies for fighting infectious diseases: From battlefields to global health. *Biomedicines*. 2022, *10*(8), 2050.
2. Muteeb, G.; Rehman, M. T.; Shahwan, M.; Aatif, M. Origin of antibiotics and antibiotic resistance, and their impacts on drug development: A narrative review. *Pharmaceuticals*. 2023, *16*(11), 1615.
3. Lewis, K. The science of antibiotic discovery. *Cell*. 2020, *181*(1), 29–45.
4. Chawla, M.; Verma, J.; Gupta, R.; Das, B. Antibiotic potentiators against multidrug-resistant bacteria: discovery, development, and clinical relevance. *Frontiers in Microbiology*. 2022, *13*, 887251.
5. Mog, M.; Ngasotter, S.; Tesia, S.; Waikhom, D.; Panda, P.; Sharma, S.; Varshney, S. Problems of antibiotic resistance associated with oxytetracycline use in aquaculture: A review. *Journal of Entomology and Zoology Studies*. 2020, *8*, 1075–1082.
6. Jamal, A. Embracing nature's therapeutic potential: Herbal medicine. *International Journal of Multidisciplinary Sciences and Arts*. 2023, *2*(1), 117–126.
7. Balunas, M. J.; Kinghorn, A. D. Drug discovery from medicinal plants. *Life Sciences*. 2005, *78*(5), 431–441.
8. Dutra, R. C.; Campos, M. M.; Santos, A. R.; Calixto, J. B. Medicinal plants in Brazil: Pharmacological studies, drug discovery, challenges and perspectives. *Pharmacological Research*. 2016, *112*, 4–29.
9. Wu, M.; Su, X.; Wu, Y.; Luo, Y.; Guo, Y.; Xue, Y. Glycosylated coumarins, flavonoids, lignans and phenylpropanoids from Wikstroemia nutans and their biological activities. *Beilstein Journal of Organic Chemistry*. 2022, *18*(1), 200–207.
10. Hao, X.; Sun, W.; Ke, C.; Wang, F.; Xue, Y.; Luo, Z.; Wang, X.; Zhang, J.; Zhang, Y. Anti-Inflammatory Activities of Leaf Oil from Cinnamomum subavenium In Vitro and In Vivo. *BioMed Research International*. 2019, *2019*(1), 1823149.
11. Praveen, N.; Preetha, R. S. T.; Pagare, V.; Devasia, J.; Nizam, A.; Mukherjee, E.; Vasantha, V. L.; Al-Khayri, J. M. Plant-based Metabolites as Source of Antimicrobial Therapeutics: Prospects and Challenges. *Antimicrobials in Pharmaceutical and Medicinal Research*. 2023, 165–201.
12. Mehrotra, N. Herbs that heal: Nature's pharmacy endowed remedies for better health. *Annals of Phytomedicine*. 2021, *10*(1), 6–22.
13. Khare, T.; Anand, U.; Dey, A.; Assaraf, Y. G.; Chen, Z.-S.; Liu, Z.; Kumar, V. Exploring phytochemicals for combating antibiotic resistance in microbial pathogens. *Frontiers in Pharmacology*. 2021, *12*, 720726.

14. Alara, J. A.; Alara, O. R. An Overview of the Global Alarming Increase of Multiple Drug Resistant: A Major Challenge in Clinical Diagnosis. *Infectious Disorders-Drug Targets*. 2024, *24*(3), 26–42.
15. Wu, L.; Chen, W.; Wang, Z. Traditional Indian medicine in China: The status quo of recognition, development and research. *Journal of Ethnopharmacology*. 2021, *279*, 114317.
16. Goyal, S.; Thirumal, D.; Yapar, E. A.; Gürer, E. S.; Kumar, A.; Babu, M. A.; Sindhu, R. K. Asian Veterinary Medicines: From the Past to the Future. *Journal of Research in Pharmacy*. 2023, *27*(4), 1313–1328.
17. Mostafa, S.; Wang, Y.; Zeng, W.; Jin, B. Floral scents and fruit aromas: Functions, compositions, biosynthesis, and regulation. *Frontiers in Plant Science*. 2022, *13*, 860157.
18. Sarkar, M.; Talukder, P. Role of Plant Metabolites in Curing Neurodegenerative Disorders. *American Journal of Applied Bio-Technology Research*. 2023, *3*(4), 34–45.
19. Wu, Y.; Su, X.; Lu, J.; Wu, M.; Yang, S. Y.; Mai, Y.; Deng, W.; Xue, Y. In Vitro and in silico analysis of phytochemicals from fallopia dentatoalata as dual functional cholinesterase inhibitors for the treatment of Alzheimer's disease. *Frontiers in Pharmacology*. 2022, *13*, 905708.
20. Chaachouay, N.; Zidane, L. Plant-derived natural products: a source for drug discovery and development. *Drugs and Drug Candidates*. 2024, *3*(1), 184–207.
21. Ashraf, M. V.; Pant, S.; Khan, M. H.; Shah, A. A.; Siddiqui, S.; Jeridi, M.; Alhamdi, H. W. S.; Ahmad, S. Phytochemicals as antimicrobials: prospecting Himalayan medicinal plants as source of alternate medicine to combat antimicrobial resistance. *Pharmaceuticals*. 2023, *16*(6), 881.
22. Napagoda, M.; Wijesundara, D. Medicinal plants as sources of novel therapeutics: The history, present, and future. *Chemistry of Natural Products, Phytochemistry and Pharmacognosy of Medicinal Plants*. 2022, *3*, 4–18.
23. Usha, P.; Jose, S.; Nisha, A. Antimicrobial Drug Resistance – A global concern. *Veterinary World*. 2010, *3*(3), 138–139.
24. Kotwal, S.; Borude, P.; Humbe, T. A review: Most permitted traditional systems of Indian culture. *World Journal of Pharmaceutical Research*. 2021, *10*(5), 1112–1124.
25. McMullen, R. L.; Dell'Acqua, G. History of natural ingredients in cosmetics. *Cosmetics*. 2023, *10*(3), 71.
26. Parveen, B.; Parveen, A.; Parveen, R.; Ahmad, S.; Ahmad, M.; Iqbal, M. Challenges and opportunities for traditional herbal medicine today, with special reference to its status in India. *Ann Phytomed*. 2020, *9*(2), 97–112.
27. Sinha, D.; Odoh, U. E.; Ganguly, S.; Muhammad, M.; Chatterjee, M.; Chikeokwu, I.; Egbuna, C., Phytochemistry, history, and progress in drug discovery, in *Phytochemistry, Computational Tools and Databases in Drug Discovery*. 2023, Elsevier. pp. 1–26.
28. Jităreanu, A.; Trifan, A.; Vieriu, M.; Caba, I.-C.; Mârțu, I.; Agoroaei, L. Current trends in toxicity assessment of herbal medicines: A narrative review. *Processes*. 2022, *11*(1), 83.
29. Liang, Z.; Hu, H.; Li, J.; Yao, D.; Wang, Y.; Ung, C. O. L. Advancing the regulation of traditional and complementary medicine products: A comparison of five regulatory systems on traditional medicines with a long history of use. *Evidence-Based Complementary and Alternative Medicine*. 2021, *2021*(1), 5833945.
30. Malik, A.; Pereira, V.; Budhwar, P.; Varma, A.; Del Giudice, M. Sustainable innovations in an indigenous Indian Ayurvedic MNE. *Journal of Business Research*. 2022, *145*, 402–413.
31. Qadir, S. U.; Raja, V., Herbal medicine: Old practice and modern perspectives, in *Phytomedicine*. 2021, Elsevier. pp. 149–180.
32. Najmi, A.; Javed, S. A.; Al Bratty, M.; Alhazmi, H. A. Modern approaches in the discovery and development of plant-based natural products and their analogues as potential therapeutic agents. *Molecules*. 2022, *27*(2), 349.
33. Nasim, N.; Sandeep, I. S.; Mohanty, S. Plant-derived natural products for drug discovery: current approaches and prospects. *The Nucleus*. 2022, *65*(3), 399–411.
34. Dzobo, K. The role of natural products as sources of therapeutic agents for innovative drug discovery. *Comprehensive Pharmacology*. 2022, 408. 10.1016/B978-0-12-820472-6.00041-4
35. Berdigaliyev, N.; Aljofan, M. An overview of drug discovery and development. *Future Medicinal Chemistry*. 2020, *12*(10), 939–947.
36. Motyka, S.; Jafernik, K.; Ekiert, H.; Sharifi-Rad, J.; Calina, D.; Al-Omari, B.; Szopa, A.; Cho, W. C. Podophyllotoxin and its derivatives: Potential anticancer agents of natural origin in cancer chemotherapy. *Biomedicine and Pharmacotherapy*. 2023, *158*, 114145.

37. Bairagi, B.; Ali, M. Z.; Tiwari, H. *Rauwolfia serpentina*: A review on biological and medicinal properties. *European Journal of Modern Medicine and Practice*. 2024, *4*(6), 523–529.
38. Deep, A.; Kumar, D.; Bansal, N.; Narasimhan, B.; Marwaha, R. K.; Sharma, P. C. Understanding mechanistic aspects and therapeutic potential of natural substances as anticancer agents. *Phytomedicine Plus*. 2023, *3*(2), 100418.
39. Mustafa, G.; Arif, R.; Atta, A.; Sharif, S.; Jamil, A. Bioactive compounds from medicinal plants and their importance in drug discovery in Pakistan. *Matrix Science Pharma*. 2017, *1*(1), 17–26.
40. Twaij, B. M.; Hasan, M. N. Bioactive secondary metabolites from plant sources: types, synthesis, and their therapeutic uses. *International Journal of Plant Biology*. 2022, *13*(1), 4–14.
41. Noman, A.; Aqeel, M.; Qasim, M.; Haider, I.; Lou, Y. Plant-insect-microbe interaction: A love triangle between enemies in ecosystem. *Science of the Total Environment*. 2020, *699*, 134181.
42. Lv, M.; Sun, D.-W.; Huang, L.; Pu, H. Precision release systems of food bioactive compounds based on metal-organic frameworks: synthesis, mechanisms and recent applications. *Critical Reviews in Food Science and Nutrition*. 2022, *62*(15), 3991–4009.
43. Khare, S.; Singh, N.; Singh, A.; Hussain, I.; Niharika, K.; Yadav, V.; Bano, C.; Yadav, R. K.; Amist, N. Plant secondary metabolites synthesis and their regulations under biotic and abiotic constraints. *Journal of Plant Biology*. 2020, *63*, 203–216.
44. Rautela, A.; Kumar, S. Engineering plant family TPS into cyanobacterial host for terpenoids production. *Plant Cell Reports*. 2022, *41*(9), 1791–1803.
45. Li, H.; Yang, Q.; Fan, N.; Zhang, M.; Zhai, H.; Ni, Z.; Zhang, Y. Quantitative trait locus analysis of heterosis for plant height and ear height in an elite maize hybrid zhengdan 958 by design III. *BMC Genet*. 2017, *18*, 1–10.
46. Wang, L.; Wu, L. M.; Greaves, I. K.; Dennis, E. S.; Peacock, W. J. In Arabidopsis hybrids and Hybrid Mimics, up-regulation of cell wall biogenesis is associated with the increased plant size. *Plant Direct*. 2019, *3*(11), e00174.
47. Elshafie, H. S.; Camele, I.; Mohamed, A. A. A comprehensive review on the biological, agricultural and pharmaceutical properties of secondary metabolites based-plant origin. *International Journal of Molecular Sciences*. 2023, *24*(4), 3266.
48. Silva, A.; Silva, V.; Igrejas, G.; Aires, A.; Falco, V.; Valentão, P.; Poeta, P. Phenolic compounds classification and their distribution in winemaking by-products. *European Food Research and Technology*. 2023, *249*(2), 207–239.
49. Prieto, J. M.; Schinella, G. R. Anti-inflammatory and antioxidant Chinese herbal medicines: Links between traditional characters and the skin lipoperoxidation “Western” model. *Antioxidants*. 2022, *11*(4), 611.
50. Hassan, M. A.; Xu, T.; Tian, Y.; Zhong, Y.; Ali, F. A. Z.; Yang, X.; Lu, B. Health benefits and phenolic compounds of Moringa oleifera leaves: A comprehensive review. *Phytomedicine*. 2021, *93*, 153771.
51. Shah, A.; Smith, D. L. Flavonoids in agriculture: Chemistry and roles in, biotic and abiotic stress responses, and microbial associations. *Agronomy*. 2020, *10*(8), 1209.
52. Dias, M. C.; Pinto, D. C.; Silva, A. M. Plant flavonoids: Chemical characteristics and biological activity. *Molecules*. 2021, *26*(17), 5377.
53. Zhuang, W.-B.; Li, Y.-H.; Shu, X.-C.; Pu, Y.-T.; Wang, X.-J.; Wang, T.; Wang, Z. The classification, molecular structure and biological biosynthesis of flavonoids, and their roles in biotic and abiotic stresses. *Molecules*. 2023, *28*(8), 3599.
54. Roufogalis, B. *Flavonoid pharmacokinetics: Methods of analysis, preclinical and clinical pharmacokinetics, safety, and toxicology*. 2012. John Wiley & Sons.
55. Shakour, Z. T. A.; Fayek, N. M.; Farag, M. A. How do biocatalysis and biotransformation affect Citrus dietary flavonoids chemistry and bioactivity? *Critical Reviews in Biotechnology*. 2020, *40*(5), 689–714.
56. Pal, D.; Sahu, P. Potential of Flavonoids in Viral Infection: From Ethnopharmacology to Clinical Trials 28. *Anti-Viral Metabolites from Medicinal Plants*. 2024, 1009. 10.1007/978-3-031-12199-9_28
57. Pal, D.; Sahu, P., *Potential of Flavonoids in Viral Infection: From Ethnopharmacology to Clinical Trials, in Anti-Viral Metabolites from Medicinal Plants*. 2023, Springer. pp. 1009–1038.
58. Alamgir, A.; Alamgir, A. Vitamins, nutraceuticals, food additives, enzymes, anesthetic aids, and cosmetics. *Therapeutic Use of Medicinal Plants and their Extracts: Volume 2: Phytochemistry and Bioactive Compounds*. 2018, 407–534.
59. Roh, C. Microbial transformation of bioactive compounds and production of ortho-dihydroxyisoflavones and glycitein from natural fermented soybean paste. *Biomolecules*. 2014, *4*(4), 1093–1101.

60. Lalouckova, K.; Mala, L.; Marsik, P.; Skrivanova, E. In vitro antibacterial effect of the methanolic extract of the Korean soybean fermented product doenjang against Staphylococcus aureus. *Animals*. 2021, *11*(8), 2319.
61. Lerman, J. S.; Haslem, J.; Kim, L.; Ponderelli, J.; Ma, M.; Felson, R.; Sanderson, K.; Murphy, K.; Hopkins, K.; Wright, E. Collected research on phytonutrients: Flavonoids. *Journal of Culinary Science and Technology*. 2015, *13*(3), 214–241.
62. Terahara, N. Flavonoids in foods: a review. *Natural Product Communications*. 2015, *10*(3), 1934578 X1501000334.
63. Khoo, H. E.; Azlan, A.; Tang, S. T.; Lim, S. M. Anthocyanidins and anthocyanins: Colored pigments as food, pharmaceutical ingredients, and the potential health benefits. *Food and Nutrition Research*. 2017. 10.1080/16546628.2017.1361779
64. Dini, C.; Zaro, M. J.; Viña, S. Z. Bioactivity and functionality of anthocyanins: A review. *Current Bioactive Compounds*. 2019, *15*(5), 507–523.
65. Paulino, B. N.; Silva, G. N.; Araújo, F. F.; Néri-Numa, I. A.; Pastore, G. M.; Bicas, J. L.; Molina, G. Beyond natural aromas: The bioactive and technological potential of monoterpenes. *Trends in Food Science and Technology*. 2022, *128*, 188–201.
66. Shin, S.-A.; Moon, S. Y.; Kim, W.-Y.; Paek, S.-M.; Park, H. H.; Lee, C. S. Structure-based classification and anti-cancer effects of plant metabolites. *International Journal of Molecular Sciences*. 2018, *19*(9), 2651.
67. Mabou, F. D.; Yossa, I. B. N. Terpenes: Structural classification and biological activities. *IOSR Journal of Pharmacy and Biological Sciences*. 2021, *16*, 25–40.
68. Ninkuu, V.; Zhang, L.; Yan, J.; Fu, Z.; Yang, T.; Zeng, H. Biochemistry of terpenes and recent advances in plant protection. *International Journal of Molecular Sciences*. 2021, *22*(11), 5710.
69. Runyoro, D.; Ngassapa, O.; Vagionas, K.; Aligiannis, N.; Graikou, K.; Chinou, I. Chemical composition and antimicrobial activity of the essential oils of four Ocimum species growing in Tanzania. *Food Chemistry*. 2010, *119*(1), 311–316.
70. Muhammed, D.; Odey, B. O.; Alozieuwa, B. U.; Alawode, R. A.; Okunlola, B.; Ibrahim, J.; Lawal, A.; Berinyuy, E. B. Azadirachtin-A a bioactive compound from Azadirachta indica is a potential inhibitor of SARS-CoV-2 main protease. *AROC in Pharmaceutical and Biotechnology*. 2021, *1*(1), 1–8.
71. Hassan, W.; Rehman, S.; Noreen, H.; Gul, S.; Neelofar, A. N. Gas chromatography mass spectrometric (GCMS) analysis of essential oils of medicinal plants. *Adv. Anim. Vet. Sci.* 2016, *4*(8), 420–437.
72. Karimkhani, M. M.; Nasrollahzadeh, M.; Maham, M.; Jamshidi, A.; Kharazmi, M. S.; Dehnad, D.; Jafari, S. M. Extraction and purification of α-pinene; a comprehensive review. *Critical Reviews in Food Science and Nutrition*. 2024, *64*(13), 4286–4311.
73. Xu, J.; Wang, S.; Feng, T.; Chen, Y.; Yang, G. Hypoglycemic and hypolipidemic effects of total saponins from Stauntonia chinensis in diabetic db/db mice. *Journal of cellular and molecular medicine*. 2018, *22*(12), 6026–6038.
74. Ahmed, E.; Arshad, M.; Khan, M. Z.; Amjad, M. S.; Sadaf, H. M.; Riaz, I.; Sabir, S.; Ahmad, N. Secondary metabolites and their multidimensional prospective in plant life. *Journal of Pharmacognosy and Phytochemistry*. 2017, *6*(2), 205–214.
75. Jan, B.; Choudhary, B.; Malik, Z.; Dar, M. I. A descriptive review on exploiting the therapeutic significance of essential oils as a potential nutraceutical and food preservative. *Food Safety and Health*. 2024, *2*(2), 238–264.
76. Banday, A. H. *Steroids and their Medicinal Potential*. 2023. Bentham Science Publishers.
77. Nattagh-Eshtivani, E.; Barghchi, H.; Pahlavani, N.; Barati, M.; Amiri, Y.; Fadel, A.; Khosravi, M.; Talebi, S.; Arzhang, P.; Ziaei, R. Biological and pharmacological effects and nutritional impact of phytosterols: A comprehensive review. *Phytother. Res*. 2022, *36*(1), 299–322.
78. Sharma, A.; Ramakrishnan, M.; Khanna, K.; Landi, M.; Prasad, R.; Bhardwaj, R.; Zheng, B. Brassinosteroids and metalloids: regulation of plant biology. *J. Hazard. Mater*. 2022, *424*, 127518.
79. Truchan, M.; Tkachenko, H.; Buyun, L.; Kurhaluk, N.; Góralczyk, A.; Tomin, V.; Osadowski, Z. J. Antimicrobial Activities of Three Commercial Essential Oils Derived from Plants Belonging to Family Pinaceae. *Agrobiodiversity for Improving Nutrition, Health and Life Quality*. 2019(3). https://agrobiodiversity.uniag.sk/scientificpapers/article/view/287

80. Karimkhani, M. M.; Nasrollahzadeh, M.; Maham, M.; Jamshidi, A.; Kharazmi, M. S.; Dehnad, D.; Jafari, S. M. Extraction and purification of α-pinene; a comprehensive review. *Critical Reviews in Food Science Nutrition & Diabetes*. 2024, *64*(13), 4286–4311.
81. Hachlafi, N. E.; Aanniz, T.; Menyiy, N. E.; Baaboua, A. E.; Omari, N. E.; Balahbib, A.; Shariati, M. A.; Zengin, G.; Fikri-Benbrahim, K.; Bouyahya, A. In vitro and in vivo biological investigations of camphene and its mechanism insights: a review. *Food Rev. Int*. 2023, *39*(4), 1799–1826.
82. Keservani, R. K.; Tung, B. T.; Kesharwani, R. K.; Ahire, E. D. *Plant Metabolites and Vegetables as Nutraceuticals*. 2024. CRC Press.
83. Luo, Z.; Wang, F.; Zhang, J.; Li, X.; Zhang, M.; Hao, X.; Xue, Y.; Li, Y.; Horgen, F. D.; Yao, G. Cytotoxic alkaloids from the whole plants of Zephyranthes candida. *J. Nat. Prod*. 2012, *75*(12), 2113–2120.
84. Daley, S.-K.; Cordell, G. A. Biologically significant and recently isolated alkaloids from endophytic fungi. *Journal of Natural Products*. 2021, *84*(3), 871–897.
85. González-Juárez, D. E.; Escobedo-Moratilla, A.; Flores, J.; Hidalgo-Figueroa, S.; Martínez-Tagüeña, N.; Morales-Jiménez, J.; Muñiz-Ramírez, A.; Pastor-Palacios, G.; Pérez-Miranda, S.; Ramírez-Hernández, A. A review of the Ephedra genus: distribution, ecology, ethnobotany, phytochemistry and pharmacological properties. *Molecules*. 2020, *25*(14), 3283.
86. Butnariu, M.; Quispe, C.; Herrera-Bravo, J.; Pentea, M.; Sarac, I.; Küşümler, A. S.; Özçelik, B.; Painuli, S.; Semwal, P.; Imran, M. Papaver plants: current insights on phytochemical and nutritional composition along with biotechnological applications. *Oxid. Med. Cell. Longev*. 2022, *2022*(1), 2041769.
87. Rui, H. Traditional Chinese medicine and herbal medicine: current research and their clinical application for the treatment and prevention of cancer. *Pharmacological Research on Traditional Herbal Medicines*. 2021, 17–30.
88. Zielińska, S.; Czerwińska, M. E.; Dziągwa-Becker, M.; Dryś, A.; Kucharski, M.; Jezierska-Domaradzka, A.; Płachno, B. J.; Matkowski, A. Modulatory effect of Chelidonium majus extract and its alkaloids on LPS-stimulated cytokine secretion in human neutrophils. *Molecules*. 2020, *25*(4), 842.
89. Gupta, I.; Adin, S. N.; Aqil, M.; Mujeeb, M.; Ahad, A. Computer-aided Box–Behnken outlook towards optimization of extraction of piperine from Piper longum L. Fruits. *World J Pharm Res*. 2022, *11*, 1439–1455.
90. Aboelnaga, S. M. H. Evaluation of the Antihyperlipidemic and antioxidant effects of Catharanthus roseus extracted from Vinca minor in diabetic rats. *J Pharm Res Int*. 2021, *33*(2), 1–13.
91. Tahir, F.; Ali, E.; Hassan, S. A.; Bhat, Z. F.; Walayat, N.; Nawaz, A.; Khaneghah, A. M.; Phimolsiripol, Y.; Khan, M. R.; Aadil, R. M. Cyanogenic glucosides in plant-based foods: Occurrence, detection methods, and detoxification strategies–A comprehensive review. *Microchemical Journal*. 2024, 110065. 10.1016/j.microc.2024.110065
92. Jha, M.; Rahman, M.; Sheikh, H. Vinpocetine: a smart drug and a smart nutrient: a review. *International Journal of Pharmaceutical Sciences and Research*. 2012, *3*(2), 346.
93. Masihuddin, M.; Jafri, M.; Siddiqui, A.; Chaudhary, S. Traditional uses, phytochemistry and pharmacological activities of papaver somniferum with special reference of unani medicine an updated review. *Journal of Drug Delivery and Therapeutics*. 2018, *8*(5-s), 110–114.
94. Nain, T.; Sharma, S.; Chawariya, N.; Yadav, J. P. Prospect of natural compounds against malaria: a review. *Bulletin of Pharmaceutical Sciences Assiut University*. 2022, *45*(2), 629–653.
95. Zhao, C.-X.; Liu, H.; Zhang, X.; Yang, M.-Y.; Wang, Y.-T.; Xing, Y.-J.; Hua, J.-X.; Zhang, Q.; Li, D.-H.; Bai, J. Cephalotaxine-type and homoerythrina-type alkaloids with antiproliferative effects from Cephalotaxus fortunei. *Org. Biomol. Chem*. 2022, *20*(35), 7076–7084.
96. Stéphane, F. F. Y.; Jules, B. K. J.; Batiha, G. E.-S.; Ali, I.; Bruno, L. N. Extraction of bioactive compounds from medicinal plants and herbs. *Natural Medicinal Plants*. 2021, *2021*, 1–39.
97. Jha, A. K.; Sit, N. Extraction of bioactive compounds from plant materials using combination of various novel methods: A review. *Trends in Food Science and Technology*. 2022, *119*, 579–591.
98. Zhu, F.; Zhao, B.; Hu, B.; Zhang, Y.; Xue, B.; Wang, H.; Chen, Q. Review of available "extraction+ purification" methods of natural ceramides and their feasibility for sewage sludge analysis. *Environmental Science and Pollution Research*. 2023, *30*(26), 68022–68053.
99. Fonmboh, D. J.; Abah, E. R.; Fokunang, T. E.; Herve, B.; Teke, G. N.; Rose, N. M.; Borgia, N. N.; Fokunang, L. B.; Andrew, B. N.; Kaba, N. An overview of methods of extraction, isolation and characterization of natural medicinal plant products in improved traditional medicine research. *Asian J Res Med Pharm Sci*. 2020, *9*(2), 31–57.

100. Kiraman, N. A. S.; Yusof, H. A mini review on the methods for the extraction, isolation, and determination of P. odorata's bioactive compounds. *Healthscope*. 2022, *5*(2), 38–44.
101. Câmara, J. S.; Perestrelo, R.; Berenguer, C. V.; Andrade, C. F.; Gomes, T. M.; Olayanju, B.; Kabir, A.; Rocha, C.M.R.; Teixeira, J. A.; Pereira, J. A. Green extraction techniques as advanced sample preparation approaches in biological, food, and environmental matrices: a review. *Molecules*. 2022, *27*(9), 2953.
102. Salmerón-Manzano, E.; Garrido-Cardenas, J. A.; Manzano-Agugliaro, F. Worldwide research trends on medicinal plants. *International journal of environmental research public health*. 2020, *17*(10), 3376.
103. Organization, W. H. *WHO global report on traditional and complementary medicine 2019*. 2019. World Health Organization.
104. Süntar, I. Importance of ethnopharmacological studies in drug discovery: role of medicinal plants. *Phytochemistry Reviews*. 2020, *19*(5), 1199–1209.
105. Prasad, C. C.; Khemchandani, N. V.; Bhagat, R. T.; Pimpale, A. D. Ethnomedicinal Plant-A Review. *Journal of pharmaceutical research international*. 2021, *33*(29B), 17–30.
106. Daulagala, P. Chitinolytic endophytic bacteria as biocontrol agents for phytopathogenic fungi and nematode pests: a review. *Asian Journal of Research in Botany*. 2021, *5*(3), 14–24.
107. Iqbal, M. S.; Ahmad, K. S.; Ali, M. A.; Akbar, M.; Mehmood, A.; Nawaz, F.; Hussain, S. A.; Arshad, N.; Munir, S.; Arshad, H. An ethnobotanical study of wetland flora of Head Maralla Punjab Pakistan. *Plos One*. 2021, *16*(10), e0258167.
108. Jadhav, C. A.; Vikhe, D. N.; Jadhav, R. Global and domestic market of herbal medicines: A review. *Research Journal of Science and Technology*. 2020, *12*(4), 327–330.
109. Avigan, M. I.; Mozersky, R. P.; Seeff, L. B. Scientific and regulatory perspectives in herbal and dietary supplement associated hepatotoxicity in the United States. *International journal of molecular sciences*. 2016, *17*(3), 331.
110. Sen, S.; Chakraborty, R.; De, B. Challenges and opportunities in the advancement of herbal medicine: India's position and role in a global context. *Journal of Herbal medicine*. 2011, *1*(3–4), 67–75.
111. Ghosh, S.; Bishal, A.; Ghosh, S. K.; Jana, K.; Gayen, B.; Sahu, S.; Debnath, B. Herbal medicines: A potent approach to human diseases, their chief compounds, formulations, present status, and future aspects. *Int J*. 2023, *10*, 442–464.
112. Singh, A. K.; Rai, S. N.; Maurya, A.; Mishra, G.; Awasthi, R.; Shakya, A.; Chellappan, D. K.; Dua, K.; Vamanu, E.; Chaudhary, S. K. Therapeutic potential of phytoconstituents in management of Alzheimer's disease. *Evidence-Based Complementary and Alternative Medicine*. 2021, *2021*(1), 5578574.
113. Parvin, S.; Reza, A.; Das, S.; Miah, M. M. U.; Karim, S. Potential Role and International Trade of Medicinal and Aromatic Plants in the World. *European Journal of Agriculture and Food Sciences*. 2023, *5*(5), 89–99.
114. Saggar, S.; Mir, P. A.; Kumar, N.; Chawla, A.; Uppal, J.; Kaur, A. Traditional and herbal medicines: opportunities and challenges. *Pharmacognosy Research*. 2022, *14*(2), 107–114.
115. Harikrishnan, R.; Balasundaram, C., *Potential of Herbal Extracts and Bioactive Compounds for Human Healthcare, in The Role of Phytoconstitutents in Health Care*. 2020, Apple Academic Press. p. 3–158.
116. Balkrishna, A.; Sharma, N.; Srivastava, D.; Kukreti, A.; Srivastava, S.; Arya, V. Exploring the Safety, Efficacy, and Bioactivity of Herbal Medicines: Bridging Traditional Wisdom and Modern Science in Healthcare. *Future Integrative Medicine*. 2024, *3*(1), 35–49.
117. Malongane, F.; McGaw, L. J.; Mudau, F. N. The synergistic potential of various teas, herbs and therapeutic drugs in health improvement: a review. *Journal of the Science of Food and Agriculture*. 2017, *97*(14), 4679–4689.
118. Moin, M. S.; Siddiqui, J. I.; Alam, M. A.; Khatoon, F.; Khan, S.; Minhajuddin, A. Ethnomedicinal potential of widely used plant Azadirachta indica A. Juss: A comprehensive review. *The Journal of Phytopharmacology*. 2021, *10*(6), 456–467.
119. Rahmi, S. L.; Supriyana, S.; Kartini, A. The Effect of Extract (Angelica Keiskei) on Reducing Blood Pressure Level among Post-Partum Period with Hypertension. *International Journal of Nursing and Health Services*. 2020, *3*(1), 192–199.
120. Kaushik, D.; Kumar, M.; Kaushik, R.; Kumar, A. Role of Bioactive Compounds of Bauhinia variegata and their Benefits. *Harvesting Food from Weeds*. 2023, 217–266. 10.1002/9781119793007.ch7
121. Ghosh, S.; Kumar, V.; Mukherjee, H.; Saini, S.; Gupta, S.; Chauhan, S.; Kushwaha, K.; Lahiri, D.; Sircar, D.; Roy, P. Assessment of the mechanistic role of an Indian traditionally used ayurvedic herb Bacopa monnieri (L.) Wettst. for ameliorating oxidative stress in neuronal cells. *Journal of Ethnopharmacology*. 2024, *328*, 117899.

122. Fuloria, S.; Mehta, J.; Chandel, A.; Sekar, M.; Rani, N. N. I. M.; Begum, M. Y.; Subramaniyan, V.; Chidambaram, K.; Thangavelu, L.; Nordin, R. A comprehensive review on the therapeutic potential of Curcuma longa Linn. in relation to its major active constituent curcumin. *Frontiers in Pharmacology*. 2022, *13*, 820806.
123. Dailah, H. G. The ethnomedicinal evidences pertaining to traditional medicinal herbs used in the treatment of respiratory illnesses and disorders in Saudi Arabia: A review. *Saudi Journal of Biological Sciences*. 2022, *29*(9), 103386.
124. Younis, T.; Jabeen, F.; Hussain, A.; Rasool, B.; Raza Ishaq, A.; Nawaz, A.; El-Nashar, H. A.; El-Shazly, M. Antioxidant and pulmonary protective potential of Fraxinus xanthoxyloides bark extract against CCl4-induced toxicity in rats. *Chemistry and Biodiversity*. 2023, *20*(3), e202200755.
125. Aldibekova, A.; Kurmanbayeva, M.; Aksoy, A.; Permitina, V.; Dimeyeva, L.; Zverev, N. Anatomical Structure and Phytochemical Composition of a Rare Species Fraxinus sogdiana Bunge (Oleaceae) Growing in Different Soils in Kazakhstan. *Diversity*. 2023, *15*(6), 769.
126. Chanmahasathien, W.; Li, Y.; Satake, M.; Oshima, Y.; Ishibashi, M.; Ruangrungsi, N.; Ohizumi, Y. Prenylated xanthones from Garcinia xanthochymus. *Chemical and pharmaceutical bulletin*. 2003, *51*(11), 1332–1334.
127. Oyeyemi, I. T.; Ojo, T. D.; Oyeyemi, O. T. Hoslundia opposita Vahl.-A Promising Source of Bioactive Compounds against Infectious and Non-infectious Diseases. *Current Traditional Medicine*. 2023, *9*(4), 127–140.
128. Ma, R.-H.; Zhang, X.-X.; Thakur, K.; Zhang, J.-G.; Wei, Z.-J. Research progress of Lycium barbarum L. as functional food: phytochemical composition and health benefits. *Current Opinion in Food Science*. 2022, *47*, 100871.
129. Balakrishnan, R.; Vijayraja, D.; Jo, S.-H.; Ganesan, P.; Su-Kim, I.; Choi, D.-K. Medicinal profile, phytochemistry, and pharmacological activities of Murraya koenigii and its primary bioactive compounds. *Antioxidants*. 2020, *9*(2), 101.
130. Farkhondeh, T.; Kianmehr, M.; Kazemi, T.; Samarghandian, S.; Khazdair, M. Toxicity effects of Nerium oleander, basic and clinical evidence: A comprehensive review. *Human and experimental toxicology*. 2020, *39*(6), 773–784.
131. Kumar, R.; Saha, P.; Lokare, P.; Datta, K.; Selvakumar, P.; Chourasia, A. A systemic review of Ocimum sanctum (Tulsi): Morphological characteristics, phytoconstituents and therapeutic applications. *International Journal for Research in Applied Sciences and Biotechnology*. 2022, *9*(2), 221–226.
132. Guo, S.; Liao, X.; Chen, S.; Liao, B.; Guo, Y.; Cheng, R.; Xiao, S.; Hu, H.; Chen, J.; Pei, J. A comparative analysis of the chloroplast genomes of four Polygonum medicinal plants. *Frontiers in Genetics*. 2022, *13*, 764534.
133. Ojeda-Ayala, M.; Gaxiola-Camacho, S. M.; Delgado-Vargas, F. Phytochemical composition and biological activities of the plants of the genus Randia. *Botanical Sciences*. 2022, *100*(4), 779–796.
134. Adnan, M.; Siddiqui, A. J.; Arshad, J.; Hamadou, W. S.; Awadelkareem, A. M.; Sachidanandan, M.; Patel, M. Evidence-based medicinal potential and possible role of selaginella in the prevention of modern chronic diseases: Ethnopharmacological and ethnobotanical perspective. *Records of Natural Products*. 2021, *15*(5), 355.
135. Bulbul, M. R. H.; Chowdhury, M. N. U.; Naima, T. A.; Sami, S. A.; Imtiaj, M. S.; Huda, N.; Uddin, M. G. A comprehensive review on the diverse pharmacological perspectives of Terminalia chebula Retz. *Heliyon*. 2022, *8*(8), e10220.
136. Taheri, Y.; Quispe, C.; Herrera-Bravo, J.; Sharifi-Rad, J.; Ezzat, S. M.; Merghany, R. M.; Shaheen, S.; Azmi, L.; Prakash Mishra, A.; Sener, B. Urtica dioica-derived phytochemicals for pharmacological and therapeutic applications. *Evidence-based Complementary and Alternative Medicine*. 2022, *2022*(1), 4024331.
137. Saleem, S.; Muhammad, G.; Hussain, M. A.; Altaf, M.; Bukhari, S. N. A. Withania somnifera L.: Insights into the phytochemical profile, therapeutic potential, clinical trials, and future prospective. *Iranian Journal of Basic Medical Sciences*. 2020, *23*(12), 1501.
138. Chavan, S.; Kulkarni, A. Morphological and Phytochemical Studies on Xanthium strumarium L. *Plantae Scientia*. 2021, *4*(6), 287–290.
139. Mohammadhosseini, M. The ethnobotanical, phytochemical and pharmacological properties and medicinal applications of essential oils and extracts of different Ziziphora species. *Industrial Crops and Products*. 2017, *105*, 164–192.

140. Pelvan, E.; Karaoğlu, Ö.; Fırat, E. Ö.; Kalyon, K. B.; Ros, E.; Alasalvar, C. Immunomodulatory effects of selected medicinal herbs and their essential oils: A comprehensive review. *Journal of Functional Foods*. 2022, *94*, 105108.
141. Debnath, S.; Chakravorty, R.; Devi, D. A Review on Role of Medicinal plants in Immune system. *Asian Journal of Pharmacy and Technology*. 2020, *10*(4), 273–277.
142. Pecora, F.; Persico, F.; Argentiero, A.; Neglia, C.; Esposito, S. The role of micronutrients in support of the immune response against viral infections. *Nutrients*. 2020, *12*(10), 3198.
143. Daëron, M. The immune system as a system of relations. *Frontiers in Immunology*. 2022, *13*, 984678.
144. Hao, X.; Zhang, F.; Yang, Y.; Shang, S. The evaluation of cellular immunity to avian viral diseases: methods, applications, and challenges. *Frontiers in Microbiology*. 2021, *12*, 794514.
145. Lee, H.-J.; Woo, Y.; Hahn, T.-W.; Jung, Y. M.; Jung, Y.-J. Formation and maturation of the phagosome: a key mechanism in innate immunity against intracellular bacterial infection. *Microorganisms*. 2020, *8*(9), 1298.
146. Andrés, C. M. C.; Pérez de la Lastra, J. M.; Juan, C. A.; Plou, F. J.; Pérez-Lebeña, E. The role of reactive species on innate immunity. *Vaccines*. 2022, *10*(10), 1735.
147. Eswar, K.; Mukherjee, S.; Ganesan, P.; Rengan, A. K. Immunomodulatory natural polysaccharides: An overview of the mechanisms involved. *European Polymer Journal*. 2023, *188*, 111935.
148. Abdel Shaheed, C.; Beardsley, J.; Day, R. O.; McLachlan, A. J. Immunomodulatory effects of pharmaceutical opioids and antipyretic analgesics: Mechanisms and relevance to infection. *British Journal of Clinical Pharmacology*. 2022, *88*(7), 3114–3131.
149. Shukla, M. K.; Singh, S. K.; Pandey, S.; Gupta, P. K.; Choudhary, A.; Jindal, D. K.; Dua, K.; Kumar, D. Potential immunomodulatory activities of plant products. *South African Journal of Botany*. 2022, *149*, 937–943.
150. Vetvicka, V.; Vannucci, L. Biological properties of andrographolide, an active ingredient of Andrographis Paniculata: A narrative review. *Annals of Translational Medicine*. 2021, *9*(14), 1186.
151. Shoba, E. R.; Sivagamasundari, U.; Devi, V. V.; Josely, A. E.; Afreen, J. I.; Valiyapurayil, K. M. Phytochemical analysis of Baliospermum montanum leaves and evaluation of in vitro anticoagulant, antioxidant and anticancer properties. *Medicinal Plants-International Journal of Phytomedicines and Related Industries*. 2023, *15*(2), 344–353.
152. Jayaraman, S.; Variyar, J. Plant metabolites as immunomodulators. *Plant Metabolites: Methods, Applications and Prospects*. 2020, 441–464.
153. Chan, S. M.; Khoo, K. S.; Sit, N. W. Interactions between plant extracts and cell viability indicators during cytotoxicity testing: implications for ethnopharmacological studies. *Tropical Journal of Pharmaceutical Research*. 2015, *14*(11), 1991–1998.
154. Gharred, N.; Ali, L. M.; Bettache, N.; Dridi-Dhaouadi, S.; Morere, A.; Menut, C. In vitro anti-inflammatory activity of three Inula species essential oils in lipopolysaccharide-stimulated RAW 264.7 macrophages. *Chemistry Africa*. 2023, *6*(4), 1933–1942.
155. Klaophimai, S.; Pouyfung, P.; Wongnoppavich, A.; Chairatvit, K. Induction of S arrest and apoptosis in human oral cancer cells by Rhinacanthin-C extracted from Rhinacanthus nasutus via modulating Akt and p38 signaling pathways. *Journal of Ethnopharmacology*. 2023, *317*, 116813.
156. Shan, Y.; Zhao, J.; Wei, K.; Jiang, P.; Xu, L.; Chang, C.; Xu, L.; Shi, Y.; Zheng, Y.; Bian, Y. A comprehensive review of Tripterygium wilfordii hook. f. in the treatment of rheumatic and autoimmune diseases: bioactive compounds, mechanisms of action, and future directions. *Frontiers in Pharmacology*. 2023, *14*, 1282610.
157. Jain, S.; Arora, P.; Popli, H. A comprehensive review on Citrus aurantifolia essential oil: its phytochemistry and pharmacological aspects. *Brazilian Journal of Natural Sciences*. 2020, *3*(2), 354–354.
158. Kenechukwu, E. C.; Chigozie, U. M.; Chinemerem, N. D.; Uchenna, E. G.; Chijioke, E. S.; Belinda, U. C.; Chiedu, O. F. B. Immunostimulatory and Antimicrobial Claims of Super-7 Herbal Mixture (SHM) and Odogwu Cleanser Herbal Mixture (OCHM) Commercially Available in Enugu State, Nigeria. *American Journal of BioScience*. 2023, *11*(1), 11–20.
159. Shin, S.-A.; Moon, S. Y.; Kim, W.-Y.; Paek, S.-M.; Park, H. H.; Lee, C. S. Structure-Based Classification and Anti-Cancer Effects of Plant Metabolites. *Int. J. Mol. Sci*. 2018, *19*(9), 2651.
160. Salehi, B.; Ata, A. V.; Anil Kumar, N.; Sharopov, F.; Ramírez-Alarcón, K.; Ruiz-Ortega, A.; Abdulmajid Ayatollahi, S.; Valere Tsouh Fokou, P.; Kobarfard, F.; Amiruddin Zakaria, Z. Antidiabetic potential of medicinal plants and their active components. *Biomolecules*. 2019, *9*(10), 551.

161. Ahmad, M. F. Ganoderma lucidum: A rational pharmacological approach to surmount cancer. *Journal of ethnopharmacology*. 2020, *260*, 113047.
162. Patel, A.; Vanecha, R.; Patel, J.; Patel, D.; Shah, U.; Bambharoliya, T. Development of natural bioactive alkaloids: anticancer perspective. *Mini Reviews in Medicinal Chemistry*. 2022, *22*(2), 200–212.
163. Melfi, F.; Carradori, S.; Mencarelli, N.; Campestre, C.; Gallorini, M.; Di Giacomo, S.; Di Sotto, A. Natural products as a source of new anticancer chemotypes. *Expert Opinion on Therapeutic Patents*. 2023, *33*(11), 721–744.
164. Pang, B.; Zhang, J.; Zhang, X.; Yuan, J.; Shi, Y.; Qiao, L. Inhibition of lipogenesis and induction of apoptosis by valproic acid in prostate cancer cells via the C/EBPα/SREBP-1 pathway. *Acta Biochimica et Biophysica Sinica*. 2021, *53*(3), 354–364.
165. Malmir, S.; Ebrahimi, A.; Mahjoubi, F. Effect of ginger extracts on colorectal cancer HCT-116 cell line in the expression of MMP-2 and KRAS. *Gene Reports*. 2020, *21*, 100824.
166. Mao, H.; Zhou, J.; Yan, L.; Zhang, S.; Yu, D.-G. Hybrid films loaded with 5-fluorouracil and Reglan for synergistic treatment of colon cancer via asynchronous dual-drug delivery. *Frontiers in Bioengineering and Biotechnology*. 2024, *12*, 1398730.
167. Ahmad, R.; Khan, M. A.; Srivastava, A.; Gupta, A.; Srivastava, A.; Jafri, T. R.; Siddiqui, Z.; Chaubey, S.; Khan, T.; Srivastava, A. K. Anticancer potential of dietary natural products: A comprehensive review. *Anti-Cancer Agents in Medicinal Chemistry*. 2020, *20*(2), 122–236.
168. Tarar, A.; Peng, S.; Cheema, S.; Peng, C.-A. Anticancer activity, mechanism, and delivery of allyl isothiocyanate. *Bioengineering*. 2022, *9*(9), 470.
169. Zhao, J.; Carbone, J.; Farruggia, G.; Janecka, A.; Gentilucci, L.; Calonghi, N. Synthesis and Antiproliferative Activity against Cancer Cells of Indole-Aryl-Amide Derivatives. *Molecules*. 2022, *28*(1), 265.
170. Şeker Karatoprak, G.; Küpeli Akkol, E.; Yücel, Ç.; Bahadır Acıkara, Ö.; Sobarzo-Sánchez, E. Advances in understanding the role of aloe emodin and targeted drug delivery systems in cancer. *Oxidative Medicine and Cellular Longevity*. 2022, *2022*(1), 7928200.
171. Chen, J.; Xu, B.; Sun, J.; Jiang, X.; Bai, W. Anthocyanin supplement as a dietary strategy in cancer prevention and management: A comprehensive review. *Critical reviews in food science nutrition*. 2022, *62*(26), 7242–7254.
172. Sabeel, Z.; Liang, Y.; Hao, M.; Ying, L.; Guo, R.; Chen, R.; Li, X.; Yu, C.; Yang, Z. A comprehensive review of antitumor properties of Angelica species and their antitumor-responsible constituents and the underlying molecular mechanisms involved in tumor inhibition. *Phytotherapy Research*. 2023, *37*(5), 2187–2211.
173. Wani, H. A.; Majid, S.; Wani, R. A.; Khan, M. S.; Qureshi, W.; Bhat, A. A.; Bhat, S. A.; Rasool, S.; Amin, H.; Masoodi, M. Phytochemicals from Honey as MAP-Kinase Inhibitors: Current Therapeutic Standing and Future Prospects. *Therapeutic Applications of Honey and its Phytochemicals: Volume* II. 2020, 141–164.
174. Emami, S. A.; Ramazani, E.; Mousavi, S. H.; Vahdati-Mashhadian, N.; Asili, J.; Parsaee, H.; Tayarani-Najaran, Z. Neobaicalein, a flavonoid from the Scutellaria litwinowii Bornm. *& Sint. ex Bornm. induced apoptosis in human leukemic cell lines. Iranian Journal of Basic Medical Sciences*. 2023, *26*(3), 269.
175. Satpathy, S.; Bhuyan, S. K.; Bhuyan, R. Targeting Autophagic Pathway in Oral Cancer Therapy Through Phytoconstituents: A Short Review. *Biomedical and Pharmacology Journal*. 2024, *17*(2). 10.13005/bpj/2890
176. Wang, T.; Lu, Z.; Qu, X.-H.; Xiong, Z.-Y.; Wu, Y.-T.; Luo, Y.; Zhang, Z.-Y.; Han, X.-J.; Xie, C.-F. Chrysophanol-8-O-glucoside protects mice against acute liver injury by inhibiting autophagy in hepatic stellate cells and inflammatory response in liver-resident macrophages. *Frontiers in Pharmacology*. 2022, *13*, 951521.
177. Ao, X.; Luo, C.; Zhang, M.; Liu, L.; Peng, S. The efficacy of natural products for the treatment of nasopharyngeal carcinoma. *Chemical Biology and Drug Design*. 2024, *103*(1), e14411.
178. Ng, C. X.; Affendi, M. M.; Chong, P. P.; Lee, S. H. The potential of plant-derived extracts and compounds to augment anticancer effects of chemotherapeutic drugs. *Nutrition and Cancer*. 2022, *74*(9), 3058–3076.
179. Hu, Q.; Li, Z.; Li, Y.; Deng, X.; Chen, Y.; Ma, X.; Zeng, J.; Zhao, Y. Natural products targeting signaling pathways associated with regulated cell death in gastric cancer: Recent advances and perspectives. *Phytotherapy Research*. 2023, *37*(6), 2661–2692.
180. Akkol, E. K.; Tatlı, I. I.; Karatoprak, G. Ş.; Ağar, O. T.; Yücel, Ç.; Sobarzo-Sánchez, E.; Capasso, R. Is emodin with anticancer effects completely innocent? Two sides of the coin. *Cancers*. 2021, *13*(11), 2733.

181. Matulja, D.; Vranješević, F.; Kolympadi Markovic, M.; Pavelić, S. K.; Marković, D. Anticancer activities of marine-derived phenolic compounds and their derivatives. *Molecules*. 2022, *27*(4), 1449.
182. Cao, L.-M.; Sun, Z.-X.; Makale, E. C.; Du, G.-K.; Long, W.-F.; Huang, H.-R. Antitumor activity of fucoidan: A systematic review and meta-analysis. *Translational Cancer Research*. 2021, *10*(12), 5390.
183. Vijayalakshmi, A.; Prabha, T.; Lalitha, V.; Hemalatha, S.; Jagadeeswaran, M.; Chaitanya, M.; Selvamani, P.; Latha, S. Dietary carotenoid fucoxanthin as a promising biomarker to target the cancer cells: A focused review. *Annals of Phytomedicine*. 2022, *11*(1), 164–174.
184. Davanso, M. R.; Crisma, A. R.; Murata, G.; Newsholme, P.; Curi, R. Impact of dietary fatty acids on macrophage lipid metabolism, signaling and function. *Immunometabolism*. 2020, *2*(1), e200008.
185. Zhao, P.; Guan, M.; Tang, W.; Walayat, N.; Ding, Y.; Liu, J. Structural diversity, fermentation production, bioactivities and applications of triterpenoids from several common medicinal fungi: Recent advances and future perspectives. *Fitoterapia*. 2023, *166*, 105470.
186. Keyvani-Ghamsari, S.; Rahimi, M.; Khorsandi, K. An update on the potential mechanism of gallic acid as an antibacterial and anticancer agent. *Food Science and Nutrition*. 2023, *11*(10), 5856–5872.
187. Jiang, M.; Sun, Y.; Song, K. Ginsenoside compound K suppresses tumour growth in the 22Rv1 xenograft model and inhibits androgenic responses via the transcriptional mechanism in human prostate cancer cells. *The FEBS Journal*. 2023, *290*(7), 1840–1854.
188. Ali, A.-S.; As, A.-D. E. Ginseng modulates steroid hormones action, production and metabolism; a narative review on androgens, estrogens and corticosteroids. *Indian Drugs*. 2020, *57*(10), 7–16.
189. Liu, Y.; Wang, L.; Zhao, L.; Zhang, Y. Structure, properties of gossypol and its derivatives from physiological activities to drug discovery and drug design. *Natural Product Reports*. 2022, *39*(6), 1282–1304.
190. Chiu, Y. J.; Hour, M. J.; Lu, C. C.; Chung, J. G.; Kuo, S. C.; Huang, W. W.; Chen, H. J.; Jin, Y. A.; Yang, J. S. Novel quinazoline HMJ-30 induces U-2 OS human osteogenic sarcoma cell apoptosis through induction of oxidative stress and up-regulation of ATM/p53 signaling pathway. *Journal of Orthopaedic Research*. 2011, *29*(9), 1448–1456.
191. Chaudhary, M., *Role of plant secondary metabolites as modulators of multidrug resistance in cancer therapy, in Plant Secondary Metabolites: Physico-Chemical Properties and Therapeutic Applications*. 2022, Springer. p. 415–435.
192. Manochkumar, J.; Doss, C. G. P.; Efferth, T.; Ramamoorthy, S. Tumor preventive properties of selected marine pigments against colon and breast cancer. *Algal Research*. 2022, *61*, 102594.
193. Umamahesh, K.; Gandhi, A. D.; Reddy, O. V. Ethnopharmacological Applications of Mango (Mangiferaindica L.) Peel-A Review. *Current Pharmaceutical Biotechnology*. 2020, *21*(13), 1298–1303.
194. Atiq, A.; Parhar, I. Anti-neoplastic potential of flavonoids and polysaccharide phytochemicals in glioblastoma. *Molecules*. 2020, *25*(21), 4895.
195. Farghadani, R.; Naidu, R. The anticancer mechanism of action of selected polyphenols in triple-negative breast cancer (TNBC). *Biomedicine and Pharmacotherapy*. 2023, *165*, 115170.
196. Gajjar, D.; Thakkar, J.; Patel, P. K.; Sagar, S. R. Phytochemistry, ethnopharmacology and novel formulations based approaches for the treatment of irritable bowel syndrome: a comprehensive review. *Phytochemistry Reviews*. 2024, 1–38. 10.1007/s11101-024-09970-8
197. Sun, Y. F.; Wink, M. Tetrandrine and fangchinoline, bisbenzylisoquinoline alkaloids from Stephania tetrandra can reverse multidrug resistance by inhibiting P-glycoprotein activity in multidrug resistant human cancer cells. *Phytomedicine*. 2014, *21*(8–9), 1110–1119.
198. Shu, B.; Duan, W.; Yao, J.; Huang, J.; Jiang, Z.; Zhang, L. Caspase 3 is involved in the apoptosis induced by triptolide in HK-2 cells. *Toxicology in vitro*. 2009, *23*(4), 598–602.
199. Salem, M. A.; Perez de Souza, L.; Serag, A.; Fernie, A. R.; Farag, M. A.; Ezzat, S. M.; Alseekh, S. Metabolomics in the context of plant natural products research: From sample preparation to metabolite analysis. *Metabolites*. 2020, *10*(1), 37.
200. Seca, A. M.; Pinto, D. C. Plant secondary metabolites as anticancer agents: successes in clinical trials and therapeutic application. *International journal of molecular sciences*. 2018, *19*(1), 263.
201. Leuci, R.; Brunetti, L.; Poliseno, V.; Laghezza, A.; Loiodice, F.; Tortorella, P.; Piemontese, L. Natural compounds for the prevention and treatment of cardiovascular and neurodegenerative diseases. *Foods*. 2020, *10*(1), 29.
202. Vacca, R. A.; Valenti, D.; Caccamese, S.; Daglia, M.; Braidy, N.; Nabavi, S. M. Plant polyphenols as natural drugs for the management of Down syndrome and related disorders. *Neurosci. Biobehav. Rev*. 2016, *71*, 865–877.

203. Shahidi, F.; Yeo, J. Bioactivities of phenolics by focusing on suppression of chronic diseases: A review. *Int. J. Mol. Sci.* 2018, *19*(6), 1573.
204. Maisam, M.; Khan, M. T.; Lodhi, M. S.; Mou, K.; Liu, Z.; Wei, D. Alzheimer's Disease; Mechanism, Mutations, and Applications of Nano-Medicine. *Frontiers in Bioscience-Landmark.* 2023, *28*(10), 258.
205. Brunetti, L.; Laghezza, A.; Loiodice, F.; Tortorella, P.; Piemontese, L. Combining fatty acid amide hydrolase (FAAH) inhibition with peroxisome proliferator-activated receptor (PPAR) activation: A new potential multi-target therapeutic strategy for the treatment of Alzheimer's disease. *Neural Regeneration Research.* 2020, *15*(1), 67–68.
206. Jia, L.; Wang, Y.; Sang, J.; Cui, W.; Zhao, W.; Wei, W.; Chen, B.; Lu, F.; Liu, F. Dihydromyricetin inhibits α-synuclein aggregation, disrupts preformed fibrils, and protects neuronal cells in culture against amyloid-induced cytotoxicity. *J. Agric. Food Chem.* 2019, *67*(14), 3946–3955.
207. Imran, M.; Irfan, A.; Ibrahim, M.; Assiri, M. A.; Khalid, N.; Ullah, S.; Al-Sehemi, A. G. Carbonic anhydrase and cholinesterase inhibitory activities of isolated flavonoids from Oxalis corniculata L. and their first-principles investigations. *Industrial Crops and Products.* 2020, *148*, 112285.
208. Karakoyun, Ç.; Bozkurt, B.; Çoban, G.; Masi, M.; Cimmino, A.; Evidente, A.; Somer, N. U. A comprehensive study on Narcissus tazetta subsp. tazetta L.: Chemo-profiling, isolation, anticholinesterase activity and molecular docking of Amaryllidaceae alkaloids. *S. Afr. J. Bot.* 2020, *130*, 148–154.
209. Chaurasiya, N. D.; León, F.; Ding, Y.; Gómez-Betancur, I.; Benjumea, D.; Walker, L. A.; Cutler, S. J.; Tekwani, B. L. Interactions of Desmethoxyyangonin, a Secondary Metabolite from Renealmia alpinia, with Human Monoamine Oxidase-A and Oxidase-B. *Evidence-Based Complementary and Alternative Medicine* 2017, *2017*(1), 4018724.
210. Augustin, N.; Nuthakki, V. K.; Abdullaha, M.; Hassan, Q. P.; Gandhi, S. G.; Bharate, S. B. Discovery of helminthosporin, an anthraquinone isolated from Rumex abyssinicus Jacq as a dual cholinesterase inhibitor. *ACS omega.* 2020, *5*(3), 1616–1624.
211. Lee, H. W.; Ryu, H. W.; Kang, M.-G.; Park, D.; Oh, S.-R.; Kim, H. Potent selective monoamine oxidase B inhibition by maackiain, a pterocarpan from the roots of Sophora flavescens. *Bioorganic and Medicinal Chemistry Letters.* 2016, *26*(19), 4714–4719.
212. Laksemi, D.; Suwanti, L. T.; Mufasirin, M.; Suastika, K.; Sudarmaja, M. Opportunistic parasitic infections in patients with human immunodeficiency virus/acquired immunodeficiency syndrome: A review. *Veterinary world.* 2020, *13*(4), 716.
213. Renault, C.; Veyrenche, N.; Mennechet, F.; Bedin, A.-S.; Routy, J.-P.; Van de Perre, P.; Reynes, J.; Tuaillon, E. Th17 CD4+ T-cell as a preferential target for HIV reservoirs. *Frontiers in Immunology.* 2022, *13*, 822576.
214. Berhan, A.; Bayleyegn, B.; Getaneh, Z. HIV/AIDS associated lymphoma. *Blood and Lymphatic Cancer: Targets and Therapy.* 2022, *2022*, 31–45.
215. Kankara, S. S.; Nuhu, A. I.; Bindawa, K. A.; Haruna, M. R. U.; Bello, A.; Abubakar, I. B. Indigenous traditional knowledge of medicinal plants used for the management of HIV/AIDS opportunistic infections in Katsina State, Nigeria. *Ethnobotany Research and Applications.* 2022, *23*, 1–17.
216. Shahrajabian, M. H.; Sun, W. The Importance of Traditional Chinese Medicine in the Intervention and Treatment of HIV While Considering Its Safety and Efficacy. *Current HIV Research.* 2023, *21*(6), 331–346.
217. Hu, L.; Zhang, Y.; Zhu, H.; Liu, J.; Li, H.; Li, X.-N.; Sun, W.; Zeng, J.; Xue, Y.; Zhang, Y. Filicinic acid based meroterpenoids with anti-Epstein–Barr virus activities from Hypericum japonicum. *Org. Lett.* 2016, *18*(9), 2272–2275.
218. Roy, A.; Roy, M.; Gacem, A.; Datta, S.; Zeyaullah, M.; Muzammil, K.; Farghaly, T. A.; Abdellattif, M. H.; Yadav, K. K.; Simal-Gandara, J. Role of bioactive compounds in the treatment of hepatitis: A review. *Frontiers in Pharmacology.* 2022, *13*, 1051751.
219. Rigopoulou, E. I.; Smyk, D. S.; Matthews, C. E.; Billinis, C.; Burroughs, A. K.; Lenzi, M.; Bogdanos, D. P. Epstein-Barr Virus as a Trigger of Autoimmune Liver Diseases. *Advances in virology.* 2012, *2012*(1), 987471.
220. Perera, W.; Liyanage, J. A.; Dissanayake, K.; Gunathilaka, H.; Weerakoon, W.; Wanigasekara, D.; Fernando, W.; Rajapaksha, R.; Liyanage, R.; Perera, B. T. Antiviral potential of selected medicinal herbs and their isolated natural products. *BioMed research international.* 2021, *2021*(1), 7872406.
221. Singh, I. P.; Bharate, S. B.; Bhutani, K. Anti-HIV natural products. *Current Science.* 2005, *2005*, 269–290.

222. Junior, V. V.; Rosas, E.; Carvalho, M.; Henriques, M.; Pinto, A. C. Chemical composition and anti-inflammatory activity of copaiba oils from Copaifera cearensis Huber ex Ducke, Copaifera reticulata Ducke and Copaifera multijuga Hayne—A comparative study. *Journal of ethnopharmacology*. 2007, *112*(2), 248–254.
223. Sillapachaiyaporn, C.; Rangsinth, P.; Nilkhet, S.; Moungkote, N.; Chuchawankul, S. HIV-1 protease and reverse transcriptase inhibitory activities of Curcuma aeruginosa Roxb. Rhizome extracts and the phytochemical profile analysis: in vitro and in silico screening. *Pharmaceuticals*. 2021, *14*(11), 1115.
224. Sharma, R.; Bhattu, M.; Tripathi, A.; Verma, M.; Acevedo, R.; Kumar, P.; Rajput, V. D.; Singh, J. Potential medicinal plants to combat viral infections: A way forward to environmental biotechnology. *Environmental Research*. 2023, *227*, 115725.
225. Nasef, A. Z.; El-Sheikh, N. A. Potential Effects of Horny Goat Weed (Epimedium grandiflorum) on The Level of Fertility in Male Rats Infected with Cadmium Chloride: Biochemical and Histopathological Study. *Alexandria Science Exchange Journal*. 2023, *44*(3), 433–443.
226. Zhang, Y.; Zhang, G.; Ling, J. Medicinal fungi with antiviral effect. *Molecules*. 2022, *27*(14), 4457.
227. Erhabor, J.; Komakech, R.; Kang, Y.; Tang, M.; Matsabisa, M. Ethnopharmacological importance and medical applications of Myrothamnus flabellifolius Welw.(Myrothamnaceae)-A review. *Journal of ethnopharmacology*. 2020, *252*, 112576.
228. Jama-Kmiecik, A.; Sarowska, J.; Wojnicz, D.; Choroszy-Król, I.; Frej-Mądrzak, M. Natural products and their potential anti-HAV activity. *Pathogens*. 2021, *10*(9), 1095.
229. Moghe, A.; Deshpande, M.; Kamyab, S.; Chunarkar-Patil, P.; Nandi, S. S.; Bhatt, N. Hepatitis C Virus (HCV) and the Role of Phytochemicals in the Antiviral Effects of Different Medicinal Plants Against Infection. *Anti-Viral Metabolites from Medicinal Plants*. 2023, 1–31. 10.1007/978-3-030-83350-3_8-1
230. Ahmed, S. R.; Rabbee, M. F.; Roy, A.; Chowdhury, R.; Banik, A.; Kubra, K.; Hassan Chowdhury, M. M.; Baek, K.-H. Therapeutic promises of medicinal plants in Bangladesh and their bioactive compounds against ulcers and inflammatory diseases. *Plants*. 2021, *10*(7), 1348.
231. Gupta, M.; Singh, N.; Gulati, M.; Gupta, R.; Sudhakar, K.; Kapoor, B. Herbal bioactives in treatment of inflammation: An overview. *South African Journal of Botany*. 2021, *143*, 205–225.
232. Tran, N.; Pham, B.; Le, L. Bioactive compounds in anti-diabetic plants: From herbal medicine to modern drug discovery. *Biology*. 2020, *9*(9), 252.
233. Fang, M.; Echouffo-Tcheugui, J. B.; Selvin, E. J. D. C. *Clinical and public health implications of 2019 endocrine society guidelines for diagnosis of diabetes in older adults*. 2020, *43*(7), 1456–1461.
234. Sivakumar, T.; Deepa, B. A critical review on Antidiabetic Potential of Herbal plants and its their bioactive components. *Journal of University of Shanghai for Science and Technology*. 2023, *25*(01), 303–314.
235. Al-Shafie, T. A.; Mahrous, E. A.; Shukry, M.; Alshahrani, M. Y.; Ibrahim, S. F.; Fericean, L.; Abdelkader, A.; Ali, M. A. A Proposed Association between Improving Energy Metabolism of HepG2 Cells by Plant Extracts and Increasing Their Sensitivity to Doxorubicin. *Toxics*. 2023, *11*(2), 182.
236. Adomas, A.; Heller, G.; Li, G.; Olson, Å.; Chu, T.-M.; Osborne, J.; Craig, D.; van Zyl, L.; Wolfinger, R.; Sederoff, R. Transcript profiling of a conifer pathosystem: response of Pinus sylvestris root tissues to pathogen (Heterobasidion annosum) invasion. *Tree Physiology*. 2007, *27*(10), 1441–1458.
237. Vrancianu, C. O.; Dobre, E. G.; Gheorghe, I.; Barbu, I.; Cristian, R. E.; Chifiriuc, M. C. Present and future perspectives on therapeutic options for carbapenemase-producing Enterobacterales infections. *Microorganisms*. 2021, *9*(4), 730.
238. Newman, D. J.; Cragg, G. M. Natural products as sources of new drugs over the nearly four decades from 01/1981 to 09/2019. *Journal of Natural Products*. 2020, *83*(3), 770–803.

6 Pharmacological Properties of Herbal Drugs

Maria Faraz, Iftikhar Ahmad, Roomah Javed, Sohail Ahmad, and Arshad Farid
Gomal Center of Biochemistry and Biotechnology, Gomal University, D.I. Khan, Pakistan

Samy Selim
Department of Clinical Laboratory Sciences, College of Applied Medical Sciences, Jouf University, Sakaka, Saudi Arabia

6.1 INTRODUCTION

6.1.1 The Development of Antibiotics

In the control of infectious diseases, one very successful and tremendous accomplishment of contemporary technology and modern science is the development and discovery of antibiotics. Nonetheless, the conventional antimicrobial agents used for the eradication of disease-causing microorganisms are no longer effective due to the increasing emergence of resistant strains of pathogenic microorganisms (Ge et al., 2002; Neogi et al., 2007). There is a global increase in the isolation of microorganisms that are resistant to conventional antibiotics as well as the resistant strains recovery during antibacterial therapy (Hancock, 2005). Additionally, the harmful side effects of antibiotics on the body of the host include the reduction of useful microorganisms on gut and mucosal surfaces, hypersensitivity allergic reactions, and immune suppression, which are other substantial problems from the use of conventional antibiotics (Al-Jabri, 2005)

6.1.2 Antibiotic-Resistant Microorganisms

Microorganisms such as bacteria are becoming less susceptible to antibiotics for some time and this resistance towards antimicrobial medicines has developed naturally. Nonetheless, over recent years, the major cause of the occurrence of pathogens resistant to various drugs has been the misuse and overuse of antibiotics. The advent of novel diseases and the occurrence of epidemics because of the emergence of drug-resistant strains of microorganisms cause many challenges to the health concerns of the public. Recently, the potential substitute for the currently used antimicrobial agents has been shown by plants. Human health is significantly influenced by medications derived from plants. A significant increase has been observed in the occurrence of multidrug-resistant bacterial strains and the emergence of strains with reduced susceptibility to antimicrobial drugs. The random utilization of broad-spectrum antibiotics, immunosuppressive agents, organ transplantation, intravenous catheters, and constant HIV infection outbreaks have all been linked to this increase in the emergence of multidrug-resistant microorganisms (Gonzalez et al., 1996). Pneumococci resistant to macrolides and penicillin, vancomycin-resistant enterococci, staphylococci resistant to methicillin, and multidrug-resistant gram-negative strains are among the examples of this phenomenon (Norrby et al., 2005)

 DOI: 10.1201/9781003544418-6

Immediate action is required for the control of antimicrobial resistance. These actions include better use of antibiotics and reduced levels of cross-infection in hospitals (Sung & Lee, 2007). Therefore, it is very important to produce new antibiotic drugs as it is very crucial to maintain the effectiveness of antimicrobial therapy (Marchese & Schito, 2000).

6.1.3 Necessity of Developing Natural Plant-Derived Drugs

Plants are one of the cornerstones for current medicines aid in attaining new ideas and principles as plants are promising agents for the development of antimicrobials that can combat a vast majority of microorganisms (Evans et al., 2002). Antimicrobials obtained from plants are an enormous untapped medicine source. The therapeutic application of antimicrobials derived from plants is great because plants can function for the intended purpose without causing any of the harmful effects on the body which can often be associated with conventional antibiotics.

Furthermore, it is very crucial to investigate and develop antimicrobials derived from plants As medication derived from plants has greatly improved the well-being and health of humans, plants have frequently served as a source of development of new drugs (Iwu et al., 1999). Both traditional and modern medicines can be effectively constituted by medicinal plants. Herbal medicines are commonly used by approximately 80% of people in rural areas as the major source of healthcare because of their tremendous efficacy in eradicating infectious diseases (Akinyemi et al., 2005). The World Health Organization (WHO) has been advocating the idea that countries should support the use of conventional medicines to identify and utilize these compounds which provide effective and safe treatment for diseases caused by both microbial and non-microbial agents (World Health Organization, 1978). In order to produce more affordable and inexpensive disease treatments for the population, the pharmaceutical industries have been devoting a great deal of time and money to the production and development of natural compounds derived from plants (Doughari, 2007). The first scientific investigations on plants' antimicrobial properties date back to the late 19th century (ZAIKA, 1988). Studies have suggested that 50% of Western drugs contain plant-derived compounds as well as serve as models (Ashutosh & Kar, 2007). Numerous commercially approved medications that are now widely utilized in contemporary medicines were initially exploited in their basic forms in folk and conventional curing practices or for some other purposes which have demonstrated possibly therapeutic biological activity. The utilization of plant-derived medications has several benefits over synthetic drugs, including significant therapeutic advantages and providing less expensive treatment.

6.2 PHARMACOLOGICAL ACTIVITIES OF MEDICINAL PLANTS

6.2.1 Antimicrobial Activity of Herbal Drugs

It has been suggested that several biological qualities, including anti-inflammatory, antimicrobial, and antioxidant properties, have been demonstrated by extracts derived from medicinal plants (Mehta et al., 2001) (Figure 6.1). The growth of microorganisms, including bacteria, viruses, fungi, and some protozoans, are significantly inhibited by the antimicrobial substances extracted from plants through the use of different mechanisms from those of currently used antimicrobials; therefore, these plant-derived compounds have potentially important clinical value in combating different diseases caused by microorganisms (Shankar et al., 2010). Some of the plant-derived compounds, when coupled with other antimicrobials, can modify antibiotic-resistant bacteria that were not eradicated by antibiotics alone, as well some of these compounds exhibit both antibiotic resistance-modifying activities and intrinsic antibacterial activity. The relatively few side effects of herbal medications, when compared to those of synthetic conventional drugs, make these compounds a good candidate for eradicating microbial infections and the chances of developing resistance also become low in the case of using herbal medicines (Ruddaraju et al., 2020). In some cases, medicinal plants having a single active component with a specific target are used may then act in a similar way to conventional antibiotics, leading

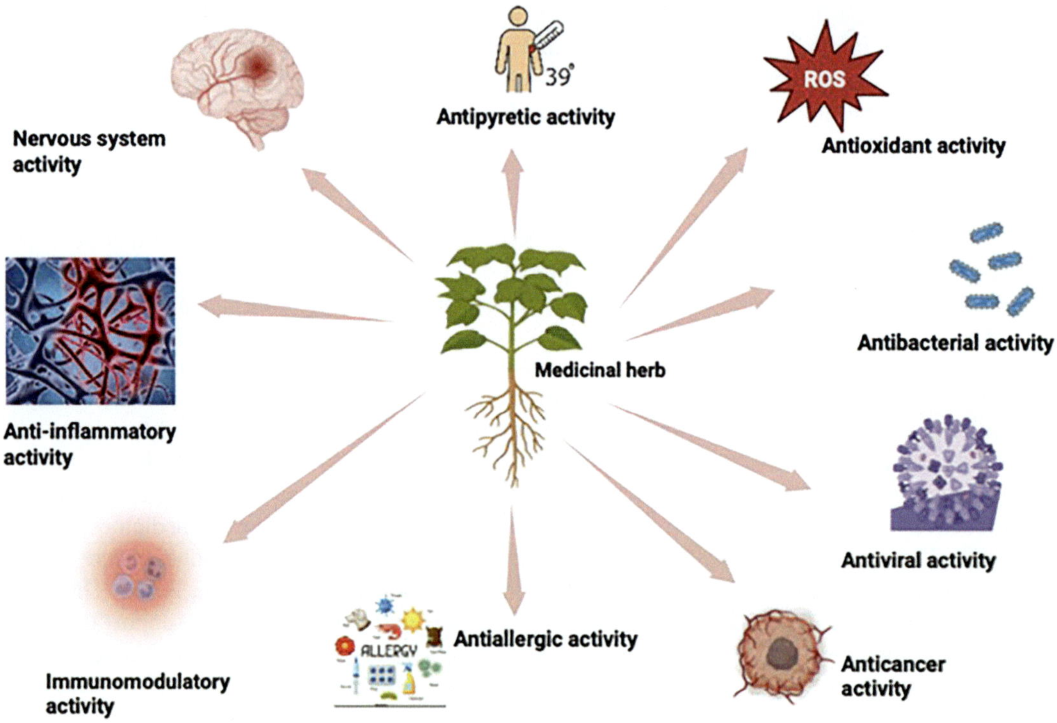

FIGURE 6.1 Pharmacological activity of medicinal herbs.

bacteria to develop resistance (Reker et al., 2014). Thus, more research is required on resistance mechanisms because, at present, there is very little research into the ways in which microorganisms develop resistance against plants (Almabruk et al., 2018). Additionally, the synergies between the active compounds of the medicinal plant extracts increases their effectiveness in the inhibition of bacterial growth (Wagner & Ulrich-Merzenich, 2009). Different types of effects are linked to the synergistic action, which includes the presence of components that are capable of suppressing the resistance mechanism of bacteria, the development of multi-target mechanism, physicochemical effects, and pharmacokinetics, resulting in an increased rate of resorption and solubility, increased bioavailability, and the increased neutralization of toxic and adverse effects (Wagner & Ulrich-Merzenich, 2009). There is a wide variety of chemical compounds that have tremendous antimicrobial properties present in medicinal plants. The mechanism of antimicrobial activity of phytocomponents derived from plants includes DNA damage, acidification of cytoplasm, iron chelation, the disruption of the bacterial cell membrane, and oxidative stress induction (Cowan, 1999) (Figure 6.2). Phytochemical investigation showed that a large variety of different components isolated from caper (*Capparis sp.*), exhibiting anti-inflammatory, antimicrobial, antiviral, and antioxidative properties, are quercetin, spermidine, carotenoids, and rutin. Glucosinolate and quaternary ammonium are the major components of *Capparis decidua* seed extracts responsible for antifungal, anti-bacterial, and anti-leishmanial properties (Tlili et al., 2011). Urinary tract infections have been reported to be effectively treated by the consumption of juice of *Vaccinium macrocarpon* and bearberry (*Arctostaphylos uraursi*), commonly known as cranberry and bearberry respectively; however, several plant species have been identified as broad-spectrum antimicrobial agents, including *Allium sativum* (garlic), *Melissa officinalis* (lemon balm), and *Melaleuca alternifolia* (tea tree) (Joshi, 2016). Cameroonian plant metabolites, such as flavonoids, triterpenes, alkaloids, phenolics, and steroids, are considered the major bioactive components with strong antimicrobial properties (Dzotam & Kuete, 2017). The major ingredient of the medicinal plant *Euphorbiaceae Croton lechleri* mostly grown in Western Amazonian areas of South America

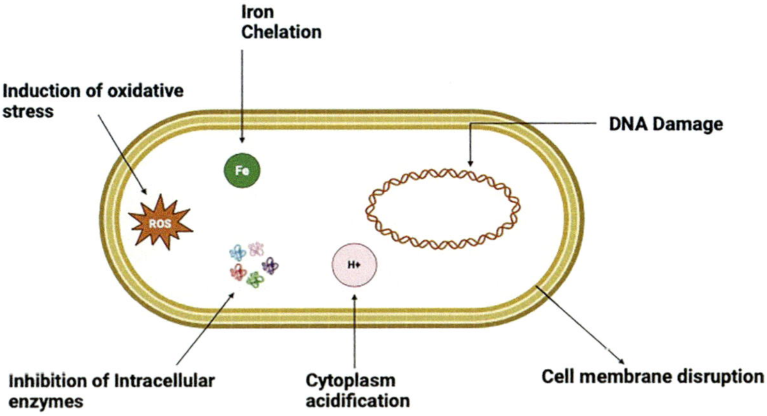

FIGURE 6.2 Antibacterial mechanism of bioactive compounds of plants.

is Fulyzaq (crofelemer, a proanthocyanidin oligomer, is responsible for notable antibacterial activity (Orozco-Topete et al., 1997)). *Escherichia coli*, *Staphylococcus aureus*, and *Salmonella typhi* are significantly eradicated by the *Myrtus communis* and *Verbena officinalis* leaf extracts. Furthermore, remarkable antibacterial activity is exhibited by *Myrtus communis* against *Pseudomonas aeruginosa*. Significant antibacterial effects against *Mycoplasma pneumoniae* and *Helicobacter pylori* were shown by seed oil of *Daucus carota* (Carrot) and *Melaleuca alternifolia* (tea tree) oil, respectively (Wangchuk et al., 2011). *Salmonella typhi*, *Citrobacter koseri*, *Klebsiella pneumoniae*, and *Staphylococcus aureus* are shown to be significantly inhibited by *Artemisia vulgaris*, *Cinnamomum tamala*, *Oxalis corniculate*, and *Ageratina adenophora* methanolic extract (Manandhar et al., 2019). Additionally, a high antibacterial effect against *Staphylococcus aureus*, *Enterococcus faecalis*, and *Enterobacter cloacae* has been demonstrated by the hydro-methanolic extracts of *Punica granatum*, *Berberis vulgaris*, and *Cistus monspeliensis* (Bereksi et al., 2018). Various components such as, hyperphorin, hypercalin B, emodin and, hyperenone A from an endophytic fungus which is isolated from *Hypericum acmosepalum*, a well-known medicinal plant, exhibited potent antibacterial efficacy against various bacteria such as *Pseudomonas aeruginosa*, *Staphylococcus aureus*, *Klebsiella pneumoniae, Escherichia coli*, *Salmonella enterica*, and *Mycobacterium tuberculosis*. These components also show antifungal property against *Candida albicans* and *Aspergillus niger* (Osman et al., 2012). Numerous essential molecules are found in Hypericum olympicum, among which the main compounds are β-farnesene, spathulenol, and E-anethole while other components are also isolated from Hypericum olympicum E-caryophyllene, a novel form of acylphloroglucinol and germacrene D. It has been suggested that Hypericum olympicum crude extract demonstrated strong antibacterial action against various resistant strains of *Klebsiella pneumoniae* and *Salmonella enteritidis* (Shiu et al., 2012). Strong antiprotozoal activity and antibacterial activity is observed in natural resins isolated from medicinal herbs and their derivatives (Paraschos et al., 2012). Specifically, propolis extract rich in flavonoid content mainly (pinocembrin and galangin) displayed greater antibacterial activity against various strains of *Streptococcus pyogenes* (Bosio et al., 2000). Studies revealed that Korean propolis demonstrates significant antimicrobial activity against *Streptococcus mutans* (Kim et al., 2011). A very strong antibacterial action against *Staphylococcus aureus* strains, including methicillin-resistant *Staphylococcus aureus (MRSA)*, was demonstrated by the compound diaporthalasin, which is isolated from *Diaporthaceae* sp., a fungus from marine sponge (Liming et al., 2016). Some viruses and both gram-positive and gram-negative bacterial isolates are significantly inhibited by the essential oil isolated from aromatic medicinal plants which include peppermint, fennel, lavender, and thyme, and these plants consist of a mixture of active compounds, such as phenylpropanoids, sesquiterpenes and, monoterpenes, which are responsible for antibacterial activity (Reichling et al., 2009; Sienkiewicz et al., 2012).

6.2.2 Anticancer Activity of Medicinal Herbs

Cancer is one of the significant causes of death globally, affecting both developed and developing countries. It is thus considered a significant public health concern. About 12.5% of the world's population dies because of cancer according to a survey published by the WHO. Cancer is a disease condition that is characterized by the abnormal and uncontrollable growth of body cells. It may form tumors and it may be metastatic and therefore move towards other body parts, where it forms secondary tumors (Akindele et al., 2015; Ochwang'i et al., 2014). Various factors that are considered to be the cause of cancer include heredity, physical inactivity, various environmental factors, and an unbalanced diet (Siddiqui et al., 2022). There are several chemical substances used to treat cancer, but their utilization is prohibited because they have adverse side effects (Kathiresan et al., 2006). Non-targeted cells/tissues are also affected by chemotherapy, immunotherapy, radiotherapy, and some surgical practices. This emphasizes the urge to employ alternative therapies and treatments against cancer (Veerakumar et al., 2016). Numerous investigations on cancer have been executed with traditional medicinal plants to find new therapeutic medication with fewer side effects than the present chemotherapeutic medicines (Shaikh et al., 2014). In the treatment of cancer, herbal treatments have been accepted worldwide to have very few adverse effects on the body (Hartwell, 1967). The majority of the world's population uses medicinal plants; these play a continuous role in human healthcare. Only a few medicinal plants around the world, including in Pakistan and India, have attracted the interest of researchers who are eager to investigate their potential use in the treatment of cancer (Shaikh et al., 2014). Medicinal plants have naturally occurring bioactive components, including flavonoids and phenols, which are responsible for protecting biological systems from adverse and harmful effects. Their proapoptotic, anti-tumor, and antiangiogenic effects have been investigated (Carocho & Ferreira, 2013). Recently anti-tumor compounds isolated from medicinal plants, including campothecin, vinblastine, flavofiridol, silvestrol, and podophyllotoxin, have been utilized worldwide (Batra & Sharma, 2013). Strong good immunomodulatory and antioxidant properties of medicinal plants result in anti-cancer effects. The cells of the body are protected from oxidative damage by the action of these antioxidant phytochemicals (Madhuri & Pandey, 2009)

Great cytotoxicity in many cancer cell lines is demonstrated by AP9-cd, which is a standardized lignin isolated from the medicinal plant *Cedrus deodara*. The mechanism by which medicinal plants stimulate the death of cells in human leukemia Molt-4 and HL-60 cells was also studied. The mechanism by which medicinal plants show anticancer activity includes the suppression of Molt-4 cell proliferation, induce the production of apoptotic body masses, and trigger the formation of a DNA ladder. Time-dependent increase and post-apoptotic necrosis were revealed by flow cytometric analysis (Shashi et al., 2006). HeLa cells were used to examine the anti-cancer cytotoxic properties of two saponins such as gymnemagenol and dasyscyphin derived from medicinal plants *Gymnema sylvestre* and *Eclipta prostrata*, respectively. A significant cytotoxic property was demonstrated by dayscyphin C and gymnemagenol. The positive control group used was 5-Fluorouracil (5-FU). According to in vitro analysis, it is determined that medicinal plant-derived saponins dayscyphin C and gymnemagenol have tremendous anticancer-cytotoxic activity (V. Khanna & Kannabiran, 2009).

The first medicinal drugs to start in clinical trials for the treatment of cancer were vinblastine, vincristine, and Vinca alkaloids derived from the medicinal herb *Catharanthus roseus*. When combined with other chemotherapeutic medicines, these herbal medicines are mainly used to treat various cancer types such as lung cancer, leukemia, breast cancer, advanced testicular cancer, lymphoma, and Kaposi's sarcoma. Likewise, the one of the great anticancer drugs, roscovitine, is a synthetic compound produced from the natural substance known as olomucine, which has been isolated from the medicinal plant *Raphanus sativus* (Meijer & Raymond, 2003; Bhushan et al., 2007). In human leukemia HL-60, the natural product Triterpenediol (TPD), derived from *Boswellia serrata*, causes apoptosis induction. The proliferation of cells is inhibited and apoptosis is induced by Triterpenediol (TPD) as confirmed by the more sub-G0 DNA fraction, increased formation of DNA ladder and increased AnnexinV-FITC binding of the cells (Bhushan et al., 2007). Herbal substance gallic acid (GA), which is considered a natural antioxidant and is derived from the medicinal plant *Phaleria macrocarpa*, leads to a considerable decrease in the proliferation of cells in several cancer cell lines and in TE-2, that is esophageal cancer cells, it triggers apoptosis, while it does not do so in CHEK-1 which is noncancerous cells (Faried et al., 2007). Silymarin, which has significant chemopreventative properties, inhibits many carcinogenic chemicals. The carcinogenic action of azoxymethane, which is responsible for colon cancer in rats, is significantly inhibited by Silymarin. Silymarin also inhibited benzoyl peroxide which causes skin carcinogenesis (Kohno et al., 2002).

6.2.3 Antiviral Activity of Medicinal Herbs

The utilization of medicinal herbs on earth began at the very beginning of human civilization. These medicinal plants have many therapeutic applications. Many of these have been employed in the treatment of viral infections. The Boots Drug Company in England launched a study which investigated the anti-influenza efficacy of 288 medicinal herbs due to which interest has been developed in plants as antiviral agents (Chantrill et al., 1952). Further studies revealed that medicinal plant extracts have significantly inhibited various virus replication. These plants have inhibitory effects against HIV (Asres & Bucar, 2005), hepatitis B virus (HBV), and Herpes simplex virus type 2 (HSV-2) (K. L. Huang et al., 2006a). Plant extract strongly inhibited severe acute respiratory syndrome (SARS) virus and Poxvirus infections (Kotwal et al., 2005). Alcoholic and water-soluble extracts of medicinal herbs have been used to study the therapeutic action of these plants. Very little effort has been made to identify the natural ingredient possessing antiviral properties. Furthermore, studies also revealed that plant extracts show potential antiviral activity against viral strains that are otherwise resistant to conventional antiviral drugs. The antiviral mechanism of the plants is different for different strains of the virus. However, plant extracts use common pathways to boost the human body's inherent defense system against viral infections, which is regulated by a complex immune system. The immunostimulatory effects of many medicinal plants with potent antiviral potential have been investigated in several studies (Webster et al., 2006)

6.2.3.1 Medicinal Plants Exhibiting Antiviral Activity

The macrophage activation assay revealed that the extract of the herb *Heracleum maximum* Bartr. (Umbelliferae) roots possess potent antiviral activity, along with antibacterial and antifungal activity. It induced interleukin 6 production, which thus confirmed the link between antiviral properties and immunostimulatory activity (Webster et al., 2006). Moreover, *P. asiatica* Linn. *(Plantaginaceae) and Plantago major* Linn., commonly known medicinal plants utilized in Taiwan folk medicine as a treatment for various infections, have been demonstrated that at low concentrations, they stimulate the proliferation of lymphocytes and interferon-gamma (IFN-γ) secretion. The increased proliferation of lymphocytes and interferon-gamma (IFN-γ) induction thus indicates the cell-mediated immune response modulation (Chiang et al., 2003). Similarly, the compounds extracted from *Sambucus nigra* L, namely Sambucol, have potent anti-influenza effects and are also known to stimulate immune system responses by the secretion of inflammatory cytokines (IL-1 beta, TNF-alpha, IL-6, and IL-8) (Barak et al., 2001). The most interesting finding about medicinal plants is their effective antiviral activity against various viral strains along with their ability to modulate the immune system (Pompei et al., 1979). This effect may be due to a particular plant component or several different constituents of plants. Different studies revealed that different infectious viruses, (i) including HIV, HCV, human papillomavirus, influenza, Marburg, and HBV, can be effectively inhibited by the root extract of *Trifollium* species Secomet-V (Kotwal et al., 2005). (ii) A significant antiviral effect against influenza virus strain H1N1 and HSV-1 has been shown by Pandanin, a lectin that is extracted from the saline *Pandanus amaryllifolius* Roxb leaves extract (Ooi et al., 2004), (iii) Hop crude extract demonstrated antiviral efficacy against many viruses, thereby indicating various plant parts contain broad-spectrum antiviral components (Buckwold et al., 2004). Medicinal herbs

TABLE 6.1
Medicinal Plants Possess Antiviral Effects against Various Viruses

Medicinal Plant	Virus	Antiviral Effect of Medicinal Plant	Reference
Geranium sanguineum L.	Influenza virus	The inactivity of various strains of influenza is considerably reduced by medicinal plants	(Pantev et al., 2006)
Elderberry extract		An inexpensive, efficient and safe treatment is exhibited by elderberry extract	(Zakay-Rones et al., 2004)
Carissa edulis Vahl	Herpes simplex virus (HSV)	Potent anti HSV-1 and -2 activity is demonstrated by *Carissa edulis Vahl extract*	(Tolo et al., 2006)
Phyllanthus urinaria L.		HSV-1 and -2 is strongly inhibited by geraniin, an active biochemical of *Phyllanthus urinaria L.*	(Yang et al., 2007)
Saxifraga melanocentra Engl.	Hepatitis C virus (HCV)	A significant anti HCV is exhibited by compound 1,2,3,4,6-penta-*O*-galloyl-beta-d-glucoside derived from *Saxifraga melanocentra Engl*	(Zuo et al., 2005)
Azadirachta indica Juss. (Neem)	Dengue virus type-2 (DEN-2)	The invitro and in vivo growth of DEN-2 is inhibited by neem leaves extract	(Parida et al., 2002)
Trichilia glabra L.	Vesicular stomatitis virus (VSV)	VSV is inhibited by *Trichilia glabra* leaves extract	(Cella et al., 2004)
Guazuma ulmifolia Lam	Poliovirus	Replication of polio virus as well as the viral antigen synthesis is inhibited by *Guazuma ulmifolia* extract	(Felipe et al., 2006)
Black soybean extract	Human adenovirus type 1	Dose dependent inhibitory effect of Black soybean extract against Human adenovirus type 1 was observed.	(Yamai et al., 2003)
Lycoris radiate	SARS-CoV	Anti SARS-CoV activity is exhibited by Lycorine, derived from *Lycoris radiate*	(Li et al., 2005)

Source: Table created from the article: Antiviral potentials of medicinal plants (Mukhtar et al., 2008).

are utilized for different purposes all around the world, but they are excessively used in countries like India, Japan, China, Pakistan, Thailand, Sri Lanka, and African countries. Developing countries are also encouraging the utilization of plant-based medicinal products for the treatment of diseases. The Canadian Natural Product Regulation established in 2004 represents an important step in the advancement of the utilization of plant-derived products in healthcare. This regulation promotes evidence-based scientific support and the use of modern technology encourages medicinal herbs and their derived substances (Siow et al., 2005). The replication of various viruses are inhibited by medicinal plants listed in Table 6.1.

6.2.4 Antioxidant Activity of Medicinal Herbs

The compounds that prevent or slow down the process of oxidation, thereby prolonging the lifespan of oxidizable substances, are known as antioxidants or inhibitors of oxidation (Panchawat et al., 2010). The species are known as oxidants or free radicals and are very highly reactive; their half-life is short and causes damage to macromolecules like DNA, lipids, and proteins. Oxidants may be produced from oxygen as reactive oxygen species (ROS) or they may originate from nitrogen as reactive nitrogen species (RNS). Hydrogen peroxide (H_2O_2), superoxide anions (O_2), reactive hydroxyl radicals (OH), and peroxyl radicals (ROO) are key reactive oxygen species. The free radicals produced from nitrogen are nitrogen dioxide (NO_2), nitric oxide (NO), dinitrogen trioxide (N_2O_3), and the peroxynitrite anion (ONOO) (Roja & Rao, 2000). Free radicals are generated continuously, resulting in severe damage to cells, tissue, and biomolecules and thereby leading to several diseases. Thus, an alternative medication to treat diseases linked to oxidative stress is medicinal plants with potent antioxidant activity (Narayanaswamy & Balakrishnan, 2011). The entity capable of slowing down or preventing other molecules from oxidation is called an antioxidant. In the process of oxidation, electrons are transferred from one component to the oxidizing agent. During this process, free radicals are produced as a result of an oxidation reaction. This starts a chain reaction, which damages the cells. The compounds known as antioxidants prevent the process of oxidation and can alleviate the harmful effects of the oxidation process in the tissue of the body. The damage caused by free radicals is prevented by antioxidants. Free radicals are molecules which act extremely erratically with unpaired electrons. They play a crucial role as intermediates in many physiological mechanisms such as neurotransmission, cytotoxicity, and the control of vascular tone. Many human diseases, such as Alzheimer's disease, cancer, kidney disease, cardiac reperfusion abnormalities, and fibrosis, are caused by free radicals. Antioxidants are crucial for many cellular functions and provide many benefits when present in food (Bharti et al., 2013).

Antioxidant activity in plants is due to the presence of nutrient components with demonstrated radical scavenging ability as well as being the result of minerals or non-vitamin substances (Sies, 1997). Medicines derived from plants consist of plants' phytochemicals such as flavonoids, polyphenols, and flavoproteins, along with ascorbate, zinc, alpha-tocopherol, and carotenoids. Additionally, some plants or specific herb combinations in particular formulations may function as antioxidants by superoxide scavenging activity or by enhancing the activity of superoxide dismutase at several tissue sites (Niwano et al., 2011). Several mechanisms are used by antioxidants to exert their cell protection effect. Many plants, like herbs, fruits, vegetables, and some commonly used spices, possess cancer-protecting factors along with anti-oxidant properties. Based on the chemical structures, these factors are divided into several different groups, for example, carotenoids, carbohydrates, retinoids, polyphenols, trace metals, terpenes, thiols, tocopherols, glucosinolates (isothiocyanates, indoles and dithiothiols), and others. The protecting effects are exerted by the group of several biochemical mechanisms. The biochemical processes responsible for carcinogenesis are still not understood and may differ according to the cancer type. Therefore, the explanation of the mechanism of carcinogenesis utilized by cancer-protecting factors must depend on the simplified carcinogenesis process. Here the model presented is a generalized initiation-promotion-conversion model. In this model, initiators are considered to be genotoxic both directly or indirectly, promotors are considered as particles that are capable of conferring a growth advantage on initiated cells, and converters, including mutagens, recombinogens, and clastogens, are considered to be genotoxic. Studies suggest that different mechanisms are used by cancer-protecting substances in fruits and vegetables against cancer initiation. These mechanisms include the polyphenol-scavenging effects on activated carcinogens and mutagens, the carotenoid's quenching effect on radicals, and singlet oxygen, additional the antioxidant effects is also possessed by several components like polyphenols and ascorbic acid. Flavonols and tannins pose inhibiting effect on activating enzymes, furthermore, the induction of enzymes involved in conjugation and oxidation by indoles, dithiothiones, and isothiocyanates, the protecting of some sensitive structures by some phenols and the DNA repair stimulation by sulphur-containing compounds (Dragland et al., 2003). The mechanism of antipromotion at the biochemical level involves the antioxidant effect of carotenoids' and polyphenols' membrane-stabilizing effects. Furthermore, the inhibition of protease by components isolated from soybeans, the induction of immune response via carotenoids and ascorbic acid and ornithine decarboxylase inhibition by carotenoids and polyphenols. Experimentation identified a few inhibitors of conversion, but it is theoretically reasonable that many initiation inhibitors may also be efficient against conversion. The effect of anticarcinogenic compounds found in vegetables and fruits is investigated in the context of cancer inhibition and prevention (Saito et al., 2008). The antioxidant substances of plants are more than supporting agents to fight against disease and cellular damage. As indicated by folklore, many plants have been demonstrated to have particular functions in disease treatment and prevention. Silymarin is a commonly known liver antioxidant derived from the medicinal plant *Silybum marianum* (milk thistle) prevents the damage of the liver by free radicals scavenging among other mechanisms (Saito et al., 2008). This effective antioxidant prevents liver damage which is caused by poisoning by highly poisonous compounds present

TABLE 6.2
Anti-Oxidant Activity of Some Medicinal Plants

Medicinal Plant	Bioactive Component of Medicinal Plant	Clinical Uses
Musa acuminate	Dietary fibers, sugar, fats, proteins	Since the earliest times this plant has been used as a source of food by humans
Theobroma cacao	Theobromine	Potent antioxidant
Olea euroapea	Phenolic compounds, Oleuropein	Excellent antioxidant and diuretic
Coriandrum sativum	Pinene, terpenes linalool	Strong antioxidant, diuretic and carminative
Origanum vulgare	Phenolic compounds, thymol, ocimene, limonene, caryphyllene	High antioxidant activity
Sasamum indicum	Essential fatty acids, lariciresinol	Strong antioxidant activity
Rubus ursinus	Salicylic acid, ellagic acid, anthocyanins	Exhibits excellent antioxidant property
Rubus occidentalis	Quercetin, pelargonidins, catechins, ellagic acid, kaempferol	Demonstrating significant antioxidant and antiproliferative functions
Piper nigrum	Selenium, piperine, beta-carotene, vitamin B	Strong anti-carcinogenic and antioxidant in nature
Arachis hypogaea	P-coumaric, Niacin, Folate fiber, phytonutrients, Vitamin E	Exhibiting strong antioxidant activities.
Allium cepa	Quercetin, 3,4-diglucoside	Anti-inflammatory and antioxidant property

Source: Table created from the article: some medicinal plants with antioxidant activity – a review (Nigam & Sodhi, 2014).

in the death cap mushroom *Amanita phalloides*, pharmaceutical drugs, or even alcohol. It is noteworthy that toxins of amanita are not easily eliminated by free radical-scavenging effects. However, it is hypothesized that the toxins of amanita and silymarin vie for the same receptors on cell membranes. Again, it is confirmed by contemporary laboratory research and thus explains the hepatic-protecting property of milk thistle, which has been utilized in folk medicines for 2000 years (Adhikari et al., 2007). Various medicinal plants with potent anti-oxidant activity are listed in Table 6.2.

6.2.5 Hepatoprotective Activity of Medicinal Herbs

The liver is considered to be one of the most important organs of the human body because of its function in the metabolism of different nutrients such as proteins, lipids, and carbohydrates, as well as its major role in the excretion of metabolic wastes. Furthermore, it also plays a crucial role in protecting foreign particles as it effectively excretes drugs and other xenobiotics from the body. The detoxification and elimination of foreign components is another major function of the liver. In addition to other functions, the liver also secretes bile which has a primary function in the digestion of food (Das et al., 2022). Hepatic disease or liver disease is a disease condition in which the liver cells, tissue, structure, and function are affected. Several types of complex functions are performed by the liver, including components detoxification, protein synthesis, and the production of biochemicals, which play an important role in the digestion of different substances. The liver also plays a function in the production and breakdown of complex and simple molecules, which are required for regulating important normal functions. In the protection of the liver, herbal medicine is preferrable to conventional allopathic medicines for several different reasons: herbal medicines are not expensive; better culture acceptability; they have greater compatibility with the human body; and they have few (or no) adverse side effects. The normal physiological function of the liver is maintained by these herbal drugs with very few side effects. An array of important functions such as maintenance, performance, and the regulation of body homeostasis is controlled by the liver. The liver plays an important function in nearly all biochemical pathways linked to growth, nutrient absorption, reproduction, disease resistance, and energy supply. Therefore, the well-being of an individual depends upon the health of the liver. Studies suggested that different types of toxicants, such as chronic alcohol consumption, microbes, chemotherapeutic agents, thioacetamide, and carbon tetrachloride, can all cause liver cell injury. Owing to their safety, affordability, and efficiency, herbal drugs become more important and popular in recent years. The use of herbal medications to treat diseases of the liver has a venerable and old history. Medicinal herbs and their derived chemical compounds are still utilized in different forms for this purpose all over the world. Scientific studies of herbs have revealed that the presence of active biochemicals is responsible for the therapeutic action of medicinal plants. These therapeutic biochemical components of the plants are utilized for a different variety of disease treatments. Various chemical constituents, such as phenols, lignans, monoterpenes, glycosides, flavonoids, essential oils, organic acids, alkaloids, carotenoids, and xanthenes, are found in liver-protective plants (Gupta & Misra, 2006). Therefore, several herbs and their mixtures have been demonstrated to exhibit hepatoprotective activity; accordingly, the production of hepatoprotective drugs from plants has gained in prominence around the globe. The family of composites has a member known as *Eclipta alba* (Bhringaraja), which is a perennial shrub that has been demonstrated to exhibit potent hepatoprotective properties and is therefore useful in the treatment of treating liver cell injuries caused by carbon tetrachloride (Chandra et al., 1987). The aromatic herb *Foeniculum vulgare Mill* is commonly known as the 'Fennel' of the Umbelliferae family. The essential oil of *Foeniculum vulgare* considerably inhibits carbon tetrachloride-induced hepatoxicity as indicated by reduced, alkaline phosphatase, serum aspartate aminotransferase, bilirubin levels, and alanine aminotransferase levels (Öbek et al., 2004). Similarly, fenugreek, which has the botanical name *Trigonella foenum graecum*, is a member of the family Leguminosae. An annual herb, it has been shown in studies that the phenolic extract of fenugreek seeds exerts a protective effect and that it protects human liver cells from the toxicity caused by ethanol. The ethanolic extract of fenugreek reduced the change liver cells growth and induced the formation of oxygen radicals, cytotoxicity, and mitochondrial dysfunction. Co-incubation of FPEt and EtOH dramatically enhances the dose-dependent increase in cell viability, decreases the leakage of lactate dehydrogenase, and restores the GSH/GSSG ratio to the normal range. *Jatropha curcas* Linn is a medicinal herb from the Euphorbiaceae family. Different compounds, including the flavonoids apigenin and its glycosides, vitexin, and isovitexin the sterols stigmasterol, α-D-sitosterol and its α-Dglucoside, have been derived from the leaves extract of *Jatropha curcas*, which possessed significant hepatoprotective activity (El-Baz et al., 2015). Methanolic extract of *Jatropha curcas* (MFJC) leaves is known to treat Aflatoxin B1 (AFB1)-induced hepatocellular carcinoma (El-Baz et al., 2015). The medicinal plant *Wedelia calendulacea* L., a member of the Asteraceae family, is considered to exhibit hepatoprotective properties. The *Wedelia calendulacea* ethanolic extract was evaluated for its hepatoprotective efficacy against CCL4-stimulated cytotoxicity in rats. The dose-dependent decrease in the elevated range of serum enzyme activities, which is induced by CCI4, is observed in rats when treated with the ethanolic extract *Wedelia calendulacea*. The proteins and bilirubin concentrations were also observed to be enhanced with the treatment of this medicinal plant. This result suggests that the extract of *Wedelia calendulacea* can restore the normal function of the liver, making it comparable to that of normal healthy rats (Murugaian et al., 2008). Other medicinal plants with potent hepatoprotective properties include *Andrographic paniculate, Solanum nigrum, Flacourtia indica, Phyllanthus emblica*, and *Sargassum polycystum*, etc. (Shirani et al., 2017).

6.2.6 Nervous System Activity of Medicinal Herbs

The brain, spinal cord, and billion neurons, which are also known as nerve cells, constitute the central nervous system. The aberration in the nervous system leads to a dysfunction of different body organs. The gradual and irreversible neuron loss in specific brain areas results in Huntington's disease (HD), Parkinson's disease (PD), and neurodegenerative disorders. An example of neurodegenerative conditions is characterized by the irregularities in the control of movement because of the neuron loss from basal ganglia. Alzheimer's disease, another neurodegenerative disease, is caused by hippocampal and cortical neuron loss and thus leads to memory impairment. Another neurodegenerative disease in muscular muscles, known as amyotrophic lateral sclerosis (ALS), becomes weak and this is caused by spinal degeneration, bulbar, and cortical and motor neuron degeneration. Nowadays, most of the pharmacological treatments for neurodegenerative disorders mainly focus on lessening the symptoms rather than altering the underlying disease progression. The activity of the nervous system is regulated by several medicinal plants. The ethanolic extract of medicinal plant leaves, known as *Vitex leucoxylon*, inhibited the spontaneous motor activity, thereby resulting in reduction of psychoactivity (Makwana et al., 1994). *Azadirachta indica* is known to possess analgesic activity in mice by regulating the neurotransmitter system in their bodies (N. Khanna et al., 1995). The hepatic microsomal enzyme system is significantly stimulated by *Pongamia pinnata* root and seed extract (R. K. Singh et al., 1996). The retention and consolidation of memory is facilitated by the alcoholic extract of *Bacopa monniera*, which mainly consists of bacosides A and B (H. K. Singh & Dhawan, 1997). *Hibiscus vitifolius* has been known for its anti-nociceptive activity, as with morphine, by regulating several neurotransmitter systems, and the main bioflavonoid compound known as gossypin present in the plant is responsible for this effect (Ramaswamy & Viswanathan, 1997). The hydroalcoholic extract of roots of the medicinal herb *Argyreia speciose* at 500 mg/kg dosage was investigated for neuropharmacological properties in mice, the measurement of pentobarbitol-induced sleeping time and spontaneous motor activity were evaluated. These results indicated the depressant activity of the central nervous system (Shen et al., 2002). The method was used to assess the depressant activity of the central nervous system of mice using *Russelia equisetiformis (REC)* and its derivatives (RE1, RE2, and RE3) crude methanol extract; the assessment involved a test for amphetamine-induced stereotypy, phenobarbitone sleeping time, and picrotoxin-induced convulsion. The phenobarbitone sleeping time is significantly enhanced by REC; it also significantly reduced the sleep latency. The RE1, RE2, and RE3 fractions considerably extended the phenobarbitone sleeping time and also delayed the onset of sleep. REC notably reduced stereotype behavior induced by amphetamine, but it does not prevent amphetamine-induced mortality. The aforementioned results indicated the central nervous system depressant activities possessed by the methanolic extract of *Russelia equisetiformis*. It is therefore a suitable choice for the treatment of disorders of the central nervous system (Kolawole et al., 2007)

6.2.7 Anti-Inflammatory Activity of Medicinal Herbs

Globally, the prevalence of inflammatory etiology or pathology is increasing daily (Iwalewa et al., 2007). These inflammatory disorders are very difficult to treat because they involve several systems, their treatment is very costly, they involve long-duration treatment if necessary, and they may have adverse side effects (Ponder & Long, 2013). In the treatment of anti-inflammatory diseases, the common anti-inflammatory drugs that are regularly applied are chemical synthetic drugs. These drugs have many drawbacks, even if they are usually effective and quick-acting (Ganesh et al., 2014). They are very cost-effective, rarely available in many countries, and pose some serious side effects (Sostres et al., 2013). By contrast, medicinal herbs are gaining more importance day by day and become beneficial for the treatment and prevention of inflammatory diseases (Gessner et al., 2017). These plant-based medications are becoming very popular because of their effectiveness, and the facts that they are both relatively inexpensive, and widely available in many regions (Yatoo et al., 2017). Inflammatory responses that create serious irregularities in the body can often be effectively treated by anti-inflammatory herbs. It should be borne in mind that the inflammatory response is a natural protective defense action of the body against injury or infection; however, it can cause some adverse effects when it becomes intensified or excessive, meaning that early intervention is necessary in order to secure effective outcomes. Beneficial properties of medicinal plants, such as reasonable potency, cost-effectiveness, safety, very little or no side effects, and easy availability make them the best candidate to treat anti-inflammatory disorders. The phytoconstituents derived from medicinal plants have anti-inflammatory properties; therefore, they have considerable potency for the prevention of adverse inflammatory processes (Adegbola et al., 2017). There are various phytoconstituents which are derived from the plants, and which are commonly known by a number of different names. These include flavonoids, steroids, phenolics, terpenoids, glycosides, polysaccharides, fatty acids, alkaloids, and cannabinoids. These active phytochemicals use different mechanisms for the prevention of inflammatory actions. In some cases, these components reduce inflammatory conditions by synergizing the enzymes, proteins and some other factors involved in the anti-inflammatory pathway. Alternatively, these phytochemicals may interfere with inflammatory protein and enzymes like lipooxygenases, interleukins, prostaglandin, tumor necrosis factors, cyclooxygenases, mitogen-activated protein, nitric oxide, and nuclear factors. Taking all these aforementioned factors into account, more molecular and cellular research is needed to gain a deeper understanding of the preventive phenomena (Yatoo et al., 2017). At present, the most commonly used anti-inflammatory medicinal herbs are *Curcuma longa*, *Urtica dioica*, *Zingiber officinale*, *Vaccinium myrtillus*, *Borago officinalis*, *Rosmarinus officinalis*, and *Vaccinium myrtillus* These plants have very minimal side effects, in contrast to synthetic chemical drugs, such as the nonsteroid anti-inflammatory drugs and immunosuppressants which are employed in the treatment of such conditions. Medicinal herbs are easily available and are inexpensive. A thorough and proper assessment of their pharmacological, phytochemical, and physiological properties is essential for the effective and safe use of these phytochemicals in anti-inflammatory treatment (Medzhitov, 2010).

6.2.7.1 Mechanism of Action

The enzyme phospholipase A2 damages phospholipids, causing them to be released from leukocytes and platelets. This leads to a condition known as inflammation. These phospholipids are activated by proinflammatory cytokines, such as IL-1 and TNF-α. Leukotrienes and prostaglandins, prostacyclins, and thromboxanes are produced by the metabolism of arachidonic acid via cyclooxygenase enzymes (COX). It has been found that different tissues and organs consist of both COX1 and COX2; however, most of the body tissues consist of COX1, whereas COX2 is present only at the inflammation site. Different mechanisms are being used by anti-inflammatory agents, including resins, alkaloids, lignans, and essential oils, to prevent inflammation. As stated in the previous section, these agents either affect the inflammatory pathways or inhibit the proteins, enzymes, and hormones involved in the inflammatory pathway (T. H. W. Huang et al., 2006b).

Medicinal plants prevent inflammation through the action of COX inhibitors, phospholipase A2 inhibitors, LOX inhibitors, and A2 inhibitors, which are the essential components of medicinal plants. The inhibition of the enzyme phospholipase A2 is achieved by phospholipase A2 inhibitors; this is considered to be the starting point of the inflammatory processes. Similarly, the formation of prostaglandins and thromboxane from arachidonic acid is prevented by COX inhibitors (Setty & Sigal, 2005). Inflammation is also prevented by leukotriene and prostaglandin inhibitors through their actions on different inflammatory pathways. Medicinal plants have been demonstrated to have significant anti-leukotriene and anti-prostaglandin effects. Oxidative stress is dramatically also reduced by medicinal plants as they can significantly reduce the levels of oxidants. Additionally, the amount of anti-oxidant is increased, which results in the prevention of inflammation (Anilkumar et al., 2017). The aggregation of platelets and the constriction of vessels is prevented by constriction, thereby inhibiting diffusion, ischemia, edema, and necrosis. A bisbenzylisoquinoline alkaloid known as Tetrandrine has also been shown to have promising antirheumatic properties, making it perhaps a suitable herbal drug for the treatment of rheumatic diseases such as silicosis. The anti-inflammatory effects of medicinal plants are attributed to their effects on a number of factors that are involved in inflammatory pathways, including the reduction of IL1-level, Platelet Activating Factors (PAF), and endothelin-1. The production of reactive oxygen species (ROS) is also reduced by medicinal plants, which is induced by polymethacrylic acid, iNOS, and NO release inhibition and the prevention of arachidonic acid synthesis. The scavenging action of the superoxide radical inhibits the photolysis of vitamin B, decreases the production of hydroxyl free radicals, and suppresses TNF-α production (Kassuya et al., 2005).

6.2.8 Antipyretic Activity of Medicinal Herbs

Fever, also known as pyrexia, is characterized by an increase in body temperature beyond the normal physiological range. Pyrexia is caused by different factors, including microbial infections, increased secretion from the thyroid, due to leukemia, excessive exercise, physiological stress such as during ovulation, and injuries to the central nervous system (Balunas & Kinghorn, 2005). Whenever the body comes in contact with any infectious particles, such as bacteria, viruses, or fungi, the body's natural defense mechanism becomes activated to make the environment unfavorable in order to make it more difficult for these pathogens to survive. Proinflammatory mediators, such as interleukin 1β, β, α, and TNF-α production, is increased in response to damaged tissue or the presence of any infectious agents (Sultana et al., 2013). The production of prostaglandin E2 (PGE2), which occurs close to the peptic hypothalamus area, is stimulated by these cytokines. Prostaglandin, in turn, acts on the hypothalamus to increase body temperature. Antipyretic drugs are used to reduce the increased body temperature; they work by inhibiting the expression of COX-2, which subsequently inhibits the synthesis of prostaglandin (Vane & Botting, 1995). Though COX-2 is inhibited by synthetic antipyretic drugs with great selectivity, there are certain limitations of synthetic antipyretic agents as they have adverse effects on other organs as well such as the cortex of the brain, glomeruli, heart muscles, and hepatic cells. Natural COX-2 inhibitors, on the other hand, have low selectivity and minimal side effects (Vane & Botting, 1995). For many centuries, fever has been treated with the help of antipyretic agents isolated from plants. It is therefore essential to look for more herbal drugs that also have significant antipyretic activity, but which are less toxic, have fewer side effects, and can be used as an alternative to synthetic drugs e.g., paracetamol (Chhetri, 2004).

6.2.8.1 Medicinal Plants Possessing Antipyretic Properties

The member of the Maliaceae family, *M. azedarach*, generally called Bakyain, is a medicinal herb and exhibits potent antipyretic properties. Studies revealed a dose-dependent decrease in elevated body temperature when treated with hydro-methanol extract of *M. azedarach*. The extract at 500 mg/kg has a similar effect as the standard prescribed medication paracetamol (Sultana et al., 2013). The medicinal plant *P. cineraria* of Leguminosae, also called Kherji, Ghaf, or Jandi, is grown in different countries in the Middle East. This plant is also famous for its significant antipyretic activity. Albino rats were used to investigate the ethanol extract of *P. cineraria* for its antipyretic effect against brewer's-yeast-induced pyrexia, which is present in its leaves and fruits. It was found that the in studies temperature of a rat was significantly lowered by the application of *P. cineraria* leaves and fruit extract in a manner comparable to that achieved by standard drugs (Joseph et al., 2011). *P. nigrum*, also known as species of king from the family *Piperaceae*, is another example of an antipyretic herb. This herb is utilized for conventional home treatment to reduce fever. Wistar albino rats were shown with an elevation of antipyretic properties of the alcoholic extracts of *P. nigrum*. A brewer's yeast suspension injection at 15% (w/v) was used to induce pyrexia in rats. *P. nigrum* with concentrations of 250 and 500 mg/kg considerably lowers the temperature of the body, when compared with the use of standard drugs (Damanhouri, 2014). The medicinal herb known as *P. rubra* L. is from the family Apocynaceae. It is commonly known as true frangipani. The ethanolic extract of *P. rubra* was evaluated for its antipyretic properties and this revealed that the *P. rubra* at 200 mg/kg considerably reduced the increased body temperature of rabbits comparable to the standard drug aspirin (Baghel et al., 2010). A three-metre-tall shrub from the Capparaceae family is known as *C. trifoliata* (Roxb.) Wt.& Arn. This medicinal plant is known to treat several ailments such as syphilis and sores, and can be utilized as an emmenagogue, deobstruent, antiphlogistic anthelmintic, and so on. The aqueous extract and ethanolic extract of *C. trifoliata* was assessed for its antipyretic activity on Wistar rats in which pyrexia was induced by yeast. The extracts of aqueous and ethanol of *C. trifoliata* at a concentration of 500 mg/kg body weights are drastically seen in the reduction of the body temperature and pyrexia induced by yeast which is comparable to the standard antipyretic drug paracetamol at 45mg/kg. This study reveals the potent antipyretic property of *C. trifolia* (Mythreyi et al., 2009).

6.2.9 Antiallergic Activity of Medicinal Herbs

An allergy is an exaggerated reaction caused by the body's immune system, whenever it comes in contact with some sort of foreign particle. This reaction can be regarded as exaggerated because in the body of non-allergic people, these foreign particles are typically recognized as harmless and the body doesn't trigger such a reaction. Herbal drugs, along with synthetic drugs, are considered suitable anti-allergens because they have minimal side effects (P. P. Gupta et al., 1995).

Ethanolic extract of medicinal plant *Vitex negundo* has been demonstrated to inhibit mast cell degranulation, which is mediated by the immune system more efficiently than the synthetic compound. Furthermore, in mice during active paw anaphylaxis, this ethanolic extract reduced oedema. The tracheal contractions, both initially and at later sustained phases, are also inhibited by this extract. The extract was found to inhibit the release of histamine, which causes the initial phase of paw anaphylaxis and the later phase is primarily by the lipid mediators release from arachidonic acid is also inhibited by this extract. The notable anti-allergic activity was found to be demonstrated by an alcoholic extract of *Nyctanthus arbortristis* and *Andrographis paniculate* seeds, *Cedrus deodara wood* extract as well as *Albizzia lebbeck* bark aqueous extract in rat anaphylaxis, and mast cell degranulation investigation (Barua et al., 1997).

6.2.10 Antidiabetic Activity of Medicinal Herbs

It has been found that natural substances from organisms (including plants, animals or microorganisms) acquire beneficial properties for humans, plants, and animals. According to the estimation of the World Health Organization, approximately 80% of the people of developing countries use folk medicines or traditional medicines. Most of these medications are prepared from plants and are used for treating various diseases. Traditional medicines have several advantages over synthetic drugs as these medicines are less toxic to the body, have very minimal side effects, are very inexpensive, and are clinically effective. The study demonstrates an increasing focus on using medicinal plant-derived phytochemicals in pharmaceutical industries. Steroids, lignin, glycosides, phenolic and carbohydrates, and so on, are small molecules (known as secondary metabolites) which are isolated from plants and these components are known to have a diverse range of biological properties for the welfare of humans (Tran et al., 2020). These properties include antibacterial, anti-allergic, antidiabetic, and anti-inflammatory ones. The metabolic disorder of the β-cells of the pancreas results in a chronic disease commonly known as diabetes mellitus. The insufficient production of insulin by β-cells of the pancreas results in Type 1 diabetes mellitus and this condition leads to hyperglycemia. This condition is also caused by inadequate production of insulin because of insulin resistance in Type 2 diabetes mellitus (Kayarohanam, 2015). The current diabetes mellitus medications have the goal of controlling and reducing the glucose level of blood back to normal levels. However, during treatment, the usage of synthetic drugs brings many side effects and thus causes many serious health problems. Therefore, traditional medicines are the best suitable alternative for traditional medicines and they have thus been used for a long time due to their importance in the treatment of the disease. Additionally, in recent years, new bioactive medicines derived from plants have exhibited significant antidiabetic activity. These components have more effective antidiabetic activity than currently used oral hypoglycemic agents in clinical treatments. Traditional medicines have been shown to be an efficient clinical practice and to hold a bright future for diabetes mellitus treatment. Many medicinal plants which have been used as traditional antidiabetic medication demonstrated a significant antidiabetic and hypoglycemic effect and the hypoglycemic mechanism of these medicinal plants has been investigated thoroughly (M. C. Sidhu, 2013).

6.2.10.1 Medicinal Plants Possessing Antidiabetic Activity

One of the most common vegetables found in the tropical areas is *Momordica charantia* (MC) also known as bitter melon. Although this is commonly used as a vegetable, it can also be used as a herbal drug and has been used in traditional medicine for many centuries. MC has been demonstrated to have many biological properties, such as antibacterial, anti-oxidant, antiviral, and anticancer effects. It has been shown that MC possesses significant antidiabetic properties. Some insulin-like protein particles are present in the callus, seeds, and fruits of *Momordica charantia* (P. Khanna et al., 1981). These protein works in a similar manner to human insulin and reduce high levels of glucose when its impacts are investigated effect in trials on rats, humans, langurs, and gerbils (Dans et al., 2007). Similarly, in Korea, Ginseng has been a widespread medicinal plant in folk medicines over many centuries. Ginseng belongs to the family Araliaceae and the genus Panax (Attele et al., 2002). The anti-diabetic properties of *Panax ginseng* roots have been investigated by many researchers through both in vitro and in vivo studies (Ratan et al., 2021). Ginseng-specific saponins and ginsenosides are the most important phytochemicals isolated from *Panax ginseng*. Among these, the most effective component in diabetic treatment in rats induced by streptozotocin is Ginsenosied Rb2, which indicates that Ginsenosied Rb2 decreases the glucose level of blood (Korayashi & Oura, 1985). A promising antidiabetic property has been demonstrated by saponins, which have been derived from ginseng (Attele et al., 2002). These phytochemicals act as anti-diabetic agents by regulating and reducing the enzymatic activity in the metabolism of glucose and thus regulating the production of insulin (Hyun et al., 2020). The onion, which is also known as *Allium cepa* Linn, belongs to the Liliaceae family. A large number of phytochemicals have been identified and isolated from this vegetable. These bioactive substances include flavonoids, phenolics, amino acids, essential oils, thiosulfinates, vitamins, and other substances. Diverse bioactivities have been exhibited by these isolated bioactive compounds such as antioxidant activity, anti-inflammatory activity anti-diabetic activity, from their onion extracts. Different studies confirmed *Allium cepa* a potent hypoglycemic agent (Galal & Gawad, 1965). S-methyl cysteine sulfoxide is an important component of the bulb part of the onion and is the source of the active components responsible for its anti-diabetic activity. These components reduce the blood glucose value and possess a significant antioxidant property. S-methylcysteine sulfoxide exerts its antidiabetic activity through three different mechanisms: (1) S-methylcysteine sulfoxide increases the pancreatic secretion of insulin and thus increases the insulin level in the blood; (2) S-methylcysteine regulates the absorption of dietary glucose; and (3) it makes effective use of insulin (Kumari & Augusti, 2002)

6.2.11 Immunomodulatory Activity of Medicinal Herbs

Immunomodulators are crucial substances that actively regulate and normalize the immune system, whether they are natural or synthetic. They can rectify weakened immune systems and balance overly active immune systems. The use of immunomodulators activates the body's natural defense systems, including cytokines, to enhance its ability to protect itself. Furthermore, the stimulation of the immune system by plants has been employed to treat autoimmune diseases. A potent immunostimulant activity has been demonstrated by *A. paniculata* and andrographolide ethanolic extract (Puri et al., 1993). Andrographolide, through the induction of macrophage, cytokines, and natural killer cells, regulates specific and nonspecific functions of the immune (Peng et al., 2002). A disease-free state of the body is regulated with the help of the stimulation and repression of the immune response. Animals in both normal and immunosuppressed states which are induced by surgery or hemi-splenectomy can be protected from infection with the aid of medicinal plants, including *Tinospora cordifolia, Asparagus racemosus*, and *Withania somnifera*. These medicinal plants are known to produce white blood cells, specifically neutrophils, and inhibit the induction of leucopenia to varying degrees. It has also been suggested that polymorphonuclear and monocyte-macrophage systems are activated by these plants (Dahanukar & Thatte, 1997). A cell-mediated response is also triggered by *A. indica*, as confirmed by the inhibition of macrophage migration and enhanced the thickness of the footpad (Muntgantiwar et al., 1997). It has been found that immunostimulatory effects are demonstrated by *Janakia arayalpathra* root suspension in mice.

Plaque-forming cells assay reveals that, following immunization with sheep erythrocytes, the level of humoral antibody is increased by *Janakia arayalpathra* and it also increases the number of spleen cells that secrete antibodies after immunization with sheep erythrocytes. It significantly raised the peritoneal macrophage number and led to a notable increase in delayed hypersensitivity reaction in mice. In suppressed mice, the suppressed humoral response is significantly reversed by the alkaloidal fraction of *Boerrhiva diffusa* (Verma & Singh, 2008).

6.3 ADVANTAGES OF MEDICINAL HERBS

Herbal medicines are considered to be more effective and potent for serious health issues than traditional medicines, which do not respond effectively to chronic diseases. Herbal medicines are safer for use than traditional medicines because they have minimal side effects. The treatment of arthritis is one of the best examples of herbal drugs and alternative medications. Viox, a well-known prescribed medication for arthritis treatment, was recalled from the market because it caused some serious cardiovascular issues. By contrast, alternative treatments, such as the herbal drugs used for arthritis treatment, have very few side effects.

Cost is another big advantage of herbal drugs. Herbal drugs are much less expensive than traditional medicines. The costs of research, testing, and marketing are the main reasons for the relatively high prices of modern medicines

Availability is another of the big advantages of herbal medicines. Some simple and potential herbs, such as chamomile and peppermint, can be easily grown at home while other medicines can be used without prescription. The only treatment used by people in some remote parts of the world is herbs.

6.4 DISADVANTAGES OF MEDICINAL HERBS

Although there are many advantages of herbal medicines, there are also some drawbacks to their use. Herbalists cannot efficiently treat serious medical conditions such as heart attacks and appendicitis, or treat traumas and fractured bones with herbal medicines. However, professional doctors can treat these conditions effectively with synthetic drugs and they have advanced diagnostic tests. They also have access to advanced surgical tools and drugs. Herbal medicines cannot effectively treat sudden illnesses and accidents; these are treated efficiently with modern medicine. Another big drawback of herbal medicines is the risk of harming oneself with improper self-dosing with herbs. Although it may be debated that accidental overdosing can also occur with medication, many herbal medicines do not have clear instructions or package inserts. Overdose is thus a very real risk. It is very risky to harvest herbs in the wild; however, some people still identify, gather, and utilize wild herbs. Due to the lack of tight regulation on herbal medicinal products, people still risk their lives consuming low-quality herbal medicines. Herbal products may differ in quality between producers, brands, and batches, meaning that the proper dosage of herbal medicine is much more difficult to prescribe (S. Singh & Sedha, 2016).

6.5 FUTURE PROSPECTS OF MEDICINAL HERBS

For a long time, the production procedures of medicines from plants have been passed only verbally by practitioners of traditional herbal. In general, the records of the production of these medicines is not commonly kept. However, the WHO has started to authenticate the use of medicinal plants for the treatment of diseases by researchers across the world. As a result, present-day developing countries are making efforts to accumulate ethnomedical information on herbs so that the pharmacological values of herbs can be easily validated scientifically. Once the preparation of the ethnomedical herbs is assessed and properly publicized, people will better understand and be satisfied with effective drug treatment, leading to enhanced health outcomes (Jahromi et al., 2021). Additionally, because of their abundance in therapeutic compounds, plants can also be used as raw materials for the manufacture of a vast variety of semisynthetic chemical compounds such as perfumes, cosmetics, and foodstuffs (Efferth et al., 2016). Products that are plant-derived are gaining in popularity day by day and gaining in acceptance for their utilization in many industries, including the global cosmetic industry. Therefore, medicinal plants are now performing crucial activities in the development of economic growth because of their double role as a source of healthcare and a source of commercial income.

The demand for herbal medicines and treatments is likely to increase in future years, making the supply of these herbal drugs a prosperous business (Rodrigues & Barnes, 2013). This implies that doctors, pharmaceutical industries, and researchers will turn to countries like India, China, and other well-developed countries to source their supplies because these countries are very diverse in terms of medicinal plant resources, and are the leading exporters of herbal materials. Plant-derived products are gaining in popularity as affordable and safer substitutes to conventional therapeutic drugs; therefore, in the coming years plants will be utilized, either in their entirety or in the form of drugs (Farah et al., 2000).

REFERENCES

Adegbola, P., Aderibigbe, I., Hammed, W., & Omotayo, T. (2017). Antioxidant and anti-inflammatory medicinal plants have potential role in the treatment of cardiovascular disease: a review. *American Journal of Cardiovascular Disease*, 7(2), 19.

Adhikari, S., Priyadarsini, K. I., & Mukherjee, T. (2007). Physico-chemical studies on the evaluation of the antioxidant activity of herbal extracts and active principles of some Indian medicinal plants. *Journal of Clinical Biochemistry and Nutrition*, 40(3). https://doi.org/10.3164/jcbn.40.174

Akindele, A. J., Wani, Z. A., Sharma, S., Mahajan, G., Satti, N. K., Adeyemi, O. O., Mondhe, D. M., & Saxena, A. K. (2015). In vitro and in vivo anticancer activity of root extracts of sansevieria liberica gerome and labroy (agavaceae). *Evidence-Based Complementary and Alternative Medicine*, 2015. https://doi.org/10.1155/2015/560404

Akinyemi, K. O., Oladapo, O., Okwara, C. E., Ibe, C. C., & Fasure, K. A. (2005). Screening of crude extracts of six medicinal plants used in South-West Nigerian unorthodox medicine for anti-methicillin resistant Staphylococcus aureus activity. *BMC Complementary and Alternative Medicine*, 5. https://doi.org/10.1186/1472-6882-5-6

Al-Jabri, A. A. (2005). Honey, milk and antibiotics. *African Journal of Biotechnology*, 4(13) SPEC.ISS.

Almabruk, K. H., Dinh, L. K., & Philmus, B. (2018). Self-Resistance of Natural Product Producers: Past, Present, and Future Focusing on Self-Resistant Protein Variants. *ACS Chemical Biology*, 13(6). https://doi.org/10.1021/acschembio.8b00173

Anilkumar, K., Reddy, G. V., Azad, R., Yarla, N. S., Dharmapuri, G., Srivastava, A., Kamal, M. A., & Pallu, R. (2017). Evaluation of Anti-Inflammatory Properties of Isoorientin Isolated from Tubers of Pueraria tuberosa. *Oxidative Medicine and Cellular Longevity*, 2017. https://doi.org/10.1155/2017/5498054

Ashutosh, & Kar. (2007). *Pharmacognosy and Pharmacobiotechnology*. In New Delhi: New Age International (P) Ltd (Vol. 2nd).

Asres, K., & Bucar, F. (2005). Anti-HIV activity against immunodeficiency virus type 1 (HIV-I) and type II (HIV-II) of compounds isolated from the stem bark of Combtetum molle. *Ethiopian Medical Journal*, 43(1).

Attele, A. S., Zhou, Y. P., Xie, J. T., Wu, J. A., Zhang, L., Dey, L., Pugh, W., Rue, P. A., Polonsky, K. S., & Yuan, C. S. (2002). Antidiabetic effects of Panax ginseng berry extract and the identification of an effective component. *Diabetes*, 51(6). https://doi.org/10.2337/diabetes.51.6.1851

Baghel, A. S., Mishra, C. K., Rani, A., Sasmal, D., & Kumar Nema, R. (2010). Antibacterial activity of Plumeria rubra Linn. plant extract. *Journal of Chemical and Pharmaceutical Research*, 2(6).

Balunas, M. J., & Kinghorn, A. D. (2005). Drug discovery from medicinal plants. *Life Sciences*, 78(5). https://doi.org/10.1016/j.lfs.2005.09.012

Barak, V., Halperin, T., & Kalickman, I. (2001). The effect of Sambucol, a black elderberry-based, natural product, on the production of human cytokines: I. *Inflammatory cytokines. European Cytokine Network*, 12(2).

Barua, C. C., Gupta, P. P., Patnaik, G. K., Kulshrestha, D. K., & Dhawan, B. N. (1997). Studies on antianaphylactic activity of fractions of Albizzia lebbeck. *Current Science*, 72(6).

Batra, P., & Sharma, A. K. (2013). Anti-cancer potential of flavonoids: recent trends and future perspectives. *3 Biotech*, 3(6). https://doi.org/10.1007/s13205-013-0117-5

Bereksi, M. S., Hassaïne, H., Bekhechi, C., & Abdelouahid, D. E. (2018). Evaluation of antibacterial activity of some medicinal plants extracts commonly used in Algerian traditional medicine against some pathogenic bacteria. *Pharmacognosy Journal*, 10(3). https://doi.org/10.5530/pj.2018.3.83

Bharti, R., Ahuja, G., Ganapathy, S., & Dakappa, S. S. (2013). A review on medicinal plants having antioxidant potential. *Indian Journal of Research in Pharmacy and Biotechnology*, 8(5), 404.

Bhushan, S., Kumar, A., Malik, F., Andotra, S. S., Sethi, V. K., Kaur, I. P., Taneja, S. C., Qazi, G. N., & Singh, J. (2007). A triterpenediol from *Boswellia serrata* induces apoptosis through both the intrinsic and extrinsic apoptotic pathways in human leukemia HL-60 cells. *Apoptosis*, 12(10). https://doi.org/10.1007/s10495-007-0105-5

Bosio, K., Avanzini, C., D'Avolio, A., Ozino, O., & Savoia, D. (2000). In vitro activity of propolis against Streptococcus pyogenes. *Letters in Applied Microbiology*, 31(2). https://doi.org/10.1046/j.1365-2672.2000.00785.x

Buckwold, V. E., Wilson, R. J. H., Nalca, A., Beer, B. B., Voss, T. G., Turpin, J. A., Buckheit, R. W., Wei, J., Wenzel-Mathers, M., Walton, E. M., Smith, R. J., Pallansch, M., Ward, P., Wells, J., Chuvala, L., Sloane, S., Paulman, R., Russell, J., Hartman, T., & Ptak, R. (2004). Antiviral activity of hop constituents against a series of DNA and RNA viruses. *Antiviral Research*, 61(1). https://doi.org/10.1016/S0166-3542(03)00155-4

Carocho, M., & Ferreira, I. (2013). The Role of Phenolic Compounds in the Fight against Cancer – A Review. *Anti-Cancer Agents in Medicinal Chemistry*, 13(8). https://doi.org/10.2174/18715206113139990301

Cella, M., Riva, D. A., Coulombié, F. C., & Mersich, S. E. (2004). Virucidal activity presence in Trichilia glabra leaves. *Revista Argentina de Microbiologia*, 36(3).

Chandra, T., Sadique, J., & Somasundaram, S. (1987). Effect of Eclipta alba on inflammation and liver injury. *Fitoterapia*, 58(1), 23–32.

Chantrill, B. H., Coulthard, C. E., Dickinson, L., Inkley, G. W., Morris, W., & Pyle, A. H. (1952). The action of plant extracts on a bacteriophage of Pseudomonas pyocyanea and on influenza A virus. *Journal of General Microbiology*, 6(1–2). https://doi.org/10.1099/00221287-6-1-2-74

Chhetri, D. R. (2004). Medicinal plants used as antipyretic agents by the traditional healers of Darjeeling Himalayas. *Indian Journal of Traditional Knowledge*, 3(3).

Chiang, L. C., Chiang, W., Chang, M. Y., & Lin, C. C. (2003). In vitro cytotoxic, antiviral and immunomodulatory effects of Plantago major and Plantago asiatica. *American Journal of Chinese Medicine*, 31(2). https://doi.org/10.1142/S0192415X03000874

Cowan, M. M. (1999). Plant products as antimicrobial agents. *Clinical Microbiology Reviews*, 12(4). https://doi.org/10.1128/cmr.12.4.564

Dahanukar, S. A., & Thatte, U. M. (1997). Current status of ayurveda in phytomedicine. *Phytomedicine*, 4(4). https://doi.org/10.1016/s0944-7113(97)80048-7

Damanhouri, Z. A. (2014). A Review on Therapeutic Potential of Piper nigrum L. (Black Pepper): The King of Spices. *Medicinal & Aromatic Plants*, 03(03). https://doi.org/10.4172/2167-0412.1000161

Dans, A. M. L., Villarruz, M. V. C., Jimeno, C. A., Javelosa, M. A. U., Chua, J., Bautista, R., & Velez, G. G. B. (2007). The effect of Momordica charantia capsule preparation on glycemic control in Type 2 Diabetes Mellitus needs further studies. *Journal of Clinical Epidemiology*, 60(6). https://doi.org/10.1016/j.jclinepi.2006.07.009

Das, R., Mitra, S., Tareq, A. M., Emran, T. Bin, Hossain, M. J., Alqahtani, A. M., Alghazwani, Y., Dhama, K., & Simal-Gandara, J. (2022). Medicinal plants used against hepatic disorders in Bangladesh: A comprehensive review. *Journal of Ethnopharmacology*, 282. https://doi.org/10.1016/j.jep.2021.114588

Doughari, J. H. (2007). Antimicrobial Activity of Tamarindus indica Linn. *Tropical Journal of Pharmaceutical Research*, 5(2). https://doi.org/10.4314/tjpr.v5i2.14637

Dragland, S., Senoo, H., Wake, K., Holte, K., & Blomhoff, R. (2003). Several culinary and medicinal herbs are important sources of dietary antioxidants. *Journal of Nutrition*, 133(5). https://doi.org/10.1093/jn/133.5.1286

Dzotam, J. K., & Kuete, V. (2017). Antibacterial and Antibiotic-Modifying Activity of Methanol Extracts from Six Cameroonian Food Plants against Multidrug-Resistant Enteric Bacteria. *BioMed Research International*, 2017. https://doi.org/10.1155/2017/1583510

Efferth, T., Romero, M., Bilia, A., Osman, A., Elsohly, M., Wink, M., Bauer, R., Khan, I., Bergonzi, M., & G. Marin, J. (2016). Expanding the therapeutic spectrum of artemisinin: Activity against infectious diseases beyond malaria and novel pharmaceutical developments. World *Journal of Traditional Chinese Medicine*, 2(2). https://doi.org/10.15806/j.issn.2311-8571.2016.0002

El-Baz, F. K., Aly, H. F., & Saad, S. A. (2015). Potential impact of Jatropha curcas in combating liver dysfunction induced by carbon tetrachloride in rats. *International Journal of Pharmaceutical Sciences Review and Research*, 34(2).

Evans, C. E., Banso, A., & Samuel, O. A. (2002). Efficacy of some nupe medicinal plants against Salmonella typhi: An in vitro study. *Journal of Ethnopharmacology*, 80(1). https://doi.org/10.1016/S0378-8741(01)00378-6

Farah, M. H., Edwards, R., Lindquist, M., Leon, C., & Shaw, D. (2000). International monitoring of adverse health effects associated with herbal medicines. *Pharmacoepidemiology and Drug Safety*, 9(2). https://doi.org/10.1002/(SICI)1099-1557(200003/04)9:2<105::AID-PDS486>3.0.CO;2-2

Faried, A., Kurnia, D., Faried, L. S., Usman, N., Miyazaki, T., Kato, H., & Kuwano, H. (2007). Anticancer effects of gallic acid isolated from Indonesian herbal medicine, Phaleria macrocarpa (Scheff.) Boerl, on human cancer cell lines. *International Journal of Oncology*, 30(3). https://doi.org/10.3892/ijo.30.3.605

Felipe, A. M. M., Rincão, V. P., Benati, F. J., Linhares, R. E. C., Galina, K. J., De Toledo, C. E. M., Lopes, G. C., De Mello, J. C. P., & Nozawa, C. (2006). Antiviral effect of Guazuma ulmifolia and Stryphnodendron adstringens on poliovirus and bovine herpesvirus. *Biological and Pharmaceutical Bulletin*, 29(6). https://doi.org/10.1248/bpb.29.1092

Galal, E. E., & Gawad, M. A. (1965). Antidiabetic activity of Egyptian onion "Allium cepa" extract. *The Journal of the Egyptian Medical Association*, 48.

Ganesh, S., Mukesh, G., Ranjan, B., & Vishal, V. (2014). A review on some plants having anti-inflammatory activity. *The Journal of Phytopharmacology*, 3(3), 241.

Ge, Y., Difuntorum, S., Touami, S., Critchley, I., Bürli, R., Jiang, V., Drazan, K., & Moser, H. (2002). In vitro antimicrobial activity of GSQ1530, a new heteroaromatic polycyclic compound. *Antimicrobial Agents and Chemotherapy*, 46(10). https://doi.org/10.1128/AAC.46.10.3168-3174.2002

Gessner, D. K., Ringseis, R., & Eder, K. (2017). Potential of plant polyphenols to combat oxidative stress and inflammatory processes in farm animals. *Journal of Animal Physiology and Animal Nutrition*, 101(4). https://doi.org/10.1111/jpn.12579

Gonzalez, C. E., Venzon, D., Lee, S., Mueller, B. U., Pizzo, P. A., & Walsh, T. J. (1996). Risk factors for fungemia in children infected with human immunodeficiency virus: A case-control study. *Clinical Infectious Diseases*, 23(3). https://doi.org/10.1093/clinids/23.3.515

Gupta, A. K., & Misra, N. (2006). Hepatoprotective Activity of Aqueous Ethanolic Extract of Chamomile capitula in Paracetamol Intoxicated Albino Rats. *American Journal of Pharmacology and Toxicology*, 1(1). https://doi.org/10.3844/ajptsp.2006.17.20

Gupta, P. P., Srimal, R. C., Srivastava, M., Singh, K. L., & Tandon, J. S. (1995). Antiallergic activity of arbortristosides from nyctanthus arbortristis. *Pharmaceutical Biology*, 33(1). https://doi.org/10.3109/13880209509088151

Hancock, R. E. W. (2005). Mechanisms of action of newer antibiotics for Gram-positive pathogens. *In Lancet Infectious Diseases*, 5(4). https://doi.org/10.1016/S1473-3099(05)70051-7

Hartwell, J. L. (1967). Plants used against cancer. *A survey. Lloydia*, 30(4). https://doi.org/10.2307/1220819

Huang, K. L., Lai, Y. K., Lin, C. C., & Chang, J. M. (2006a). Inhibition of hepatitis B virus production by Boehmeria nivea root extract in HepG2 2.2.15 cells. *World Journal of Gastroenterology*, 12(35). https://doi.org/10.3748/wjg.v12.i35.5721

Huang, T. H. W., Tran, V. H., Duke, R. K., Tan, S., Chrubasik, S., Roufogalis, B. D., & Duke, C. C. (2006b). Harpagoside suppresses lipopolysaccharide-induced iNOS and COX-2 expression through inhibition of NF-κB activation. *Journal of Ethnopharmacology*, 104(1–2). https://doi.org/10.1016/j.jep.2005.08.055

Hyun, S. H., Kim, S. W., Seo, H. W., Youn, S. H., Kyung, J. S., Lee, Y. Y., In, G., Park, C. K., & Han, C. K. (2020). Physiological and pharmacological features of the non-saponin components in Korean Red Ginseng. *In Journal of Ginseng Research*, 44(4). https://doi.org/10.1016/j.jgr.2020.01.005

Iwalewa, E. O., McGaw, L. J., Naidoo, V., & Eloff, J. N. (2007). Inflammation: The foundation of diseases and disorders. A review of phytomedicines of South African origin used to treat pain and inflammatory conditions. *African Journal of Biotechnology*, 6(25). https://doi.org/10.5897/ajb2007.000-2457

Iwu, M. W., Duncan, A. R., & Okunji, C. O. (1999). New antimicrobials of plant origin. *Perspectives on new crops and new uses*. ASHS Press, Alexandria, 03(01).

Jahromi, B., Pirvulescu, I., Candido, K. D., & Knezevic, N. N. (2021). Herbal medicine for pain management: Efficacy and drug interactions. *In Pharmaceutics*, 13(2). https://doi.org/10.3390/pharmaceutics13020251

Joseph, L., George, M., Sharma, A., & Gopal, N. (2011). Antipyretic and analgesic effects of the aqueous extract of the Prosopis cineraria. *Global Journal of Pharmacology*, 5(2).

Joshi, R. K. (2016). A Perspective on the Phytopharmaceuticals Responsible for the Therapeutic Applications. https://doi.org/10.4018/978-1-5225-0754-3.ch008

Kassuya, C. A. L., Leite, D. F. P., De Melo, L. V., Rehder, V. L. C., & Calixto, J. B. (2005). Anti-inflammatory properties of extracts, fractions and lignans isolated from *Phyllanthus amarus*. *Planta Medica*, 71(8). https://doi.org/10.1055/s-2005-871258

Kathiresan, K., Sithranga Boopathy, N., & Kavitha, S. (2006). Coastal vegetation – an underexplored source of anticancer drugs. *Indian Journal of Natural Products and Resources*, 5(2).

Kayarohanam, S. (2015). Current Trends of Plants Having Antidiabetic Activity: A Review. *Journal of Bioanalysis & Biomedicine*, 07(02). https://doi.org/10.4172/1948-593x.1000124

Khanna, N., Goswami, M., Sen, P., & Ray, A. (1995). Antinociceptive action of Azadirachta indica (Neem) in mice: Possible mechanisms involved. *Indian Journal of Experimental Biology*, 33(11).

Khanna, P., Jain, S. C., Panagariya, A., & Dixit, V. P. (1981). Hypoglycemic activity of polypeptide-p from a plant source. *Journal of Natural Products*, 44(6). https://doi.org/10.1021/np50018a002

Khanna, V., & Kannabiran, K. (2009). Anticancer-cytotoxic activity of saponins isolated from the leaves of gymnema sylvestre and eclipta prostrata on HeLa cells. *International Journal of Green Pharmacy*, 3(3). https://doi.org/10.4103/0973-8258.56280

Kim, M. J., Kim, C. S., Kim, B. H., Ro, S. B., Lim, Y. K., Park, S. N., Cho, E., Ko, J. H., Kwon, S. S., Ko, Y. M., & Kook, J. K. (2011). Antimicrobial effect of Korean propolis against the mutans streptococci isolated from Korean. *Journal of Microbiology*, 49(1). https://doi.org/10.1007/s12275-011-1002-8

Kohno, H., Tanaka, T., Kawabata, K., Hirose, Y., Sugie, S., Tsuda, H., & Mori, H. (2002). Silymarin, a naturally occurring polyphenolic antioxidant flavonoid, inhibits azoxymethane-induced colon carcinogenesis in male F344 rats. *International Journal of Cancer*, 101(5). https://doi.org/10.1002/ijc.10625

Kolawole, O. T., Makinde, J. M., & Olajide, O. A. (2007). Central nervous system depressant activity of Russelia equisetiformis. *Nigerian Journal of Physiological Sciences : Official Publication of the Physiological Society of Nigeria*, 22(1–2). https://doi.org/10.4314/njps.v22i1-2.54871

Korayashi, T., & Oura, H. (1985). Studies on the Mechanism of the Hypoglycemic Activity of Ginsenoside-Rb2 in Streptozotocin-Diabetic Rats. *Chemical and Pharmaceutical Bulletin*, 33(2). https://doi.org/10.1248/cpb.33.869

Kotwal, G. J., Kaczmarek, J. N., Leivers, S., Ghebremariam, Y. T., Kulkarni, A. P., Bauer, G., De Beer, C., Preiser, W., & Mohamed, A. R. (2005). Anti-HIV, anti-poxvirus, and anti-SARS activity of a nontoxic, acidic plant extract from the Trifollium species Secomet-V/anti-Vac suggests that it contains a novel broad-spectrum antiviral. *Annals of the New York Academy of Sciences*, 1056. https://doi.org/10.1196/annals.1352.014

Kumari, K., & Augusti, K. T. (2002). Antidiabetic and antioxidant effects of S-methyl cysteine sulfoxide isolated from onions (Allium cepa Linn) as compared to standard drugs in alloxan diabetic rats. *Indian Journal of Experimental Biology*, 40(9), 1005–1009.

Li, S. Y., Chen, C., Zhang, H. Q., Guo, H. Y., Wang, H., Wang, L., Zhang, X., Hua, S. N., Yu, J., Xiao, P. G., Li, R. S., & Tan, X. (2005). Identification of natural compounds with antiviral activities against SARS-associated coronavirus. *Antiviral Research*, 67(1). https://doi.org/10.1016/j.antiviral.2005.02.007

Liming, J., Chunshan, Q., Xiyan, H., & Shengdi, F. (2016). Potential pharmacological resources: Natural bioactive compounds from marine-derived fungi. *Marine Drugs*, 14(4). https://doi.org/10.3390/md14040076

Madhuri, S., & Pandey, G. (2009). Some anticancer medicinal plants of foreign origin. *In Current Science*, 96(6).

Makwana, H. G., Ravishankar, B., Shukla, V. J., Bhaskaran Nair, R., Vijayan, N. P., Sasikala, C. K., Saraswathy, V. N., & Bhatt, S. V. (1994). General pharmacology of Vitex leucoxylon Linn leaves. *Indian Journal of Physiology and Pharmacology*, 38(2), 95.

Manandhar, S., Luitel, S., & Dahal, R. K. (2019). In Vitro Antimicrobial Activity of Some Medicinal Plants against Human Pathogenic Bacteria. *Journal of Tropical Medicine*, 2019. https://doi.org/10.1155/2019/1895340

Marchese, A., & Schito, G. C. (2000). Resistance patterns of lower respiratory tract pathogens in Europe. *International Journal of Antimicrobial Agents*, 16(SUPPL. 1). https://doi.org/10.1016/s0924-8579(00)00302-2

Medzhitov, R. (2010). Inflammation 2010: New Adventures of an Old Flame. *Cell*, 140(6). https://doi.org/10.1016/j.cell.2010.03.006

Mehta, S. R., Yusuf, S., Peters, R. J. G., Bertrand, M. E., Lewis, B. S., Natarajan, M. K., Malmberg, K., Rupprecht, H. J., Zhao, F., Chrolavicius, S., Copland, I., & Fox, K. A. A. (2001). Effects of pretreatment with clopidogrel and aspirin followed by long-term therapy in patients undergoing percutaneous coronary intervention: The PCI-CURE study. *Lancet*, 358(9281). https://doi.org/10.1016/S0140-6736(01)05701-4

Meijer, L., & Raymond, E. (2003). Roscovitine and other purines as kinase inhibitors. From starfish oocytes to clinical trials. *Accounts of Chemical Research*, 36(6). https://doi.org/10.1021/ar0201198

Mukhtar, M., Arshad, M., Ahmad, M., Pomerantz, R. J., Wigdahl, B., & Parveen, Z. (2008). Antiviral potentials of medicinal plants. *Virus Research*, 131(2). https://doi.org/10.1016/j.virusres.2007.09.008

Muntgantiwar, A. A., Nair, A. M., Shinde, U. A., & Saraf, M. N. (1997). Effect of stress on plasma and adrenal cortisol levels and immune responsiveness in rats: Modulation by alkaloidal fraction of *Boerhaavia diffusa. Fitoterapia*, 68(6).

Murugaian, P., Ramamurthy, V., & Karmegam, N. (2008). Hepatoprotective Activity of Wedelia calendulacea L. Against Acute Hepatotoxicity in Rats. *Research Journal of Agriculture and Biological Sciences*, 4(6).

Mythreyi, R., Sasikala, E., Geetha, A., & Madhavan, V. (2009). Antibacterial activity of leaves of *Cadaba trifoliata* (Roxb.) Wt. Arn. *Indian Journal of Pharmaceutical Sciences*, 71(2). https://doi.org/10.4103/0250-474X.54271

Narayanaswamy, N., & Balakrishnan, K. P. (2011). Evaluation of some medicinal plants for their antioxidant properties. *International Journal of Pharm. Tech. Research*, 3(1).

Neogi, U., Saumya, R., Mishra, R. K., & Raju, K. C. (2007). Lipid Content and in vitro Antimicrobial Activity of Oil Seeds of Some Indian Medicinal Plants. *Current Research in Bacteriology*, 1(1). https://doi.org/10.3923/crb.2008.1.6

Nigam, V., & Sodhi, J. S. (2014). Some medicinal plants with antioxidant activity- A review. *International Journal of Pharmacy and Biological Sciences*, 4(1), 173–178.

Niwano, Y., Saito, K., Yoshizaki, F., Kohno, M., & Ozawa, T. (2011). Extensive screening for herbal extracts with potent antioxidant properties. *Journal of Clinical Biochemistry and Nutrition*, 48(1). https://doi.org/10.3164/jcbn.11-013FR

Norrby, S. R., Nord, C. E., & Finch, R. (2005). Lack of development of new antimicrobial drugs: a potential serious threat to public health. *The Lancet Infectious Diseases*, 5(2). https://doi.org/10.1016/s1473-3099(05)01283-1

Öbek, H., Uğraş, S., Bayram, I., Uygan, I., Erdoğan, E., Öztürk, A., & Huyut, Z. (2004). Hepatoprotective effect of Foeniculum vulgare essential oil: A carbon-tetrachloride induced liver fibrosis model in rats. *Scandinavian Journal of Laboratory Animal Science*, 31(1), 9–17.

Ochwang'i, D. O., Kimwele, C. N., Oduma, J. A., Gathumbi, P. K., Mbaria, J. M., & Kiama, S. G. (2014). Medicinal plants used in treatment and management of cancer in Kakamega County, Kenya. *Journal of Ethnopharmacology*, 151(3). https://doi.org/10.1016/j.jep.2013.11.051

Ooi, L. S. M., Sun, S. S. M., & Ooi, V. E. C. (2004). Purification and characterization of a new antiviral protein from the leaves of Pandanus amaryllifolius (Pandanaceae). *International Journal of Biochemistry and Cell Biology*36(8). https://doi.org/10.1016/j.biocel.2004.01.015

Orozco-Topete, R., Sierra-Madero, J., Cano-Dominguez, C., Kershenovich, J., Ortiz-Pedroza, G., Vazquez-Valls, E., Garcia-Cosio, C., Soria-Cordoba, A., Armendariz, A. M., Teran-Toledo, X., Romo-Garcia, J., Fernandez, H., & Rozhon, E. J. (1997). Safety and efficacy of Virend® for topical treatment of genital and anal herpes simplex lesions in patients with AIDS. *Antiviral Research*, 35(2). https://doi.org/10.1016/S0166-3542(97)00015-6

Osman, K., Evangelopoulos, D., Basavannacharya, C., Gupta, A., McHugh, T. D., Bhakta, S., & Gibbons, S. (2012). An antibacterial from Hypericum acmosepalum inhibits ATP-dependent MurE ligase from Mycobacterium tuberculosis. *International Journal of Antimicrobial Agents*, 39(2). https://doi.org/10.1016/j.ijantimicag.2011.09.018

Panchawat, S., Rathore, K. S., & Sisodia, S. S. (2010). A review on herbal antioxidants. *International Journal of Pharm. Tech. Research*, 2(1), 232–239.

Pantev, A., Ivancheva, S., Staneva, L., & Serkedjieva, J. (2006). Biologically active constituents of a polyphenol extract from Geranium sanguineum L. with anti-influenza activity. Zeitschrift Fur Naturforschung - Section C *Journal of Biosciences*, 61(7–8). https://doi.org/10.1515/znc-2006-7-807

Paraschos, S., Mitakou, S., & Skaltsounis, A.-L. (2012). Chios Gum Mastic: A Review of its Biological Activities. *Current Medicinal Chemistry*, 19(14). https://doi.org/10.2174/092986712800229014

Parida, M. M., Upadhyay, C., Pandya, G., & Jana, A. M. (2002). Inhibitory potential of neem (Azadirachta indica Juss) leaves on Dengue virus type-2 replication. *Journal of Ethnopharmacology*, 79(2). https://doi.org/10.1016/S0378-8741(01)00395-6

Peng, G. Y., Zhou, F., Ding, R. L., Li, H. Di, & Yao, K. (2002). Modulation of lianbizi jnjection (andrographolide) on some immune functions. *Zhongguo Zhongyao Zazhi*, 27(2), 147–150.

Pompei, R., Flore, O., Marccialis, M. A., Pani, A., & Loddo, B. (1979). Glycyrrhizic acid inhibits virus growth and inactivates virus particles. *Nature*, 281(5733). https://doi.org/10.1038/281689a0

Ponder, A., & Long, M. D. (2013). A clinical review of recent findings in the epidemiology of inflammatory bowel disease. *Clinical Epidemiology*, 5(1). https://doi.org/10.2147/CLEP.S33961

Puri, A., Saxena, R., Saxena, R. P., Saxena, K. C., Srivastava, V., & Tandon, J. S. (1993). Immunostimulant agents from *andrographis paniculata. Journal of Natural Products*, 56(7). https://doi.org/10.1021/np50097a002

Ramaswamy, S., & Viswanathan, S. (1997). Influence of gossypin on the development of acute tolerance to morphine induced antinociception. *Indian Journal of Experimental Biology*, 35(4), 413–414.

Ratan, Z. A., Haidere, M. F., Hong, Y. H., Park, S. H., Lee, J. O., Lee, J., & Cho, J. Y. (2021). Pharmacological potential of ginseng and its major component ginsenosides. *In Journal of Ginseng Research*, 45(2). https://doi.org/10.1016/j.jgr.2020.02.004

Reichling, J., Schnitzler, P., Suschke, U., & Saller, R. (2009). Essential oils of aromatic plants with antibacterial, antifungal, antiviral, and cytotoxic properties - An overview. *In Forschende Komplementarmedizin*, 16(2). https://doi.org/10.1159/000207196

Reker, D., Perna, A. M., Rodrigues, T., Schneider, P., Reutlinger, M., Mönch, B., Koeberle, A., Lamers, C., Gabler, M., Steinmetz, H., Müller, R., Schubert-Zsilavecz, M., Werz, O., & Schneider, G. (2014). Revealing the macromolecular targets of complex natural products. *Nature Chemistry*, 6(12). https://doi.org/10.1038/nchem.2095

Rodrigues, E., & Barnes, J. (2013). Pharmacovigilance of herbal medicines: The potential contributions of ethnobotanical and ethnopharmacological studies. *Drug Safety*, 36(1). https://doi.org/10.1007/s40264-012-0005-7

Roja, G., & Rao, P. S. (2000). Anticancer compounds from tissue cultures of medicinal plants. *Journal of Herbs, Spices and Medicinal Plants*, 7(2). https://doi.org/10.1300/J044v07n02_08

Ruddaraju, L. K., Pammi, S. V. N., Guntuku, G. S., Padavala, V. S., & Kolapalli, V. R. M. (2020). A review on anti-bacterials to combat resistance: From ancient era of plants and metals to present and future perspectives of green nano technological combinations. *Asian Journal of Pharmaceutical Sciences*, 15(1). https://doi.org/10.1016/j.ajps.2019.03.002

Saito, K., Kohno, M., Yoshizaki, F., & Niwano, Y. (2008). Extensive screening for edible herbal extracts with potent scavenging activity against superoxide anions. *Plant Foods for Human Nutrition*, 63(2). https://doi.org/10.1007/s11130-008-0071-2

Setty, A. R., & Sigal, L. H. (2005). Herbal medications commonly used in the practice of rheumatology: Mechanisms of action, efficacy, and side effects. *Seminars in Arthritis and Rheumatism*, 34(6). https://doi.org/10.1016/j.semarthrit.2005.01.011

Shaikh, R., Pund, M., Dawane, A., & Iliyas, S. (2014). Evaluation of anticancer, antioxidant, and possible anti-inflammatory properties of selected medicinal plants used in Indian traditional medication. *Journal of Traditional and Complementary Medicine*, 4(4). https://doi.org/10.4103/2225-4110.128904

Shankar, S. R., Rangarajan, R., Sarada, D. V. L., & Kumar, C. S. (2010). Evaluation of antibacterial activity and phytochemical screening of wrightia tinctoria l. *Pharmacognosy Journal*, 2(14). https://doi.org/10.1016/S0975-3575(10)80066-5

Shashi, B., Jaswant, S., Madhusudana, R. J., Kumar, S. A., & Nabi, Q. G. (2006). A novel lignan composition from Cedrus deodara induces apoptosis and early nitric oxide generation in human leukemia Molt-4 and HL-60 cells. *Nitric Oxide - Biology and Chemistry*, 14(1). https://doi.org/10.1016/j.niox.2005.09.009

Shen, Y. C., Chen, C. F., & Chiou, W. F. (2002). Andrographolide prevents oxygen radical production by human neutrophils: Possible mechanism(s) involved in its anti-inflammatory effect. *British Journal of Pharmacology*, 135(2). https://doi.org/10.1038/sj.bjp.0704493

Shirani, M., Raeisi, R., Heidari-Soureshjani, S., Asadi-Samani, M., & Luther, T. (2017). A review for discovering hepatoprotective herbal drugs with least side effects on kidney. *Journal of Nephropharmacology*, 6(2). https://doi.org/10.15171/npj.2017.03

Shiu, W. K. P., Rahman, M. M., Curry, J., Stapleton, P., Zloh, M., Malkinson, J. P., & Gibbons, S. (2012). Antibacterial acylphloroglucinols from hypericum olympicum. *Journal of Natural Products*, 75(3). https://doi.org/10.1021/np2003319

Siddiqui, A. J., Jahan, S., Singh, R., Saxena, J., Ashraf, S. A., Khan, A., Choudhary, R. K., Balakrishnan, S., Badraoui, R., Bardakci, F., & Adnan, M. (2022). Plants in Anticancer Drug Discovery: From Molecular Mechanism to Chemoprevention. *BioMed Research International*, 2022. https://doi.org/10.1155/2022/5425485

Sidhu, T. S. (2013). Medicinal Plants From Twelve Families Having Antidiabetic Activity: A Review. *Am. J. Pharm. Tech. Res.*, 3(5), 779.

Sienkiewicz, M., Lysakowska, M., Ciecwierz, J., Denys, P., & Kowalczyk, E. (2012). Antibacterial Activity of Thyme and Lavender Essential Oils. *Medicinal Chemistry*, 7(6). https://doi.org/10.2174/157340611797928488

Sies, H. (1997). Oxidative stress: Oxidants and antioxidants. *Experimental Physiology*, 82(2). https://doi.org/10.1113/expphysiol.1997.sp004024

Singh, H. K., & Dhawan, B. N. (1997). Neuropsychopharmacological effects of the ayurvedic nootropic *Bacopa monniera* Linn. (Brahmi). *Indian Journal of Pharmacology*, 29(5), 359–365.

Singh, R. K., Joshi, V. K., Goel, R. K., Gambhir, S. S., & Achaiya, S. B. (1996). Pharmacological actions of Pongamia pinnata seeds - A preliminary study. *Indian Journal of Experimental Biology*, 34(12), 1204–1207.

Singh, S., & Sedha, S. (2016). Medicinal Plants and Their Pharmacological Aspects. *Food & Pharma International*, 1(4). https://doi.org/10.5530/fpi.1.4.1

Siow, Y. L., Gong, Y., Au-Yeung, K. K. W., Woo, C. W. H., Choy, P. C., & Karmin, O. (2005). Emerging issues in traditional Chinese medicine. *In Canadian Journal of Physiology and Pharmacology*, 83(4). https://doi.org/10.1139/y05-029

Sostres, C., Gargallo, C. J., & Lanas, A. (2013). Nonsteroidal anti-inflammatory drugs and upper and lower gastrointestinal mucosal damage. *In Arthritis Research and Therapy*, 15(3). https://doi.org/10.1186/ar4175

Sultana, S., Akhtar, N., & Asif, H. M. (2013). Phytochemical screening and antipyretic effects of hydromethanol extract of Melia azedarach leaves in rabbits. *Bangladesh Journal of Pharmacology*, 8(2). https://doi.org/10.3329/bjp.v8i2.14708

Sung, W. S., & Lee, D. G. (2007). In vitro antimicrobial activity and the mode of action of indole-3-carbinol against human pathogenic microorganisms. *Biological and Pharmaceutical Bulletin*, 30(10). https://doi.org/10.1248/bpb.30.1865

Tlili, N., Elfalleh, W., Saadaoui, E., Khaldi, A., Triki, S., & Nasri, N. (2011). The caper (Capparis L.): Ethnopharmacology, phytochemical and pharmacological properties. *Fitoterapia*, 82(2). https://doi.org/10.1016/j.fitote.2010.09.006

Tolo, F. M., Rukunga, G. M., Muli, F. W., Njagi, E. N. M., Njue, W., Kumon, K., Mungai, G. M., Muthaura, C. N., Muli, J. M., Keter, L. K., Oishi, E., & Kofi-Tsekpo, M. W. (2006). Anti-viral activity of the extracts of a Kenyan medicinal plant Carissa edulis against herpes simplex virus. *Journal of Ethnopharmacology*, 104(1–2). https://doi.org/10.1016/j.jep.2005.08.053

Tran, N., Pham, B., & Le, L. (2020). Bioactive compounds in anti-diabetic plants: From herbal medicine to modern drug discovery. *Biology*, 9(9). https://doi.org/10.3390/biology9090252

Vane, J. R., & Botting, R. M. (1995). New insights into the mode of action of anti-inflammatory drugs. *Inflammation Research*, 44(1). https://doi.org/10.1007/BF01630479

Veerakumar, S., Amanulla, S. S. D., & Ramanathan, K. (2016). Anti-cancer efficacy of ethanolic extracts from various parts of Annona squamosa on MCF-7 cell line. *Journal of Pharmacognosy and Phytotherapy*, 8(7). https://doi.org/10.5897/JPP2016.0398

Verma, S., & Singh, S. P. (2008). Current and future status of herbal medicines. *In Veterinary World*, 1(11). https://doi.org/10.5455/vetworld.2008.347-350

Wagner, H., & Ulrich-Merzenich, G. (2009). Synergy research: Approaching a new generation of phytopharmaceuticals. *In Phytomedicine*, 16(2–3). https://doi.org/10.1016/j.phymed.2008.12.018

Wangchuk, P., Keller, P. A., Pyne, S. G., Taweechotipatr, M., Tonsomboon, A., Rattanajak, R., & Kamchonwongpaisan, S. (2011). Evaluation of an ethnopharmacologically selected Bhutanese medicinal plants for their major classes of Phytochemicals and biological activities. *Journal of Ethnopharmacology*, 137(1). https://doi.org/10.1016/j.jep.2011.06.032

Webster, D., Taschereau, P., Lee, T. D. G., & Jurgens, T. (2006). Immunostimulant properties of Heracleum maximum Bartr. *Journal of Ethnopharmacology*, 106(3). https://doi.org/10.1016/j.jep.2006.01.018

World Health Organization. (1978). The promotion and development of traditional medicine : report of a WHO meeting [held in Geneva from 28 November to 2 December 1977]. *In World Health Organization technical report series* (Issue 622).

Yamai, M., Tsumura, K., Kimura, M., Fukuda, S., Murakami, T., & Kimura, Y. (2003). Antiviral activity of a hot water extract of black soybean against a human respiratory illness virus. *Bioscience, Biotechnology and Biochemistry*, 67(5). https://doi.org/10.1271/bbb.67.1071

Yang, C. M., Cheng, H. Y., Lin, T. C., Chiang, L. C., & Lin, C. C. (2007). The in vitro activity of geraniin and 1,3,4,6-tetra-O-galloyl-β-d-glucose isolated from Phyllanthus urinaria against herpes simplex virus type 1 and type 2 infection. *Journal of Ethnopharmacology*, 110(3). https://doi.org/10.1016/j.jep.2006.09.039

Yatoo, M. I., Saxena, A., Gopalakrishnan, A., Alagawany, M., & Dhama, K. (2017). Promising antidiabetic drugs, medicinal plants and herbs: An update. *In International Journal of Pharmacology*, 13(7). https://doi.org/10.3923/ijp.2017.732.745

Zaika, L. L. (1988). Spices and herbs: Their antimicrobial activity and its determination. *Journal of Food Safety*, 9(2). https://doi.org/10.1111/j.1745-4565.1988.tb00511.x

Zakay-Rones, Z., Thom, E., Wollan, T., & Wadstein, J. (2004). Randomized study of the efficacy and safety of oral elderberry extract in the treatment of influenza A and B virus infections. *Journal of International Medical Research*, 32(2). https://doi.org/10.1177/147323000403200205

Zuo, G. Y., Li, Z. Q., Chen, L. R., & Xu, X. J. (2005). In vitro anti-HCV activities of Saxifraga melanocentra and its related polyphenolic compounds. *Antiviral Chemistry and Chemotherapy*, 16(6). https://doi.org/10.1177/095632020501600606

7 Biotechnological Approaches for Herbal Drug Discovery

Sohail Ahmad, Maria Faraz, Iftikhar Ahmad, and Arshad Farid
Gomal Center of Biochemistry and Biotechnology, Gomal University, D.I. Khan, Pakistan

Samy Selim
Department of Clinical Laboratory Sciences, College of Applied Medical Sciences, Jouf University, Sakaka, Saudi Arabia

7.1 INTRODUCTION TO HERBAL DRUG DISCOVERY

Herbs are employed globally in the form of drugs to cure or prevent diseases. Historical evidence of the usage of medicinal plants can be traced to around 5,000 years ago in China and India. The evidence shows that plants have been an essential component of medicine [1]. The employment of plants as drugs started at earliest Paleolithic, considering that early people have been exploiting most of the existing plant species for medication for not less than sixty thousand years. It has been estimated that approximately 80% of the entire global population depends on plant-based medication for their primary healthcare needs and also that medicinal plants provide about 80% of the raw materials used in the traditional medical system [2]. Secondly, the public's need for, interest in, and the use of plant-based remedies, is increasing on a daily basis. The market of natural products formulated with herbs has increased due to side effects of allopathic medicines [3, 4]. Concerning the use of medicinal plants, [5] posited that people, especially those in developing countries, prefer to use plants to treat illnesses and disorders because such practices are culturally acceptable. They are approaching the plants as a source of new solutions for health needs and saving people's lives from incurable infectious diseases. As noted from the yearly statistics, the usage of medicinal plants is rising; recent studies have recorded that around one-third of the America's population have adapted to using herbal products [6]. Drug discovery can be defined as the efforts to identify and develop new drug molecules from natural sources or through processes of synthesis. When compounds produce pharmacological activity with low toxicity in preclinical studies, they are put through clinical trial as new drugs [7].

Plants constitute the richest and most widely available source for drug discovery. From 94 species, for example, Fabricant and Farnsworth isolated 122 structurally defined compounds [8]. According to the most recent taxonomic revisions, angiosperms now are considered to consist of about 295,383 species, while global terrestrial vascular plants including angiosperms, gymnosperms, ferns, and lycophytes are inferred to be about 308,312 species, which may be regarded as the efficient resource for finding lead molecules/drugs [9]. The global population is expanding; in order to fulfill this demand, scientists are finding for clues in our planet's supply, plants. A decrease in biological diversity may lead to increases in specific plant species due to changes in the environment. The red list of the International Union for Conservation of Nature contains data on over 79,800 species. Of these, more than 23,000 species are at risk of extinction status; this accounts for 13% of bird species, 33% of reef-building corals 25% of mammals, 34% conifers, and 41% of amphibians [10].

DOI: 10.1201/9781003544418-7

7.1.1 History of Herbal Drug Discovery

The history of discovering plant-based compounds is deeply rooted in the traditional use of herbal medicines across various cultures throughout history [11]. Herbal medicine have been empirically utilized for years in the treatment of diseases, showcasing their longstanding presence in healthcare practices [12]. These herbal remedies have been recognized for their ability to maintain health and treat various ailments since ancient times [13]. Despite the advancements in modern medicine, traditional herbal remedies have continued to attract attention as potential alternative therapies for a wide range of diseases [14]. The process of herbal drug discovery has evolved with the integration of novel technologies and approaches. The development of novel therapeutic agents now involves concepts such as ADMET (absorption, distribution, metabolism, excretion, and toxicity) and considers interaction profiles between herb–herb and herb–synthetic compounds [15]. Additionally, computational frameworks have been developed to explore the contributions of herbal ingredients in treating conditions such as cancer by modulating immune responses and metabolic processes [16]. These advancements highlight the integration of modern scientific methodologies in the exploration of herbal medicine. The popularity of herbal medicines results from increased concerns about the efficacy and safety of conventional allopathic drugs, leading to the increased usage of herbal remedies for both mild and severe illnesses [17]. The therapeutic value of herbal medicines has been acknowledged by both healthcare providers and patients, with a recognition of their potential to offer superior benefits with fewer side effects compared to modern medications [18]. The cartilage-protective, anti-inflammatory, or antioxidant effects of herbal products use has shown promise in the treatment of conditions such as osteoarthritis [19]. The efficacy and safety of herbal medicine is ensured by its quality control and standardization. The development of techniques such as fingerprint analysis is helpful in standardizing and controlling the quality of herbal medicines, thereby enhancing their reliability and consistency [20]. Furthermore, the evaluation of standardized herbal extracts has demonstrated potential in terms of enhancing the effectiveness of chemotherapy while reducing its associated side effects, thereby demonstrating the incorporation of herbal medicine in advanced treatment approaches. With regard to major world health challenges such as the recent COVID-19 pandemic, there is a call for precision herbal medicine that integrates local community knowledge and omics systems science technologies to enable more targeted and effective herbal treatments. This approach emphasizes the importance of leveraging traditional wisdom alongside modern scientific advancements to enhance the discovery and application of herbal remedies in addressing contemporary health issues.

7.2 CURRENT TRENDS IN HERBAL DRUG DISCOVERY

The field of herbal drug discovery is currently experiencing a significant surge in research and interest on a global basis. The field is revolutionized through the use of new technologies in research. Different kinds of new techniques and other isolation and extraction techniques came into play which have changed the scenario of conventional drug discovery. Below are some of the details of the modern technologies that have played a crucial role in herbal drug discovery.

7.2.1 Molecular and Genetic Study Levels

Even if there is little in the way of a written account detailing the medicinal importance of several plants, it is still necessary to investigate, record, and organize the conventional wisdom on herbal medicine. Metabolites and bioactivities encoding by genes can be explored, as can the investigation of a particular plant metabolite and the evaluation of its treatment efficacy, the interaction of a plant bioactive with a drug target interaction in the human body and its effect on relevant biological pathways, disease-modifying potential, and related toxic effects, using contemporary genomics, proteomics, bioinformatics, and metabolomics approaches [21, 22]. Using these contemporary methods, scientists are trying to confirm the conventional perception of herbal therapy and reported influence of sickness on it. Initiatives of various genome sequencing and transcriptional profiling studies pertaining to herbal plants are underway, offering genetic and molecular data for metabolite profiling and investigating potential synthetic pathways associated with plant products [23].

7.2.2 Molecular Pharmacognosy

Molecular pharmacognosy is the study of how to produce effective components at the molecular level and how to classify, identify, cultivate, and safeguard crude pharmaceuticals. It involves a number of stages: (a) Sorting the real from the fake to resolve the variety confusion issue: as medical applications and dosages expand, homonyms of plant and animal and materials with identical appearances are emerging. The same medications in different areas are mistaken by these materials, causing variety and confusion. Because of this, it's important to separate the real from the fake when it comes to their places of origin and dissemination. The only way to assure quality is to do this. (b) Quality assessment: To confirm standard varieties and the aspects that may have an influence on them, crude drugs study with numerous origins and genuine quality should be conducted. This study should address a range of factors: place of origin, harvesting, processing, storage, and the influence of transportation upon active ingredients. Furthermore, in order to fulfill the growing need for medication, superior cultivars should be studied and cultivated to ensure quick development, good quality, and high production.

Dendrobium catenatum Lindl., *Orchidaceae*, *Panax ginseng* C.A. Mey, Araliaceae, and *Bupleurum chinense* DC, Apiaceae, like medications, have been subjected to DNA barcoding techniques in recent years. To prevent adulteration and harmful situations, a precise and prompt scientific identification of the herbal plant is essential [24]. Plant species identification using taxonomy is conventional, time-consuming, and sometimes imprecise; by contrast, the contemporary DNA barcode method is quick and accurate. Herbal plants are identified using DNA barcodes such as matK, rbcL, trnH-psbA, ITS, trnL-F, 5S-rRNA, and 18S-rRNA [25].

For the effective confirmation of herbal products, conjunction with transcriptomics utilization can be achieved by DNA barcoding, metabolomics, and proteomics. The main source of bioactive substances and metabolites is a vast number of plant species belonging to the Lamiaceae family. Because there are a few raw plants in this species, products are frequently tampered with. The examination of proposed DNA barcode loci (matK, trnH-psbA, and trnL) for their PCR amplification in order to identify several Lamiaceae species revealed that matK locus reliably identifies all selected species, followed by trnH-psbA and trnL [26]. For Mentha, Ocimum, and Plectranthus medicinal plants, DNA barcode-based verification may lessen associated unfair trades and adulterations.

For the purpose of addressing the quality, efficacy, safety concerns, and precise identification of the medicinal plants species of a traditional multi-ingredient herbal Chinese medicine can be possible by shotgun metabolomic sequencing. Microscopy, thin-layer chromatography, and high-performance liquid chromatography [27] was undertaken as a complementary method. Combinations of ten herbal raw materials make up Longdan Xiegan Wan (LDXGW) a conventional Chinese medicine derived from a herbal prescription from the Qing Dynasty (17th century). These materials include Gentianaceae (roots); *Gentiana crassa* subsp. Rigescens (Franch. ex Hemsl.) Halda, *Bupleurum chinense* DC., Apiacea (roots); *Akebia trifoliata* (Thunb.) Koidz., Alismataceae (rhizoma and roots); Lardizabalaceae (stem); *Alisma plantago-aquatica* subsp. orientale (Sam.) Sam., *Plantago asiatica* L., Plantaginaceae (stir-fried seeds with salt solution); *Angelica sinensis* (Oliv.) Diels, Apiaceae (stir-fried roots with yellow rice wine); *Scutellaria baicalensis* Georgi, Lamiaceae (roots); *Gardenia jasminoides, Rehmannia glutinosa* (Gaertn.) DC, Plantaginaceae (roots); *Glycyrrhiza uralensis Fisch.*, Fabaceae (roots stir-fried with honey); J. Ellis, Rubiaceae (stir-fried fruits).

7.2.3 Combination Therapy

A single metabolite or family of chemicals is not present in herbal extracts made from plant parts. It is nonetheless a complicated blend of several bioactive substances. Consequently, it becomes imperative to investigate the primary ingredients in the blend that have a therapeutic effect. Understanding the therapeutic properties of each bioactive metabolite in a herbal combination would increase our comprehension of its therapeutic efficacy and lower the possibility of adverse effects brought on by the presence of additional bioactive specialized compounds [28]. Herbal remedies are thought to be safer and less likely to have negative effects than manufactured medications. However, taking it without a prescription and in an unfavourable combination can potentially prove fatal. It is expected that the use of comparative genomics to identify a synthetic pathway linked to a bioactive agent would be easy once the genomic and proteome information of the majority of herbal plants have been screened and made available in biological databases [29, 30].

For a very long time, people have had a firm belief in the traditional herbal method. To assess the efficiency and safety of herbal medicine, it is essential for clinical studies to be carried out. The World Health Organization published the recommendations for these tests and is in favor of conducting herbal medicine clinical trials. Numerous instances bolster the idea of combination treatment, indicating that two herbs taken together may have a synergistic impact on the illness [31]. However, the interaction of different compounds present in these two herbs may confer negative health effects. Additionally, the effectiveness of herbal treatments varies depending on the person or group. Consequently, the pharmacogenomics principles have to be consistently implemented for herbal remedies as well as their phytopharmaceutical derivatives. The complex and varied components found in herbal remedies are affected by several factors: the type of plant used, when it was harvested, and whether or not it was contaminated or adulterated with microorganisms. However, following clinical examination, high-throughput investigation and contemporary biotechnology technologies have made it feasible to assess the efficacy of herbal medication and suggest appropriate treatments.

Traditional medicine has a long history in China and India in the use of herbal medicine for therapeutic purposes. With a combined history spanning over two millennia, the two oldest medical systems are traditional Chinese medicine (TCM) and Ayurveda. These two systems have a number of similarities. The main therapeutic ingredients utilized in each of these systems are medicinal plants. Indians have long been aware of the benefits of herbal remedies in treating a variety of skin conditions. Numerous similar herbal remedies are also suggested by the Ayurvedic and Unani medical systems for the treatment of various illnesses. For instance, in India, there are a total of just 119 herbal plants that can be used to treat 39 skin conditions [32]. Many medicinal plants are processed using methods provided by Ayurveda and TCM, and in order to make full use traditional knowledge, more research on the extraction and preparation of herbal extracts is needed [33].

Additionally, it's necessary to look for related plant species that can serve as high-quality substitutes for endangered or expensive herbal extract sources. There are several instances of herbal formula creation in the traditional medical system which incorporate extracts from numerous herbal plants. When two plant compounds or drugs are combined, they may work synergistically to enhance therapeutic outcomes. For example, Xiaozhang Tie is a herbal remedy which is used to treat ascites related to cirrhosis. Various methods, such as proteomics, biochemistry, histopathology, and immunohistochemistry, were employed to determine the potential therapeutic targets linked to this herbal remedy [34]. By interfering with the l-arginine and nitric oxide pathways, Xiaozhang Tie raises serum arginine levels and lowers the levels of nitric oxide in the bloodstream.

7.2.4 Conservation and Propagation Strategies

Research has indicated that the pace of extinction of plant species is far higher than anticipated. We will eventually run out of many significant medical plants if this rate continues. Data from the World

Wildlife Fund and the International Union for Conservation of Nature indicate that between 50,000 and 80,000 kinds of flowering plants are used for medicinal purposes. Due to human population growth destroying their natural habitats, the overutilization of plant species, and unfavorable climate changes, some 15,000 plants are currently in danger of going extinct [35]. One of the primary causes of medicinal plant extinction in countries such as India, Kenya, China, and Nepal is habitat loss.

Regarding the preservation, propagation, inventorying, and status monitoring of medicinal plants, several guidelines and recommendations have been evaluated. Globally, the rate at which medicinal plants are disappearing can be slowed down by the maintainable utilization of natural resources. Brazil, China, India, and South Africa have witnessed declines in their flora and fauna because of the high demand brought on by the increases in population. Enhancing the quantity, caliber, and effectiveness of pharmaceuticals can be achieved through the use of plant tissue culture, micropropagation, and synthetic seed development techniques [35, 36]. The mass production of desired bioactive compounds can be achieved through the fermentation of tissue culture and of medicinal plants. Tissue culture can also be used to generate a large number of secondary metabolites and to multiply uncommon medicinal plants more quickly. Cultivating the herbal plant in vitro or ex vitro can be achieved by synthetic seed technology when regular seeds are unable to germinate. It is also possible to shorten the time needed for large-scale manufacturing to breed the desired herbal plant using genetic modification [37, 38].

It is necessary to conduct exploratory research on medicinal plants and find novel bioactive molecules in order to evaluate and characterize possible lead compounds for drug development. Using plant cell cultures, heterologous biosynthesis, and synthetic biotechnology techniques the large-scale, economical synthesis of secondary metabolites can be achieved [39]. The functional expression of plant biosynthetic cascades requires a deeper comprehension of the genetic control of pathways and protein production. Understanding the information of pathway reconstruction and the gene involvement in synthesis are important for the de novo manufacture of a plant compound.

Various in-vitro studies have been conducted for the culturing of medicinal plants using the technique of micropropagation. Different techniques of micropropagation are available for culturing the plants. The synthesis of secondary metabolites is beneficial in rare conservation of endangered and vulnerable specie of medicinal plants can be achieved by in vitro culture procedures. Numerous medicinal plants, such as like *Aloe vera* (L.) Burm.f., Xanthorrhoeaceae; *Artemisia annua* (L.), Asteraceae; *Catharanthus roseus* (L.) G. Don, Apocynaceae; *Withania somnifera* (L.) Dunal, Solanaceae; and *Rauvolfia serpentina* Benth. ex Kurz, Apocynaceae, have all been successfully propagated through the use of effective protocols. Secondary metabolites from medicinal plants are produced by callus induction and growth, shoot proliferation, cell culture, and the use of transgenic techniques (Agrobacterium rhizogenes). Sanguinarine from *Papaver somniferum* (L.), Papaveraceae, shikonin from cell cultures of *Lithospermum erythrorhizon* Siebold & Zucc., Boraginaceae, and berberine from *Coptis japonica* (Thunb.) Makino, Ranunculaceae, and berberine from *Coptis japonica* (Thunb.) Makino are examples of such techniques. Conservation of therapeutic plants including *Saussurea costus* (Falc.) Lipsch., Asteraceae; *Ginkgo biloba* (L.), Ginkgoaceae; *Gymnema sylvestre* (Retz.) R.Br. ex Sm., Apocynaceae; *Tinospora sinensis* (Lour.) Merr., Menispermaceae; and *Oroxylum indicum* (L.) Kurz, Bignoniaceae is carried out through the use of tissue culture techniques [40]. Normally, a plant creates secondary metabolites in response to a variety of adverse environmental stressors, including nutrition deficiency, predator contact, and pathogenic interaction. One substantial, environmentally acceptable, and vital method for conserving medicinal plants and their germplasm is tissue culture. It is a process for producing secondary metabolites in plants [41].

7.2.5 Pharmacogenomics

Since they have been used for a considerable period, many herbs that we use as spices and condiments in our daily diet do not need to be under medical care. Before receiving permission and license to be sold, herbal medications must also adhere to a few regulatory requirements. It is necessary to outlaw the use of unauthorized herbal treatments in order to prevent poisoning and other major health problems [42, 43].

When using herbs conventionally, side effects from herb–herb and herb–drug interactions pose a serious risk to public health. Numerous chemical components included in herbal mixtures may target various hormones, enzymes, receptors, and other molecules in our biological systems, potentially producing a wide range of pharmacological responses [44, 45]. When a herb has the potential to impact the distribution, metabolism, excretion, absorption, or absorption of concurrently administered herbs or medications, this can have a negative effect.

Pharmacogenomics is study of how a person's genes affects and how they react to medications, which is the combination of genomics (the study of genes and their function), and the science of pharmaceuticals, to create safe, effective treatments and dosages that are specific to an individual's genetic composition. By using pharmacogenomics, one may forecast possible adverse consequences of herb–drug interactions based on an individual's profile of absorption, distribution, metabolism, and excretion [45, 46].

A deeper comprehension of the interactions between herbs and drugs, as well as genetic differences, is essential for the safe use of herbal drugs. Herbs and CYP450 enzymes, including CYP3A4, CYP1A2, CYP2C9, and CYP2C19, can interact to either favorably or unfavorably affect a drug's metabolism [47]. Numerous CYP450 enzyme-related polymorphisms have been documented in the human population. There are people who have vast metabolisms and others who have inadequate metabolisms. If a herb has a negative effect on a medicine's metabolism, a drug with a poor metabolizer may have hazardous effects because of inadequate drug elimination. Analogously, a herb's beneficial effects on medication metabolism may lessen the therapeutic response. Grapefruit juice is one of the strong inhibitors of the cytochrome P450 CYP3A4 enzyme. While grapefruit juice can sometimes interact fatally with medications like terfenadine or astemizole, it also boosts the bioavailability of pharmaceuticals by influencing drug metabolism [48]. Herb–drug interactions are becoming more common; however, it's still unclear how they work in various genotypes of many plants.

For more efficient and successful therapy, it is important to evaluate the genetic basis of the various medication responses in different genotypes. Additionally, pharmacokinetic pathways and other polymorphisms associated with the transporter gene are how herbs interact with transporters [47]. The metabolism of warfarin in people with the CYP1A1 and CYP2B1 genotype strongly induces the Danshen–Gegan formula (DGF). This results an increase in the absorption of warfarin in the intestine because of this decrease of warfarin binding to plasma protein [49]. Another example is berberine [6], which is easily obtained from medicinal plants belonging to various plant families that grow mostly at high altitudes. These families include the Annonaceae (e.g., Xylopia L.), Berberidaceae (e.g., Berberis L.), Menispermaceae (e.g., Tinospora Miers), Papaveraceae (e.g., Argemone L.), Ranunculaceae (e.g., Coptis Salisb.), and Rutaceae (e.g., Zanthoxylum L.). Some of these plants are used in several traditional folklore formulations in Ayurvedic and Chinese medicine to treat hyperglycemia, high blood pressure, and hyperlipidemia. Berberine can interact negatively with cyclosporine A and enhance its bioavailability due to suppression of the CYP3A4 enzyme, requiring a lower dose. Moreover, berberine may interact negatively with tolbutamide, thiopental, and warfarin, worsening blood toxicity. Azithromycin and clarithromycin are examples of macrolide antibiotics that may interact with berberine and cause cardiac problems [50].

7.2.6 Computational Resources for Drug Discovery

Over the past few years, the field of computing hardware, software, and algorithm development has developed very significantly. These initiatives have had a noteworthy influence in creating databases comprised of biogenic and synthetic drug-like compounds as well as the computer-aided drug discovery process [51]. Numerous databases, including PubChem, ChEMBL, and ZINC, offer details on hundreds of chemical compounds that are sourced from various plants or natural sources [52]. Detailed information of herbal compounds, including compound name and source, IUPAC name, chemical composition, molecular weight, lipophilicity, hydrogen bond donor and acceptor, biological targets, bioactivity assay, efficacy, toxicity, related literature, and other datasets are available in databases that have free access [53, 54]. In the creation of new drugs, these plant metabolites can be applied as a lead molecule in order to target traditional medical situations.

Herbs or medication molecules act on hormones, enzymes, receptors, lipids, carbohydrates, DNA, and RNA in order to produce therapeutic effects. Proteins are typically employed as possible pharmacological targets [55]. Rational medication design benefits greatly from the traditional understanding of herbs and their pharmacological uses. Algorithms and computational tools have been crucial in the drug design process. Currently, a number of servers and software tools are available for modeling and validating the 3D structure of the target protein [56]. The calculation of the extent and content of the binding site or cavity in the target protein can be achieved using theoretical computing. Targeting the protein can also be done using the analogue of a substrate molecule. Information on the protein–ligand complex is available in Protein data bank database for a large number of proteins, which may be observed to determine the binding-related details of a substrate, cofactor, previously identified inhibitor, or antibody [57, 58].

The structural distinction seen in herbal substances is considered to be a valuable starting points for the progress of new drugs. Most of the lead compounds come from natural sources. Lead compounds do not meet the requirements for medications, notwithstanding the possibility of there being some therapeutic benefits against illness [59, 60]. The recognition of old herbal components is a source for the discoveries of new medication. Binding interaction, specificity, selectivity, absorption, distribution, metabolism, and toxicity are factors for the optimization of herbal lead compounds. Several chemical changes occurs in the lead molecule during the lead optimization process in order to meet the various requirements to be considered as a potential medication [59]. A herbal compound needs to be chemically modified to make it easier for liver enzymes to metabolize it, especially if it has excellent absorption, distribution, and binding affinity for the drug target but has a poor metabolism and toxic substructure. Removing or substituting harmful groups with alternative groups can reduce the toxicity of herbal substances [61, 62]. An anti-inflammatory medication called lumiracoxib has serious liver toxicity. Diclofenac, another safe and efficient medication, was created by substituting chlorine for fluorine and eliminating the methyl group from the meta position of phenylacetic acid in lumiracoxib [63]. Aromatic nitro, aromatic amines, bromoarenes, hydrazines, polyhalogenated groups, and hydroxylamine like toxic groups are not added to candidate medications throughout the drug discovery process. If they already existed in the original lead compounds, they are eliminated. Several theoretical guidelines to structure-property and structure-activity help in determining what modifications to the lead molecule are necessary to improve effectiveness, pharmacokinetics, and pharmacodynamics [62]. According to the binding energy of the docked complex, molecular docking software is particularly useful in sifting through a large number of compounds to find those that may be potentially effective against a target [64]. Predicting the compound's binding posture and the associated interaction with the target protein's amino acids is now feasible thanks to docking and structural visualization technologies. This method is being used by pharmaceutical corporations for the creation of new drugs, and there are several examples of successful computer-aided drug design [23, 65]. Computational approaches for the development of medications include some examples, such as saquinavir for the target HIV-1 and HIV-2 protease (AIDS), zanamivir for the target neuraminidase (influenza), aliskiren for the target renin (hypertension, high blood pressure), and captopril for the target angiotensin-converting enzyme (hypertension) [66] (Figure 7.1).

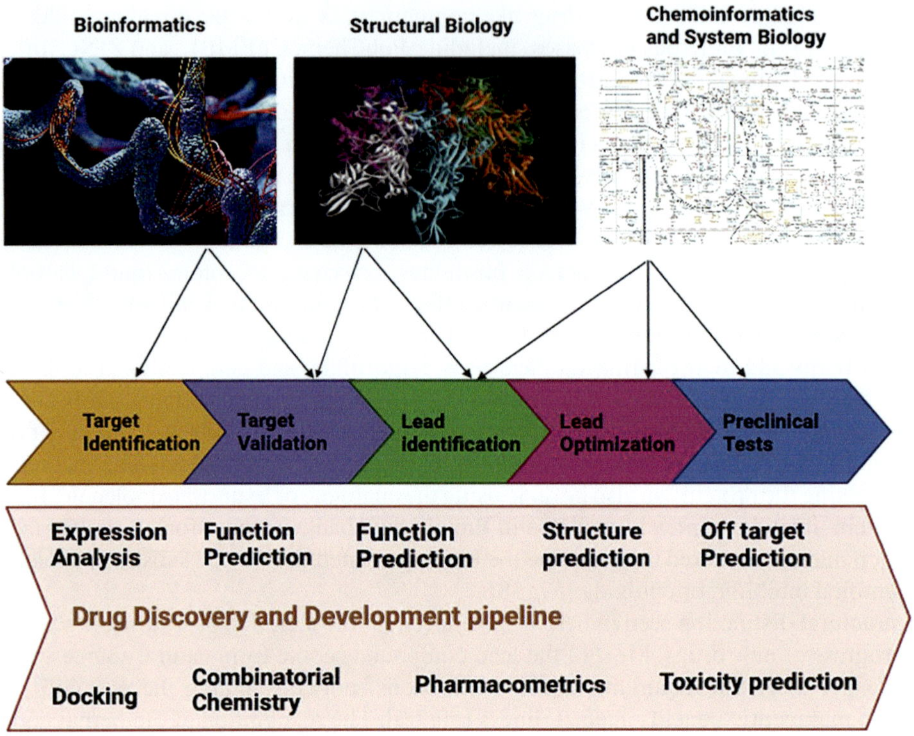

FIGURE 7.1 Drug discovery and development pipeline.

7.3 CONNECTION BETWEEN HERBAL DRUG DISCOVERY AND BIOTECHNOLOGY

The relationship between herbal drug discovery and biotechnology is a multifaceted and evolving field that encompasses various aspects of research, development, and application. Herbal medicine, derived from plants and natural sources, has been a significant part of traditional medicine systems globally and continues to gain recognition for its therapeutic potential [67]. The integration of biotechnological interventions in herbal medicine plays a vital role in improving the conservation, cultivation, and utilization of medicinal plants [68]. Biotechnology offers tools and techniques that can aid in the sustainable production, standardization, and quality assessment of herbal products, thereby contributing to the advancement of herbal drug discovery [69]. The emergence of network pharmacology as a research paradigm has provided a systematic approach to exploring the complex interactions between herbal compounds and biological systems, facilitating evidence-based drug discovery [70]. This approach allows for a comprehensive understanding of the mode of action of herbal remedies, supporting identification of potential therapeutic targets for various conditions, including obesity, neurodegenerative diseases, and cancer [71]. Thus, using the network pharmacology approach, it is possible to reveal the interactions of several compounds in the complex of herbal medicine and develop new targeted therapies. Considering the fact that precision medicine—the use of the genomic information to personalize the medical treatment—has become a trendy topic, more and more attention is paid to the prospects of using herbal medicine for personalized medicine. When using concepts and data from omics technologies, including genome and metabolomics, with traditional knowledge on natural products comprising herbs, the quality and safety of traditional cured herbs can be boosted, especially when applied to current global concerns like COVID-19 [20]. This approach emphasized the positive way of synergy between the old knowledge of herbs and the new biotechnological approach to the solution of present-day health complications. Additionally, the incorporation of herbal drugs with orthodox practice has been testified to help in the control of many diseases such as cancer. That being said, paradoxical to its immunosuppressive effects, herbal medicine has been recognized to have a positive influence on the immune microenvironment alongside the metabolic pathways that relate to cancer development [72]. The inclusion of herbal medicine into conventional health systems, in other centres of academic oncology, shows that the concepts offered by traditional herbal practices can be incorporated into existing medical practices [73]. The use of herbal medicine to cure such ailments as mouth ulcers, gynecologic cancer, and Behcet's disease demonstrates how—and in what ways—some Internet-sourced herbal remedies were used in other branches of medicine [74, 75]. Most patients use herbal treatment to manage symptoms in addition to conventional medical treatment, showing why it is essential for healthcare workers to understand and have a basic knowledge of herbal medicine. The general patterns and temporal trends of herbal medicine use among patients with different diseases can help healthcare providers design individual patient-centered treatment plans that combine American Medical Association (AMA) and Western biomedical medications [76].

7.4 APPLICATIONS OF BIOTECHNOLOGICAL TOOLS FOR HERBAL DRUG DISCOVERY

Thus, the application of biotechnological methods to study herbal drugs is an innovative approach to studying natural compounds through the synergistic blend of traditional practices and modern discoveries. When the two fields of technology are combined with herbal medicine, new opportunities, such as the development of new drugs, the quality control of medicines, and the use of individualized patient-orientated medicines can be enhanced by researchers. This integration also applies the use of artificial intelligence and machine learning algorithms in quickening the process of drug discovery [77]. These technologies allow the forecasting of molecular interactions, the potential drug identification targets and the fine-tuning of an herb's chemical composition, effectively making the process of developing new herbal medicines more efficient [78]. In addition, network pharmacology studies has become a viable method in integrating herbal drugs and natural products in the processes of complex drug discovery [79]. Applying network pharmacology to analyze and comprehend the multitarget nature of herbal ingredients and their biological effect, it is possible to discover new targets of disease interventions, like cancer, neurodegenerative diseases, and other diseases and infections. This IT approach facilitates the study of herb–medicine relations along with the simultaneous action of numerous compounds in the formulation, and lays down the framework for improving present-day herbal drugs. In addition to artificial intelligence and network pharmacology, computational technologies are considered a major determinant of enhancing herbal drug discovery [80]. The uses of these tools are unlimited, encompassing the virtual screening of phytochemicals with pharmacokinetics of herbal compounds [81]. Thus, computational models provide an opportunity to speed up the process of selection of bioactive compounds, to fine-tune a drug delivery system, and to the assessment of the safety and efficacy of the components of herbal products. This computational method is beneficial not only in accelerating the drug discovery process but also in increasing the accuracy and speed of the formulation of herbal medicines. Furthermore, the use of bio-modern techniques, including metabolic profiling, genomic sequencing and the bioactivity screening of the natural product, has been enhanced to boost drug discovery and development [82]. They help the researchers to replicate, discover the chemical constitution of bioactive organisms, and identify drug leads from natural products. Thus, integrating such biotechnological tools with the conventional system of using herbal medicine will enable the researchers to trace new bioactive molecules of pharmaceutical importance and foster more development on the use of herbal medicine as drugs.

7.4.1 Metabolomics Approaches in Herbal Drug Discovery

The study of small molecules, which includes drug metabolites, is a comparatively recent branch of science which entails the characterization of small molecules with the help of various physical methods including nuclear magnetic resonance (NMR), mass spectrometry, along with chromatography [83]. It also allows a special type of 'top-down' examination of other complex biological structures, such as cells [84]. The flexibility and the broad spectrum of the analysis methods, which comprise mass spectrometry [85–90], NMR spectroscopy [91–96], and infrared spectrophotometry [97], integrated in the metabolomics has boosted the field's growth. With the help of the available instruments, thousands of metabolites in complex biological samples might be evaluated [98]. There is no one analytical tool or technique that can be used to detect every single metabolite in a biological model because of the complexity of the biological system and the many substances in the metabolome. These more recent techniques are being combined to increase the diagnostic effectiveness and dereplication procedures, hence broadening the range of metabolites that are being identified. The primary analytical methods that have been often used to produce metabolomics data are MS and NMR. However, for the purposes of metabolite separation, identification, and quantification, various analytical hyphenated platforms—such as gas chromate-MS (GC-MS), HPLC, UPLC, and compound systems of those mentioned—are being employed more appropriately [99]. Although NMR may yield precise and reliable results quickly, making it suitable for examining bulk materials, mass spectrometry's great sensitivity makes it possible to analyze many metabolites both qualitatively and quantitatively.

7.4.2 Genomic Approaches

'Genome mining' is the potential function and chemical connections between bioactive natural compounds and the bioinformatics inquiry that identifies the biosynthesis route of those compounds [100]. More precisely, genome mining involves discovering earlier unidentified gene clusters responsible for the biosynthesis of natural products within the genomes of sequenced organisms. This process includes examining the enzymes encoding these gene clusters by sequencing, and experimentally identifying the result of the resulting products (Figure 7.2) [101].

The majority of clinically authorized medications, including antibiotics, are derived from natural materials. Combinatorial biosynthesis approaches is the novel technique for the identification of naturally occurring compounds with biotechnological uses, having been made possible by the development of new bioinformatics, genetics, and analytical tools [102, 103]. These methods have demonstrated the capability of organisms, especially microorganisms generating bioactive natural compounds, has been undervalued, in conjunction with bioinformatic methodologies [104]. The genomes of these creatures have been sequenced, and genome mining techniques have been used to further investigate them [105]. Biosynthetic gene clusters that have an unknown origin are referred to as orphan and cryptic and may be involved in the production of additional secondary metabolites. Their origins have been revealed by genome analysis [104]. Genome mining uses various bioinformatic procedures to find gene that code for novel natural compounds of biological benefits [106]. The growing prevalence of serious diseases (like cancer) and the declining effectiveness of current medications have made it imperative to find novel therapeutic entities, which is why genome mining is so important [107]. Additionally, by encouraging cost-cutting and their commercial application, the components contained in the biosynthetic gene clusters help enhance the production of both natural and engineered products [105].

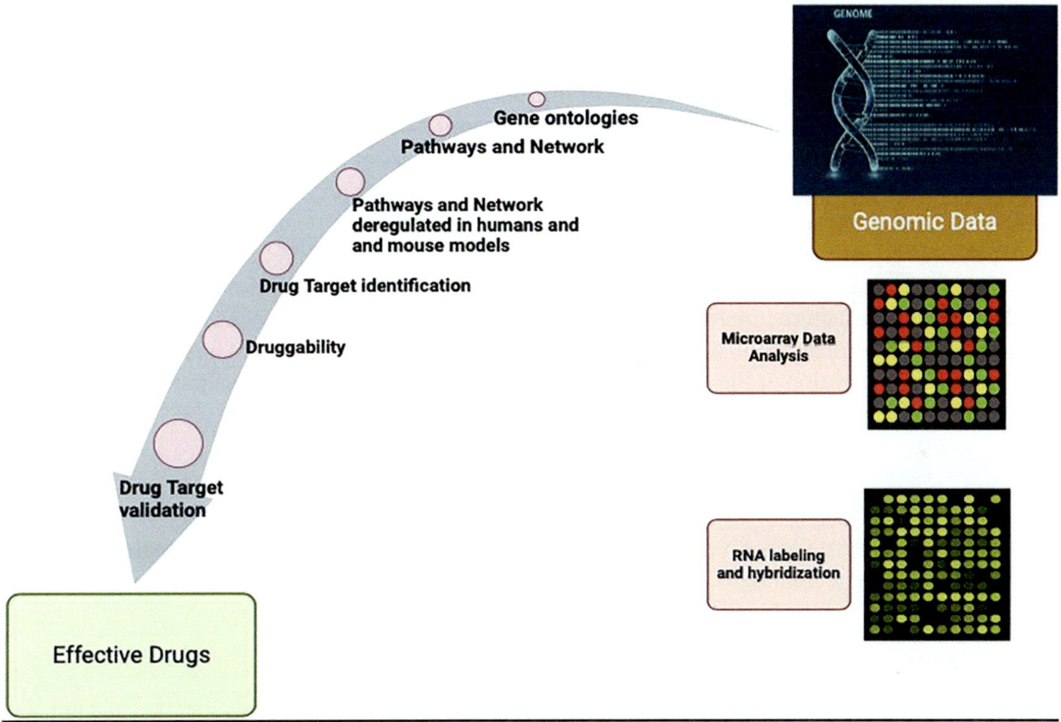

FIGURE 7.2 Genomic approach for the discovery of drugs.

7.5 CASE STUDIES TREATED WITH BIOTECHNOLOGICALLY-EXTRACTED HERBAL DRUGS

Many of the diseases that have previously been treated with pharmacological drugs are today treated with herbal extracts. The herbal compounds are readily available and have either no or very rare side effects. Here are some of the case studies now treated with biotechnologically extracted herbal sources.

7.5.1 Quinine for Malarial Treatment

Malaria, a very fatal disease caused by the Plasmodium parasite, has plagued humankind for generations. The bark of the cinchona tree (*Cinchona officinalis*) is widely used by Native tribes of South America as a fever treatment [108]. Back in the 17th century, quinine, a bioactive substance present in cinchona bark, was discovered to be a very effective treatment for malaria [109]. This discovery marked a major breakthrough in the treatment of the illness; quinine remained a vital anti-malarial drug for decades until more modern treatments were developed [110].

The significance of plant-based medicine identifying the healing potential of the natural environment is the persistent understanding of ancient civilization [111]. These early methods contributed to the advancement of healthcare by laying the foundation for the development of modern pharmacology [112]. In addition, they stress the need to protect and honor our natural resources while also looking into the potential medical benefits of natural products based on plants in the contemporary field of pharmaceutical study and development.

7.5.2 Aspirin for Pain and the Treatment of Inflammation

The development of modern pharmacology has been greatly influenced by the historical usage of herbal treatments. One important example of this phenomena is aspirin, which is a common analgesic used globally. The willow tree, which is the commonly known specie of the *Salix alba* species, is the original source of aspirin, which was initially isolated from its bark. This natural therapy was transformed into a synthetic version through a synthesis procedure. Consequently, it has gained widespread recognition and is employed globally as an essential element in pain management. As a result, the variety of medicinal compounds found in nature has had a big impact on the present-day pharmaceutical business. One example of this is the transition from traditional herbal remedies to synthetically manufactured pharmaceutical molecules, which is what happened with aspirin [113]. Willow bark was utilized as an analgesic by ancient civilizations such as the Greeks and Egyptians. The successful extraction of the active component, acetylsalicylic acid, throughout the 19th century led to a dramatic change in the management of pain and inflammation [114].

7.6 LIMITATIONS IN HERBAL DRUG DISCOVERY

Herbal drug discovery faces several limitations, including regulatory hurdles and challenges associated with emerging technologies.

7.6.1 Regulatory Hurdles

Regulatory frameworks for herbal medicines vary globally, with disparities in policies and enforcement mechanisms affecting the herbal products in terms of the quality, safety, and efficacy of herbal products. In many countries, such as those in sub-Saharan Africa, national regulations regarding the use of herbal medicines remain scarce; this makes it possible for counterfeit herbal remedies to flood the market and therefore is a potentially grave source of danger to the health of its consumers. This is due to the fact that there appears to be no recognized protocol at both national and international levels for the formulation of policies and the implementation of quality control on herbal drugs [115]. Confirmation of in vitro bioassay techniques is important in the investigation of most oriental medicinal compounds. It is , however, faced with several difficulties as to the accuracy of subsequent findings. Any validation of the bioassay method is crucial for the identification of the pharmacological activity of the compounds derived from a herb. Nevertheless, variability in the validation procedures can lead to variations in efficacies' estimates. Such an absence of the uniform validation process is one of the factors to complicate the process of identifying adequate experimental models for discovering herbal drugs, not to mention translating in vitro data into clinical reality [116]. Another crucial factor of herbal product commercialization and safe use is the lack of standard policies and quality control. The spiking of herbal drugs with synthetic substances increases the danger to consumers and demonstrates the need for quality control measures. However, there are technical difficulties in identifying and controlling some of these impurities in the final herbal products with regards to quality and safety. Thus, there is a need to establish efficient analytical methods and product regulation [117]. The major problem of notably various multiple-herb formulae is in pointing to which of the numerous compounds have therapeutically active components. Many herbal medicines are reported to consist of several bioactive compounds so there is a problem of analysing the individual constituents for pharmacodynamics studies. Lack of standardized validation protocols results in the limitation of this whole process up to in vitro tests and the herbal compounds cannot be clinically practiced [118]. Market factors that flank the production and sale of herbal supplements include statutes and rules by federal agencies such as the US FDA, which inhibit the marketing of substandard supplements. The current regulation of herbal products relies on the extent to which the authorities may choose to enforce standards with regard to the products' quality and performance. Their failure to prescribe improved instrumentalities of regulation, also making it difficult to assure safety and the sound profile of herbal drugs in the market [119]. Regulatory problems in developing countries, as well as worldwide issues regarding the quality of herbal medicine and compliance to good standards, are among the issues that slow down the development of the herbal business. The policies of the FDA and other related bodies have introduced measures that limit market access for substandard and adulterated herbal products, thereby calling for good and established policies that will protect society. As a result of diverse regional policies that are not in harmony, there is a major challenge in the formulation of the herbal drugs and selling them around the world [14]. Hence with regard to herbal wound care products, regulatory and legal hurdles are the key barriers to the creation and commercialization of safe and efficacious remedies. Due to the factors such as product registration, product quality standards, and requirements needed to meet the regulatory agencies' standards present some hurdles regarding the entry of herbal wound care products into the market. There are issues with the regulation behavior that slow down the process of commercialization of herbal therapies, and its usage in healthcare systems [120]. The use of chemicals and other substances not approved by doctors and other medical practitioners on patients remains extremely rampant in regions such as Nigeria explains the difficulties involved in regulating the usage of herbal products. The easy access to these herbal remedies from sources other than the formal pharmacies and the absence of measures being taken to ensure that the herbs meet the specified quality catches a lot of blame as far as consumption of the herbal medicines are concerned. Lack of and/or insufficient regulation of traditional herbal remedies for sale thus contributes to public health issues and implies the necessity of the improved regulation of these types of medications [121]. Challenges of pharmaceutical companies get into the development of orphan drugs mainly because the patient pool for these diseases is limited which brings the issue of finance into play. The speciality market of rare disease therapies eliminates the profit-making potential of mainstream research approaches and hinders the funding of further exploration on herbal drugs for rare diseases. Regulatory issues influence herbal medicines' accessibility and acceptance in medical settings and hinder their timely and efficient commercialization [122]. Among the challenges that are encountered during the development of herbal compounds to treat inflammatory bowel disease (IBD) include: Regulatory frameworks limiting the use of safety and efficacies of the herbal medicine. Another limitation is the requirement to conform to strict preclinical and clinical testing procedures that are entailed by compliance with regulatory measures, which increases the drug development process's complexity and timeframe. The stringent regulation regarding the formulation of herbal drugs complicates safety and efficacy concerns, which hamper research in the development of new treatments for IBD [123]. Nonetheless, medicinal plants represent promising reservoirs of novel therapeutic leads in the process of drug discovery. Current issues include the difficulties in identifying appropriate high-throughput screening bioassays and how to increase the production of bioactive molecules for translational purposes. The number of screening assays has not stabilized which significantly hinders the recognition of molecules contained in herbs that can be therapeutic. Furthermore, difficulties in the procurement and increasing the production of plant-derived bioactive molecules limit the conversion of folk medicines into drugs [124]. While attempting to establish ways of excluding herbal ingredients that hit specific types of pathways, it is a challenge to do so because herbs are not as straightforward as one-targeted or one-pathway affected compounds. Herbal ingredients work through the combined action of multiple fractions and influences different bio-pathways. This is why the straightforward drug target hunting technology does not work well in this context. Due to the diverse composition of herbal preparations, different strategies are used to discover and validate the molecules possessing required pharmacological effects [15]. Imbalances in the global accessibility of drugs concerns the opportunities to use therapeutic monoclonal antibodies and their availability in pediatric populations. The shortage of treatment antibodies restricts the healthcare available to children with different diseases, thereby making a statement about healthcare inequalities. These problems around accessing therapeutic monoclonal antibodies speak of the need to have policies that will allow the availability of vital medications in the market [125]. Evaluating the adverse drug reactions (ADRs) linked with herbal medicines suggests issues in the legal standing, risk assessment, and quality standards of herbs and supplements. The low awareness among the healthcare workers regarding herbal products, together with poor efficacy, an absence of regulatory and the lack of quality control measures are just some of the factors which are attributed to the hazards related to the usage of herbal medicines. To tackle these challenges, more stringent regulatory measures at the administrative level and safety control of the herbal compounds are mandatory in order to protect the health of the community [126].

7.6.2 Emerging Technologies

Although innovative technologies have the potential to radically enhance drug discovery, their application in the study of herbal medicine is still limited by a number of difficulties. The incorporation of Computer-Aided Design and Drafting (CADD) and other related tools in the development of herbal drugs entails aspects of overcoming technical difficulties in adopting these novel technologies in coping with the complications that surround the structures of most natural products. The weaknesses that have been observed during the application of these advanced technologies on herbal drug discovery therefore call for an invention of strategies that will favor the discovery of natural products [127]. These confusing single botanical drugs could be due to the multifaceted medical applications of single botanical drugs that Ayurveda and other traditional medical practices have helped in using, thereby messing up the concept of precise herbal medicinal usage. The literature review followed this topical approach because the medicinal plants are generally complex biological products used in the treatment of diverse conditions whose pharmacology cannot easily be standardized. These findings around herbal medicine usage show that there was a dire need for intensive and exhaustive research coupled with positive indicators to support the healing prowess of those products [128]. Metabolomics based on mass spectrometry is, and will become, one of the most efficient techniques for the analysis of metabolic profiles, targets, and the effectiveness of herbal medicine. It has been found that metabolomics in herbal drug discovery helps the researchers to analyse the metabolites of the herb and the working of the drug [129]. Integrated use of the metabolomics in diagnosing and assessing the impact of herbal remedies improves the methods' accuracy and efficiency, most notably in terms of evaluating the metabolic changes induced by natural substances [129]. The process of collection, identification, and cataloging of reference DNA barcodes for goal identification of recipes and herbs reveals the significance of quality measures for protecting the conventional and modern herbal medicines' industrial markets [130]. Implementation of the genetic body reference standards for plant products in herbal products helps to prevent the sale of counterfeit products and to meet the requirement of the regulatory agencies. The utilization of DNA barcoding technologies solves some of the issues related to the identification of the botanical source for herbal products and eliminates the aspects of addition and misidentification of the plant species. Information regarding the efficacy and therapeutic achievements of complex herbal formulas in the context of herbal medication combinations can be acquired with the use of knowledge graph-based methodologies. The use of knowledge graphs in the research of herbal medicines helps in the determination of combined effects and the role of multiple compounds present in a drug formulation. This approach is vital in overcoming issues related to the safety and effectiveness of herbal formulations in relation to their combination form, which is useful in generating important information concerning improvement in the treatment regimen of the patients [131]. The relevance of toxicity and standardization concerns in herbal treatment during the COVID-19 pandemic highlight some of the problems that are encased by the regulation of the use of herbs. Possible issues of quality control, interaction of herbal and drug and, generally, the absence of adequate regulatory policies add challenges to the processes of safety and efficacy of herbal items in public health emergencies. These regulatory barriers show that there is a gap in monitoring herbal therapies with subsequent need of quality assurance of the drugs, and standardized methods of using herbal medicines [132]. A regulator has a significant duty to ensure the quality and safety of herbal medicines, but there are difficulties in the effective management of herbal products. Thus, the development of new forms of healthcare calls for the elaboration of more detailed regulatory measures that contain references to the specifics of herbal preparations. Balancing the need for regulatory oversight with the promotion of herbal medicine innovation requires collaborative efforts between regulatory agencies and industry stakeholders [133]. Difficulties in ensuring the safety of herbal medicines underscore the importance of robust pharmacovigilance systems and the measurement of quality control. The lack of standardized protocols for assessing the efficacy and safety of herbal products hinders the identification of potential risks and adverse reactions. Enhancing pharmacovigilance practices and regulatory oversight is essential for confirming the responsible use of herbal drugs and safeguarding public health [134].

REFERENCES

1. Rivera, J. O., Loya, A. M., & Ceballos, R. (2013). Use of herbal medicines and implications for conventional drug therapy medical sciences. *Altern. Integ. Med.*, *2*(6), 1–6.
2. Mishra, P., Kumar, A., Nagireddy, A., Mani, D.N., Shukla, A.K., Tiwari, R., et al. DNA barcoding: An efficient tool to overcome authentication challenges in the herbal market. *Plant Biotechnol. J.* 2016 Jan 1;14(1):8–21.
3. Gurib-Fakim, A.. Medicinal plants: traditions of yesterday and drugs of tomorrow. *Mol. Asp. Med.*, 2006;*27*(1):1–93.
4. Rout, J., Sajem, A.L., & Nath, M. Medicinal plants of North Cachar Hills district of Assam used by the Dimasa tribe. *Indian J. Tradit. Knowl.* 2012 [cited 2024 Jul 20];11(3):520–7. Available from: https://nopr.niscpr.res.in/handle/123456789/14395
5. Ekor, M. The growing use of herbal medicines: Issues relating to adverse reactions and challenges in monitoring safety. *Front. Neurol.* 2014;4 Jan:177.
6. Hussin, A. H.. Adverse effects of herbs and drug-herbal interactions. *Malays. J. Pharm. Sci.* 2002;1(2):39–44.
7. Lahlou, M. The success of natural products in drug discovery. 2013 [cited 2024 Jul 20]; Available from: https://www.scirp.org/html/3-2500287_33502.htm
8. Fabricant, D.S., & Farnsworth, N.R. The value of plants used in traditional medicine for drug discovery. *Environ. Health Perspect.*. 2001 [cited 2024 Jul 20];109(Suppl. 1):69–75. Available from: http://ehpnet1.niehs.nih.gov/docs/2001/suppl-1/69-75fabricant/abstract.html
9. Christenhusz, M., & Phytotaxa, J.B.. The number of known plants species in the world and its annual increase. *Phytotaxa*. 2016 [cited 2024 Jul 20];261(3):201–17. Available from: https://phytotaxa.mapress.com/pt/article/view/16325
10. Parasuraman, S. Herbal drug discovery: challenges and perspectives. *Curr. Pharmacogenomics Pers. Med.* 2018;*16*(1):63–68.
11. Noviana, E., Indrayanto, G., & Rohman, A. Advances in fingerprint analysis for standardization and quality control of herbal medicines. *Front. Pharmacol.* 2022 Jun 2;13:853023.
12. Roy, D., Chakraborty, A., Mukherjee, P., Ghosh, B., & Chakraborty, M. Recent advances in herbal medicines – An updated review. *Asian J. Pharm. Res. Dev.* 2022 Feb 15;10(1):45–8.
13. Batiha, G. E. S., Beshbishy, A. M., El-Mleeh, A., Abdel-Daim, M. M., & Devkota, H. P.. Traditional uses, bioactive chemical constituents, and pharmacological and toxicological activities of Glycyrrhiza glabra L. (Fabaceae). *Biomolecules* 2020;10(3):352.
14. Hossain, C. M., Gera, M. E. E. T. A., & Ali, K. A.. Current status and challenges of herbal drug development and regulatory aspect: a global perspective. *Asian J. Pharm. Clin. Res.* 2022;15:31–41.
15. Tian, S., Li, Y., Xu, J., Zhang, L., Zhang, J., Lu, J., … & Zhang, W.. COIMMR: a computational framework to reveal the contribution of herbal ingredients against human cancer via immune microenvironment and metabolic reprogramming. *Brief. Bioinform.* 2023;24(6): bbad346.
16. Atoe, K., Idu, M., Ikhajiagbe, B., & Bakre, A. G.. The Influence of Methanol Extracts of Some Plant Species Used in the Management of Pregnancy-Related Symptoms on the Reproductive Parameters of Pregnant Wistar Rats. *Trop. J. Nat. Prod. Res.* 2023;*7*(12):5658–5663.
17. Darji, S., Rani, K.S., & Patani, P. A Review on Herbal Nanoparticles. *J. Pharm. Negat. Results*. 2022 Nov 7;1990–2002.
18. Li, X.Z., & Zhang, S.N. Recent advance in treatment of osteoarthritis by bioactive components from herbal medicine. *Chin. Med.*. 2020 Aug 1;15(1).

19. Manmuan, S., Manmuan, P., Yoykaew, P., Thuetong P., Asipong, P., Riantong, N., et al. Evaluation of standardized extract of centella asiatica on cell viability and repressive cancer migration in metastatic colorectal cancer cells in vitro. *Walailak J. Sci. Technol.* 2021;18(5):1–18.
20. Thomford, N.E., Yahaya, E.S., Ekor, M., & Awortwe, C. Turning up the Volume for Precision Herbal Medicine in Africa in an Era of COVID-19 and Planetary Biodiversity Loss. *OMICS* 2020 Dec 1;24(12): 682–4.
21. Rinschen, M. M., Ivanisevic, J., Giera, M., & Siuzdak, G. (2019). Identification of bioactive metabolites using activity metabolomics. *Nat. Rev. Mol. Cell Biol.*, 20(6), 353–367.
22. Salem, M. A., Perez de Souza, L., Serag, A., Fernie, A. R., Farag, M. A., Ezzat, S. M., & Alseekh, S.. Metabolomics in the context of plant natural products research: From sample preparation to metabolite analysis. *Metabolites* 2020;*10*(1):37.
23. Pathak, R.K., Singh, D.B., Sagar, M., Baunthiyal, M., & Kumar, A. Computational approaches in drug discovery and design. *Comput Aided Drug Des.* 2020 Jan 1;2020:1–21.
24. Gesto-Borroto, R., Medina-Jiménez, K., Lorence, A., & Villarreal, M.L. Application of DNA barcoding for quality control of herbal drugs and their phytopharmaceuticals. *Rev. Bras* 2021 Apr 1;31(2):127–41.
25. Mishra, P., Kumar, A., Nagireddy, A., Mani, D.N., Shukla, A.K., Tiwari, R., et al. DNA barcoding: An efficient tool to overcome authentication challenges in the herbal market. *Plant Biotechnol. J.* 2016 Jan 1;14(1):8–21.
26. Thakur, V.V., Tripathi, N., & Tiwari, S. DNA barcoding of some medicinally important plant species of Lamiaceae family in India. *Mol. Biol. Rep.* 2021 Apr 1;48(4):3097–106.
27. Xin, T., Su, C., Lin, Y., Wang, S., Xu, Z., Phytomedicine, J.S. et al. *Precise species detection of traditional Chinese patent medicine by shotgun metagenomic sequencing*. Elsevier [Internet]. [cited 2024 Jul 18]; Available from: https://www.sciencedirect.com/science/article/pii/S094471131830148X
28. Sasidharan, S., Chen, Y., Saravanan, D., Sundram, K. M., & Latha, L. Y.. Extraction, isolation and characterization of bioactive compounds from plants' extracts. *Afr. J. Tradit. Complement. Altern. Med.* 2011;8(1). 10.4314/ajtcam.v8i1.60483
29. Karimi, A., Majlesi, M., & Rafieian-Kopaei, M. Herbal versus synthetic drugs; beliefs and facts. *J. Nephropharmacol.* 2015 [cited 2024 Jul 18]; Available from: https://www.ncbi.nlm.nih.gov/pmc/articles/PMC5297475/
30. Ekor, M. The growing use of herbal medicines: Issues relating to adverse reactions and challenges in monitoring safety. *Front. Neurol.* 2014 4 Jan;4:177.
31. Ventola, C. L. Current issues regarding complementary and alternative medicine (CAM) in the United States: part 1: the widespread use of CAM and the need for better-informed health care professionals to provide patient counseling. *Pharmacy and Therapeutics*, 2010;*35*(8):461.
32. Anand, U., Tudu, C. K., Nandy, S., Sunita, K., Tripathi, V., Loake, G. J., … & Proćków, J. (2022). Ethnodermatological use of medicinal plants in India: From ayurvedic formulations to clinical perspectives–A review. *J. Ethnopharmacol.*, *284*, 114744.
33. Jaiswal, Y., Liang, Z., & Zhao, Z. Botanical drugs in Ayurveda and traditional Chinese medicine. *J. Ethnopharmacol.* 2016. [cited 2024 Jul 18]; Available from: https://www.sciencedirect.com/science/article/pii/S0378874116304123
34. Zhang, K., Zhang, Y., Li, N., Xing, F., Zhao, J., Yang, T., … & Feng, N.. An herbal-compound-based combination therapy that relieves cirrhotic ascites by affecting the L-arginine/nitric oxide pathway: A metabolomics-based systematic study. *J. Ethnopharmacol.* 2019;241:112034.
35. Chen, S.L., Yu, H., Luo, H.M., Wu, Q., Li, C.F., & Steinmetz, A. Conservation and sustainable use of medicinal plants: Problems, progress, and prospects. *Chin. Med.* 2016 Jul 30;11(1):1–10.
36. Chokheli, V., Dmitriev, P., Rajput, V., & Plants, S.B. Recent development in micropropagation techniques for rare plant species. *Plants.* 2020. [cited 2024 Jul 18]; Available from: https://www.mdpi.com/2223-7747/9/12/1733
37. Hussain, M., Fareed, S., … Saeed, M. Current approaches toward production of secondary plant metabolites. *J. Pharm. Bioallied Sci.*. 2012. [cited 2024 Jul 18]; Available from: https://journals.lww.com/jpbs/fulltext/2012/04010/Current_approaches_toward_production_of_secondary.3.aspx
38. Chandran, H., Meena, M., Barupal, T., & Sharma, K.. Plant tissue culture as a perpetual source for production of industrially important bioactive compounds. *Biotechnol. Rep.*. 2020;26:e00450.
39. Kayser, O. Ethnobotany and medicinal plant biotechnology: from tradition to modern aspects of drug development. *Planta Med.* 2018 Aug 1;84(12–13):834–8.

40. Singh, S., & Singh, V.K. Molecular dynamics simulation: methods and application. *Frontiers in Protein Structure, Function, and Dynamics* 2020;213–38.
41. Patil, V.N., & Shahnawaz, M. Tissue culture approaches to enhance plant secondary metabolites production. *Biotechnological Approaches to Enhance Plant Secondary Metabolites: Recent Trends and Future Prospects*. 2021 Jan 1;89–97.
42. Corns, C.M. Herbal remedies and clinical biochemistry. *Ann. Clin. Biochem.* 2003 Sep;40(5):489–507.
43. Prasad, S., & Aggarwal, B. B. Turmeric, the golden spice. *Biomolecular and Clinical Aspects*. 2011. [cited 2024 Jul 18]; Available from: https://www.ncbi.nlm.nih.gov/books/n/herbalmed/ch13/?report=reader
44. Fasinu, P.S., Bouic, P.J., & Rosenkranz, B. An overview of the evidence and mechanisms of herb-drug interactions. *Front. Pharmacol.* 2012;3 Apr:69.
45. Borse, S. P., Singh, D. P., & Nivsarkar, M. Understanding the relevance of herb–drug interaction studies with special focus on interplays: a prerequisite for integrative medicine. *Porto Biomed. J.*. 2019. [cited 2024 Jul 18]; Available from: https://journals.lww.com/pbj/fulltext/2019/04000/Understanding_the_relevance_of_herb_drug.1.aspx
46. Thomford, N. E., Dzobo, K., Chopera, D., Wonkam, A., Skelton, M., Blackhurst, D., … & Dandara, C.. Pharmacogenomics implications of using herbal medicinal plants on African populations in health transition. *Pharmaceuticals* 2015;8(3):637–663.
47. Liu, M.Z., Zhang, Y.L., Zeng, M.Z., He, F.Z., Luo, Z.Y., Luo, J.Q., et al. Pharmacogenomics and Herb-Drug Interactions: Merge of Future and Tradition. *Evidence-Based Complementary and Alternative Medicine*. 2015 [cited 2024 Jul 18];2015. Available from: https://onlinelibrary.wiley.com/doi/abs/10.1155/2015/321091
48. Khan, S. L., Sonwane, G. M., Siddiqui, F. A., Jain, S. P., Kale, M. A., & Borkar, V. S. Discovery of naturally occurring flavonoids as human cytochrome P450 (CYP3A4) inhibitors with the aid of computational chemistry. *Indo Glob J Pharm Sci.* 2020. [cited 2024 Jul 19]; Available from: https://www.academia.edu/download/78872062/09.-Khan-et-al.-2020.pdf
49. Zhang, Z., Ge, B., Zhou, L., Lam, T. N., & Zuo, Z. Induction of liver cytochrome P450s by Danshen–Gegen formula is the leading cause for its pharmacokinetic interactions with warfarin. *J. Ethnopharmacol.*. 2014. [cited 2024 Jul 18]; Available from: https://www.sciencedirect.com/science/article/pii/S0378874114003353
50. Khashayar, A., Bahari, Z., Elliyeh, M., & Ghasemi, M. Therapeutic Effects of Berberine in Metabolic Diseases and Diabetes Mellitus. *Rev. Bras* 2021 Jun 1;31(3):272–81.
51. Stanzione, F., Giangreco, I., & Cole, J. C. Use of molecular docking computational tools in drug discovery. *Prog. Med. Chem.*. 2021. [cited 2024 Jul 18]; Available from: https://www.sciencedirect.com/science/article/pii/S0079646821000047
52. Zhao, L., Ciallella, H. L., Aleksunes, L. M., & Zhu, H. Advancing computer-aided drug discovery (CADD) by big data and data-driven machine learning modeling. *Drug Discov. Today*. 2020. [cited 2024 Jul 18]; Available from: https://www.sciencedirect.com/science/article/pii/S1359644620302646
53. Ramsay, R. R., Popovic-Nikolic, M. R., Nikolic, K., Uliassi, E., & Bolognesi, M. L. A perspective on multi-target drug discovery and design for complex diseases. *Clin. Transl. Med.*. 2018 Dec [cited 2024 Jul 18];7(1):3. Available from: https://link.springer.com/article/10.1186/s40169-017-0181-2
54. Rifaioglu, A. S., Atas, H., Martin, M. J., Cetin-Atalay, R., Atalay, V., & Doğan, T. Recent applications of deep learning and machine intelligence on in silico drug discovery: methods, tools and databases. *Brief. Bioinform.*. 2019. [cited 2024 Jul 19]; Available from: https://academic.oup.com/bib/article-abstract/20/5/1878/5062947
55. Schenone, M., Dančík, V., Wagner, B. K., & Clemons, P. A. Target identification and mechanism of action in chemical biology and drug discovery. *Nat. Chem. Biol.*. 2013. [cited 2024 Jul 18]; Available from: https://www.nature.com/articles/nchembio.1199
56. Gupta, R., Srivastava, D., Sahu, M., Tiwari, S., Ambasta, R.K., & Kumar, P. Artificial intelligence to deep learning: machine intelligence approach for drug discovery. *Mol. Divers.* 2021 Aug 1;25(3):1315–60.
57. Sliwoski, G., Kothiwale, S., Meiler, J., & Lowe, E.W. Computational methods in drug discovery. *Pharmacol. Rev.*. 2014. [cited 2024 Jul 18]; Available from: https://pharmrev.aspetjournals.org/content/66/1/334.short
58. Burley, S. K., Bhikadiya, C., Bi, C., Bittrich, S., Chen, L., Crichlow, G. V., … & Zhuravleva, M. RCSB Protein Data Bank: powerful new tools for exploring 3D structures of biological macromolecules for basic and applied research and education in fundamental. *Nucleic Acids Res.*. 2021. [cited 2024 Jul 18]; Available from: https://academic.oup.com/nar/article-abstract/49/D1/D437/5992282

59. Thomford, N. E., Senthebane, D. A., Rowe, A., Munro, D., Seele, P., Maroyi, A., & Dzobo, K. Natural products for drug discovery in the 21st century: innovations for novel drug discovery. *Int. J. Mol. Sci.*. 2018. [cited 2024 Jul 18]; Available from: https://www.mdpi.com/1422-0067/19/6/1578
60. Batool, M., Ahmad, B., & Choi, S. A structure-based drug discovery paradigm. *Int. J. Mol. Sci.*. 2019. [cited 2024 Jul 18]; Available from: https://www.mdpi.com/1422-0067/20/11/2783
61. Gertsch, J. undefined. Botanical drugs, synergy, and network pharmacology: forth and back to intelligent mixtures. *Planta Med.*. 2011 [cited 2024 Jul 18];77(11):1086–98. Available from: https://www.thieme-connect.com/products/ejournals/html/10.1055/s-0030-1270904
62. Singh, D. B., & Pathak, R. K. Computational approaches in drug designing and their applications. *Exp. Protoco. Biotechnol.*. 2020 [cited 2024 Jul 18];95–117. Available from: https://link.springer.com/protocol/10.1007/978-1-0716-0607-0_6
63. Selg, E., Buccellati, C., Andersson, M., Rovati, G.E., Ezinga, M., Sala, A., et al. Antagonism of thromboxane receptors by diclofenac and lumiracoxib. *Br. J. Pharmacol.* 2007 [cited 2024 Jul 18];152(8):1185–95. Available from: https://bpspubs.onlinelibrary.wiley.com/doi/abs/10.1038/sj.bjp.0707518
64. Pathak, R. K., Gupta, A., Shukla, R., & Baunthiyal, M. Identification of new drug-like compounds from millets as Xanthine oxidoreductase inhibitors for treatment of Hyperuricemia: A molecular docking and simulation study. *Comput. Biol. Chem.* 2018 [cited 2024 Jul 18]; Available from: https://www.sciencedirect.com/science/article/pii/S1476927118300173
65. Agnihotry, S., Pathak, R.K., Srivastav, A., Shukla, P.K., & Gautam, B. Molecular docking and structure-based drug design. *Comput Aided Drug Des.* 2020 Jan 1;115–31.
66. Pathak, R.K., Singh, D.B., Sagar, M., Baunthiyal, M., & Kumar, A. Computational approaches in drug discovery and design. *Comput Aided Drug Des.* 2020 Jan 1;1–21.
67. Sharma, A., Kumar, L., Malhotra, M., Singh, A.P., & Singh, A.P. Ananas comosus (Pineapple): A comprehensive review of its medicinal properties, phytochemical composition, and pharmacological activities. *J. Drug Deliv. Ther.* 2024 May 15;14(5):148–57.
68. Asase, A. Ghana's herbal medicine industry: prospects, challenges and ways forward from a developing country perspective. *Front. Pharmacol.* 2023;14:1267398.
69. Wang, H., Chen, Y., Wang, L., Liu, Q., Yang, S., & Wang, C. Advancing herbal medicine: enhancing product quality and safety through robust quality control practices. *Front. Pharmacol.* 2023;14:1265178.
70. Wang, Y., Yang, S.H., Zhong, K., Jiang, T., Zhang, M., Kwan, H.Y., et al. Network Pharmacology-Based Strategy for the Investigation of the Anti-Obesity Effects of an Ethanolic Extract of Zanthoxylum bungeanum Maxim. *Front. Pharmacol.* 2020 Nov 13;11:572387.
71. Guan, Y., Tang, G., Li, L., Shu, J., Zhao, Y., Huang, L., & Tang, J. Herbal medicine and gut microbiota: exploring untapped therapeutic potential in neurodegenerative disease management. *Arch. Pharm. Res.*. 2024 Feb 1 [cited 2024 Jul 19];47(2):146–64. Available from: https://link.springer.com/article/10.1007/s12272-023-01484-9
72. Tian, S., Li, Y., Xu, J., Zhang, L., Zhang, J., Lu, J., et al. COIMMR: a computational framework to reveal the contribution of herbal ingredients against human cancer via immune microenvironment and metabolic reprogramming. *Brief. Bioinform.* 2023 Nov 1;24(6):bbad346.
73. Hou, Y. N., Chimonas, S., Gubili, J., Deng, G., & Mao, J. J. Integrating herbal medicine into oncology care delivery: development, implementation, and evaluation of a novel program. *Support. Care Cancer.* 2023 Feb 1 [cited 2024 Jul 18];1(2):3. Available from: https://link.springer.com/article/10.1007/s00520-023-07577-x
74. Shahare, N., Chouhan, S., & Darwhekar, G.N. Herbs used in treatment of mouth ulcer- a review. *Int. J. Pharmacogn. Chem.* 2021 Jul 29;68–74. https://doi.org/10.46796/ijpc.v2i3.212
75. Jun, J.H., Ang, L., Choi, T.Y., Lee, H.W., & Lee, M.S. Integrative medicine (Herbal medicine combined with drug therapy) for behcet's disease: A systematic review and meta-analysis of randomized controlled trials. *Pharmaceutics.* 2021;13(4):476.
76. Theuser, A.K., Hack, C.C., Fasching, P.A., Antoniadis, S., Grasruck, K., Wasner, S., et al. Patterns and Trends of Herbal Medicine Use among Patients with Gynecologic Cancer. *Geburtshilfe Frauenheilkd.* 2021 Jun 1;81(6):699–707.
77. David, L., Thakkar, A., Mercado, R., & Engkvist, O. Cheminformatics OEJ of, 2020 undefined. Molecular representations in AI-driven drug discovery: a review and practical guide. *J.Cheminform.*, 2020 Sep 17 [cited 2024 Jul 18];12(1). Available from: https://link.springer.com/article/10.1186/s13321-020-00460-5
78. Paul, D., Sanap, G., Shenoy, S., Kalyane, D., Kalia, K., & Tekade, R.K. Artificial intelligence in drug discovery and development. *Drug Discov. Today* 2021 Jan 1;26(1):80–93.

79. Park, M.N. The Therapeutic Potential of a Strategy to Prevent Acute Myeloid Leukemia Stem Cell Reprogramming in Older Patients. *Int. J. Mol. Sci.* 2023 Aug 1;24(15):12037.
80. Cox, P.B., & Gupta, R. Contemporary computational applications and tools in drug discovery. *ACS Med. Chem. Lett.* 2022 Jul 14;13(7):1016–29.
81. Li, C., Jia, W. W., Yang, J. L., Cheng, C., & Olaleye, O. E. Multi-compound and drug-combination pharmacokinetic research on Chinese herbal medicines. *Acta Pharmacol. Sin.*. 2022 Dec 1 [cited 2024 Jul 18];43(12):3080–95. Available from: https://www.nature.com/articles/s41401-022-00983-7
82. Gaudêncio, S.P., Bayram, E., Lukić Bilela, L., Cueto, M., Díaz-Marrero, A.R., Haznedaroglu, B.Z., et al. Advanced methods for natural products discovery: bioactivity screening, dereplication, metabolomics profiling, genomic sequencing, databases and informatic tools, and structure elucidation. *Mar. Drugs* 2023 May 1;21(5):308.
83. Wishart, D.S. Applications of metabolomics in drug discovery and development. *Drugs R&D* 2008;9(5): 307–22.
84. Kosmides, A. K., Kamisoglu, K., Calvano, S. E., Corbett, S. A., & Androulakis, I. Metabolomic fingerprinting: challenges and opportunities. *Crit. Rev. Biomed. Eng.*. 2013 [cited 2024 Jul 18]; Available from: https://www.dl.begellhouse.com/journals/4b27cbfc562e21b8,4863d6072e2ac4d2,7fd840d0527bd32e.html
85. Chiu, H. H., & Kuo, C. H. Gas chromatography-mass spectrometry-based analytical strategies for fatty acid analysis in biological samples. *J. Food Drug Anal.* 2020. [cited 2024 Jul 18]; Available from: https://www.sciencedirect.com/science/article/pii/S1021949819300985
86. Chiu, H. H., & Kuo, C. H. Mass spectrometry-based forest tree metabolomics. *Mass Spectrom. Rev.*. 2019 Mar 1 [cited 2024 Jul 18];40(2):126–57. Available from: https://analyticalsciencejournals.onlinelibrary.wiley.com/doi/abs/10.1002/mas.21603?casa_token=j2ZND51fsdEAAAAA:vGpAqg6Dl70a_cQrGyyTVvW8I1rwcQ3XkWMy7Q0LCPtzI55YFIlBkMB1YgO1i0-n00jAVDPF-1Pppp_nkQ
87. Wang, R., Yin, Y., & Zhu, Z.J. Advancing untargeted metabolomics using data-independent acquisition mass spectrometry technology. *Anal. Bioanal. Chem.* 2019 Jul 19]411(19):4349–57.
88. Wang, S., Blair, I.A., & Mesaros, C. Analytical methods for mass spectrometry-based metabolomics studies. *Adv. Exp. Med. Biol.* 2019;1140:635–47.
89. Duncan, K. D., Fyrestam, J., & Lanekoff, I. Advances in mass spectrometry based single-cell metabolomics. *Analyst.* 2019 [cited 2024 Jul 18];144:782. Available from: https://pubs.rsc.org/en/content/articlehtml/2019/an/c8an01581c
90. Diez-Simon, C., Mumm, R., & Hall, R. D. Mass spectrometry-based metabolomics of volatiles as a new tool for understanding aroma and flavour chemistry in processed food products. *Metabolomics.* 2019. 123AD Mar 1 [cited 2024 Jul 18];1(3):41. Available from: https://link.springer.com/article/10.1007/s11306-019-1493-6
91. Vignoli, A., Ghini, V., Meoni, G., Licari, C., Takis, P.G., Tenori, L., et al. High-throughput metabolomics by 1D NMR. *Angew. Chem. Int. Ed.* 2019 Jan 21;58(4):968–94.
92. Ranjan, R, & Sinha, N. Nuclear magnetic resonance (NMR)-based metabolomics for cancer research. *NMR Biomed.*. 2018 Oct 1 [cited 2024 Jul 18];32(10). Available from: https://analyticalsciencejournals.onlinelibrary.wiley.com/doi/abs/10.1002/nbm.3916?casa_token=UnmadNHf_WEAAAAA:52708Uji6g_LVZ-uHX_UxgD4sp05axyVBaWVe0AslHOePOUhJdj6hy1s0Gj_28x_kl-6WQi_xjb3Ghe78w
93. Lucas-Torres, C., & Wong, A. Current developments in μMAS NMR analysis for metabolomics. *Metabolites.* 2019 [cited 2024 Jul 18]; Available from: https://www.mdpi.com/2218-1989/9/2/29
94. Consonni,R.,&Cagliani,L.R.ThepotentialityofNMR-basedmetabolomicsinfoodscienceandfoodauthentication assessment. *Magn. Reson. Chem.* 2019. 2018 Sep 1 [cited 2024 Jul 18];57(9):558–78. Available from: https://analyticalsciencejournals.onlinelibrary.wiley.com/doi/abs/10.1002/mrc.4807?casa_token=F6zAVMHagCAAAAAA:8V-RpHPM77ucKTwmL_3jytoMnsG4KdJq0Z-5LN1POf0MDh-Var3eeyZ2NoIfVWPW_SLHNxHSQXprcn1S_w
95. Song, Z., Wang, H., Yin, X., Deng, P., & Jiang, W. Application of NMR metabolomics to search for human disease biomarkers in blood. *Clin. Chem. Lab. Med.* 2019 Apr 1;57(4):417–41.
96. De Castro, F., Benedetti, M., Del Coco, L., & Fanizzi, F. P. NMR-based metabolomics in metal-based drug research. *Molecules.* 2019. [cited 2024 Jul 19]; Available from: https://www.mdpi.com/1420-3049/24/12/2240
97. Volpe, M.G., Costantini, S., Coccia, E., Parrillo, L., & Paolucci, M. Evaluation of metabolic changes induced by polyphenols in the crayfish Astacus leptodactylus by metabolomics using Fourier transformed infrared spectroscopy. *J. Biosci.* 2018 Sep 1;43(4):585–96.

98. Alonso, A., Marsal, S., Julià, A. Analytical methods in untargeted metabolomics: State of the art in 2015. *Front Bioeng Biotechnol* 2015;3(MAR):23.
99. Zhang, A., Sun, H., Wang, P., Han, Y., & Wang, X. Modern analytical techniques in metabolomics analysis. *Analyst*. 2012. [cited 2024 Jul 19]; Available from: https://pubs.rsc.org/en/content/articlehtml/2012/an/c1an15605e
100. Ziemert, N., Alanjary, M., & Weber, T. The evolution of genome mining in microbes–a review. *Nat. Prod. Rep.*. 2016. [cited 2024 Jul 19]; Available from: https://pubs.rsc.org/en/content/articlehtml/2016/np/c6np00025h
101. Trivella, D. B., & de Felicio, R. The tripod for bacterial natural product discovery: genome mining, silent pathway induction, and mass spectrometry-based molecular networking. *System*. 2018 Apr 24 [cited 2024 Jul 19];3(2). Available from: https://journals.asm.org/doi/abs/10.1128/msystems.00160-17
102. Helfrich, E. J., Reiter, S., & Piel, J. Recent advances in genome-based polyketide discovery. *Curr. Opin. Biotechnol.*. 2014. [cited 2024 Jul 19]; Available from: https://www.sciencedirect.com/science/article/pii/S0958166914000676
103. Olano, C., Méndez, C., & Salas, J.A. Molecular insights on the biosynthesis of antitumour compounds by actinomycetes. *Microb. Biotechnol.* 2011 Mar;4(2):144–64.
104. Nett, M., Ikeda, H., & Moore, B. S. Genomic basis for natural product biosynthetic diversity in the actinomycetes. *Nat. Prod. Rep.*. 2009. [cited 2024 Jul 19]; Available from: https://pubs.rsc.org/en/content/articlehtml/2009/np/b817069j
105. Olano, C., Méndez, C, & Salas, J.A. Strategies for the design and discovery of novel antibiotics using genetic engineering and genome mining. *Antimicrobial Compounds: Current Strategies and New Alternatives*. 2013 Dec 1;1–25.
106. Machado, H., Tuttle, R. N., & Jensen, P. R. Omics-based natural product discovery and the lexicon of genome mining. *Curr. Opin. Microbiol.*. 2017. [cited 2024 Jul 19]; Available from: https://www.sciencedirect.com/science/article/pii/S136952741730108X
107. Haldar, S., & Mody, K.H. Genome mining for bioactive compounds. *Springer Handbook of Marine Biotechnology*. 2015 Jan 1;531–9.
108. Packard, R.M. The Making of a Tropical Disease: A… - Google Scholar. [cited 2024 Jul 19]. Available from: https://scholar.google.com/scholar?hl=en&as_sdt=0%2C5&q=Packard%2C+R.M.+The+Making+of+a+Tropical+Disease%3A+A+Short+History+of+Malaria%3B+JHU+Press%3A+Baltimore%2C+MD%2C+USA%2C+2021%3B+Available+online%3A+&btnG=
109. Capasso, F. Phytotherapy: a quick reference to herbal medicine. 2003 [cited 2024 Jul 19]. Available from: https://books.google.com/books?hl=en&lr=&id=ejHpcJXzPTsC&oi=fnd&pg=PA1&dq=Capasso,+F.+Phytotherapy:+A+Quick+Reference+to+Herbal+Medicine%3B+Springer+Science+%26+Business+Media:+Berlin/Heidelberg,+Germany,+2003%3B+Available+online:+&ots=sWfdgrXe8j&sig=XL0j9tx3mH0szOZAixn_HGjt7xY
110. Dagen, M. History of malaria and its treatment. *Antimalarial agents*. Elsevier. 2020. [cited 2024 Jul 19]; Available from: https://www.sciencedirect.com/science/article/pii/B9780081012109000019
111. Rogers, K. Out of Nature: Why Drugs from Plants Matter to the Future of Humanity. 2012 [cited 2024 Jul 19]. Available from: https://books.google.com/books?hl=en&lr=&id=HTbUdjdUBy8C&oi=fnd&pg=PR5&dq=+Rogers,+K.+Out+of+Nature:+Why+Drugs+from+Plants+Matter+to+the+Future+of+Humanity%3B+University+of+Arizona+Press:+Tucson,+AZ,+USA,+2012%3B+Available+online:+&ots=4t766LDZWL&sig=5243GQFTtnY5VXvWLYw2-W1SOmI
112. David, B., Wolfender, J.L., & Dias, D.A. The pharmaceutical industry and natural products: historical status and new trends. *Phytochem. Rev*. 2015 Apr 1;14(2):299–315.
113. Ugurlucan, M., Caglar, I.M., Turhan Caglar, F.N., Ziyade, S., Karatepe, O., Yildiz, Y., et al. Aspirin: from a historical perspective. *Recent Pat. Cardiovasc. Drug Discov.*. 2012 [cited 2024 Jul 19];7:71–6. Available from: https://www.ingentaconnect.com/content/ben/prc/2012/00000007/00000001/art00009
114. Levesque, H., & Lafont, O. Aspirin throughout the ages: a historical review. *Rev. Médecine Interne.*. [cited 2024 Jul 19]; Available from: https://europepmc.org/article/med/10763200
115. El Hajj, M., & Holst, L. Herbal medicine use during pregnancy: A review of the literature with a special focus on Sub-Saharan Africa. *Front. Pharmacol.* 2020 Jun 9;11.
116. Indrayanto, G., Putra, G.S., & Suhud, F. Validation of in-vitro bioassay methods: Application in herbal drug research. *Profiles Drug Subst. Excip. Relat. Methodol.* 2021 Jan 1;46:273–307.
117. Nunavath, R.S., & Nagappan, K. Identification of synthetic impurities and their significance in natural products quality control. *Nat. Prod. Commun.* 2024 May 1;19(5).

118. Li, C., Jia, W. W., Yang, J. L., Cheng, C., & Olaleye, O. E.. Multi-compound and drug-combination pharmacokinetic research on Chinese herbal medicines. *Acta Pharmacol. Sin.*. 2022 Dec 1 [cited 2024 Jul 19];43(12):3080–95. Available from: https://www.nature.com/articles/s41401-022-00983-7
119. Steenkamp, V., Parkar, H., & Dasgupta, A. Utility of therapeutic drug monitoring in identifying clinically significant interactions between St. John's Wort and Prescription Drugs. *Ther. Drug Monit.* 2023 Feb 1;45(1):35–44.
120. Jayram, J., Kondaveeti, S.S., Gnanaraj Johnson, C., Sampath, P.J., Kalachaveedu, M. Challenges and prospects of development of herbal biomaterial based ethical wound care products — A scoping review. Int J Low Extrem Wounds 2024 Jun 1;23(2):291–305.
121. Adeoye, I., & Etuk, V. Prevalence, predictors and pregnancy outcomes of unprescribed and herbal medicine use in Ibadan, Nigeria. *BMC Complement. Med. Ther.*. 2023 Dec 1 [cited 2024 Jul 19];23(1). Available from: https://link.springer.com/article/10.1186/s12906-023-03838-8
122. Ko, J.M. Advancing orphan drug development for rare diseases. *Clin Exp Pediatr*. 2023 Jul 15;67:356
123. Dubey, R. K., & Shukla, S. Exploring Novel Herbal Compounds and Formulations for Inflammatory Bowel Disease (IBD) Management. *Chin. J. Appl. Physiol.*. 2023 Dec 16 [cited 2024 Jul 19];39:2023–35. Available from: http://47.94.85.12:8889/resources/uploads/20240508/1715153350607052306.pdf
124. Sezer, F., Deniz, S., Sevim, D., Chaachouay, N., & Zidane, L. Plant-derived natural products: a source for drug discovery and development. *Drugs Drug Candidates*. 2024 Feb 19 [cited 2024 Jul 18];3(1):184–207. Available from: https://www.mdpi.com/2813-2998/3/1/11
125. Larrosa, C., Mora, J., Cheung, N.K. Global Impact of Monoclonal Antibodies (mAbs) in children: A focus on Anti-GD2. Cancers (Basel) 2023 Jul 1;15(14).
126. van Hunsel, F. P., van der Kooi, D., van de Koppel, S., Kroes, B. H., & Woerdenbag, H. J. Analysis of reports on adverse drug reactions related to herbal medicinal products and herbal supplements in the Netherlands received by the national. *Drug Saf.*. 2022 Jun 1 [cited 2024 Jul 19];45(6):651–61. Available from: https://link.springer.com/article/10.1007/s40264-022-01180-5
127. Niazi, S., & Pharmaceuticals, Z.M. Computer-aided drug design and drug discovery: a prospective analysis. *Pharmaceuticals*, [cited 2024 Jul 19];17(1):22. Available from: https://www.mdpi.com/1424-8247/17/1/22
128. Yao, R., Heinrich, M., Zhang, B., Wei, X., Qi, Y., & Gao, W. Single botanical drugs in the Ayurvedic Pharmacopoeia of India—A quantitative ethnobotanical analysis. *Front. Pharmacol.* 2023;14:1136446.
129. Lin, C., Tian, Q., Guo, S., Xie, D., Cai, Y., Wang, Z., et al. Metabolomics for clinical biomarker discovery and therapeutic target identification. *Molecules* 2024 May 1;29(10):2198.
130. Kumar, S., Urumarudappa, J., Tungphatthong, C., Jaipaew, J., Pornputtapong, N., Pakdeesattayapong, D., et al. Development of a DNA barcode library of plants in the Thai Herbal Pharmacopoeia and Monographs for authentication of herbal products. *Sci. Rep.*, [cited 2024 Jul 19];12(1). Available from: https://www.nature.com/articles/s41598-022-13287-x
131. Liu, C., Yu, H., Chen, G., Yang, Q., Wang, Z., Niu, N., et al. An herbal drug combination identified by knowledge graph alleviates the clinical symptoms of plasma cell mastitis patients: A nonrandomized controlled trial. *elife*, [cited 2024 Jul 19]; Available from: https://elifesciences.org/articles/84414
132. Komolafe, K., Komolafe, T.R., Fatoki, T.H., Akinmoladun, A.C., Brai, B.I.C., Olaleye, M.T., et al. Coronavirus Disease 2019 and Herbal Therapy: Pertinent Issues Relating to Toxicity and Standardization of Phytopharmaceuticals. *Rev. Bras* 2021 Apr 1;31(2):142–61.
133. Eruaga, M.A., Itua, E.O., & Bature, J.T. The role of regulatory authorities in the regulation and control of herbal medicines: A case study of NAFDAC. *Int. J. Sci. Res. Arch.*. 2024 Mar 30 [cited 2024 Jul 19];2024(02):207–11. Available from: https://ijsra.net/content/role-regulatory-authorities-regulation-and-control-herbal-medicines-case-study-nafdac
134. Bhat, M.D.A., & Malik, R. Challenges in monitoring the safety of herbal medicines-an appraisal. *J. Pharmacovigil. Drug Res.* 2022 Dec 1;3(4):3–8.

8 Herbal Formulation Development and Standardization

Rabiya Rashid
Ayush Medical officer, Community health Centre Sankoo Kargil, Ladakh, India

Sanjeev Kumar, Ab Waheed Wani, and Anis Ahmad Mirza
Department of Horticulture, School of agriculture, Lovely Professional University, Phagwara, India

Bilal Ah Bhat and Zarina
Ayush Medical officer, Primary health Centre Shargole, Kargil, Ladakh, India

8.1 INTRODUCTION

Herbal medicine has been an integral part of human healthcare for thousands of years, with its roots in various conventional practices like traditional Chinese medicine (TCM) and Ayurveda, and Indigenous knowledge systems across the globe. These systems have established the framework for the creation of herbal formulations, which are used to treat a wide array of health conditions. The past few years have seen a resurgence in the popularity of herbal products, driven by the global shift towards natural and holistic approaches to health and wellness. The increasing demand for herbal products, coupled with the advancements in scientific research, has brought the development and standardization of herbal formulations to the forefront of modern pharmaceutical practices (Patel et al., 2022).

Despite the rich history and widespread use of herbal medicine, the development and standardization of herbal formulations pose significant challenges. These challenges stem from the inherent variability in the chemical composition of medicinal plants, which is affected by elements like geographic location, climate, cultivation practices, and harvesting techniques. Additionally, the lack of standardized protocols for the extraction, formulation, and quality control of herbal products has led to inconsistencies in their therapeutic efficacy and safety. This has underscored the need for rigorous standardization processes that ensure the consistency, safety, and efficacy of herbal formulations (Singh et al., 2021).

Standardization of herbal formulations involves the establishment of consistent manufacturing processes, including the use of standardized extracts with well-defined chemical profiles, the implementation of quality control measures throughout the production process, and the adherence to regulatory guidelines. This chapter delves into the various aspects of herbal formulation

DOI: 10.1201/9781003544418-8

development, from traditional approaches to modern extraction techniques, and explores the critical role of standardization in guaranteeing the effectiveness and quality of herbal products. Through a series of case studies and discussions on current challenges and future directions, this chapter seeks to offer a thorough summary of the strategies employed in the standardization of herbal formulations.

8.2 TRADITIONAL APPROACHES TO HERBAL FORMULATION

Herbal medicine has been practiced for millennia, with each traditional system developing its own methods for formulating herbal remedies. These traditional approaches to herbal formulation are deeply rooted in the cultural and historical context of the societies that developed them. For instance, one of the oldest medical systems, Ayurveda, dates back more than 5,000 years to India and is predicated on the balancing of the three doshas: Pitta, Kapha, and Vata. Ayurvedic formulations often consist of multiple herbs, combined in specific proportions to achieve a synergistic effect that balances these doshas and restores health (Mukherjee et al., 2015).

In traditional Chinese medicine (TCM), herbal formulations are also a cornerstone of treatment, with over 13,000 medicinal herbs and more than 100,000 recipes documented in ancient texts. TCM formulations are based on the principles of Yin and Yang, and the Five Elements, which govern the balance and harmony within the body. Herbal combinations are meticulously crafted to address the underlying causes of disease by harmonizing the body's internal environment. Commonly used TCM herbs include ginseng (*Panax ginseng*), astragalus (*Astragalus membranaceus*), and licorice (*Glycyrrhiza glabra*), each selected for its unique properties and its role within the larger formulation (Wang et al., 2019).

While these traditional approaches have provided effective treatments for centuries, they are often based on empirical knowledge and lack the rigorous scientific validation required by modern pharmaceutical standards. The challenge lies in translating this traditional knowledge into standardized formulations that meet contemporary expectations for safety, efficacy, and quality. This requires a deep understanding of the pharmacological properties of individual herbs, as well as the interactions between them within a formulation.

8.3 PHYTOCHEMICAL CONSTITUENTS IN HERBAL FORMULATIONS

Phytochemicals are naturally occurring compounds found in plants which are essential to their medicinal properties. These bioactive compounds are the cornerstone of many herbal formulations, contributing to their therapeutic effects. Understanding the types of phytochemicals present in herbs and their biological activities is essential for the development and standardization of herbal formulations. This section delves into the common phytochemical groups, their specific examples, biological activities, and sources, as outlined in Table 8.1 and Figure 8.1.

8.3.1 Alkaloids

Alkaloids are a broad class of nitrogen-containing substances with strong biological effects. Examples include morphine, quinine, and berberine. Morphine, derived from the poppy plant (*Papaver somniferum*), is widely recognized for its analgesic properties, making it one of the most effective pain relievers available. Quinine, extracted from *Cinchona* species, has been used for centuries as an antimalarial agent. Finally, berberine, found in *Berberis* species, exhibits antimicrobial properties, making it useful in treating infections (Heinrich et al., 2009).

TABLE 8.1
Phytochemical Constituents Commonly Found in Herbal Formulations

Phytochemical Group	Examples	Biological Activity	Common Sources	References
Alkaloids	Morphine, Quinine, Berberine	Analgesic, antimalarial, antimicrobial	Poppy (*Papaver somniferum*), Cinchona, Berberis	Heinrich et al., 2009; Gibson et al., 2019
Flavonoids	Quercetin, Kaempferol, Rutin	Antioxidant, anti-inflammatory, cardioprotective	Citrus fruits, onions, tea	Mukherjee et al., 2012; Li et al., 2008
Terpenoids	Menthol, Limonene, Carotenoids	Anti-inflammatory, antimicrobial, antioxidant	Mint (*Mentha*), Citrus peels, Carrots	Verpoorte et al., 2005; Sarker & Nahar, 2012
Glycosides	Digoxin, Salicin, Saponins	Cardiotonic, anti-inflammatory, expectorant	Foxglove (*Digitalis*), Willow bark, Soapwort	Heinrich et al., 2009; Bent, 2008
Tannins	Ellagic acid, Tannic acid	Astringent, antimicrobial, antioxidant	Oak (*Quercus*), Pomegranate, Tea	Sarker & Nahar, 2012; Patra et al., 2018
Phenolic acids	Caffeic acid, Gallic acid	Antioxidant, anti-inflammatory, antimicrobial	Berries, Coffee, Spices	Li et al., 2008; Mahapatra et al., 2021
Saponins	Dioscin, Ginsenosides	Antioxidant, immunomodulatory, anti-inflammatory	Ginseng, Soybeans, Yucca	Ekor, 2013; Heinrich et al., 2009

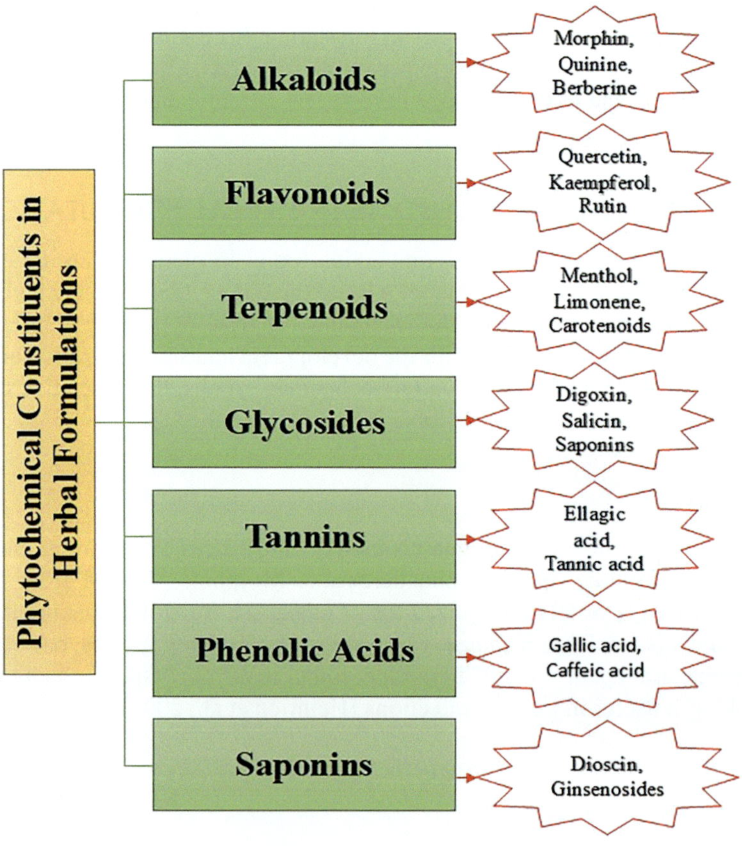

FIGURE 8.1 Commonly occurring phytochemical components in herbal formulations.

These alkaloids are often the active ingredients in traditional herbal remedies and are critical in modern pharmacology due to their efficacy in treating various conditions. However, their potent effects also necessitate careful standardization and dosage control to avoid toxicity (Gibson et al., 2019).

8.3.2 Flavonoids

Flavonoids, such as quercetin, kaempferol, and rutin, are widely dispersed polyphenolic chemicals found in the kingdom of plants. These compounds are renowned for their antioxidant properties, which aid in the body's defence against oxidative stress and the neutralization of free radicals. Quercetin, found in onions and apples, also exhibits anti-inflammatory and cardioprotective effects. Kaempferol and rutin, present in various fruits and vegetables, contribute to vascular health by strengthening capillaries and reducing inflammation (Mukherjee et al., 2012).

The inclusion of flavonoid-rich herbs in formulations is common due to their broad spectrum of health benefits. Their role in preventing chronic diseases, such as cardiovascular diseases and cancer, highlights the importance of maintaining high levels of these compounds in herbal products through proper extraction and formulation techniques (Li et al., 2008).

8.3.3 Terpenoids

Terpenoids, including menthol, limonene, and carotenoids, are another significant group of phytochemicals. Menthol, derived from mint species (*Mentha*), is renowned for providing a cooling feeling and is widely used in products for respiratory relief. Limonene, found in citrus peels, has anti-inflammatory and antimicrobial properties, while carotenoids, like beta-carotene, are potent antioxidants that contribute to skin health and immune function (Verpoorte et al., 2005).

Terpenoids are often incorporated into formulations for their medicinal qualities, such as antioxidant, antibacterial, and anti-inflammatory capabilities. Their natural aroma and flavor also enhance the sensory appeal of herbal products (Sarker & Nahar, 2012).

8.3.4 Glycosides

A sugar molecule linked to a non-carbohydrate component is called a glycoside. Notable examples include digoxin, salicin, and saponins. Digoxin, extracted from *Digitalis* species (foxglove), is an effective cardiotonic medication used to treat arrhythmias and heart failure. Salicin, found in willow bark, serves as a precursor to aspirin and has analgesic and anti-inflammatory properties. Saponins, present in soapwort and ginseng, exhibit expectorant and immune-boosting activities (Heinrich et al., 2009).

The presence of glycosides in herbal formulations contributes to their efficacy in treating cardiovascular, inflammatory, and respiratory conditions. Standardizing these compounds is crucial to ensure consistency and safety in herbal remedies (Bent, 2008).

8.3.5 Tannins

Tannins, like tannic and ellagic acids, are astringent compounds that precipitate proteins and other organic compounds. These phytochemicals are commonly found in oak (*Quercus*), pomegranate, and tea. Tannins possess antimicrobial and antioxidant properties, making them useful in wound healing, gastrointestinal disorders, and as natural preservatives (Sarker & Nahar, 2012).

Their astringent action is particularly beneficial in formulations intended for treating diarrhoea, bleeding, and other conditions requiring tissue tightening and protection. The standardization of tannin content is vital to avoid excessive astringency, which can lead to adverse effects (Patra et al., 2018).

8.3.6 Phenolic Acids

Gallic acid and caffeic acid are examples of phenolic acids, which are secondary metabolites with strong anti-inflammatory and antioxidant characteristics. These substances are abundant in berries, coffee, and spices. Free radical scavenging is the well-known property of gallic acid, whereas the anti-inflammatory and antibacterial properties of caffeic acid (Li et al., 2008).

Herbal formulations containing phenolic acids are often targeted at reducing oxidative stress and inflammation, which are underlying factors in many chronic diseases. Ensuring an adequate concentration of these compounds in the final product is essential for achieving the desired therapeutic outcomes (Mahapatra et al., 2021).

8.3.7 Saponins

Saponins, including dioscin and ginsenosides, are glycosides with soap-like properties. They are found in plants like ginseng and soybeans. Saponins have immunomodulatory and anti-inflammatory activities, making them valuable in formulations aimed at boosting immune function and reducing inflammation. The special qualities of saponins, namely their capacity to create emulsions and stable foams, are also leveraged in the formulation of topical and oral products. The standardization of saponin content is necessary to balance their therapeutic benefits with potential irritant effects (Ekor, 2013; Heinrich et al., 2009).

8.4 MODERN EXTRACTION TECHNIQUES IN HERBAL FORMULATION

The development of herbal formulations has greatly benefited from advancements in extraction techniques, which make it possible to separate bioactive substances from plant materials in a targeted manner. These modern extraction methods have been instrumental in enhancing the potency and consistency of herbal products, while also enabling the production of standardized extracts with well-defined chemical profiles. Some of the most commonly used extraction techniques in herbal formulation include solvent extraction, enzyme-assisted extraction (EAE), microwave-assisted extraction (MAE), ultrasonic extraction, and supercritical fluid extraction (SFE) (Figure 8.2).

8.4.1 Solvent Extraction

Solvent extraction is one of the most traditional and often-utilized techniques for extracting bioactive compounds from medicinal plants. Using solvents like water, ethanol, or methanol, the targeted chemicals are dissolved from the plant matrix using this approach. The content of the final extract and the effectiveness of the extraction process are both influenced by the choice of solvent. For instance, polar chemicals like flavonoids and glycosides are often extracted using water, whereas less polar compounds like terpenoids and alkaloids are best extracted using ethanol and methanol (Dhanani et al., 2017).

Solvent extraction is a valuable technique that may be applied to both large-scale industrial production and small-scale laboratory investigations due to its scalability and simplicity. However, this method also has limitations, including the potential for solvent residues in the final product, the degradation of thermolabile compounds during the extraction process, and the extraction of unwanted impurities. To overcome these challenges, researchers have developed optimized extraction protocols that involve the careful selection of solvents, extraction temperatures, and extraction times to maximize the yield and purity of the desired compounds (Li et al., 2020).

For instance, the extraction of curcumin from turmeric (*Curcuma longa*) typically involves the use of ethanol or acetone as solvents, followed by purification steps to isolate curcumin from other curcuminoids and impurities. The resulting standardized extract, containing a high concentration of curcumin, is widely used in herbal formulations for its anti-inflammatory and antioxidant properties (Aggarwal et al., 2013).

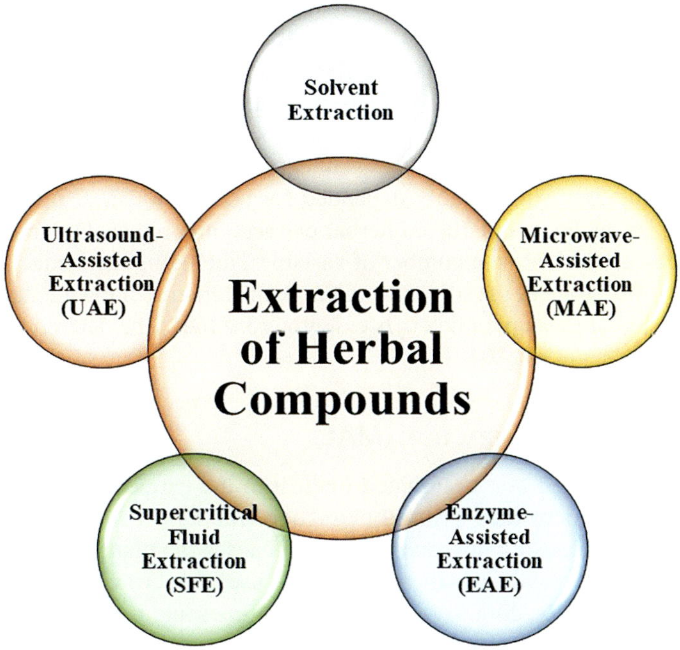

FIGURE 8.2 Contemporary methods of extraction in herbal formulation.

8.4.2 Supercritical Fluid Extraction (SFE)

Supercritical fluid extraction (SFE) is a modern technique that uses supercritical fluids, typically using carbon dioxide (CO_2) to remove bioactive substances from plant matter. In its supercritical state, CO_2 exhibits both liquid-like solvating power and gas-like diffusivity, allowing it to efficiently penetrate the plant matrix and dissolve target compounds. The use of CO_2 as a solvent is particularly advantageous because it is non-toxic, non-flammable, and easily removed from the final product by simple depressurization (Khaw et al., 2017).

SFE is especially effective for extracting thermolabile and non-polar substances, such as fatty acids, carotenoids, and essential oils, which may be degraded or lost during traditional solvent extraction. Additionally, by modifying the extraction process' pressure and temperature, SFE enables the selective extraction of particular chemicals. For instance, it has been demonstrated that employing SFE to extract the essential oils of lavender (*Lavandula angustifolia*) results in a greater concentration of volatile components, such as linalool and linalyl acetate, compared to traditional hydro distillation methods (Pourmortazavi & Hajimirsadeghi, 2007).

Despite its advantages, SFE is a relatively complex and expensive technique, needing specific tools and knowledge. Consequently, its main application in the pharmaceutical industry is the extraction of high-value chemicals, cosmetic, and food industries. However, ongoing research is focused on improving the cost-effectiveness and scalability of SFE, making it a promising technique for the production of standardized herbal extracts in the future (Lang & Wai, 2021).

8.4.3 Ultrasonic Extraction

High-frequency sound waves are used in ultrasonic extraction (sometimes referred to as ultrasonication or sonication) to improve the extraction of bioactive chemicals from plant materials. The extraction solvent experiences cavitation bubbles due to the mechanical vibrations produced by ultrasonic waves. These bubbles burst violently, breaking down the walls of the plant and allowing intracellular chemicals to be released into the solvent (Tiwari, 2015).

Compared to conventional procedures, ultrasonic extraction has a number of benefits, such as reduced solvent consumption, increased extraction efficiency, and shorter extraction timeframes. This method works especially well for removing flavonoids, polyphenols, and other phenolic chemicals from plant sources. For example, the extraction of catechins and other polyphenols from green tea (*Camellia sinensis*) using ultrasonic extraction has been shown to yield higher concentrations of these compounds compared to conventional solvent extraction methods (Wang et al., 2016).

Moreover, ultrasonic extraction readily expanded for use in industrial production, making it a practical choice for producing standardized herbal extracts on a big scale. However, the effectiveness of this method is dependent on a number of variables, including the frequency and intensity of the ultrasonic waves, the nature of the solvent, and the characteristics of the plant material. Therefore, the careful optimization of these parameters is essential to achieve the desired extraction efficiency and product quality (Zhang et al., 2018).

8.4.4 Microwave-Assisted Extraction (MAE)

Another contemporary method that has gained interest in the extraction of bioactive components from medicinal plants is microwave-assisted extraction (MAE). In MAE, the plant material and extraction solvent are heated using microwaves, which causes the plant cell walls to break quickly and release intracellular chemicals into the solvent. The use of microwaves allows for the selective heating of the plant material, leading to faster extraction times and higher yields compared to conventional heating methods (Eskilsson & Björklund, 2000).

It works especially well for extracting substances that are thermally stable, such as alkaloids, terpenoids, and flavonoids, which require higher temperatures for efficient extraction. For instance, phenolic chemicals are extracted from medicinal herbs like sage (*Salvia officinalis*) and rosemary (*Rosmarinus officinalis*), using MAE has been shown to yield higher concentrations of these compounds compared to traditional solvent extraction methods (Xiao et al., 2015).

One of MAE's main benefits is its capacity to reduce the use of solvents, making it an environmentally friendly extraction technique. Additionally, MAE can be easily integrated into existing production lines, making it a cost-effective option for the extensive manufacturing of herbal extracts. However, like ultrasonic extraction, the effectiveness of MAE depends on various factors, such as the microwave power, extraction time, and solvent type. Therefore, careful optimization of these parameters is necessary to achieve the desired extraction efficiency and product quality (Routray & Orsat, 2012).

8.4.5 Enzyme-Assisted Extraction (EAE)

By using particular enzymes to break down plant materials' cell walls, the process known as 'enzyme-assisted extraction' (EAE) makes it easier for bioactive chemicals to be released into the extraction solvent. In EAE, enzymes like cellulases, pectinases, and proteases are frequently employed to target certain plant cell wall constituents including cellulose, pectin, and proteins. The use of enzymes in the extraction process not only improves the yield of target compounds but also reduces the extraction time and the need for harsh solvents (Rai et al., 2016).

EAE is particularly effective for extracting complex polysaccharides, glycosides, and other macromolecules that are difficult to extract using traditional methods. For example, the extraction of saponins from ginseng (*Panax ginseng*) using EAE has been shown to yield higher concentrations of these compounds in contrast to traditional solvent extraction methods (Zheng et al., 2019).

Enhancing the extraction of bioactive chemicals from plant materials that would otherwise be challenging to handle is one of the main benefits of EAE. Additionally, EAE is a gentle extraction method that preserves the integrity of thermolabile compounds, making it appropriate for removing delicate bioactive substances. However, the effectiveness of EAE depends on the selection of appropriate enzymes, as well as the optimization of parameters for the extraction process, including temperature, pH, time, and enzyme concentration (Ibrahim et al., 2018).

TABLE 8.2
Common Techniques for the Extraction of Herbal Compounds

Extraction Technique	Description	Advantages	Disadvantages	References
Maceration	Plant material is soaked for a while at room temperature in a solution.	Simple, cost-effective, no specialized equipment needed.	Time-consuming, less efficient extraction.	Verpoorte et al., 2005
Percolation	Continuous flow of solvent through a column of powdered plant material.	Faster than maceration, higher yield.	Requires more solvent, may extract unwanted components.	Heinrich et al., 2009
Soxhlet Extraction	Continuous extraction process using a reflux condenser, where solvent is repeatedly vaporized and condensed.	Efficient extraction with minimal solvent use.	Requires special equipment, not suitable for thermolabile compounds.	Sarker & Nahar, 2012
Supercritical Fluid Extraction (SFE)	Uses supercritical fluids (usually CO_2) to extract compounds at moderate temperatures.	High selectivity, environmentally friendly, solvent-free.	Expensive equipment, not suitable for all types of compounds.	Li et al., 2008
Ultrasound-Assisted Extraction (UAE)	Uses ultrasonic waves to enhance solvent penetration and improve extraction efficiency.	Faster extraction, energy-efficient, uses less solvent.	Potential degradation of sensitive compounds, requires specialized equipment.	Mukherjee et al., 2012

8.4.6 Comparative Analysis of Extraction Techniques

The selection of an extraction process is contingent upon a number of criteria, such as the target compounds, the nature of the plant material, and the intended use of the extract. Each of the extraction techniques previously covered has pros and cons of its own. For instance, solvent extraction is a straightforward and affordable method for removing a variety of bioactive substances from a mixture; but it could call for the use of organic solvents, which may leave residues in the finished product. On the other hand, SFE is a very effective and eco-friendly method of extracting non-polar chemicals, but it needs specific tools and knowledge (Table 8.2).

8.5 ADVANCED FORMULATION STRATEGIES

The development of herbal formulations has evolved significantly with the integration of advanced technologies that increase the stability, bioavailability, and therapeutic efficacy of herbal compounds. These advanced formulation strategies include the use of nanotechnology, encapsulation techniques, standardized extracts, and synergistic combinations of herbs. In this section, we will explore these strategies in detail, highlighting their role in conquering the difficulties posed by conventional herbal remedies and their capacity to completely transform the herbal medicine industry.

8.5.1 Nanotechnology in Herbal Formulations

Nanotechnology has emerged as a powerful tool in the formulation of herbal products, offering solutions to some of the key challenges in herbal medicine, such as poor bioavailability, the instability of active compounds, and variable therapeutic efficacy. By decreasing the size of the particles of herbal compounds to the nanoscale, nanotechnology enhances their solubility, absorption, and bioavailability, leading to improved therapeutic outcomes.

8.5.1.1 Nanoemulsions

Surfactants stabilize the colloidal dispersions of two immiscible liquids, usually water and oil, to create nanoemulsions with droplet sizes in the nanometre range (20–200 nm). In herbal formulations, nan emulsions are used to enhance the herbal components that are hydrophobic and their bioavailability the herbal components that are hydrophobic and their bioavailability, such as curcumin, quercetin, and essential oils. The small droplet size in nanoemulsions leads to a larger surface area, which facilitates the absorption of the active compounds in the gastrointestinal tract (Shakeel et al., 2010).

For example, curcumin, a hydrophobic polyphenol derived from turmeric (*Curcuma longa*), has poor bioavailability due to its low solubility in water and rapid metabolism in the body. However, when formulated as a nanoemulsion, curcumin shows significantly enhanced bioavailability, leading to better therapeutic outcomes in conditions such as inflammation, cancer, and neurodegenerative diseases (Gutte et al., 2018).

8.5.1.2 Liposomes

Phospholipid bilayers form the spherical vesicles known as liposomes, which are capable of encasing both hydrophilic and hydrophobic substances. In herbal formulations, liposomes are used to improve the delivery of herbal compounds by protecting them from degradation, enhancing their absorption, and targeting them to specific tissues or cells. Liposomal encapsulation also helps in controlling the release of active compounds, thereby prolonging their therapeutic effects (Akbarzadeh et al., 2013).

For example, it has been demonstrated that the liposomal encapsulation of quercetin, a flavonoid present in numerous fruits and vegetables, enhances its antioxidant and anti-inflammatory properties as well as its stability and bioavailability. This indicates that liposomal quercetin is a viable option for the management of a number of inflammatory and oxidative stress-related illnesses (Zhang et al., 2019).

8.5.1.3 Solid Lipid Nanoparticles (SLNs) and Nanostructured Lipid Carriers (NLCs)

Lipid-based nanocarriers known as solid lipid nanoparticles (SLNs) and nanostructured lipid carriers (NLCs) are employed to improve the transport of herbal components that are poorly soluble. Whereas NLCs contain a mixed lipid core made up of both liquid and solid lipids, SLNs are made up of a solid lipid core stabilized by surfactants. These nanocarriers provide a number of benefits, such as increased stability, regulated release, and focused delivery of active ingredients (Müller et al., 2000).

SLNs and NLCs have been successfully used for the purpose of improving the therapeutic efficacy of various herbal compounds. For example, SLNs loaded with resveratrol, a polyphenol found in grapes and berries, have shown improved antioxidant activity and photostability compared to free resveratrol. Similarly, NLCs have been used to deliver berberine, an alkaloid from the plant *Berberis vulgaris*, with enhanced bioavailability and anti-inflammatory effects (Yadav et al., 2013).

8.5.2 Encapsulation Techniques

Encapsulation techniques play a vital part in the growth of herbal formulations by protecting the active compounds from degradation, masking unpleasant tastes or odors, and controlling the release of the active ingredients. Various encapsulation methods, such as microencapsulation, coacervation, and spray drying, are used in herbal formulations to enhance the stability and bioavailability of herbal compounds.

8.5.2.1 Microencapsulation

Microencapsulation involves the encapsulation of active compounds within a protective coating, typically made of polymers, lipids, or proteins, to form micro-sized particles. This technique is widely used in herbal formulations to improve the stability, bioavailability, and controlled release of active compounds. Microencapsulation also helps in masking the bitter taste or unpleasant odor of certain herbal extracts, making them more palatable (Jyothi et al., 2010). For example, microencapsulation of bitter gourd (*Momordica charantia*) extract using alginate and chitosan as encapsulating agents has been shown to improve the stability and controlled release of its bioactive compounds, such as charantin and polypeptide-p, which are known for their antidiabetic properties (Kumar et al., 2014).

8.5.2.2 Coacervation

Coacervation is an encapsulation technique that involves the phase separation of a polymer solution to form a coacervate, which can then encapsulate the active compounds. Coacervation is particularly useful for the encapsulation of delicate plant substances, as it provides a protective barrier against environmental factors, such as light, heat, and oxygen. This technique is commonly used in the food and pharmaceutical industries for the encapsulation of flavors, vitamins, and enzymes (Ribeiro et al., 1999).

In herbal formulations, coacervation has been used to encapsulate essential oils, such as lavender and peppermint oil, to enhance their stability and control their release. The encapsulated essential oils can be incorporated into various dosage forms, such as capsules, tablets, or topical formulations, for improved therapeutic efficacy (Varona et al., 2011).

8.5.2.3 Spray Drying

One popular encapsulation method is spray drying, which entails atomizing a liquid solution or suspension containing the active compound into a hot drying chamber, where the solvent evaporates, leaving behind dry powder particles. This is an economical and expandable technique that can be used in industrial settings to encapsulate plant extracts (Patel et al., 2015).

For instance, spray drying has been used to encapsulate turmeric extract, rich in curcumin, to improve its solubility, stability, and bioavailability. The spray-dried turmeric powder can be used in various dosage forms, such as capsules, tablets, or functional foods, for enhanced therapeutic effects (Bagchi et al., 2012).

8.5.3 Standardized Extracts

Standardization is a critical aspect of herbal formulation development, as it ensures the consistency, safety, and efficacy of the final product. Standardized extracts are herbal extracts that have been processed to contain a specific concentration of one or more bioactive compounds, which are used as markers for quality control. The use of standardized extracts in herbal formulations helps to overcome the variability in the chemical composition of medicinal plants, which can result from differences in cultivation practices, harvesting times, and environmental conditions (Bauer, 1998).

8.5.3.1 Methods of Standardization

Several methods are used to standardize herbal extracts, including the use of reference compounds, chromatographic techniques, and spectrophotometric assays. These methods allow in order to precisely quantify bioactive substances in the extract, ensuring that each batch of the product contains the same concentration of these compounds. For the purpose of standardizing herbal extracts, one of the most popular methods is high-performance liquid chromatography (HPLC). HPLC allows for the separation, identification, and quantification of individual bioactive compounds in the extract, making it a powerful tool for quality control (Agarwal et al., 2012).

For example, HPLC is used to standardize ginkgo biloba extract to contain a specific concentration of flavonoids (24%) and terpenoids (6%), which are the primary active compounds responsible for its therapeutic effects. The use of standardized ginkgo biloba extract ensures consistent product quality and effectiveness in treating cognitive illnesses like Alzheimer's (DeFeudis & Drieu, 2000).

8.5.3.2 Challenges in Standardization

Despite the importance of standardization in herbal formulation development, several challenges remain. One of the primary challenges is the complexity of herbal extracts, which often contain hundreds of bioactive compounds that can vary in concentration depending on various factors. Standardizing these complex mixtures requires the identification and quantification of multiple marker compounds, which can be time-consuming and expensive.

The absence of established procedures for the extraction and processing of herbal materials presents another difficulty and may result in product diversity. International guidelines, including those issued by the European Medicines Agency (EMA) and the World Health Organization (WHO), have been devised to address this issue and standardize the development of herbal medicines (WHO, 2011).

8.5.4 Synergistic Formulations

Herbal formulations often consist of multiple herbs combined in specific proportions to achieve a synergistic effect, where the combined medicinal impact of the herbs is more than the sum of their separate effects. The use of synergistic compounds is fundamental to traditional medical systems like TCM and Ayurveda, where herbs are carefully selected and combined based on their complementary actions and interactions.

8.5.4.1 Mechanisms of Synergy

The synergistic effects of herbal formulations can result from various mechanisms, including the enhancement of bioavailability, the modulation of pharmacokinetics, and the targeting of multiple pathways involved in disease. For example, certain herbs may enhance the absorption of other herbs by inhibiting their metabolism or increasing their solubility. Additionally, herbs with complementary actions, such as anti-inflammatory and antioxidant effects, can work together to provide a more comprehensive therapeutic effect (Wagner & Ulrich-Merzenich, 2009).

For example, the mixture of black pepper (*Piper nigrum*) and turmeric (*Curcuma longa*) in a synergistic formulation has been shown to enhance the bioavailability of curcumin, the active compound in turmeric, by inhibiting its metabolism in the liver. This synergistic effect results in a more potent anti-inflammatory and antioxidant activity, making the combination more effective than turmeric alone (Shoba et al., 1998).

8.5.4.2 Examples of Synergistic Formulations

Several herbal formulations that leverage synergistic effects have been developed and are widely used in traditional and modern medicine. One example is the Ayurvedic formulation Triphala, which is made up of three fruits: Bibhitaki (*Terminalia bellirica*), Haritaki (*Terminalia chebula*), and Amla (*Emblica officinalis*). Triphala's beneficial effects on digestion, immunity, and antioxidants are a result of the harmonious interactions between the several herbs that make up the blend (Naik et al., 2012).

Another example is the Chinese herbal remedy Buxue Tang Danggui, which consists of two herbs: Danggui (*Angelica sinensis*) and Huangqi (*Astragalus membranaceus*). This formulation is used to treat anaemia and improve blood circulation, with the two herbs working synergistically to enhance haematopoiesis and increase blood flow (Zhang et al., 2008).

8.5.5 Personalized Herbal Formulations

The creation of customized herbal formulas based on a person's genetic composition, state of health, and way of life is becoming increasingly popular as genomics and personalized medicine evolve. With a lower chance of side effects and a higher possibility for therapeutic results, customized herbal formulations may offer more focused and efficient treatments.

8.5.5.1 Role of Genomics in Personalized Herbal Medicine

Because it sheds light on how a person's genetic composition affects how they respond to herbal remedies, genomics is essential to the creation of customized herbal formulations. For example, genetic variations in drug-metabolizing enzymes, such as cytochrome P450, can affect the metabolism and bioavailability of herbal compounds, leading to differences in therapeutic efficacy and safety (Nebert & Russell, 2002).

By analysing an individual's genetic profile, it is possible to identify the most suitable herbal treatments for their specific condition, as well as the appropriate dosage and formulation. This approach allows for the customization of herbal formulations according to each person's distinct genetic and metabolic composition, leading to more personalized and effective treatments.

8.5.5.2 Challenges in Personalized Herbal Formulations

Despite the potential benefits of personalized herbal formulations, several challenges remain. One of the primary challenges is the complexity of herbal medicine, which involves multiple bioactive compounds with diverse pharmacological actions. Identifying the most suitable combination of herbs and dosages for an individual requires a deep understanding of the interactions between herbs, as well as the individual's genetic and metabolic profile. Another challenge is the lack of standardized protocols for the personalization of herbal formulations, which can lead to variability in the final product. Additionally, the high cost of genomic testing and the complexity of interpreting genetic data may limit the widespread adoption of personalized herbal medicine.

8.6 QUALITY CONTROL AND STANDARDIZATION IN HERBAL FORMULATION DEVELOPMENT

To guarantee the security, effectiveness, and uniformity of herbal products, quality control and standardization are crucial elements of the formulation process. The different facets of quality control and standardization will be discussed in this section. These include the use of reference standards, the identification and authentication of herbal materials, and the application of good manufacturing principles (GMP).

8.6.1 Recognition and Verification of Herbal Materials

To guarantee the efficacy and security of herbal formulations, it is imperative to identify and authenticate botanical components. Misidentification or adulteration of herbal materials can lead to variations in the chemical composition of the final product, resulting in reduced efficacy or even adverse effects. Therefore, accurate identification and authentication of herbal materials are essential to maintaining the integrity of herbal formulations.

8.6.1.1 Morphological Identification

Morphological identification involves the examination of the physical characteristics of herbal materials, such as the shape, size, color, and texture of the plant parts. This method is commonly used in the initial stages of herbal material identification, particularly for whole plants, leaves, roots, and seeds. However, morphological identification has its limitations, as it may not be reliable for closely related species or for processed herbal materials, such as powders or extracts (Li et al., 2010).

8.6.1.2 Microscopic Identification

Microscopic identification involves the examination of the internal structure of herbal materials using a microscope. This method is particularly useful for identifying the cellular and anatomical features of plant parts, such as trichomes, stomata, and vascular bundles. Microscopic identification is often used in conjunction with morphological identification to provide a more accurate and reliable means of identifying herbal materials (Liu et al., 2012).

8.6.1.3 Chemical Identification

Chemical identification involves employing analytical methods like chromatography and spectroscopy to identify the chemical constituents of herbal materials. This method is particularly useful for the identification of specific bioactive compounds that serve as markers for quality control. Chemical identification provides a high level of accuracy and reliability, making it an essential tool in the standardization of herbal formulations (Wang et al., 2009).

8.6.1.4 DNA Barcoding

A contemporary method called DNA barcoding makes use of brief, standardized DNA sequences to identify and authenticate herbal materials at the species level. DNA barcoding provides a rapid and accurate means of identification, particularly for processed herbal materials where traditional morphological and microscopic methods may be unreliable. This technique is increasingly being used in the quality control of herbal medicines to prevent misidentification and adulteration (Chen et al., 2013).

8.6.2 Use of Reference Standards

Reference standards are well-characterized chemical compounds that are used as benchmarks for the identification, quantification, and quality control of herbal formulations. The use of reference standards is essential for ensuring the consistency and accuracy of herbal products, as they provide a basis for comparing the chemical composition of different batches of the product.

8.6.2.1 Primary and Secondary Reference Standards

Primary reference standards are highly purified chemical compounds that have been thoroughly characterized and are used as the definitive benchmark for quality control. Secondary reference standards, on the other hand, are less pure and are calibrated against primary reference standards. Both types of standards are used in the standardization and quality control of herbal formulations, depending on the specific requirements of the product (USP, 2021).

8.6.2.2 Development of Reference Standards

The isolation, purification, and characterization of certain bioactive chemicals from herbal materials is necessary for the formation of reference standards. To guarantee the correctness and purity of the reference standard, this procedure calls for the application of sophisticated analytical tools such as nuclear magnetic resonance (NMR) spectroscopy, mass spectrometry, and HPLC. Reference standards are used to determine the identity, potency, and quality of herbal formulations once they are created (Zhang et al., 2010).

8.6.3 Good Manufacturing Practices (GMP)

A collection of rules known as good manufacturing practices, or GMPs, guarantee the uniformity, quality, and safety of herbal products at every stage of production. GMP addresses every facet of manufacturing, from the sourcing and handling of raw materials to the final packaging and distribution of the product. The implementation of GMP is essential for maintaining the integrity of herbal formulations and ensuring that they meet regulatory standards (WHO, 2007).

8.6.3.1 Sourcing and Handling of Raw Materials

The quality of herbal formulations begins with the sourcing and handling of raw materials. GMP guidelines emphasize the importance of sourcing herbal materials from reputable suppliers and ensuring that they are properly identified, authenticated, and stored. Proper handling of raw materials, including their storage under appropriate conditions to prevent contamination and degradation, is critical to maintaining the quality of the final product (Brinckmann, 2011a).

8.6.3.2 Manufacturing Processes

GMP guidelines also cover the manufacturing processes used to produce herbal formulations, including extraction, processing, and packaging. These processes must be carried out under controlled conditions to ensure the consistency and quality of the product. GMP requires that all equipment and facilities used in the manufacturing process are regularly maintained, cleaned, and validated to prevent contamination and ensure product safety (EMA, 2007).

8.6.3.3 Quality Control Testing

This is an integral part of GMP and involves the testing of raw materials, intermediates, and final goods to make sure they comply with established specifications. Testing of chemical, physical, and microbiological factors is included in this, as well as the use of reference standards for the quantification of bioactive compounds. Quality control testing ensures that each batch of the product meets the required standards for safety, efficacy, and consistency (ICH, 2005).

8.6.3.4 Documentation and Record-Keeping

GMP calls for meticulous record-keeping and documentation at every stage of the production process. This includes the documentation of all procedures, protocols, and test results, as well as the maintenance of records for each batch of the product. Proper documentation is essential for ensuring traceability, accountability, and compliance with regulatory requirements (FDA, 2016).

8.7 CHALLENGES IN HERBAL FORMULATION DEVELOPMENT

To fully achieve the potential of herbal medicine, a number of obstacles still need to be overcome, notwithstanding the advancements in the production of herbal formulations. Among these difficulties are the complicated herbal extracts and the inconsistent chemical composition of herbal components, the standardization of formulations, and the regulatory hurdles associated with the approval of herbal products.

8.7.1 Variability in Chemical Composition

The variable chemical makeup of plant resources poses a significant problem in the development of herbal formulations. Environmental variables like soil type, climate, and altitude, as well as variations in the production, harvesting, processing, and storage of medicinal plants, can all contribute to this variability. It might be challenging to guarantee uniformity in the finished product since variations in the chemical makeup of herbal formulations can affect their performance and safety (Verpoorte et al., 2005).

8.7.1.1 Factors Affecting Chemical Composition

Numerous elements may impact the chemical makeup of herbal materials, including the species and variety of the plant, the part of the plant used, and the stage of growth or maturity at which the plant is harvested. Additionally, post-harvest processing, such as drying, grinding, and extraction, can affect the concentration of bioactive compounds in the final product (Wu et al., 2007a, 2007b).

8.7.1.2 Strategies to Address Variability

To address the variability in chemical composition, several strategies can be employed, including the use of standardized extracts, the implementation of good agricultural and collection practices (GACP), and the development of robust analytical methods for quality control. Standardized extracts, as discussed earlier, provide a consistent concentration of bioactive compounds, ensuring the quality and efficacy of herbal formulations. GACP guidelines, developed by the WHO, provide recommendations for the cultivation, harvesting, and processing of medicinal plants to minimize variability in chemical composition (WHO, 2003, 2003b).

8.7.2 Complexity of Herbal Extracts

The complexity of herbal extracts, which often contain hundreds of bioactive compounds, poses a significant challenge in the development of herbal formulations. The interactions between these compounds, as well as their individual and collective pharmacological effects, are not fully understood, making it difficult to predict the therapeutic outcomes of herbal formulations. Additionally, the presence of multiple compounds can complicate the standardization and quality control of herbal products (Sarker & Nahar, 2012).

8.7.2.1 Analytical Challenges

The complexity of herbal extracts presents several analytical challenges, including the process of identifying, measuring, and characterizing bioactive substances. Extensive analytical methods, including mass spectrometry, NMR spectroscopy, and HPLC, are necessary to precisely profile the chemical makeup of plant extracts. However, the development of robust and reproducible analytical methods for complex mixtures remains a challenge (Heinrich et al., 2009).

8.7.2.2 Formulation Challenges

The complexity of herbal extracts also poses challenges in the formulation of herbal products, particularly in ensuring the stability, bioavailability, and consistency of the final product. The interactions between different compounds in the extract can affect the solubility, absorption, and metabolism of the active ingredients, leading to variability in therapeutic outcomes. Addressing these challenges requires a deep understanding of the pharmacokinetics and pharmacodynamics of herbal compounds, as well as the development of advanced formulation strategies, such as nanotechnology and encapsulation (Li et al., 2008).

8.7.3 Standardization of Herbal Formulations

Standardization is a critical aspect of herbal formulation development, but it remains a significant challenge due to the complexity and variability of herbal materials. Ensuring the consistency, quality, and efficacy of herbal products requires the identification and quantification of specific bioactive compounds, as well as the establishment of standardized protocols for extraction, processing, and quality control (Ekor, 2014).

8.7.3.1 Challenges in Standardization

One of the primary challenges in standardization is the identification of suitable marker compounds that can be used to assess the quality and efficacy of herbal formulations. Marker compounds should be specific to the plant species, pharmacologically active, and present in sufficient quantities to allow for accurate quantification. However, the selection of marker compounds can be complicated by the presence of multiple bioactive compounds in the extract, as well as by the variability in their concentration due to environmental factors and processing methods (Bilia et al., 2014).

Another challenge is the lack of standardized protocols for the extraction and processing of herbal materials, which can lead to variability in the final product. The development of standardized protocols requires a thorough understanding of the pharmacognosy and phytochemistry of the medicinal plant, as well as the optimization of extraction and processing methods to maximize the yield and consistency of bioactive compounds (Mukherjee et al., 2012).

8.7.3.2 Advances in Standardization

Despite these challenges, significant advances have been made in the standardization of herbal formulations. The use of modern analytical techniques, such as HPLC, mass spectrometry, and DNA barcoding, has improved the accuracy and reliability of herbal product identification and quantification. Additionally, the development of standardized extracts and reference standards has provided a basis for ensuring the consistency and quality of herbal formulations. The adoption of international guidelines, such as those provided by the WHO and the EMA, has also contributed to the standardization of herbal products (EMA, 2016).

8.7.4 Regulatory Hurdles

The regulatory approval of herbal products is a complex and challenging process, as herbal medicines are often subject to different regulatory requirements than conventional pharmaceuticals. The lack of harmonized regulations across different countries and regions can create barriers to the global marketing and distribution of herbal products. Additionally, the requirement for scientific evidence of safety and efficacy, as well as the need for quality control and standardization, poses significant challenges for the regulatory approval of herbal formulations (Ekor, 2013).

8.7.4.1 Regulatory Requirements

Regulatory requirements for herbal products vary widely depending on the country or region. Certain nations regulate herbal products as dietary supplements or traditional medicines, with less stringent requirements for safety and efficacy compared to conventional pharmaceuticals. In other countries, herbal products are subject to the same regulatory standards as conventional drugs, requiring extensive clinical trials and scientific evidence of safety and efficacy (Bent, 2008).

For instance, the Dietary Supplement Health and Education Act (DSHEA) of 1994 governs herbal items in the United States as dietary supplements. Manufacturers must guarantee the safety of their goods under the DSHEA, but they are exempt from having to submit proof of scientific efficacy before to marketing. On the other hand, herbal products are subject to regulation in the European Union under the Traditional Herbal Medicinal Products Directive (THMPD), which calls for quality control and standardization in addition to scientific proof of safety and efficacy (EMA, 2007).

8.7.4.2 Challenges in Meeting Regulatory Requirements

Meeting regulatory requirements for herbal products can be challenging due to the complexity of herbal extracts, the variability in chemical composition, and the lack of standardized protocols for quality control. Additionally, the requirement for scientific evidence of safety and efficacy can be difficult to fulfil, particularly for traditional herbal formulations that have been used for centuries but have not undergone modern clinical trials. The high cost and time required for conducting clinical trials and obtaining regulatory approval can also be a barrier for small and medium-sized herbal companies (Barnes et al., 2015).

8.7.4.3 Strategies to Overcome Regulatory Hurdles

To overcome regulatory hurdles, several strategies can be employed, including the use of standardized extracts, the implementation of GMP, and the conduct of well-designed clinical trials. Standardized extracts provide a consistent and reliable product that meets regulatory requirements for quality and efficacy. GMP ensures that herbal products are manufactured to the highest standards of quality and safety. Conducting clinical trials, particularly randomized controlled trials (RCTs), provides the scientific evidence needed to support the safety and efficacy of herbal formulations (Izzo & Ernst, 2009).

Additionally, engaging with regulatory authorities early can assist in identifying and addressing any regulatory problems during the product development process. Collaboration with academic institutions and research organizations can also provide the expertise and resources needed to conduct high-quality clinical trials and meet regulatory requirements (Brinckmann, 2011b).

8.8 FUTURE DIRECTIONS IN HERBAL FORMULATION DEVELOPMENT

The field of herbal formulation development is rapidly evolving, with new technologies and approaches being explored to enhance the efficacy, safety, and consistency of herbal products. This section will cover some of the major future directions in the development of herbal formulations, such as the application of machine learning and artificial intelligence (AI), the incorporation of omics technologies, and the investigation of novel delivery systems.

8.8.1 Artificial Intelligence and Machine Learning

More and more, machine learning (ML) and artificial intelligence (AI) are used in the development of herbal formulations to analyse complex datasets, predict therapeutic outcomes, and optimize formulation strategies. These technologies have the potential to revolutionize the field of herbal medicine by providing insights into the interactions between herbal compounds, predicting the effects of different formulations, and identifying novel therapeutic targets (Topol, 2019a, 2019b).

8.8.1.1 Applications in Herbal Formulation Development

AI and ML can be applied at various stages of herbal formulation development, from the identification of bioactive compounds to the optimization of delivery systems. For example, ML algorithms can be used to analyse large datasets of phytochemical profiles, identifying patterns and correlations that can be used to predict the therapeutic effects of different herbal formulations. AI can also be used to design and optimize innovative methods of administration, like liposomes and nanoparticles, to enhance the bioavailability and targeted delivery of herbal compounds (Wang et al., 2021).

8.8.1.2 Challenges and Opportunities

The use of AI and ML in herbal formulation development presents both challenges and opportunities. One of the primary challenges is the availability of high-quality, standardized data on herbal compounds and their pharmacological effects. Additionally, the complexity of herbal medicine, with its multiple bioactive compounds and diverse mechanisms of action, can make it difficult to develop accurate predictive models. However, the integration of AI and ML with other advanced technologies, such as omics and high-throughput screening, offers significant opportunities for the discovery of new herbal formulations and the optimization of existing products (Sarker & Fricker, 2021a, 2021b).

8.8.2 Integration of Omics Technologies

Genomics, proteomics, metabolomics, and transcriptomics are examples of omics technologies that offer deep insights into the molecular mechanisms behind the actions of herbal compositions. The integration of omics technologies in herbal formulation development allows for the identification of novel bioactive compounds, the elucidation of their mechanisms of action, and the development of more targeted and effective formulations (Wishart, 2016).

8.8.2.1 Applications in Herbal Medicine

Omics technologies can be applied at various stages of herbal formulation development, from the screening of medicinal plants for bioactive compounds to the evaluation of their effects on gene expression, protein function, and metabolic pathways. Transcriptomics can be used to examine the effects of herbal formulations on gene expression, offering insights into their mechanisms of action and possible therapeutic targets. For instance, metabolomics can be used to profile the chemical composition of herbal extracts, identifying key metabolites that contribute to their therapeutic effects (Gibson et al., 2019).

8.8.2.2 Challenges and Opportunities

The integration of omics technologies in herbal formulation development presents several challenges, including the complexity of data analysis, the need for standardized protocols, and the high cost of omics studies. However, the use of omics technologies offers significant opportunities for the discovery of new herbal compounds, the identification of biomarkers for quality control, and the development of personalized herbal formulations. The integration of omics with other advanced technologies, such as AI and ML, further enhances the potential for innovation in herbal medicine (Cimino et al., 2018a, 2018b).

8.8.3 Novel Delivery Systems

The development of novel delivery systems for herbal formulations is a key area of research aimed at improving the bioavailability, stability, and targeted delivery of herbal compounds. Traditional herbal formulations, such as powders, teas, and tinctures, often have limitations in terms of bioavailability and stability. Novel delivery technologies, such as nanoparticles, liposomes, and hydrogels, offer innovative answers to these issues, increasing the medicinal efficacy of herbal mixtures (Patra et al., 2018).

8.8.3.1 Nanotechnology in Herbal Medicine

One of the most promising methods for enhancing the transport of herbal components is nanotechnology. Solid lipid nanoparticles (SLNs), polymeric nanoparticles, and liposomes are examples of nanoparticles which can encapsulate herbal compounds, protecting them from degradation and enhancing their absorption and bioavailability. Additionally, nanoparticles can be engineered to target specific tissues or cells, allowing for more precise and effective delivery of herbal compounds (Gul et al., 2019).

8.8.3.2 Other Novel Delivery Systems

In addition to nanotechnology, other novel delivery systems, such as hydrogels, micelles, and transdermal patches, are being explored for the delivery of herbal formulations. Hydrogels, for example, can provide a controlled release of herbal compounds over an extended period, improving the duration of therapeutic effects. Transdermal patches offer a non-invasive and convenient method of delivering herbal compounds directly through the skin, bypassing the digestive system and enhancing bioavailability (Sharma et al., 2020).

8.8.3.3 Challenges and Opportunities

The development of innovative delivery mechanisms for herbal formulations presents several challenges, including the complexity of formulation, the need for biocompatibility and safety, and the regulatory requirements for approval. However, the use of advanced delivery systems offers significant opportunities for enhancing the therapeutic efficacy of herbal medicine, improving patient compliance, and expanding the range of conditions that can be treated with herbal formulations. The combination of novel delivery systems with personalized medicine approaches further enhances the potential for innovation in herbal formulation development (Mahapatra et al., 2021) (Table 8.3).

TABLE 8.3
Challenges and Solutions in Herbal Formulation Development

Challenge	Description	Proposed Solution	References
Variability in Chemical Composition	Herbal extract composition might differ depending on plant species, growth conditions, and processing techniques.	Use of standardized extracts and application of chromatographic techniques to ensure consistency.	Ekor, 2013; WHO, 2003a, 2003b
Complexity of Herbal Extracts	Herbal extracts contain a complex mixture of compounds, making it difficult to isolate active components	Application of advanced analytical techniques such as HPLC, GC-MS, and NMR for profiling and standardization.	Heinrich et al., 2009; Gibson et al., 2019
Poor Bioavailability of Active Compounds	Many herbal compounds have low bioavailability, reducing their therapeutic efficacy.	Creation of cutting-edge delivery methods like liposomes, transdermal patches, and nanoparticles to increase bioavailability.	Patra et al., 2018; Mahapatra et al., 2021
Lack of Regulatory Standards	There is a lack of standardized regulatory guidelines for the production and marketing of herbal formulations.	Harmonization of regulatory frameworks, adoption of GMP, and establishment of international standards.	EMA, 2016; Ekor, 2013
Safety and Efficacy Concerns	Potential adverse effects and drug-herb interactions are not always well-documented.	Conducting thorough preclinical and clinical studies, proper labeling, and public education on usage.	Bent, 2008; Izzo & Ernst, 2009

8.9 CONCLUSION

The process of creating herbal formulations is intricate and multidimensional, requiring the careful selection of therapeutic plants as well as the extraction and purification of their bioactive ingredients, and the formulation of these compounds into safe, effective, and standardized products. Advances in science and technology have significantly enhanced our understanding of herbal medicine, leading to the development of more sophisticated and targeted herbal formulations. However, several challenges remain, including the variations in chemical makeup, the intricacy of plant extracts, and the formulation standards, and the regulatory hurdles associated with the approval of herbal products.

The development of new herbal compounds, customized formulations, and delivery system optimization are all made possible by the combination of modern technologies like AI, ML, and omics with traditional knowledge of herbal medicine. Strict quality control measures like GMP adherence, reference standard usage, and herbal material identification and authentication are also necessary to guarantee the efficacy, safety, and consistency of herbal products. This is why more research and innovation are needed to overcome these obstacles and fully realize the potential of herbal medicine.

A balance between innovation and tradition is crucial as the field of developing herbal formulations continues to develop, ensuring that the therapeutic benefits of herbal medicine are preserved while embracing the advances in science and technology that can enhance its efficacy and safety. With continued research, collaboration, and regulatory support, herbal medicine has the potential to become more significant in the global healthcare system, providing safe, effective, and natural solutions to a wide range of health conditions.

REFERENCES

Agarwal, A., Singh, S. K., & Arora, S. (2012). Standardization of herbal formulations: A review. *International Journal of Pharmaceutical Sciences and Research*, *3*(9), 2232–2244.

Aggarwal, B. B., Gupta, S. C., & Kim, J. H. (2013). Historical perspectives on tumor necrosis factor and its superfamily: 25 years later, a golden journey. *Blood*, *121*(2), 439–444.

Akbarzadeh, A., Rezaei-Sadabady, R., & Davaran, S. (2013). Liposome: Classification, preparation, and applications. *Nanoscale Research Letters*, *8*(1), 102.

Bagchi, D., Preuss, H. G., & Bagchi, M. (2012). Phytochemicals in cancer chemoprevention. *Toxicology*, *302*(1), 23–31.

Barnes, J., Anderson, L. A., & Phillipson, J. D. (2015). *Herbal Medicines* (4th ed.). Pharmaceutical Press.

Bauer, R. (1998). Quality criteria and standardization of phytopharmaceuticals: Can acceptable drug standards be achieved? *Drug Information Journal*, *32*(1), 101–110.

Bent, S. (2008). Herbal medicine in the United States: Review of efficacy, safety, and regulation. *Journal of General Internal Medicine*, *23*(6), 854–859.

Bilia, A. R., Santomauro, F., Sacco, C., Bergonzi, M. C., & Donato, R. (2014). Essential oil of Artemisia annua L.: An extraordinary component with numerous antimicrobial properties. *Evidence-Based Complementary and Alternative Medicine*, *2014*, 159819.

Brinckmann, J. A. (2011a). *Guide to Medicinal and Aromatic Plants: Their Cultivation and Processing*. CAB International.

Brinckmann, J. A. (2011b). Medicinal plants and the law. *Herbal Gram*, *90*, 56–63.

Chen, Y., Feng, L., & Wei, Y. (2013). Phytochemical and pharmacological properties of *Polygonum cuspidatum*: A review. *Journal of Ethnopharmacology*, *148*(2), 729–745.

Cimino, F., Speciale, A., Anwar, S., Canali, R., Ricciardi, R., Virgili, F., & Trombetta, D. (2018a). The emerging role of omics approaches in herbal medicine and traditional uses. *Journal of Ethnopharmacology*, 226, 48–63.

Cimino, M., Brown, B., & Daly, L. (2018b). The role of herbal medicines in managing diabetes: A systematic review. *Journal of Herbal Medicine*, *8*(2), 10–20.

DeFeudis, F. V., & Drieu, K. (2000). Ginkgo biloba extract (EGb 761) as a cognitive enhancer. *Drugs of the Future*, *25*(8), 741–761.

Dhanani, T., Shah, S., & Sharma, S. (2017). A comprehensive review on extraction techniques, phytochemical characterization, and biological activities of plant extracts. *Journal of Pharmacognosy and Phytochemistry*, *6*(2), 32–39.

Ekor, M. (2013). Regulatory control of herbal medicines: Lessons from the global situation. *African Journal of Traditional, Complementary, and Alternative Medicines*, 10(4), 58–69.

Ekor, M. (2014). The growing use of herbal medicines: Issues relating to adverse reactions and challenges in monitoring safety. *Frontiers in Pharmacology*, 4, 177.

Eskilsson, C. S., & Björklund, E. (2000). Supercritical fluid extraction for the determination of pesticide residues in plant matrices. *Journal of Chromatography A*, *902*(1), 227–244.

European Medicines Agency. (2007). *Reflection paper on markers used for quantitative and qualitative analysis of herbal medicinal products*. EMA.

European Medicines Agency. (2016). *Guideline on quality of herbal medicinal products/traditional herbal medicinal products*. EMA.

European Medicines Agency (EMA). (2016). Guideline on quality of herbal medicinal products/traditional herbal medicinal products. *EMA/CPMP/QWP/2819/00 Rev*. 2.

Food and Drug Administration. (2016). *Guidance for Industry: Botanical Drug Development*. FDA.

Gibson, G., & Wieser, M. J. (2019). Translating traditional herbal medicine into novel therapeutics: An integrative approach. *Bioorganic & Medicinal Chemistry Letters*, *29*(9), 126–135.

Gul, P., & Khanum, A. (2019). Nanotechnology in the treatment of herbal medicine. *Pakistan Journal of Pharmaceutical Sciences*, *32*(1), 305–315.

Gul, R., Jan, S. U., & Farooq, U. (2019). Review on the extraction techniques and pharmacological studies of bioactive compounds from medicinal plants. *Current Pharmaceutical Biotechnology*, *20*(6), 456–464.

Gutte, N. R., Kulkarni, A. S., & Kulkarni, A. V. (2018). Recent advances in herbal drug delivery systems and its applications. *Journal of Pharmacy and Bioallied Sciences*, *10*(4), 208–217.

Heinrich, M., Barnes, J., Gibbons, S., & Williamson, E. M. (2009). *Fundamentals of pharmacognosy and phytotherapy* (2nd ed.). Elsevier Health Sciences.

Ibrahim, M. A., Koorbanally, N. A., & Islam, M. S. (2018). Anti-diabetic effects of naturally occurring phenolic compounds: Recent advances. *Natural Product Research*, *32*(10), 1–19.

International Council for Harmonisation of Technical Requirements for Pharmaceuticals for Human Use. (2005). *ICH Harmonised Tripartite Guideline: Quality Guidelines Q7-Q10*. ICH.

Izzo, A. A., & Ernst, E. (2009). Interactions between herbal medicines and prescribed drugs: A systematic review. *Drugs*, *69*(13), 1777–1798.

Khaw, K. Y., Parat, M. O., Shaw, P. N., & Falconer, J. R. (2017). Solvent supercritical fluid technologies to extract bioactive compounds: A review. *Molecules*, *22*(7), 1186.

Kumar, A., Verma, S., & Jha, K. K. (2014). Nanotechnology: A tool to improve the bioavailability of herbal drugs. *Journal of Medicinal Plant Research*, *8*(24), 836–846.

Lang, Q., & Wai, C. M. (2021). Recent advances in extraction and sample preparation for natural products analysis. *Journal of Chromatography A*, *1020*(2), 63–70.

Li, F., Wang, Z., & Dong, Y. (2020). Comprehensive evaluation of *Curcuma longa* extracts for their therapeutic potential against metabolic syndromes. *Phytotherapy Research*, *34*(4), 733–742.

Li, S., Yuan, W., Deng, G., Wang, P., Yang, P., & Aggarwal, B. B. (2008). Chemical composition and product quality control of turmeric (Curcuma longa L.). *Pharmaceutical Crops*, *1*(1), 28–54.

Li, X. J., & Jiang, Y. M. (2010). Molecular mechanisms of action of extracts from medicinal plants. *Phytomedicine*, *17*(1), 1–7.

Mahapatra, S., Mishra, S. R., & Pradhan, R. (2021). Nanotechnology in herbal medicine: Opportunities and challenges. *Journal of Applied Pharmaceutical Science*, *11*(2), 112–121.

Mukherjee, P. K., Harwansh, R. K., Bahadur, S., Banerjee, S., Kar, A., & Chanda, J. (2015). Development of Ayurveda -Tradition to trend. *Journal of Ethnopharmacology*, *169*, 10–18.

Mukherjee, P. K., Kumar, V., Mal, M., & Houghton, P. J. (2012). Evaluation of free-radical scavenging potential of some Indian medicinal plants. *Journal of Ethnopharmacology*, *84*(2–3), 195–201.

Müller, R. H., Radtke, M., & Wissing, S. A. (2000). Nanostructured lipid matrices for improved microencapsulation of drugs. *International Journal of Pharmaceutics*, *242*(1–2), 121–128.

Naik, S. R., & Panda, V. S. (2012). Phytochemicals as antioxidants in chemoprevention of cancer. *Journal of Clinical Biochemistry and Nutrition*, *51*(2), 131–140.

Nebert, D. W., & Russell, D. W. (2002). Clinical importance of the cytochrome P450 gene family. *Lancet*, *360*(9340), 1155–1162.

Patel, K., Patel, D. K., & Patel, N. K. (2015). Evaluation of bioactive components and medicinal properties of *Eclipta alba*. *Journal of Pharmacognosy and Phytochemistry*, *4*(3), 40–43.

Patel, S., Verma, A., & Gupta, A. (2022). Phytochemical evaluation and antioxidant activity of medicinal plants used in traditional medicine. *Journal of Ethnopharmacology*, *274*, 114015.

Patra, J. K., Das, G., Fraceto, L. F., Campos, E. V. R., del Pilar Rodríguez-Torres, M., Acosta-Torres, L. S., & Shin, H. S. (2018). Nano based drug delivery systems: Recent developments and future prospects. *Journal of Nanobiotechnology*, 16(1), 71.

Pourmortazavi, S. M., & Hajimirsadeghi, S. S. (2007). Supercritical fluid extraction in plant essential and volatile oil analysis. *Journal of Chromatography A*, *1163*(1–2), 2–24.

Rai, A., Bhattacharyya, D., & Mandal, N. (2016). Potential role of bioactive compounds from medicinal plants in the management of cancer: A review. *Journal of Traditional and Complementary Medicine*, *6*(1), 1–16.

Ribeiro, A. C., Rodrigues, R. A., & da Silva, S. A. (1999). The role of phytochemicals in disease prevention: A review. *Brazilian Journal of Medical and Biological Research*, *32*(8), 101–111.

Routray, W., & Orsat, V. (2012). Microwave-assisted extraction of flavonoids: A review. *Food and Bioprocess Technology*, *5*(2), 409–424.

Sarker, S. D., & Fricker, S. P. (2021a). *Medicinal Chemistry of Anticancer Drugs* (3rd ed.). Elsevier.

Sarker, S. D., & Fricker, S. P. (2021b). *Medicinal Chemistry of Bioactive Natural Products*. Wiley.

Sarker, S. D., & Nahar, L. (2012). Natural products isolation: principles and practice. *Methods in Molecular Biology*, *864*, 1–25.

Shakeel, F., Shafiq, S., & Haq, N. (2010). Nanoemulsions as drug delivery vehicles for bioavailability enhancement: A review. *Phytomedicine*, *17*(8–9), 654–664.

Sharma, A., Khuller, G. K., & Bhaskar, S. (2020). Novel carriers for herbal formulations: Recent trends and advancements. *Journal of Controlled Release*, *327*, 69–90.

Shoba, G., Joy, D., Joseph, T., Majeed, M., Rajendran, R., & Srinivas, P. S. (1998). Influence of piperine on the pharmacokinetics of curcumin in animals and human volunteers. *Planta Medica*, *64*(4), 353–356.

Singh, R., Kumar, V., & Kumar, S. (2021). Pharmacological potential of bioactive compounds from medicinal plants: Therapeutic aspects. *Journal of Molecular and Cellular Biochemistry*, *476*(3), 1063–1075.

Tiwari, P. (2015). Phytochemical screening and extraction: A review. *International Journal of Pharmacy and Bio Sciences*, *6*(1), 66–75.

Topol, E. J. (2019a). High-performance medicine: The convergence of human and artificial intelligence. *Nature Medicine*, *25*(1), 44–56.

Topol, E. J. (2019b). The future of medicine in the digital age: Harnessing technology to improve healthcare. *Journal of Medical Internet Research*, *21*(1), e10816.

United States Pharmacopeia. (2021). *USP 44-NF 39, Supplement 2: United States Pharmacopeia National Formulary*. United States Pharmacopeia Convention.

Varona, S., Rojo, I., & Peinado, I. (2011). Extraction and functional properties of essential oils from different parts of *Zingiber officinale*. *Journal of Food Engineering*, *103*(1), 1–8.

Verpoorte, R., Choi, Y. H., & Kim, H. K. (2005). Ethnopharmacology and systems biology: A perfect holistic match. *Journal of Ethnopharmacology*, *100*(1–2), 53–56.

Wagner, H., & Ulrich-Merzenich, G. (2009). Synergy research: Approaching a new generation of phytopharmaceuticals. *Phytomedicine*, *16*(2–3), 97–110.

Wang, S., He, G., & Zhu, W. (2016). Therapeutic approaches of phytochemicals to cancer chemoprevention. *Journal of Cancer Prevention*, *21*(1), 1–14.

Wang, S., Zhang, Q., Yang, X., Zhang, H., & Zhu, W. (2021). Application of artificial intelligence and machine learning in the pharmaceutical sector: Practical considerations. *Pharmaceutical Research*, *38*, 1653–1663.

Wang, W., Zhang, P., & Li, Y. (2009). Phytochemicals and biological activities of *Salvia miltiorrhiza*. *Phytochemistry Reviews*, *8*(1), 27–52.

Wang, Y., Wang, H., Zhao, Z., & Ma, L. (2019). Phytochemical profile and therapeutic potential of *Curcuma longa* in cancer prevention. *Nutrients*, *11*(10), 2376.

Wishart, D. S. (2016). Emerging applications of metabolomics in drug discovery and precision medicine. *Nature Reviews Drug Discovery*, *15*(7), 473–484.

World Health Organization (WHO). (2003a). *WHO guidelines on good agricultural and collection practices (GACP) for medicinal plants*. WHO.

World Health Organization (WHO). (2003b). *Good agricultural and collection practices (GACP) for medicinal plants*. WHO Press.

World Health Organization. (2007). *Guidelines for assessing the quality of herbal medicines with reference to contaminants and residues*. WHO Press.

World Health Organization. (2011). *Quality control methods for herbal materials*. WHO Press.

Wu, C., Chen, Y., & Yang, L. (2007a). Botanical extracts and standardization: A comprehensive review of scientific validation. *Journal of Food and Drug Analysis*, *15*(3), 1–15.

Wu, L., Chen, L., & Tzeng, C. (2007b). The effect of medicinal plant extractions on the properties of polymer membranes and the resulting antibacterial activity. *International Journal of Pharmaceutics*, *334*(1–2), 61–68.

Xiao, J., Chen, Y., & Huang, Q. (2015). Phytochemical and pharmacological properties of *Solanum nigrum*: A review. *Journal of Ethnopharmacology*, *163*, 247–259.

Yadav, H. K., Patel, M. P., & Raval, J. A. (2013). Nanotechnology: A safe and effective drug delivery system for herbal medicines. *International Journal of Pharmaceutical Research and Bio-Science*, *2*(1), 1–25.

Zhang, H., Tang, G., & Xue, H. (2010). Extraction and bioactive components of medicinal plants: New aspects for pharmacology. *Journal of Ethnopharmacology*, *131*(2), 1–12.

Zhang, H., Xie, X., & Yang, D. (2018). Phytochemical composition and bioactivity of extracts from Chinese medicinal plants. *Journal of Ethnopharmacology*, *221*, 193–201.

Zhang, L., Li, J., & He, Y. (2019). Recent advances in plant-derived natural products for inflammatory diseases. *Phytochemistry Letters*, *31*, 86–94.

Zhang, Y., Gan, R. Y., & Li, H. B. (2008). Antioxidant activity and phenolic content of extracts from Chinese medicinal plants. *Journal of Agricultural and Food Chemistry*, *56*(10), 3969–3976.

Zheng, X., Zhao, C., & Yang, C. (2019). Recent advances in supercritical fluid extraction of bioactive compounds from medicinal plants. *Critical Reviews in Food Science and Nutrition*, *59*(7), 1118–1131.

9 Emerging Trends in Herbal Nanotechnology

Muhammad Sirab Khan
Department of Botany, University of Lakki Marwat, Lakki Marwat, Pakistan

Muhammad Mudasar Aslam and Nida Khan
Department of Botany, University of Science and Technology Bannu, Bannu, Pakistan

Maha Rehman and Nazneen Akhtar
Department of Biotechnology and Genetic Engineering, Kohat University of Science and Technology, Kohat, Pakistan

9.1 INTRODUCTION

The emergence of nanomaterial sciences has altered every aspect of human life. Nanomaterial science is a field that encompasses various disciplines and focuses on the study of particles with nanostructures and their practical application across multiple dimensions. Nanoparticles (NPs) are materials that can be organic, inorganic, or hybrid in nature, and have a minimum of one dimension that lies between 1 and 100 nm. NPs can have a variety of dimensions depending on how they are manipulated. They can be zero-dimensional, meaning all dimensions are at the nanoscale; one-dimensional, which is rod-shaped; two-dimensional, which are ultra-thin films; or three-dimensional, which can be in any shape [1, 2]. NPs are classified into three groups according to their source: natural, engineered, and incidental NPs [3]. Naturally generated NPs have been present since the earth's inception and can be found in various sources such as weathering, celestial dust, volcanic dust, terrestrial storms that produce dust, photochemical processes, mineral material made of composites, forest fire, etc. Incidentally generated NPs primarily originate from anthropogenic activities such as burning coal, engine exhaust, industrial fumes, and welding gases [4]. Engineering NPs can be classified into five distinct categories: NPs composed of carbon, NPs made of metals, magnetic NPs, composite NPs, and dendrimers [5].

Nanotechnology is the domain that focuses on the synthesis, manipulation, and use of components at extremely small scales, such as the atomic, molecular, or supramolecular levels [6]. Richard P. Feynman, the winner of the Nobel Prize in Physics, advanced the idea for nanotechnology in 1959, which involves using larger things and mechanical instruments on lower scales [7]. Norio Taniguchi first introduced the word 'nanotechnology' in 1960, a decade after the initial concept was conceived. Yet the development of cutting-edge nanotechnology started to form in 1981 when the scanning tunneling microscope was invented [8]. Nanotechnology is the most rapidly expanding technology globally, and has sometimes been referred to as the technological revolution of the 21st century [9]. In recent times, nanotechnology has been employed in a wide array of technical procedures. These include the characterization, production, and control of different materials, with the aim of creating innovative materials that can be used in fields such as biology, chemistry, bioengineering, pharmaceuticals, medicine, agriculture, and electronics [10–13]. For almost twenty years,

DOI: 10.1201/9781003544418-9

there has been a consistent rise in the quantity of scholarly articles concerning nanotechnology and nanotechnology-based products. It is anticipated that the commercialization of nanotechnology will exceed a trillion dollars within a timeframe of 10 to 15 years. Hence this technology will profoundly transform the fields of academia, science, industries, and the way humans live worldwide [9]. Nowadays, the utilization and advancement of herbal items through the application of nanotechnology are rising fields [14]. This chapter looks at these applications and developments by exploring phytonanomedicines' functions in human healthcare, nanoparticles as protectants, and nanocarriers for plant disease control.

9.2 HERBAL NANOTECHNOLOGY AND PHYTONANOMEDICINES

Phytonanomedicines, which refers to plant chemical-enriched nanoformulations for individual health assistance, has recently received significant attention [15]. Phytonanomedicines are small-sized medicinal substances that consist of plant extracts or powder enclosed in nanocarriers. Phytomedicines, often known as herbal medicines, are a combination of plant metabolites that contain medicinally active chemicals and possess therapeutic and medicinal properties. For many centuries, herbal medications have been utilized as therapeutic agents for many disorders due to their advantages, including reduced unwanted and inexpensive costs. Furthermore, it is worthy of note that more than one-third of all newly identified molecules authorized by the US Food and Drug Administration (FDA) are derived from natural sources [16, 17]. Although herbal drugs have significant pharmacological effects in treating different diseases, they also face several obstacles that can hinder their effectiveness when taken orally. These challenges include low bioavailability, reduced water and lipid solubility, limited adsorption, instability, and substantial molecular weight [18, 19]. To overcome the said barriers and improve patient satisfaction, an efficient system of drug discovery is required that reduces the need for continued administration [20].

In recent years, nanotechnology-based delivery in herbal medicines has been significantly investigated. Enclosing herbal medicines in nanocarriers and addressing the restrictions outlined above, Figure 9.1 highlighted the advantages of phytonanomedicines such as enhanced solubility, enhanced

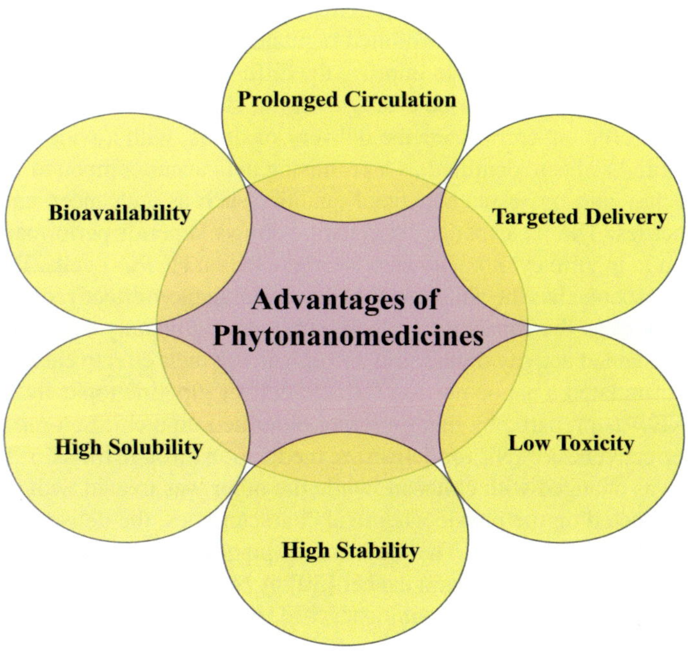

FIGURE 9.1 Advantages of phytonanomedicines.

circulation time, safeguarding against degradation, decreased side effects, regulated release, and, thereby, therapeutic effectiveness and ideal bioavailability [21–23].

9.2.1 Role of Phytonanomedicines in Disease Management

Phytonanomedicines have been crucial in delivering advanced medical treatments to fight against diseases. This chapter emphasizes the use of phytonanomedicines in treating cancer, diabetes mellitus, and neurodegenerative and cardiovascular diseases.

9.2.1.1 Cancer

Cancer is a diverse disease that could be affected by genetic and ecological factors. It leads to the uncontrolled development of cells and disturbances in cellular processes, especially metabolism and defense [24]. Cancer symptoms can appear as structural, such as losing weight or tiredness, or affect specific organs, including coughs or pain [25]. The most recent global cancer statistics report indicates that around 19.3 million fresh cases of cancer and almost 1 crore deaths from cancer were recorded globally in 2020 [26]. Skin cancer is a particularly prevalent kind of cancer, impacting millions of individuals annually, and is widely recognized as the most widespread kind of cancer. Annually, almost three and a half million individuals in the United States are affected by skin cancer [27]. Various drug delivery techniques have been examined and studied to achieve desired results in controlling subcutaneous drug distribution. These are NPs that utilize nanotechnology, including phytosomes, dendrimers, micelles, carbon nanotubes, nanosponges nanostructured lipid carriers (NLCs), solid lipid nanoparticles (SLNs), magnetic NPs, mesoporous NPs, polymeric NPs, and vesicular systems. These nanocarriers enhance the therapeutic efficacy. The skin cancer is an exceptionally highly organized physiological hurdle and is used to assess the therapeutic efficacy of topical herbal medicines. Nevertheless, a range of solutions are employed to surmount the skin cancer obstacle by the customization of appropriate formulations. Hence, it is crucial to have a well-designed formulation on order to ensure the effective dispersion of medicinal molecules. Chitosan and β-cyclodextrin nanosponges exhibit promising capabilities for medication delivery. Their research found that sesamol β-cyclodextrin nanosponges medication delivery charged with a substance had increased cytotoxic effects against B16F12 melanoma cell lines, had anti-inflammatory properties, as well as exhibited regulated release of drugs when applied topically. Chitosan nanosponges were discovered to improve the diffusion of medications over the skin without causing any harm and effectively distribute drugs via the skin. The results emphasize the capability of nanosponges as efficient carriers for the delivery of drugs, with a wide range of uses [28]. In recent years, curcumin has been identified as a promising anticancer compound. Resveratrol belongs to the polyphenol class and possesses beneficial qualities such as anticancer, antioxidant, and anti-inflammatory properties. The nanosponge formulation shows superior performance compared with the plain medication in in vitro cytotoxicity assays conducted on HCPC-1 cells. This suggests that the formulation of a nanosponge has the ability to enhance the anticancer effectiveness of resveratrol [29]. These nanocarriers possess the capacity to release substances gradually over a prolonged period of time, allowing for sustained activity on the skin and enhancing their effectiveness in treatment.

Giacone et al. formulated a nanoemulsion (NE) to deliver piperine topically for the treatment of skin cancer. The NE was prepared by incorporating piperine into a chitosan-modified NE. Giacone et al. evaluate the effectiveness of NE for delivering medication through the skin. Two varieties of NE were created: one was changed with chitosan, while the other was treated with sodium alginate. An assessment was conducted on the physicochemical characteristics, the delivery of piperine, and the effectiveness of the Nes' formulation. The inclusion of piperine in chitosan-modified NE shows an effective approach for locally treating skin cancer [30]. A NE with a low hydrophilic-lipophilic balance surface-active agent was created to administer 5-FU topically and reduce skin cancer. The drug-based NE showed significantly higher and more effective in vitro penetration over rat skin compared to free 5-FU. This study demonstrated the capability of the tested substance to kill SK-MEL-5 cancer cells in a laboratory setting. This suggests that it could be a promising treatment for the control of skin cancer. Yet additional investigation is required to assess its effectiveness and safety in humans [31]. In 2021, Asasutjarit et al. aimed to create NE-silver NPs and assessing their efficacy in the treatment of skin cancer. The results demonstrated notable cytotoxic effects on A-431 and A-375 cells through the induction of selective apoptosis.AG-NE inhibited the tyrosinase activity in A-375 cells, suggesting a potential therapeutic approach for skin cancer [32]. Falamas et al. investigate the chemopreventative characteristics of a betulin NE product [33]. Kaplan et al. formulated NE and NE-based gels with daidzein to be applied topically for the treatment of melanoma with significant positive outcomes being observed in relation to melanoma cells [34]. A research experiment was conducted to modify the physicochemical parameters of an NE to enhance the penetration of apigenin-loaded NE through rat skin. The study also evaluated the toxicological capability of the NE against HaCaT and A-431 cells and found significant results. The study examined the potential of using curcumin, a widely used spice, to enhance penetration. This was done by utilizing a nanocarrier called NE, which consisted to ethanol, lecithin, labral, and transcutol. The aim was to explore its effectiveness in treating skin cancer and psoriasis [35]. Mukherjee et al. developed a NE utilizing the essential oil of usnic acid and cinnamon by the process of ultrasonic emulsion. Usnic acid and cinnamon diminished tumors by reinstating the antioxidant capacity of intrinsic enzymes found in mice skin. The skin sample study showed a notable decrease in the quantity and size of acanthosis and keratinized pearls when contrasted with the results that were not expected [36]. A study conducted a comparative evaluation of different 5-FU incorporated nanoemulsion-based gels to determine their effectiveness in treating skin cancer [37]. An experiment was conducted to evaluate the effects of NNE loaded with dacarbazine on skin cancer compared to a suspension [38]. Andrographolide (AG) is a promising plant-based medication that has a low ability to dissolve in water. The substance was administered intranasally in order to evaluate its efficacy against non-melanoma skin cancer using HFE-1 and A-431 cell lines. Both AG-NE and AG were nontoxic to HFF-1 cells, while they caused apoptosis in A-431 cells [39]. A recent study investigated the use of a combination of lower laser therapy and cellulose nanocrystals/nanofibrils carrying NE to treat skin cancer. This dual approach improved the effectiveness of NE by controlling the signaling routes within cancer cells [40].

Chitosan-based NPs are frequently employed for the purpose of delivering substances to the skin surface. This is because they possess advantageous characteristics such as biodegradability, antibacterial, anti-inflammatory, and antioxidant capabilities. The extensively studied biodegradable NPs 'PLGA' was utilized for the delivery of 5-aminolevulinic acid (ALA) to enhance cellular absorption, regulate drug delivery, and exhibit comparatively high cell toxicity against SCC cells in comparison to ALA. The result obtained from the experiment conducted on hairless male SKH-1 mice shown significant enhancement against cancers [41, 42]. Another study investigated the use of PLGA-based NPs for regulated drug administration and improved retention of drugs in the epidermal and dermal layers of rat skin. The researcher encapsulated protoporphyrin IX (PpIX) in PLGA-nanoparticles and achieved a regulated release of the medicine over a 10-day period using a laboratory release model. The ex vivo drug absorption investigation demonstrated a 23-fold increase in drug deposition inside the skin cancer layer and a 10-fold increase in the dermal and epidermal layers contrasted to free PpIX following topical administration [43].

In 2013, researchers used chitin-derived natural NPs measuring 120 to 140 nm to deliver the drug 5-FU for the purpose of controlling skin cancer. They achieved this by using a pH-sensitive nanogel that allowed regulated drug release. The NPs had a high drug retention rate. Additionally, the NPs demonstrated cell toxicity against A375 melanoma cell lines at a concentration of 0.4-2 mg/mL, while having less adverse impacts on human dermal fibroblast [44]. A commercially available nanosphere called 'tyroSpheres™', made from a special polymer generated from tyrosine, was used to administer indocyanine green for controlling skin cancer in an experimental model using CD1 mice [45]. TyroSpheres™ utilized lipophilic medicines, including cholcalciferol and paclitaxel, to potentially treat skin cancer and psoriasis and provide protection against photodegradation [46]. NPs were utilized to administer zinc phthalocyanine and dacarbazine for the purpose of managing melanoma. The in vitro study did not demonstrate any notable toxicity, whereas the in vivo experiments revealed considerable toxicity [47]. Das et al. (2013) did a study in which they enclosed apigenin within

PLGA NPs to examine the improved anticarcinogenic properties against BaP- and UVB-induced skin cancers and improper mitochondrial function in mice [48]. His findings indicated that the use of apigenin-loaded NPs reduces the severity of skin cancer. This suggests that this approach was effective [48]. A study investigated a new way of treating melanoma cancer by synthesizing PLGA-PEG NPs in a single-step process. These NPs were used to deliver chrysin and curcumin simultaneously. The combination of chrysin and curcumin enhanced its efficacy as a co-delivery strategy for the cure of melanoma. This technique has the ability to serve as an appealing and easy therapy alternative for people diagnosed with melanoma. However, extensive investigation is needed to better examine the effectiveness of this medication [49].

The use of elastic liposomes and NEs was investigated for the management of cutaneous problems such as skin-related keratoses and cancer. The researcher has discovered many methods for delivering herbal medicines, such as apigenin, luteolin and gefitinib. These methods include NE, cationic NE, solid dispersion, and elastic liposomes [50, 51]. Transfersomes prolong the release of drugs, enabling sustained drug administration. Waheed et al (2022) utilized the quality-by-design method to create lyotropic liquid crystalline NPs (LLC-NPs) that was loaded with apigenin for skin application. Their objective was to boost the penetration of apigenin, leading to improved bioavailability. The apigenin LLC-NPs exhibited the ability to specifically reach the innermost parts of the skin, indicating their potential as a nanocarrier for delivering drugs by topical application in managing skin cancer [52]. The phytonanomedicines and their efficacy in managing cancer have been arranged in Table 9.1.

TABLE 9.1
Phytonanomedicines for the Treatment of Cancer

Nanoform/Nanocarrier	Phytonanomedicine Function	Reference
Cyclodextrin-based nanosponges	The simultaneous administration of curcumin and resveratrol	[53]
NLCs, SLNs and nanoemulsion	Utilizing topical administration of lutein for the purpose of managing skin deterioration. Functions as a stress-reducing agent and antioxidant. Preserved lutein against deterioration caused by ultraviolet radiation	[54]
Dacarbazine and eugenol in liposomes	The objective was to use hyaluronic acid-loaded liposomes to simultaneously distribute dacarbazine and eugenol, with the aim of controlling resistant metastatic melanoma model (MM)	[55]
Chitosan-coated liposomes loaded with Indocyanine green	Photodynamic treatment of melanoma. Enhanced and optimized medication penetration via the skin. Enhanced cellular absorption and increased photo-cellular toxicity of IC in B16F10 MM cellular lineages	[56]
Hyaluronic acid and oleic acid-loaded gold nanoparticles	Sustainable gold NPs exhibited enhanced cellular toxicity towards two specific cellular lineages, namely B16F10 MM and keratinocytes (HaCat). Effective toward a non-melanoma model	[57]
Ethosomes and transfersome-coated sulforaphane	The non-proliferation method of natural isothiocyanates promoted MM along with the additional skin cancers in the laboratory (SK MEL 28 cell models)	[58]
Carbon nanotubes conjugated with doxorubicin	Enhanced cellular toxicity towards B16-F10 MM achieved by fast absorption using lysozymes from MM cells	[59]
Carboxymethylcellulose-caped silver nanoparticles	Doxorubicin-loaded silver NPs encapsulated with carboxymethylcellulose showed a combined and enhanced efficacy toward a non-melanoma model	[60]
Delivery of trametinib and doxorubicin using microneedles with dextran methacrylate hydrogel	Mice Xenograft B16 types were tested for anticancer activity. Drugs work together to fight skin tumors	[61]
Delivery of immunoadjuvant and doxorubicin using cationic dendrimer	Metastasized MM was managed. pH and cationic-sensitive loading of drugs inhibited RES removal for optimum cancer drug buildup	[62]

9.2.1.2 Diabetes Mellitus

Diabetes mellitus (DM) is a long-term medical disorder characterized by a high level of sugar in the blood, known as hyperglycemia. This condition occurs due to a lack of insulin, either relative or absolute [63], reduced responsiveness of cells to insulin, and disruptions in the metabolism of proteins and glycolipids [64]. Individuals with DM often have four health concerns: obesity [65], enhanced glucose production [66], aberrant insulin function, and failure in secretion [67, 68]. In DM types, type 2 DM affects almost 90% of individuals diagnosed with DM [69]. DM poses a significant risk to both individuals and society due to its high death rate [70, 71]. The International Diabetes Federation (IDF) has just released data indicating that the worldwide incidence of DM has reached 10.5% in 2021. Out of all the instances, there are currently 537 million adults who have DM, representing a 16% rise (74 million) compared to 2019. Nevertheless, a significant proportion of adults (44.7%) remain undiagnosed. According to the IDF's forecast, the number of adults with DM is estimated to reach 784 million by 2045, that is greater than twice the predicted population rate of 20% for the same time [72, 73].

Emerging research on traditional medications substantiates the claim that herbal products are efficient in a significant number of diabetic individuals. The data reveals that more than 50% of the already marked pharmaceuticals are derived from herbal products extracted using herbal products [74]. The study developed a nanodelivery system for treating insulin resistance in type 2 DM. This system involved loading the oleanolic acid (OA) around the polygalacturonic acid (PGA) acting as a natural, building oneself. The resulting nano-based drug, called polygalacturonic acid-loaded oleanolic acid (PGAOA), was designed for oral administration and showed biocompatibility with the body. Previous literature has demonstrated that PGAOA micelles loaded with OA have enhanced stability in traversing the gastrointestinal barriers and improved intestinal uptake of the drug. Additionally, these micelles exhibit exceptional ability to sustain plasma drug levels over an extended period. Therefore, nano-formulated PGAOA micelles in the rat model of type 2 DM were used. Furthermore, it was observed that this treatment had a lasting effect on controlling glucose levels regardless of drug was discontinued [75]. In Indonesia, fruits of *Zanthoxylum acanthopdium* and leaves of *Rhodomyrtus tomentosa* are widely found plants known for their antioxidant capabilities [76]. Elevated glucose levels in patients with DM impact the process of angiogenesis, hence influencing the time it takes for wounds to heal. The study utilized Andaliman and Haramonting phytonanomedicines to assess the histological alterations in wound healing caused by diabetes in the dermal tissue of rats, specifically focusing on the fibroblast growth factor. The clinical investigation demonstrated that the process of epithelialization had entirely enveloped the epidermis, organized the robust basal membrane in an orderly manner, and resulted in rich collagen-connected tissue covering the skin, hence increasing the abundance of fibroblast cells. The phytonanomedicines Andaliman and Haramonting stimulated cell proliferation in the damaged skin layer, leading to differentiation and the development of cells, ultimately facilitating the healing of the damaged tissues [77]. The silver nanoparticle methanolic extract from *Costus pictus* D. Don, generally known as the insulin plant, was used to treat diabetes. It effectively suppressed that α-glucosidase inhibitory action, impeding the degradation of sugars into glucose. Furthermore, it had a notable impact on DM in comparison to the acarbose drug [78]. A solid lipid nanoparticle based on the flame flower, which is scientifically known as *Talinum portulacifolium*, exhibits a potent antidiabetic effect in comparison to *T. portulacifolium* treatment for diabetes by preventing abnormalities in the breakdown of lipids caused by high blood sugar levels, which leads to higher lipids and the development of hyperlipidemia, ultimately resulting in cardiovascular disease [79]. The usefulness of curcumin (CUR) loaded poly(caprolactone) nanofiber carriers and CUR-loaded CSNPs coupled and constructed with collagen-alginate scaffolds has been demonstrated in the treatment of diabetic ulcers [80]. Additionally, these scaffolds have shown efficacy in promoting the healing of diabetic wounds and lowering inflammation [81].

The combined use of gold NPs with plant extract of *Bauhinia variegata* enhanced the effectiveness of *B. variegata* by improving its characteristics, including the contents of polyphenols and antioxidant potential against free radicals. These enhancements were observed when contrasted to using the extract alone. The *B. variegate*-mediated Au-NPs was demonstrated to be more effective against diabetes through restoring the normal structure of pancreatic ß-cells and providing hypolipidemic and antioxidant potential towards streptozotocin-mediated DM in rates. This effect was not observed with the plant extract alone, indicating that the therapeutic effect of the plant extract of *B. variegata* was improved after the addition of Au-NPs [82]. Quercetin is a chemical that may be dissolved in lipids. It has been shown to reduce the production of lipid hydroperoxide [83] and has the ability to avoid damage to lipids. In addition, it can scavenge free radicals, protect biomolecules from oxidation, and affect the routes involved in the antioxidant system. Nevertheless, the effectiveness of quercetin was enhanced by administering quercetin NPs (QUNPs) to rats with DM. The blood biochemical and morphological observations indicated that QUNPs had a beneficial impact on repairing damaged cells in the kidney. QUNPs have been found to possess therapeutic properties in treating streptozotocin (STZ)-mediated DM in rats. These compounds can serve as a phytomedicine to protect pancreatic cells and the kidney [84]. The laboratory experiments on quercetin contained poly-D, I-lactide (PLA) nanoparticles, averagely sized 250 nm, demonstrated the highest quercetin intake in the intestines. These findings align with the findings reported in previous literature [85, 86]. PLA nanoparticles demonstrated efficacy in delivering quercetin orally, enhancing cellular absorption and maintaining a suitable presence in the bloodstream. In addition, the use of *Catharanthus roseus*, commonly known as rosy periwinkle alba extract, coupled with chitosan-based nanoparticles [87] and nano-emulsified ethanol-based extracts of *E. Littorale* was found to be more beneficial in managing DM [88]. The phytonanomedicines and their efficacy in managing diabetes mellitus have been arranged in Table 9.2.

TABLE 9.2
Phytonanomedicines for the Treatment of Diabetes Mellitus

Nanoform/Nanocarrier	Phytonanomedicine Function	Reference
Polymer polygalacturonic acid	Oleanolic acid, a herbal stimulator, was incorporated into polygalactuonic acid, a naturally occurring polymer, to overcome gastrointestinal obstacles and enhanced drug intake in the intestine. This formulation also has a sustained impact on glucose levels over an extended time	[75]
Solid lipid nanoparticles	*Talinum portulacifolium*-based solid lipid NPs were used to treat diabetes, effectively reducing lipid metabolic abnormalities	[79]
Silver nanoparticles	The *Costus pictus* D. Don-derived silver NPs were utilized for the management of diabetes. It hinders the function of α-glucosidase suppression	[78]
Gold nanoparticles	The antidiabetic properties of gold NPs conjugated with *Bauhinia variegate*-derived drugs were investigated. It leads to the regulations of the buildings of the β-cells and exhibits lipid-lowering and antioxidant activities	[82]
Poly (caprolactone) nanofibers	A carrier consisting of nanofibers made from poly (caprolactone) and mediated with Cur and Cur-mediated CSNPs was inserted into collagen-alginate scaffolds. This therapy was shown to be helpful in reducing inflammatory rate, wound recovery, time, and diabetic lesions in patients with diabetes	[80, 81]
Chitosan nanoparticles	The antidiabetic effect of extract from *Catharanthus roseus* loaded chitosan-mediated nanoparticles was investigated. It has the ability to maintain pancreatic β-cells, have antidiabetic actions, and exhibit antioxidant activity	[87]
Nanoparticles, poly-D, L-lactide (PLA) nanoparticles	The effectiveness of quercetin was enhanced by providing quercetin NPs to rats with diabetes. The outcomes were the repair of injured kidney cells and the improvement of intestinal utilization of quercetin	[84, 85]
Nanocapsules	The antidiabetic potential of *Enicostemma littorale* was evaluated using nanoemulsion. The result was a potent antidiabetic impact that was both prolonged and maintained over an extended duration	[88]

9.2.1.3 Neurodegenerative Diseases (NDDs)

Neurodegenerative diseases (NDDs), including Parkinson's disease (PD), Alzheimer's disease (AD), Huntington's disease (HD), and amyotrophic lateral sclerosis (ASL), are a diverse set of diseases characterized by gradual deterioration and specific loss of neurons that are physiologically or anatomically connected. This loss severely impairs cognitive behavioral function [89–91]. NDDs are characterized by the buildup of improperly folded proteins in the central nervous system (CNS), that form insoluble aggregation or inclusions. This leads to the gradual degeneration of neurons in the affected areas [92]. The rates of morbidity and mortality of these illnesses are experiencing a significant increase due to the aging of the worldwide population [93]. NDDs are projected to surpass cancer as the second most prevalent death cause worldwide by 2024, as stated by the World Health Organization. Due to the limited regenerating capacity of the human neural system, the disease's course is irreversible [89]. Nanotechnology is an emerging method for delivering drugs that can potentially overcome obstacles in the CNS such as the blood-brain barrier (BBB) [94]. The therapeutic benefits of drug release rely on their ability to effectively avoid the immune system, traverse BBB, and specifically accumulate in the tissues of interest, any form of obstruction during the administration of drugs reduces the effectiveness of medicinal treatments. Over the past few years, numerous nanodelivery technologies have been created for diverse applications in the pharmaceutical sector [95]. Interest in using polyphenolic compounds is growing for medical purposes to prevent and treat many degenerative and long-term illnesses, such as neurological disorders, cardiovascular problems, and cancer [94]. The extract of *Ginkgo biloba* leaves has a significant concentration of polyphenolic components such as myricetin, quercetin and kaempferol [96]. The article states that extract has neuroprotective benefits in the recombinant mouse model (G93A) of amyotrophic lateral sclerosis (ALS). The administration of the extract orally significantly reduced the aberration in motor function and increased the duration of survival [97]. Moreover, research findings demonstrated a significant reduction in the degeneration of motor neuron cells in the spinal cord within the ALS mice model when treated with the *Gingko biloba* plant extract. Additionally, there are several polyphenolic substances that show beneficial benefits in ALS, including ginseng [98], genistein [99], epigallocatechin gallate [100], and resveratrol [101]. Currently, there are numerous unsolved issues about the effective transportation of drugs into the CNS. The specific difficulties can be attributed to various reasons, such as the existence of the BBB, the blood-spinal cord barrier, and the inherent characteristics of the medications themselves, such as insufficient biostability, low solubility, and undesired effects among others [102]. Due to low bioavailability, the traditional medication cannot effectively reach the area of interest in individuals who have ALS to promote a full recovery, However, the nanotechnological methods employ specially designed nanocarriers that show great promise in delivering individual or combined drugs that can eliminate biological obstacles, enable real-time monitoring while minimizing widespread adverse consequences, enhance interaction with specific sites, and enhance drug stability and availability [103].

Mathew et al. (2011) discovered that hydrophilic Cur-loaded PLGA nanoparticles, in a laboratory model of AD, were able to bind to Amyloid beta (Aβ) assembles and facilitate their separation. This suggests that Cur-loaded PLGA nanoparticles possess the capability to specifically target the brain through the BBB and successfully interrupt the formation of amyloid plaques in AD [104]. Furthermore, the finding also showed that by stimulating the wnt/β-catenin route in a laboratory setting, Cur-PLGA nanoparticles led to a substantial rise in the development of neural stem cells and neurons, as opposed to the free Cur, in the subventricular region and hippocampus of adult rats. A separate investigation utilizing Cur-loaded selenium PLGA nanospheres demonstrated its efficacy in treating AD deficits. The introduction of Cur-loaded nanoparticles had a dual effect on the brain, both improving memory and spatial identification and also reducing behavioral abnormalities. This was observed in the context of brain neuropathy [105]. The inclusion of Cur in alginate nanoparticles showed significant neuroprotective effects against Parkinson's disease in Drosophila (insects). This was attributed to the significant potential for the absorption of Cur in nanoplatforms, which reduced the oxidative harm and death of brain cells in a recombinant Drosophila PD model [106]. In their study, Bollimpelli et al. (2016) found that Cur-loaded lactoferrin (LF) nanoparticles had neuroprotective benefits in the SK-N-SH cell line. These effects were witnessed when the NPs were tested on rotenone, which increased cell toxicity. Additionally, the NPs were found to boost the expression of Lf receptors in individuals with Parkinson's disease [107].

Naringenin (NRG) is a flavonoid that belongs to a class of polyphenolic phytochemicals. It is commonly found in citrus fruits, as well as in other fruits, including coca. bergamot, tomatoes, and cherries. NRG possesses remarkable therapeutic qualities, including anticancer, antioxidant and anti-inflammatory effects [108]. NRG bioavailability in the brain for treating cerebral ischemia was enhanced when administered in the form of nanoemulsion [109]. The research study also demonstrated that NRG-loaded nanoparticles had a positive impact on SH-SY5Y cells by enhancing their potential to protect against neurotoxicity generated by 6-hydroxydopamine (6-OHDA). This effect was attributed to the nanoparticles' neuroprotective and antioxidant properties [110]. Ghaffari et al. states that quercetin and its nanosized crystal molecules exhibit neuroprotective qualities against 6-OHDA-induced damage in rat models. Quercetin nanocrystals provided considerable protection against neurobehavioral impairments and oxidative harm in the hippocampus regions. The dosage control study of quercetin and its nanosized crystals at doses of 10 and 25 mg/kg effectively prevented memory disruption, reduced the quantity of MDA in the hippocampus, and enhanced the action of the antioxidant enzyme. Furthermore, quercetin nanosized crystals showed a substantial increase in bioavailability and established superior efficacy compared to pure quercetin in managing Parkinson's disease in a rate model [111]. Gold nanocluster (AuNCs) enhanced the efficacy of herbal medicines, such as berberine [44], diosgenin (DIO), and Astragalus polysaccharide (APS), in treating nerve damage caused by spinal cord injury. The three consecutive phytonanomedicines, BRB-AuNCs, APS-AuNCs and DIO-AuNCs, effectively reduced the excessive activation of inflammation-promoting M1 type microglia and inhibited the expression of domain-containing protein (3NLRP3), most likely through the nuclear factor kappa B (NF-κB) signals route. The AuNC5 carrier offers a potential method for improving the effectiveness of phytomedicines in the treatment of NDDs in living organisms [112]. The process of nanoencapsulation of the combination of the phytomedicine piperine and Cur in the lipid glyceryl monooleate was discovered to be successful in treating Parkinson's disease. These nanoparticles possess a tendency to inhibit the formation of alpha-synuclein material, enhance the absorption of Cur, decrease oxidative harm, and improve autophagic operation when compared to the drugs in their original form [113]. The phytonanomedicines and their efficacy in managing neurodegenerative diseases have been arranged in Table 9.3.

9.2.1.4 Cardiovascular Diseases (CVD)

Cardiovascular diseases (CVD) are a group of disorders that disturb the heart and blood vessels and have the greatest rates of hospitalizations and death [114]. CVDs are the prevailing noncontiguous diseases on a global scale, responsible for over 33% of global deaths [115]. Flexible risk indicators, including diabetes, blood pressure, body-mass index, lower density lipoprotein (LDL) cholesterol, and smoking cause the occurrence and growth of CVD. However, the specific percentage of their impact depends on the population being studied and the approaches used [116]. Nanotechnology's significance in pharmaceuticals could present a new and innovative method for resolving CVD [117]. Unmodified therapeutic drugs lack the ability to particularly target the heart and are recurrently deposited in the lungs, spleen, and liver. Therefore, the incorporation of drugs into NPs enhances their longevity and duration in the bloodstream in comparison to their original state. In addition, a number of medications have been produced through natural plant compounds in order to explore commercial treatments for controlling and preventing CVD [118, 119]. Therefore, the advancement of nanomedicines and techniques for delivering drugs has enhanced both the security and effectiveness of phytonanomedicines for CVD [120, 121]. The methanol-based extract produced from seeds of *S. cumini* decreased the amount of collagen and disputed the membrane of the mitochondria. It also triggered a significant increase in the generation of reactive oxygen species (ROS) in the H9C2

TABLE 9.3
Phytonanomedicines for the Treatment of Neurodegenerative Diseases

Nanoform/Nanocarrier	Phytonanomedicine Function	Reference
Selenium PLGA nanospheres	A research investigation used selenium PLGA nanosphere coated with quercetin. It was found that brain inflammation has a positive impact on consciousness and spatial awareness	[105]
PLGA nanoparticles	Using curcumin-loaded PLGA NPs disrupts amyloid plaque during Alzheimer's disease, acting as the Wnt/B-catenin route	[104]
Nanoemulsion	NRG-loaded nanoparticles enhanced brain-protecting capacity and antioxidant activity in SH-SY5Y cells	[110]
Lactoferrin /alginate nanoparticles	The Drosophila was used to coat curcumin in alginate NPs, whereas the SK-N-SH cellular model in individuals with Parkinson's disease was used to incorporate curcumin-loaded lactoferrin (Lf) NPs. Results showed an effective protective response against neurodegeneration in a Drosophila model of Parkinson's disease	[106, 107]
Gold nanoclusters	Three phytonanomedicines, berberine, astragalus polysaccharides and diosgenin, were administered for the treatment of harm caused by spinal cord injury. The stimulation of inflammation-inducing MI-type microglia has been reduced, and the oligomeric assembly of nucleotide-binding sites has been hindered	[112]
Nanocrystal	The potential of quercetin and associated nanocrystals enhanced and safeguarded neurocognitive harms and triggered oxidative harm in the hippocampus regions	[111]
Glyceryl monooleate lipid	Suppression of α-synuclein clusters in Parkinson's disease. Decreasing oxidative harm and improving autophagic activity	[113]

cardiomyocyte cell line, effectively decreasing the stress caused by elevated glucose levels [122]. In an in vitro study to investigate the potential cardioprotective effect of silver nanoparticles derived from *S. cumini* seeds (SmSNPs) in glucose-mediated cardiac damage, it was shown that the therapeutic efficacy of *S. cumini* against cardiomyopathy due to diabetes was augmented by incorporating the extract alongside silver nanoparticles, resulting in improved safety and increased delivery. The SmSNPs phytonanomedicines exhibited stability, crystallinity, purity, and a most dispersed nanosized structure. When exposed to sugar-stressed H9C2 cardiac cells, it successfully restored the size of a cell, lipid peroxide generation and nuclear form [123]. Resveratrol is a natural polyphenol molecule that is utilized for the cure of CVD. Despite its unfavorable pharmacokinetics features, including sluggish photostability, low solubility, and extensive first passage metabolism, its limited bioavailability significantly impairs its therapeutic potential [124, 125]. Lipid nanoparticles can serve as carriers, providing an opportunity to develop novel therapies to address this problem. Furthermore, solid lipid nanoparticles show a pronounced ability in attaining the goal of regulated and site-specific administration of drugs [126]. The research utilized F127 micelles to encapsulate resveratrol and curcumin, resulting in enhanced solubility of both compounds in water. Additionally, this encapsulation technique boosted cardiovascular safety in H9C2 cells [127]. The heart rate, fractional shortness, and ejection fractions in doxorubicin-mediated cardiac toxicity mice were considerably improved by encapsulating resveratrol in solid lipid nanoparticles. Furthermore, following the administration of medication, myocardial fiber exhibited a normal arrangement and a slight degradation of vacuoles was seen in the cells of the myocardium [128].

Nanocarriers made of polyvinylpyrrolidone-amphiphilic carboxymethyl-hexanoyl chitosan (CHC) range between 90 and 170 nm in size have been employed to transport the herbal medicine demethoxycumin. CHC nanocarriers boosted the release of drugs, facilitated rapid incorporation into cells, and increased the growth inhibition and movement of smooth muscle cells. Furthermore, the in vitro studies indicated prolonged therapeutic benefits and the incorporation of medications into nanocarriers could potentially provide an alternative therapy approach for CVD and demonstrate exceptional stability [129]. *Panax notoginsenoside saponin* (PNS) is the main bioactive constituent. of *P. notoginseng*, which was utilized for medical intentions [130]. PNS exhibits low bioavailability upon oral ingestion [131]. According to Deng et al. (2015), salvianolic acid B (Sal B), an active compound found in the *Salvia miltiorrhiza* roots, has the ability to safeguard the heart and the arteries. The RGD-S/P-LPNs system, consisting of Arg-Gly Asp (RGD) attached lipid polymer hybrid nanoparticles, was developed to evaluate the impact of co-delivering SalB and PNS on the combined treatment of acute myocardial ischemia [132]. The laboratory-based studies showed that tailored lipid-polymer, hybrid NPs (LPNPs) were highly effective nanocarriers. These LPNPs exhibited superior stability in serum and provided extended drug administration, making them appropriate for long-term use. Moreover, the in vivo dual drug investigations indicated that the RGD-S/P-LPNPs might enhance the effectiveness of the medications in terms of cardiac dissemination, pharmacokinetics, and therapy of infarcts [133]. *Dracocephalum moldavica* includes the natural flavonoid component tilianin, which has been utilized for the management of CVD such as coronary heart disorder, myocardial ischemia, atherosclerosis and high blood pressure [88–90]. The study examined the therapeutic impact of tilianin-loaded micelles (TLMs) on H9C2 cardiomyocytes subjected to hypoxia-reoxygenation. The use of TLMs resulted in a reduction of LDH level, the preservation of ROS generation, an enhancement in cell survival, and a substantial decrease in the level of malondialdehyde in the cardiomyocytes. PNMs were found to decrease the level of IL-1 and TNF-α and lower the rate of apoptosis in H9C2 cardiomyocytes [134]. The phytonanomedicines and their efficacy in managing neurodegenerative diseases have been arranged in Table 9.4.

TABLE 9.4
Phytonanomedicines for the Treatment of Cardiovascular Diseases

Nanoform/Nanocarrier	Phytonanomedicine Function	Reference
Poly [lactic-co-glycolic] acid (PLGA) nanoparticles	The application of nanoemulsion-based quercetin led to the promotion of antioxidant activity and attraction, adhesion, proliferation and expression of cardiac protein within the myocardium	[135, 136]
Silver nanoparticles	Laboratory research was conducted to assess the effects of *Syzygium cumini* synthesized silver NPs on glucose-mediated heart problems. As a consequence, there was an improvement in the treatment of CVDs caused by diabetes. When exposed to high glucose levels. H9C2 cells of the heart exhibited a recovery in their size, nucleus structure and generation of lipids peroxides	[123]
Polyvinylpyrrolidone amphiphilic carboxymethyl hexanoyl chitosan (CHC)	CHC nanocarriers that were employed in the transportation of the phytomedicine demethoxycurcumin demonstrated greater delivery of the drug, increased restriction of the movement, and the growth of smooth muscle cells in the vascular system	[129]
F127 micelles	The study evaluated the use of polymeric micelles for delivering curcumin and resveratrol together to reduce the heart problems caused by doxorubicin in a laboratory setting. The results demonstrated enhanced solubility in a water-based solution and better cardiovascular protection in H9C2 cells	[127, 128]
Arginylglycylaspartic acid (RGD) lipid polymer	Salvianolic acid B and *Panax notoginsenoside saponin* on the combined treatment of acute ischemia of the myocardium using a lipid polymeric NPs system coupled with Arg-Gly Asp (RGD) was assessed. As a consequence, the phytomedicine's performance in heart dispersion was enhanced	[133]
Micelles	The amounts of LDH were reduced, the health of cells was improved, ROS production was preserved and the amount of MDA was reduced in heart muscle cells.	[134]

9.3 NANOPARTICLES FOR PLANT DISEASE MANAGEMENT

Nanotechnology has brought creative solutions to a variety of industries, including agriculture, where it offers tremendous potential for plant disease management [137]. Nanoparticles (NPs) have distinct qualities such as high surface area-to-volume ratio, increased reactivity, and the potential to interact at the molecular level with pathogens [138]. This chapter looks at how silver (Ag), gold (Au), zinc (Zn), palladium (Pd), titanium (Ti), iron (Fe), selenium (Se) and copper (Cu) nanoparticles can be used to manage plant disease (Figure 9.2).

9.3.1 Role of Silver Nanoparticles (AgNPs) in Plant Disease Management

Silver nanoparticles are recognized for their powerful antibacterial characteristics, making them an effective tool for plant disease management [139]. AgNPs produce antibacterial activity in a variety of ways. They impair the integrity of microbial cell membranes, producing structural damage and increased permeability, which finally leads to cell death [140]. Furthermore, AgNPs cause the creation of reactive oxygen species (ROS), which harm essential cellular components like proteins, lipids and DNA. AgNPs impair important biological functions in pathogens by interacting with proteins and containing sulphur and DNA-carrying phosphorus. The practical applications of AgNPs to manage plant disease are numerous. They can be applied as foliar sprays to control bacterial and fungal infections such as Xanthomonas, Pseudomonas and Botrytis [141]. When applied to soil, AgNPs aid in treating root infections caused by fungi like Fusarium and Rhizoctonia. Furthermore, covering seeds with AgNPs increases germination rates and protects seedlings from seed-borne diseases, resulting in healthier plant growth from the start [142].

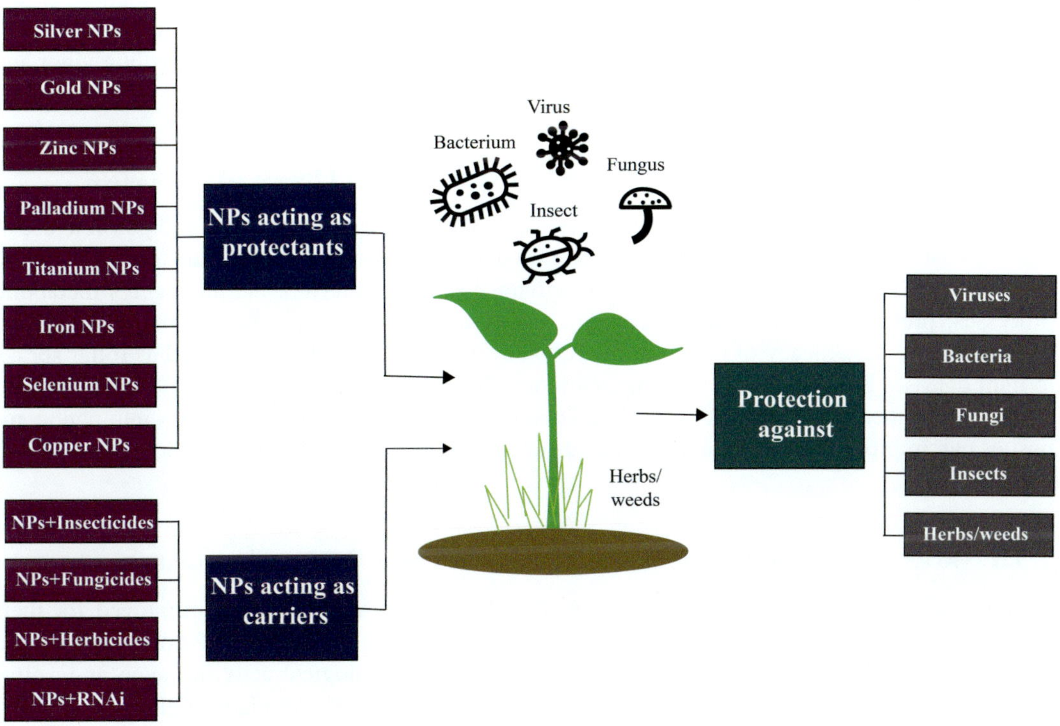

FIGURE 9.2 A plant protection method that uses nanoparticles as protectants or carriers.

9.3.2 Role of Gold Nanoparticles (AuNPs) in Plant Disease Management

Gold nanoparticles are valued for their stability and biocompatibility, and their synthesis may be precisely controlled to obtain specific shapes and sizes, which influence their interactions with pathogens [143]. AuNPs attack plant pathogens using a variety of ways. They interrupt the electron transport pathway in microbial cells, causing ATP depletion [144]. AuNPs can also alter the genes' expression linked with stress response and pathogenicity in pathogens, decreasing their virulence [145]. AuNPs can also elicit systemic acquired resistance (SAR) in plants, which improves their overall resistance to a varied range of pathogens. AuNPs are applied in agriculture for anno-priming seeds, which improves germination, growth, and disease resistance. Foliar sprays of AuNPs act as plant immunity boosters, fortifying plant defenses, and making them more resistant to diseases [146].

9.3.3 Role of Zinc Nanoparticles (ZnNPs) in Plant Disease Management

Zinc nanoparticles (ZnNPs) show a significant part in enzyme function and protein synthesis, making them crucial for plant nutrition and disease management [147]. ZnNPs supply zinc ions that are required for plant growth and development. They possess antibacterial activities by producing reactive oxygen species (ROS) and damaging microbial membranes. Furthermore, ZnNPs promote the expression of genes related to defense in plants, hence increasing their innate immunity to pathogens [148]. In practice, ZnNPs can be utilized as foliar sprays to control diseases like powdery mildew and leaf blight. Zinc is commonly administered to prevent plant disease, particularly fungal diseases, though, zinc oxide nanoparticles (ZnO-NPs) are more effective at inhibiting the development of fungal pathogens in plants. Plant fungal diseases like *Fusarium oxysporum* ZnO-NPs are controlled by distorting growing mycelia, eliminating mycotoxins like fusaric acid, and altering the membrane integrity and morphology of macroconidia via increased lipid peroxidation, ergosterol content, and ROS levels [149]. When ZnNPs are absorbed into the soil, they increase nutrient availability and decrease soil-borne disease. Seed coating with ZnNPs improves seedling vigor and disease resistance, resulting in healthier crop development [150].

9.3.4 Role of Palladium Nanoparticles (PdNPs) in Plant Disease Management

Palladium nanoparticles are gaining popularity for their catalytic characteristics and possible use in plant disease management [151]. PdNPs have antibacterial activity principally due to their catalytic reduction of ROS, which is harmful to microorganisms. They interrupt the pathogen's metabolic processes, preventing their development and proliferation. PdNPs also have synergistic effects, which increase the efficacy of other antimicrobial drugs [152]. PdNPs can be used as foliar sprays to fight diseases like Altermaria and Phytophthora. PdNPs, when employed as soil treatments, aid in the management of fungal and bacteria-related root illness. Incorporating PdNPs into nanocomposites can improve antibacterial activity and stability, offering long-term protection against plant diseases [153].

9.3.5 Role of Titanium Nanoparticles (TiNPs) in Plant Disease Management

Titanium nanoparticles, specifically titanium dioxide (TiO_2), are commonly employed for their photocatalytic capabilities and stability [154]. When exposed to light, TiO_2 NPs produce reactive oxygen species (ROS), which harm bacteria. They interrupt biological activities such as respiration and DNA replication, causing pathogen death. TiO_2 NPs can trigger plant's defense pathways, increasing resistance to pathogens. TiO_2 NPs can be used to create photocatalytic coatings on plant surfaces that defend against infections when exposed to light. Adding TiO_2 NPs to soil improves plant health by suppressing infections [155, 156]. Titanium nanoparticles activate plant defense mechanisms, offering a substitute to treating fungal infections. Findings support the promising role of biogenic TiO_2 NPs in the up and down-regulation of proteins that improve wheat plant defense and disease resistance to the biotic stress caused by *Puccinia striiformis* [157].

9.3.6 Role of Iron Nanoparticles (FeNPs) in Plant Disease Management

Iron nanoparticles (FeNPs) have appeared as a viable tool in plant disease management owing to their distinct characteristics and multifunctional activities [150]. These nanoparticles can successfully battle numerous plant diseases, including bacteria, fungi and viruses, thanks to their high redox potential and capability to produce reactive oxygen species (ROS). FeNPs antimicrobial action damages pathogens' cellular integrity, causing inactivation and death. Furthermore, iron is an essential micronutrient for plants, playing a significant role in a diversity of physiological processes such as chlorophyll synthesis and enzyme activity. The use of FeNPs not only kills pathogens and bacteria but also improves plant development and stress resistance by increasing iron availability in soil and plant tissues. This dual role of FeNPs makes them an appealing alternative for sustainable agriculture, as they provide an integrated strategy for pest control and nutrient management. However, the environmental impact and potential toxicity of FeNPs on non-target organisms should be cautiously evaluated to certify their harmless and successful application in agriculture techniques [158].

9.3.7 Role of Copper Nanoparticles (CuNPs) in Plant Disease Management

Copper nanoparticles (CuNPs) are becoming widely recognized for their powerful antibacterial characteristics, making them an important tool in plant disease management. CuNPs have high bactericidal and fungicidal properties due to their capacity to release copper ions (Cu2+), which disrupt pathogens' critical functions such as enzyme function and cell membrane integrity. These nanoparticles can efficiently target a varied range of plant pathogens, including fungi, bacteria, and viruses, preventing diseases such as blight, rust and mildew in crops. In addition to their antibacterial properties, CuNPs can enhance plant development by acting as a micronutrient, increasing photosynthesis and strengthening plant defense mechanisms [159]. The use of CuNPs in agriculture is a possible alternative to typical chemical pesticides, minimizing environmental contamination, and the development of resistant pathogens. However, the dosage and potential phytotoxicity of CuNPs must be carefully considered as excessive copper accumulation can be harmful to plants and beneficial soil microbes. Thus, optimizing CuNPs application tactics is critical for ensuring their long-term and successful use in plant disease management [160].

9.3.8 Role of Selenium Nanoparticles (SeNPs) in Plant Disease Management

Selenium nanoparticles have appeared as a new and influential technique for plant disease management because of their unique antibacterial characteristics and minimal toxicity. SeNPs can prevent the development of a diversity of plant pathogens, including bacteria, fungi and viruses by producing ROS, which cause oxidative stress and damage to the pathogen's cellular structures. This activity disturbs their metabolic process, eventually leading to their demise. Along with their direct antibacterial effects, SeNPs can improve plant defense systems by activating antioxidant enzymes and boosting the plants' immunological response. The use of SeNPs is especially useful because selenium is an important element for plants, which aids growth and development. By introducing SeNPs into agriculture methods, it is possible to lessen reliance on traditional chemical pesticides, lowering environmental pollution and the possibility of generating resistant disease strains. However, the use of SeNPs must be carefully controlled to avoid selenium accumulation in soil and plats, which could result toxicity. Overall, selenium provides a promising and long-term method for enhancing plant health and safeguarding crops from disease [158, 161].

9.4 NANOPARTICLES AS CARRIERS

9.4.1 Nanoparticles as Carriers for Insecticides

Insecticides are chemicals that are used to kill insects while they have the potential to contaminate the ecosystem and kill non-target animals like bees and other useful insects. It may be possible to get around some of these problems with nanoparticles. Nanotechnology has demonstrated recently that it can outperform traditional pesticides in terms of adherence to crop foliage, solubility, stability, targeted distribution, and other aspects. Nano-pesticides can promote higher crop yields and lay the foundation for sustainable agriculture and worldwide food security [162]. Nanoparticles as a carrier can be useful for many reasons such as they possess the ability to augment the efficacy of delivering insecticides more precisely to their intended targets. These nanoparticles can boost their potency and minimize the amount required. These nanoparticles have lessened influence on the environment such as they are more effective at targeting insects, there will likely be less insecticide released into the environment. Nanoparticles can be engineered to lower the chance of exposure to creatures other than those intended for them, increased safety such as by releasing insecticides gradually over time [163].

It was reported by [164] that nanomaterials can withstand insects and transport medications to boost agent efficacy, and the usage of nano-agents ensures the control of vector-borne illness. [165] reported that stakeholders and farmers introduce the latest developments of nanomaterials for usage in agriculture, particularly in the fight against plant diseases and pests. It is critical to examine the benefits and drawbacks of nano-agrochemicals before using them in sustainable agriculture. Zinc oxide nanoparticles made with a non-biogenic (sol–gel) method showed two functions: they acted as a growth stimulant in maize seeds and as an insecticide against *S. Oryzae* [166]. CuO-NP is a reasonably priced and ecologically beneficial way to control mosquito larvae. It is necessary to conduct additional research to examine the long-term consequences, environmental effects, and possible field use of CuO-NPs [163]. The assessment of phytotoxicity made it abundantly evident that nanoparticles had no effect on the corn plants' morphology. The results of the nutrient analysis of soil indicated that, in comparison to the control treatments, neither the pH nor the nutrients of the soil changed. The results of the investigation unequivocally demonstrated that *S. frugiperda* larvae are toxically affected by nanoparticles [167].

Liposomes, polymer nanoparticles, dendrimers, nanoemulsions, and nanocapsules are examples of nanoparticles that can be released directly onto crops and pests in a targeted and controlled manner. This lessens toxicity and leaks into soil and water systems while increasing pesticide efficacy. Furthermore, functionalizing nanoparticles can enhance their absorption, targeting, and controlled release into the environment [168]. Two examples include (i) the use of pH-responsive nanoparticles that release pesticides when consumed by pests and (ii) ligands that bind to plant surfaces. Nanotechnology reduces the risk of exposure and environmental pollution by reducing the demand for pesticides. There are still issues to be resolved, like increased prices, accelerating production, evaluating environmental effects, and a lack of uniform laws. The development of nano-enabled pesticide delivery and additional research may open the door to next-generation crop protection strategies [169]. Using nanoparticles as insecticide carriers have some drawbacks to such as the production of nanoparticles might be costly. There is significant ambiguity around the safety of nanoparticles, and regulations pertaining to nanotechnology are still developing. Although the effects of continuous exposure to these nanoparticles on the environment are still unknown, they do have an effect.

9.4.2 Nanoparticles as Carriers for Fungicides

Similar to pesticides, fungicide-chemical nanoparticles have the potential to be utilized to treat fungal diseases in plants. The following are some ways in which fungicide applications could be revolutionized using nanoparticles. Nanoparticles as fungicides can enhance the effectiveness, even at a lower overall dosage, and have the ability to deliver them straight to fungal cells. Nano-fungicides have better solubility and are challenging due to their poor water solubility [170]. Nano-fungicides have limited water solubility and it might be difficult to apply them consistently. These fungicides can be encapsulated by nanoparticles, which will increase their solubility and allow for more effective application techniques like spraying. Certain fungal infections are targeted by nano-fungicides, which also reduce the harm to beneficial microbes and the likelihood that non-targeted fungus

The most potent fungicide was AgNPs. *A. alternata* was the most susceptible, whereas *M. phaseolina* had the highest statistical resistance. This is the first report that we are aware of about *R. stricta* antifungal activity against these species [171] . Since SLNs exhibit promise against other fungal plant diseases, they are highly intriguing as biodegradable transport agents that can be injected into the trunk to prevent harmful fungal illnesses [172]. The antifungal analysis showed that there were severe ultra-structural changes, together with a considerable inhibition of the mycelial growth of all species. AgNPs were the most successful fungicide among all the treatments. M. phaseolina exhibited the highest statistical resilience, whereas A. alternate was the most vulnerable. This is the first study on the antifungal activity of R. stricta against these species. [171]. The literature emphasizes the use of nanotechnology to enable transdermal administration of antifungal ointments based on NPs. It also explains the new transdermal method by using a variety of nanoparticles that effectively deliver medication to the intended spot. This study includes a summary of earlier research and developments along with the current nanoparticle-based ointments [173].

It was suggested by [174] that the mechanism of fungicide-loaded MSN regulates the amino acid metabolic pathways to protect plants from the harmful effects of fungicides. Numerous metallic nanoparticles (MNPs) have already been used to show that they can act as fungicidal agents in an efficient and different way. MNPs have a lot of promise due to their inherent antifungal qualities and the transport ability of antifungal medications. For example, gold nanoparticles (Au-NPs) can cause the Au-NP-mediated death of cells in *Candida albicans* by disrupting the calcium homeostasis in the mitochondria. Additionally, there were notable inhibitory effects of copper oxide nanoparticles against pathogenic fungi. Strong antifungal effects were demonstrated by silver nanoparticles against a variety of pathogenic fungi, including, *C. tropicalis, A. fumigatus, Penicillium brevicompactum Trichophyton rubrum, Cladosporium cladosporioides*, and *C. albicans*. When applied to *A. niger* and *P. chrysogenum*, iron oxide nanoparticles had strong antifungal properties [175].

9.4.3 Nanoparticles as Carriers for Herbicides

Herbicide delivery using nanoparticles is being investigated as a unique approach. Herbicides can be encapsulated in these minuscule particles, which are thousands of times smaller than the breadth of a human hair. Farmers may be able to use fewer herbicides overall if they utilize nanoparticles to prevent the herbicide from degrading. This may lessen the expense of managing weeds and the harm that pesticides cause to the environment [178]. Nanoparticles can be made to specifically target weeds, such as those that have developed resistance to conventional pesticides. This may lessen the demand for broad-spectrum pesticides and increase the efficiency of weed control. PCL nanoparticles may find some uses in agriculture as herbicide nano-carriers due to their low toxicity, high loading capacity, ability to dissolve herbicides, and large-scale synthesis potential from low-cost materials. They may also help to reduce the chemical breakdown of herbicides. In agricultural nanotechnology, the creation of sustainable nanomaterials is promoted in an effort to decrease the usage of traditional pesticides, herbicides, and fertilizers. Metal-organic Frameworks (MOFs) are porous materials composed of metal nodes or clusters joined by covalent bonds and functional organic ligands. The use of nano-agrochemicals has the ability to revolutionize agriculture by boosting crop yields while lowering environmental pollution. It offers a succinct synopsis of the synthesis of green nano-agrochemicals, their agricultural uses, and their effectiveness against pests, including weeds and insects. Nano-agrochemical pesticides are being researched because of their extraordinary performance benefits over traditional pesticides and their distinctive size [165].

NPs can enter a plant by its seeds, roots, or leaves. The apoplastic and simplistic routes are primarily responsible for NP translocation. The kind, charge, size, and concentration of NPs all have a significant impact on how they affect plant growth with respect to the particular plant species. The intrinsic photoluminescence of NPs or their combination with fluorescent dyes allows them to emit fluorescence, making it possible to track their supply throughout the crop. The advantages and difficulties of NPs are explored by researchers and further study will allow NPs to support targeted and sustainable agriculture [179]. The herbicide's genotoxicity was shown to be lessened by the nanoparticle systems through the use of the *Allium cepa* chromosomal aberration assay. The created formulations provide an effective way to manage agricultural weeds while lowering the possibility of harmful effects on the environment and public health [180]. By tackling these issues and encouraging interdisciplinary research, the field of nanoparticle transport in plants is well-positioned to significantly advance sustainable agriculture and environmental management [181].

9.4.4 Role of Nanoparticles and RNAi in Plant Disease Management

Research is being done on the powerful combination of RNA and nanoparticles for plant protection. RNA interference (RNAi) is a natural defense mechanism found in plants. The method to silence undesirable genes makes use of certain RNA molecules. Scientists can target genes critical to pests or pathogens with small RNA molecules by nanoparticles and protecting plants [182]. These particles function as microscopic transporters by encasing the RNA molecules and shielding them from deterioration. Researchers are always working to improve the properties of these nanoparticles, which can be created from a variety of materials. Additionally, this evaluation offers helpful advice on selecting the best nanoparticle-based technology for efficiently and sustainably delivering dsRNA for pest control. Additionally, suggests future lines of inquiry for improving pest management techniques by utilizing dsRNA and nanoparticles [183].

A detailed and comprehensive knowledge of the connection between nanoparticles and plants will provide genetic engineers and plant scientists with a roadmap for creating plant biomarkers and investigating their possible uses in crop development [184]. The study shows that plant endogenous genes can be transiently silenced using polymer-encapsulated dsRNA to provide long-lasting resistance against pests. This could be a useful technique for protecting crops and maintaining productivity [185]. Chitosan/SPc complex (CSC) may increase the stability of dsRNA and result in a 7% decrease in fluorescence intensity by nuclease treatment. Pathogens' efficiency of dsRNA absorption was effectively increased by CSC and chitosan (CS). Moreover, CSC may lessen pathogen invasion and increase the duration of dsRNA's protection up to twenty days. This work offers a fresh and successful SIGS-based method for creating RNA-based fungicides [186].With the advancement of phytonanotechnology, modern sustainable green agriculture is now feasible which is used to protect and cultivate plants while anticipating phytopathogens and boosting the immune system beforehand. Additionally, the use of phytonanotechnology in genetic engineering to improve plants' resilience and nutritional status, by boosting agricultural yields and crop quality. Furthermore, the benefits of edible nanocoatings formed by phytonanotechnology in the post-harvest preservation of crops and other aspects are appearing rapidly. Research is being done on the powerful combination of RNA and nanoparticles for plant protection [182]. RNAi offers a more focused and eco-friendly option which denotes the temporary silencing of disease- or insect-deadly genes for agricultural protection. The capacity of various nanoparticle-mediated delivery materials to distribute double-stranded RNA (dsRNA) more effectively than traditional methods has been developed over time [187].

REFERENCES

1. Bernhardt, E.S., et al., An ecological perspective on nanomaterial impacts in the environment. *Journal of Environmental Quality*, 2010. 39(6): 1954–1965.
2. Tiwari, J.N., R.N. Tiwari, and K.S. Kim, Zero-dimensional, one-dimensional, two-dimensional and three-dimensional nanostructured materials for advanced electrochemical energy devices. *Progress in Materials Science*, 2012. 57(4): 724–803.

3. Monica, R.C. and R. Cremonini, Nanoparticles and higher plants. *Caryologia*, 2009. 62(2): 161–165.
4. Buzea, C., I.I. Pacheco, and K. Robbie, Nanomaterials and nanoparticles: sources and toxicity. *Biointerphases*, 2007. 2(4): MR17–MR71.
5. Banerjee, J. and C. Kole, Plant nanotechnology: an overview on concepts, strategies, and tools. *Plant nanotechnology: Principles and practices*, 2016: Springer, Cham: pp. 1–14.
6. Drexler, E., *Engines of creation: The coming era of nanotechnology* 1987: Anchor.
7. Thakur, A., P. Thakur, and S.P. Khurana, *Synthesis and applications of nanoparticles*. Vol. 1. 2022: Springer.
8. Hulla, J., S.C. Sahu, and A.W. Hayes, Nanotechnology: History and future. *Human & Experimental Toxicology*, 2015. 34(12): 1318–1321.
9. Khan, W.S. and R. Asmatulu, *Nanotechnology emerging trends, markets, and concerns*, in *Nanotechnology safety* 2013, Elsevier. p. 1–16.
10. Ansari, M.A., et al., Sol–gel synthesis of dy-substituted Ni0.4Cu0.2Zn0.4 (Fe2-xDyx) O4 nano spinel ferrites and evaluation of their antibacterial, antifungal, antibiofilm and anticancer potentialities for biomedical application. *International Journal of Nanomedicine*, 2021. 16: pp. 5633–5650.
11. Ray, S.S. and J. Bandyopadhyay, Nanotechnology-enabled biomedical engineering: Current trends, future scopes, and perspectives. *Nanotechnology Reviews*, 2021. 10(1): pp. 728–743.
12. Mazayen, Z.M., et al., Pharmaceutical nanotechnology: from the bench to the market. *Future Journal of Pharmaceutical Sciences*, 2022. 8(1): p. 12.
13. Gupta, A., et al., Nanoemulsions: formation, properties and applications. *Soft Matter*, 2016. 12(11): pp. 2826–2841.
14. Tiwari, A., et al., *Nanophytomedicine: nanotechnology for herbal product development and value addition*, in *Phytopharmaceuticals and Herbal Drugs*2023, Elsevier. p. 197–212.
15. Mohapatra, P., P. Singh, and S.K. Sahoo, Phytonanomedicine: A novel avenue to treat recurrent cancer by targeting cancer stem cells. *Drug Discovery Today*, 2020. 25(8): pp. 1307–1321.
16. Hafez, D.A., et al., Nanomedicine-based approaches for improved delivery of phyto-therapeutics for cancer therapy. *Expert Opinion on Drug Delivery*, 2020. 17(3): pp. 279–285.
17. Sharma, R., J. Hazra, and P. Prajapati, Nanophytomedicines: A novel approach to improve drug delivery and pharmacokinetics of herbal medicine. *Bio Bull*, 2017. 3(1): pp. 132–135.
18. Kumar, K. and A. Rai, Miraculous therapeutic effects of herbal drugs using novel drug delivery systems. *International Research Journal of Pharmacy*, 2012. 3(2): pp. 27–30.
19. Gunasekaran, T., et al., Nanotechnology: an effective tool for enhancing bioavailability and bioactivity of phytomedicine. *Asian Pacific Journal of Tropical Biomedicine*, 2014. 4: pp. S1–S7.
20. Sandhiya, V. and U. Ubaidulla, A review on herbal drug loaded into pharmaceutical carrier techniques and its evaluation process. *Future Journal of Pharmaceutical Sciences*, 2020. 6: pp. 1–16.
21. Rahman, H.S., et al., Novel drug delivery systems for loading of natural plant extracts and their biomedical applications. *International Journal of Nanomedicine*, 2020. 15: pp. 2439–2483.
22. Aqil, F., et al., Bioavailability of phytochemicals and its enhancement by drug delivery systems. *Cancer Letters*, 2013. 334(1): pp. 133–141.
23. Thakur, L., et al., Novel approaches for stability improvement in natural medicines. *Pharmacognosy Reviews*, 2011. 5(9): p. 48.
24. Dobhal, C., et al., *Multi-omics Approach Towards Cancer Therapy*, in *Personalized and Precision Nanomedicine for Cancer Treatment*2024, Springer. p. 313–338.
25. Olver, I.N., Cancer symptoms and side effects of treatment. *The MASCC Textbook of Cancer Supportive Care and Survivorship*, 2011: pp. 3–7.
26. Sung, H., et al., Global cancer statistics 2020: GLOBOCAN estimates of incidence and mortality worldwide for 36 cancers in 185 countries. *CA: A Cancer Journal for Clinicians*, 2021. 71(3): pp. 209–249.
27. Balaha, H.M. and A.E.-S. Hassan, Skin cancer diagnosis based on deep transfer learning and sparrow search algorithm. *Neural Computing and Applications*, 2023. 35(1): pp. 815–853.
28. Lee, J.S., et al., A novel chitosan nanosponge as a vehicle for transepidermal drug delivery. *Pharmaceutics*, 2021. 13(9): pp. 1329.
29. Ansari, K.A., et al., Cyclodextrin-based nanosponges for delivery of resveratrol: in vitro characterisation, stability, cytotoxicity and permeation study. *Aaps Pharmscitech*, 2011. 12: pp. 279–286.
30. Giacone, D.V., et al., Effect of nanoemulsion modification with chitosan and sodium alginate on the topical delivery and efficacy of the cytotoxic agent piplartine in 2D and 3D skin cancer models. *International Journal of Biological Macromolecules*, 2020. 165: pp. 1055–1065.

31. Shakeel, F., et al., Chemoprevention of skin cancer using low HLB surfactant nanoemulsion of 5-fluorouracil: A preliminary study. *Drug Delivery*, 2015. 22(4): pp. 573–580.
32. Asasutjarit, R., et al., Optimization of production parameters for andrographolide-loaded nanoemulsion preparation by microfluidization and evaluations of its bioactivities in skin cancer cells and uvb radiation-exposed skin. *Pharmaceutics*, 2021. 13(8): pp. 1290.
33. Falamas, A., C.A. Dehelean, and S.C. Pinzaru, Monitoring of betulin nanoemulsion treatment and molecular changes in mouse skin cancer using surface enhanced Raman spectroscopy. *Vibrational Spectroscopy*, 2018. 95: pp. 44–50.
34. Kaplan, A.B.U., et al., Formulation and in vitro evaluation of topical nanoemulsion and nanoemulsion-based gels containing daidzein. *Journal of Drug Delivery Science and Technology*, 2019. 52: pp. 189–203.
35. Yousef, S.A., et al., Mechanistic evaluation of enhanced curcumin delivery through human skin in vitro from optimised nanoemulsion formulations fabricated with different penetration enhancers. *Pharmaceutics*, 2019. 11(12): pp. 639.
36. Mukerjee, A., et al., Development, characterization and evaluation of cinnamon oil and usnic acid blended nanoemulsion to attenuate skin carcinogenicity in swiss albino mice. *Biocatalysis and Agricultural Biotechnology*, 2019. 20: pp. 101227.
37. Ahmad, N., et al., A comparative ex vivo permeation evaluation of a novel 5-Fluorocuracil nanoemulsion-gel by topically applied in the different excised rat, goat, and cow skin. *Saudi Journal of Biological Sciences*, 2020. 27(4): pp. 1024–1040.
38. Kakumanu, S., et al., A nanoemulsion formulation of dacarbazine reduces tumor size in a xenograft mouse epidermoid carcinoma model compared to dacarbazine suspension. *Nanomedicine: Nanotechnology, Biology and Medicine*, 2011. 7(3): pp. 277–283.
39. Sooksai, N., et al., Andrographolide-loaded nanoemulsion and its activity against non-Melanoma skin cancer cells. *Key Engineering Materials*, 2019. 819: pp. 139–144.
40. Shanmugapriya, K., et al., Cellulose nanocrystals/nanofibrils loaded astaxanthin nanoemulsion for the induction of apoptosis via ROS-dependent mitochondrial dysfunction in cancer cells under photobiomodulation. *International Journal of Biological Macromolecules*, 2020. 149: pp. 165–177.
41. Shi, L., et al., In vitro evaluation of 5-aminolevulinic acid (ALA) loaded PLGA nanoparticles. *International Journal of Nanomedicine*, 2013. 8: pp. 2669–2676.
42. Wang, X., et al., Topical 5-aminolaevulinic acid-photodynamic therapy for the treatment of urethral condylomata acuminata. *British Journal of Dermatology*, 2004. 151(4): pp. 880–885.
43. Da Silva, C.L., et al., Improved in vitro and in vivo cutaneous delivery of protoporphyrin IX from PLGA-based nanoparticles. *Photochemistry and Photobiology*, 2013. 89(5): pp. 1176–1184.
44. Sabitha, M., et al., Development and evaluation of 5-fluorouracil loaded chitin nanogels for treatment of skin cancer. *Carbohydrate Polymers*, 2013. 91(1): pp. 48–57.
45. Gamal-Eldeen, A.M., et al., Photodynamic therapeutic effect of indocyanine green entrapped in polymeric nanoparticles and their anti-EGFR-conjugate in skin cancer in CD1 mice. *Photodiagnosis and Photodynamic Therapy*, 2013. 10(4): pp. 446–459.
46. Ramezanli, T., et al., Polymeric nanospheres for topical delivery of vitamin D3. *International Journal of Pharmaceutics*, 2017. 516(1–2): pp. 196–203.
47. Do Reis, S.R.R., et al., Dual encapsulated dacarbazine and zinc phthalocyanine polymeric nanoparticle for photodynamic therapy of melanoma. *Pharmaceutical Research*, 2021. 38: pp. 335–346.
48. Das, S., et al., Efficacy of PLGA-loaded apigenin nanoparticles in Benzo [a] pyrene and ultraviolet-B induced skin cancer of mice: Mitochondria mediated apoptotic signalling cascades. *Food and Chemical Toxicology*, 2013. 62: pp. 670–680.
49. Tavakoli, F., et al., Effects of nano-encapsulated curcumin-chrysin on telomerase, MMPs and TIMPs gene expression in mouse B16F10 melanoma tumour model. *Artificial Cells, Nanomedicine, and Biotechnology*, 2018. 46(sup2): pp. 75–86.
50. Hussain, A., et al., Elastic liposome-based gel for topical delivery of 5-fluorouracil: in vitro and in vivo investigation. *Drug Delivery*, 2016. 23(4): pp. 1115–1129.
51. Hussain, A., et al., Optimized permeation enhancer for topical delivery of 5-fluorouracil-loaded elastic liposome using Design Expert: part II. *Drug Delivery*, 2016. 23(4): pp. 1242–1253.
52. Waheed, A., et al., Engineering of QbD driven and ultrasonically shaped lyotropic liquid crystalline nanoparticles for Apigenin in the management of skin cancer. *European Journal of Pharmaceutics and Biopharmaceutics*, 2022. 180: pp. 269–280.

53. Pushpalatha, R., S. Selvamuthukumar, and D. Kilimozhi, Cyclodextrin nanosponge based hydrogel for the transdermal co-delivery of curcumin and resveratrol: Development, optimization, in vitro and ex vivo evaluation. *Journal of Drug Delivery Science and Technology*, 2019. 52: pp. 55–64.
54. Mitri, K., et al., Lipid nanocarriers for dermal delivery of lutein: preparation, characterization, stability and performance. *International Journal of Pharmaceutics*, 2011. 414(1–2): pp. 267–275.
55. Mishra, H., et al., Co-delivery of eugenol and dacarbazine by hyaluronic acid-coated liposomes for targeted inhibition of survivin in treatment of resistant metastatic melanoma. *Pharmaceutics*, 2019. 11(4): pp. 163.
56. Lee, E.-H., S.-J. Lim, and M.-K. Lee, Chitosan-coated liposomes to stabilize and enhance transdermal delivery of indocyanine green for photodynamic therapy of melanoma. *Carbohydrate Polymers*, 2019. 224: pp. 115143.
57. Lopes, J., et al., Preliminary assays towards melanoma cells using phototherapy with gold-based nanomaterials. *Nanomaterials*, 2020. 10(8): pp. 1536.
58. Cristiano, M.C., et al., Sulforaphane-loaded ultradeformable vesicles as a potential natural nanomedicine for the treatment of skin cancer diseases. *Pharmaceutics*, 2019. 12(1): pp. 6.
59. Chaudhuri, P., S. Soni, and S. Sengupta, Single-walled carbon nanotube-conjugated chemotherapy exhibits increased therapeutic index in melanoma. *Nanotechnology*, 2009. 21(2): p. 025102.
60. Capanema, N.S., et al., Hybrid hydrogel composed of carboxymethylcellulose–silver nanoparticles–doxorubicin for anticancer and antibacterial therapies against melanoma skin cancer cells. *ACS Applied Nano Materials*, 2019. 2(11): pp. 7393–7408.
61. Huang, S., et al., Dextran methacrylate hydrogel microneedles loaded with doxorubicin and trametinib for continuous transdermal administration of melanoma. *Carbohydrate Polymers*, 2020. 246: p. 116650.
62. Muller, H., Shegokar, R. R., and Keck, C. M., 20 years of lipid nanoparticles (SLN & NLC): present state of development & industrial applications. *Current Drug Discovery Technologies*, 2011. 8(3): pp. 207–227.
63. DiMeglio, L.A., C. Evans-Molina, and R.A. Oram, Type 1 diabetes. *The Lancet*, 2018. 391(10138): pp. 2449–2462.
64. Russo, S., et al., Meta-inflammation and metabolic reprogramming of macrophages in diabetes and obesity: the importance of metabolites. *Frontiers in Immunology*, 2021. 12: pp. 746151.
65. Malone, J.I. and B.C. Hansen, Does obesity cause type 2 diabetes mellitus (T2DM)? Or is it the opposite? *Pediatric Diabetes*, 2019. 20(1): pp. 5–9.
66. Alsahli, M. and J.E. Gerich, Renal glucose metabolism in normal physiological conditions and in diabetes. *Diabetes Research and Clinical Practice*, 2017. 133: pp. 1–9.
67. Chen, L., et al., PAQR3 regulates phosphorylation of FoxO1 in insulin-resistant HepG2 cells via NF-κB signaling pathway. *Experimental Cell Research*, 2019. 381(2): pp. 301–310.
68. Schwartsburd, P., Glucose-lowering Strategies in Diabetes: Pharmacological Development of New Anti-diabetic Drugs. *Current Pharmaceutical Design*, 2018. 24(9): pp. 1007–1011.
69. Ma, Q., et al., Research progress in the relationship between type 2 diabetes mellitus and intestinal flora. *Biomedicine & Pharmacotherapy*, 2019. 117: pp. 109138.
70. Liu, M., et al., Burden of diabetes, hyperglycaemia in China from to 2016: findings from the 1990 to 2016, global burden of disease study. *Diabetes & Metabolism*, 2019. 45(3): pp. 286–293.
71. Lovic, D., et al., The growing epidemic of diabetes mellitus. *Current Vascular Pharmacology*, 2020. 18(2): pp. 104–109.
72. Saeedi, P., et al., Global and regional diabetes prevalence estimates for 2019 and projections for 2030 and 2045: Results from the International Diabetes Federation Diabetes Atlas. *Diabetes Research and Clinical Practice*, 2019. 157: pp. 107843.
73. Sun, H., et al., IDF Diabetes Atlas: Global, regional and country-level diabetes prevalence estimates for 2021 and projections for 2045. *Diabetes Research and Clinical Practice*, 2022. 183: p. 109119.
74. National Research Council et al., From monsoons to microbes: Understanding the ocean's role in human health. 1999.
75. Zhang, Y., et al., Natural plant-derived polygalacturonic acid-oleanolic acid assemblies as oral-delivered nanomedicine for insulin resistance treatment. *Chemical Engineering Journal*, 2020. 390: p. 124630.
76. Manurung, R.D., et al., Diabetic Wound Healing in IL-1β expression by Nano Herbal of Zanthoxylum acanthopodium and Rhodomyrtus tomentosa. *Research Journal of Pharmacy and Technology*, 2022. 15(5): pp. 2041–2046.

77. Ilyas, S., et al., Diabetic wound healing in FGF expression by nano herbal of Rhodomyrtus tomentosa L. and Zanthoxylum acanthopodium fruits. *Pakistan Journal of Biological Sciences: PJBS*, 2021. 24(3): pp. 401–408.
78. Aruna, A., et al., Comparative anti-diabetic effect of methanolic extract of insulin plant (Costus pictus) leaves and its silver nanoparticle. *Indo American Journal of Pharmaceutical Research*, 2014. 4(7): pp. 3217–3230.
79. Bindu, R.H., et al., Formulation characterization and antidiabetic evaluation of Talinum portulacifolium (Forssk.) loaded solid lipid nanoparticles in Streptozotocin and high fat diet induced diabetic rats. *Journal of Global Trends in Pharmaceutical sciences*, 2014. 5(4): pp. 2108–2114.
80. Merrell, J.G., et al., Curcumin loaded poly (ε-caprolactone) nanofibers: diabetic wound dressing with antioxidant and anti-inflammatory properties. *Clinical and Experimental Pharmacology & Physiology*, 2009. 36(12): p. 1149.
81. Karri, V.V.S.R., et al., Curcumin loaded chitosan nanoparticles impregnated into collagen-alginate scaffolds for diabetic wound healing. *International Journal of Biological Macromolecules*, 2016. 93: pp. 1519–1529.
82. Abdel-Halim, A.H., et al., Assessment of the anti-diabetic effect of Bauhinia variegata gold nano-extract against streptozotocin induced diabetes mellitus in rats. *Journal of Applied Pharmaceutical Science*, 2020. 10(05): pp. 77–91.
83. Coldiron Jr, A.D., R.A. Sanders, and J.B. Watkins III, Effects of combined quercetin and coenzyme Q10 treatment on oxidative stress in normal and diabetic rats. *Journal of Biochemical and Molecular Toxicology*, 2002. 16(4): pp. 197–202.
84. Al-Jameel, S.S. and S.N. Abd El-Rahman, Effect of quercetin nanoparticles on the kidney of the streptozotocin-induced diabetes in male rats: A histological study and serum biochemical alterations. *African Journal of Biotechnology*, 2017. 16(39): pp. 1944–1952.
85. Kumari, A., et al., Nanoencapsulation and characterization of Albizia chinensis isolated antioxidant quercitrin on PLA nanoparticles. *Colloids and Surfaces B: Biointerfaces*, 2011. 82(1): pp. 224–232.
86. Kumari, A., V. Kumar, and S.K. Yadav, Plant extract synthesized PLA nanoparticles for controlled and sustained release of quercetin: a green approach. *PLoS One*, 2012. 7(7): p. e41230.
87. Barkat, M.A. and M.M. Mohd. Mujeeb, Comparative study of catharanthus roseus extract and extract loaded chitosan nanoparticles in alloxan induced diabetic rats. 2013.
88. Deepa, V. and R. Sridhar, Development, characterization, efficacy and repeated dose toxicity of nanoemulsified ethanolic extract of Enicostemma littorale in Streptozotocin-induced diabetes rats. *International Journal of Phytomedicine*, 2012. 4(1): p. 70.
89. Heemels, M.-T., Neurodegenerative diseases. *Nature*, 2016. 539(7628): pp. 179–180.
90. Taylor, J.P., R.H. Brown Jr, and D.W. Cleveland, Decoding ALS: from genes to mechanism. *Nature*, 2016. 539(7628): pp. 197–206.
91. Saudou, F. and S. Humbert, The biology of huntingtin. *Neuron*, 2016. 89(5): pp. 910–926.
92. Dugger, B.N. and D.W. Dickson, Pathology of neurodegenerative diseases. *Cold Spring Harbor Perspectives in Biology*, 2017. 9(7): pp. a028035.
93. Ferri, C.P., et al., Global prevalence of dementia: a Delphi consensus study. *The Lancet*, 2005. 366(9503): pp. 2112–2117.
94. Babazadeh, A., F.M. Vahed, and S.M. Jafari, Nanocarrier-mediated brain delivery of bioactives for treatment/prevention of neurodegenerative diseases. *Journal of Controlled Release*, 2020. 321: pp. 211–221.
95. Babazadeh, A., B. Ghanbarzadeh, and H. Hamishehkar, Formulation of food grade nanostructured lipid carrier (NLC) for potential applications in medicinal-functional foods. *Journal of Drug Delivery Science and Technology*, 2017. 39: pp. 50–58.
96. Singh, B., et al., Biology and chemistry of Ginkgo biloba. *Fitoterapia*, 2008. 79(6): pp. 401–418.
97. Ferrante, R.J., et al., Therapeutic efficacy of EGb761 (Gingko biloba extract) in a transgenic mouse model of amyotrophic lateral sclerosis. *Journal of Molecular Neuroscience*, 2001. 17: pp. 89–96.
98. Jiang, F., S. DeSilva, and J. Turnbull, Beneficial effect of ginseng root in SOD-1 (G93A) transgenic mice. *Journal of the Neurological Sciences*, 2000. 180(1–2): pp. 52–54.
99. Trieu, V.N. and F.M. Uckun, Genistein is neuroprotective in murine models of familial amyotrophic lateral sclerosis and stroke. *Biochemical and Biophysical Research Communications*, 1999. 258(3): pp. 685–688.
100. Koh, S.-H., et al., Epigallocatechin gallate prevents oxidative-stress-induced death of mutant Cu/Zn-superoxide dismutase (G93A) motoneuron cells by alteration of cell survival and death signals. *Toxicology*, 2004. 202(3): pp. 213–225.

101. Mancuso, R., et al., Resveratrol improves motoneuron function and extends survival in SOD1G93A ALS mice. *Neurotherapeutics*, 2014. 11(2): pp. 419–432.
102. Kabu, S., et al., Drug delivery, cell-based therapies, and tissue engineering approaches for spinal cord injury. *Journal of Controlled Release*, 2015. 219: pp. 141–154.
103. Wang, G., et al., Advances in nanotechnology-based strategies for the treatments of amyotrophic lateral sclerosis. *Materials Today Bio*, 2020. 6: pp. 100055.
104. Mathew, A., et al. *Curcumin nanoparticles-a gateway for multifaceted approach to tackle Alzheimer's disease*. in *2011 11th IEEE International Conference on Nanotechnology*. 2011. IEEE.
105. Huang, N., et al., PLGA nanoparticles modified with a BBB-penetrating peptide co-delivering Aβ generation inhibitor and curcumin attenuate memory deficits and neuropathology in Alzheimer's disease mice. *Oncotarget*, 2017. 8(46): pp. 81001.
106. Ganesan, P., et al., Recent trends in the development of nanophytobioactive compounds and delivery systems for their possible role in reducing oxidative stress in Parkinson's disease models. *International Journal of Nanomedicine*, 2015. 10: pp. 6757–6772.
107. Bollimpelli, V.S., et al., Neuroprotective effect of curcumin-loaded lactoferrin nano particles against rotenone induced neurotoxicity. *Neurochemistry international*, 2016. 95: pp. 37–45.
108. Bhia, M., et al., Naringenin nano-delivery systems and their therapeutic applications. *Pharmaceutics*, 2021. 13(2): pp. 291.
109. Ahmad, N., et al., Poloxamer-chitosan-based Naringenin nanoformulation used in brain targeting for the treatment of cerebral ischemia. *Saudi Journal of Biological Sciences*, 2020. 27(1): pp. 500–517.
110. Md, S., et al., Neuroprotective and antioxidant effect of naringenin-loaded nanoparticles for nose-to-brain delivery. *Brain Sciences*, 2019. 9(10): pp. 275.
111. Ghaffari, F., A.H. Moghaddam, and M. Zare, Neuroprotective effect of quercetin nanocrystal in a 6-hydroxydopamine model of Parkinson disease: biochemical and behavioral evidence. *Basic and Clinical Neuroscience*, 2018. 9(5): pp. 317.
112. Zhou, Z., et al., Gold nanoclusters for optimizing the general efficacies of herbal medicines on nerve repair after spinal cord injury. *Materials & Design*, 2022. 215: p. 110465.
113. Kundu, P., et al., Delivery of dual drug loaded lipid based nanoparticles across the blood–brain barrier impart enhanced neuroprotection in a rotenone induced mouse model of Parkinson's disease. *ACS Chemical Neuroscience*, 2016. 7(12): pp. 1658–1670.
114. Gaidai, O., Y. Cao, and S. Loginov, Global cardiovascular diseases death rate prediction. *Current Problems in Cardiology*, 2023. 48(5): pp. 101622.
115. Joseph, P., et al., Reducing the global burden of cardiovascular disease, part 1: the epidemiology and risk factors. *Circulation Research*, 2017. 121(6): pp. 677–694.
116. Yusuf, S., et al., Modifiable risk factors, cardiovascular disease, and mortality in 155 722 individuals from 21 high-income, middle-income, and low-income countries (PURE): a prospective cohort study. *The Lancet*, 2020. 395(10226): pp. 795–808.
117. Radomska, A., J. Leszczyszyn, and M.W. Radomski, The nanopharmacology and nanotoxicology of nanomaterials: new opportunities and challenges. *Advances in Clinical and Experimental Medicine*, 2016. 25(1): pp. 151–162.
118. Lin, M.C., et al., State of complementary and alternative medicine in cardiovascular, lung, and blood research: executive summary of a workshop. *Circulation*, 2001. 103(16): pp. 2038–2041.
119. Sobhani, Z., et al., Medicinal plants targeting cardiovascular diseases in view of Avicenna. *Current Pharmaceutical Design*, 2017. 23(17): pp. 2428–2443.
120. Jérôme, C. and P. Lecomte, Recent advances in the synthesis of aliphatic polyesters by ring-opening polymerization. *Advanced Drug Delivery Reviews*, 2008. 60(9): pp. 1056–1076.
121. Koehn, F.E. and G.T. Carter, The evolving role of natural products in drug discovery. *Nature Reviews Drug Discovery*, 2005. 4(3): pp. 206–220.
122. Atale, N., K. Gupta, and V. Rani, Protective effect of Syzygium cumini against pesticide-induced cardiotoxicity. *Environmental Science and Pollution Research*, 2014. 21: pp. 7956–7972.
123. Atale, N., et al., Synthesis and characterization of Sygyzium cumini nanoparticles for its protective potential in high glucose-induced cardiac stress: a green approach. *Applied Biochemistry and Biotechnology*, 2017. 181: pp. 1140–1154.
124. Summerlin, N., et al., Resveratrol nanoformulations: challenges and opportunities. *International Journal of Pharmaceutics*, 2015. 479(2): pp. 282–290.

125. Shahraki, A., et al., Resveratrol nanocapsule as an efficient tool for blood pressure regulation: a study on metabolic syndrome induced mice. *Bioscience Biotechnology Research Communications*, 2017. 10(4): pp. 623–630.
126. Hesari, M., et al., Current advances in the use of nanophytomedicine therapies for human cardiovascular diseases. *International Journal of Nanomedicine*, 2021. 16: pp. 3293–3315.
127. Carlson, L.J., et al., Polymeric micellar co-delivery of resveratrol and curcumin to mitigate in vitro doxorubicin-induced cardiotoxicity. *Journal of Pharmaceutical Sciences*, 2014. 103(8): pp. 2315–2322.
128. Zhang, L., et al., Resveratrol solid lipid nanoparticles to trigger credible inhibition of doxorubicin cardiotoxicity. *International Journal of Nanomedicine*, 2019. 14: pp. 6061–6071.
129. Wang, Y.-J., et al., Forming of demethoxycurcumin nanocrystallite-chitosan nanocarrier for controlled low dose cellular release for inhibition of the migration of vascular smooth muscle cells. *Molecular Pharmaceutics*, 2012. 9(8): pp. 2268–2279.
130. Chen, Z.-H., et al., Saponins isolated from the root of Panax notoginseng showed significant anti-diabetic effects in KK-Ay mice. *The American Journal of Chinese Medicine*, 2008. 36(05): pp. 939–951.
131. Lee, J., et al., Studies on absorption, distribution and metabolism of ginseng in humans after oral administration. *Journal of Ethnopharmacology*, 2009. 122(1): pp. 143–148.
132. Deng, Y., et al., Combined salvianolic acid B and ginsenoside Rg1 exerts cardioprotection against ischemia/reperfusion injury in rats. *PLoS One*, 2015. 10(8): p. e0135435.
133. Qiu, J., et al., αvβ3 integrin receptor specific peptide modified, salvianolic acid B and panax notoginsenoside loaded nanomedicine for the combination therapy of acute myocardial ischemia. *Biomedicine & Pharmacotherapy*, 2017. 96: pp. 1418–1426.
134. Wang, Y., et al., Tilianin-loaded reactive oxygen species-scavenging nano-micelles protect H9c2 cardiomyocyte against hypoxia/reoxygenation-induced injury. *Journal of Cardiovascular Pharmacology*, 2018. 72(1): pp. 32–39.
135. Lozano, O., et al., Nanoencapsulated quercetin improves cardioprotection during hypoxia-reoxygenation injury through preservation of mitochondrial function. *Oxidative Medicine and Cellular Longevity*, 2019. 2019(1): p. 7683051.
136. Lim, C.L., et al., Precision and advanced nano-phytopharmaceuticals for therapeutic applications. *Nanomaterials*, 2022. 12(2): p. 238.
137. Fu, L., et al., Nanotechnology as a new sustainable approach for controlling crop diseases and increasing agricultural production. *Journal of Experimental Botany*, 2020. 71(2): pp. 507–519.
138. Ahmad, F., et al., Unique properties of surface-functionalized nanoparticles for bio-application: functionalization mechanisms and importance in application. *Nanomaterials*, 2022. 12(8): p. 1333.
139. Tariq, M., et al., Biological synthesis of silver nanoparticles and prospects in plant disease management. *Molecules*, 2022. 27(15): p. 4754.
140. Vazquez-Muñoz, R., et al., Enhancement of antibiotics antimicrobial activity due to the silver nanoparticles impact on the cell membrane. *PloS One*, 2019. 14(11): p. e0224904.
141. Mikhailova, E.O., Silver nanoparticles: Mechanism of action and probable bio-application. *Journal of Functional Biomaterials*, 2020. 11(4): p. 84.
142. Dutta, P., et al., Silver nanoparticles and attainment through their application in soil-borne disease management. *Journal of Mycology and Plant Pathology*, 2023. 53(3): pp. 233–247.
143. Hammami, I. and N.M. Alabdallah, Gold nanoparticles: Synthesis properties and applications. *Journal of King Saud University-Science*, 2021. 33(7): p. 101560.
144. Jeevanandam, J., et al., Green approaches for the synthesis of metal and metal oxide nanoparticles using microbial and plant extracts. *Nanoscale*, 2022. 14(7): pp. 2534–2571.
145. Elshaer, S.L. and M.I. Shaaban, Inhibition of quorum sensing and virulence factors of Pseudomonas aeruginosa by biologically synthesized gold and selenium nanoparticles. *Antibiotics*, 2021. 10(12): p. 1461.
146. Kongala, S.I., S.R. Nadendla, and P. Mamidala, Internalization and induction of defense responses in tobacco by harpinPss conjugated gold nanoparticles as a foliar spray. *Colloid and Interface Science Communications*, 2021. 43: p. 100438.
147. Thounaojam, T.C., et al., Zinc oxide nanoparticles (ZnO-NPs): a promising nanoparticle in renovating plant science. *Acta Physiologiae Plantarum*, 2021. 43: pp. 1–21.
148. Rani, S., et al., Biogenic synthesis of zinc nanoparticles, their applications, and toxicity prospects. *Frontiers in Microbiology*, 2022. 13: p. 824427.

149. Abdelaziz, A.M., et al., Potential of biosynthesized zinc oxide nanoparticles to control Fusarium wilt disease in eggplant (Solanum melongena) and promote plant growth. *BioMetals*, 2022. 35(3): pp. 601–616.
150. Kumar, A., et al., Potential applications of engineered nanoparticles in plant disease management: a critical update. *Chemosphere*, 2022. 295: p. 133798.
151. Vinodhini, S., B.S.M. Vithiya, and T.A.A. Prasad, Green synthesis of palladium nanoparticles using aqueous plant extracts and its biomedical applications. *Journal of King Saud University-Science*, 2022. 34(4): p. 102017.
152. Saleh, E.A.M., et al., Phytoassisted synthesis and characterization of palladium nanoparticles (PdNPs); with enhanced antibacterial, antioxidant and hemolytic activities. *Photodiagnosis and Photodynamic Therapy*, 2021. 36: p. 102542.
153. Yadav, S.A., et al., Fungal-derived nanoparticles for the control of plant pathogens and pests. *Fungal cell factories for sustainable nanomaterials productions and agricultural applications*, 2023: pp. 755–784.
154. Ziental, D., et al., Titanium dioxide nanoparticles: prospects and applications in medicine. *Nanomaterials*, 2020. 10(2): pp. 387.
155. Dalai, S., et al., A comparative cytotoxicity study of TiO2 nanoparticles under light and dark conditions at low exposure concentrations. *Toxicology Research*, 2012. 1(2): pp. 116–130.
156. Liao, C., Y. Li, and S.C. Tjong, Visible-light active titanium dioxide nanomaterials with bactericidal properties. *Nanomaterials*, 2020. 10(1): p. 124.
157. Satti, S.H., et al., Plant-based titanium dioxide nanoparticles trigger biochemical and proteome modifications in Triticum aestivum L. under biotic stress of Puccinia striiformis. *Molecules*, 2022. 27(13): p. 4274.
158. Vijayaram, S., et al., Applications of green synthesized metal nanoparticles—a review. *Biological Trace Element Research*, 2024. 202(1): pp. 360–386.
159. Antonio-Pérez, A., et al., Biosynthesis of copper nanoparticles with medicinal plants extracts: from extraction methods to applications. *Micromachines*, 2023. 14(10): p. 1882.
160. Liu, Y., et al., Agricultural applications and potential risks of copper-based nanoagrochemicals in crop cultivation. *Reviews of Environmental Contamination and Toxicology*, 2022. 260(1): p. 20.
161. Nikam, P.B., et al., *Selenium Nanomaterials: Contribution Toward Crop Development*, in *Nanofertilizers for Sustainable Agroecosystems: Recent Advances and Future Trends* 2023. Springer. pp. 317–342.
162. Dong, W., Y. Ren, and H. Xue, Fabrication and application of carrier-free and carrier-based nanopesticides in pest management. *Archives of Insect Biochemistry and Physiology*, 2024. 116(2): p. e22124.
163. Riaz, M.A., et al., Assessing the Effectiveness of Copper Oxide Nanoparticles (Cuo-Nps) as Larvicidal Agents against Aedes Aegypti Larvae in a Laboratory Setup. *Journal of Bioresource Management*, 2024. 11(1): pp. 14.
164. Nie, D., et al., Nanoparticles: A Potential and Effective Method to Control Insect-Borne Diseases. *Bioinorganic Chemistry and Applications*, 2023. 2023(1): p. 5898160.
165. Ali, S., et al., Nano-agrochemicals as substitutes for pesticides: prospects and risks. *Plants*, 2023. 13(1): p. 109.
166. Aisvarya, S., et al., Comparative analysis of the insecticidal activity against Sitophilus oryzae (L.) and agro-morphological characteristics of maize using non-biogenic and biogenic ZnO nanoparticles. *Environmental Science: Nano*, 2024. 11(5): pp. 2173–2187.
167. Pittarate, S., et al., Insecticidal efficacy of nanoparticles against Spodoptera frugiperda (JE Smith) larvae and their impact in the soil. *Heliyon*, 2023. 9(5): p. e16133.
168. Thabet, A.F., et al., Silica nanoparticles as pesticide against insects of different feeding types and their non-target attraction of predators. *Scientific Reports*, 2021. 11(1): p. 14484.
169. Rathore, A., et al., Nanotech for Crop Protection: Utilizing Nanoparticles for Targeted Pesticide Delivery. *Uttar Pradesh Journal of Zoology*, 2024. 45(6): p. 46–71.
170. Almawash, S., Solid lipid nanoparticles, an effective carrier for classical antifungal drugs. *Saudi Pharmaceutical Journal*, 2023. 31(7): pp. 1167–1180.
171. Al-Sahli, S.A., et al., Silver nanoparticles improve the fungicidal properties of Rhazya stricta decne aqueous extract against plant pathogens. *Scientific Reports*, 2024. 14(1): p. 1297.
172. Machado, T.O., et al., Bio-based lignin nanocarriers loaded with fungicides as a versatile platform for drug delivery in plants. *Biomacromolecules*, 2020. 21(7): pp. 2755–2763.
173. Tarannum, N., et al., Nanoparticles assisted intra- and transdermic delivery of antifungal ointment: an updated review. *Discover Nano*, 2024. 19(1): p. 11.

174. Zhao, P., et al., Fungicide-loaded mesoporous silica nanoparticles promote rice seedling growth by regulating amino acid metabolic pathways. *Journal of Hazardous Materials*, 2022. 425: p. 127892.
175. Madkhali, O.A., A comprehensive review on potential applications of metallic nanoparticles as antifungal therapies to combat human fungal diseases. *Saudi Pharmaceutical Journal*, 2023: p. 101733. https://doi.org/10.1016/j.jsps.2023.101733
176. Bratovcic, A., et al., Application of Nanotechnology in Agroecosystems: Nanoparticles for Improving Agricultural Production. *Reviews in Agricultural Science*, 2023. 11: pp. 291–309.
177. Malandrakis, A., N. Kavroulakis, and C. Chrysikopoulos. Nano-fungicides against plant pathogens: Copper, silver and zinc NPs. In *Geophysical Research Abstracts*. 2019.
178. Zargar, M., et al., New Advances in Nano-Enabled Weed Management Using Poly (Epsilon-Caprolactone)-Based Nanoherbicides: A Review. *Agriculture*, 2023. 13(10): p. 2031.
179. Zhang, H.-Y. and W.-H. Su, Classification, uptake, translocation, and detection methods of nanoparticles in crop plants: a review. *Environmental Science: Nano*, 2024. https://doi.org/10.1039/d4en00059e
180. Pereira, A.E., et al., Application of poly (epsilon-caprolactone) nanoparticles containing atrazine herbicide as an alternative technique to control weeds and reduce damage to the environment. *Journal of Hazardous Materials*, 2014. 268: pp. 207–215.
181. Sembada, A.A. and I.W. Lenggoro, Transport of nanoparticles into plants and their detection methods. *Nanomaterials*, 2024. 14(2): p. 131.
182. Jiang, M., et al., Phytonanotechnology applications in modern agriculture. *Journal of Nanobiotechnology*, 2021. 19: pp. 1–20.
183. Arjunan, N., V. Thiruvengadam, and S. Sushil, Nanoparticle-mediated dsRNA delivery for precision insect pest control: A comprehensive review. *Molecular Biology Reports*, 2024. 51(1): p. 355.
184. Daniel, A.I., et al., Application of nanotechnology and proteomic tools in crop development towards sustainable agriculture. *Journal of Crop Science and Biotechnology*, 2024. 27: pp. 1–21.
185. Pal, G., et al., Exogenous application of nanocarrier-mediated double-stranded RNA manipulates physiological traits and defence response against bacterial diseases. *Molecular Plant Pathology*, 2024. 25(1): p. e13417.
186. Wang, Y., et al., Nanoparticle carriers enhance RNA stability and uptake efficiency and prolong the protection against Rhizoctonia solani. *Phytopathology Research*, 2023. 5(1): p. 2.
187. Verma, K. and M. Modgil, RNA interference (RNAi) mediated technique for combating plant diseases: Harnessing nanoparticles for effective delivery and enhanced efficacy. *Plant Cell, Tissue and Organ Culture (PCTOC)*, 2024. 157(3): p. 53.

10 Applications of Nanotechnology in Herbal Pharmacology

Jyoti Rani, Deepak Beniwal, Sahil Dhull, and Vibhuti Gulia
Department of Botany, Chaudhary Devi Lal University, Sirsa, India

Mukul Machhindra Barwant
Department of Botany, Sanjivani Rural Education Society's (SRES), Sanjivani Arts, Commerce and Science College, Kopargaon, India

Balwant Singh
Department of Botany, Dr. Ram Manohar Lohia Avadh University, Ayodhya, India

10.1 INTRODUCTION

The rapid evolution of nanotechnology has revolutionized numerous fields, from electronics and materials science to medicine and environmental science. Among the myriad applications of nanotechnology, one of the most promising and impactful areas is its integration into herbal pharmacology (Ansari et al., 2012). For millennia, herbal medicine has been a cornerstone of healthcare. Its roots are deeply embedded in traditional practices across cultures (Mishra et al., 2022). Nanotechnology has opened up a new way to improve the efficacy, bioavailability, and therapeutic potential of herbal components, bridging the gap between old wisdom and modern scientific breakthroughs (Chakraborty et al., 2016).

In 1959, Richard Feynman's visionary lecture, 'There's Plenty of Room at the Bottom,' laid the framework for future discoveries in nanotechnology. Feynman envisioned manipulating individual atoms and molecules (Vaibhav et al., 2024). Advances in microscopy, materials science, and quantum mechanics have moved nanotechnology from theoretical concepts to practical applications during recent decades. Today, nanotechnology involves the characterization, design, and use of structure and systems by amendable size and shape on the nanoscale (Sharma, 2014). Herbal pharmacology has been practiced for millennia, drawing on traditional medicine systems like traditional Chinese medicine (TCM), Ayurveda, and Native knowledge. Herbs have been utilized to cure a variety of disorders by exploiting their complex phytochemical combinations (Chakraborty et al., 2016). These natural substances, which include alkaloids, flavonoids, terpenoids, and phenolics, have an extensive variety of biological functions, such as anti-inflammatory, antioxidant, antibacterial, and anticancer characteristics. Despite their medicinal promise, herbal medications encounter problems relating to stability, solubility, and absorption, which restrict their clinical usefulness (Sharma, 2014; Vaibhav et al., 2024).

DOI: 10.1201/9781003544418-10

Nanotechnology is the manipulation of matter at the atomic and molecular levels, usually at scales smaller than 100 nanometers. At this size, materials possess different properties, biological, physical and chemical, that differ dramatically from their bulk counterparts. These qualities can be used to enhance the administration and efficacy of herbal medications (Harika et al., 2021; Mishra et al., 2022). Uniting nanotechnology with herbal pharmacology is a process to generate nanoformulations that improve the medicinal effects of herbal substances (Ansari et al., 2012). These nanoformulations can comprise nanoparticles, nanospheres, nanoemulsions, liposomes, and dendrimers, among others. Such formulations can improve the solubility and stability of herbal substances, preserve them from degradation, increase their absorption and bioavailability, and allow targeted distribution to specific tissues or cells. (Harika et al., 2021).

Nanotechnology, a revolutionary discipline at the crossroads of physics, chemistry, biology, and engineering, has transformed several scientific fields since its inception. The ability to modify materials on the nanoscale (1 to 100 nanometers) has created new opportunities for invention and application, particularly with regard to medicine and pharmacology (Gavhane et al., 2021). Herbal pharmacology, which relies on the medicinal characteristics of plants and their derivatives, appears to benefit greatly from nanotechnology advances (Mishra et al., 2022). This chapter investigates the synergistic potential of incorporating nanotechnology into herbal pharmacology, with the objective of refining the efficacy, bioavailability, and delivery of herbal medications (Harika et al., 2021). The intersection of nanotechnology and herbal pharmacology offers hope for addressing these obstacles. Nanotechnology can increase the transport, efficacy, and safety of herbal medicines by creating nanoscale formulations (Gopi et al. 2016). Nanoparticles, nanocapsules, nanoemulsions, and nanogels can encapsulate herbal extracts or active chemicals, keeping them from degradation, increasing solubility, and allowing for regulated and targeted release (Rushikesh et al., 2021).

Many phytochemicals have low water solubility and are rapidly metabolized in the body, limiting their therapeutic efficacy. Nanotechnology can help solve this problem by enhancing the solubility and stability of these chemicals (Afroj et al., 2021). Curcumin, which is derived from turmeric, has limited bioavailability due to its low solubility and quick metabolism. Nanoparticle formulations of curcumin have demonstrated higher bioavailability, longer circulation duration, and improved therapeutic efficiency in numerous disease types (Ruturaj et al., 2023).

Nanotechnology is used for the delivery of herbal medications to precise cells and thereby improving therapeutic outcomes (Bruna et al., 2014). This is especially important in cancer treatment, as targeted delivery might boost anticancer chemical accumulation in tumour tissues while preserving healthy cells (Ruturaj et al., 2023). Nanoparticles loaded with herbal extracts, such as paclitaxel from the Pacific yew tree, have shown specific targeting and significant anticancer action against a variety of cancer cell lines (Bruna et al., 2014). Nanotechnology enables the creation of delivery systems that provide continuous and controlled release of herbal ingredients (Sahu, 2013).

This can keep therapeutic concentrations of the medicine in the body for longer periods of time, lowering the frequency of delivery and enhancing patient compliance (Rushikesh et al., 2021). Nanocapsules and nanogels, for example, can encapsulate herbal extracts and slowly release them over time, maintaining a consistent supply of active components (Sharma, 2014).

The human body contains various biological barriers that limit the efficacy of herbal medications. These include the gastrointestinal system, the blood–brain barrier, and cell membranes. Nanotechnology can aid herbal substances in overcoming these obstacles (Arifin et al., 2019). Nanoemulsions, for example, can improve the absorption of herbal components in the gastrointestinal tract, whilst liposomes can help transfer medicinal molecules across the blood–brain barrier (Vaibhav et al., 2020).

Numerous case studies show the successful use of nanotechnology in herbal pharmacology (Bruna et al., 2014). Nanoformulations of ginseng, a traditional herbal treatment, have demonstrated improved pharmacokinetics and therapeutic efficacy in alleviating exhaustion and improving immunological function. Similarly, green tea polyphenol nanoparticles have shown increased anticancer and anti-inflammatory activity (Afroj et al., 2021; Mishra et al., 2022).While using nanotechnology into herbal medicine has numerous advantages, it also presents safety and regulatory concerns (Ansari et al., 2012). Nanoparticles' unique features, such as their small size and large surface area, can cause unexpected toxicological effects (Chakraborty et al., 2016). Rigorous testing and evaluation are required to assure the safety and efficacy of nano-herbal compositions. Regulatory systems must develop to handle the problems posed by these innovative products, ensuring that they meet safety requirements and are supported by scientific evidence (Rupali, 2022).

The future of nanotechnology in herbal pharmacology seems bright, with continuing research and development targeted at improving nanoformulations and discovering new therapeutic applications (Hiwa et al., 2020). Nanotechnology advancements are projected to result in more complex and multifunctional delivery methods, such as stimuli-responsive nanoparticles that release their payload in reaction to specific physiological situations (Ansari et al., 2012). In addition, integrating nanotechnology with other cutting-edge sciences, such as biotechnology and synthetic biology, could further boost the medicinal potential of herbs (Hiwa et al., 2020).

10.2 TYPES OF NANOMATERIALS UTILIZED IN HERBAL PHARMACEUTICALS

Employing nanotechnology revolutionized the contemporary and improvised approach of pharmaceutical distribution and reaches peak efficacy with minimal toxicity used conventional herbal products. The nanomaterials were utilized to enhance properties such as solubility, bioavailability, and the targeting ability of herbal substances (also described in Figure 10.1). The types of these engineered nanomaterials are nanoparticles, nanolayers, micelles, nanocapsules, and carbon-nanotubes. That constitutes some nanomaterials explaining their characteristics and uses, mainly in the field of herbal drugs (Table 10.1).

10.2.1 Nanoparticles

Nanoparticles are solid colloidal particles that range in size from 1 to 100 nanometers. They improve the solubility, stability, and bioavailability of medicines by acting as carriers. Nanoparticles could better encapsulate poorly soluble plant extracts, perhaps improving the therapeutic efficacy of herbal medications (Sandhiya & Ubaidulla, 2020). Coupling is thus considered one of the chief advantages

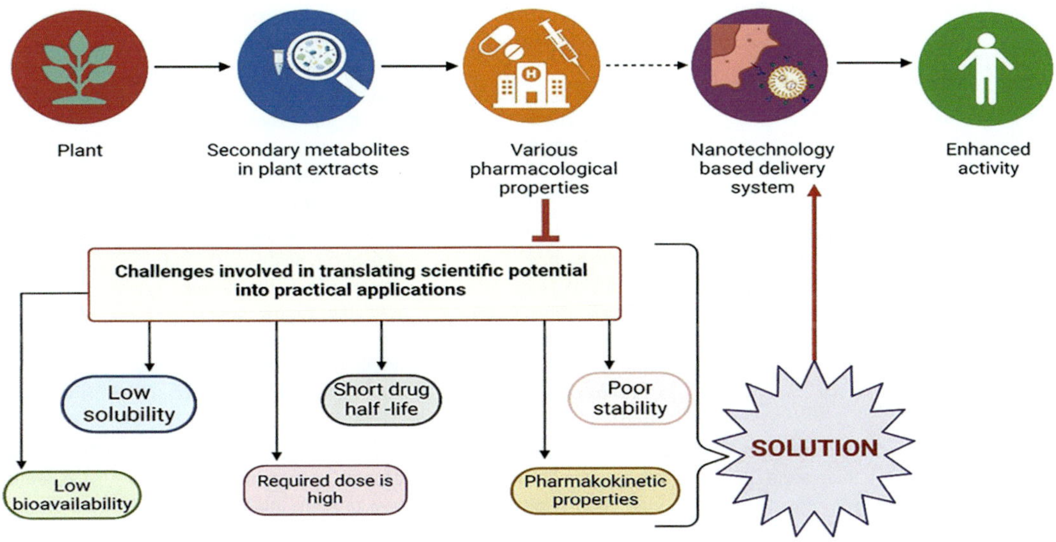

FIGURE 10.1 Innovative pharmaceutical solutions for improved treatment.

TABLE 10.1
Different Nanomaterial Types, Their Attributes, and Applications in Herbal Medicines

Type of Nanomaterials	Properties	Herbal Pharmaceuticals	Application	References
Nanoparticles	A greater dose proportionality, increased bioavailability, reduced dose form, Decreased toxicity Diminished fed or rapid variability	Curcumin, Paclitaxel, Berberin, Camptothecin, Ginkgo biloba, Triptolide, Salvia miltiorrhiza, Quercetin, Breviscapine, Naringenin, Dodder, Silymarins, Genistein, Centella asiatica, Annual mugwort	Actions to counteract cognitive decline, including memory loss, impaired reasoning, language difficulties, and behavioral changes, Actions targeting the prevention of memory loss, cognitive decline, linguistic impairment, behavioral changes, as well as inflammatory and immunological conditions. Medical conditions, particularly rheumatoid arthritis, are often associated with illnesses. Antioxidant, anxiolytic, and antimalarial.	(Chakraborty et al., 2016)
Nanoemulsions	Regulating the gradual release of active substances while ensuring their safeguarding.	Furocoumarin Psoralen, Curcumin, Triptolide	This medication is used to treat autoimmune disorders, including rheumatoid arthritis, psoriasis, leukaemia, and to exhibit antineoplastic effects.	(Thapa et al., 2013)
Nanocapsules	Protection and controlled release of active substances	Dutasteride, Olanzapine, Clarithromycin, Bedaquiline, Quercetin, Carvedilol	Used to produce a significant amount of drug content for both water-soluble and oil-soluble herbal drugs.They increase the efficacy of herbal treatments by employing precise and sustained delivery techniques.	(Deng et al., 2020)
Nanoliposomes	Efficient encapsulation of drugs using biocompatible materials.	Curcumin, Catechins constituents (catechin, epicatechin, epigallocatechin-3-gallate), SilymarinConstituents (silybin,taxifolin, isosilybin,silydianin, silychristin)	The results include chemoprevention, anticarcinogenesis, antiviral effects, antioxidation, anti-obesity effects, anti-inflammatory effects, antidiabetic effects, antimutagenesis, antiangiogenesis, antibacterial effects, and anti-aging effects.	(Thapa et al., 2013)
Nanofibers	High porosity, controlled release of drugs	*Nigella sativa* (seed), *Satureja mutica* (seed), Flax (seed), Turmeric (root), *Cissus quadrangularis* (stem), Aloe vera (leave), *Citrullus colocynthis* (fruit).	antioxidant, anti-inflammatory, anticancer, analgesic, blood pressure-lowering, anxiety-relieving, angiogenesis, nerve repair, Insecticide, analgesic, anesthetic, fungicide, antioxidant, and antibiosis	(Liu et al., 2023)
Quantum Dots	High stability and fluorescent	Quantum dots, PEG-encapsulated QDs, QDs encapsulated in phospholipid micelles.	Use in immunity tests to evaluate changes in protein structure and how proteins interact with one another.	(Sandhiya & Ubaidulla, 2020)

(*Continued*)

TABLE 10.1 (CONTINUED)
Different Nanomaterial Types, Their Attributes, and Applications in Herbal Medicines

Type of Nanomaterials	Properties	Herbal Pharmaceuticals	Application	References
Dendrimers	High loading capacity, branched	Isothiocyanate, Triglyceride of docosahexaenoic acid, Glucosamine, Capsaicin, N-Acetylcysteine	With their unique polyvalency and nanoencapsulation feature, dendrimers provide a novel approach to solubilizing these components and facilitating the transport of active biomolecules.Dendrimers regulate the release of loaded phytochemicals and shield their contents from harsh environments like pH fluctuations, enzymatic destruction, etc.	(Yousefi et al., 2020)
Solid Lipid Nanoparticles	Biocompatible, biodegradable, and capable of encasing lipophilic medications	Triptolide, Podophyllotoxin, Curcuminoids, Tetrandrine, Cryptotanshinone	Cytotoxic, antibacterial, antiparasitic, anti-inflammatory, antioxidant, and anti-parasitic.	(Thapa et al., 2013)
Polymeric Nanoparticles	Enhanced surface area, higher bioavailability	Triptolide, Curcumin, Camptothecin, Hypericin	These substances have properties that inhibit tumour growth, reduce oxidative stress, prevent the formation of amyloid plaques, inhibit the aggregation of platelets, and reduce inflammation.	(Thapa et al., 2013)
Carbon Nanotubes	High strength and ability to conduct electricity	Oridonin, Camptothecin, Gallic acid derivatives, Vinblastine, Betulinic acid	Surface functionalization of carbon nanotubes with polymers and other biocompatible materials has significantly enhanced their inherent characteristics. Furthermore, the shape of carbon nanotubes enables them to be combined with other nanosystems and used to create hybrid nanocarriers. These nanocarriers may then be used as theranostic modalities for the diagnosis and treatment of cancer and other diseases.	(Jogi et al., 2018; Saliev, 2019)

nanoparticles have since their surface area allows them to hold an amount of medication. This is particularly useful for delivering remedies that require concentrations for their medicinal effects. Again, the controlled release of drugs enclosed in nanoparticles can be adjusted thus facilitating delivery (Wahab et al., 2022). For instance, when administered orally, curcumin – an extract from turmeric – is poorly water-soluble and bioavailable. Scientists have discovered that by delivering the drug encapsulated in nanoparticles, its solubility can be enhanced while avoiding metabolic degradation of curcumins, the efficacy of the drug with respect to diseases like cancer and inflammatory disorders increased hundredfold (Sohn et al., 2021).

10.2.2 Nanocapsules

In these systems, the drug is tightly caged in a core which is, in turn, covered by a membrane. This protective construct shields ingredients from degradation. It can allow for the controlled release of contents. Treatment including nanocapsules brings benefits associated with better drug stability, targeted distribution, and reduced side effects (Kaur et al., 2022). To transfer the substance to the desired location of action, the nanocapsules associate the release of their payload with signals such as pH fluctuation or the presence of an enzyme. Essential oils, for example, can be successfully contained within nanocapsules, despite their susceptibility to oxidation and disintegration. Within such encapsulation, their therapeutic capabilities are kept, and a controlled release is made possible; therefore, they exhibit improved multifunctionality in the treatment of numerous ailments (Elmarzugi et al., 2023).

10.2.3 Nanospheres

The drug, in nanosphere matrix systems, is uniformly distributed over the polymer matrix. Herbal ingredients, for which these solid colloidal particles are particularly useful, lead to a sustained therapeutic effect from the long-term, steady release. Nanosphere herbal medicines enhance the bioavailability and raise resistance to the environmental deterioration of poorly soluble herbal chemicals. Additionally, there is a possible creation of nanospheres programmed to release their content in a controlled manner, which enhances the overall effect of the herbal medicine (Dubey et al., 2022). For example, the alkaloid berberine, which has numerous therapeutic benefits, has low solubility and quick metabolism, reducing its bioavailability. These studies have revealed that berberine has great therapeutic potential in the treatment of cardiovascular and diabetes problems. Berberine is encapsulated in nanospheres, which slows the release of the substance and boosts bioavailability (Mirhadi et al., 2018).

10.2.4 Nanotubes

A nanotube is a cylindrical nanostructure with special thermal, electrical, and mechanical characteristics. Nanotubes have the potential to serve as vehicles for medication delivery in herbal medicine by bypassing biological barriers and covering huge surface areas. Carbon nanotubes (CNTs) have sparked widespread interest due to their possible applications in medicine administration. Their vast surface area allows for effective medication loading, while the capacity to pass through biomembranes allows them to carry herbal ingredients to some specified cells or tissues, if needed. The herbal medicine nanotubes increase the bioavailability and efficiency of some herbal constituents (Waris et al., 2022). For instance, resveratrol has been studied in transport mechanisms as it is associated with CNTs for its anti-inflammatory and antioxidant properties. The loading of the polyphenol compound into carbon nanotubes increases the solubility and bioavailability of resveratrol by many folds, thereby increasing its therapeutic effect properly (Kale, 2023).

10.3 INNOVATIVE APPLICATIONS OF NANOTECHNOLOGY

Nanoscience, which investigates structures and materials at the nanoscale, has brought about significant advancements in other disciplines via its creative implementations. Table 10.2 is a comprehensive summary of some crucial domains in which nanoscience is exerting a substantial influence.

10.3.1 Anti-Cancer Herbal Nanomedicine

It has been specifically shown that nanotechnology increases the potential of herbal medicines in treating cancer. Figure 10.2 illustrates how herbal nanomedicines for cancer treatment use

TABLE 10.2
Key Attributes and Outcomes of Nano Formulation with Pharmaceutically Active Compounds

Biological Activity	Nano Formulation	Pharmaceutical Active Compound	Major Outcomes	References
Anti-inflammatory activity	Nanocarrier transdermalgel	Diclofenac diethylamine with curcumin.	Enhanced biological activity. Targeted medication delivery	(Chaudhary et al., 2014)
	Nanotransfersomes	Diclofenac, diethylamine, and curcumin	High bioavailability.	(Chaudhary et al., 2013)
	Gold metallic nanoparticles	Quercetin	Enhance Bioavailability	(Ozdal et al., 2019)
	Nanoemulsion	Capsaicin	Reduce adverse effects while boosting anti-inflammatory properties	(Ghiasi et al., 2019)
	PLGA/ Polymeric nanoparticles	Capsaicin	persistent impact and focus on the action's site	(Baskaran et al., 2017)
Anticancer	Magnetic Nanoparticles	Curcumin and temozolomide	A dual drug delivery system combining curcumin and temozolomide is enhancing anticancer activity through stimulation.	(Dilnawaz & Sahoo, 2013)
	Silver Nanoparticles	*Bauhinia tomentosa* Linn Leaves extract	Effective drug for treating cancer	(Mukundan et al., 2015)
	Liposome based gold nanoparticles	Curcumin	When tried on the B16 F10 (melanoma) cell type, it caused damage to cells that could not be fixed.	(Nakamura et al., 2014)
	Zein nanocapsules	Resveratrol	An increase in efficacy was seen for Ehrlich ascites mammary tumors.	(Elzoghby et al., 2017)
	Albumin nanoparticles	Berberine	Effects that work together in lung cancer better internalization of cells higher levels of caspase-3 and lower levels of VEGF.	(Elgohary et al., 2018)
Antidiabetic Activity	Polymeric Nanoparticles	*Syzygium cumini* extract	A big improvement in the long-term problems caused by diabetes mellitus.	(Bitencourt et al., 2016)
	Zinc oxide nanoparticles	*Hibiscus subdariffa* leaf extract	Makes Th1 and Th2 cells work more Increases the production of insulin receptors Mice's blood sugar levels were raised again after they were given.	(Bala et al., 2015)
	Liposomes	Sterols	Blood sugar level drops by 50%. More insulin can pass through Caco-2 monolayers.	(Cui et al., 2015)
	Gold nanoparticles	Guavanoic acid	In vivo tests show that it has effective anti-diabetic ability in L6 rat skeletal muscle cell lines.Better intake of glucose that depends on insulin	(Govindaraju & Suganya, 2020)

(Continued)

TABLE 10.2 (CONTINUED)
Key Attributes and Outcomes of Nano Formulation with Pharmaceutically Active Compounds

Biological Activity	Nano Formulation	Pharmaceutical Active Compound	Major Outcomes	References
Antifungal Activity	Liposome	Garlic extract	Liposomes made with garlic extract showed a promise as an organic method to kill fungi in baked products.	(Pinilla et al., 2019)
	Micelles	Capsofungin + Amphotericin B	Small drugs that do not dissolve in micelles should be used more often to treat diseases in the gall bladder or bile duct.	(Hsieh et al., 2017)
	Solid lipid Nanoparticles	Terbinafine hydrochloride	The gel formulation of Terbinafine hydrochloride, which utilizes solid lipid nanoparticles, exhibited superior efficacy against *Candida albicans* in comparison to the commercially available traditional preparation.	(S et al., 2014)
	Silver, titanium dioxide, cobalt (II) hydroxide and cobalt (II,III) oxide nanomaterials	*Spirulina platensis* extract	The silver nanoparticles and cobalt (II) hydroxide nanomaterials exhibited potent antifungal activity at a concentration of 50 μg/mL, as shown by their minimum inhibitory concentration (MIC) values.	(Sidorowicz et al., 2022)
Antibacterial	Nanosuspension	Zerumbone	Formulations including particles with a size of 200 nm were created, resulting in a substantial ($p < 0.05$) increase in both the saturation solubility and dissolving rate by a factor of 2.	(Md et al., 2018)
	Nanoemulsion	Curcumin	The formulation had a droplet size of 196 nm, resulting in a significant improvement in solubility by up to 95% and an eight-fold increase in bioavailability.	(Onoue et al., 2010)
	Solid lipid nanoparticles	Triptolide	The lymphatic system efficiently absorbed SLN containing triptolide, resulting in little damage to the liver and kidneys. Enhanced anti-inflammatory efficacy was reported because of increased oral bioavailability and sustained plasma drug levels.	(MEI et al., 2005)
	Silver nanoparticles	Cinnamon, clove	The silver nanoparticles produced from cinnamon and clove displayed bigger regions of inhibition (10 mm) than amoxycillin (8 mm), indicating a high level of antibacterial efficacy.	(Cinthura & Rajaseka, 2020)
	Liposomes	Garlic oil	SLN were manufactured with an entrapment effectiveness of more than 90%. The formulation considerably increased garlic oil solubility, as evidenced by drug release studies in a phosphate-buffered media (11% in 17 hours).	(Wencui et al., 2015)
	Microspheres	Camptothecin	Camptothecin is pH sensitive inside the human body. Encapsulation in PLGA microspheres provided stability in an acidic microenvironment. The microspheres (1.3 μm) had a stronger anticancer effect when cancer cells absorbed them more effectively.	(Tong et al., 2003)

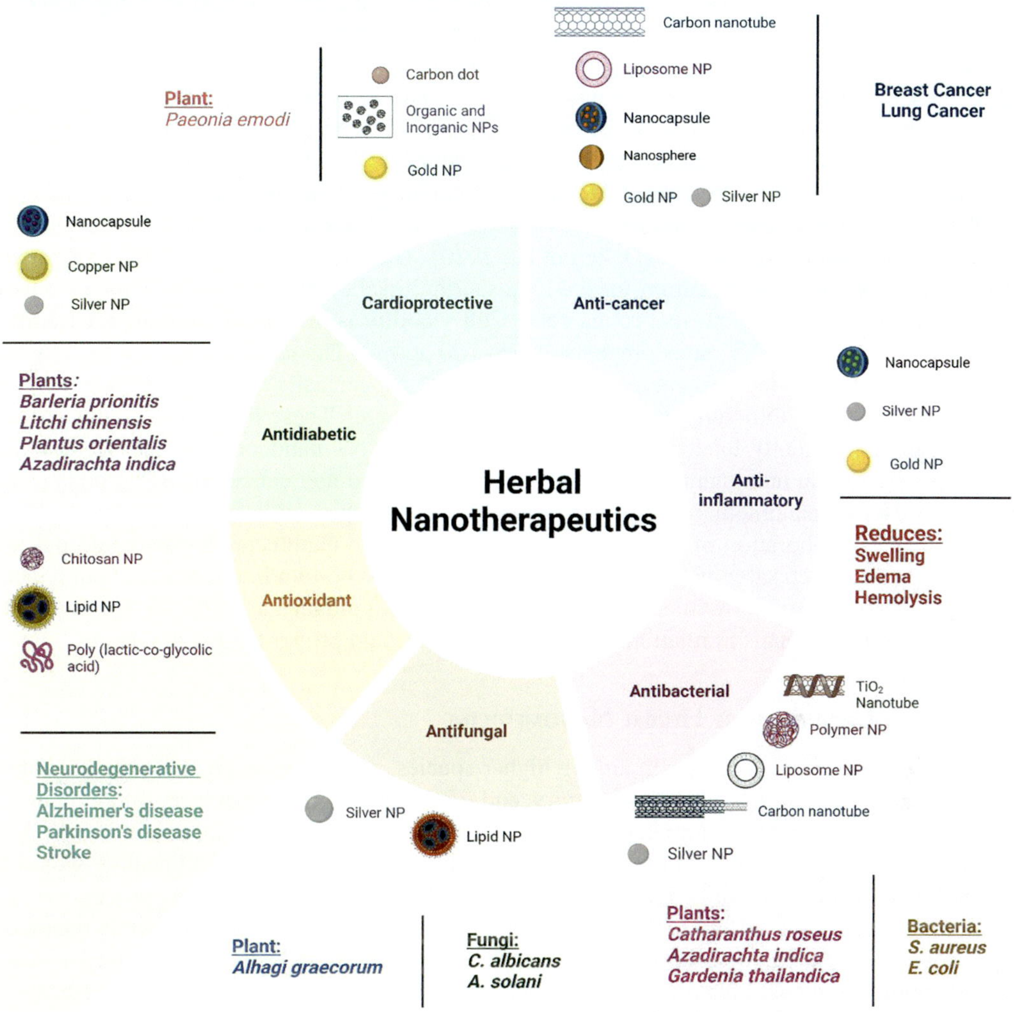

FIGURE 10.2 Herbal Nanotherapeutics: Applications and diverse nanomaterials employed in nanomedicine.

nanocarriers to improve the absorption and accurate distribution of medicinal herbal elements. Nanocarriers, such as nanoparticles, liposomes, and dendrimers, might overcome the limitations of traditional herbal therapy by shielding active components from degradation, increasing absorption, and ensuring effective targeting of cancer cells (Hare et al., 2017). All preclinical studies using in vitro and in vivo models of breast cancers grown on mice have demonstrated the efficacy of Nano Swarna Bhasma (NSB). Ayurvedic drug containing gold nanoparticles is a highly potent agent in preventing the growth of breast tumors. During the animal trials, the NSB medication showed great efficiency in regulating the growth. The NSB drug effectively inhibited tumor growth in untreated controls at doses Sof 3–7 mg per 30 kg mouse, twice a week. The findings provide strong empirical evidence supporting the therapeutic use of NSB medicine in humans for the treatment of breast cancer, as well as the potential to increase patient duration and quality of life (Khoobchandani et al., 2020). Researchers found that AgNPs derived from *Dimocarpus longan* leaf extract effectively inhibited non-small cell lung cancer (NSCLC) cell proliferation in H1299 lines by targeting NFκB, bcl-2, caspase-3, and survivin. The AgNPs significantly suppressed tumor formation when administered to SCID mice using the mouse H1299 xenograft tumor model. AgNPs are effective alternatives to chemotherapy for preventing lung cancer in living creatures. Further study is needed to assess the possible use of AgNPs in NSCLC therapy (He et al., 2016). In vitro investigations on the MCF-7 tumor cell line exposed that nanocapsules with no load (CNs) had extremely low levels of cell toxicity, indicating good biocompatibility. The cytotoxicity of 5-fluorouracil-loaded nanocapsules (CN-5-FU) was shown to be dosage-dependent, with cell viability reduced by up to 61.33%. However, the cytotoxic effect of hybrid nano capsules (CNM-4-5-FU) was less potent. The uncharged treatment for the 5-fluorouracil therapy resulted in a cell viability of 58.24% at maximum dose. The outcomes reveled that nanocapsules loaded with 5-fluorouracil efficiently decreased the cell viability in a controlled manner (Dellali et al., 2020). The MTT test was used to examine and compare the cytotoxicity of curcumin-loaded chitosan/multiwalled carbon nanotube nanocomposite on breast cancer cells and normal retina cells. Cell viability is measured 48 hours after adding curcumin, Cs, or Cs/MWCNT nanocomposite (0.78–100 µg/ml). The results show that there would be a larger reduction in the percentage of viable MCF-7 cancer cells compared to normal cells RPE1, indicating that cytotoxicity against cancer cells is stronger. These findings thus indicate that curcumin can be efficiently loaded within a Cs or a Cs/MWCNT nanocomposite for the purpose of preferentially killing malignant MCF-7 cells in comparison to non-cancerous RPE1 cells (Sobh et al., 2019). Herbal nanomedicine shows potential as a new approach in cancer treatment. Close interdisciplinary collaboration of experts in herbal medicine and nanotechnologists could pull out strategies to create much safer and more effective therapies. The bottom line is that, although nanomedicine confers good opportunities, each case needs correct guidance and specific recommendations by health professionals in relation to time-tested advice and proper treatment choices.

10.3.2 Anti-Inflammatory Herbal Nanomedicine

Inflammation is a defence mechanism among higher species, occurring as a response to hazardous stimuli such as microbial infection, tissue injury, and other types of tissue damage. Inflammation is a necessary aspect of the host's immunological response to tissue damage or infection. If the physiological response involved in inflammation were absent, then the rate of wound healing would be faster, with fatal results from infections. During the inflammatory response, healing processes begin to replace damaged tissue and fill in the gaps with fibrous tissue. As shown in Figure 10.2, nanotechnology is just one of the new technologies being used to numerous industries in a variety of ways. Different nanotechnological pharmaceutical formulations have been created to deliver medications to damaged organs while reducing unwanted drug effects caused by medication accumulation in areas not targeted (Abo-Zeid et al., 2021). Kedi et al. (2018) found that oral treatment of silver nanoparticles significantly reduced carrageenan-induced paw edema in rats. Reduction rates corresponded to the three dosages in the 1h group, which were at a 44.30% reduction, the 5h group at a 57.60% reduction, and the 5h group at a 60.50% reduction. These were inferior to the effects of the classic medicine, indomethacin, with an inhibition of 71.50% obtained at the 5th hour (Kedi et al., 2018). Silver nanoparticles (AgNPs) were tested for anti-inflammatory efficacy using an in vivo experimental model of inflammation generated by carrageenan in Wistar rats. The Animal Ethics Board at the 'Iuliu Haţieganu' University approved this work. The rats were kept in good condition before being divided into four groups: two experimental groups, one treated with the extract from the leaves of *Viburnum opulus*, one treated with AgNPs, and a positive control and a negative control that received NaCl treatment. Induction of inflammation was implemented by injecting the right hind footpads of the rats with carrageenan. The measurements for the edema of the rat paws were done in the paw at 2, 24, and 48 hours following the induction of inflammations (Moldovan et al., 2017). Levels of the cytokines in the soft paw tissues were accessed using the multiplex cytokine kit to assess inflammation. Units of various therapies used to render paw oedema using rats that had induced arthritis. On th first day after Complete Freund's adjuvant (CFA) administration, edema was found, and the control groups, which were only treated with saline, continued to show this characteristic. Resveratrol, curcumin, and combined solution administration revealed no statistically significant decrease in edema with respect to the control. Nanoencapsulation favored anti-edema action of polyphenols. Nanoencapsulated curcumin (C-LNC) significantly reduced edema on the 16th and 22nd days whereas Coencapsulated polyphenols (R-LNC) showed significant from days 15 to 20 and on day 22. RC-LNC reduced edema during days 16 to 22 with percentage reductions of 37–55%, which is very superior when compared with the solution form control (Coradini et al., 2015). Sulaiman et al. (2020) confirmed the biocompatibility of Hsp-AuNPs, which is validated by earlier research of comparable nanoparticle systems, indicating that they are safe to use in an organism. With the global population increasingly resorting to medicinal plants to manage inflammation, this discovery will pave the way for the development of new pharmaceuticals that harness the therapeutic benefits of Hsp and its active variants (Sulaiman et al., 2020). Many plants were tested for anti-inflammatory properties, but only a few have advanced to clinical studies. This is primarily due to issues with the bioavailability and solubility of herbal extracts/rutin. Nanoparticles have been utilized as herbal medication carriers, as the objective has been to improve the generally poor bioavailability of herbals or to improve tissue distribution of treatments. It also provides prolonged release while protecting the pharmaceuticals from physicochemical degradation.

10.3.3 Antibacterial Herbal Nanomedicine

Since ancient times, organic goods and remedial plants have played an important part in providing healthcare and promoting the health of various populations. Natural origin pharmaceuticals, such as medical plants, herbal remedies, animals, fungi, and marine creatures, account for approximately 50% of the market. Some of the biologically active compounds present in extracts show poor absorption and dispersion, as depicted in Figure 10.2. This may lead to reduced bioavailability and efficiency, thereby limiting their practical applications in a clinical setting. Most herbal compositions consist of biologically active ingredients with poor absorption and poor dispersion that together reduce their bioavailability and effectiveness. This can prevent applicability in the clinical setting. Several nanomedical techniques have been proposed to address this issue. These proposed techniques include polymeric nanoparticles, solid lipid nanoparticles, liposomes, microemulsions, liquid crystals, and transparent precursor carriers for liquid crystals. Overall, various sorts of nanotechnological systems, applied alone or in combination, enhance the potential of herbal prescription formulations for greater absorption and biological action (Ahmed et al., 2021). An experiment was planned to investigate the potential for wound healing following treatment with silver nanoparticles derived from two major medicinal plants, *Catharanthus roseus* and *Azadirachta indica* leaves, on female BALB/c mice in an excision wound model. Both C Ag NPs and A Ag NPs demonstrated greater wound constriction than the control and positive-control groups. C AgNPs healed the wound by 94%, while A AgNPs closed it by 87%, as opposed to silica gel, which only closed it by 74%. This faster-healing tendency was attributed to the antibacterial character, which prevented the bacterium from causing irritation and inflammation. Overall, C Ag NPs displayed greater healing than A Ag NPs (Lakkim et al., 2020). The study examined by Attallah et al. (2022) measured macroscopic wound healing rates for the various groups on days 0, 3, and 7. The treatments used either Betadine™ or *Gardenia thailandica* Tirveng silver nanoparticles (AgNPs). By the end of the experiment, leaves treated with either of the two treatments had completely healed compared to the control. By the third day, the Betadine and AgNPs groups had significantly healed their wounds, at 90.19% and 92.3%, respectively, compared to the control group. On the seventh day, the Betadine™ and AgNP groups saw nearly complete healing, with rates of 99.02% and 99.23%, respectively, compared to the control group. Furthermore, colony-forming units per milliliter (CFU/mL), as the count for the presence of bacteria, reflects a significant decrease in both groups compared with the control group. Results also reveal the fact that Betadine and AgNPs have drastically affected the healing process of the wound and have synergistically reduced the bio-burden (Attallah et al., 2022). In a *S. aureus* burn infection model, the combination of multi-walled carbon nanotube tensor and levofloxacin (MWCNT-LVX) was found to kill or limit bacterial growth. At doses of 312.5,

156.25, and 78.125 μg/mL, there were no visible bacterial colonies during the exposure time. The descriptive analysis was followed by a statistical analysis, which showed that there was a significant reduction of bacterial numbers in the MWCNT-LVX treatment groups in comparison to the control groups, where $p<0.02$, indicating that the medication through the MWCNT-LVX delivery system was better able to convey. The nontreated group had the most bacterial colonies per milliliter, while those treated with MWCNT-LVX had significantly fewer colonies ($p<0.005$) (Hassani et al., 2022). Titanium alloys have recently received a lot of attention for their use in bone implants. For these reasons, the use of titanium alloys in bone implants is hampered by various constraints, such as metal incompatibility, callus formation, and delayed fracture healing. It developed anodic oxidation with a TiO_2 nanotubular layer for enhanced surface biocompatibility. In this, an antimicrobial characteristic was given by coating it with curcumin. A curcumin drug loading on the TiO_2 surface was prepared by the direct dropping method. The results from the antibacterial activity confirm the report that the *E. coli* and *S. aureus* bacteria populations are cut back by 43% and 38%, respectively, in a 24-hour period. This research showed that the curcumin coat did not promote the rates at which the hMSCs had proliferated relative to a polished surface. It was, however, more favourable for the attachment of stem cells (Saha et al., 2021). Medicinal nanotechnology with an antibacterial property holds great prospects for the fight against bacterial infections. It will provide more effective and targeted, safe antibacterial therapeutics if developed to exploit the multiple cutting-edge advantages that only nanotechnology can offer in the future. It is along these lines that future studies are likely to greatly benefit to bringing significant advancement in the management of bacterial diseases and help combat the growing menace of antibiotic resistance.

10.3.4 Antifungal Herbal Nanomedicine

Fungal infection has emerged as a significant public health threat among patients whose immunity has been compromised, hospitalized, and chronic patients, and patients undergoing organ transplantation. These diseases are increasingly prevalent and fatal, all because of the potent property of the microorganism due to which resistant strains have developed against current treatments. The main factors responsible for the resistance include the emergence of new strains and unrestricted usage of antifungal drugs. For this reason, the medical fraternity must devise new mechanisms that assist in fighting such infections. Natural products, of course, have the greatest potential for developing new drugs for fungal infections due to their biocompatibility and low toxicology. Antibacterial activities and natural products are mainly aimed at flavonoids, terpenes, and quinones, accounting for antifungal activities. These are studied to increase the pharmaceutical drug's transportation, specificity, and medical effectiveness. Nanotechnological systems can provide a secure and specific environment for different chemicals, including natural products, thereby enhancing their antifungal efficacy (Marena et al., 2022). The efficiency of hydrogel containing terbinafine hydrochloride-loaded solid lipid nanoparticles was evaluated in vivo in a rat mycosis model generated by *C. albicans*. The results revealed a substantial reduction in Candida infection with the use of this hydrogel. In the culture test, just two of the seven animals tested positive and were infected, whereas the controls had higher rates. The hydrogel containing solid lipid nanoparticles (SLN) with terbinafine (TH) demonstrated efficacy comparable to commercially available terbinafine formulations, while requiring less frequent applications than the latter. The hydrogel can be repurposed as a sustained release treatment, requiring fewer doses ($p < 0.05$).(Rarokar et al., 2022). The antifungal activity of AgNPs was produced from *Alhagi graecorum* plant leaves and evaluated against several Candida species using the well diffusion method. The results showed inhibitory zones of 14–22 mm at a concentration of 0.01 mmol/ml and 16–27 mm at a concentration of 0.02 mmol/ml. The findings were significantly superior than those obtained with the $AgNO_3$ solution alone, 8–11 mm, and fluconazole, whose zone of inhibition is within the range 3–5 mm (Hawar et al., 2022). The in vitro experiment demonstrated that pomegranate and orange peel extracts, as well as their biosynthesized AgNPs, effectively inhibited *A. solani* mycelia at all concentrations tested. AgNPs inhibited mycelial development more effectively than peel extracts and $AgNO_3$. At a concentration of 100 μg/mL, pomegranate peel extract reduced mycelial growth by 6.12 cm, AgNPs by 16.14 cm, and $AgNO_3$ by 14.12 cm. Similarly, the orange peel extract reduced mycelial development by 5.21 cm, the AgNPs by 12.52 cm, and the $AgNO_3$ by 10.61 cm. Among the various treatments, the pomegranate peel extract and its AgNPs showed better efficacy compared to the orange peel extract and its AgNPs (Mostafa et al., 2021). Currently, the rapid increase in the number of multidrug resistant fungal strains necessitates the development of new natural antifungal categories as well as a unique drug delivery mechanism. Medical plant research has recently shown tremendous pharmacological usefulness due to the existence of active phytoconstituents. Efforts to create new drug delivery methods, including plant-based antifungals, are underway to improve efficacy against invasive mycoses and reduce high mortality rates; such a system must aid in reducing the time and cost of therapy. Antifungal chemicals can now be carried and released via a wide range of nano-drug delivery systems constructed from lipids, polymers, and metals thanks to advancements in encapsulation and materials science. The possibility of tuning nanostructures having different compositions or surface properties with distinct membrane fluidity enables one to utilize them as efficient vehicles for the transport of natural antifungal compounds (Yadav et al., 2022).

10.3.5 Antioxidant Neuroprotective Herbal Nanomedicine

In the medical field, managing central nervous system (CNS) illnesses, particularly those involving neurodegenerative processes, has traditionally been one of the most difficult challenges. The current drugs available have limited use and are not neuroprotective; in fact, the majority of them have other unwanted side effects. In this setting, there is a growing interest in the research of natural compounds that could serve as neuroprotective treatments for such illnesses, owing to their antioxidant activity. Flavonoid polyphenols such as quercetin, as well as nonflavonoids like resveratrol and curcumin, have been shown to be neuroprotective. Despite their many benefits, these chemicals have limited bioavailability and transport into the CNS due to their inability to penetrate the blood–brain barrier. An encouraging approach to counteract these limitations is to employ nanotechnology for the preparation of nanoparticles with features that can improve the efficiency of these neuroprotectors (Hort et al., 2024). Artichoke agricultural waste has revealed a comprehensive and varied herbal remedy for Alzheimer's disease (AD). This revealed a high concentration of caffeoylquinic acids and flavonoid glycosides in the artichoke bracts extract, including apigentin, luteolin, kaempferol, and quercetin. It has a significant amount of total phenolic and flavonoid chemicals. Antioxidant activity were determined in vitro using a variety of approaches. Artichoke bract extracts and chitosan-coated solid lipid nanoparticles were tested in vivo for their ability to treat Alzheimer's disease using a streptozotocin-induced mice model. Mice treated with artichoke extract or chitosan-coated artichoke-loaded solid lipid nanoparticles showed improved memory and cognitive functions, as well as down-regulation of inflammatory markers and key proteins involved in the progression of Alzheimer's disease, such as β-amyloid and tau proteins. Furthermore, neuroprotective activity was witnessed in the dentate Gyrus areas of the brain only by histopathological examination (El-Nashar et al., 2022). Zhang et al. found that administering curcumin (CUR) via intranasal route improves its transport into the brain. Curcumin-encapsulated chitosan-coated poly (lactic-co-glycolic acid) nanoparticles (CUR-CS-PLGA-NPs) and hydroxypropyl-β-cyclodextrin-encapsulated curcumin complexes (CUR/HP-β-CD) inclusion complexes improved brain delivery after intranasal administration at the same dose. CUR/HP-β-CD inclusion complexes had greater bioavailability and brain distribution compared to CUR-CS-PLGA-NPs in the same settings. CUR/HP-β-CD is easier to prepare than CUR-CS-PLGA-NPs. Hence, the former approach is more practicable for industrial magnification (Zhang et al., 2020). Because of their intrinsic potential to boost free radical scavenging activity, antioxidants will be more effective when combined as nanoparticles. Moving further, the integration of material science with nanobiotechnology has already been shown to prevent oxidative damage caused by free radicals in living biological systems more effectively. Antioxidant nanoparticles are expected to outperform all standard antioxidant therapy in terms of improving quality of life and increasing life span. Nanotechnology refers to materials that are specifically created and developed to interact with biological systems on a molecular level. This technology has the potential to profoundly change the management of neurodegenerative disorders since it stimulates, responds to, and interacts with any place within the body in such a way as to elicit a required response with minimized unwanted side effects. Without a single doubt, over the past decade, nanotechnology has been hailed and recognized worldwide as a very promising tool for not only the diagnosis but also the treatment of neurodegenerative diseases. Thus, NPs have the potential to revolutionize the course of treatment for diseases like Alzheimer's disease, Parkinson's disease, and stroke (Sandhir et al., 2015).

10.3.6 Anti-Diabetic Herbal Nanomedicine

Diabetes is a chronic metabolic illness that affects millions of people worldwide and, as a result, contributes significantly to human mortality. Plant-based therapies have long been used to treat diabetes mellitus around the world. Herbal medicines are one of many medications and alternative therapies that have been shown to be effective in treating and maintaining diabetes without causing side effects. Researchers are seeking for innovative drugs that have maximal efficacy while causing the fewest side effects. Herbal drugs are widely utilized in daily life, are easily accessible, and have few side effects. Nanoscience and nanotechnology are used extensively in a variety of fields, including green chemistry and herbal medicine research. In herbal medicine, the drug will take a longer amount of time. Because herbal-mediated nanoparticles have smaller particles, a bigger surface area, and enhanced solubility, they may be able to deliver an appropriate therapeutic dosage that can swiftly reach the systemic circulation and the intended location of action. Herbal plant extracts for nanoparticle production are an emerging area of research with enormous potential for the advancement and development of nanomedicine through innovative approaches (Shanker Kalakotla et al., 2015). Bhagwat et al. (2018) demonstrated the manufacture of copper nanoparticles (CuNPs) utilizing extracts from the leaves and peels of the *Barleria prionitis*, *Litchi chinensis*, and *Platanus orientalis* plants. The bioreduction process was quick and efficient, resulting in homogenous, tiny, and stable copper nanoparticles, which are critical for prospective nanomedicine applications. CuNPs show significant antidiabetic activity by inhibiting α-amylase and α-glucosidase enzymes. CuNPs produced from *Platanus orientalis* leaf extract demonstrated the highest level of α-glucosidase inhibition (89.31±0.17%). The CuNPs produced by the *Barleria prionitis* leaf extract had a slightly lower inhibition of 80.09±1.41%. CuNPs produced from *Litchi chinensis* peel extract had the strongest inhibition, which was as high as 25.92±1.9%. In contrast, CuNPs generated by *Barleria prionitis* leaf extract and *Platanus orientalis* leaf extract showed inhibition at 14.28±1.58% and 13.22±0.52%, respectively. As a result, phytogenic CuNPs have great promise as a therapeutic nanomaterial for the treatment of type 2 diabetes mellitus (Bhagwat et al., 2018). *Azadirachta indica* silver nanoparticles and crude extract can suppress α-amylase activity. AI-AgNPs had higher percent inhibition and a lower IC50 value (48.26 μg/mL) compared to the crude extract (68.37 μg/mL). Furthermore, diabetic mice's blood glucose levels were significantly reduced after 30 days of therapy with AI-AgNPs. The dosage of AI-AgNPs-dependent decrease of hyperglycemia has been identified. The studies indicate that AI-AgNPs are more powerful than the crude extract and can potentially be used for the management of diabetes (Rehman et al., 2023). The study looked at the effect of combination nanoformulations (NFs), including glycyrrhizin (GL)-loaded nanoparticles (NPs) and thymoquinone (TQ)-loaded nanocapsules (NCs), on diabetic rats during a 21-day period. An investigation was carried out to contrast the pure and NF forms, as well as the individual and combined applications. The combined nutraceutical formulations (NFs) of GL and TQ were significantly more beneficial than the separate NFs in terms of lowering fasting blood glucose levels, glycated hemoglobin, enhancing lipid profiles, and reducing body weight. Although the pure drug components were reduced, mixed nanofibers (NFs) outperformed all examined measures, implying increased efficacy and reduced toxicity. These data imply that combining bioactive substances at lower concentrations is more effective than using individual components alone. This highlights the need for more investigation into the synergistic effects of these compounds for the treatment of diabetes (Rani et al., 2019). Several antidiabetic phytochemicals have been produced in nanoscale formulations in this context of medicinal chemistry, with the potential to improve adherence and clinical success. These formulations address concerns with medication transportation and distribution in the body, as well as drug release properties. As a result, the creation of therapeutic nanoformulations, including antidiabetic medicines, has enormous potential for increasing their clinical efficacy. However, more research is needed to develop an efficient nanoformulation that succeeds in controlling diabetes and related complications (Dewanjee et al., 2020).

10.3.7 Cardioprotective Herbal Nanomedicine

Since the incidence of cardiovascular diseases (CVD) has been excessively high and a main cause of mortality over the last several decades, there is a crucial need to develop methods for preventing and treating CVD. Because of the severe side effects of current treatments, other alternative therapeutic options, such as medicinal herbs and natural products, are being promoted. As a result, nanoformulation is a pragmatic approach to treatment that tries to reduce therapeutic side effects while also improving drug delivery. As a result, the specific characteristics of the targeted organs serve as recommendations for the sort of nanostructures that must be generated in order to treat diseases. This high order includes nanostructures of organic and inorganic nature designed for the delivery of phytochemicals such as berberine, tilianin, naringenin, gymnemic acid, quercetin, puerarin, scutellarin, magnolol, breviscapine, resveratrol, and others. Nevertheless, some of its therapeutic potentials have been elaborated with regard to nano-curcumin or nano-resveratrol. Nevertheless, available data are not sufficient in the case of CVDs (Hesari et al., 2021). In this context, a dose of 500 mg/kg body weight of *Paeonia emodi*'s methanolic extract is thought to offer cardioprotective properties. Among the fractions, the ethyl acetate fraction of *P. emodi* was the most powerful and is hence chosen for further separation. Pe.EA40 was found to be the most potent subfraction after a scaled-up screening at 100 mg/kg dosage. Pe.EA40 decreases Lactate Dehydrogenase (LDH), Aspartate Amino Transferase (AST), Creatine Phosphokinase (CPK) and Alanine Amino Transferase (ALT) levels in a dose-dependent manner across all doses ranging from 20 mg/kg to 120 mg/kg. Compared to Pe.EA40, Pe.EA40-AuNPs displayed superior cardio-protective properties, since the above biomarkers have already shown a substantial drop even at a dose of 40 mg/kg, which is half of the dose used in the treatment of Pe.EA40 alone. Better solubility, permeability, and tissue penetration provided by AuNPs could be accountable for enhanced drug delivery and therapeutic efficiency (Muhammad et al., 2018). Carbon dots (CDs) were extracted from Curcumae Radix Carbonisata (CRC) hydrolysates after it was observed that the pre-administration of CRC-CDs lowered serum levels of cardiac enzymes and increased the antioxidant capacity of myocardial tissues in rats. One of them may be an excess oxidative stress reducer and a reduced inducer of cardiomyocyte apoptosis in cardiac tissues. CRC-CDs work in tandem with a promising medication candidate to treat myocardial ischemia, giving supporting evidence for the expanded use of cardiovascular disease treatment and nanomedicine applications in treating difficult conditions (Dong et al., 2024). Cardiovascular disease remains a leading cause of death around the world. Currently, extensive research is being conducted around the world to discover innovative alternative treatments for cardiovascular disease. Curcumin has demonstrated exceptional therapeutic potential in cardiovascular disease preventive studies, ranging from the preclinical to the clinical stages. It reduces hypercholesterolemia and atherosclerosis. Furthermore, it protects against cardiac ischemia and reperfusion. These properties scientifically validate curcumin (CUR) as an effective treatment for cardiovascular illnesses. However, clinical investigations of curcumin administration revealed a variety of effects on cardiovascular parameters at dosages ranging from 20 to 4,000 mg. One of the most significant constraints to using CUR as a medical medication is its low absorption. In addition to better targeting, pharmacokinetics, and efficacy other than cellular absorption, nanomedicine-based formulations of CUR are being developed to overcome this problem. The nanoformulations are representing a new generation of medicinal treatment studies. Further studies should direct new CUR nanomedicines towards intensive clinical testing (Salehi et al., 2020).

10.4 ADVANTAGES OF NANOTECHNOLOGY IN HERBAL PHARMACOTHERAPY

10.4.1 Combining Nanotechnology and Herbal Pharmacotherapy

Combining nanotechnology and herbal pharmacotherapy is a feasible solution to a number of problems that traditional herbal medicine faces. These advancements help to advance integrative and customized healthcare practices by improving the safety and efficacy of herbal medications and expanding their use to a broader spectrum of medical concerns.

10.4.2 Enhanced Bioavailability

The low solubility and absorption of active compounds in herbal medicine raises serious concerns about their bioavailability. Nanotechnology has the potential to make herbal extracts more stable and soluble by forming them into small particles or transport systems (such as liposomes or nanoparticles). This enhancement leads to better absorption and utilization of herbal substances in the body. These carriers improve absorption across biological barriers, promote bioavailability, and protect herbal compounds from deterioration in the gastrointestinal system. (Das & Sharangi, 2020).

10.4.3 Targeted Delivery

It is possible to design nanocarriers that specifically target organelles, tissues, or cells in the body. The systemic negative effects of herbal medications are decreased and their therapeutic effectiveness is increased by this tailored distribution. For instance, in cancer treatment, nanoparticles may directly carry herbal ingredients to tumor cells, and in inflammatory illnesses, they can deliver them to inflamed tissues. Personalized treatment and improved adherence by patients are possible benefits of using nanotechnology into herbal preparations (Javed et al., 2020).

10.4.4 Improved Stability or Shelf Life

Herbal extracts often degrade quickly, leading to decreased shelf life particularly in the presence of light, air, or changes in temperature. By encapsulating these substances in nanoparticles or covering them with protective layers, nanotechnology may safeguard these substances. The stabilization process prolongs the shelf life and guarantees the continuous effectiveness of herbal products. Because of this stability, herbal formulations have a longer shelf life and are thus more suited for commercial application (Dewi et al., 2022).

10.4.5 Synergistic Effects and Combination Therapies

Combining several plant extracts or herbs with traditional medications in a single composition is made possible by nanotechnology. This method maximizes therapeutic results by enabling synergistic effects and the tailoring of therapy to meet the demands of specific patients. The combined use of certain herbal preparations may have synergistic effects that increase their overall medicinal value. Enhanced therapies and decreased medication resistance may result from such combinations (Anwar et al., 2021).

10.4.6 Reduced Dosage and Toxicity

Nanotechnology improves bioavailability and targets specific regions, allowing herbal medicines to be administered at a lower dose without losing potency. Patients are more likely to stick to their treatment plans when the risk of side effects is reduced. This dose reduction may help lessen the risk of toxicity associated with utilizing high doses of herbal treatments. Herbal compounds can be gently released from nanoparticles, keeping medicinal concentrations for a long period. This delayed release profile decreases dosing frequency and improves patient compliance. (Ahmed et al., 2021).

10.4.7 Crossing Biological Barriers

Some herbal compounds have difficulties crossing biological barriers, such as the intestinal epithelium or the blood–brain barrier. By inventing delivery techniques that make it easier for these drugs to pass over these impediments, nanotechnology may open up new therapeutic paths for maladies including gastrointestinal or neurological problems, as well as prenatal conditions. This concentrated distribution minimizes potential side effects, reduces the required dose, and boosts the absorption of herbal components. Furthermore, the regulated release kinetics enabled by nanotechnology ensure long-term therapeutic benefits. As a result of breaking through biological boundaries, nanotechnology is vital to maximizing the safety and effectiveness of herbal medicine, bringing up new possibilities for personalized and successful medical care (Makeen & Barik, 2016).

10.5 CHALLENGES AND LIMITATIONS

Even though nanoparticle-based plant products might have some benefits, they also have some problems and safety issues. To make sure that these nanocarriers are safe, more study needs to be done on their toxicity, possible protection, and long-term effects on human health. Challenges related to quality and regulatory aspects hinder the widespread industrialization and adoption of plant-based nanotechnology products.

10.5.1 Complexity of Herbal Systems

One of the main obstacles to using nanotechnology in herbal pharmacology is the inherent high complexity of herbal medications. Herbal medications, in contrast to manufactured treatments, often include a large variety of bioactive substances with distinct physicochemical characteristics. Indeed, molecular size, solubility, stability, and interaction with biological systems may vary significantly. Hence, the complexity that has to be bargained in order that delivery systems satisfactorily encapsulate, stabilize, and deliver herbal ingredients to respective targets lies with nanotechnology (Barkat et al., 2020).

10.5.2 Bioavailability Enhancement

Herbal nanomedicines are very popular in the search for new drugs. Nanotechnology has profited much to the healthcare industry, churning out a good number of innovative nanocarriers that enhance the clinical efficacy and herbal medication bioavailability (Shree et al., 2024). The herbal components ought to be more bioavailable in order to maximize their therapeutic power. Among the more interesting ways in which nano-science is helping herbal medications to become poorly soluble, permeable, and stable are strong lipid nanoparticles, polymeric nanoparticles, and nanoemulsions. However, adequate bioavailability continues to be elusive due to variety factors such as quick clearance from the circulatory system, initial metabolism, and gastric degradation. The complex interaction between biological barriers and nanoparticles requires thorough investigation to ensure safe and effective therapeutic outcomes (Teja et al., 2022).

10.5.3 Regulatory and Ethical Considerations

Regarding herbal medicine goods, several regulatory frameworks exist around the world. For example, many of these items lack explicit laws covering formulations including nanotechnology. Herbal nano-pharmaceuticals pose a challenge to regulatory bodies that seek to assess the quality, safety, and efficacy of these products since particular properties and intricate interactions may be unique and complex in this heterogeneous group. As a result, harmonizing the regulatory framework and establishing exact criteria is critical for clinical development, market approval, and, particularly, post-marketing monitoring of herbal nanoparticle products. Furthermore, there are moral concerns about intellectual property rights over these discoveries, fair access to cutting-edge medical technology, and the preservation of traditional herbal knowledge and practices that are passively passed down or diffused within families or ethnic social groups (Mohi-Ud-Din et al., 2020).

10.5.4 Cost and Scalability

Significant amounts of money are spent on research and development to create and promote herbal nanotechnology-based products, including the purchase of specialized instruments, nanomaterials, and research. Enhanced product production is always coupled with increasing costs, which poses a difficulty in resource-constrained contexts and markets. It appears that financial challenges have a bearing on issues relating to pricing and access; therefore, this may undermine the successful application of new nanotechnology solutions to herbal pharmacology (Kumar, 2023).

10.5.5 Safety and Toxicity Issues

Nanoparticles utilized in drug delivery systems may be toxic, immunogenic, or accumulate in some essential organs, etc. in a time-dependent manner, some of the associated safety problems. In herbal nanopharmaceuticals, these issues are multiplied manifold due to their interactions with the endogenous biological molecules and the bioactive phytochemicals. Proper pre-clinical studies will help to establish the safety of nanotechnology-based herbal formulations regarding their pharmacological biological compatibility and possible undesirable side effects. The authorities must come up with strictly regulated guidelines that consider testing the safety profile of such novel formulations prior to their approval for clinical use and commercialization (Bandakinda et al., 2024). Herbal medications and derivatives have long been used in a variety of global therapeutic systems, including African, Chinese, Indian, and Sowa-Rigpa. Because of their importance in fundamental medical treatment, conventional medications have remained advantageous in drug research (Naik et al., 2023).

10.5.6 Standardization and Quality Control

Herbal formulations are already extremely challenging to standardize because of variation in plant species, growth environment, techniques of harvesting, and extraction procedures. Nanoparticles can modify stability and bioactivity of herbal components during formulation and storage; therefore, adding nanotechnology into the mix increases the variability still further. There is a need to develop the standard procedure for the manufacturing, analysis, and quality check of nanotechnology herbal products to ensure medicinal value as well as protection for patients from sample to sample for stability (Wang et al., 2023). Otherwise, standardization of herbal medicines will be purposeless with the growing interest in natural therapies if not directed towards the treatment of disorders caused by the current lifestyles that have become a rampant problem. Diseases in the form of obesity, diabetes, cardiovascular problems, and hypertension are global health problems rooted in poor lifestyle. Herbal therapy is sought after by many as an alternative to conventional pharmaceutical therapies because of their drawbacks (Paul & Kumar, 2023).

10.6 FUTURE PROSPECTS AND TRENDS

Nanotechnology holds potential for herbal medicine if regulatory issues are to be sorted out, and safety must be assured. In addition, the development in nanomaterials and production processes makes feasible the formulation of compositions with better and more specific efficacy from plants. Herein, co-operation would be needed in a big way from pharmaceutical researchers, herbal experts, and nanotechnology experts to be able to fully harness and exploit nanotechnology for the betterment of herbal medicine (Kumar, 2023). It is here with nanotechnology that one can enhance the delivery, effectiveness, and safety profiles of herbal medications. The current study of the pharmacological activities of Berberine (BBR) is also active, with several findings published in journals or presented at international congresses. One potential future application is the isolation or neosynthesis of BBR compounds with higher bioavailability. Another example is that some studies have employed acute, subacute, and subchronic toxicities to evaluate the medical therapeutic efficacy of BBR. This means that an important application is the increased bioavailability achieved through nanoformulations by herbal extracts encapsulation, together with chemicals. It is enveloped to bypass gastrointestinal tract degradation and enhance the process of absorption for systemic circulation. The development removes a long-standing problem in conventional herbal treatment where low bioavailability has too often limited therapeutic effectiveness (Behl et al., 2022). Nanotechnology also makes site-specific transport easier, which means that plant bioactive can only affect the cells or tissues they are supposed to reach, reducing any effects that happen elsewhere. This makes therapy more effective. Improving the physical and chemical qualities of plant extract nanoencapsulation could help control how quickly bioactive ingredients are released, allowing for a steady medicinal dose (Teja et al., 2022). Furthermore, nanocarriers have demonstrated the ability to cross the blood–brain barrier; as a result, herbal remedies pass through it to treat neurological problems. Aside from improved distribution, nanotechnology provides instruments of herbal product characterization and standardization that provide quality control and batch-to-batch consistency, which is a very significant component in regulatory clearance and consumer confidence (Wahab et al., 2022). The stabilizing feature that functionalization at a nanoscale level imparts is responsible for increasing the shelf life of herbal formulations from dependence on hard chemical preservatives Predictable future developments in herbal pharmacology and nanotechnology could establish new biochemical combination treatments that take advantage of synergistic interactions between herbal extracts and nanoparticles, thereby elevating therapeutic efficacy while reducing total dosage requirements and associated side effects (Anand et al., 2024). However, there is an undeniable need for rigorous scientific control and regulatory administration in areas such as scalability, cost-effectiveness, and safety concerns about the long-term effects of nanomaterial exposure. As this field expands, herbalists, clinicians, pharmacologists, and nanotechnologists will need to collaborate more than ever to fully realize nanotechnology's potential to cause a paradigm shift in herbal pharmacology, resulting in a whole new generation of treatments that are personalized, targeted, and effective, tailored to each patient's needs and way of life. Herbal medicine has finally discovered a new way to work in nanotechnology, making it a strong challenger as an alternative or additional treatment to traditional medications in new healthcare settings (Barkat et al., 2020).

10.7 CONCLUSION

Nanoscience application in herbal medicine has been very promising across various sectors, including neuroprotection, care for diabetes, antifungal, antibacterial, antioxidant, and cancer treatment. Nanotechnology increases the possibilities of herbal components to be absorbed by the body, delivered target-specific, and protected from breakage, hence increasing their efficacy. Effective delivery of active components from herbal therapies to cancer cells is one of the issues of the conventional form of herbal therapies. This can be achieved through nanocarriers, for example, liposomes, nanoemulsions, nanocapsules, nanofibers, dendrimers, and nanoparticles. Among the flavonoid, polyphenols, furocoumarin, psoralen, curcumin, and triptolide has evidence for neuroprotective properties shown in neurodegenerative diseases, as also with non-flavonoids such as resveratrol and curcumin. Their inability to pass the blood–brain barrier restricts their effectiveness. Nanotechnology will be able to navigate this by creating nanoparticles that successfully carry chemicals into the central nervous system, hence increasing their therapeutic power against diseases such as Alzheimer's. This facilitates the use of nano herbal medicines in the treatment of diabetes without any difficulties that may come with pharmaceuticals as a result of limited solubility, thereby boosting their anti-diabetic qualities. This promotes a more effective and constant distribution of active substances, hence leading to improved control over blood glucose levels and fewer side effects about classic therapies. In general, nanoscience combined with herbal medicine shows potential for creating improved and safer choices of medication. Nanotechnology breakthroughs improve the pharmacological features of herbal medicines and open further possibilities for other innovative therapies, which may make up a principal influence on patient care and results. Simultaneously, further research and the promotion of multidisciplinary collaboration are required to realize the full potential of such innovative therapeutic applications.

REFERENCES

Abo-Zeid, Y., Diab, R., Sanad, R., & Skran, W. (2021). Recent advances in herbal-based nanomedicine for anti-inflammatory purposes. *Journal of Advanced Pharmacy Research*, *5*(4), 387–397. https://doi.org/10.21608/aprh.2021.79946.1135

Afroj, S. C., Babar, A. S., Dahiwade, L. K., Payghan, S. A.. (2021). Recent advances in herbal nanotechnology. *International Journal of Creative Research Thoughts* *9*(5), 312–321.

Ahmed, H. M., Nabavi, S., & Behzad, S. (2021). Herbal drugs and natural products in the light of nanotechnology and nanomedicine for developing drug formulations. *Mini-Reviews in Medicinal Chemistry*, *21*(3), 302–313. https://doi.org/10.2174/1389557520666200916143240

Anand, A., Gautam, P., & Ojha, S. (2024). Application of nanotechnology for herbal medicine development: A review. *Letters in Drug Design & Discovery*, *21*(8), 1325–1333. https://doi.org/10.2174/1570180820666230308105723

Ansari, S.H., Islam, F., Sameem, M. (2012). Influence of nanotechnology on herbal drugs: A Review. *Journal of Advanced Pharmaceutical Technology & Research 3*, 142–146.

Anwar, D. M., El-Sayed, M., Reda, A., Fang, J.-Y., Khattab, S. N., & Elzoghby, A. O. (2021). Recent advances in herbal combination nanomedicine for cancer: delivery technology and therapeutic outcomes. *Expert Opinion on Drug Delivery*, *18*(11), 1609–1625. https://doi.org/10.1080/17425247.2021.1955853

Arifin, S.F., AlShami, A., Omar, S.S.S., Jalil, M.A.A., Khalid, K.A., Hadi, H. (2019). Impact of modern technology on the development of natural-based products. *International Journal of Ayurvedic And Herbal Medicine* *5*(4), 133–142.

Attallah, N. G. M., Elekhnawy, E., Negm, W. A., Hussein, I. A., Mokhtar, F. A., & Al-Fakhrany, O. M. (2022). In vivo and in vitro antimicrobial activity of biogenic silver nanoparticles against staphylococcus aureus clinical isolates. *Pharmaceuticals*, *15*(2), 194. https://doi.org/10.3390/ph15020194

Bala, N., Saha, S., Chakraborty, M., Maiti, M., Das, S., Basu, R., & Nandy, P. (2015). Green synthesis of zinc oxide nanoparticles using Hibiscus subdariffa leaf extract: Effect of temperature on synthesis, anti-bacterial activity and anti-diabetic activity. *RSC Advances*, *5*(7), 4993–5003. https://doi.org/10.1039/C4RA12784F

Bandakinda, M., Kumar, A., & Mishra, A. (2024). Toxicity issues of nanoparticles in the delivery of phytoconstituents and cosmeceuticals. In *Nanotechnology Based Delivery of Phytoconstituents and Cosmeceuticals* (pp. 329–342). Springer Nature Singapore. https://doi.org/10.1007/978-981-99-5314-1_11

Barkat, M. A., Harshita, D. S. S., Beg, S., & Ahmad, F. J. (2020). Nanotechnology-based phytotherapeutics: Current status and challenges. In *Nanophytomedicine* (pp. 1–17). Springer Singapore. https://doi.org/10.1007/978-981-15-4909-0_1

Baskaran, M., Baskaran, P., Arulsamy, N., & Thyagarajan, B. (2017). Preparation and evaluation of PLGA-coated capsaicin magnetic nanoparticles. *Pharmaceutical Research*, *34*(6), 1255–1263. https://doi.org/10.1007/s11095-017-2142-2

Behl, T., Singh, S., Sharma, N., Zahoor, I., Albarrati, A., Albratty, M., Meraya, A. M., Najmi, A., & Bungau, S. (2022). Expatiating the pharmacological and nanotechnological aspects of the alkaloidal drug berberine: Current and future trends. *Molecules*, *27*(12), 3705. https://doi.org/10.3390/molecules27123705

Bhagwat, T. R., Joshi, K. A., Singh Parihar, V., Asok, A., Bellare, J., & Ghosh, S. (2018). Biogenic copper nanoparticles from medicinal plants as novel antidiabetic nanomedicine. *World Journal of Pharmaceutical Research*, *7*, 183. https://doi.org/10.20959/wjpr20184-10773

Bitencourt, P. E. R., Ferreira, L. M., Cargnelutti, L. O., Denardi, L., Boligon, A., Fleck, M., Brandão, R., Athayde, M. L., Cruz, L., Zanette, R. A., Alves, S. H., & Moretto, M. B. (2016). A new biodegradable polymeric nanoparticle formulation containing Syzygium cumini: Phytochemical profile, antioxidant and antifungal activity and in vivo toxicity. *Industrial Crops and Products*, *83*, 400–407. https://doi.org/10.1016/j.indcrop.2016.01.007

Borase, V.A., Kapadnis, R. S., Chaudhari, A. Y., Bachhav, R. S. (2020). A review on novel drug delivery system. *Journal of Emerging Technologies and Innovative Research* *7*(9), 19–27.

Bruna, Vidal Bonifácio et al. (2014). Nanotechnology-based drug delivery systems and herbal medicines: a review. *International Journal of Nanomedicine*. *9*, 1–15.

Chakraborty, K., Shivakumar, A., Ramachandran, S. (2016). Nano-technology in herbal medicines: A review. *International Journal of Herbal Medicine* *4*(3), 21–27.

Chaudhary, H., Kohli, K., & Kumar, V. (2013). Nano-transfersomes as a novel carrier for transdermal delivery. *International Journal of Pharmaceutics*, *454*(1), 367–380. https://doi.org/10.1016/j.ijpharm.2013.07.031

Chaudhary, H., Kohli, K., & Kumar, V. (2014). A novel nano-carrier transdermal gel against inflammation. *International Journal of Pharmaceutics*, *465*(1–2), 175–186. https://doi.org/10.1016/j.ijpharm.2014.02.023

Cinthura, C., & Rajasekar, A. (2020). Antibacterial activity of cinnamon-Clove mediated silver nanoparticles against Streptococcus mutans. *Plant Cell Biotechnology and Molecular Biology* *21*, 11–17.

Coradini, K., Friedrich, R. B., Fonseca, F. N., Vencato, M. S., Andrade, D. F., Oliveira, C. M., Battistel, A. P., Guterres, S. S., da Rocha, M. I. U. M., Pohlmann, A. R., & Beck, R. C. R. (2015). A novel approach to arthritis treatment based on resveratrol and curcumin co-encapsulated in lipid-core nanocapsules: In vivo studies. *European Journal of Pharmaceutical Sciences*, *78*, 163–170. https://doi.org/10.1016/j.ejps.2015.07.012

Cui, M., Wu, W., Hovgaard, L., Lu, Y., Chen, D., & Qi, J. (2015). Liposomes containing cholesterol analogues of botanical origin as drug delivery systems to enhance the oral absorption of insulin. *International Journal of Pharmaceutics*, *489*(1–2), 277–284. https://doi.org/10.1016/j.ijpharm.2015.05.006

Das, S., & Sharangi, A. B. (2020). Nanotechnology: A potential tool in exploring herbal benefits (pp. 27–46). https://doi.org/10.1007/978-3-030-41464-1_2

Dellali, K. Z., Rata, D. M., Popa, M., Djennad, M., Ouagued, A., & Gherghel, D. (2020). Antitumoral drug: Loaded hybrid nanocapsules based on chitosan with potential effects in breast cancer therapy. *International Journal of Molecular Sciences*, *21*(16), 5659. https://doi.org/10.3390/ijms21165659

Deng, S., Gigliobianco, M. R., Censi, R., & Di Martino, P. (2020). Polymeric nanocapsules as nanotechnological alternative for drug delivery system: Current status, challenges and opportunities. *Nanomaterials*, *10*(5), 847. https://doi.org/10.3390/nano10050847

Dewanjee, S., Chakraborty, P., Mukherjee, B., & De Feo, V. (2020). Plant-based antidiabetic nanoformulations: The emerging paradigm for effective therapy. *International Journal of Molecular Sciences*, *21*(6), 2217. https://doi.org/10.3390/ijms21062217

Dewi, M. K., Chaerunisaa, A. Y., Muhaimin, M., & Joni, I. M. (2022). Improved activity of herbal medicines through nanotechnology. *Nanomaterials*, *12*(22), 4073. https://doi.org/10.3390/nano12224073

Dilnawaz, F., & Sahoo, S. K. (2013). Enhanced accumulation of curcumin and temozolomide loaded magnetic nanoparticles executes profound cytotoxic effect in glioblastoma spheroid model. *European Journal of Pharmaceutics and Biopharmaceutics*, *85*(3), 452–462. https://doi.org/10.1016/j.ejpb.2013.07.013

Dong, L., Zhao, Y., Luo, J., Li, X., Wang, S., Li, M., Zou, P., Kong, H., Wang, Q., Zhao, Y., & Qu, H. (2024). Carbon dots derived from curcumae radix and their heartprotective effect. *International Journal of Nanomedicine*, *19*, 3315–3332. https://doi.org/10.2147/IJN.S444125

Dubey, S. K., Parab, S., Achalla, V. P. K., Narwaria, A., Sharma, S., Jaswanth Gowda, B. H., & Kesharwani, P. (2022). Microparticulate and nanotechnology mediated drug delivery system for the delivery of herbal extracts. *Journal of Biomaterials Science. Polymer Edition*, *33*(12), 1531–1554. https://doi.org/10.1080/09205063.2022.2065408

Elgohary, M. M., Helmy, M. W., Abdelfattah, E.-Z. A., Ragab, D. M., Mortada, S. M., Fang, J.-Y., & Elzoghby, A. O. (2018). Targeting sialic acid residues on lung cancer cells by inhalable boronic acid-decorated albumin nanocomposites for combined chemo/herbal therapy. *Journal of Controlled Release*, *285*, 230–243. https://doi.org/10.1016/j.jconrel.2018.07.014

Elmarzugi, N., Amara, R., Eshmela, M., & Eid, A. (2023). An overview of nanocapsule and lipid nanocapsule: Recent developments and future prospects. *Palestinian Medical and Pharmaceutical Journal*, *8*(3). https://doi.org/10.59049/2790-0231.1244

El-Nashar, H. A. S., Abbas, H., Zewail, M., Noureldin, M. H., Ali, M. M., Shamaa, M. M., Khattab, M. A., & Ibrahim, N. (2022). Neuroprotective effect of artichoke-based nanoformulation in sporadic Alzheimer's disease mouse model: Focus on antioxidant, anti-inflammatory, and amyloidogenic pathways. *Pharmaceuticals*, *15*(10), 1202. https://doi.org/10.3390/ph15101202

Elzoghby, A. O., El-Lakany, S. A., Helmy, M. W., Abu-Serie, M. M., & Elgindy, N. A. (2017). Shell-crosslinked Zein nanocapsules for oral codelivery of exemestane and resveratrol in breast cancer therapy. *Nanomedicine*, *12*(24), 2785–2805. https://doi.org/10.2217/nnm-2017-0247

Gavhane, S. A., & Gade, Aditi T. (2021). The novel drug delivery system. *International Journal of Creative Research Thoughts* 9(9), 373–385.

Ghiasi, Z., Esmaeli, F., Aghajani, M., Ghazi-Khansari, M., Faramarzi, M. A., & Amani, A. (2019). Enhancing analgesic and anti-inflammatory effects of capsaicin when loaded into olive oil nanoemulsion: An in vivo study. *International Journal of Pharmaceutics*, *559*, 341–347. https://doi.org/10.1016/j.ijpharm.2019.01.043

Gopi, S., Amalraj, A., Haponiuk, J.T., & Thomas, S. (2016). Introduction of nanotechnology in herbal drugs and nutraceutical: A review. *Journal of Nanomedicine and Biotherapeutic Discovery* 6(2), 1000143.

Govindaraju, K., & Suganya, K. S. U. (2020). In vitro anti-diabetic assessment of guavanoic acid functionalized gold nanoparticles in regulating glucose transport using L6 rat skeletal muscle cells. *RSC Medicinal Chemistry*, *11*(7), 814–822. https://doi.org/10.1039/D0MD00125B

Hare, J. I., Lammers, T., Ashford, M. B., Puri, S., Storm, G., & Barry, S. T. (2017). Challenges and strategies in anti-cancer nanomedicine development: An industry perspective. *Advanced Drug Delivery Reviews*, *108*, 25–38. https://doi.org/10.1016/j.addr.2016.04.025

Harika, P., Deepthi, B.V.P., Vinitha, B., Baherji, R., Ali, J., Sharma, J.V.C.. (2021). Herbal Nanoparticles. *World Journal of Pharmaceutical and Medical Research* 7(3), 127–130.

Hassani, M., Tahghighi, A., Rohani, M., Hekmati, M., Ahmadian, M., & Ahmadvand, H. (2022). Robust antibacterial activity of functionalized carbon nanotube- levofloxacine conjugate based on in vitro and in vivo studies. *Scientific Reports*, *12*(1), 10064. https://doi.org/10.1038/s41598-022-14206-w

Hawar, S. N., Al-Shmgani, H. S., Al-Kubaisi, Z. A., Sulaiman, G. M., Dewir, Y. H., & Rikisahedew, J. J. (2022). Green synthesis of silver nanoparticles from Alhagi graecorum leaf extract and evaluation of their cytotoxicity and antifungal activity. *Journal of Nanomaterials*, *2022*. https://doi.org/10.1155/2022/1058119

He, Y., Du, Z., Ma, S., Liu, Y., Li, D., Huang, H., Jiang, S., Cheng, S., Wu, W., Zhang, K., & Zheng, X. (2016). Effects of green-synthesized silver nanoparticles on lung cancer cells in vitro and grown as xenograft tumors in vivo. *International Journal of Nanomedicine*, *11*, 1879–1887. https://doi.org/10.2147/IJN.S103695

Hesari, M., Mohammadi, P., Khademi, F., Shackebaei, D., Momtaz, S., Moasefi, N., Farzaei, M. H., & Abdollahi, M. (2021). Current advances in the use of nanophytomedicine therapies for human cardiovascular diseases. *International Journal of Nanomedicine 16*, 3293–3315). https://doi.org/10.2147/IJN.S295508

Hiwa, M. Ahmed, Nabavi, Seyed, Behzad, Sahar. (2020). Herbal drugs and natural products in the light of nanotechnology and nanomedicine for developing drug formulations. *Mini-Reviews in Medicinal Chemistry 20*, 1–12.

Hort, M. A., Vaz, G. R., Marques, M. Da, S., Schimith, L. E., Dos Santos, M. G., Arbo, B. D., Horn, A. P., & Dora, C. L. (2024). Natural antioxidant nanoparticles in neuroprotection. In *Natural Molecules in Neuroprotection and Neurotoxicity* (pp. 1905–1934). Elsevier. https://doi.org/10.1016/B978-0-443-23763-8.00081-6

Hsieh, S.-H., Brunke, S., & Brock, M. (2017). Encapsulation of antifungals in micelles protects candida Albicans during gall-bladder infection. *Frontiers in Microbiology*, *8*. https://doi.org/10.3389/fmicb.2017.00117

Javed, M. N., Dahiya, E. S., Ibrahim, A. M., Alam, Md. S., Khan, F. A., & Pottoo, F. H. (2020). Recent advancement in clinical application of nanotechnological approached targeted delivery of herbal drugs. In *Nanophytomedicine* (pp. 151–172). Springer Singapore. https://doi.org/10.1007/978-981-15-4909-0_9

Jogi, H., Maheshwari, R., Raval, N., Kuche, K., Tambe, V., Mak, K. K., Pichika, M. R., & Tekade, R. K. (2018). Carbon nanotubes in the delivery of anticancer herbal drugs. *Nanomedicine*, *13*(10), 1187–1220. https://doi.org/10.2217/nnm-2017-0397

Kalakotla, S., Sandhya Rani, M., Gottumukkala, K.M., & Pravallika, P. L. (2015). Herbal drugs and herbal mediated silver nano particles as anti diabetics: A new horizon. *International Journal of Pharmaceutical Sciences Review and Research*, *31*, 142–148.

Kale, N. (2023). Treatment of cancer using combination of herbal and novel drug delivery system. In *Bioprospecting of Tropical Medicinal Plants* (pp. 1177–1197). Springer Nature Switzerland. https://doi.org/10.1007/978-3-031-28780-0_49

Kaur, A., Kaur, L., Singh, G., Dhawan, R. K., & Mahajan, A. (2022). Nanotechnology-based herbal formulations: A survey of recent patents, advancements, and transformative headways. *Recent Patents on Nanotechnology*, *16*(4), 295–307. https://doi.org/10.2174/1872210515666210428135343

Kedi, P. B. E., Meva, F. E., Kotsedi, L., Nguemfo, E. L., Zangueu, C. B., Ntoumba, A. A., Mohamed, H. E. A., Dongmo, A. B., & Maaza, M. (2018). Eco-friendly synthesis, characterization, in vitro and in vivo anti-inflammatory activity of silver nanoparticle-mediated Selaginella myosurus aqueous extract. *International Journal of Nanomedicine*, *13*, 8537–8548. https://doi.org/10.2147/IJN.S174530

Khoobchandani, M., Katti, K. K., Karikachery, A. R., Thipe, V. C., Srisrimal, D., Mohandoss, D. K. D., Darshakumar, R. D., Joshi, C. M., & Katti, K. V. (2020). New approaches in breast cancer therapy through green nanotechnology and nano-ayurvedic medicine – pre-clinical and pilot human clinical investigations. *International Journal of Nanomedicine*, *15*, 181–197. https://doi.org/10.2147/IJN.S219042

Kumar, R. (2023). Nanotechnology in herbal medicine: Challenges and future perspectives. In *Nanotechnology in Herbal Medicine* (pp. 515–548). Elsevier. https://doi.org/10.1016/B978-0-323-99527-6.00008-2

Lakkim, V., Reddy, M. C., Pallavali, R. R., Reddy, K. R., Reddy, C. V., Inamuddin, B. A. L., & Lomada, D. (2020). Green synthesis of silver nanoparticles and evaluation of their antibacterial activity against multidrug-resistant bacteria and wound healing efficacy using a murine model. *Antibiotics*, *9*(12), 902. https://doi.org/10.3390/antibiotics9120902

Liu, H., Bai, Y., Huang, C., Wang, Y., Ji, Y., Du, Y., Xu, L., Yu, D.-G., & Bligh, S. W. A. (2023). Recent progress of electrospun herbal medicine nanofibers. *Biomolecules*, *13*(1), 184. https://doi.org/10.3390/biom13010184

Makeen, H. A., & Barik, B. (2016). Nanotechnology and traditional medicine: Challenges and opportunities. *Therapeutic Delivery*, *7*(3), 157–162. https://doi.org/10.4155/tde.15.97

Marena, G. D., Ramos, M. A. Dos, S., Carvalho, G. C., Junior, J. A. P., Resende, F. A., Corrêa, I., Ono, G. Y. B., Sousa Araujo, V. H., De Camargo, B. A. F., Bauab, T. M., & Chorilli, M. (2022). Natural product-based nanomedicine applied to fungal infection treatment: A review of the last 4 years. *Phytotherapy Research*, *36*(7), 2710–2745. https://doi.org/10.1002/ptr.7460

Mathpati, R. V., Harangule, Y. R., Bavage, S. B., Bavage, N. B. (2021). Nano-technology in herbal medicines: Advancements in herbal treatment. *International Journal of Innovative Research in Technology* *8*(3), 116–122.

Md, S., Kit, B. C. M., Jagdish, S., David, D. J. P., Pandey, M., & Chatterjee, L. A. (2018). Development and in vitro evaluation of a zerumbone loaded nanosuspension drug delivery system. *Crystals*, *8*(7), 286. https://doi.org/10.3390/cryst8070286

Mei, Z., Li, X., Wu, Q., Hu, S., & Yang, X. (2005). The research on the anti-inflammatory activity and hepatotoxicity of triptolide-loaded solid lipid nanoparticle. *Pharmacological Research*, *51*(4), 345–351. https://doi.org/10.1016/j.phrs.2004.10.007

Mirhadi, E., Rezaee, M., & Malaekeh-Nikouei, B. (2018). Nano strategies for berberine delivery, a natural alkaloid of Berberis. *Biomedicine & Pharmacotherapy*, *104*, 465–473. https://doi.org/10.1016/j.biopha.2018.05.067

Mishra, Y. et al.. (2022). Application of nanotechnology to herbal antioxidants as improved phytomedicine: An expanding horizon. *Biomedicine & Pharmacotherapy* *153*(2022), 113413.

Mohi-Ud-Din, R., Mir, R. H., Pottoo, F. H., Sawhney, G., Masoodi, M. H., & Bhat, Z. A. (2020). Nanophytomedicine ethical issues, regulatory aspects, and challenges. In *Nanophytomedicine* (pp. 173–192). Springer Singapore. https://doi.org/10.1007/978-981-15-4909-0_10

Moldovan, B., David, L., Vulcu, A., Olenic, L., Perde-Schrepler, M., Fischer-Fodor, E., Baldea, I., Clichici, S., & Filip, G. A. (2017). In vitro and in vivo anti-inflammatory properties of green synthesized silver nanoparticles using Viburnum opulus L. fruits extract. *Materials Science and Engineering: C*, *79*, 720–727. https://doi.org/10.1016/j.msec.2017.05.122

Mostafa, Y. S., Alamri, S. A., Alrumman, S. A., Hashem, M., & Baka, Z. A. (2021). Green synthesis of silver nanoparticles using pomegranate and orange peel extracts and their antifungal activity against Alternaria solani, the causal agent of early blight disease of tomato. *Plants*, *10*(11), 2363. https://doi.org/10.3390/plants10112363

Muhammad Ibrar, M. I., Khan, M. A., Abdullah, A. & Muhammad Imran, M. I (2018). Evaluation of Paeonia emodi and its gold nanoparticles for cardioprotective and antihyperlipidemic potentials. *Journal of Photochemistry and Photobiology B: Biology*, *189*, 5–13. https://doi.org/10.1016/j.jphotobiol.2018.09.018

Mukundan, D., Mohankumar, R., & Vasanthakumari, R. (2015). Green synthesis of silver nanoparticles using leaves extract of bauhinia tomentosa linn and its invitro anticancer potential. *Materials Today Proceedings*, *2*(9), 4309–4316. https://doi.org/10.1016/j.matpr.2015.10.014

Naik, G. G., Sahu, A. N., Kaushik, V., Kaushik, A., & Sarkar, B. K. (2023). Phytopharmaceuticals and herbal drugs: Prospects and safety issues in the delivery of natural products. In *Phytopharmaceuticals and Herbal Drugs* (pp. 215–248). Elsevier. https://doi.org/10.1016/B978-0-323-99125-4.00006-8

Nakamura, O., Hitora, T., Yamagami, Y., Mori, M., Nishimura, H., Horie, R., Yamaguchi, K., & Yamamoto, T. (2014). The combination of rapamycin and MAPK inhibitors enhances the growth inhibitory effect on Nara-H cells. *International Journal of Molecular Medicine*, *33*(6), 1491–1497. https://doi.org/10.3892/ijmm.2014.1715

Narwade, V.V., Game, M. D., Kadam, S.G. (2024). A Review on Nanotechnology in Herbal Medicine. *International Journal of Pharmaceutical Sciences* 2(4), 411–416.

Onoue, S., Takahashi, H., Kawabata, Y., Seto, Y., Hatanaka, J., Timmermann, B., & Yamada, S. (2010). Formulation design and photochemical studies on nanocrystal solid dispersion of curcumin with improved oral bioavailability. *Journal of Pharmaceutical Sciences*, *99*(4), 1871–1881. https://doi.org/10.1002/jps.21964

Ozdal, Z. D., Sahmetlioglu, E., Narin, I., & Cumaoglu, A. (2019). Synthesis of gold and silver nanoparticles using flavonoid quercetin and their effects on lipopolysaccharide induced inflammatory response in microglial cells. *3 Biotech*, *9*(6), 212. https://doi.org/10.1007/s13205-019-1739-z

Paul, T., & Kumar, K. J. (2023). Standardization of herbal medicines for lifestyle diseases. In *Role of Herbal Medicines* (pp. 545–557). Springer Nature Singapore. https://doi.org/10.1007/978-981-99-7703-1_27

Pinilla, C. M. B., Thys, R. C. S., & Brandelli, A. (2019). Antifungal properties of phosphatidylcholine-oleic acid liposomes encapsulating garlic against environmental fungal in wheat bread. *International Journal of Food Microbiology*, *293*, 72–78. https://doi.org/10.1016/j.ijfoodmicro.2019.01.006

Rani, R., Dahiya, S., Dhingra, D., Dilbaghi, N., Kaushik, A., Kim, K. H., & Kumar, S. (2019). Antidiabetic activity enhancement in streptozotocin + nicotinamide-induced diabetic rats through combinational polymeric nanoformulation. *International Journal of Nanomedicine*, *14*, 4383–4395. https://doi.org/10.2147/IJN.S205319

Rarokar, N. R., Menghani, S. S., Kerzare, D. R., Khedekar, P. B., Bharne, A. P., Alamri, A. S., Alsanie, W. F., Alhomrani, M., Sreeharsha, N., & Asdaq, S. M. B. (2022). Preparation of terbinafin-encapsulated solid lipid nanoparticles containing antifungal Carbopol® hydrogel with improved efficacy: in vitro, ex vivo and in vivo study. *Pharmaceutics*, *14*(7), 1393. https://doi.org/10.3390/pharmaceutics14071393

Rehman, G., Umar, M., Shah, N., Hamayun, M., Ali, A., Khan, W., Khan, A., Ahmad, S., Alrefaei, A. F., Almutairi, M. H., Moon, Y.-S., & Ali, S. (2023). Green Synthesis and characterization of silver nanoparticles using Azadirachta Indica seeds extract: In vitro and in vivo evaluation of anti-diabetic activity. *Pharmaceuticals*, *16*(12), 1677. https://doi.org/10.3390/ph16121677

Rupali, S. P. 2022, Nanotechnology in herbal drug: A review. *International Journal of Scientific Development and Research*. *7*(12), 127–135.

Ruturaj, D. D., Jatale, A. P., Deshmukh, S. P., Hire, M. K. (2023) Nanotechnology for Enhancing the Bioavailability of Herbal Drug. *International Journal of Creative Research Thoughts*. *11*(10), 790–801.

Saha, S., Pramanik, K., & Biswas, A. (2021). Antibacterial activity and biocompatibility of curcumin/TiO2 nanotube array system on Ti6Al4V bone implants. *Materials and Technologies*, *36*(4), 221–232. https://doi.org/10.1080/10667857.2020.1742984

Sahu, A. N. (2013). Nanotechnology in herbal medicines and cosmetics. *International Journal of Research in Ayurveda and Pharmacy* 4(3), 472–474.

Salehi, B., Del Prado-Audelo, M. L., Cortés, H., Leyva-Gómez, G., Stojanović-Radić, Z., Singh, Y. D., Patra, J. K., Das, G., Martins, N., Martorell, M., Sharifi-Rad, M., Cho, W. C., & Sharifi-Rad, J. (2020). Therapeutic applications of curcumin nanomedicine formulations in cardiovascular diseases. In *Journal of Clinical Medicine* (Vol. 9, Issue 3). MDPI. https://doi.org/10.3390/jcm9030746

Saliev, T. (2019). The advances in biomedical applications of carbon nanotubes. *C*, *5*(2), 29. https://doi.org/10.3390/c5020029

Sandhir, R., Yadav, A., Sunkaria, A., & Singhal, N. (2015). Nano-antioxidants: An emerging strategy for intervention against neurodegenerative conditions. *Neurochemistry International*, *89*, 209–226. https://doi.org/10.1016/j.neuint.2015.08.011

Sandhiya, V., & Ubaidulla, U. (2020). A review on herbal drug loaded into pharmaceutical carrier techniques and its evaluation process. *Future Journal of Pharmaceutical Sciences*, *6*(1), 51. https://doi.org/10.1186/s43094-020-00050-0

Sharma, M.. (2014). Applications of nanotechnology based dosage forms for delivery of herbal drugs. *Research and Reviews: Journal of Pharmaceutics and Nanotechnology* 2(1), 23–30.

Shree, D., Patra, C. N., & Sahoo, B. M. (2024). Applications of nanotechnology-mediated herbal nanosystems for ophthalmic drug. *Pharmaceutical Nanotechnology*, *12*(3), 229–250. https://doi.org/10.2174/2211738511666230816090046

Sidorowicz, A., Margarita, V., Fais, G., Pantaleo, A., Manca, A., Concas, A., Rappelli, P., Fiori, P. L., & Cao, G. (2022). Characterization of nanomaterials synthesized from Spirulina platensis extract and their potential antifungal activity. *PLoS One*, *17*(9), e0274753. https://doi.org/10.1371/journal.pone.0274753

Sobh, R. A., Nasr, H. E. S., & Mohamed, W. S. (2019). Formulation and in vitro characterization of anti-cancer drugs encapsulated chitosan/multi-walled carbon nanotube nanocomposites. *Journal of Applied Pharmaceutical Science*, *9*(8), 32–40. https://doi.org/10.7324/JAPS.2019.90805

Sohn, S.-I., Priya, A., Balasubramaniam, B., Muthuramalingam, P., Sivasankar, C., Selvaraj, A., Valliammai, A., Jothi, R., & Pandian, S. (2021). Biomedical applications and bioavailability of curcumin—An updated overview. *Pharmaceutics*, *13*(12), 2102. https://doi.org/10.3390/pharmaceutics13122102

Sulaiman, G. M., Waheeb, H. M., Jabir, M. S., Khazaal, S. H., Dewir, Y. H., & Naidoo, Y. (2020). Hesperidin loaded on gold nanoparticles as a drug delivery system for a successful biocompatible, anti-cancer, anti-inflammatory and phagocytosis inducer model. *Scientific Reports*, *10*(1), 9362. https://doi.org/10.1038/s41598-020-66419-6

Teja, P. K., Mithiya, J., Kate, A. S., Bairwa, K., & Chauthe, S. K. (2022). Herbal nanomedicines: Recent advancements, challenges, opportunities and regulatory overview. *Phytomedicine*, *96*, 153890. https://doi.org/10.1016/j.phymed.2021.153890

Thapa, R. K., Khan, G. M., Parajuli-Baral, K., & Thapa, P. (2013). Herbal medicine incorporated nanoparticles: Advancements in herbal treatment. *Asian Journal of Biomedical and Pharmaceutical Sciences*, *3*(24), 7–14.

Tiwari, S., Mistry, P., & Patel, V. (2014). SLNs based on co-processed lipids for topical delivery of terbinafine hydrochloride. *Journal of Pharmaceutics and Drug Development*, *1*(6). https://doi.org/10.15744/2348-9782.1.604

Tong, W., Wang, L., & D'Souza, M. J. (2003). Evaluation of PLGA microspheres as delivery system for antitumor agent-camptothecin. *Drug Development and Industrial Pharmacy*, *29*(7), 745–756. https://doi.org/10.1081/DDC-120021774

Wahab, S., Ahmad, Md. P., Hussain, A., & Qadir, S. F. A. (2022). Nanomaterials for the Delivery of Herbal Bioactive Compounds. *Current Nanoscience*, *18*(4), 425–441. https://doi.org/10.2174/1573413717666211004090341

Wang, H., Chen, Y., Wang, L., Liu, Q., Yang, S., & Wang, C. (2023). Advancing herbal medicine: enhancing product quality and safety through robust quality control practices. *Frontiers in Pharmacology*, *14*. https://doi.org/10.3389/fphar.2023.1265178

Waris, A., Ali, A., Khan, A. U., Asim, M., Zamel, D., Fatima, K., Raziq, A., Khan, M. A., Akbar, N., Baset, A., & Abourehab, M. A. S. (2022). Applications of Various Types of Nanomaterials for the Treatment of Neurological Disorders. *Nanomaterials*, *12*(13), 2140. https://doi.org/10.3390/nano12132140

Wencui, Z., Qi, Z., Ying, W., Di, W., Wencui, Z., & Qi, Z. (2015). Preparation of solid lipid nanoparticles loaded with garlic oil and evaluation of their in vitro and in vivo characteristics, *European Review for Medical and Pharmacological Sciences*, *19*(19), 3742–3750.

Yadav, R., Pradhan, M., Yadav, K., Mahalvar, A., & Yadav, H. (2022). Present scenarios and future prospects of herbal nanomedicine for antifungal therapy. *Journal of Drug Delivery Science and Technology*, *74*, 103430. https://doi.org/10.1016/j.jddst.2022.103430

Yousefi, M., Narmani, A., & Jafari, S. M. (2020). Dendrimers as efficient nanocarriers for the protection and delivery of bioactive phytochemicals. *Advances in Colloid and Interface Science*, *278*, 102125. https://doi.org/10.1016/j.cis.2020.102125

Zhang, L., Yang, S., Wong, L. R., Xie, H., & Ho, P. C.-L. (2020). *In vitro* and *in vivo* comparison of curcumin-encapsulated chitosan-coated poly (lactic-*co*-glycolic acid) nanoparticles and curcumin/hydroxypropyl-β-cyclodextrin inclusion complexes administered intranasally as therapeutic strategies for Alzheimer's disease. *Molecular Pharmaceutics*, *17*(11), 4256–4269. https://doi.org/10.1021/acs.molpharmaceut.0c00675

11 Nanoparticle Synthesis and Characterization for Herbal Drug Delivery

Kashmala Khalid
Department of Plant Sciences, Quaid-i-Azam University, Islamabad, Pakistan

Muhammad Abdullah Aziz
Institute of Soil and Environmental Sciences, Pir Mehr Ali Shah Arid Agriculture University, Rawalpindi, Pakistan

Sana Saleem
Department of Microbiology, Quaid-i-Azam University, Islamabad, Pakistan

Arshad Farid
Gomal Center of Biochemistry and Biotechnology, Gomal University, D.I. Khan, Pakistan

Fati Ullah Khan and Hafiza Sehrish Kiani
Department of Biochemistry, Quaid-i-Azam University, Islamabad, Pakistan

11.1 INTRODUCTIONS

Recently, there has been significant focus on the construction of a novel drug delivery system (NDDS) specifically designed for herbal medications. Traditional dosage forms, such as those with a prolonged release, are unable to meet the needs of the body for both retaining the drug component at a specific rate during the course of treatment and delivering the phytoconstituents to the intended target site for the best possible therapeutic effect (Aminu et al., 2020). Prior to reaching the bloodstream, numerous phytochemicals present in herbal medications undergo decomposition when they pass through the very acidic pH of the stomach and are acted upon by liver enzymes. Therefore, the optimal quantity of herbal medications may not reach the bloodstream. For the drug to have the desired effect, it must reach the affected region in the optimal amount necessary to produce a therapeutic impact. If it fails to do so, it will not be powerful enough (Jain, 2020). Nanoparticles employed in herbal medicines unequivocally possess the capability to transport an optimal quantity of the medication to the targeted area of effect, thereby bypassing obstacles such as, for instance, the acidic pH of the stomach, liver processing, and increased systemic drug circulation, which is due to their diminutive size (Devi et al., 2010). Phytoformulation research focuses on creating small-sized forms of herbal drugs, such as polymeric nanoparticles, liposomes, phytosomes, nanocapsules, solid lipid nanoparticles, nanoemulsions,

DOI: 10.1201/9781003544418-11

and many others. These various forms offer several benefits, including the increased solubility and bioavailability of the drugs, improved pharmacological activity, protection against toxicity, enhanced stability, better distribution in body tissues, sustained release, and safeguarding against chemical as well as physical degradation (Amol & Pratibha, 2014). Therefore, these studies show that the nano-sized nanodelivery systems of herbal therapeutics possess promising prospects for augmenting the efficacy and surmounting challenges linked to plant-based medications (Devi et al., 2010).

11.2 BACKGROUND AND LITERATURE REVIEW

11.2.1 Historical Overview and Present Trends in Herbal Medicine

Treatment for human illness has always relied on natural resources like plants. The primary idea behind the evolution of contemporary medicine can be linked to conventional therapies and medicine. Since prehistoric times, plants have been utilized for medical purposes in several regions of the world, including Africa, ancient China, America, Egypt, and India, (Darji et al., 2022). Plants produce a diverse array of bioactive compounds from various components. These include, among others, latex, stems, leaves, bark, seeds, roots, and flowers. These substances, which are known as secondary metabolites, play a crucial role in the defense mechanism of plants against bacteria, insects, and herbivores. Plants contain various bioactive substances, such as flavonoids, terpenoids, saponins, steroids, alkaloids, cardiac glycosides, tannins, phenolic compounds, and others (Sendker & Sheridan, 2017). Even today, herbal remedies and natural products remain essential for treating various illnesses. Contrary to the commonly used allopathic method, herbal treatments contain several phytochemicals that act collectively in combating ailments (Afzal et al., 2022). The availability of chemical analysis emerged in the early 19th century, marking the commencement of the extraction and alteration of herbal extracts. Herbal medications were formerly disregarded for creating new formulations because of insufficient scientific validation and challenges in processing, extracting, standardizing, and identifying the different components of these complex mixtures (Darji et al., 2022). Due to their possible therapeutic efficacy and reduced side effects when compared to other medications, herbal medicines have gained recognition from both doctors and patients. In addition, they boost the medicine's bioavailability (Sendker & Sheridan, 2017).

11.2.2 Overview of Nanoparticles in Drug Delivery

Conventional drug dosage forms often struggle to deliver the intended dose to specific areas of the body due to the tendency of drugs to disperse throughout the body based on their biochemical as well as physicochemical properties (Jain, 2020). The novel drug delivery system (NDDS) for herbal medicines is designed to address this issue by focusing on targeted drug delivery, which can reduce the frequency of dosage and elimination while also improving solubility and absorption (Dongare et al., 2021). Among them, nanoparticles (NPs), that is, those with a diameter of 10–100nm, play an essential role in which they are recognized as the versatile nanoscale drug delivery system to modulate pharmacodynamic and pharmacokinetic properties for the distribution of both small molecules and large molecules (Hsu et al., 2023). The initial observation of nanoparticles revealed that they were composed of non-biodegradable polymers (polymethyl-methacrylate, polyacrylamide, and polystyrene). These polymeric nanoparticles demonstrate the ability to load drugs and proteins, allowing bioactive substances to be incorporated within the polymer matrix through chemical or physical attachment to the particle surface (Idrees et al., 2020). Enzyme activity and nanoparticle solubilization in the blood vessels within the target tissue is proven to be helpful in raising the retention and stability of the drug. During nanoparticle design, it is essential to carefully consider surface properties and dimensions as well as release patterns in order to establish the accurate interaction, rate, and dosage with the specific locations in the body (Chenthamara et al., 2019).

11.2.3 Advantages of Nanoparticle-Based Drug Delivery Systems

Some of the benefits associated with nanoparticles are their site-targeting properties, biodegradability, and non-toxicity with a shelf life of at least one year (Afzal et al., 2022). Magnetic guidance or attachment of specific ligands on particle surfaces can help in targeting a medicine to some particular location within the body. A nano-sized delivery system was chosen based on its ability to target disease sites effectively. They are nano-sized, so they have a high loading efficiency and, therefore, offer the advantage of delivering a high concentration of drugs at the targeted site. It provides controllable drug release rates and customizable particle degradation behavior by selecting specific matrix components (Chakraborty et al., 2016). This is due to the ability of the nano-scaled drug delivery system to release drugs in a controlled manner and has a tendency of maintaining drug concentration at the disease sites for longer durations. They also manifest the EPR effect – increased permeation and retention –, meaning they can easily permeate barriers due to their small size and are retained in the body due to limited the drainage of lymphatic fluid, especially in the case of cancer. Since the particle size and surface properties of the nanoparticles can be personalized, it can easily reach passive and active pharmaceutical sites after injection. This enables the introduction of drugs into the systems without any chemical reaction due to high loading of medication, which is essential in the safe operation of drugs. Moreover, nanoparticles are used for passive targeting of diseases, and there is no need to attach specific ligands to drugs and, therefore, causing less harm and requiring lower drug formulation concentrations (Afzal et al., 2022).

11.3 TYPES OF NANOPARTICLES EMPLOYED IN THE DELIVERY OF HERBAL DRUGS

11.3.1 Polymer Nanoparticle

Polymer nanoparticles are solid colloidal particles that vary between 10 and 1000 nm in size. The most commonly physical systems of polymer nanoparticles are called nanospheres and nanocapsules (Chakraborty et al., 2016). These can be made either directly from monomers through polymerization or from preformed polymers (Afzal et al., 2022). A variety of techniques are employed, including dialysis, supercritical fluid evaporation, solvent evaporation, salting out, and quick supercritical solution expansion. Several factors, such as, for instance, the type of polymeric system, the intended application area, and the desired size specifications, determine the selection of the preparation method (Chakraborty et al., 2016). These techniques have proven successful in creating polymeric nanoparticles with demonstrated effective medicinal properties.

11.3.2 Metallic Nanoparticles

Metal nanoparticles, which typically measure between 1 and 100 nm in length, width, and thickness, are commonly produced using liquid phase techniques – for instance, chemical reduction, reversed micelle methods as well as sol gel. These nanoparticles are prized for their large surface area, their unique electronic properties which bridge molecular and metallic states, and the presence of multiple low coordination sites. They have diverse applications, including magnetic separation for isolating labeled cells, along with other biological entities, gene and drug delivery, as well as in radio frequency techniques for hyperthermia-based tumor breakdown and as contrast enhancement agents in magnetic resonance imaging (Chakraborty et al., 2016).

11.3.3 Magnetic Nanoparticles

Various compositions and phases of magnetic nanoparticles have been synthesized, for instance, pure metals such as iron, copper, and nickel as well as alloys of metals, such as iron platinum and copper platinum (Chakraborty et al., 2016). Magnetic nanoparticles can be utilized in producing particles as small as 3 nm in size. Magnetic nanoparticles (MNPs) have found a significant application in bioseparation. By modifying MNPs with specific receptors and linking them to target biomolecules, complexes are formed. The aforementioned complexes can be separated out from the original mixture via magnetic field application. Thus, this approach proves to be an efficient one when compared with approaches such as filtration as well as centrifugation (Chakraborty et al., 2016).

11.3.4 Liposomes

These are tiny sacs that are spherical in nature and have size ranges from 0.05 to 5.0 μm in diameter and they are observed to be formed when the polar lipids undergo self-organization and become hydrated in aqueous media (Nalla & Chinnala, 2017). Such colloidal particles have the ability to encapsulate a solvent portion and allow it to move freely in its central sac (Afzal et al., 2022). The absolute benefits of liposomes include their chemical stability as they have ability to encapsulate various compounds, such as hydrophilic, amphiphilic, and lipophilic ones. It is also easy for them to be prepared and exhibit higher compatibility. In addition, it is easy to alter liposomes' pharmacokinetic properties through changing its chemical composition (Tiwari et al., 2020).

11.3.5 Dendrimers

These kinds of nanomaterials are considered to be highly branched and exhibit unique features – for instance, small size, large surface area, high biocompatibility, lower side effects, and the ability to encapsulate a larger drug quantity in its core, as well as being non-toxic, allowing for sustained and controlled release and being biodegradable. Their functionalization can be enhanced when it is combined with targeting agents – for instance, antibodies or peptides – which could lead to increased specificity of dendrimers for specific regions in the body (Tiwari et al., 2020).

11.3.6 Niosomes

Multilamellar vesicles, also known as niosomes, are formed by the combination of cholesterol and nonionic surfactants from the alkyl or dialkylpolyglycerol ether class. Previous research suggests that niosomes, like liposomes, have the potential to act as effective drug carriers. Niosomes offer several advantages over liposomes, including their relatively low cost, chemically stable against oxidation and temperature, longer shelf life, better skin penetration, as well as higher encapsulation efficiency, (Nalla & Chinnala, 2017).

11.3.7 Proniosomes

The proniosome gel system represents a progression from niosomes and offers a versatile approach to delivering actives to specific locations. Proniosomal gels are formulations that, upon hydration with skin water, transform into niosomes. Proniosomes, which are carrier particles soluble in water and coated with surfactant, can be hydrated to form a niosomal dispersion before use by agitating them briefly in hot water (Nalla & Chinnala, 2017) (Figure 11.1).

11.3.8 Phytosomes

Flavonoids, the main bioactive components of phytomedicines, have low oral bioavailability. Phytosomes are molecular complexes formed from water-soluble phytoconstituent molecules, primarily polyphenols, and are lipid-compatible (Nalla & Chinnala, 2017). Due to their ability to navigate through lipid-rich biomembranes and reach their source, phytosomes have higher bioavailability compared to simple herbal extracts. These complexes are prepared by encapsulating phytoconstituents in soy-derived phospholipids phosphatidyl choline to make them lipid-compatible (Tiwari et al., 2020).

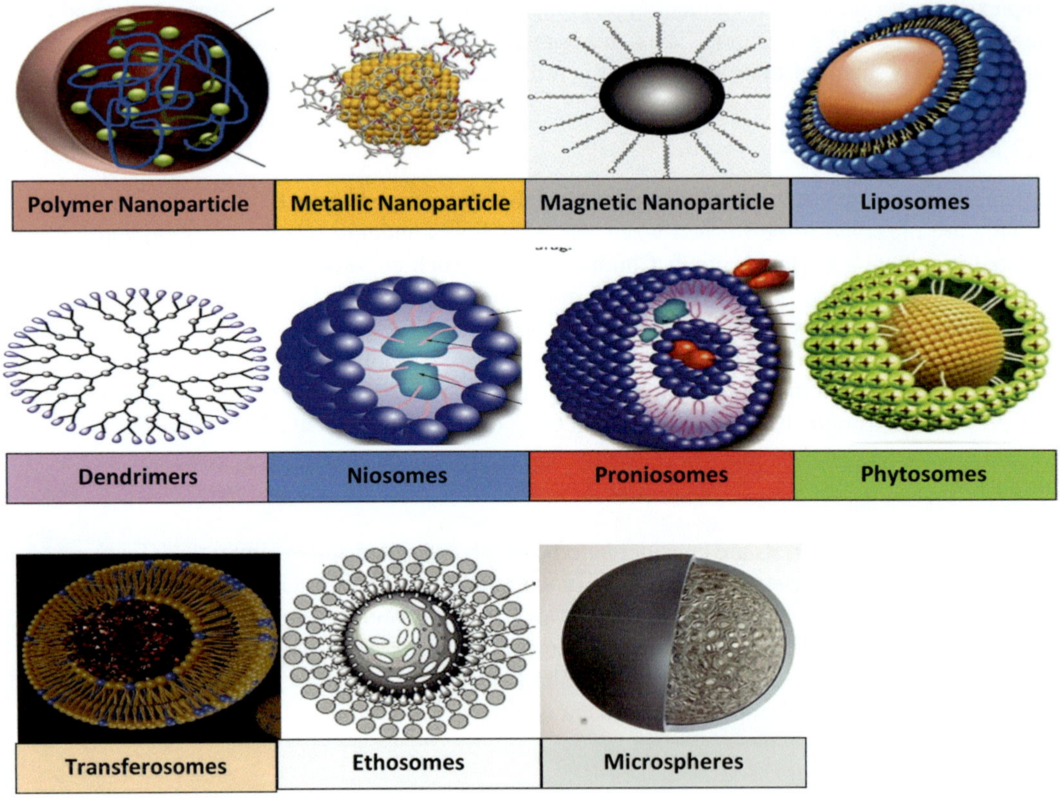

FIGURE 11.1 Types of nanoparticles employed in herbal drug delivery.

11.3.9 Transfersomes

Transfersomes can be described as vesicles of remarkable design capable of transformation of speed at breakneck speed, and within a relatively affordable process, when exposed to stress on the outside. The advancement of efficient strategies like transfersomes has reasonably reduced problems found in transdermal drug delivery systems like the stratum corneum barrier, the transport of large molecules and the attributes of physical and chemical of drugs that hinder their transport through the skin. In addition to their suitability for larger molecules, these adjustable vesicles can pass through skin pores much smaller than the size of the vesicles. Transfersomes have the ability to penetrate the skin's initial barrier, also known as the *stratum corneum*, to deliver nutrition that is vital in carrying out skin functions, thereby supporting skin health (Nalla & Chinnala, 2017).

11.3.10 Microspheres

Microspheres are defined as small spherical beads which have a diameter of between 1μ and 50 μ. At the moment, there is an extensive uninterrupted study on these particles mainly regarding their efficiency in targeting a certain area in the body, and retaining the medication in that particular region without provoking adverse effects (Tiwari et al., 2020). Of the technologies relevant to this field, microencapsulation is known to greatly prolong the effectiveness of the administered drugs and increase patient compliance with the prescribed medication. Currently, microspheres are being used to encapsulate various bioactive substances extracted from plants such as zedoary oil, camptothecin, tetrandrine, quericin, Cynara scolymus extract, and rutin (Nalla & Chinnala, 2017).

11.3.11 Ethosomes

Ethosomal patches are the most recent innovation in the field of patches. These reagents are prepared from soya phosphatidylcholine, ethanol and water, and incorporate multilamellar vesicles that are capable of capturing particles with variable lipophilic properties. Ethosomes' encapsulation efficiency and deformability improve drug penetration into the skin because of their ability to get deeper into the skin layers. The ethosomes possess characteristics pertaining to physical and chemical profile that enable enhanced diffusion across the skin layers and bloodstream; thus, ethosomes can play a key role in applications in transdermal delivery systems and for topical drug carriers (Tiwari et al., 2020). Ethosomes can accommodate hydrophilic as well lipophilic drugs and can deliver these drugs into the cells efficiently. This property increases the percutaneous penetration of other enterprises such as matrine, which is an herbal anti-inflammatory drug and also increases the penetration of antibacterial peptides into fibrocytes (Nalla & Chinnala, 2017).

11.4 NANOPARTICLE SYNTHESIS TECHNIQUES

The preparation techniques used have considerable influence on the shape, size, architecture and efficiency of the nanoparticles. Additionally, the electrical, optical physiochemical, and electrochemical properties of the nanoparticles are also influenced. In some cases, the aim of coating of nanoparticles is to maintain their properties while they form a sediment in the suspension solution. Two conventional methods exist for creating nanoparticles, namely the top-down and bottom-up approaches (Kumari et al., 2023).

11.4.1 Top-Down Approach

The top-down approach is a technique that fragments big molecules into smaller pieces before turning them into the appropriate nanoparticles. This technique has been used to create carbon, dichalcogenides, iron oxide, and also cobalt (III) oxide nanoparticles. Some of the common methods used to achieve synthesis using a top-down strategy include thermal decomposition, mechanical milling, nanolithography, and laser ablation (Kumari et al., 2023). The production of nanoparticles can also be achieved by adopting other strategies as physical, chemical, or biological means.

11.4.2 Bottom-Up Approach

The process entails the systematic production of nanoparticles from basic raw materials. It is feasible, inexpensive, less toxic, and environmentally beneficial. Examples of the bottom-up approach, according to this definition, include laser pyrolysis, sol-gel processing, chemical vapor deposition and so on (Kumari et al., 2023).

Typically, there are four major methods (chemical, physical, biological and mechanical) employed for synthesizing nanoparticles, which are discussed below:

11.4.3 Chemical Methods

The synthesis of nanoparticles often involves the use of various chemical techniques – for instance, the microemulsion technique, hydrothermal synthesis, the sol-gel method, polyol synthesis, chemical vapor synthesis and also plasma-enhanced chemical vapor deposition (Dhand et al., 2015).

11.4.3.1 Sol-Gel Method

The sol-gel method is the most popular bottom-up method because it is easy to use and can make a variety of nanoparticles (Ealia & Saravanakumar, 2017). The precursor materials are uniformly mixed in a solvent to create a gel, which is then heated to generate the desired nanoparticles. The first step is to prepare a sol, which is then gelled, and the solvent is removed. 'Sol' and 'gel' are the two components via which the sol-gel method is named (Khan et al., 2022). It has various names in the following sorted according to how the solids are contained. Sol is an instance in which solids are dispersed in a liquid in a colloidal fashion. On the other hand, solid polymers dispersed in a liquid are called gel. There are essentially two steps in the sol-gel process: as in the case of hydrolysis and condensation reactions. The hydrolysis process involves the use of water to cause break down in the precursor's bonds, which is the first step towards forming the gel phase. Subsequent to the hydrolysis, the condensation makes it possible to generate nanomaterials. The excess water is removed in order to give the formed material its ultimate structure. Normally, the sol-gel technique employs chlorides and metal oxides in precursors as they are used commonly (Ealia & Saravanakumar, 2017). The precursor is mixed or shaken in a liquid so that we obtain a system with a liquid and suspended solid phase.

11.4.3.2 Spinning

Spinning is one of the techniques to produce nanoparticles in a reactor that is known as a spinning disc reactor (SDR) (Khan et al., 2022). This method uses a rotating disc to be fitted in a reactor or chamber. This allows the control of other physical characteristics such as temperature. Normally, in practice, the reactor is purged with nitrogen or with some other inert gas in order to minimize the formation of chemical reactions (Dhand et al., 2015). Water and the precursor are placed in the reactor; as the disc rotates at different speeds, spinning occurs., The atoms/molecules then precipitate, accumulate, and then desiccate and fuse. The nature of nanoparticles synthesized in the spinning disc reactor depends on the various operating parameters, such as the flow rate of liquid, the position of feed, the ratio of liquid to precursor, the speed of rotation of the disc, and the nature of the disc surface, as stated in the work of Ealia and Saravanakumar (2017).

11.4.3.3 Microemulsion Technique

A microemulsion technique comprises of a minimum of three phases: the water constitutes the polar phase; the oil or hydrocarbon liquid constitutes the non-polar phase; and the surfactant, which acts as the surface-active agent. It is an optically transparent, macroscopically homogenous, thermally stable as well as isotropic dispersion. The surfactant acts as a barrier to stop the droplets from coalescing (Solanki & Murthy, 2011). The microemulsion system comprises spherical droplets of water-in-oil (W/O) and oil-in-water (O/W) that are evenly spread out and have dimensions between 600 nm and 8000 nm (Dhand et al., 2015). The reverse micellar system with water in oil is a good reaction site for nanoparticle synthesis. Simply put, micellas are lipid molecules arranged spherically in water-containing solutions. The amphipathic nature of fatty acids leads to the formation of a micelle. This indicates that the long hydrophobic chain (hydrophobic regions) and the polar head groups, which are hydrophilic regions, are both present in micelles. The outermost parts of the micelles are hydrophilic regions, which are polar and drawn to water. Because the micelles are non-polar, their hydrophobic tails are located away from water. The hydrophilic head of the surfactant, responsible for forming the aqueous core, is situated within the reverse micelle, while the lipophilic tail points outward, creating a water-in-oil (W/O) microemulsion. In a W/O microemulsion, metallic nanoparticles are produced by combining two microemulsions that contain a reducing agent as well as a metal salt. Typically, sodium borohydride serves as the reducing agent, and the surfactant used is often Triton X-100 (Solanki & Murthy, 2011).

11.4.3.4 Hydrothermal Synthesis

The hydrothermal synthesis method utilized high temperature as well as pressure, which causes a heterogeneous reaction to take place in a water-based environment. Hence, the conditions, such as the temperature, pH and pressure, purely impacts the properties that are required to synthesize nanoparticles. This type of surface hydrophilicity makes these nanoparticles appropriate for biotechnological applications. The process can be carried out in two kinds of systems: continuous hydrothermal or batch hydrothermal (Gan et al., 2020). While the continuous hydrothermal makes it possible to achieve a higher rate of reaction in a shorter amount of time, the batch hydrothermal can operate a system with the intended ratio phases. When two or more phases of matter, such as gels and foams, are combined under carefully regulated pressure and temperature, a colloidal system is created in a chemical solution. This method can produce nanoparticles. One advantage of using this method is that it produces many NPs in the right shape, composition, size, and surface chemistry at an affordable price (Abedini et al., 2013).

11.4.3.5 Electrochemical Synthesis

Electrochemical synthesis typically entails constructing chemical compounds inside an electrochemical cell. This technique has merits, including the elimination from consideration of half-reactions that could be wastage-prone and the required potential can be controlled to the desired value (Sajid & Płotka-Wasylka, 2020). In recent times, there has been thorough research on producing silver nanoparticles through electrochemical synthesis methods. The cutting-edge method required dissolving a metallic anode in an aprotic solvent for generating the silver nanoparticles. However, these particles fall into the size range of 2–7 nm, while through the management of current density, the size can be accurately regulated. Counter electrodes provide an excellent possibility for changing different electrochemical parameters and studying their effects on final particle size. The electrochemical generation of silver nanoparticles in an aqueous solution of polyvinyl alcohol (PVA) was studied. PVA is a cheap man-made polymer which is safe, soluble in water, and biodegradable (Nešović et al., 2020).

11.4.3.6 Polyol Synthesis

The polyol process is a transformative way to obtain metal-containing compounds when poly(ethylene glycols) are used like the reaction medium. This serves not only as a solvent but also a reducing agent wherein it is combined with other dissolved stabilizing agents (Rahman & Green, 2009). Besides, this method synthesizes numerous types of nanoparticles including; metal oxide nanoparticles such as copper(I) oxide and zinc oxide, magnetic nanoparticles, and metal-based nanoparticles such as silver, copper platinum (Pt) or palladium (Pd). In one case, polycrystalline copper(I) oxide nanoparticles could have been formed using a copper nitrate precursor along with a poly(vinyl pyrrolidone) stabilizing agent followed by an ethylene glycol reducing agent. Such an approach presents exciting possibilities for the synthesis of many different kinds of metal-containing compounds in a fast, efficient, and controlled manner all at once (Dhand et al., 2015).

11.4.3.7 Thermal Decomposition

Thermal decomposition is an endothermic reaction triggered by the application of heat, resulting in the disruption of chemical bonds within a compound. If this decomposition leads to the release of heat from the compound's structure, it establishes a positive feedback loop and can cause thermal runaway (Khan et al., 2022). Moreover, the infrared spectra and thermal decompositions of the metal acetates and dicarboxylates were studied with respect to the metal-acetate compound bonding. Thermal decomposition with copper, lead, and rare earth acetates was investigated using thermogravimetry and differential thermal analysis, and desirable results were obtained (Ealia & Saravanakumar, 2017).

11.4.3.8 Chemical Vapor Deposition & Chemical Vapor Synthesis

To get solid films from the vapor phase, Chemical Vapor Deposition (CVD) is a method that relies on chemical reactions at extremely high thermodynamic conditions (Khan et al., 2022). Under certain conditions, thin films produced using the CVD process also contain ultrafine particles. By subjecting the system to elevated temperatures, high supersaturations, longer residence periods, and compact substrates among other conditions, CVD can facilitate production of nanoparticles (Ealia & Saravanakumar, 2017). The chemical vapor deposition technique, whose operating conditions have been altered for nanoparticle generation instead of a thin solid film, is called chemical vapor synthesis (CVS). Vapor precursors, which can be solid, liquid, or gas, are generated inside a reactor under specific conditions, initiating particle formation through a nucleation process and enabling the production of multi-component or doped nanomaterials (Hachem et al., 2022). In the course of chemical vapor synthesis, gaseous molecules undergo a chemical reaction to form a condensing phase that allows the growth of particles. As soon as the temperature increases, so also does the rate of particle formation. This method effectively produces nanoparticles of carbon, titania, and silica (Dhand et al., 2015).

11.4.3.9 Plasma-Enhanced Chemical Vapor Deposition

Plasma-enhanced/plasma-assisted chemical vapor deposition (PACVD/PECVD) is a widely employed technique for depositing thin films. Plasma unequivocally amplifies the chemical reactions required for generating thin films and nanoparticles, as its name signifies (Khan et al., 2022). The PECVD unit consists of four main components: a power supply (either AC or DC), gaseous precursors, a vacuum processing system, and a heater. Not only the nanoparticles but also thin film production occurs using plasma-enhanced chemical vapor deposition (PECVD) at relatively lower temperatures compared to typical chemical vapor deposition (CVD). This process involves the participation of ionized species and radicals due to the partially ionized nature of plasma. The process of synthesizing nanoparticles (NPs) from various materials was successfully demonstrated using plasma-enhanced chemical vapor deposition (PECVD) (Dhand et al., 2015).

11.4.4 Physical Methods

In order to produce nanoparticles, physical methods can use electrical energy, mechanical pressure, thermal energy, or high-energy radiation to induce evaporation, material abrasion, melting, or condensation. Typically, top-down operation is used by physical approaches. This method offers the benefit of producing nanoparticles that are uniform and monodisperse, without any contamination from solvents. However, given the substantial volume of waste generated during the synthesis process, the physical processes are considered to be less cost-effective. Physical methods such as high-energy ball milling, laser ablation, electrospraying, inert gas condensation, laser pyrolysis, and melt mixing can be employed to produce nanoparticles.

11.4.4.1 High-Energy Ball Milling Process

Due to the exceptionally high temperatures and pressures involved, the high-energy ball milling process is classified as a mechanochemical synthesis process. High-energy ball milling was being introduced by John Benjamin in 1970, and he declared this method as a promising approach to produce nanoparticles of various sizes and shapes (Xing et al., 2013). In this process, moving balls produce kinetic energy that directly influences the material which is being milled, which further causes the breakage of the chemical bonds hence leading to smaller particles from the milled substances. This all leads to exposure to new surfaces (Dhand et al., 2015).

11.4.4.2 Physical Vapor Deposition (PVD)

To create NPs and thin material layers, physical vapor deposition (PVD) methods are often used. These methods typically produce particles ranging in size from a few nanometers to several micrometers. PVD, considered an eco-friendly vacuum deposition method, consists of three basic steps: (1) vaporizing the material from a solid source, (2) nucleating and growing to obtain thin films and NPs, and (3) transporting the vaporized material. Common PVD techniques used for creating NPs include sputtering, pulsed laser deposition, electron beam evaporation, and vacuum arc (Dhand et al., 2015).

11.4.4.3 Pyrolysis

Pyrolysis involves the thermal breakdown of a precursor material using a flame. The precursor is subjected to high pressure as it enters the furnace and combusts in a liquid or vapor form through a small opening (Khan et al., 2022). The resulting combustion gases, or byproducts, are categorized as air to facilitate the recovery of nanoparticles. Some furnaces utilize plasma as well as lasers, instead of flames, to achieve the high temperatures required for evaporation. Pyrolysis offers advantages such as ease of use, cost-effectiveness, continuous operation, and high yield.

11.4.4.4 Melt Mixing

In the process of melt mixing, polymer and modified nanofillers are homogenized using mechanical mixing techniques, such as extrusion, kneading and occasionally injection molding (Kamal et al., 2022). This can be regarded as the most applied mechanical process because it is easily integrated with existing industrial processes while also being environmentally friendly. It is one of the initial approaches to the preparation of polymer composites with NPs as fillers for the creation of desired material properties (Dhand et al., 2015).

11.4.4.5 Laser Ablation (LA) and Pulse Laser Deposition (PLD)

One of the most frequently used methods for the generation of nanoparticles is Laser Ablation Synthesis in Solution (LASiS); this technique uses various solvents to create a vapor of a material from its solid source using a high-powered laser beam to produce nanoparticles, excluding any other chemicals or stabilizing agents (Khan et al., 2022). It is depicted that a laser beam (either continuous or pulsed) could be focused on a metal which is being submerged in an aqueous solution, leading to the formation of nanoparticles and plasma plumes (Dhand et al., 2015). This approach can synthesized to produce a broader range of nanomaterials, for instance metallic nanoparticles, oxide composites, and carbon nanomaterials (Ealia & Saravanakumar, 2017).

Colloidal nanoparticles in aqueous media could be generated by pulse laser ablation without the involvement of any surfactants, and these are considered to be uniform particles. All the parameters, for instance, ablation time, laser wavelength and fluence as well as liquid environment are adjustable to produce the nanoparticles of desired properties. Another vacuum-based technique, Pulsed Laser Deposition (PLD), can be used to extract a substance from the target with the help of laser energy. This causes reactions such as ionization, evaporation, and melting, resulting in ablation and the deposition of material onto a substrate (Radičić et al., 2022).

11.4.4.6 Electron Beam Evaporation (EBE)

Electron beam evaporation (EBE) is another PVD process carried out in vacuum conditions. It produces thin films and nanoparticles (Dhand et al., 2015). The EBE system consists of three components, namely electron beam source, a vacuum unit and target materials. The current that passes through the filament causes it to heat up and thereafter produce an electron beam. Magnets further direct the focused electron beam to the target material. Regarding the Electron Beam Evaporation method, the electron beam is responsible in heating the target material, which would later evaporate when its boiling point is reached at a certain temperature. The evaporated material is then conveyed and spread over the substrate. EBE has rapid deposition rates; hence, it can be used for a wide range of materials that have varying electrical conductivities and deposits materials with higher boiling points compared to thermal evaporation. NPs with different sizes and shapes can be deposited through the EBE process by the selection of proper process parameters and conditions (Bello et al., 2014).

11.4.4.7 Inert Gas Condensation (IGC)

One of the oldest methods of NP synthesis involves the technique termed inert gas condensation (IGC), which relies on cooled liquid nitrogen substrate holders and uses inert gases such as helium or argon for NP production (Kumari et al., 2023). Inert gases are used to carry the evaporated materials, and liquid nitrogen is used to condense them onto the substrate. Hybrid nanoparticles can be synthesized by atomic vapor phase condensation under high-pressure application by sputtering multiple targets for instance silver, Iron and silicon (Benelmekki et al., 2015). In the process of DC sputtering – metal atom vapors are deliberately generated in close proximity to the target surface. These vapors are then intentionally condensed into nanoclusters as they precisely move through the aggregation zone. The nanoclusters are subsequently extracted and propelled by the significant pressure disparity between the aggregation zone (10 mbar) and the deposition unit (10^{-5} mbar), before being firmly placed onto the substrate (Dhand et al., 2015).

11.4.4.8 Flame Spray Pyrolysis (FSP)

At the time of writing, the flame spray process (FSP) technology is the most recent of all the flame aerosol technologies (Teoh et al., 2010). The combustion process is a single-step process in which the precursor is in a liquid state. The combustion enthalpy is significantly higher, accounting for more than 50% of the total energy released during combustion. Typically, an organic solvent is used in this process. The production of nanoparticles necessitates the subsequent series of steps: (1) The initial spray undergoes evaporation or decomposition, resulting in the formation of metal vapors. (2) Nucleation arises because of supersaturation. (3) Growth occurs through the processes of sintering and coalescence. (4) Particle aggregation is the result of chemical bonding, while agglomeration is caused by physical interactions. It is a highly utilized method for creating complicated and effective nanoparticles. Liquid precursors can form particles by either the gas-to-particle or droplet-to-particle pathway; however, the gas-to-particle route leads to more uniform particle sizes and morphologies (Dhand et al., 2015).

11.4.4.9 Laser Pyrolysis

Vapor phase synthesis is the method used in CO_2 laser pyrolysis (D'Amato et al., 2013). The process produces condensable products as a result of chemical reactions induced by the laser at the interface between the laser beam as well as the molecular flow of reactants in the gaseous and vapor phases. One requirement for achieving energy coupling in the reaction system is that one of the reactants or precursors must be capable of absorbing infrared (IR) CO_2 laser radiation through resonant vibrational modes (Kozuch et al., 2023). In addition, inert photo-sensitizers such as sulphur hexafluoride (SF_6), ammonia (NH_3), ethylene gas (C_2H_4), and so on can be added to the reactants to facilitate the energy transfer between precursors and the laser light. In contrast to alternative vapor phase techniques, laser pyrolysis enables more focused and swift heating, resulting in quick nucleation and quicker quenching of particle growth (in a matter of milliseconds). In light of this, this process shows promise for producing NPs with a narrow size distribution in the hot zone and an average diameter of between 5 and 60 nm. Unfortunately, the nanoparticles attempt to aggregate as soon as they leave the heated chamber, which leads to the creation of chains of nanoparticles (Dhand et al., 2015).

11.4.4.10 Nanolithography

Nanolithography involves the precise design of nanostructures, typically with at least one dimension ranging from 1 to 100 nm (Khan et al., 2022). It encompasses various processes such as optical, multiphoton, electron-beam, nanoimprint, and scanning probe lithography. Lithography, a key technique in nanolithography, involves printing specific shapes or structures onto light-sensitive materials by selectively removing portions of the material to achieve the desired shape as well as structure. Nanolithography provides unparalleled precision in controlling the size along with shape of nanoparticles and clusters, ranging from single particles to desired clusters. However, it requires sophisticated equipment and entails significant costs (Ealia & Saravanakumar, 2017).

11.4.4.11 Electrospraying Technique

The electrospraying method utilizes an electromechanical device to generate charged droplets. This is achieved by taking a solution that contains the desired polymer as well as solvent into a syringe, and applying a high voltage to the tip of the capillary (Anu Bhushani & Anandharamakrishnan, 2014). The solvent undergoes evaporation during its journey toward the counter electrode, resulting in the collection of particles or fibers as the final product. With the use of various solvents, it is possible to create different types and sizes of nanoparticles (NPs) through the electrospraying technique. These kinds of nanoparticles can be applied to treat many different serious diseases. The capacity to produce uniform nanoparticle sizes, a quick preparation process, and bulk NP assembly are just a few of the benefits of using the electrospraying technique. However, due to thermal or shear stresses in the syringe and drying process, this method may also result in the degradation of some macromolecules (Dawadi et al., 2021). It is also possible to create engineered water nano-structures (EWNS) from atmospheric water vapor by using the electrospraying technique. With a low toxicity level and an impressive array of chemical, physical, biological, as well as morphological properties, the 25 nm-sized EWNS has a remarkable mechanism of action (Pyrgiotakis et al., 2014).

11.4.5 Biosynthesis of Nanoparticles

An environmentally friendly, low-toxic, economical, effective, as well as biodegradable method for creating nanoparticles is biosynthesis, also known as green synthesis or bio-assisted synthesis (Dhand et al., 2015). The formation of metal oxide nanoparticles and metal nanoparticles can be achieved by using biological systems such as actinomycetes, plant extracts, fungi, bacteria viruses, yeast and yeast-like species The formation of nanoparticles by chemical means carries certain risks – for example, carcinogenic, genotoxic and cytotoxic risks. However, pressure, high energy and temperature are employed in the process of physical synthesis. The biosynthesis of nanoparticles offers significant advantages over physical and chemical methods due to its non-toxic nature, simplicity, cost-effectiveness, ability to detoxify heavy metals, and environmental friendliness. It eliminates the need for pressure, high temperature, and energy, along with toxic chemicals (Karunakaran et al., 2023). Three steps are typically involved in biosynthesis: choosing a stabilizing agent, choosing a reducing agent, and choosing a reaction medium. Biosynthesis could be categorized into three main groups on the basis of the origin of raw materials, namely: biomolecules, microorganisms, and extracts derived from plants.

11.4.5.1 Utilizing Biomolecules as Templates for Synthesis

To create nanoparticles, different biomolecules, including viruses, membranes, diatoms, and nucleic acids, were used as templates. It is well known that transition metal ions have a strong affinity for DNA, a biomolecular template. Nanoparticles of gold were prepared by cross-linking of DNA hydrogel and by incorporating transition metal ions into the DNA macromolecules such as Au (III) ions. The decrease of Au (III) also leads to the formation of single gold (Au) atoms and metal clusters to form gold nanoparticles adjacent to the DNA chain (Morikawa et al., 2021).

11.4.5.2 Microbial Synthesis

Most of the nanoparticles are produced within bioreactors that include algae, actinomycetes, yeasts, fungi and the prokaryotic bacteria. Extensive scientific efforts have been channelled to the advancement of this method for the synthesis of gold, cadmium sulfide, palladium, silver, titanium dioxide, and others. Microbes uptake target ions from the surroundings and transform them into element metals through the enzymes released by their cellular actions (Zhang et al., 2011). On the basis of the formation site, nanoparticle synthesis can be categorized into extracellular synthesis and intracellular synthesis. Another method of nanoparticle formation is intracellular biosynthesis in which metal ions are taken into microorganism cells and then used with the aid of enzymes to form nanoparticles. Otherwise, it can happen extracellularly when enzymes bind metal ions to the surfaces of microbial cells. Metal ions are present in the bacteria cell by using the enzymes and anionic functional groups that minimize interactions (Dhand et al., 2015).

11.4.5.3 Utilizing Botanical Extracts for Synthesis

Using plant extracts to produce nanoparticles through a green synthesis process is a highly efficient, non-toxic, rapid, and environmentally friendly method. It has proven possible to create metal oxide

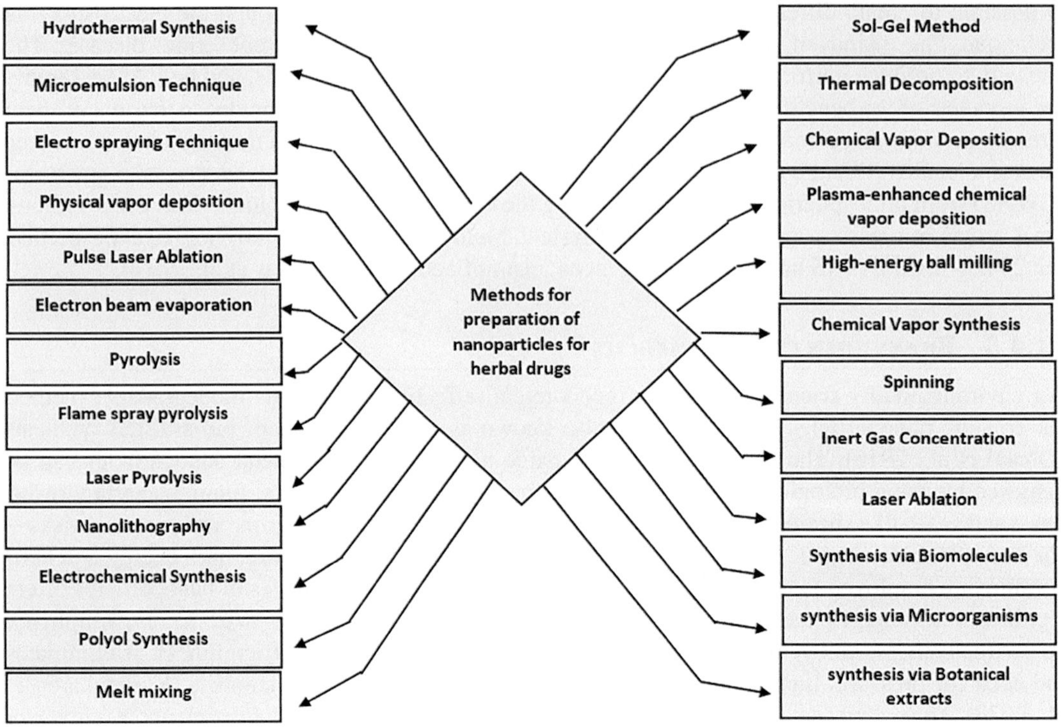

FIGURE 11.2 Methods for preparation of nanoparticles.

and noble metal nanoparticles through green synthesis. Due to their roles as capping and reducing agents, various plant biometabolites may contribute to the synthesis of nanoparticles. Terpenoids, amino acids, hydrogenase, tartaric acid, quercetin, sequiterpenes, tannic acid, saponin, peptide, citric acid, secondary metabolites, flavonoids, protein, phenolics, heterocyclic compounds, as well as functional groups (alcohols, sulfhydryl, ketones, amines as well as carboxyl acids) are among the reducing agents. The capping agents include pralines, peptides, extracellular proteins, tannic acid, enzymes, tartaric acid, functional groups (such as ketones, carboxylic acid, aldehydes, alcohols, and amines), and citric acid (Adeyemi et al., 2022).

11.4.6 Mechanical Techniques

Nanoparticles can also be produced through mechanical means such as milling, mechanical alloying, as well as mechanochemical processes (Tulinski & Jurczyk, 2017). At low temperatures, the milling method restores chemical interactions between surfaces. The mechanochemical technique is a continuous welding process that minimizes agglomerations and appropriately selects milling materials. To achieve efficient production, it is of utmost importance to thoroughly analyze the thermal treatment, stoichiometry of source materials, reaction pathways, as well as milling conditions (Figure 11.2).

11.5 CHARACTERIZATION OF NANOPARTICLES

Characterization of materials entails studying materials with an aim of identifying their properties, structures, and compositions by looking at chemical as well as the physical properties. The most common classification of nanoparticles encompasses dimension, structure, as well as electric charge that is determined by analysis with the aid of instruments such as Transmission electron microscopy

(TEM), Atomic force microscopy (AFM) as well as Scanning electron microscopy (SEM). Chemical properties of nanoparticles include size, shape, color, refractive index, fluorescence and others, while physical properties include size, density, surface area, and others (Khan et al., 2022).

11.5.1 Chemical

It is crucial to note that the applications of nanoparticles are closely coupled to their chemical nature with regard to reactivity to target, and sensitivity to factors such as light, heat, moisture, atmosphere and stability. Among the useful properties inherent to the nanoparticles are fungicidal, decontaminating, bactericidal, and anti-toxic effects, which is why the nanoparticles can be effectively used in biomedical and ecological applications (Ealia & Saravanakumar, 2017). The following techniques can be used to determine the stability of nanoparticles – for example, Ultraviolet-visible spectroscopy (UV-Vis), Transmission Electron Microscopy (TEM), dynamic light scattering, and zeta potential (Sandhiya & Ubaidulla, 2020).

11.5.2 Physical

Optical properties – for instance, coloration, ability to transmit, ability to absorb and ability to reflect light – are aspects of the physical characteristics of nanoparticles. These nanoparticles are also effective in absorbing and reflecting ultra violet (UV) light when they are incorporated in a solution or used in forming a coating layer. Moreover, their mechanical properties, including malleability, strength and flexibility, elasticity or tensile strength affect their usefulness (Ealia & Saravanakumar, 2017). In addition, the magnetic and electrical characteristics, including resistivity, conductivity, and semi-conductivity, have made it possible to utilize nanoparticles in current electronics and renewable power sources.

11.5.2.1 Particle Size Analyzer

The analysis of synthesized nanoparticles is heavily dependent on their morphology and particle size distribution (Al-Gebory & Mengüç, 2018). Nanoparticles are primarily utilized for drug delivery and targeted drug administration. The size of the particles significantly affects drug release; smaller particles with larger surface areas lead to faster drug release as a greater proportion of the loaded drugs are exposed to the surface. Conversely, drugs diffuse more slowly within larger particles. However, one of the disadvantages of smaller particles is their tendency to become aggregated during nanoparticle dispersion storage and transportation. Therefore, there is a trade-off between smaller particle size and the stability of nanoparticle dispersion (Chakraborty et al., 2016).

11.5.2.2 Surface Area Analysis

The surface area of the particles is calculated by adding up the areas of the surfaces that are exposed on the particles, and then dividing that by the mass of the particles. The relationship between surface area and particle size is an inverse one. The surface area of a powder can be determined by utilizing nitrogen adsorption. The Brunauer, Emmett, and Teller (BET) method is frequently employed for the quantification of the overall surface area. Assuming that the particles are perfectly spherical as well as having a narrow size distribution, the specific surface area can be used to calculate the average particle diameter in nanometers using the formula below: dBET =6000/ñs. S represents the specific surface area measured in square meters per gram, while ñ represents the theoretical density measured in grams per cubic centimeter (Chakraborty et al., 2016).

11.5.2.3 Zeta Potential

As for the measure of the electrical charge in the surface of the nanoparticle, zeta potential is used. The term zeta potential defined the ability of the measuring of the effective charge on the surface of a nanoparticle that plays the most vital role in controlling the motion of the nanoparticle. This charge is protected by the bringing of ions of opposite charges close to the surface of the nanoparticle. The above-described layers of ions run separately and jointly with the nanoparticle. There are claims that the value of zeta potential has a straight link to the stability of the particles. The greater the value of the potential, for example, the magnitude of the wave function is small and the electrostatic repulsion is larger leading to increase stability (Khatak & Dureja, 2015; Sandhiya & Ubaidulla, 2020).

11.5.2.4 Thermogravimetric Analysis (TGA)

Thermogravimetric analysis (TGA) is an analytical technique employed to investigate the weight changes of a sample as a function of temperature and time at a well-defined thermal history (Kharisov et al., 2014). This tool can be used for the characterization of inorganic as well as organic solid substances. Non-isothermal transitions of a substance can be understood with the help of differential thermal analysis, which is a method of calorimetry that measures temperature and heat flux. This technique also enables the selection of the right transition points, such as melting point and glass transition, as well as crystallization point (Sandhiya & Ubaidulla, 2020).

11.5.2.5 Dynamic Light Scattering

Dynamic light scattering or photon correlation spectroscopy (also called PCS) is the most widely applied technique in particle size analysis. Namely, if the precise set of conditions is applied, such as an angle and optimal temperature in the case of the current experiment, dynamic light scattering can serve as the means for the determination of the particle size and the polydispersity index of the given material. The technique is useful in determining the surface charge and the physical state of the emulsion (Sandhiya & Ubaidulla, 2020). DLS is highlighted as the most efficient technique for quantifying the Particle Size Distribution, particularly for the Brownian particles within the nano and submicron range in colloidal media (Chakraborty et al., 2016).

11.5.2.6 Scanning Electron Microscopy (SEM)

The magnification power of the scanning electron microscope (SEM) means its images have a high depth of focus, which provide a realistic three-dimensional view of surface topography: SEMs are therefore better suited to looking at the surface details of samples (Kumar & Seth, 2021). Electrons emitted from a source are concerned in an electric field gradient in a vacuum environment. It allowed for the direct visualization and evaluation of the samples' morphological features and was employed in this work for morphological and sizing assessment. It has, however, proved of some use in the determination of size distribution, but falls short in offering all the rounded details. When using SEM to characterize samples, nanoparticles have to be dried, ground into a fine powder, placed evenly on a sample stub, and spray-coated with conductive metal (usually gold) through the sputter coater. The scanned sample done with high current density and narrow electron probe and information about the features of sample surface is extracted from the secondary electrons emitted from the sample surface. This is of great concern because the nanoparticles have to withstand the electron beams and the vacuum conditions that are likely to degrade polymers (Dawadi et al., 2021).

11.5.2.7 Nuclear Magnetic Resonance

Characterization of nanoparticles in terms of size and composition can be done using nuclear magnetic resonance (NMR). NMR offers a number of cues about the physicochemical properties of the constituents in nanoparticles (Sandhiya & Ubaidulla, 2020).

11.5.2.8 Transmission Electron Microscopy (TEM)

Transmission electron microscopy TEM) is a high-amplification technique used to study the analysis of internal structures of materials using an electron beam that has to be passed through the sample. It allows the examination of factors such as the structural surface of the carriers, their shape, and formulations (Kumar & Seth, 2021). TEM images have clear and detailed information on the size and dispersal intensities. The technique involves use of an electron beam through an ultra-thin sample to obtain surface properties of the sample. Before subjecting a sample to TEM it has to be diluted with distilled water, have a droplet placed on a carbon-stained copper grid and be possibly stained for visibility. To enhance the stability of nanoparticles, they can either be coated with negative staining substances or the nanoparticles can be embedded in plastic to make them more manageable. The other technique is to remove the sample, before placing it on vitreous ice and then exposing it to liquid nitrogen (Sandhiya & Ubaidulla, 2020).

11.5.2.9 X-Ray Powder Diffraction (XRD)

X-ray diffraction (XRD) is a technique that studies the crystal structure of the powdered material by observing the diffraction pattern, which is caused by X-ray pass through a powdered material. It rapidly assesses the properties of crystalline materials and, if desired, produces the unit cell, and atomic spacing data (Sandhiya & Ubaidulla, 2020). It is produced using an X-ray tube that has been designed in a very elaborate manner to produce only one wavelength of radiation, something known as monochromatic radiation. This radiation is deliberately beamed on the sample subject of analysis (Sandhiya & Ubaidulla, 2020).

11.5.2.10 Evaluation of Recovery and Encapsulation Performance

Encapsulation efficiency and recovery involve assessing the ability by which a drug is entrapped in a carrier. After that, a volume of the sample is mixed with the organic solvent and then placed in an ultrasonic bath for 30 minutes to extract the drug. The mixture is then spun at a fixed rpm/min for 10 min. and could be best estimated by High-Performance Liquid Chromatography (HPLC) or ultraviolet (UV) spectroscopy (Sandhiya & Ubaidulla, 2020).

11.5.2.11 Atomic Force Microscopy

Atomic Force Microscopy (AFM) can be described as a complex imaging technique that employs the contact modes of scanning at below the micrometer scale using a probe with atomic scale dimensions to measure the size of particles with great precision. By operating at contact and noncontact modes, AFM forms topographical images with a high degree of resolution (Yadav et al., 2014). Its capability to image non-conducting samples without special treatment makes it invaluable in visualizing complex biological as well as polymeric nano- and microstructures. Additionally, AFM stands out by providing the most accurate description of particle size and size distribution without the need for mathematical processing (Chakraborty et al., 2016).

11.5.2.12 UV-Visble Spectroscopy

An experiment sample is positioned halfway between the light source and a photodetector – a device used in detecting and measuring the intensity of light striking (Kumar et al., 2012). In the UV-visible absorption spectroscopy, the amount of intensity of a UV-visible light beam is determined before and after passing through the sample. From these measurements, one can compare the HWHM at two different wavelengths and thereby deduce the spectrum of the sample as a function of wavelength. As a rule, this information is presented in the form of an absorption spectrum where the ordinate is the absorbance and the abscissa is the wavelength (Sandhiya & Ubaidulla, 2020).

11.5.2.13 Surface Plasmon Resonance

Nanoparticles exhibit specific absorption resonance wavelengths based on the natural frequency of surface electrons interacting with light photons. This interaction involves oscillation against the restoring force of positive nuclei, establishing a resonance condition (Sandhiya & Ubaidulla, 2020). Nanoparticles at the nanoscale display unique optical properties not found in bulk materials or individual atoms. The optical characteristics of nanoparticles are heavily influenced by particle size and the surrounding medium. Additionally, UV light can be utilized to observe particle agglomeration, as the movement of nanoparticles towards each other changes their pH, leading to their aggregation (Al-Gebory & Mengüç, 2018).

11.5.2.14 Acoustic Methods

The method involves measuring sound wave attenuation and applying a physical equation to calculate particle size (Devi et al., 2010). When a charged particle interacts with the oscillating electric field, it causes the field to rotate in the direction of the acoustic energy. This rotational movement yields valuable information about the surface charge (Sandhiya & Ubaidulla, 2020).

11.6 CONCLUSION

Nanoparticles offer a promising drug delivery system with potential for site-specific targeting, enhanced efficacy, and reduced side effects. Conventional drug delivery systems encounter challenges such as non-targeted delivery, drug wastage, and adverse effects. Nanotechnology is expected to address these issues. Various nanoparticle production techniques have been explored, each with its own advantages and disadvantages. Notably, biosynthesis or green synthesis has emerged as a highly advantageous method due to its simplicity, cost-effectiveness, availability, and environmentally friendly nature. However, thorough investigation of potential nanoparticle toxicity is imperative.

REFERENCES

Abedini, A., Daud, A. R., Abdul Hamid, M. A., Kamil Othman, N., & Saion, E. (2013). A review on radiation-induced nucleation and growth of colloidal metallic nanoparticles. *Nanoscale Res Lett*, *8*(1), 474. https://doi.org/10.1186/1556-276x-8-474

Adeyemi, J. O., Oriola, A. O., Onwudiwe, D. C., & Oyedeji, A. O. (2022). Plant extracts mediated metal-based nanoparticles: synthesis and biological applications. *Biomolecules*, *12*(5), 627.

Afzal, O., Altamimi, A. S. A., Nadeem, M. S., Alzarea, S. I., Almalki, W. H., Tariq, A., Mubeen, B., Murtaza, B. N., Iftikhar, S., & Riaz, N. (2022). Nanoparticles in drug delivery: From history to therapeutic applications. *Nanomaterials*, *12*(24), 4494.

Al-Gebory, L., & Mengüç, M. P. (2018). The effect of pH on particle agglomeration and optical properties of nanoparticle suspensions. *Journal of Quantitative Spectroscopy and Radiative Transfer*, *219*, 46–60. https://doi.org/10.1016/j.jqsrt.2018.07.020

Aminu, N., Bello, I., Umar, N. M., Tanko, N., Aminu, A., & Audu, M. M. (2020). The influence of nanoparticulate drug delivery systems in drug therapy. *Journal of Drug Delivery Science and Technology*, *60*, 101961.

Amol, K., & Pratibha, P. (2014). Novel drug delivery system in herbal's. *International Journal of Pharmaceutical, Chemical & Biological Sciences*, *4*(4).

Anu Bhushani, J., & Anandharamakrishnan, C. (2014). Electrospinning and electrospraying techniques: Potential food based applications. *Trends in Food Science & Technology*, *38*(1), 21–33. https://doi.org/10.1016/j.tifs.2014.03.004

Bello, A., Fabiane, M., Dodoo-Arhin, D., Ozoemena, K. I., & Manyala, N. (2014). Silver nanoparticles decorated on a three-dimensional graphene scaffold for electrochemical applications. *Journal of Physics and Chemistry of Solids*, *75*(1), 109–114. https://doi.org/10.1016/j.jpcs.2013.09.006

Benelmekki, M., Vernieres, J., Kim, J.-H., Diaz, R.-E., Grammatikopoulos, P., & Sowwan, M. (2015). On the formation of ternary metallic-dielectric multicore-shell nanoparticles by inert-gas condensation method. *Materials Chemistry and Physics*, *151*, 275–281. https://doi.org/10.1016/j.matchemphys.2014.11.066

Chakraborty, K., Shivakumar, A., & Ramachandran, S. (2016). Nano-technology in herbal medicines: A review. *International Journal of Herbal Medicine*, *4*, 21–27. https://doi.org/10.22271/flora.2016.v4.i3.05

Chenthamara, D., Subramaniam, S., Ramakrishnan, S. G., Krishnaswamy, S., Essa, M. M., Lin, F. H., & Qoronfleh, M. W. (2019). Therapeutic efficacy of nanoparticles and routes of administration. *Biomater Res*, *23*, 20. https://doi.org/10.1186/s40824-019-0166-x

D'Amato, R., Falconieri, M., Gagliardi, S., Popovici, E., Serra, E., Terranova, G., & Borsella, E. (2013). Synthesis of ceramic nanoparticles by laser pyrolysis: From research to applications. *Journal of Analytical and Applied Pyrolysis*, *104*, 461–469. https://doi.org/10.1016/j.jaap.2013.05.026

Darji, S., Rani, K. S., & Patani, P. (2022). A review on herbal nanoparticles. *Journal of Pharmaceutical Negative Results*, 13(3), 1990–2002.

Dawadi, S., Katuwal, S., Gupta, A., Lamichhane, U., Thapa, R., Jaisi, S., Lamichhane, G., Bhattarai, D. P., & Parajuli, N. (2021). Current research on silver nanoparticles: synthesis, characterization, and applications. *Journal of Nanomaterials*, *2021*(1), 6687290.

Devi, V. K., Jain, N., & Valli, K. S. (2010). Importance of novel drug delivery systems in herbal medicines. *Pharmacognosy Reviews*, *4*(7), 27.

Dhand, C., Dwivedi, N., Loh, X. J., Jie Ying, A. N., Verma, N. K., Beuerman, R. W., Lakshminarayanan, R., & Ramakrishna, S. (2015). Methods and strategies for the synthesis of diverse nanoparticles and their applications: a comprehensive overview [10.1039/C5RA19388E]. *RSC Advances*, *5*(127), 105003–105037. https://doi.org/10.1039/C5RA19388E

Dongare, P. N., Motule, A. S., Dubey, M. R., More, M. P., Patinge, P. A., Bakal, R. L., & Manwar, J. V. (2021). Recent development in novel drug delivery systems for delivery of herbal drugs: An updates. *GSC Advanced Research and Reviews*, *8*(2), 8–18.

Ealia, S. A. M., & Saravanakumar, M. P. (2017). A review on the classification, characterisation, synthesis of nanoparticles and their application, IOP Conf. *Series: Materials Science and Engineering* , 263 (3), 032019

Gan, Y. X., Jayatissa, A. H., Yu, Z., Chen, X., & Li, M. (2020). Hydrothermal synthesis of nanomaterials. *Journal of Nanomaterials*, 2020, 3

Hachem, K., Ansari, M. J., Saleh, R. O., Kzar, H. H., Al-Gazally, M. E., Altimari, U. S., Hussein, S. A., Mohammed, H. T., Hammid, A. T., & Kianfar, E. (2022). Methods of chemical synthesis in the synthesis of nanomaterial and nanoparticles by the chemical deposition method: a review. *Bionanoscience*, *12*(3), 1032–1057.

Hsu, C.-Y., Rheima, A. M., Kadhim, M. M., Ahmed, N. N., Mohammed, S. H., Abbas, F. H., Abed, Z. T., Mahdi, Z. M., Abbas, Z. S., & Hachim, S. K. (2023). An overview of nanoparticles in drug delivery: properties and applications. *South African Journal of Chemical Engineering*, 46, 233–270.

Idrees, H., Zaidi, S. Z. J., Sabir, A., Khan, R. U., Zhang, X., & Hassan, S. U. (2020). A Review of Biodegradable Natural Polymer-Based Nanoparticles for Drug Delivery Applications. *Nanomaterials (Basel)*, *10*(10). https://doi.org/10.3390/nano10101970

Jain, K. K. (2020). An overview of drug delivery systems. *Methods in Molecular Biology*, 2059, 1–54.

Kamal, A., Ashmawy, M., Algazzar, A. M., & Elsheikh, A. H. (2022). Fabrication techniques of polymeric nanocomposites: A comprehensive review. *Proceedings of the Institution of Mechanical Engineers, Part C: Journal of Mechanical Engineering Science*, *236*(9), 4843–4861.

Karunakaran, G., Sudha, K. G., Ali, S., & Cho, E. B. (2023). Biosynthesis of Nanoparticles from Various Biological Sources and Its Biomedical Applications. *Molecules*, *28*(11). https://doi.org/10.3390/molecules 28114527

Khan, Y., Sadia, H., Ali Shah, S. Z., Khan, M. N., Shah, A. A., Ullah, N., Ullah, M. F., Bibi, H., Bafakeeh, O. T., Khedher, N. B., Eldin, S. M., Fadhl, B. M., & Khan, M. I. (2022). Classification, Synthetic, and Characterization Approaches to Nanoparticles, and Their Applications in Various Fields of Nanotechnology: A Review. *Catalysts*, *12*(11), 1386. https://www.mdpi.com/2073-4344/12/11/1386

Kharisov, B. I., Dias, H. V. R., Kharissova, O. V., Vázquez, A., Pena, Y., & Gomez, I. (2014). Solubilization, dispersion and stabilization of magnetic nanoparticles in water and non-aqueous solvents: recent trends. *RSC Advances*, *4*(85), 45354–45381.

Khatak, S., & Dureja, H. (2015). Recent techniques and patents on solid lipid nanoparticles as novel carrier for drug delivery. *Recent Patents on Nanotechnology*, *9*(3), 150–177.

Kozuch, J., Ataka, K., & Heberle, J. (2023). Surface-enhanced infrared absorption spectroscopy. *Nature Reviews Methods Primers*, *3*(1), 70.

Kumar, D., & Seth, C. S. (2021). Green-synthesis, characterization, and applications of nanoparticles (NPs): a mini review. *International Journal of Plant and Environment*, *7*(01), 91–95.

Kumar, S., Dilbaghi, N., Saharan, R., & Bhanjana, G. (2012). Nanotechnology as emerging tool for enhancing solubility of poorly water-soluble drugs. *Bionanoscience*, *2*, 227–250.

Kumari, S., Raturi, S., Kulshrestha, S., Chauhan, K., Dhingra, S., András, K., Thu, K., Khargotra, R., & Singh, T. (2023). A comprehensive review on various techniques used for synthesizing nanoparticles. *Journal of Materials Research and Technology*, *27*, 1739–1763. https://doi.org/10.1016/j.jmrt.2023.09.291

Morikawa, K., Masubuchi, Y., Shchipunov, Y., & Zinchenko, A. (2021). DNA-chitosan hydrogels: Formation, properties, and functionalization with catalytic nanoparticles. *ACS Applied Bio Materials*, *4*(2), 1823–1832.

Nalla, A., & Chinnala, K. M. (2017). Novel herbal drug delivery system-an overview. *WJPPS*, *6*(8), 369–395.

Nešović, K., Janković, A., Radetić, T., Perić-Grujić, A., Vukašinović-Sekulić, M., Kojić, V., Rhee, K. Y., & Mišković-Stanković, V. (2020). Poly (vinyl alcohol)/chitosan hydrogels with electrochemically synthesized silver nanoparticles for wound dressing applications. *Journal of Electrochemical Science and Engineering*, *10*(2), 185–198.

Pyrgiotakis, G., McDevitt, J., Bordini, A., Diaz, E., Molina, R., Watson, C., Deloid, G., Lenard, S., Fix, N., Mizuyama, Y., Yamauchi, T., Brain, J., & Demokritou, P. (2014). A chemical free, nanotechnology-based method for airborne bacterial inactivation using engineered water nanostructures [10.1039/C3EN00007A]. *Environmental Science: Nano*, *1*(1), 15–26. https://doi.org/10.1039/C3EN00007A

Radičić, R., Maletić, D., Blažeka, D., Car, J., & Krstulović, N. (2022). Synthesis of silver, gold, and platinum doped zinc oxide nanoparticles by pulsed laser ablation in water. *Nanomaterials*, *12*(19), 3484.

Rahman, P., & Green, M. (2009). The synthesis of rare earthfluoride based nanoparticles [10.1039/B9NR00089E]. *Nanoscale*, *1*(2), 214–224. https://doi.org/10.1039/B9NR00089E

Sajid, M., & Płotka-Wasylka, J. (2020). Nanoparticles: Synthesis, characteristics, and applications in analytical and other sciences. *Microchemical Journal*, *154*, 104623.

Sandhiya, V., & Ubaidulla, U. (2020). A review on herbal drug loaded into pharmaceutical carrier techniques and its evaluation process. *Future Journal of Pharmaceutical Sciences*, *6*, 1–16.

Sendker, J., & Sheridan, H. (2017). History and current status of herbal medicines. *Toxicology of herbal products*, 11–27. https://doi.org/10.1007/978-3-319-43806-1_2

Solanki, J. N., & Murthy, Z. V. P. (2011). Controlled Size Silver Nanoparticles Synthesis with Water-in-Oil Microemulsion Method: A Topical Review. *Industrial & Engineering Chemistry Research*, *50*(22), 12311–12323. https://doi.org/10.1021/ie201649x

Teoh, W. Y., Amal, R., & Mädler, L. (2010). Flame spray pyrolysis: An enabling technology for nanoparticles design and fabrication [10.1039/C0NR00017E]. *Nanoscale*, *2*(8), 1324–1347. https://doi.org/10.1039/C0NR00017E

Tiwari, S., Talreja, M. S., & Pandey, M. S. (2020). A review on use of novel drug delivery systems in herbal medicines. *Science and Engineering Journal*, *24*(8), 190–197.

Tulinski, M., & Jurczyk, M. (2017). Nanomaterials Synthesis Methods. *Metrology and Standardization of Nanotechnology*. https://doi.org/10.1002/9783527800308.ch4

Xing, T., Sunarso, J., Yang, W., Yin, Y., Glushenkov, A. M., Li, L. H., Howlett, P. C., & Chen, Y. (2013). Ball milling: a green mechanochemical approach for synthesis of nitrogen doped carbon nanoparticles [10.1039/C3NR02328A]. *Nanoscale*, *5*(17), 7970–7976. https://doi.org/10.1039/C3NR02328A

Yadav, M., Bhatia, V. J., Doshi, G., & Shastri, K. (2014). Novel techniques in herbal drug delivery systems. *International Journal of Pharmaceutical Sciences Review and Research*, *28*, 83–89.

Zhang, X., Yan, S., Tyagi, R. D., & Surampalli, R. Y. (2011). Synthesis of nanoparticles by microorganisms and their application in enhancing microbiological reaction rates. *Chemosphere*, *82*(4), 489–494. https://doi.org/10.1016/j.chemosphere.2010.10.023

12 Enhanced Bioavailability of Herbal Extracts using Nanocarriers

Marij Noor, Amina Khalid, Muhammad Umair, Tooba Khan, and Muhammad Imran Khan
Department of Biomedical Sciences, Pak Austria Fachhochschule: Institute of Applied Sciences and Technology, Haripur, Pakistan

12.1 INTRODUCTION

The healing power of nature has always been a consistent thread within the vast tapestry of human history, weaving through prehistoric and modern societies alike. Herbal extracts, with their potent properties and rich heritage, have always been appreciated as nature's pharmacy. Yet, despite all their benefits the body's own defenses act as a formidable barrier against them. The journey from plant to pill is challenged by the interaction of plants' bioactive compounds and biological barriers, compromising the overall efficacy. With the advent of nanotechnology, scientists have now unlocked a way to enhance the bioavailability of herbal medicines. This involves the synthesis of ancient traditions to modern innovation, changing the dynamics of nature's healing power. Imagine a tiny, engineered carrier that will directly deliver the bioactive compounds to their targets, while navigating through the complex terrain of the human body, with unprecedented efficiency.

12.1.1 Challenges of Herbal Extracts in Traditional Medicine

Herbal extracts are used for their medicinal properties and are derived from plants or plant parts. Since the time of early humankind, they have been acting as biosynthetic chemical laboratories, often containing several constituents that act together synergistically. 'Active ingredients' are the chemical constituents that provide medicinal benefits. However, the use of herbal extracts remains challenging due to several factors. These factors are poor solubility, instability, low bioavailability, and variability of the constituents due to environmental and genetic factors, and so on [1].

Herbal extracts may consist of either a single or multiple active compounds. Unlike other drugs, however, they are thermolabile to volatile, leading to the degradation of active compounds during storage and processing. Temperature, moisture, light, oxygen, microbial, and trace metal contamination are among the environmental factors that influence the stability of herbal products [2].

Almost every herbal extract, despite their extraordinary benefits in vitro, shows less to no in vivo results because of their inappropriate molecular size and poor solubility in lipids, which is a major cause of low absorption and low bioavailability. Procedures such as purification can lead to the loss of certain chemically related substances involved in synergistic action, thereby compromising the active compounds' desired activity. Additionally, water-soluble components cannot cross the lipid membranes in the intestines, further compromising bioavailability. Scientists have now found a solution to tackle this problem. As a result, bioavailability is increased using novel delivery systems for drugs like micelles, liposomes, and polymeric nanoparticles to improve movement across the barriers and drug release [3].

DOI: 10.1201/9781003544418-12

12.1.2 Importance of Bioavailability in Therapeutic Efficacy

The concept of bioavailability is a cornerstone of therapeutic efficacy. It determines the efficiency with which a bioactive compound reaches its site of action in the body. If the therapeutic agent available produces an intended effect, it is considered to have high bioavailability. By contrast, drugs with suboptimal levels of bioactive element at the target site will be described as having poor bioavailability. This will result in reduced efficacy, and an increased likelihood of adverse effects, necessitating higher or more frequent doses, and compromising patient adherence. Hence, the optimization of bioavailability is crucial in developing drugs from herbal extracts to facilitate precise dosage adjustments, enhanced patient outcomes, and better pharmacodynamic and pharmacokinetic profiles.

Bioavailability is defined as the measure absorption level of an active constituent present in a drug and its absorbed rate in the bloodstream from the administered dosage in a patient, reaching the targeted location to achieve the desired therapeutic outcome. A drug's bioavailability dictates the active ingredient percentage present in the organism in comparison with the concentration of drug administered [4].

For accurate dosage adjustment, knowledge regarding the absorption, transport mechanism, metabolism, and elimination of the Active Pharmaceutical Ingredient (API) is necessary. Conversely, the administration of improper medication contributes to the unnecessary utilization of drugs. Hence, for the patient's safety and treatment efficiency, bioavailability is significant. To maximize the bioavailability of medication, scientists utilize certain protocols such as appropriate drug delivery techniques, identifying and managing factors that hinder bioavailability, adjusting drug formulations, optimizing dosages, and monitoring blood levels of drug along with adjustment in the dosage.

Once the dose is optimized with improved bioavailability, this will result in a reduced required dose and frequency of dose administration, and therefore improved patient compliance and reduced chances of side effects [4].

It is necessary to optimize herbal extracts' pharmacokinetic and pharmacodynamic profiles. The pharmacokinetic profile shows how a drug is absorbed in the bloodstream, distributed, metabolized, and eliminated from the body. This will help to determine the amount of active ingredients in the bloodstream at any particular time. Poor bioavailability may be a result of inadequate absorption, rapid metabolism, or premature elimination, leading to insufficient levels of drug at the targeted site and reduced therapeutic efficiency. Conversely, the pharmacodynamic profile determines the therapeutic effects, harmful side effects, and the mechanism of action. High bioavailability will ensure an adequate amount of active compound at the target site which will assist in producing the desired therapeutic effect. It is bioavailability that influences the onset, duration of action, and intensity. A drug with poor bioavailability may not elicit a significant pharmacodynamic response, causing inadequate efficacy of the drug [5].

12.1.3 The Role of Nanotechnology in Addressing Bioavailability Issues

A nanotechnology-based drug delivery system was introduced to overcome the dilemmas related to herbal medicines. This acts as a drug carrier, increasing bioavailability and therapeutic efficacy. This nanoscale system, based on particle size of 0.1μm, has been developed to enhance activity, minimize dosage requirement, and mitigate adverse effects. The potential of herbal medicines has significantly increased using nanotechnology-based drug delivery systems, including the capacity to convert less soluble, less absorbed, and unstable constituents into effective drugs [6].

Consider a nanobiologist-designed system in a laboratory setting that can perform targeted delivery. It consists of an imaging molecule, ligands for site-specific targeting, an active constituent for delivery, a destabilizing lipid for drug release at the required site, and, finally, a sensor to probe the efficacy in real time. The fabrication of this delivery vehicle's structural components is based on structural motifs that undergo self-assembly, targeting ligands and pharmaceutical agents. This is the vision behind the creation of drug-loaded nanocarriers, which are tuned to trigger the release of the drug in order to significantly improve the efficacy of herbal extracts, which will potentially overcome drug resistance. In the case of liposomes, the basic underlying principle revolves around creating defects in the liposome membrane. Membranes sensitive to mild hyperthermia are potential methods for triggered drug delivery. Additionally, by utilizing external triggering of liposomes, this method exploits the abnormalities of diseased cells in tumors, upregulated enzymes can cleave the membranes' lipids and create defects [7].

12.2 PRINCIPLES OF BIOAVAILABILITY ENHANCEMENT

12.2.1 Understanding ADME Profiles

Pharmacokinetics refers to the study of the drug's kinetics, or its active constituent once it enters the body. It refers to the transient changes that occur in the drug and its metabolites when they get into the plasma, serum, or whole blood, and, over time, within target tissues and organs. The human body is a very intricate system; once a drug has entered, it experiences a series of processes: absorption, distribution, metabolism, and, finally, excretion (ADME). Below we shall discuss each in detail (Ruiz-Garcia, Bermejo et al. 2008) (Figure 12.1).

12.2.1.1 Absorption

The term absorption is used to describe the process which involves the movement of a certain amount of drug from the administration site to the main compartment. It depends on the route of administration. In the case of an orally administered drug, it is first absorbed from the gastrointestinal tract, and the net absorption is limited by factors including the drug's physicochemical properties and the dosage form. Common routes of drug administration are mentioned in Table 12.1. Before

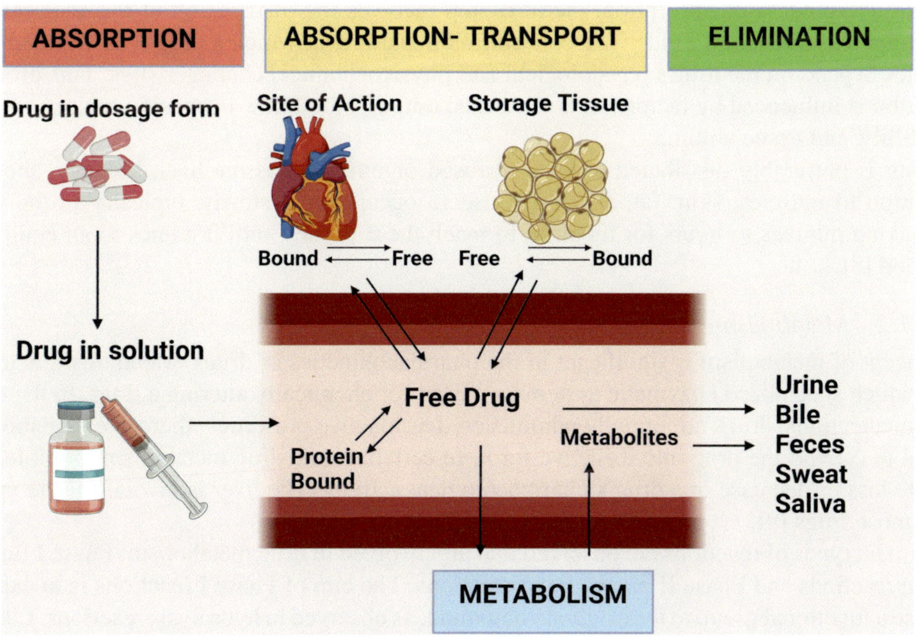

FIGURE 12.1 This diagram illustrates the complex processes of absorption, distribution, metabolism and excretion which determine the pharmacokinetics of the drugs.

TABLE 12.1
Common Routes of Drug Administration and Characteristics

Route Of Administration	Absorption Pattern	Advantages	Limitations and Precautions
Intravenous	Bypass absorption Rapid onset of action Ideal for administering irritating substances, large volumes or complex mixtures upon dilution	Used in emergencies Allows for dose adjustment Typically, necessary for large molecular weight proteins and peptide drugs	Potential risk of adverse effects Slow administration Not preferable in case of oily and poorly soluble drugs
Subcutaneous	Quick action from aqueous solution Gradual and prolonged from respiratory preparation	Suitable for poorly soluble suspension and slow-release implants	Not suggested in case of large volumes potential pain or necrosis from irritating substances
Intramuscular	Quick action from aqueous solution Gradual and prolonged from respiratory preparation	It is ideal for moderate volumes, oily vehicles, and certain irritants Can be self-administered	Not allowed during anticoagulant therapy May affect the results of certain diagnostic test like LFT
Oral ingestion	Depends on different factors	Safe, economical, and convenient	Requires patient compliance bioavailability, inconsistent and incomplete

entering the systemic circulation, the drug must pass through the liver where metabolism and biliary excretion are likely to occur. Therefore, out of the total concentration of administered and absorbed dose, a fraction will be diverted or inactivated before reaching the general circulation, followed by distribution to the site where needed.

12.2.1.1.1 Distribution

Following the process of absorption, the next step involves the distribution of the drug into both intracellular and interstitial fluids. The concentration and rate at which a drug is distributed within the tissues depend on the drug's physiological and physicochemical characteristics. This process of distribution is influenced by factors such as cardiac output, blood flow to specific regions, capillary permeability, and tissue volume.

A drug is preferably distributed to well-perfused organs such as the liver, kidneys, and brain. Distribution to muscles, skin, fat, and most viscera occurs more slowly, typically in the second phase, taking minutes to hours for the drug to reach these tissues until it comes to an equilibrium with blood [8].

12.2.1.1.2 Metabolism

The concept of metabolism is significant in the pharmacokinetics of drugs. Metabolism is the process in which specialized enzymatic systems are used for chemically altering a drug. In the case of certain medications, drugs are initially administered as inactive pro-drugs; therefore, metabolism is essential to convert the drug into its active form. In certain cases, drug metabolism might lead to a complete loss or decrease in a drug's pharmacological activity. The liver is pivotal for the metabolism of most drugs [9].

Two main types of reactions are observed that are involved in drug metabolism: Phase I functionalization reactions and Phase II biosynthetic reactions. The aim of Phase I reactions is to expose or introduce a functional group to the original compound, as observed in hydrolysis reactions. Likewise, Phase II conjugation reactions facilitate the creation of covalent bonds between metabolites from Phase I or a functional group on the parent compound and endogenously produce sulfate, glutathione, glucuronic acid, amino acids, or acetate.

The system of enzyme engaged in the drugs biotransformation is likely to be seen positioned firstly in the liver, however some metabolic activity is also detected in other tissues. Organs which have significant metabolic capacity include the kidneys, lungs and the GI tract. Drug-metabolizing activity is predominantly found in the cytosol (Phase II conjugation enzyme systems) and the smooth ER (prominent location for enzyme systems involved in Phase I), but it can also occur in the mitochondria, plasma membrane and nuclear envelope.

12.2.1.1.3 Excretion

Drugs are removed from the body through excretion in their original form or as metabolites. Organs other than the lungs, such as the kidneys and liver, exhibit more effective elimination of polar compounds compared to those with high lipid solubility. Consequently, lipid-soluble drugs undergo metabolism to become more polar compounds for elimination. The kidney is crucial in regarding the removal of drugs and their respective metabolites. Three processes are involved in renal excretion: (i) glomerular filtration, (ii) active tubular secretion, and (iii) passive tubular reabsorption.

Regarding the biliary and fecal excretion, there are transporters located in the hepatocytes' canalicular membrane, facilitating the active secretion of drugs and metabolites into bile. Ultimately, drugs and its metabolites which reside in bile are discharged into the GI tract during digestion. Given that there are secretory transporters expressed on the enterocyte's apical membrane, they may be involved in directing the drug's secretion and their metabolites into the intestinal lumen from the systemic circulation. Subsequently, reabsorption can occur in the intestine, particularly taking into consideration the conjugated metabolites such as glucuronides, which may necessitate hydrolysis by enzymes via intestinal flora [8]. The different routes involved in the excretion are mentioned in Figure 12.2.

12.2.2 Factors Affecting the Bioavailability of Herbal Compounds

Despite the therapeutic and phytochemical significance of herbal medicine in improving human health, its wider application is hindered by the issue of low bioavailability. A compounds

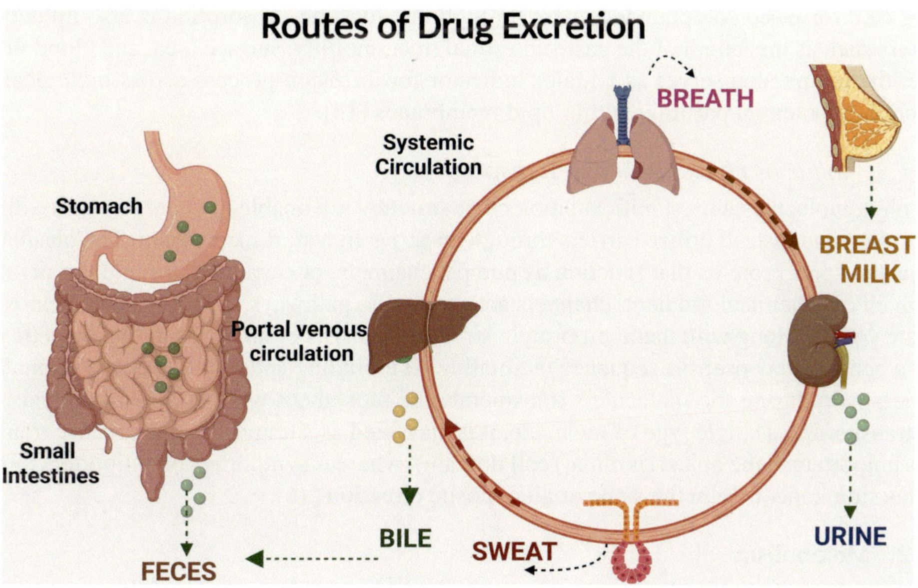

FIGURE 12.2 A diagram of the human body shows the various routes involved in drug excretion from the body. They include renal clearance, biliary excretion, pulmonary excretion, fecal excretion, breast milk excretion, sweat excretion, and skin excretion.

characteristics are crucial for improving the rate and extent of its absorption, regardless of the route of administration. Challenges arise when poorly lipid-soluble compounds face restricted permeability across membranes [10]. Due to solubility issues, many herbal products exhibit limited therapeutic efficacy leading to suboptimal bioavailability regardless of their remarkable potential [3].

The effectiveness of any herbal extract depends upon achieving adequate levels of its therapeutically active compound at the desired site [11]. Unfortunately, most bioactive substances designed for oral intake are predominantly hydrophobic, with compromised water solubility and inadequate bioavailability. Additionally, poor solubility contributes to reduced absorption in the GI tract, thus limiting therapeutic effectiveness [12].

It is not possible to accurately predict the bioavailability of a compound; however, Lipinski's 'rule of five' analysis offers a valuable insight. According to this rule, a compound's bioavailability will be better when it possesses no more than five hydrogen bond donors, ten hydrogen bond acceptors, and a molecular mass not exceeding 500 Da, log p less than or equal to 5, and a maximum of 10 rotatable bonds [13].

12.2.2.1 Absorption within the GI Lumen

Absorption involves the transferring a drug from the administration site into the general circulation. In the scenario of drugs that are orally administered, the absorption of molecules from the GI tract relies on the dissolution and permeation across membranes.

12.2.2.1.1 The Solubility of the Herbal Products

A drug's dissolution is influenced by several factors which include the compound's log p value, particle size, pKa, pH levels of gastrointestinal and intestinal fluids, and, last but not least, the surface area. For instance, ellagic acid (EA) [10] resveratrol [8], and curcumin [1] exhibit poor bioavailability as a result of limited solubility in aqueous media [14].

12.2.2.1.2 Absorption via Passive Diffusion

There are various physicochemical factors which are important for the absorption of phytochemicals via passive diffusion. The unionized molecules tend to transverse the GI lipid barriers; however, this is no big deal for basic compounds to pass through the intestine. Absorption is also influenced by parameters such as the length of the gastrointestinal tract, motility, surface area, and blood flow. For instance, the log p value serves as a major indicator for diffusion process across biological membrane and their potential partition within lipid membranes [15].

12.2.2.1.3 Influx or Efflux via Active Transport

Certain phytomolecules with significant molecular structure are unable to undergo passive diffusion for absorption and instead utilize carriers through an active transport mechanism. The plasma membrane has transport proteins that function as pumps, channels, or carriers. For rapid ion permeation along the electrochemical gradient, channels act as specific pathways, and they need adenosine triphosphate (ATP). Along with that their protein structure consists of an ATP-binding cassette (ABC) that has a conserved consensus sequence facilitating ATP binding and hydrolysis. Additionally, solute carriers help during the molecule's transmembrane movement without ATP hydrolysis. In the case of transporting a single type of molecule, it is classified as a uniporter (e.g. glucose transporter 2 acts as uniporters in the apical (luminal) cell domain), whereas symporters or antiporters move two molecules simultaneously in the same or an opposite direction [16].

12.2.2.2 Metabolism

The process of metabolism in living organisms involves a series of enzymatic pathways that are supposed to modify the chemical structure of foreign compounds. This process focuses on detoxifying phytochemicals, converting them into more hydrophilic from lipophilic compounds. Metabolism may take place either before (in the gastrointestinal tract) or after absorption (within the hepatic microsomal system).

12.2.2.2.1 Metabolism Prior to Absorption

The metabolism occurs prior to absorption in the GI tract and is crucial for the compound's bioavailabilty. The process involves digestive enzymes (such as glycosidase and protease) or gut microflora at the colonic level. For instance, flavonoids undergo hydrolysis in the small intestine via glycosidase [17]. Chlorogenic acid (CA) cannot be released by esterase in humans; instead, it is metabolized by the microflora of colon for absorption [18].

12.2.2.2.2 Metabolism Post-Absorption

After entering the bloodstream, a drug undergoes first-pass metabolism in the hepatic microsomal system, generating metabolites with increased polarity that are more readily excreted in bile or urine. A common oxidation mechanism is mediated by cytochromeP450 (CYP450)-dependent mixed function oxidase system. This system introduces an atom of oxygen into a non-activated hydrocarbon, leading to the incorporation of hydroxyl groups or N-, O-, and S-dealkylation of substrates [19].

12.2.2.3 Mechanisms of Action for Nanocarriers

Nanotechnology has significantly emerged across different scientific fields in the recent decades. It involves the manipulation of minute particles at a nanoscale. At nanoscales, atoms and molecules exhibit distinctive behaviors, offering a wide array of intriguing applications. To overcome the constraints of conventional formulations, drug delivery scientists have embraced this paradigm shift into pharmaceutical nanocarriers. They are submicron-sized drug carriers (greater than 200 nm in diameter) either with or without biodegradable properties. The broad category of nanocarriers includes nanospheres, nanocapsules, nanoparticles, nanoemulsions, and nano-sized vesicular carriers such as niosomes and liposomes [20].

The pharmacological activity of drugs is dependent not only on their therapeutic efficacy but also on their bioavailability. Most of the phytopharmaceuticals exhibit limited water solubility, leading to reduced permeability across membranes and, consequently, diminished oral bioavailability. The promotion of innovative technologies is creating a vast platform for developing novel delivery systems for herbal medicine to enhance both therapeutic efficacy and the bioavailability of compounds with low aqueous solubility [21]. Biomedical applications of nanocarriers are mentioned in Figure 12.3.

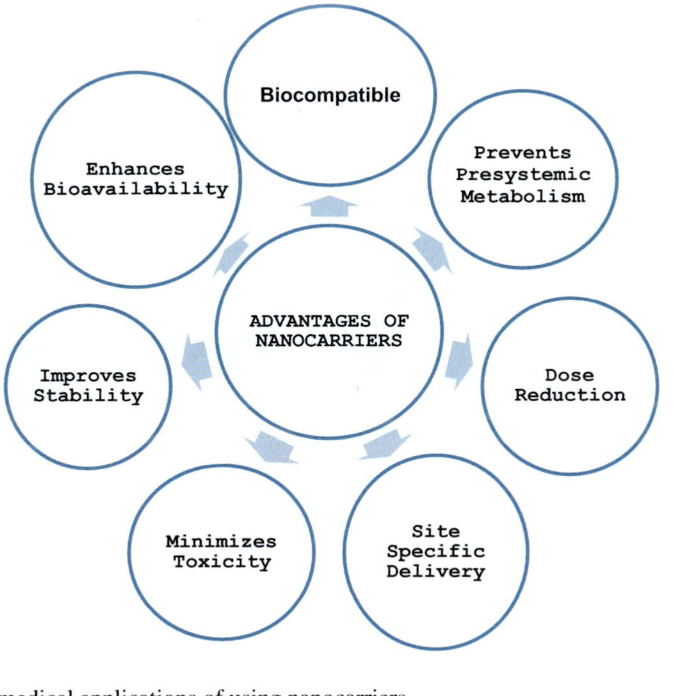

FIGURE 12.3 Biomedical applications of using nanocarriers.

12.3 TYPES OF NANOCARRIERS AND THEIR APPLICATIONS

In recent years, the submicron-system, i.e. nanosystem, has become popular in medicine. This is due to these systems having advantages over traditional approaches. The latter have limitations such as off-target effect, rapid degeneration and bioavailability. As a result of these shortcomings, new therapeutic intervention tools have been introduced, i.e. nanocarriers. In this section, we will explore several types of nanocarriers based on their structure, properties, and functionality, and their applications in medicine.

12.3.1 Liposomes: Structure, Function, and Applications

Liposomes have become popular and often used nanocarriers for different hydrophobic and hydrophilic molecules due to their properties like high compatibility, biodegradability, ease of synthesis, high loading efficiency, and low immunogenicity.

12.3.1.1 Structure

Liposomes were discovered by Bengham in 1960 [22]. Liposomes are spherical and their size typically ranges from 50–500nm in diameter. The emulsification of lipids in aqueous media leads to the formation of lipid bilayers [23]. Liposomes are generally composed of two molecules, i.e. phospholipids and cholesterol (although this is not always present). Phospholipids are major molecules composing liposomes. Phospholipids are amphiphilic, meaning that they have both a hydrophobic region (i.e. a tail) and a hydrophilic region (i.e. a head). The structure of phospholipids consists of two fatty acid chains (with 10–24 carbon atoms in each chain) and the hydrophilic head composed of phosphoric acid and water-soluble molecules. Water can be excluded from the hydrophobic domain by orienting the phospholipid tails, allowing the hydrophilic heads to be exposed to water. The structure can be seen in Figure 12.4.

The lipids commonly used in the synthesis of liposomes are either natural (phosphatidylcholine) or synthetic (dialkyl or trialkyl lipids, 1,2-dioleoyl-sn-glycero-3-phosphocholine).

While cholesterol is not always present, it is often incorporated because of its properties like modulating membrane permeability, changing fluidity, and improving the stability of the bilayer membrane [24].

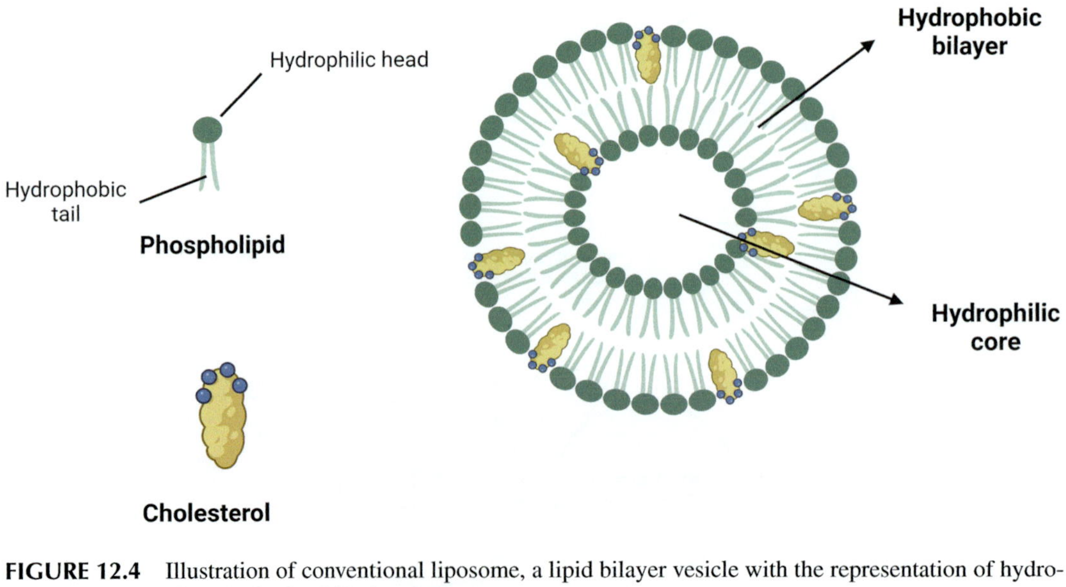

FIGURE 12.4 Illustration of conventional liposome, a lipid bilayer vesicle with the representation of hydrophobic tails as well as the hydrophilic head groups.

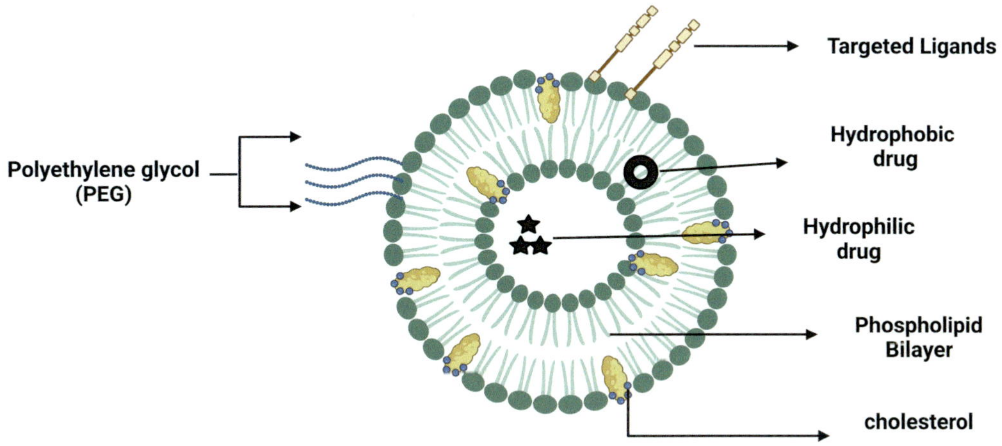

FIGURE 12.5 Schematic representation of bilaminar liposome. For active targeting, the liposomal surface can be functionalized with ligands or PEGylated. Drugs can be packed into liposomes during preparation.

Polymers and membrane proteins can also be added to liposome formulations as shown in Figure 12.5 to enhance the liposome circulation half-life, enhance the biodistribution of liposomes, and increase the effectiveness of the medicine contained [25].

12.3.1.2 Function

Most active constituents of extracts are extremely hydrophilic and have high solubility in water, but their absorption is low because they cannot pass through the plasma membrane due to hydrophobicity of the plasma membrane and excessive molecular size, resulting in poor bioavailability and a loss of efficacy. To overcome this hurdle, it has been recommended that herbal medicines should be combined with nanotechnology. Liposomes can be used as nanocarriers for hydrophilic substances.

Drugs can be loaded into the engineered liposome so that an aqueous environment encapsulates the hydrophilic substances and adsorbed hydrophobic molecules are introduced into the membrane. In this way, both types of substances can be loaded into the liposome [26]. Methods to achieve drug loading into the liposome are the PH gradient technique, utilizing organic solvents, and the solvent exchange mechanism.

Liposomes are first engineered by adding different molecules (targeting molecules, functionalized imaging agents) onto the exposed surface of the lipid bilayer. Due to this manipulation, liposomes can target specific tissues both actively and passively.

12.3.1.3 Applications

High systemic absorption of medicines leads to off-target effects on tissue other than the intended target tissue, which can cause unwanted side effects. Using liposome-encapsulated medicines and their controlled release rate reduces the adverse effects and incompatibility resulting from high systemic absorption [27].

1. Since liposomes have high substantive rates of interaction with biological membranes, this makes them capable of easily fusing with target cell membranes, releasing the encapsulated drug directly inside the cell [27].
2. Liposomes also enhance the stability of the herbal extract by acting as a protective barrier against light, enzymes, and heat. Liposomes improve the stability of herbal extract during storage and delivery.
3. For the improvement of bioavailability and to lessen the adverse reaction of a medication, liposomes are designed so they can discharge the encapsulated drug in a regulated way [28].

4. Liposomes can also be helpful in delivering drugs across the blood–brain barrier and the blood–cochlear barrier [29].
5. Liposomes are used to lower the clearance of medication and extend its half-life.

12.3.2 Polymeric Nanoparticles: Design and Delivery Mechanisms

Polymeric nanoparticles are typically ranging from 10 to 100nm in size. These polymeric nanosystems are formed by the polymerization reaction of many monomer units and under specific circumstances they can self-assemble and organize into ananometric sizes [30].

12.3.2.1 Design

This section of the chapter focuses on design considerations for polymeric nanocarriers.

12.3.2.1.1 Polymeric Material

In designing a polymeric nanocarrier, the right choice of polymer is important as this determines the properties of the nanocarrier. Polymer carriers vary in their chemical structures and functions. The types of polymers that can be used in nanocarrier design are described in Figure 12.6.

12.3.2.1.2 Drug Encapsulation Methods

A suitable transport device should have desirable characteristics i.e. small particle size, bioavailability, high drug-loading rate, and efficient drug encapsulation and entrapment. There are two ways to encapsulate drugs: (1) the incorporation of the drug during the nanocarrier formulation; and (2) the uptake of the drug after the formulation. Loading of drugs into the nanocarrier is achieved through different methods. Among the most common are those discussed below:

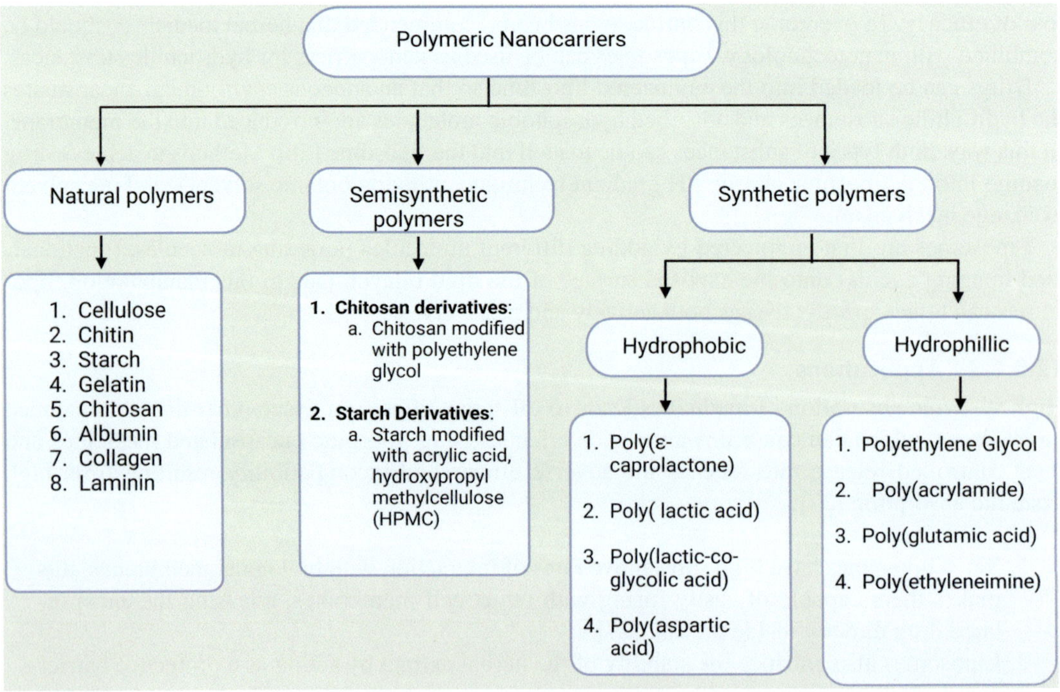

FIGURE 12.6 Types of polymeric nanocarriers based on their origin.

12.3.2.1.2.1 Solvent Evaporation The solvent evaporation technique involves both the polymer and the drug dissolving in an organic volatile solvent. The solvent is evaporated, leaving the encapsulated drug in the self-assembled nanocarrier [31].

12.3.2.1.2.2 Nano-Precipitation In the nano-precipitation technique, the polymer can be dissolved in a water-soluble solvent and then it is quickly added to water with a strong shear force, leading to rapid precipitation of the polymer as a drug-encapsulated nanoparticle [32].

12.3.2.1.2.3 Emulsion-Based Techniques A stabilizing agent is used to emulsify an oil phase containing the drug and polymer in an aqueous phase. And the nanoparticles dispersed in the aqueous phase are left with the encapsulated drug after the organic solvent evaporates [31].

12.3.2.2 The Delivery Mechanism of the Drug

12.3.2.2.1 Route of Delivery

There are various routes for the administration of the nanocarriers inside the body, each having its advantages and limitations. The delivery route can be oral, nasal, pulmonary, transdermal, or subcutaneous. For polymeric nanoparticles, intravenous injection is the main route of administration [1, 33].

12.3.2.2.2 Targeting Strategies

The site of administration of drugs is often far from pathological or targeted sites. So, the nanoparticle has to reach pathological sites and release the drug. Thus, there has been a significant study related to targeted drug delivery. Among the important targeting strategies are:

12.3.2.2.2.1 Passive Targeting Due to rapid angiogenesis in tumors, fenestrated blood vessels are formed which make tumors more permeable than normal tissues to nanoparticles. Due to delayed lymphatic drainage, passive targeted nanoparticles penetrate the fenestrated structure of blood vessels which leads to significant accumulations of the drug. This process is therefore called an enhanced permeation and retention effect [34].

12.3.2.2.2.2 Active Targeting Active targeting refers to the interactions between receptors and ligands. Different cell lines express different types of cell surface receptors. As a result, conjugating nanoparticles with molecules that bind to the receptors can enhance the adherence or absorption of nanoparticles into target cells. Several molecules can be utilized for this purpose, such as antibodies, antibody fragments, DNA/RNA aptamers, peptides and so on [34, 35].

12.3.2.2.2.3 Stimuli-Responsive Targeting Due to disease or inflammation, the microenvironment of tissues undergoes different chemical or biological changes. The pH, oxygen concentration, temperature, and enzyme levels are all typical indicators of an abnormal state. Sensitive nanocarriers can be developed that can sense stimuli and release the drug upon stimulation. The composition or structural conformation of nanocarriers can be influenced by chemical, biochemical, or physical stimuli, resulting in the release of drugs into a targeted environment.

12.3.2.2.3 Drug Release

Different factors are seen to influence the drug release from the nanocarrier, i.e. drug composition (polymer's type etc.), chemical and physical interaction among various components of the nanocarrier and drug, composition ratio, and manufacturing methods. There are four categories that a drug release mechanism falls into: (i) Diffusion-controlled release; (ii) Solvent control release; (iii) Chemical interaction-based release; and (iv) Stimulated release.

12.3.2.2.3.1 Diffusion-Controlled Release Due to the change in concentration gradient, the drug is diffused across the membrane. Thus, in the systems mentioned below, the dissolved or dispersed drug is diffused across the membrane and reaches the target site [36]. The two kinds of nanosystems that use diffusion-controlled release are

- **Reservoir systems**: The drug is encapsulated inside a central core of nanocarrier.
- **Matrix systems**: In this polymer matrix the drug is dispersed [37].

12.3.2.2.3.2 Solvent-Controlled Release There are two approaches involved in solvent control release, i.e., osmotic-controlled release and swelling-controlled release [38].

a. **Osmotic-controlled release**: This is based on the osmosis principle in which the drug and environment are separated by a membrane that is semipermeable. This allows water to move inside, but restricts the movement of drug molecules. When a nanoparticle is internalized by the cell exposed to the environment where a high concentration of external solvent is present, water starts moving inside the nanoparticle. This influx of solvent leads to high pressure inside the nanoparticle and the release of the drug through a semipermeable membrane [36].
b. **Swelling-controlled release**: In the case of high solvent concentration polymers, they absorb water and swell. As a result of this swelling, a large space is created inside the matrix allowing for the release of the drug [38].

12.3.2.2.3.3 Chemical Interaction-Based Release In this method, biodegradable polymers are used. These can be degraded and release the encapsulated drug. The degradation process can be simultaneous or enzymatic, depending on the type of polymer used.

If polymers like polylactic acid (PLA), polycaprolactone (PCL), or polylactic-co-glycolic acid (PLGA) are used for matrix preparation then the matrix will degrade simultaneously without the need for enzymatic degradation. However, if polymers such as polysaccharides, polyamides, and polyesters are used then they release the drug through enzymatic degradation [36].

12.3.2.2.3.4 Temperature-Controlled Release Heat-responsive polymers are used in nanocarrier synthesis. These polymers release the drug upon the slightest change in temperature since in response to heat these polymers can change their chemical and physical properties [39]. These thermo-responsive nanocarriers have gained popularity in recent years, as studies show that inflammatory diseases and tumors show abnormal temperatures due to high metabolic activity [40].

Besides temperature changes brought on by disease tissue, the temperature can also be changed by extrinsic factors in particular tissues. When subjected to temperature change, the polymer undergoes go a structural change which results in a shift in hydrophobicity or solubility Below a certain temperature, they are hydrophilic; above that, they are hydrophobic [41, 42].

12.3.3 Nanoemulsions: Formulation and Stability

As the name suggests, nanoemulsions are two insoluble liquids (either oil in water droplets or water in oil), stabilized by an amphiphilic surfactant. There are different techniques, such as high-pressure homogenization and ultrasonication, for rupturing large microscale droplets into nanoscale droplets [43]. Typically, a mean droplet diameter attained is <500 nm [44]. Nanoemulsions come under a broad class of multiphase colloidal dispersion. Nanoemulsion is sometimes also called submicron emulsion or mini emulsion, but it is entirely different from microemulsion in terms of structure and thermodynamic stability [45].

12.3.3.1 Formulation

Major components of nanoemulsion include: (1) lipids/oils; (2) surfactants and co-surfactants; and (3) preservatives, chemo-protectants, and antioxidants.

i. **Oil/lipid**: In the case of oil/water emulsion, the percentage of oil/lipid droplets it contains is 5–20% generally, but sometimes it can increase up to 70%. Long-chain triglyceride, medium-chain triglyceride, and short-chain triglycerides like safflower oil and soyabean oil are used to formulate nanoemulsions. They can be used alone or can be combined to formulate nanoemulsions [46, 47]. A carrier in the nanoemulsion D-α-Tocopherol (vitamin E) family can be used [48].
ii. **Surfactants and co-surfactants**: Surfactants have polar heads and non-polar tails, and they should be highly soluble in one liquid. Because of the difference in attractive interaction between molecules of two liquids, in all situations where a liquid phase contacts another liquid phase, there is an interfacial tension σ [43]. Nanoemulsions are stabilized by surfactants since they have ability to lessen the interfacial tension and prevent droplets from aggregating. Lecithin is a common surfactant used in nanoemulsion [49].
iii. **Preservatives, chemo-protectants, and antioxidants**: We use preservatives to inhibit the growth of microorganisms. Preservatives should have low toxicity, chemical and physical compatibility, affordable cost, acceptable odor, color and taste, stability to heat and storage, and a broad anti-microbial spectrum. Quaternary ammonium compounds and phenolics can be used as broad-spectrum preservatives [50].

12.3.3.1.1 The Generation of Nanoemulsion

The stability of nanoemulsion depends on many factors that need to be controlled. These factors include choosing the appropriate composition, effective shear application to rupture droplets effectively, and sequence of component addition. There are three types of tailoring techniques for forming nanoemulsions: high-energy emulsification, low-energy emulsification, and a combination of low and high energy.

12.3.3.1.1.1 Emulsification by High-Energy Method In this method, mechanical devices are used to generate strong disruptive forces for size reduction. Mechanical apparatus such as ultrasonicators, high-pressure homogenizers, and microfluidizers. For example, **u**ltrasonication uses high-frequency sound waves (20khz and up). Nanoemulsions can be formed in situ or preformed emulsions can be reduced in size using ultrasonication. Cavitation bubbles form when ultrasonic waves are dipped in samples; these bubbles continue to grow until they reach their limit and implode. This implosion produces a shock wave. As a result, a jet stream of liquid is created in the surrounding area, decreasing the size of dispersed droplets by pressurizing them [51]. The simple process involves the adding oil phase to the aqueous phase with continuous shaking to produce a coarse emulsion. Afterwards, the prepared emulsion is exposed to ultrasonication at different amplitudes until the desired characteristics are obtained [50].

12.3.3.1.1.2 Emulsification by Low-Energy Method

This process involves spontaneous emulsification [52] and the phase inversion method. Spontaneous emulsification is particularly used for the synthesis of polymeric nanoparticles. The overall process involves the preparation of two phases: an aqueous phase, which contains hydrophilic surfactant; and a second phase containing , oil . The forming of a nanoscale emulsion requires the continuous stirring of the aqueous phase and the dropwise addition of oil or the organic phase. Oil nanoemulsions can form spontaneously because the increase in entropy (disorder) from spreading the oil into tiny droplets outweighs the energy needed to break them up. The overall process can be initiated by itself or might require a little external energy (which can be supplied by a magnetic stirrer) [50].

12.3.3.2 Stability

12.3.3.2.1 Physical Stability

Nanoemulsions have a small size due to which they have advantages such as better physical stability against droplet aggregation, high optical clarity, and enhanced bioavailability of drugs. A low-viscosity oil emulsion produces significantly smaller droplets than a high-viscosity oil emulsion. In the same study, it is found that increased concentration of alcohol up to 10 wt.% leads to decreased droplet size. However, with a further increase in alcohol concentration, the droplet size begins to increase. This shows that at optimum concentrations, alcohol enhanced solvent performance [53]. In addition, natural stabilizers such as eugenol are also useful in obtaining more stable nanoemulsions [54].

12.3.3.2.2 Chemical Stability

To avoid chemical degradation, it is advisable to design nanoemulsions to avoid light-catalyzed reactions that may occur due to the transparency of nanoemulsion and surface-catalyzed reactions, i.e. lipid oxidation may occur because of high interfacial areas [55]. Also, bioactive lipids such as carotenoids, conjugated linoleic acid and so on, which are incorporated into nanoemulsions are unstable [56, 57]. To tackle these problems with solutions a number of steps can be taken, such as maintaining low storage temperature, incorporating appropriate antioxidants, controlling droplet interfacial properties, and chelating transition metal catalysts [55].

12.3.4 Micelles: Enhancing Solubility and Bioavailability

Micelles are colloidal structures with a diameter of between 5 and 200nm which are amphiphilic. Critical micellar temperature and critical micelle concentration (CMC) are the specific temperature and concentration at which molecules aggregate to form a micelle. By forming the hydrogen bonds in water and removing the micelle's hydrophobic fragments from the aqueous environment, amphiphilic molecules aggregate into micelles, reducing the free energy of the system [58]. Micelles can be polymeric micelles (poly ethylene oxide-poly(propylene oxide) [59] and triblock copolymer micelles (poly (ethylene oxide)-poly (aspartic acid) block copolymer (PEO/PASP (ADR)) and so on [60]. Figure 12.7 shows polymeric micelles as hydrophobic drug carrier.

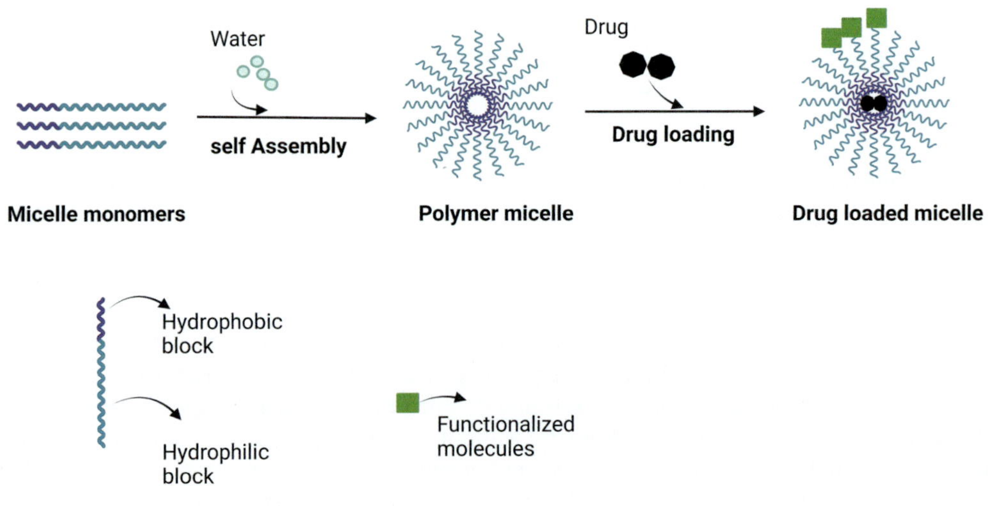

FIGURE 12.7 Polymeric micelles as hydrophobic drug delivery nanocarriers. Polymeric micelle with a hydrophobic core and hydrophilic shell, allowing for the encapsulation and delivery of hydrophobic drugs.

12.3.4.1 Enhancing Solubility and Bioavailability

The factors affecting the amount and speed of absorption of a drug include its solubility and gastrointestinal permeability [61]. The aqueous solubility of therapeutics plays an important role in absorption after drug administration [62]. The biopharmaceutics' classification indicates that Class II and IV drugs (APIs) are poorly soluble, bioavailable, and dissolvable.

In this section we will discuss micelles and polymeric micelles [63]. Nanocarriers can enhance the solubility and bioavailability of hydrophobic drugs. Of these, the most promising nanocarrier is the micelles system which can improve the drug's solubility due to its core–shell structure and can also control drug release. Two of the prominent ways to increase the solubility of drug-using micelles systems are the following:

12.3.4.1.1 Micellar Solubilization

This approach involves the incorporation of components into or onto the micelles. One of the most important properties of the micelle is that it can enhance the water solubility of compounds. Plotting the solubility of a substance with low water solubility as a function of surfactant concentration typically leads to the conclusion that drug solubility is significantly lower until the surfactant concentration approaches the CMC. Solubility increases linearly with surfactant concentration after it reaches concentrations higher than the CMC, suggesting that solubilization is important for micellization [63].

12.3.4.1.2 Polymeric Micellar Nanocarriers

Another approach to improve the solubility of weekly soluble drugs is through their incorporation into surface active agents. Amphiphilic copolymers with hydrophobic and hydrophilic building blocks dissolve in an aqueous solution to form micelles [64, 65]. Hydrophobic domains constitute the core of the micelles while the hydrophilic copolymer tails constitute the outer shell. Lipophilic drugs are entrapped in the core. The solubility of lipophilic medicine can be enhanced by incorporating micellar carriers within the micellar core [66]. Amphiphilic block copolymers offer a promising alternative to traditional delivery systems for hydrophobic drug delivery with increased bioavailability.

12.4 NANOCARRIERS AND SOLUBILITY ENHANCEMENT

12.4.1 Techniques for Improving the Solubility of Hydrophobic Compounds

The use of nanocarriers has emerged as the most promising method for improving the solubility of hydrophobic drugs and increasing bioavailability. Some nanocarriers are employed often to enhance hydrophobic drug solubility:

12.4.1.1 Lipid Dispersion Techniques

The lipid dispersion method is the synthesis of nanoparticles using lipid excipients. Dispersions of lipids are more suitable for the drug to be encapsulated as this process has features such as being low water-soluble, and having high lipid solubility, and a low melting point.

1. **Solid lipid nanoparticles**
   Solid lipid nanoparticles (SLNs) developed from oil/water emulsions [67]. SLNs solidify at room temperature and body temperature. SLNs offer several benefits as drug carriers, such as facilitating the absorption of drugs, high drug loading and regulated drug release. Because of these features, in order to enhance the oral bioavailability of readily soluble drugs, solid lipid nanoparticles are more suitable for formulation. SLNs can be produced by the high-pressure homogenization technique [68]. SLNs are now considered a possible way to increase the bioavailability of hydrophobic drugs.

2. **Nanostructured lipid carriers**
   Solid lipids have a lower ability to dissolve poorly soluble drugs when compared with liquid lipids. To overcome this limitation of SLNs, nanostructured lipid carriers (NLCs) have been developed. NLCs are created simply by adding liquid lipids to the core of the SLNs. Due to their exceptional solubilizing and dispersion abilities, NLCs have emerged as a potential nanocarrier for poorly soluble drugs [69, 70].
3. **Nanoemulsions**
   Pharmaceutical nanoemulsions are made up of the water phase, the emulsifier co-emulsifier phase, and the oil phase. Liquid lipid oil is employe to formulate nanoemulsions, which offer a high capacity for readily soluble drugs. Because of their particle size, which is less than 100nm, they have a large surface area for high drug dispersibility and absorption [71].
4. **Liposomes**
   Of all the liposome preparation methods, the film hydration method is the most developed and widely used. In this procedure, the drug is completely dispersed in the lipid mixture.
5. **Micellar solubilization**
   By lowering the surface tension in aqueous solution, surfactants efficiently increase the solubility of hydrophobic drugs. Drug suspensions can also be stabilized by surfactants. Micelle formation happens when the concentration of surfactants exceeds their critical micelle concentration, which is typically between 0.05 and 0.10%, entrapping the hydrophobic drug inside the micelle in a process known as micellization. Micellization enhances the solubility of drugs that are readily soluble in aqueous media. Micellar solubilization is the preferred substitute for dissolving poorly soluble medications [72].
6. **Polymeric composite (solid dispersions)**
   Polymeric composite is not a lipid dispersion technique. Solid dispersions involve the incorporation of not readily soluble drugs in a hydrophilic polymeric matrix, which may improve the solubility of drugs and the rate of dissolution in aqueous media. After reaching the target site, the hydrophilic polymer dissolves, releasing the fine particles of the drug [73]. Solid dispersion has proved a familiar method for improving the aqueous solubility and dissolution rate of BSC II drugs (poorly soluble but highly permeable, limiting their dissolution and absorption). Hydrophilic polymeric matrices such as poloxamer 407, carboxymethylcellulose, poloxamer 188, polyvinylpyrrolidone etc. can be used for solid dispersion preparation. A variety of techniques, including melting, kneading, solvent-evaporation, and lyophilization, can be used to create solid dispersions.

The solvent-evaporated dispersion is one of the most effective techniques to enhance the aqueous solubility and dissolution rate of BSC class II drugs. A clear and transparent solution is produced by completely dissolving the drug and polymeric matrix in a solvent system. The dried product is then obtained by evaporating this clear solution. Drug molecules are dispersed evenly or trapped within the polymeric compound in a solvent-evaporated solid dispersion [74].

a. Case Studies of Solubility Enhancement Using Nanocarriers
   i. Case study 1: Enhancement of Silymarin solubility using PVP-PEG polymeric composite
      Silybum marianum belongs to the Asteraceae family and is used to treat hepatic diseases. Silymarin is an extract derived from the fruits and seeds of *Silybum marianum*. There are four isomeric flavonoids in silymarin: silybinin, isosilybinin, silydianin, and silychristin. In the pharmaceutical industry, silybinin is of the utmost importance. Silymarin is well known for itsanticarcinogenic, hepatoprotective, and anti-inflammatory properties. It also acts as a strong antioxidant, restores damaged hepatocytes, detoxifies harmful substances, and stabilizes cell membranes. It is considered an effective decongestant for the liver and kidneys.

Challenge: Oral drug administration is the most practical and safe way. Solubility of the drug in gastrointestinal (GIT) fluid is crucial in the absorption of orally administered drugs. Silymarin belongs to the BSC II class and such compounds are only poorly soluble in aqueous solutions. The low solubility of silymarin hinders its absorption and reduces its effectiveness as a therapeutic agent.

Proposed solution: Researchers developed a novel approach in which they used a polymeric composite made of polyvinylpyrrolidone (PVP) and polyethylene glycol (PEG) to increase the solubility of silymarin [74].

Method

- Silymarin was incorporated into the PVP-PEG composite via the solvent evaporation method.
- PVP K-30 and PEG 6000 ratios were experimented with to maximize the composite's capacity to improve silymarin's solubility.

Results: The study shows that the polymeric composite increases the solubility of silymarin by 1150 folds. The drug's maximum aqueous solubility was shown by the ratio 0.25/1.5/1.5 (w/w/w) for PVP K-30, and PEG 6000.

The study suggests that solubility increases due to the increased surface area, increased interaction between PVP-PEG with water and the formation of a silymarin amorphous state facilitated by composite [74].

ii. Case Study ii: Improving Curcumin Solubility and Delivery using Solid Lipid Nanoparticles

Curcuma longa belongs to the Zingiberaceae family, and its active compound is curcumin, which has antibacterial, antidiabetic, anti-inflammatory, antioxidant and antitumor properties. Several studies show that curcumin is effective in the treatment of GIT disorders, liver disorders, and inflammatory conditions. It also shows effectiveness in cancer management by reducing protein levels like Cyclin D1 and CDK4, which are involved in cell proliferation.

Challenges: Curcumin has multiple uses, but due to its low bioavailability, rapid metabolism, almost complete insolubility in water (about 11 ng/ml), and physiological instability, it has not been effectively used as a therapeutic drug. All these factors lead to its poor absorption and reduced effectiveness as a therapeutic agent.

Proposed solution: Scientists proposed SLNs as drug delivery vehicles to overcome the above-mentioned problems. SLNs increase the drug's solubility, provide protection against degradation, and make the drugs more bioavailable.

Method: The high-pressure hot homogenization technique was used to prepare the curcumin-loaded SLNs technique. Curcumin was encapsulated within a lipid core and the lipid used was CompritolR888 ATO and GMS (4:1). To attain a 1.5% w/v curcumin concentration in the SLN dispersion, various formulations were prepared using a variety of types and concentrations of lipids and phospholipon 90G.

Results: In this study, researchers developed a high-drug loading CLEN dispersion that contains 15 mg of curcumin per ml, along with a high drug loading of 15%. According to reports, there is the largest increase in solubility of curcumin in aqueous solution (1.4×10^6 times higher than 11ng/ml in water for free curcumin) combined with high drug loading so far [75].

12.5 STABILITY OF HERBAL EXTRACTS IN NANOCARRIER SYSTEMS

The primary issue of using herbal extracts is their sensitivity to ultraviolet light, oxygen, enzymes, and chemicals. They also undergo a degradation process in the GI tract before entering the circulation. These drawbacks restrict the number of applications for these herbal extracts. Consequently, an encapsulation system is required to optimize the potential therapeutic advantages of herbal extracts. Sensitive chemicals found in plant extracts are well protected by encapsulation against oxidation and dehydration processes, which cause the degradation of these extracts. This chapter focuses on nanocarrier systems to improve the stability of extracts.

12.5.1 Protection against Degradation and Oxidation

Nanocarriers prevent oxidation and degradation using different methods like: (1) acting as a physical barrier; (2) inhibiting enzymatic degradation by developing a diffusion barrier; and (3) controlling the release of drug

i. **Physical barrier**: Nanocarriers encapsulate the herbal extract inside its core and act as a physical barrier against external factors like light, heat, PH change, oxygen etc., thereby preventing the degradation of the herbal extract. If we take the example of micelles, they incorporate the drug inside the hydrophobic core and shield it from environmental factors while the hydrophilic shell helps in water solubility [76].
ii. **Controlled release**: Nanocarrier systems are designed to deliver the controlled release of encapsulated extracts. In this way, the bioactives are slowly released within the target tissue, where absorption is more efficient. As a result, the extract is exposed to a harsh stomach environment for a shorter period, which further enhances its stability [76].
iii. **Diffusion barrier**: The major challenge involved in the administration of drugs orally is the degradation of the drug by the stomach and intestinal enzymes. To overcome this problem, researchers demonstrate a promising approach using ionic gelation to form peptide antibiotic-polyphosphate nanoparticles (NPs). In this method, the drug is present in the core shielded by a shell made of polyphosphate. This shell, comprised of polyphosphate, functions as a diffusion barrier that hinders enzymatic movement. In addition to slowing enzyme diffusion, NPs make it more difficult for enzymes to reach and degrade the drugs encapsulated within them [77].

12.5.2 Example of Stability Improvement in Herbal Extracts

12.5.2.1 Example 1: Stability and Bioavailability Enhancement of *Hypericum perforatum* using Micelles (Made up of ASC8 Surfactant) and Vehicles

Hypericum perforatum, commonly called St. John's Wort (SJW), is a medicinal herb employed to treat depression. Active components of SJW are lipophilic and poorly absorbed in the body. In addition, these active components are susceptible to enzymatic degradation in GIT.

Challenges: The major challenges related to SJW extracts are poor bioavailability and degradation by stomach acid and digestive enzymes. In addition, it is sensitive to oxidation and light.

Proposed solution: Researchers proposed utilizing micelles and vesicles to overcome these problems [78].

Process: Micelles and vesicles can stabilize the SJW extract by using two methods:

Physical barrier: This barrier can shield the SJW extract from a harsh GIT environment that otherwise might lead to degradation. Niosomes and liposomes are common types of vehicles. They have a closed bilayer structure that resembles the plasma membrane. A bilayer can provide a more robust barrier to prevent the encapsulated extract from degradation.

Controlled release: In the GIT, the carrier system can offer a controlled release of extract. As a result, the active components can be gradually released, potentially reducing their exposure to degradation factors such as stomach acid and enzymes [78].

Photostability and antioxidant properties: According to the study, ASC8 can self-assemble into micelles in water. These micelles might enclose SJW components, protecting them from exposure to light and oxygen. ASC8 possesses antioxidant properties. Micelles made of ASC8 can scavenge free radicals and prevent the oxidative degradation of SJW active components [79].

12.5.2.2 Example 2: Enhancing Curcumin Stability and Bioavailability using SLNs

Curcumin is an active component of turmeric and possesses various benefits as have been detailed earlier.

Challenges: Curcumin has low solubility in water and susceptibility to rapid enzymatic breakdown in the digestive tract. Both of these challenges lead to the poor bioavailability of curcumin and limit its therapeutic benefits.

Possible solutions: Researchers examined the effects of soy lecithin SLNs on curcumin's oral bioavailability. SLNs have several advantages, which mean that they can enhance the aqueous solubility of curcumins. They also protect curcumin from degradation.

Results: Studies show that C-SLNs achieved high drug encapsulation efficiency (92.33%). Also, SLNs enhanced the stability of the drug, i.e. after storage at 5°C for a year, the C-SLNs exhibited only a 9% decrease in encapsulation efficiency. This demonstrates that formulation will be stable for a long time [80].

12.6 TARGETED DELIVERY AND CONTROLLED RELEASE

Targeted drug delivery in nanocarrier systems involves the precise delivery of medications to specific areas, enhancing effectiveness while minimizing side effects. Nanotechnology has revolutionized drug delivery through the use of various nanoparticle carriers such as liposomes, dendrimers, and polymeric nanoparticles, each with unique properties and applications. These systems are crucial in treating conditions like tumors and colon cancer, where targeted delivery is essential for efficient therapy. By leveraging nanocarriers, drugs can be delivered directly to tumor cells, improving treatment outcomes while reducing harm to healthy tissues. Furthermore, in prostate cancer, functionalized nanocarriers are designed to target specific cell membrane proteins, enhancing the precision and sensitivity of drug delivery.

12.6.1 Key Principles

12.6.1.1 Selectivity and Specificity

The foundation of targeted delivery lies in achieving high selectivity and specificity for the target site. This can be accomplished through either passive targeting or active targeting.

12.6.1.2 Design and Composition of Nanocarriers

The structural and compositional attributes of nanocarriers are crucial in determining their targeting efficiency:

1. **Size and Shape**: Optimal sizes (typically 10–200 nm) and shapes (spherical, rod-like) are selected to enhance circulation time and evade immune detection.
2. **Surface Properties**: Surface modifications such as PEGylation (the attachment of polyethylene glycol) can improve the nanocarrier's stability and circulation time by reducing protein adsorption and immune recognition [81].

12.6.1.2.1 Integration and Optimization

The successful implementation of targeted delivery systems requires the meticulous optimization of various parameters, including the density and orientation of targeting ligands, the physicochemical properties of the nanocarrier, and the selection of appropriate stimuli for controlled release. Continuous research and development in this field aim to refine these parameters to enhance the specificity, efficacy, and safety of nanocarrier-based drug delivery systems [82].

12.6.1.2.2 Advantages of Controlled Release Systems

Controlled release systems offer several advantages over conventional drug delivery methods:

1. **Prolonged Therapeutic Effect**: By maintaining consistent drug levels, these systems ensure prolonged efficacy.
2. **Reduced Dosing Frequency**: Sustained release reduces the need for frequent dosing, improving patient compliance.
3. **Minimized Side Effects**: Controlled release minimizes peak–trough fluctuations, reducing the risk of side effects associated with high or low drug concentrations [83].

12.6.1.2.3 Applications in Medicine

Controlled release mechanisms are widely used across various medical fields:

1. **Cancer Therapy**: Controlled release of chemotherapeutics can reduce systemic toxicity and improve treatment efficacy.
2. **Chronic Diseases**: Conditions like diabetes and hypertension benefit from sustained drug release, ensuring stable levels of medication in the bloodstream.
3. **Pain Management**: Long-acting analgesics provide consistent pain relief without frequent dosing.

12.7 PHARMACOKINETICS AND PHARMACODYNAMICS

The pharmacokinetic and pharmacodynamic factors influencing the ADME profiles of nanocarriers include the nature of the drug delivery carrier chosen, the properties of the nanomedicines, and individual variations in organ size and body fat index. Physicochemical properties such as composition, particle size, shape, surface charge, and modifications like PEGylation and functionalization play a crucial role in determining the pharmacokinetics and pharmacodynamics of nanocarriers, affecting their therapeutic performance. Additionally, changes in the backbone and sugar chemistry of antisense oligonucleotides (ASOs), conjugation approaches, and routes of administration significantly impact their ADME and pharmacokinetic profiles, ultimately influencing efficacy and safety. The unique dispositional characteristics of nanocarriers, such as solubilizing platforms, controlled release mechanisms, and equilibrium binding properties, also contribute to the complexity of their pharmacokinetics, diverging from traditional drug-protein binding paradigms [84].

12.7.1 Enhancing Therapeutic Efficacy through Pharmacokinetic Modulation

Pharmacokinetic modulation using nanocarriers can significantly enhance the therapeutic efficacy of drugs by optimizing their absorption, distribution, metabolism, and excretion (ADME) profiles.

12.7.1.1 Sustained Release and Targeted Delivery

Nanocarriers can provide sustained release of drugs, maintaining therapeutic concentrations over an extended period. This can lead to improved efficacy, reduced dosing frequency, and enhanced patient compliance. Targeted delivery of nanocarriers to specific sites or cells can further enhance therapeutic efficacy by increasing local drug concentrations [76].

12.7.1.2 Improved Bioavailability and Reduced Inter-Individual Variability

Nanocarriers can improve the bioavailability of drugs by protecting them from degradation, enhancing solubility, and increasing permeability across biological membranes. This can reduce inter-individual variability in pharmacokinetic responses, leading to more consistent and predictable therapeutic outcomes [85].

12.7.1.3 Enhanced Pharmacodynamic Effects

The altered pharmacokinetic profiles of drugs achieved through nanocarrier modulation can lead to enhanced pharmacodynamic effects, such as improved efficacy, duration of action, and potency. This can be particularly beneficial for drugs with narrow therapeutic indices or those requiring precise dosing [86].

12.7.1.4 Reduced Adverse Effects and Toxicity

By optimizing the pharmacokinetic profiles of drugs, nanocarriers can reduce the risk of adverse effects and toxicity. This can be achieved by minimizing peak plasma concentrations, reducing off-target effects, and improving the safety profile of the drug [87].

12.7.1.5 Opportunities for Personalized Medicine

Nanocarrier-based pharmacokinetic modulation offers opportunities for personalized medicine by allowing for tailored design of nanocarriers to individual patient needs. This can involve optimizing nanocarrier properties, such as size, surface charge, and targeting ligands, to achieve improved therapeutic outcomes in specific patient populations [88].

12.7.2 Clinical Implications of Improved Pharmacodynamics

The utilization of herbal extracts as nanocarriers can lead to enhanced pharmacodynamics with important implications for clinical use. By using this approach, the bioavailability and therapeutic effectiveness of herbal medicines can be improved, possibly reducing the negative effects linked to traditional drug delivery techniques.

1. **Personalized Medicine**: Utilizing herbal extracts can be personalized based on individual patient requirements, taking into account factors like genetic composition and particular health issues. This individualized method has the potential to improve patient results and compliance with treatment plans.
2. **Safety and Efficacy**: The safety of herbal supplements is commonly assumed, but their interaction with traditional medications can present dangers. It is essential for healthcare providers to grasp the pharmacodynamics of these interactions in order to avoid negative effects and maintain patient safety. Healthcare professionals need thorough training and awareness of these interactions to manage patients effectively [89].
3. **Regulatory Considerations**: The herbal supplement industry's current absence of regulation is a cause for worry regarding the quality and safety of products. It is crucial to implement stricter regulations and standardized testing for herbal products to guarantee their effectiveness and safety in clinical use [90].
4. **Research and Development**: Researching the pharmacokinetic and pharmacodynamic profiles of herbal extracts is crucial. The emphasis should be on understanding the mechanisms of action, possible interactions, and creating standardized formulations that can be safely incorporated into conventional therapies [89].

12.8 CLINICAL APPLICATIONS AND CASE STUDIES

Nanotechnology has revolutionized cancer therapy and the treatment of chronic diseases by enabling the development of nanocarrier-based herbal drugs. These innovative drug delivery systems enhance bioavailability, reduce toxicity, and improve therapeutic efficacy. Clinical applications and case studies have showcased the successful implementation of nanocarrier-based herbal drugs in cancer therapy, diabetes, and other chronic illnesses. By encapsulating herbal extracts within nanoparticles, these nanocarriers offer targeted delivery, increased stability, and enhanced pharmacological activity, overcoming the limitations of traditional herbal medicine. The amalgamation of nanotechnology with herbal remedies presents promising prospects for personalized medicine and tailored therapies, demonstrating the transformative potential of nanocarrier-based herbal drugs in improving patient outcomes and advancing the field of healthcare.

12.8.1 Successful Implementations of Nanocarrier-Based Herbal Drugs

12.8.1.1 Curcumin-Loaded Nanoparticles

Application: Curcumin, derived from turmeric, is known for its anti-inflammatory and anti-cancer properties but suffers from poor bioavailability. Researchers developed curcumin-loaded solid lipid nanoparticles (SLNs) that significantly enhanced its solubility and stability.

Outcome: Clinical studies indicated improved absorption and therapeutic effects in patients with chronic inflammatory conditions, showcasing the potential of SLNs in delivering curcumin effectively [91].

12.8.1.2 Quercetin-Loaded Liposomes

Application: Quercetin has antioxidant and anti-inflammatory properties but is poorly soluble in water. Liposomal formulations of quercetin were developed to improve its bioavailability.

Outcome: In clinical trials, patients receiving quercetin liposomes showed enhanced antioxidant activity and reduced oxidative stress markers compared to those receiving standard quercetin supplements. This illustrates the effectiveness of liposomes in enhancing the therapeutic potential of herbal extracts [92].

12.8.1.3 Ginger Extract Nanocarriers

Application: Ginger extract, known for its anti-nausea and anti-inflammatory effects, was encapsulated in polymeric nanoparticles to enhance its delivery.

Outcome: Studies demonstrated that patients receiving ginger extract via nanoparticle formulations experienced significant reductions in nausea and vomiting, particularly in chemotherapy-induced cases, compared to those taking conventional ginger supplements [93].

12.8.1.4 Green Tea Extract Nanocarriers

Application: Green tea polyphenols, particularly epigallocatechin gallate (EGCG), were encapsulated in chitosan nanoparticles to enhance their stability and bioavailability.

Outcome: Clinical studies indicated that patients consuming EGCG-loaded nanoparticles experienced improved metabolic parameters and reduced oxidative stress, demonstrating the potential of chitosan nanoparticles in delivering herbal extracts effectively [94].

12.8.2 Challenges and Limitations in Clinical Settings

12.8.2.1 Quality Control and Standardization

One of the primary challenges is ensuring the quality and consistency of herbal extracts used in nanocarrier formulations. The rising demand for phytopharmaceuticals has led to concerns about adulteration and misidentification of raw materials, which can compromise the efficacy and safety of herbal medicines. Regulatory authorities face difficulties in finding high-quality phytopharmaceuticals due to interspecies diversity and variations in vernacular nomenclature, leading to potential health risks for consumers [91, 95].

12.8.2.2 Limited Encapsulation Capacity

Nanocarriers often have limitations in the amount of herbal active compounds they can encapsulate. This restriction can hinder the formulation of effective therapeutic doses, necessitating larger quantities of herbal extracts to achieve desired treatment durations. The manufacturing of functionalized nano-medicinal formulations that can effectively deliver these compounds remains a significant obstacle to commercialization [91].

12.8.2.3 Pharmacokinetic and Pharmacodynamic Variability

Herbal extracts often exhibit variability in pharmacokinetics and pharmacodynamics due to differences in plant sources, extraction methods, and individual patient responses. This variability can complicate the prediction of therapeutic outcomes and the establishment of standardized dosing regimens for nanocarrier formulations [91].

12.8.2.4 Manufacturing Challenges

The production of nanocarrier systems that are both effective and scalable poses significant technical challenges. The complexity involved in synthesizing and characterizing nanoparticles, along with ensuring their stability and compatibility with herbal extracts, can hinder the development of commercially viable [95, 96].

12.9 FUTURE PERSPECTIVES

12.9.1 Advancing Nanocarrier Design and Engineering

1. Exploring new materials and compositions for nanocarriers to further improve stability, drug-loading capacity, and targeted delivery.
2. Developing stimuli-responsive nanocarriers that can release the herbal extracts in a controlled manner based on specific environmental cues.
3. Incorporating multiple functionalities within a single nanocarrier to achieve synergistic effects, such as combining targeting ligands, imaging agents, and drug payloads.
4. Optimizing the manufacturing processes to enhance the scalability and reproducibility of nanocarrier production.

12.9.2 Expanding the Diversity of Herbal Extracts Formulated with Nanocarriers

1. Investigating the applicability of nanocarrier technology to a broader range of herbal extracts, including those with poor solubility, stability, or bioavailability.
2. Evaluating the performance of nanocarriers in delivering synergistic combinations of phytochemicals from herbal extracts.
3. Exploring the potential of nanocarriers to improve the delivery of herbal extracts for specific therapeutic areas, such as cancer, inflammation, or neurological disorders.

12.9.3 Advancing Preclinical and Clinical Evaluation

1. Conducting comprehensive in vitro and in vivo studies to assess the pharmacokinetics, pharmacodynamics, and safety profiles of nanocarrier-based herbal extract formulations.
2. Designing robust and standardized protocols for the evaluation of nanocarrier-based herbal extracts to ensure consistent and reliable data across different research groups.

3. Initiating clinical trials to evaluate the efficacy and tolerability of nanocarrier-based herbal extract formulations in human subjects, with a focus on demonstrating improved therapeutic outcomes compared to conventional herbal preparations.

12.9.4 Addressing Regulatory and Commercialization Challenges

1. Engaging with regulatory authorities to establish clear guidelines and frameworks for the development and approval of nanocarrier-based herbal extract products.
2. Addressing scalability and cost-effectiveness challenges to facilitate the translation of nanocarrier-based herbal extract formulations from the laboratory to the market.
3. Exploring strategies for intellectual property protection and effective commercialization of nanocarrier-based herbal extract products.

12.9.5 Exploring Synergies with Other Emerging Technologies

1. Integrating nanocarrier technology with modern advanced techniques, such as AI-driven drug design, 3D printing, and personalized medicine, to further enhance the effectiveness of herbal extract delivery.
2. Investigating the potential of combining nanocarrier-based herbal extracts with other therapeutic modalities, such as photodynamic therapy or gene therapy, for improved clinical outcomes.
3. Exploring the application of advanced analytical techniques, such as multi-omics approaches, to obtain a better understanding of the mechanisms of action and optimization of nanocarrier-based herbal extract formulations.

12.10 NANOTECHNOLOGY WITH HERBAL MEDICINE: OPPORTUNITIES AND CHALLENGES

12.10.1 Opportunities

Pharmaceutical scientists are increasingly focusing on developing drug delivery systems for herbal medicines using scientific methods. *Cuscuta chinensis*, a widely used traditional Chinese medicine, contains key components like flavonoids and lignans that have limited water solubility. This poor solubility can hinder their absorption when administered orally. To improve this, researchers have explored the use of nanoparticles [97, 98]. Recent studies have successfully employed the precipitation method to create polylactic acid nanoparticles for lipophilic anti-cancer compounds such as cucurbitacins and curcuminoids [99]. Additionally, there has been significant research into solid lipid nanoparticles (SLNs) aimed at enhancing the targeted delivery and bioavailability of traditional Chinese herbal medicines [100]. Various nanostructured carrier systems, including polymeric nanoparticles, liposomes, SLNs, polymeric micelles, and nanoemulsions, have also been investigated for their potential to deliver anticancer drugs via oral routes [101].

Some difficulties, such as converting the formulation of bioactive compounds into conventional dosage forms in herbal remedies, are faced due to their poor solubility or stability. These challenges could be resolved as nanoparticles provide a diverse platform to incorporate these compounds and enhance their properties, Marking a convergence of traditional wisdom with cutting-edge technology. The therapeutic potential of natural compounds could be enhanced in an efficient, targeted, and reliable manner by the encapsulating nanoparticles into herbal remedies. As advances in nanotechnology research continues to progress, the interconnection between nanoparticles and herbal medicine holds a great potential for developing innovative, standardized, and effective therapeutic solutions with diverse applications in healthcare and wellness [102].

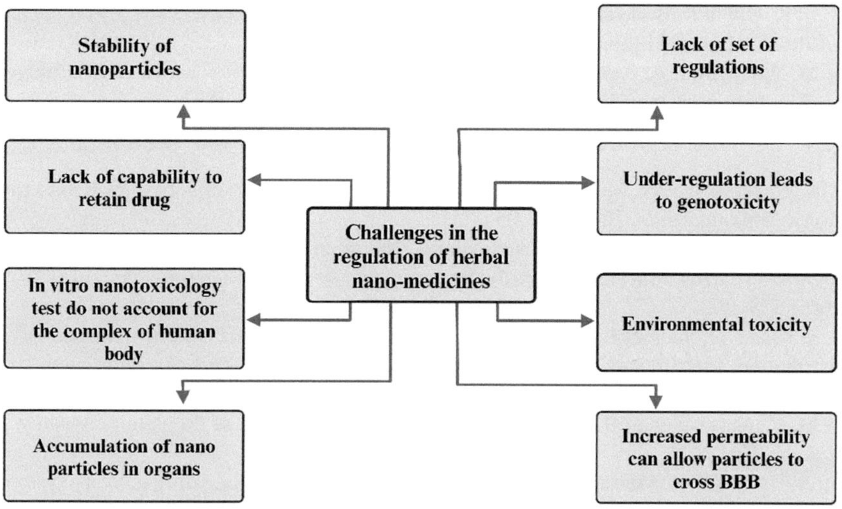

FIGURE 12.8 Challenges faced by herbal nanotechnology-based medicines.

12.10.2 Challenges

Nanopharmaceuticals may offer countless possibilities for drug delivery to diagnose and treat different conditions, but their safety profile should also be considered. The toxic effects of engineered nano-sized materials could be a result of the change in the physicochemical and structural properties and decreased size, leading to several material interactions. Since the data on safety and toxicity is insufficient, scientists must acknowledge that the toxicological evaluation for nanomaterials and nanomedicines accept that it is still very early in its development [101]. Some of the key challenges are described in Figure 12.8.

12.11 CONCLUSION

The integration of nanotechnology into herbal medicine represents a significant step forward in enhancing the therapeutic effectiveness and bioavailability of herbal extracts. These advancements address longstanding challenges associated with their stability and absorption. This chapter has explored how nanotechnology can overcome the challenges faced by traditional herbal extracts, such as poor solubility, instability, and low bioavailability. Nanocarriers like liposomes and micelles have shown great potential in improving the delivery and therapeutic efficacy of phytochemicals. By enhancing absorption, distribution, and controlled release, these nanocarrier systems can significantly boost the bioavailability of herbal active compounds. As the field progresses, future advancements in nanocarrier engineering, integration with emerging technologies, and addressing regulatory hurdles will further unlock the full potential of herbal medicines. The synergistic combination of traditional herbal knowledge and modern nanotechnology holds promise to revolutionize integrative healthcare and improve patient outcomes.

REFERENCES

1. Kunle OF, Egharevba HO, Ahmadu PO. Standardization of herbal medicines – A review. *International Journal of Biodiversity and Conservation*. 2012;4(3):101–12.
2. Sachan AK, Kumar A. Stability testing of herbal products. *Journal of Chemical and Pharmaceutical Research*. 2015;7(12):511–4.

3. Kesarwani K, Gupta R. Bioavailability enhancers of herbal origin: An overview. *Asian Pacific Journal of Tropical Biomedicine*. 2013;3(4):253–66.
4. Stielow M, Witczyńska A, Kubryń N, Fijałkowski Ł, Nowaczyk J, Nowaczyk A. The bioavailability of drugs — The current state of knowledge. *Molecules*. 2023;28(24):8038.
5. Fotherby K. Pharmacokinetics and pharmacodynamics in drug development. *Advances in Contraception*. 1991;7:107.
6. Dewi MK, Chaerunisaa AY, Muhaimin M, Joni IM. Improved activity of herbal medicines through nanotechnology. *Nanomaterials*. 2022;12(22):4073.
7. Puri A, Loomis K, Smith B, Lee J-H, Yavlovich A, Heldman E, et al. Lipid-based nanoparticles as pharmaceutical drug carriers: From concepts to clinic. *Critical Reviews in Therapeutic Drug Carrier Systems*. 2009;26(6):523–80.
8. Benet LZ, Kroetz D, Sheiner L, Hardman J, Limbird L. Pharmacokinetics: The dynamics of drug absorption, distribution, metabolism, and elimination. *Goodman and Gilman's the Pharmacological Basis of Therapeutics*. 1996;3:e27.
9. Feghali M, Venkataramanan R, Caritis S, editors. Pharmacokinetics of drugs in pregnancy. *Seminars in perinatology*; 2015: Elsevier.
10. Rahman A, Harwansh R, Mirza A, Hussain S, Hussain A. Oral lipid based drug delivery system (LBDDS): Formulation, characterization and application: A review. *Current Drug Delivery*. 2011;8(4):330–45.
11. Mukherjee PK, Venkatesh M, Maiti K, Mukherjee K, Saha B. Value added herbal drug delivery systems-perspectives and developments. *Indian Journal of Pharmaceutical Education and Research*. 2009;43(4):329–37.
12. Li Y, Zheng J, Xiao H, McClements DJ. Nanoemulsion-based delivery systems for poorly water-soluble bioactive compounds: Influence of formulation parameters on polymethoxyflavone crystallization. *Food hydrocolloids*. 2012;27(2):517–28.
13. Lipinski CA, Lombardo F, Dominy BW, Feeney PJ. Experimental and computational approaches to estimate solubility and permeability in drug discovery and development settings. *Advanced Drug Delivery Reviews*. 1997;23(1–3):3–25.
14. Aqil F, Munagala R, Jeyabalan J, Vadhanam MV. Bioavailability of phytochemicals and its enhancement by drug delivery systems. *Cancer Letters*. 2013;334(1):133–41.
15. Brown EJ, Khodr H, Hider CR, Rice-Evans CA. Structural dependence of flavonoid interactions with Cu2+ ions: Implications for their antioxidant properties. *Biochemical Journal*. 1998;330(3):1173–8.
16. Passamonti S, Terdoslavich M, Franca R, Vanzo A, Tramer F, Braidot E, et al. Bioavailability of flavonoids: A review of their membrane transport and the function of bilitranslocase in animal and plant organisms. *Current Drug Metabolism*. 2009;10(4):369–94.
17. Hollman PC, Bijsman MN, Van Gameren Y, Cnossen EP, De Vries JH, Katan MB. The sugar moiety is a major determinant of the absorption of dietary flavonoid glycosides in man. *Free Radical Research*. 1999;31(6):569–73.
18. Scalbert A, Williamson G. Dietary intake and bioavailability of polyphenols. *The Journal of nutrition*. 2000;130(8):2073S–85S.
19. Walle T, Hsieh F, DeLegge MH, Oatis JE, Walle UK. High absorption but very low bioavailability of oral resveratrol in humans. *Drug metabolism and disposition*. 2004;32(12):1377–82.
20. Ravichandran R. Nanoparticles in drug delivery: Potential green nanobiomedicine applications. *International Journal of Green Nanotechnology: Biomedicine*. 2009;1(2):B108–B30.
21. Mishra B, Patel BB, Tiwari S. Colloidal nanocarriers: A review on formulation technology, types and applications toward targeted drug delivery. *Nanomedicine: Nanotechnology, biology and medicine*. 2010;6(1):9–24.
22. Kulkarni SB, Betageri GV, Singh M. Factors affecting microencapsulation of drugs in liposomes. *Journal of Microencapsulation*. 1995;12(3):229–46.
23. Sun T, Shrike Zhang Y, Bo P, Hyun DC, Yang M, Xia Y. *Engineered Nanoparticles for Drug Delivery in Cancer Therapy**. Jenny Stanford Publishing; 2021. p. 31–142.
24. Nsairat H, Khater D, Sayed U, Odeh F, Al Bawab A, Alshaer W. Liposomes: Structure, composition, types, and clinical applications. *Heliyon*. 2022;8(5):e09394.
25. Maja L, Željko K, Mateja P. Sustainable technologies for liposome preparation. *The Journal of Supercritical Fluids*. 2020;165:104984.
26. Emami S, Azadmard-Damirchi S, Peighambardoust SH, Valizadeh H, Hesari J. Liposomes as carrier vehicles for functional compounds in food sector. *Journal of Experimental Nanoscience*. 2016;11(9):737–59.

27. Salunkhe VR, Patil PS, Wadkar GH, Bhinge SD. Herbal liposomes: Natural network for targeted drug delivery system. *Journal of Pharmaceutical Research International*. 2021;33:31–41.
28. Bibi S, Lattmann E, Mohammed AR, Perrie Y. Trigger release liposome systems: Local and remote controlled delivery? *Journal of Microencapsulation*. 2012;29(3):262–76.
29. Agrawal M, Ajazuddin TDK, Saraf S, Saraf S, Antimisiaris SG, et al. Recent advancements in liposomes targeting strategies to cross blood-brain barrier (BBB) for the treatment of Alzheimer's disease. *Journal of Controlled Release*. 2017;260:61–77.
30. Calzoni E, Cesaretti A, Polchi A, Di Michele A, Tancini B, Emiliani C. Biocompatible polymer nanoparticles for drug delivery applications in cancer and neurodegenerative disorder therapies. *Journal of Functional Biomaterials*. 2019;10(1):4.
31. Lv S, Sylvestre M, Prossnitz AN, Yang LF, Pun SH. Design of polymeric carriers for intracellular peptide delivery in oncology applications. *Chemical Reviews*. 2021;121(18):11653–98.
32. Ding S, Anton N, Vandamme TF, Serra CA. Microfluidic nanoprecipitation systems for preparing pure drug or polymeric drug loaded nanoparticles: An overview. *Expert Opinion on Drug Delivery*. 2016;13(10):1447–60.
33. Kallinteri P, Garnett MC. Polymeric Nanoparticles for Drug Delivery. *Nanotechnologies for the Life Sciences*. 2007.
34. Koo H, Huh MS, Sun I-C, Yuk SH, Choi K, Kim K, et al. In vivo targeted delivery of nanoparticles for theranosis. *Accounts of Chemical Research*. 2011;44(10):1018–28.
35. Yoo J, Park C, Yi G, Lee D, Koo H. Active targeting strategies using biological ligands for nanoparticle drug delivery systems. *Cancers*. 2019;11(5):640.
36. Son G-H, Lee B-J, Cho C-W. Mechanisms of drug release from advanced drug formulations such as polymeric-based drug-delivery systems and lipid nanoparticles. *Journal of Pharmaceutical Investigation*. 2017;47(4):287–96.
37. Fundamentals and Applications of Controlled Release Drug Delivery. 2012.
38. Peppas NA, Hilt JZ, Khademhosseini A, Langer R. Hydrogels in biology and medicine: From molecular principles to bionanotechnology. *Advanced Materials*. 2006;18(11):1345–60.
39. Devalapally H, Shenoy D, Little S, Langer R, Amiji M. Poly(ethylene oxide)-modified poly(beta-amino ester) nanoparticles as a pH-sensitive system for tumor-targeted delivery of hydrophobic drugs: Part 3. Therapeutic efficacy and safety studies in ovarian cancer xenograft model. *Cancer Chemotherapy and Pharmacology*. 2007;59(4):477–84.
40. Torchilin VP. Multifunctional, stimuli-sensitive nanoparticulate systems for drug delivery. *Nature Reviews Drug Discovery*. 2014;13(11):813–27.
41. Kamaly N, Yameen B, Wu J, Farokhzad OC. Degradable controlled-release polymers and polymeric nanoparticles: Mechanisms of controlling drug release. *Chemical Reviews*. 2016;116(4):2602–63.
42. Castro KCD, Costa JM, Campos MGN. Drug-loaded polymeric nanoparticles: A review. *International Journal of Polymeric Materials and Polymeric Biomaterials*. 2022;71(1):1–13.
43. Mason TG, Wilking JN, Meleson K, Chang CB, Graves SM. Nanoemulsions: Formation, structure, and physical properties. *Journal of Physics: Condensed Matter*. 2007;19(7):079001.
44. Aboofazeli R. Nanometric-scaled emulsions (nanoemulsions). *Iranian Journal of Pharmaceutical Sciences* 2010;9(4):325–6.
45. McClements DJ. Nanoemulsions versus microemulsions: Terminology, differences, and similarities. *Soft Matter* 2012;8(6):1719–29.
46. Anuchapreeda S, Fukumori Y, Okonogi S, Ichikawa H. Preparation of lipid nanoemulsions incorporating curcumin for cancer therapy. *Journal of Nanotechnology*. 2012;2012:1–11.
47. Hwang YY, Ramalingam K, Bienek DR, Lee V, You T, Alvarez R. Antimicrobial activity of nanoemulsion in combination with cetylpyridinium chloride in multidrug-resistant acinetobacter baumannii. *Antimicrobial Agents and Chemotherapy*. 2013;57(8):3568–75.
48. Pawar VK, Panchal SB, Singh Y, Meher JG, Sharma K, Singh P, et al. Immunotherapeutic vitamin E nanoemulsion synergies the antiproliferative activity of paclitaxel in breast cancer cells via modulating Th1 and Th2 immune response. *Journal of Controlled Release*. 2014;196:295–306.
49. Klang V, Valenta C. Lecithin-based nanoemulsions. *Journal of Drug Delivery Science and Technology*. 2011;21(1):55–76.
50. Singh Y, Meher JG, Raval K, Khan FA, Chaurasia M, Jain NK, et al. Nanoemulsion: Concepts, development and applications in drug delivery. *Journal of Controlled Release*. 2017;252:28–49.

51. Gogate PR, Kabadi AM. A review of applications of cavitation in biochemical engineering/biotechnology. *Biochemical Engineering Journal*. 2009;44(1):60–72.
52. Bouchemal K, Briançon S, Perrier E, Fessi H. Nano-emulsion formulation using spontaneous emulsification: Solvent, oil and surfactant optimisation. *International Journal of Pharmaceutics*. 2004;280(1–2):241–51.
53. Zeeb B, Herz E, McClements DJ, Weiss J. Impact of alcohols on the formation and stability of protein-stabilized nanoemulsions. *Journal of Colloid and Interface Science*. 2014;433:196–203.
54. Guan Y, Wu J, Zhong Q. Eugenol improves physical and chemical stabilities of nanoemulsions loaded with β-carotene. *Food Chemistry*. 2016;194:787–96.
55. McClements DJ. Nanoemulsion-based oral delivery systems for lipophilic bioactive components: Nutraceuticals and pharmaceuticals. *Therapeutic Delivery*. 2013;4(7):841–57.
56. Lee SJ, Choi SJ, Li Y, Decker EA, McClements DJ. Protein-stabilized nanoemulsions and emulsions: Comparison of physicochemical stability, lipid oxidation, and lipase digestibility. *Journal of Agricultural and Food Chemistry*. 2011;59(1):415–27.
57. Qian C, Decker EA, Xiao H, McClements DJ. Physical and chemical stability of β-carotene-enriched nanoemulsions: Influence of pH, ionic strength, temperature, and emulsifier type. *Food Chemistry*. 2012;132(3):1221–9.
58. Milovanovic M, Arsenijevic A, Milovanovic J, Kanjevac T, Arsenijevic N. *Nanoparticles in Antiviral Therapy*. Elsevier; 2017. p. 383–410.
59. Bose A, Roy Burman D, Sikdar B, Patra P. Nanomicelles: Types, properties and applications in drug delivery. *IET Nanobiotechnology*. 2021;15(1):19–27.
60. Kazunori K, Glenn S K, Masayuki Y, Teruo O, Yasuhisa S. Block copolymer micelles as vehicles for drug delivery. *Journal of Controlled Release*. 1993;24(1–3):119–32.
61. Amidon GL, Lennernäs H, Shah VP, Crison JR. A theoretical basis for a biopharmaceutic drug classification: The correlation of in vitro drug product dissolution and in vivo bioavailability. *Pharmaceutical Research*. 1995;12(3):413–20.
62. Williams HD, Trevaskis NL, Charman SA, Shanker RM, Charman WN, Pouton CW, et al. Strategies to address low drug solubility in discovery and development. *Pharmacological Reviews*. 2013;65(1):315–499.
63. Bhalani DV, Nutan B, Kumar A, Singh Chandel AK. Bioavailability enhancement techniques for poorly aqueous soluble drugs and therapeutics. *Biomedicines*. 2022;10(9):2055.
64. Chandel AKS, Nutan B, Raval IH, Jewrajka SK. Self-Assembly of Partially Alkylated Dextran-*graft*-poly[(2-dimethylamino)ethyl methacrylate] Copolymer Facilitating Hydrophobic/Hydrophilic Drug Delivery and Improving Conetwork Hydrogel Properties. *Biomacromolecules*. 2018;19(4):1142–53.
65. Chandel AKS, Bhingradiya N. Therapeutic efficacy of herbal formulations through novel drug delivery systems. *IGI Global*; 2021. p. 1–42.
66. Gaucher G, Dufresne M-H, Sant VP, Kang N, Maysinger D, Leroux J-C. Block copolymer micelles: Preparation, characterization and application in drug delivery. *Journal of Controlled Release*. 2005;109(1–3):169–88.
67. Xing H, Wang H, Wu B, Zhang X. Lipid nanoparticles for the delivery of active natural medicines. *Current Pharmaceutical Design*. 2018;23(43):6705–13.
68. Muchow M, Maincent P, Müller RH. Lipid Nanoparticles with a Solid Matrix (SLN®, NLC®, LDC®) for Oral Drug Delivery. *Drug Development and Industrial Pharmacy*. 2008;34(12):1394–405.
69. Andrade LM, De Fátima Reis C, Maione-Silva L, Anjos JLV, Alonso A, Serpa RC, et al. Impact of lipid dynamic behavior on physical stability, in vitro release and skin permeation of genistein-loaded lipid nanoparticles. *European Journal of Pharmaceutics and Biopharmaceutics*. 2014;88(1):40–7.
70. Poonia N, Kharb R, Lather V, Pandita D. Nanostructured lipid carriers: Versatile oral delivery vehicle. *Future Science OA*. 2016;2(3):135.
71. Neslihan Gursoy R, Benita S. Self-emulsifying drug delivery systems (SEDDS) for improved oral delivery of lipophilic drugs. *Biomedicine & Pharmacotherapy*. 2004;58(3):173–82.
72. Vemula VR, Lagishetty V, Lingala S. Solubility enhancement techniques. *International Journal of Pharmaceutical Sciences Review and Research*. 2010;5(1):41–51.
73. Chiou WL, Riegelman S. Pharmaceutical applications of solid dispersion systems. *Journal of Pharmaceutical Sciences*. 1971;60(9):1281–302.
74. Yousaf AM, Malik UR, Shahzad Y, Mahmood T, Hussain T. Silymarin-laden PVP-PEG polymeric composite for enhanced aqueous solubility and dissolution rate: Preparation and in vitro characterization. *Journal of Pharmaceutical Analysis*. 2019;9(1):34–9.

75. Gupta T, Singh J, Kaur S, Sandhu S, Singh G, Kaur IP. Enhancing bioavailability and stability of curcumin using solid lipid nanoparticles (CLEN): A covenant for its effectiveness. *Frontiers in Bioengineering and Biotechnology*. 2020;879.
76. Harwansh RK, Deshmukh R, Rahman MA. Nanoemulsion: Promising nanocarrier system for delivery of herbal bioactives. *Journal of Drug Delivery Science and Technology*. 2019;51:224–33.
77. Saleh A, Akkuş-Dağdeviren ZB, Haddadzadegan S, Wibel R, Bernkop-Schnürch A. Peptide antibiotic–polyphosphate nanoparticles: A promising strategy to overcome the enzymatic and mucus barrier of the intestine. *Biomacromolecules*. 2023;24(6):2587–95.
78. Bilia AR, Bergonzi MC, Guccione C, Manconi M, Fadda AM, Sinico C. Vesicles and micelles: Two versatile vectors for the delivery of natural products. *Journal of Drug Delivery Science and Technology*. 2016;32:241–55.
79. Bergonzi MC, Isacchi B, Bilia AR. Octanoyl-6–O-ascorbic acid: An efficient antioxidant with potent solubilising properties The case of active constituents from Saint John's wort. *Journal of Drug Delivery Science and Technology*. 2013;23(5):505–9.
80. Kakkar V, Singh S, Singla D, Kaur IP. Exploring solid lipid nanoparticles to enhance the oral bioavailability of curcumin. *Molecular Nutrition & Food Research*. 2011;55(3):495–503.
81. Suk JS, Xu Q, Kim N, Hanes J, Ensign LM. PEGylation as a strategy for improving nanoparticle-based drug and gene delivery. *Advanced Drug Delivery Reviews*. 2016;99:28–51.
82. Misra SK, Pathak K. Targeted drug delivery approaches for the management of tumors. *Tumor Discovery*. 2023;2(3):1356.
83. Chakraborty G, Meher M, Dash S, Rout RN, Pradhan S, Sahoo D. Strategies for targeted delivery via structurally variant polymeric nanocarriers. *ChemistrySelect*. 2023;8(29):e202301626.
84. Zeeshan F, Dua K. *Pharmacokinetics and pharmacodynamics of the advanced drug delivery systems*. Elsevier; 2020. p. 551–71.
85. McDonald TO, Giardiello M, Martin P, Siccardi M, Liptrott NJ, Smith D, et al. Antiretroviral solid drug nanoparticles with enhanced oral bioavailability: Production, characterization, and in vitro–in vivo correlation. *Advanced Healthcare Materials*. 2014;3(3):400–11.
86. Le H, Karakasyan C, Jouenne T, Le Cerf D, Dé E. Application of polymeric nanocarriers for enhancing the bioavailability of antibiotics at the target site and overcoming antimicrobial resistance. *Applied Sciences*. 2021;11(22):10695.
87. Lim SB, Banerjee A, Önyüksel H. Improvement of drug safety by the use of lipid-based nanocarriers. *Journal of Controlled Release*. 2012;163(1):34–45.
88. Alghamdi MA, Fallica AN, Virzì N, Kesharwani P, Pittalà V, Greish K. The promise of nanotechnology in personalized medicine. *Journal of Personalized Medicine*. 2022;12(5):673.
89. Rombolà L, Scuteri D, Marilisa S, Watanabe C, Morrone LA, Bagetta G, et al. Pharmacokinetic interactions between herbal medicines and drugs: Their mechanisms and clinical relevance. *Life*. 2020;10(7):106.
90. Hassen G, Belete G, Carrera KG, Iriowen RO, Araya H, Alemu T, et al. Clinical implications of herbal supplements in conventional medical practice: A US perspective. *Cureus*. 2022;14(7), e26893.
91. Chavda VP, Nalla LV, Balar P, Bezbaruah R, Apostolopoulos V, Singla RK, et al. Advanced phytochemical-based nanocarrier systems for the treatment of breast cancer. *Cancers*. 2023;15(4):1023.
92. Alam MK. Nanocarrier-based drug delivery systems using microfluidic-assisted techniques. *Advanced NanoBiomed Research*. 2023;3(11):2300041.
93. Alshawwa SZ, Kassem AA, Farid RM, Mostafa SK, Labib GS. Nanocarrier drug delivery systems: Characterization, limitations, future perspectives and implementation of artificial intelligence. *Pharmaceutics*. 2022;14(4):883.
94. Salih WA, Al-Taie AA. Strategies of nanocarrier-based therapeutic drug delivery. *Al-Nahrain Journal for Engineering Sciences*. 2023;26(4):365–73.
95. Kumar R. Nanotechnology in herbal medicine: Challenges and future perspectives. *Nanotechnology in Herbal Medicine*: Elsevier; 2023. p. 515–48.
96. Setia A, Vallamkonda B, Challa RR, Mehata AK, Badgujar P, Muthu MS. Herbal theranostics: Controlled, targeted delivery and imaging of herbal molecules. *Nanotheranostics*. 2024;8(3):344.
97. Yadav D, Suri S, Choudhary AA, Sikender M, Hemant B, Beg N. Novel approach: Herbal remedies and natural products in pharmaceutical science as nano drug delivery systems. *Int J Pharm Tech*. 2011;3(3): 3092–116.

98. Sharma A, Mitkare S, Moon R. Multicomponent herbal therapy: A review. *International Journal of Pharmaceutical Sciences Review and Research* 2011;6:185–7.
99. Zhang J-F, Hou S-X, Liu H-L. Comparison of preparing two polylactide nanoparticles loaded lipophilic anti-cancer herb drug by nanoprecipitation method. *Zhongguo Zhong yao za zhi= Zhongguo Zhongyao Zazhi= China Journal of Chinese Materia Medica*. 2007;32(4):303–6.
100. Li Y, Dong L, Jia A, Chang X, Xue H. Preparation and characterization of solid lipid nanoparticles loaded traditional Chinese medicine. *International Journal of Biological Macromolecules*. 2006;38(3–5):296–9.
101. Ansari S, Islam F, Sameem M. Influence of nanotechnology on herbal drugs: A Review. *Journal of Advanced Pharmaceutical Technology & Research*. 2012;3(3):142–6.
102. Shah N, Kumari M, Sadhu P, Talele C, Rajput HS, Sapra R. Nano-herbal innovations: Precision in therapeutic delivery. *Journal of Advanced Zoology*. 2024;45(1), 576–581.

13 Herbal Medicine and Neurological Diseases

Sunil Kumar Kadiri and Prashant Tiwari
Department of Pharmacology, College of Pharmaceutical Sciences, Dayananda Sagar University, Bengaluru, India

Samaresh Pal Roy
Department of Pharmacology, Sardar Patel College of Pharmacy, Vidyanagar, Gujarat, India

Suchismita Bhowmik
Department of Community Health Nursing, Manikaka Topawala Institute of Nursing, Charotar University of Science and Technology, Changa, Gujarat, India

13.1 INTRODUCTION TO HERBAL MEDICINE AND NEUROLOGICAL DISEASES

Herbal medicine has been traditionally utilized for centuries across diverse cultures to address a wide assortment of health conditions [1]. The employment of plants and plant-derived substances has coevolved alongside conventional medicine, offering choice or complementary alternatives for managing diseases, like those impacting the neurological system. Neurological afflictions encompass an expansive spectrum of disorders influencing the brain, spinal cord, and peripheral nerves [2–4]. These circumstances can vary from neurodegenerative maladies such as Alzheimer's and Parkinson's to neurological issues – for example epilepsy, migraines, and multiple sclerosis [5]. Standard treatments regularly involve pharmaceuticals targeting symptoms or slowing the progression of these diseases. Nevertheless, these treatments may come with side effects, leading numerous patients and healthcare providers to explore herbal remedies as supplementary or substitute therapies [6–8]. Herbs such as *Ginkgo biloba*, St. John's Wort, and Valerian root have been researched for their potential neuroprotective impacts and their ability to improve cognitive function, mood, and sleep. *Ginkgo biloba*, for instance, is believed to enhance cognitive function by boosting blood circulation in the brain, while St. John's Wort is commonly utilized for its antidepressant impacts. The therapeutic possibility of herbal medicine in neurological diseases lies in the intricate interplay of various active substances found in plants. These compounds may offer antioxidant, anti-inflammatory, and neuroprotective consequences, which are crucial in managing the oxidative stress and swelling that are frequently related to neurological conditions [9]. Despite the promising impacts of herbal medicine, it is essential to approach its employment with caution. The efficacy of many herbal remedies is still under investigation, and their interactions with conventional medications are not fully comprehended [10]. Additionally, the standardization of herbal products and the variability in their active components pose substantial challenges to their consistent use in clinical settings.

As interest in herbal medicine continues to grow, further research is necessary to establish the safety, efficacy, and mechanisms of action of these natural treatments in the context of neurological diseases. Healthcare providers should be informed about the potential benefits and risks of herbal therapies to guide patients in making well-informed decisions about their treatment options [11].

DOI: 10.1201/9781003544418-13

13.1.1 Overview of Herbal Medicine

Herbal medicine, also known as phytotherapy, involves harnessing nature's healing gifts from plants and their extracts for therapeutic purposes. It has been practiced for millennia across diverse cultures and remains integral to systems like Ayurveda, traditional Chinese medicine (TCM), and Native American healing practices [12].

13.1.1.1 Key Aspects of Herbal Medicine

Plant sources provide remedies as fresh, dried, or extracted leaves, roots, flowers, bark, seeds, and fruits [13]. Active compounds confer benefits through alkaloids, flavonoids, terpenes, and glycosides working anti-inflammatory, antimicrobial, antioxidant, and immune-modulating effects. Herbal medicine often adopts a holistic stance to health, focusing on equilibrium of body, mind, and spirit. It considers an individual's overall wellness, rather than solely treating symptoms [14–18].

Commonly used herbs include: immune-bolstering Echinacea reducing cold duration; St. John's Wort, which has anti-depressant and mood-stabilizing actions; energy-enhancing, cognitive-function-boosting ginseng; and anti-inflammatory, antioxidant turmeric (Figure 13.1). Similarly, peppermint frequently eases digestion and headaches.

Safety and efficacy matter: while many herbs are safe with proper use, some may interact dangerously with medications or cause side effects. Seek qualified guidance, especially with regard to health issues or other medications. Modern integration increasingly merges herbal medicine into mainstream care as complementary therapies alongside conventional treatments. This stimulated scientific research into various remedies' effectiveness and risks. Regulation varies between designating herbal medicines as dietary supplements or classifying some as drugs requiring stricter testing and approval [19]. Herbal medicine continues valuably blending antiquity's wisdom with modern science for health and wellness.

13.1.2 Scope of Neurological Diseases

Neurological diseases encompass a broad spectrum of disorders that affect the brain, spinal cord, and peripheral nervous system. These conditions can range from common ailments like migraines and epilepsy to more complex and debilitating diseases such as Alzheimer's, Parkinson's, multiple

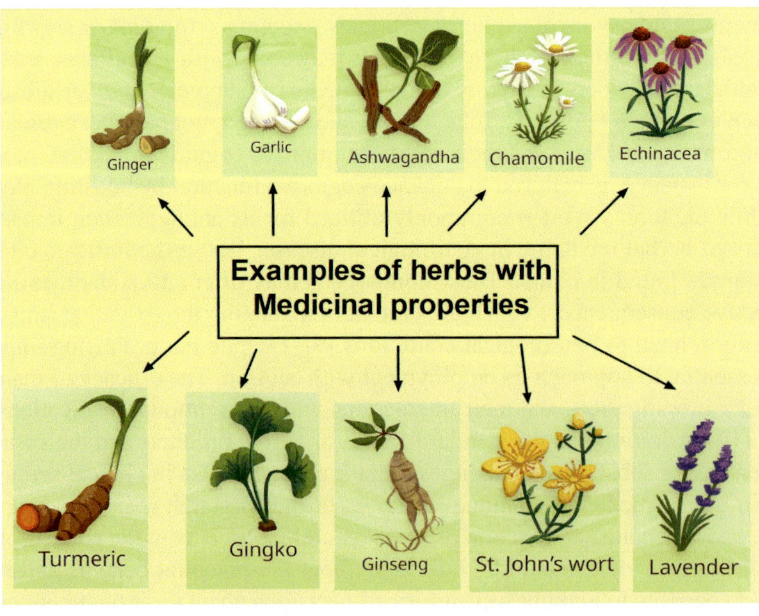

FIGURE 13.1 Some prominent herbs with Medicinal properties.

sclerosis, and the motor neuron disease known as amyotrophic lateral sclerosis [20]. The scope of neurological diseases is vast due to the immense complexity of the human nervous system, which intricately controls and coordinates each and every one of the body's innumerable functions [21]. These diseases hold the potential to impact motor skills, sensory awareness, cognition, behavior, and autonomic functions. Genetic predispositions, environmental exposures, infectious agents, physical traumas, or even etiologies that remain unknown – any of these triggers may underlie the origins of specific neurological disorders [22]. Treating and managing neurological diseases frequently demands an interdisciplinary clinical strategy, incorporating pharmaceutical interventions, physical and occupational therapies, psychological support services, and, in some cases, surgical options. Continual advances in scientific research serve to expand our fundamental understanding of these conditions, leading to enhanced diagnostic techniques, refined treatment protocols, and, in a few instances, even prospective preventative measures.

13.1.3 Rationale for Exploring Herbal Remedies

Exploring herbal remedies is a tradition rooted in humanity's innate drive to heal itself through natural means whenever possible. This pursuit of gentler options includes several compelling rationales [23]. The long cultural pedigree of many plants' medical usages, as generations shared which species soothed what suffering. That enduring ethnobotanical paper trail hints some herbs warrant careful studies into whether tradition translates to treatment. Herbs are widely perceived as wielding fewer potential side effects than manufactured pharmaceuticals, a quality valued by those navigating delicate conditions or desiring delicate approaches [24–29]. Gentleness also means gentler remedies may complement conventions where those fall short. Herbalism often sees the person, rather than just the pathology, prioritizing overall wellness – a perspective increasingly popular as prevention, not just reaction, gains favor. Treating the whole being aligns with this holistic health model.

Bacteria's accelerating resistance and the downsides of long-term medication lead to an urgent need for alternatives. Nature's plants, whose compounds work in complex concert, may offer novel ways to manage persistent infections and persistent patients [30].

Modern technology enables scientists to extract and analyze the active elements of herbs, providing valuable insights that can be applied to other fields such as medicine research and supplement formulation. This scientific validation has the potential to greatly enhance accessibility.

Herbal medicines are often more economically efficient and environmentally sensitive, which is particularly important when infrastructure is unreliable. Botanical medicine can be a crucial solution for populations that have limited access to healthcare services [31–33].

Although herbalism shows promise, it is crucial to prioritize rigorous study in order to establish the safety, effectiveness, and consistent benefits of medicines.

13.2 NEUROPROTECTIVE EFFECTS OF HERBAL COMPOUNDS

Neuroprotective properties of herbal compounds pertain to the capacity of specific molecules produced from plants to shield brain cells (neurons) against harm, deterioration, or demise [34]. The impact of these consequences is highly important in the context of neurological disorders such as Alzheimer's, Parkinson's, stroke, and multiple sclerosis. In these conditions, oxidative stress, inflammation, and excitotoxicity play a crucial role in causing damage to neurons.

13.2.1 Mechanisms of Neuroprotection

13.2.1.1 Antioxidant Activity

Compounds abundant in antioxidants, such as curcumin from turmeric and the flavonoids found in an assortment of fruits and vegetables, have demonstrated an ability to shelter neurons by scavenging damaging free radicals, which balances oxidative stress, a principal factor in neurodegeneration. Many plant extracts contain antioxidants that counteract free radicals and decrease oxidative stress [35–39].

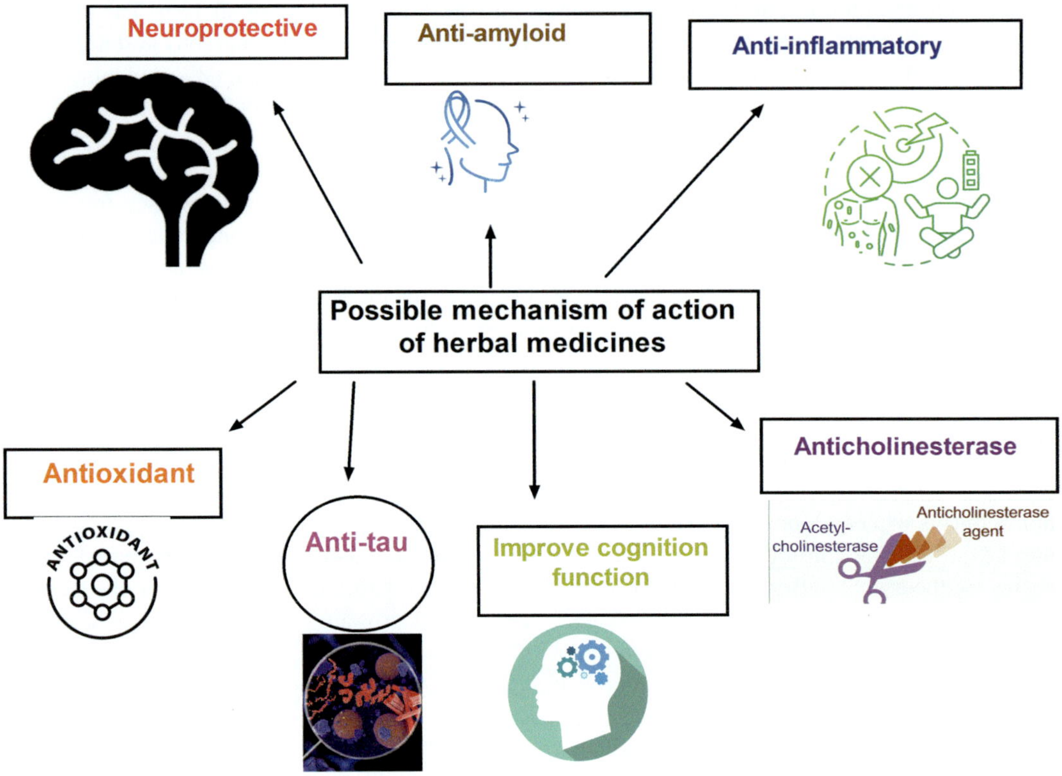

FIGURE 13.2 Potential mechanism of action of medicinal herbs.

13.2.1.2 Anti-inflammatory Properties

Persistent irritation within the mind is a hallmark of many neurological issues. Certain herbal extracts, including resveratrol present in grapes and ginsenosides discovered in ginseng, exhibit anti-inflammatory results through inhibiting the manufacturing of pro-inflammatory cytokines and modulating immune responses, thereby decreasing neuronal injury [40]. Herbal compounds have also been proven to curb inflammation in the mind, a hallmark of neurological issues (Figure 13.2).

13.2.1.3 Inhibition of Excitotoxicity

Excitotoxicity occurs when neurons are excessively stimulated by neurotransmitters such as glutamate, leading to cellular damage and death. Certain herbs include chemicals that can help control neurotransmitter levels, thereby reducing excitotoxic damage. For instance, *Ginkgo biloba* has components that help regulate neurotransmitter levels [41].

13.2.1.4 Enhancement of Neurogenesis and Synaptic Plasticity

Herbal extracts with positive effects promote the formation of new neurons, a process called neurogenesis, and enhance synaptic plasticity, which is crucial for learning, memory, and overall brain function. An example is the presence of bacosides in *Bacopa monnieri*, which have been examined for their capacity to improve cognitive functions by promoting neurogenesis and synaptic communication [42–49].

13.2.1.5 Mitochondrial Protection

Neurodegenerative disorders have a particular impact on mitochondria, which are the cellular powerhouses responsible for generating energy. Research indicates that herbal substances such as berberine found in goldenseal and quercetin found in various fruits and vegetables have the capacity to enhance mitochondrial activity and protect against energy deficiencies in neurons.

13.2.2 Role of Oxidative Stress in Neurological Diseases

Oxidative stress is a pivotal factor in the initiation and advancement of several neurological disorders. It arises from a disparity between the generation of reactive oxygen species (ROS) and the body's capacity to counteract them with antioxidants [50]. This imbalance results in the buildup of reactive oxygen species (ROS), which can harm cellular structures, including DNA, proteins, and lipids. Neurons, in particular, are prone to oxidative damage because of their high oxygen consumption and inadequate antioxidant defenses.

13.2.2.1 Essential Components of Oxidative Stress in Neurological Disorders

13.2.2.1.1 Impaired Functioning of Mitochondria

Neurons are highly dependent on mitochondria for their energy needs. Oxidative stress-induced damage to mitochondria results in decreased energy production and further amplifies the creation of reactive oxygen species (ROS), establishing a detrimental loop of damage that contributes to dementia. Oxidative stress has the potential to induce genetic mutations in DNA and alter proteins, thereby interfering with their regular functioning. This damage might result in cellular impairment and demise, hence exacerbating the advancement of neurological disorders [51–54]. Lipid peroxidation occurs when ROS target and damage lipids found in cell membranes, especially in the brain where there is a high concentration of polyunsaturated fatty acids. Lipid peroxidation is the process that causes damage to the cell membrane, resulting in the loss of cell integrity and ultimately leading to the death of neurons.

13.2.2.1.2 Neurological Disorders Linked to Oxidative Stress

Alzheimer's disease: Oxidative stress has a role in the development of beta-amyloid plaques and tau tangles, which are characteristic features of Alzheimer's disease and are linked to the loss of neurons and deterioration in cognitive function.

Parkinson's disease: Oxidative stress contributes to the deterioration of dopaminergic neurons in the substantia nigra in Parkinson's disease, primarily as a result of the oxidative breakdown of dopamine [55].

Amyotrophic lateral sclerosis (ALS): The breakdown of motor neurons is associated with oxidative stress, in which mutations in antioxidant enzymes such as superoxide dismutase 1 (SOD1) have a notable impact.

Multiple sclerosis (MS): Oxidative stress worsens inflammation and demyelination in MS, causing additional harm to neurons and advancing the disease.

Stroke: The occurrence of a stroke leads to a significant increase in ROS due to the damage caused by the interruption and restoration of blood flow. This surge in ROS results in substantial damage to the neurons, which worsens the severity of the injury [56].

Therapeutic implications: The comprehension of oxidative stress in neurological disorders has generated curiosity in antioxidant treatments designed to diminish oxidative harm and decelerate the advancement of the condition. Nevertheless, the intricate nature of oxidative stress pathways in the brain poses difficulties, necessitating further research to devise efficacious treatments.

13.2.3 Anti-Inflammatory Properties of Herbal Compounds

The varied anti-inflammatory capacities of herbal substances have long been acknowledged for their notable part in managing and forestalling diverse inflammatory issues, including those impacting the sensory system [57]. These chemicals can impede irritation through different instruments, giving a common elective or expansion to conventional anti-inflammatory medications.

13.2.3.1 Anti-Inflammatory Activity Systems

Suppression of pro-inflammatory cytokines: Certain herbs like curcumin derived from turmeric and resveratrol found in grapes have appeared capable of diminishing the combination of pro-inflammatory cytokines like TNF-α, IL-1β, and IL-6. By suppressing these signaling particles, they help in diminishing cellular irritation [58].

The NF-κB pathway plays a key job in controlling the inflammatory reaction. A few herbal substances, like gingerol found in ginger and epigallocatechin gallate (EGCG) from green tea, have the capacity to block the initiation of NF-κB, bringing about a decrease in the articulation of qualities related with swelling.

Antioxidant effects: Oxidative pressure and aggravation are firmly intertwined, as responsive oxygen types (ROS) frequently start inflammatory pathways. Herbal substances containing intense antioxidants, like quercetin present in different natural products and vegetables and flavonoids, help in diminishing oxidative pressure, subsequently diminishing irritation.

Suppression of COX and LOX enzymes: The inflammatory process is impacted by the creation of pro-inflammatory mediators, for example, prostaglandins and leukotrienes, which are created by enzymes called cyclooxygenase (COX) and lipoxygenase (LOX) [59]. Blends like boswellic corrosive, obtained from frankincense, and salicin, extracted from willow bark, have the capacity to impede these enzymes, along these lines diminishing irritation.

Immune system modulation: Certain herbs have an inherent ability to govern the immune reaction by upholding equilibrium between pro- and anti-inflammatory signals. For example, ginsenosides derived from ginseng and complex sugars obtained from fungi such as reishi have the potential to better immunological regulation, thereby facilitating a more balanced inflammatory feedback.

13.2.3.2 Uses and Advantages

Chronic Inflammatory Illnesses: Herbal anti-inflammatory agents are commonly applied to manage persistent inflammatory conditions – for example, rheumatoid arthritis, inflammatory bowel disease, and asthma [60–64]. They have the possibility to diminish the necessity for the extended application of standard anti-inflammatory medications, which can result in notable adverse effects.

Neurological Disorders: Inflammation plays a crucial part in neurodegenerative diseases such as Alzheimer's, Parkinson's, and multiple sclerosis. Herbal ingredients with anti-inflammatory qualities own the ability to retard the progression of illnesses and provide relief from symptoms. Herbal compounds are often employed to treat inflammatory skin disorders, such as eczema, psoriasis, and dermatitis, due to their relaxing and anti-inflammatory properties.

Cardiovascular Health: Enduring inflammation notably contributes to the evolution of cardiovascular sicknesses. Herbal ingredients possess the ability to mitigate inflammation in blood vessels, hence decreasing the prospect of atherosclerosis and related ailments.

The therapeutic potential of herbal components is substantiated by both ancestral usage and modern scientific investigation, owing to their anti-inflammatory traits. Although they have possible healing benefits, it is crucial to approach their application with an appreciation of suitable dosage, potential interactions with other medications, and the necessity for extra analysis to substantiate effectiveness and safety in specific circumstances.

13.2.4 Regulation of Neuronal Apoptosis by Herbal Remedies

Promising research has also been carried out on the ability of certain herbal remedies to regulate neuronal apoptosis (programmed cell death) in respect of neurodegenerative diseases where excessive or dysregulated apoptosis plays a part in the attrition of neurons [65]. Indeed, some herbal compounds have been found to positively modulate apoptotic pathways, leading to neuronal survival, and may be useful for promoting brain health (Figure 13.3).

For example, herbal compounds such as curcumin (from turmeric) and ginsenosides (ginseng) are found to inhibit pro-apoptotic proteins such as Bax, caspases, cytochrome c which, upon inhibition, reduce the markers of the apoptosis pathway, and hence prohibiting triggering an event to prevent the death of neurons.

Activation of Anti-apoptotic Pathways: Some herbal remedies increase the levels of anti-apoptotic proteins such as Bcl-2 (together with decreasing, for example, Bak), which stabilize mitochondria and inhibit apoptosis signals. Compounds like resveratrol (found in grapes) and epigallocatechin gallate, or EGCG for short (from green tea), have been shown to turn on these protective pathways.

Reducing Oxidative Stress: In the brain, one of the main stimulants for neuronal apoptosis is oxidative stress. The strong antioxidant properties of some herbal compounds like quercetin, flavonoids neutralize the ROS and thus reduced oxidative damage preventing apoptosis.

Modulation of Signaling Pathways: Herbal compounds can modulate important signaling pathways responsible for apoptosis, e.g., MAPK/ERK pathway, PI3/Akt and JNK. For example, various phytochemicals (baicalin from Sc), and honokiol (from Mo) were shown to act on this pathway protecting neurons by activation of survival genes and preventing the induction of death cell osteosis [66–68].

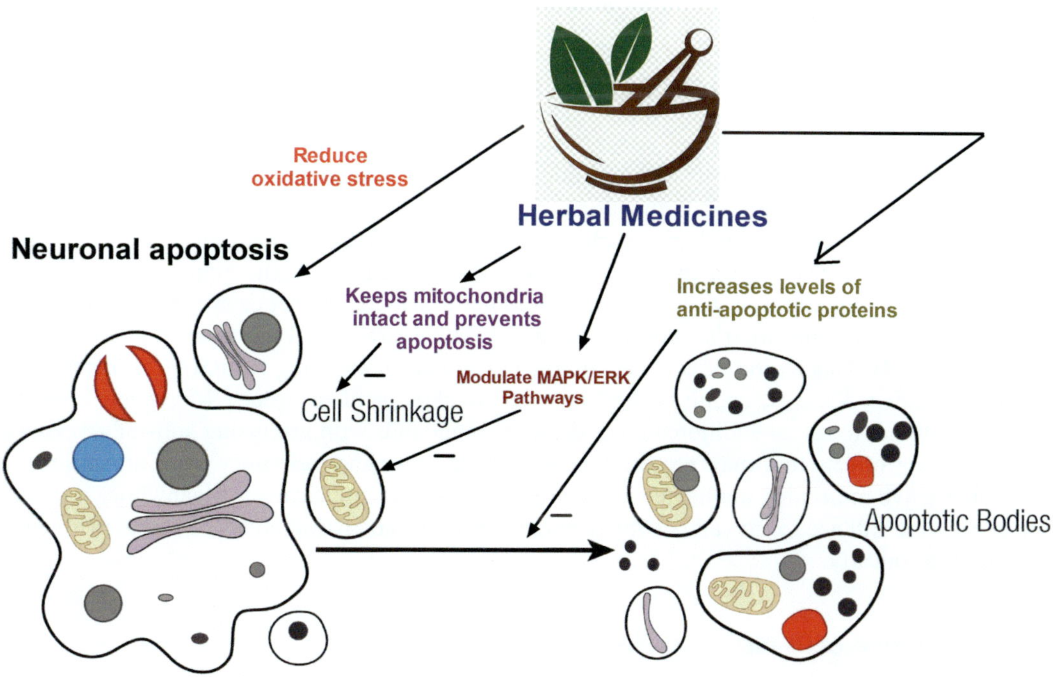

FIGURE 13.3 Mechanisms involved in the regulation of neuronal apoptosis by herbal medicines.

Mitochondrial Function Maintenance: Apoptosis is regulated through release of pro-apoptotic factors in response to stress, an important pathway controlled by proper functioning mitochondria. For example, specific plant compounds such as Berberine (via goldenseal) protect neurons by keeping the mitochondria intact and preventing damage that could lead to their dysfunctions of releasing apoptotic factors.

13.2.4.1 Neurological Diseases Applications

Neurodegenerative diseases, where neuronal apoptosis is widely believed to cause the loss of locomotor or cognitive functions like Alzheimer's, Parkinson's and ALS. Pharmacological modulation of apoptosis may protect neurons and reduce disease progression by means of regulation programmed cell death.

Stroke and Ischemic Injury: This is due to the apoptosis of neurons if widespread, resulting from an ischemic injury following a stroke. The herbal compounds with anti-apoptotic activity could result in decreasing the brain injury volume and hence improving functional recovery.

13.2.4.2 Future Scope and Challenges of Therapy

Although it sounds an interesting therapeutic approach to regulate neuronal apoptosis by herbal remedies, one must be vigilant while using them. These might be only the tip of a proverbial iceberg, as each will need to undergo evaluation in clinical trials in terms of both efficacy and safety, while potential drug–drug interactions will also have to be taken into consideration [69].

Thus, the potential of herbal remedies to modulate neuronal apoptosis through multiple mechanisms can serve as a natural way to rescue neurons under several neurological conditions. The capacity of mitochondria to affect both prosurvival and proapoptotic pathways place them at the forefront as additional candidates for extended research and therapeutic advancement.

13.3 HERBAL MODULATION OF NEUROGENESIS AND SYNAPTIC PLASTICITY

Herbal modulation of neurogenesis and synaptic plasticity is an emerging area that examines how plant constituents can influence brain functionality, specifically producing new neurons and improving synaptic connections. Certain herbs, such as *Ginkgo biloba*, Ashwagandha, and *Curcuma longa*, have demonstrated probability in fostering neurogenesis, particularly in the hippocampus, a brain region crucial for remembrance and studying. These herbs also may enhance synaptic plasticity by supporting the neurotransmitter's role, decreasing oxidative stress, and modulating inflammatory pathways [70–74]. This region of analysis retains likelihood for evolving natural therapies to fight neurodegenerative diseases, cognitive decline, and psychological disorders. Moreover, ongoing research continues to reveal additional plant-derived compounds with promising anti-inflammatory and antioxidant effects that may protect synaptic health and neural circuitry from an overabundance of reactive molecules implicated in age-related cognitive impairment. Meanwhile, rigorous human trials are still needed to validate many putative neuroprotective activities of herbal medicines observed preclinically.

13.3.1 Importance of Neurogenesis in Brain Repair

The power of self-renewal resides within our minds, as new neurons emerge to replace those that have been lost. Neurogenesis, an astounding phenomenon, continues to occur in the brain even after it has been injured. Regardless of the cause, whether it is trauma, stroke, or the gradual decline of

Alzheimer's, damage occurs, but nature initiates the process of healing through the use of newly formed cells. Newborn neurons primarily establish themselves and extend dendrites within the hippocampus, where they restore and rearrange memories that have been disrupted and thoughts that have become twisted, returning them to their original patterns. Neurogenesis repairs what was damaged and allows for the acquisition of new knowledge. When properly directed, medical techniques can assist nature by stimulating dormant precursor cells to contribute to the body's existing rescue mechanisms. By doing this, the results are more inclined towards healing rather than being controlled by impairment, which brings hope to the troubled mind. Enhancing neurogenesis offers promise in the pursuit of treatments that support the inherent abilities of self-healing present in every individual.

13.3.2 Effects of Herbal Extracts on Neurogenesis

Herbal extracts have piqued interest in their potential to spur neurogenesis, the process that leads to the development of novel brain cells. Neurogenesis underpins brain plasticity, learning, recollection, and overall neural welfare. A variety of herbal extracts have been examined for their neurogenic impacts, and the following passage summarizes some notable conclusions:

> While neurogenesis maintains brain adaptability, acquiring, and condition, specific herbal extricates have demonstrated potential in actuating the arrangement of youthful neurons. Studies have concentrated on extricates, for example, *Ginkgo biloba*, which may advance cell endurance and division in the hippocampus. Other normal treatments, for example, turmeric, rosemary, and devil's claw, have indicated neuroplastic impacts on rodent models and hold guarantees for further research in people. All in all, ongoing explores increasingly recognize the brain-aiding impacts of various home-developed medications and their components in upholding and potentially invigorating the mind's normal ability to make and incorporate new neurons throughout everyday life.

13.3.3 Enhancement of Synaptic Plasticity by Herbal Compounds

Synaptic plasticity, the brain's remodeling through synaptic strengthening and weakening enabling learning and memory, can be enhanced by herbal modulation. Curcumin, ginseng and *Bacopa monnieri* enhance it, upregulating neurotrophins like the brain-derived neurotrophic factor (BDNF) supporting synaptic growth (Figure 13.4). Curcumin and ginseng boost BDNF while Bacopa heightens acetylcholine, which is vital for hippocampal plasticity and memory. Chronic inflammation and oxidative stress impair plasticity, and green tea and curcumin's antioxidant and anti-inflammatory actions shield synapses during adaptation. Herbal extracts influence proteins like synapsins, regulating neurotransmitter discharge and synaptic strength plasticity.

Neurotrophins increased: Herbs often boost BDNF and related molecules which maintain synapses. Ginseng and curcumin specifically upregulate BDNF critical for plasticity and neuron survival.

Neurotransmission modulated: Some herbs influence key transmitters in synaptic signaling. Bacopa augments cholinergic activity, which is crucial where memory resides.

Inflammation and stress prevented: Oxidative and inflammatory insults disrupt plasticity. Green tea and curcumin counteract these with potent antioxidant and anti-inflammatory effects.

Protein expression altered: Compounds modify the levels of structure and function proteins such as synapsins, governing release and the modulation of synaptic efficacy.

Cognitively, improved plasticity is associated with enhanced learning, memory and functioning, thereby lessening neurodegenerative disease risk. Incorporating select herbs in diets or supplements may benefit lifelong mental resilience, though research elucidating long-term impacts and usage is still emerging.

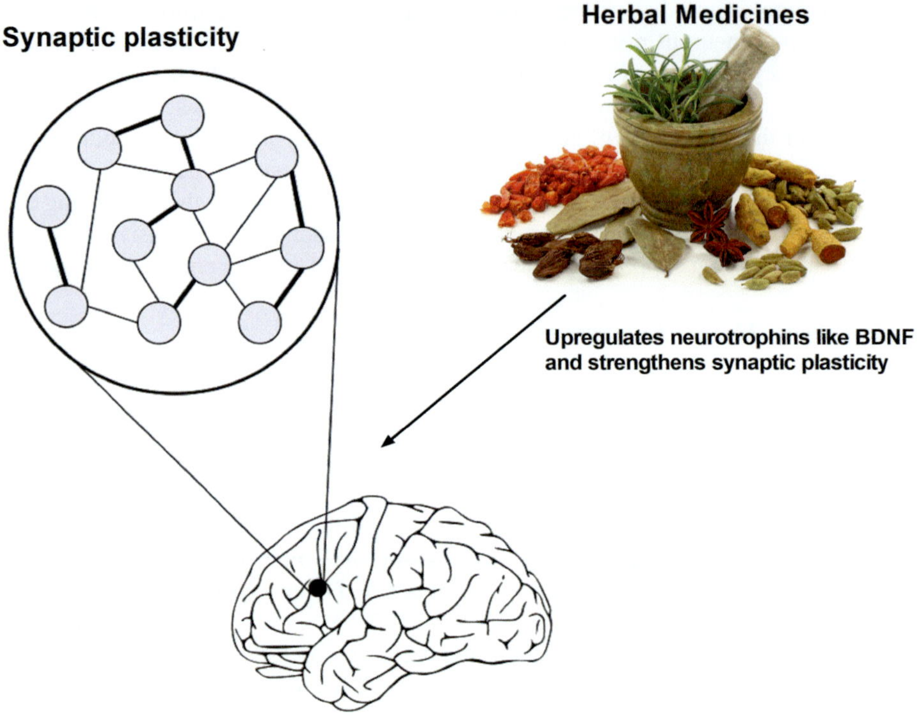

FIGURE 13.4 Regulation of synaptic plasticity by herbal compounds.

13.4 HERBAL MEDICINE AS ADJUNCTIVE THERAPY

Herbal medicine shows increasing recognition as a valuable additional therapy, complementing standard clinical treatments to boost overall wellness outcomes. As an additional therapy, herbal medicine can aid diverse health aspects, from bolstering the immune system to reducing the side effects of standard treatments and improving the quality of life [75]. For example, plants like Ashwagandha and *Rhodiola rosea* are often employed alongside standard treatments for stress and anxiety, helping to diminish signs without the harsh side effects sometimes related to modern drugs.

In cancer treatment, certain herbal compounds, for instance curcumin from turmeric, are analyzed for their potential to enhance the efficacy of chemotherapy and defend against its side consequences, including inflammation and oxidative stress. Similarly, St. John's Wort is sometimes utilized together with antidepressants to boost mood and lessen the dosage of the modern drugs necessary.

While herbal medicine offers promising advantages as an additional therapy, it's crucial for patients to seek healthcare providers before combining it with standard treatments, as interactions between herbs and medications can arise. When utilized suitably, herbal medicine can complement traditional treatments, offering a holistic method to health and wellbeing.

13.4.1 Synergistic Effects of Herbal Compounds with Conventional Treatments

The synergistic interactions between herbs and conventional medicines have become an important consideration in integrative healthcare. When combined appropriately, herbal remedies can enhance clinical outcomes, lower side effects, and improve overall well-being [76]. Here are some notable examples of herbal synergies:

13.4.1.1 Cancer Care

Curcumin boosts the potency of chemotherapy and radiation while easing inflammation and oxidative stress, helping healthy cells withstand toxic treatments. Green tea polyphenols like EGCG simultaneously inhibit tumor growth and strengthen chemotherapy's impact while potentially shielding healthy tissues from harm [77].

13.4.1.2 Depression Relief

St. John's Wort augments antidepressants by fine-tuning neurotransmitter levels, possibly allowing lower drug dosages that pose less risk. Saffron also lifts mood when paired with selective serotonin reuptake inhibitors(SSRIs), relieving symptoms with reduced side effects.

13.4.1.3 Heart Health

Garlic's blood-thinning properties complement anticoagulants' prevention of heart attacks and strokes [78]. It further lowers blood pressure and cholesterol. Omega-3 fatty acids from fish oil and flaxseed strengthen statins' cholesterol-lowering impact, further decreasing cardiovascular damage

13.4.1.4 Diabetes Management

Goldenseal's berberine improves insulin sensitivity and glucose control when allied with metformin, lessening the need for high doses of modern drugs. Fenugreek seeds do similarly when combined with insulin or oral hypoglycemic drugs.

13.4.1.5 Pain Relief

Willow bark, like aspirin from willows, improves NSAIDs' pain-relieving capacities. Chili pepper capsaicin blocks pain signals where creams are applied, synergizing with oral analgesics.

13.4.2 Mitigation of Drug-Induced Side Effects

Combining herbal compounds with conventional remedies can provide a comprehensive strategy for managing diverse medical conditions. By boosting drug efficacy and mitigating side effects, herbal additions are pivotal for integrative care plans. Still, patients must consult healthcare professionals before blending herbal and pharmaceutical therapies to avoid interactions and guarantee safety. Here are some examples of how herbal compounds can mitigate drug-induced side effects:

13.4.2.1 Digestive Challenges

Ginger: Renowned for reducing nausea, ginger is commonly applied to alleviate chemotherapy, anesthesia, and certain prescription side effects such as nausea and vomiting. It can also calm the abdomen and relieve indigestion from nonsteroidal anti-inflammatory drugs.

Peppermint: Peppermint essential oil can relieve signs of irritable bowel syndrome and drug-induced swelling and cramping, especially in patients taking treatments affecting gastrointestinal motility.

13.4.2.2 Liver Safeguarding

Milk Thistle: Milk thistle is frequently utilized to protect the liver from harm brought on by prescriptions, particularly those broken down by the liver or known to be toxic to the liver, for example, acetaminophen and specific chemotherapy medications. The dynamic compound, silymarin, has cell reinforcement and mitigating properties that back liver capacity and regeneration

Turmeric: Curcumin has liver-safeguarding properties that can help reduce liver swelling and oxidative stress induced by long-haul medication use.

13.4.2.3 Kidney Protection

Astragalus: Astragalus root is regularly utilized to ensure the kidneys from harm, particularly in patients going through treatments like chemotherapy or the long-term use of nonsteroidal anti-inflammatory drugs (NSAIDs) which can hurt renal capacity. It has mitigating and cell reinforcement impacts that help maintain kidney wellbeing.

Cordyceps: This restorative mushroom is known for its potential to uphold kidney capacity and is utilized in conventional medication to mitigate renal harm brought about by medications, particularly in patients with constant kidney illness.

13.4.2.4 Neurotoxicity

Ginkgo biloba: *Ginkgo biloba* may help reduce side effects on the mind and cognitive impacts related with chemotherapy, antipsychotics, or other neurotoxic medications. Its neuroprotective effects are generally credited to its cell reinforcement properties and capacity to increase cerebral blood flow.

Ashwagandha: Ashwagandha is an adaptogen that can mitigate the intellectual decay and uneasiness brought about by long-term utilization of corticosteroids or other stress-inducing drugs.

13.4.2.5 Cardiotoxicity

Hawthorn (*Crataegus* spp.): Native to Europe and North America, hawthorn has long been revered for its heart-supportive properties. Modern research confirms that it improves circulation, strengthens the heart, and lowers blood pressure – effects that counteract cardiotoxic side effects from some drugs. By enhancing blood flow and heart function, hawthorn helps mitigate toxicity risks [79].

Coenzyme Q10 (CoQ10): Not technically a herb, but a vital nutrient, CoQ10 supports mitochondrial energy production within heart cells. This enables it to reduce cardiotoxicity related to statins and certain chemotherapies. By maintaining cellular energy levels, CoQ10 protects the heart from drug-induced damage.

13.4.2.6 Bone Marrow Suppression

Echinacea: A popular immune booster, echinacea contains compounds that may stimulate white blood cell development. This allows it to counter bone marrow suppression from chemotherapy and other immunosuppressive medications. By improving white cell counts, echinacea reduces infection risk during treatment.

Ginseng (Panax ginseng): Ginseng has long been used for its immune-enhancing properties. Modern research confirms that it can also help to counteract immunosuppression, lowering chances of illness. For patients undergoing bone marrow suppressive regimens, ginseng offers protective effects against infection and enhances overall vitality during trying times.

13.4.2.7 Managing Fatigue

Rhodiola rosea: This adaptogenic herb enhances physical and mental endurance. For those undergoing chronic treatments, Rhodiola helps combat debilitating fatigue. It improves stamina during lengthy or recurring regimens.

Eleuthero (Siberian Ginseng): Another adaptogen, Eleuthero revitalizes energy levels that decrease due to medication use. It counters weariness and replenishes strength in patients enduring long-term therapies.

Nature provides herbs that mitigate drug side effects, enhancing treatment adherence and comfort. While offering benefits, herbs require guidance from healthcare professionals to ensure safe, targeted use, especially regarding potential interactions. When herbs are carefully monitored and guided by knowledge, they offer an efficient and natural approach to enhancing medicinal results.

13.4.3 Enhancement of Therapeutic Outcomes

While herbs can work in tandem with traditional therapies to improve healing outcomes, taking supplements always warrants medical supervision. Echinacea, *ginkgo biloba*, and curcumin have shown promise in improving standard treatments, strengthening immune defenses, reducing inflammation, and mitigating side effects, leading to quicker recoveries and increased wellness [80]. For example, combining curcumin with chemotherapy may heighten cancer treatment's effectiveness while sparing healthy cells harm, and supplementing *ginkgo biloba* along with cognitive aids can potentially boost memory and mental strength. By thoughtfully incorporating herbs under a provider's guidance, patients' prognoses and quality of life could be favourably impacted. However, it is imperative that any adjunctive natural remedies are thoroughly discussed with healthcare experts first to ensure they are administered prudently and generate only intended effects.

13.5 FUTURE DIRECTIONS AND CHALLENGES

Future directions and challenges with herbal medicine involve addressing both the potential and limitations of integrating herbal therapies into modern healthcare. Key areas that warrant exploration include research and evidence, standardization and quality control, regulation and safety, integration with conventional treatments, patient education, and personalization of therapy.

13.5.1 Opportunities for Further Research

Opportunities for additional research within the field of herbal medicine are abundant and essential to move the discipline forward in a meaningful manner. Key areas requiring focus encompass investigating the mechanisms by which herbal compounds act to understand their impacts on human biology, performing rigorous clinical tests to validate effectiveness and safety, and cultivating strategies for standardizing and ensuring the quality of herbal products. Furthermore, exploring interactions between herbs and drugs, studying how the body handles and reacts to components over time and space, and tailoring herbal treatments based on genetics and lifestyle habits has potential to optimize outcomes for patients. Studies of native plant use can uncover fresh therapeutic applications, while innovations in formulation and delivery mechanisms can strengthen the impact of herbal medicines. Addressing these open research areas will bolster the scientific foundation for herbal medicine and aid its integration into modern healthcare.

13.5.2 Challenges in Herbal Medicine Research

The issues facing herbal medicine research are extensive and multidisciplinary in nature. A predominant challenge is necessity for standardization and quality assurance, which can induce inconsistency in the potency and purity of herbal goods. Furthermore, numerous herbal therapies lack robust clinical substantiation owing to inadequately planned studies and a reliance on traditional use rather than stringent scientific testing. The potential for herb–drug interactions also introduces risks, as knowledge about how herbal remedies may influence or be influenced by standard medications is commonly limited. Additionally, the regulatory environment governing herbal medicines is less stringent relative to pharmaceuticals, leading to anxieties regarding safety and efficacy. Overcoming these difficulties necessitates rigorous scientific investigation, improved regulatory frameworks that are more stringent, and amplified collaboration between herbalists and conventional healthcare providers to guarantee the protected and effectual use of herbal therapies.

13.5.3 Integration of Traditional Knowledge with Modern Science

Integrating ancient wisdom with cutting-edge science is a promising approach for advancing herbal medicine. Traditional knowledge, accumulated painstakingly over centuries, provides invaluable insights regarding the uses and advantages of herbal remedies that can steer contemporary empirical research. By combining empirical sagacity with modern analytic techniques, investigators can substantiate and refine ancestral practices, identify bioactive ingredients, and comprehend their mechanisms of action. This integration can lead to the evolution of evidence-based herbal treatments that are scientifically validated while still culturally pertinent. Collaborations between traditional healers, ethnobotanists, and scientists can bridge the chasm between ancestral customs and scientific inquiry, fostering medical breakthroughs and ensuring that ancestral knowledge contributes to effective and safe therapeutic solutions.

13.6 CASE STUDIES AND CLINICAL APPLICATIONS

13.6.1 Illustrative Case Studies

The study of herbal medical practices sheds light on the practical applications and benefits of traditional remedies in contemporary healthcare. For example, studies have proven that curcuminoids from *Curcuma longa* can boost the potency of chemotherapy and shorten side effects in cancer sufferers. *Ginkgo biloba* has exhibited improvements in cognitive abilities and daily living activities in individuals with mild cognitive impairment. *Hypericum perforatum* has surfaced as useful in managing melancholy with less secondary effects than conventional antidepressants [81]. *Echinacea purpurea* has displayed potential in reducing the frequency and severity of respiratory infections. Last but not least, *Silybum marianum* has aided in improving liver functions in patients with liver ailments. These exploratory analyses underscore the importance of combining herbal medicine with conventional remedies, provided they are supported by rigorous scientific research.

13.6.2 Clinical Applications of Herbal Medicine in Neurological Diseases

Herbal remedies have shown promising results for neurological diseases. *Ginkgo biloba* and curcumin may aid those with Alzheimer's, due to antioxidant and anti-inflammatory effects. For Parkinson's patients, *Mucuna pruriens* contains natural levodopa. St. John's Wort and kava have assisted many dealing with depression and anxiety. Cannabidiol from cannabis demonstrates ability in managing intractable epilepsy and multiple sclerosis symptoms. Alternatives like feverfew and butterbur are also utilized for migraine prevention. Capsaicin brings relief from neuropathic pain [82, 83]. However, ensuring safety, standardizing treatments, avoiding drug interactions and navigating regulations pose challenges to fully integrating herbal medicine. With addressing such issues, herbal therapies may offer effective, side effect-friendly options for neurological disorders.

13.7 CONCLUSION

13.7.1 Summary of Key Findings

Herbal medicine is a discipline steeped in tradition, centering on the use of herbs and plants for preventive care and cures. For ages, cultures have drawn from the healing powers of nature, relying on botanical remedies to treat afflictions and promote wellness. However, interacting effects with modern drugs warrant caution when self-administering remedies. Globally, ongoing scientific inquiries explore herbal medicine's applications and benefits, seeking to integrate ancient wisdoms with modern methods. Herbal formulations come in diverse forms for different needs, so consulting your physician is imperative to avoid unintended outcomes or contradictions with existing treatments.

In summary, while herbal medicine has deep roots in harnessing nature's curative forces, taking full account of possible conflicts or risks is critical for safety when availing these complementary therapies.

13.7.2 Future Outlook for Herbal Medicine in Neurology

There is a hopeful future for herbal therapy in the field of neurology, as there is an increasing interest in incorporating these treatments into mainstream medical practice. Research and technological advancements are anticipated to enhance the standardization, quality control, and bioavailability of herbal products, thereby resolving existing restrictions. Expanded clinical trials will contribute to the validation of the effectiveness and safety of herbal treatments, potentially resulting in their broader acceptance and utilization for the treatment of neurological disorders such as Alzheimer's, Parkinson's, epilepsy, and multiple sclerosis.

Furthermore, as our comprehension of the molecular mechanisms underlying the effects of herbal remedies becomes more profound, it is possible that new therapeutic molecules could be created from traditional herbs. The application of personalized medicine, which involves customizing treatments based on an individual's genetic and biochemical characteristics, has the potential to improve the efficacy of herbal remedies in the field of neurology. Nevertheless, it is crucial to tackle regulatory obstacles and guarantee patient safety by closely monitoring the interactions between herbal medicine and conventional pharmaceuticals as the area of herbal medicine progresses.

REFERENCES

1. Lavretsky H (2009) Complementary and alternative medicine use for treatment and prevention of late-life mood and cognitive disorders. *Aging Health* 5: 61–78
2. Finkbeiner S (2011) Huntington's disease. *Cold Spring Harb. Perspect. Biol.* 3: 1–24.
3. Brodtkorb E, Torbergsen T, Nakken KO, Andersen K, Gimse R, et al. (1994) Epileptic seizures, arthrogryposis, and migrational brain disorders: a syndrome? *Acta Neurol. Scand.* 90: 232–240.
4. Ríos J, Onteniente M, Picazo D, Montesinos M (2016) Medicinal plants and natural products as potential sources for anti-Parkinson drugs. *Planta Med.* 82: 942–951.
5. DeLisi LE, Szulc KU, Bertisch HC, Majcher M, Brown K (2006) Understanding structural brain changes in schizophrenia. *Dialogues Clin. Neurosci.* 8: 71–78.
6. Curatolo P, D'Agati E, Moavero R (2010) The neurobiological basis of ADHD. *Ital. J. Pediatr.* 36: 79–85.
7. Misra LN (2013) Traditional phytomedicinal systems, scientific validations and current popularity as nutraceuticals. *Int. J. Trad. Nat. Med.* 2: 27–75.
8. Balkrishna A (2008) *Secrets of Indian herbs for good health.* Divya Prakashan, Patanjali Yogpeeth, Haridwar, India pp: 1–420.
9. Balkrishna A (2014) *Ayurved jadi-buti rahasya* (Vol. 1–3). Divya Prakashan, Patanjali Yogpeeth, Haridwar, India pp: 1–1650.
10. Anonymous (2008) *The side effects of common psychiatric drugs-A report by the citizens commission on human rights international.*
11. Pandey SK, Jangra MK, Yadav AK (2014) Herbal and synthetic approaches for the treatment of epilepsy. *Intern. J. Nutr. Pharmacol. Neurol. Dis.* 4: 43–52.
12. Parasuraman S, Thing GS, Dhanaraj SA (2014) Polyherbal formulation: Concept of ayurveda. *Pharmacogn. Rev.* 8: 73–80.
13. Husain A, Virmani OP, Popli SP, Misra LN, Gupta MM, et al. (1992) *Dictionary of Indian Medicinal Plants*, CIMAP, Lucknow, India pp: 1–546.
14. Balkrishna A, Misra LN (2017) Brief chemo-botanical account of some Ayurvedic plants useful in mental health. *Nat. Prod. J. Commun.*. 6(2), 221.
15. Kapahi, M; Sachdeva, S Mycoremediation potential of Pleurotus species for heavy metals: A review. *Bioresour. Bioprocess.* 2017, 4, 32.
16. Vetter, J Arsenic content of some edible mushroom species. *Eur. Food Res. Technol.* 2004, 219, 71–74.
17. Martino, E; Perotto, S Mineral transformations by mycorrhizal fungi. *Geomicrobiol J.* 2010, 27, 609–623.
18. Gadd, GM Interactions of Fungi with Toxic Metals. In *The Genus Aspergillus*; Powell, KA, Renwick, A, Peberdy, JF, Eds.; Springer: Boston, MA, USA, 1994; Volume 69.
19. Giller, KE; Witter, E; McGrath, SP Heavy metals and soil microbes. *Soil Biol. Biochem.* 2009, 41, 2031–2037.

20. Kuehnelt, D; Goessler, W; Irogolic, KJ Arsenic compounds in terrestrial organisms II: Arsenocholine in the mushroom Amanita muscaria. *Appl. Organomet. Chem.* 1997, 11, 459–470.
21. Slejkovec, Z; Byrne, AR; Smodis, B; Rossbach, M Preliminary studies on arsenic species in some environmental samples. *Anal. Bioanal. Chem.* 1996, 354, 592–595.
22. Melgar, MJ; Alonso, J; García, MA Total contents of arsenic and associated health risks in edible mushrooms, mushroom supplements and growth substrates from Galicia (NW Spain). *Food Chem. Toxicol.* 2014, 73, 44–50.
23. Rashid, MH; Rahman, MM; Correll, R; Naidu, R Arsenic and other elemental concentrations in mushrooms from Bangladesh: Health risks. *Int. J. Environ. Res. Public Health* 2018, 15, 919.
24. Giannaccini, G; Betti, L; Palego, L; Mascia, G; Schmid, L; Lanza, M; Mela, A; Fabbrini, L; Biondi, L; Lucacchini, A The trace element content of top-soil and wild edible mushroom samples collected in Tuscany, *Italy. Environ. Monit. Assess.* 2012, 184, 7579–7595.
25. Chen, XH; Zhou, HB; Qiu, GZ Analysis of several heavy metals in wild edible mushrooms from regions of China. *Bull. Environ. Contam. Toxicol.* 2009, 83, 280–285.
26. Tyler, G Metals in sporophores of basidiomycetes. *Trans. Br. Mycol. Soc.* 1980, 74, 41–49.
27. Mendil, D; Uluözlü, ÖD; Tüzen, M; Hasdemir, E; Sarı, H Trace metal levels in mushroom samples from Ordu, Turkey. *Food Chem.* 2005, 91, 463–467.
28. Küpeli Akkol, E, Güragaç, F T, and Ilhan, M (2019). Assessment of Antidepressant Effect of the Aerial Parts of Micromeria Myrtifolia Boiss. & Hohen on Mice. *Molecules* 24 (1869), 1–13.
29. Laurent, C, Eddarkaoui, S, Derisbourg, M, Leboucher, A, Demeyer, D, Carrier, S, et al. (2014). Beneficial Effects of Caffeine in a Transgenic Model of Alzheimer's Disease-like Tau Pathology. *Neurobiol. Aging* 35 (9), 2079–2090.
30. Leahy, RL, Holland, S J, and McGinn, L K (2011). *Treatment Plans and Interventions for Depression and Anxiety Disorders*. New York: Guilford Press. Lee, H. J., Cho, H. S., Park, E., Kim, S., Lee, S. Y., Kim, C. S., et al. (2008). Rosmarinic Acid Protects Human Dopaminergic Neuronal Cells against Hydrogen Peroxide-Induced Apoptosis. *Toxicology* 250 (2–3), 109–115.
31. Lee, M-H, Lin, R-D, Shen, L-Y, Yang, L-L, Yen, K-Y, and Hou, W-C (2001). Monoamine Oxidase B and Free Radical Scavenging Activities of Natural Flavonoids in Melastoma candidumD. *Don. J. Agric. Food Chem.* 49 (11), 5551–5555.
32. Lee, S, Kim, H-B, Hwang, E-S, Kim, E-S, Kim, S-S, Jeon, T-D, et al. (2018). Antidepressant-like Effects of P-Coumaric Acid on LPS-Induced Depressive and Inflammatory Changes in Rats. *Exp. Neurobiol.* 27, 189–199.
33. Leentjens, AFG (2012). Epidemiology of Psychiatric Symptoms in Parkinson's Disease. *Adv. Biol. Psychiatry* 27, 1–12.
34. Lenzi, J, Rodrigues, AF, Rós, ADS, De Castro, BB, De Lima, DD, Magro, DDD, et al. (2015). Ferulic Acid Chronic Treatment Exerts Antidepressant-like Effect: Role of Antioxidant Defense System. *Metab. Brain Dis.* 30, 1453–1463.
35. Lépine, JP, and Briley, M (2011). The Increasing Burden of Depression. *Neuropsychiatr. Dis. Treat.* 7 (1), 3–7.
36. Leyhe, T, Reynolds, CF, 3rd, Melcher, T, Linnemann, C, Klöppel, S, Blennow, K, et al. (2017). A Common Challenge in Older Adults: Classification, Overlap, and Therapy of Depression and Dementia. *Alzheimers Dement.* 13 (1), 59–71.
37. Li, F, Han, G, and Wu, K(2016). Tanshinone IIA Alleviates the AD Phenotypes in APP and PS1 Transgenic Mice. *Biomed. Res. Int.* 2016, 1–8.
38. Nalini, K, Aroor, A, Karanth, K, and Rao, A (1992). Centella Asiatica Fresh Leaf Aqueous Extract on Learning and Memory and Biogenic Amine Turnover in Albino Rats. *Fitoterapia* 63, 232–237.
39. Nehlig, A (2016). Effects of Coffee/caffeine on Brain Health and Disease: What Should I Tell My Patients? *Pract. Neurol.* 16 (2), 89–95.
40. Miyazawa, M, and Yamafuji, C (2005). Inhibition of Acetylcholinesterase Activity by Bicyclic Monoterpenoids. *J. Agric. Food Chem.* 53, 1765–1768.
41. Moloudizargari, M, Mikaili, P, Aghajanshakeri, S, Asghari, M, and Shayegh, J (2013). Pharmacological and Therapeutic Effects of Peganum Harmala and its Main Alkaloids. *Phcog. Rev.* 7 (14), 199–212.
42. Nakashi, K (2005). Terpene Trilactones from Ginkgo Biloba: From Ancient Times to the 21st Century. *Bioorg. Med. Chem.* 13, 4987–5000.
43. Roodenrys, S, Booth, D, Bulzomi, S, Phipps, A, Micallef, C, and Smoker, J (2002). Chronic Effects of Brahmi (Bacopa Monnieri) on Human Memory. *Neuropsychopharmacolog* 27 (2), 279–281.

44. Ruipérez, V, Darios, F, and Davletov, B (2010). Alpha-synuclein, Lipids and Parkinson's Disease. *Prog. Lipid Res.* 49 (4), 420–428.
45. Ryu, S-H, Jung, H-Y, Lee, KJ, Moon, SW, Lee, DW, Hong, N, et al. (2017). Incidence and Course of Depression in Patients with Alzheimer's Disease. *Psychiatry Investig.* 14 (3), 271–280.
46. Nakazawa, T, Yasuda, T, Ueda, J, and Ohsawa, K (2003). Antidepressant-like Effects of Apigenin and 2,4,5-trimethoxycinnamic Acid from Perilla Frutescens in the Forced Swimming Test. *Biol. Pharm. Bull.* 26 (4), 474–480.
47. More, SV, Kumar, H, Kim, IS, Song, S-Y, and Choi, D-K (2013). Cellular and Molecular Mediators of Neuroinflammation in the Pathogenesis of Parkinson's Disease. *Mediat. Inflamm.* 2013, 1–12.
48. Moss, M, and Oliver, L (2012). Plasma 1,8-cineole Correlates with Cognitive Performance Following Exposure to Rosemary Essential Oil Aroma. *Ther. Adv.* 2, 103–113.
49. Rickards, H (2006). Depression in Neurological Disorders: an Update. *Curr. Opin. Psychiatry* 19 (3), 294–298. doi:10.1097/01.yco.0000218601.17722.5b
50. Rickards, H (2005). Depression in Neurological Disorders: Parkinson's Disease, Multiple Sclerosis, and Stroke. *J. Neurol. Neurosurg. Psychiatry* 76 (1), i48–i52. doi:10.1136/jnnp.2004.060426
51. Rickels, K, Shiovitz, TM, Ramey, TS, Weaver, JJ, Knapp, LE, and Miceli, JJ (2012). Adjunctive Therapy with Pregabalin in Generalized Anxiety Disorder Patients with Partial Response to SSRI or SNRI Treatment. *Int. Clin. Psychopharmacol.* 27 (3), 142–150.
52. Rodríguez-Landa, JF, Cueto-Escobedo, J, Puga-Olguín, A, Rivadeneyra DomínguezBernal-Morales, EB, Bernal-Morales, B, Herrera-Huerta, EV, et al. (2017). The Phytoestrogen Genistein Produces Similar Effects as 17β Estradiol on Anxiety-like Behavior in Rats at 12 Weeks after Ovariectomy. *Biomed. Res. Int.* 2017, 1–10.
53. Sakina, MR, and Dandiya, PC (1990). A Psycho-Neuropharmacological Profile of Centella Asiatica Extract. *Fitoterapia* 61, 291–296.
54. Santabárbara, J, Lipnicki, D, Bueno-Notivol, J, Olaya-Guzmán, B, Villagrasa, B, and López-Antón, R (2020). Updating the Evidence for an Association between Anxiety and Risk of Alzheimer's Disease: A Meta-Analysis of Prospective Cohort Studies. *J. Affect. Disord.* 262, 397–404.
55. Rodríguez-Landa, JF, Hernández-López, F, Cueto-Escobedo, J, Herrera-Huerta, EV, Rivadeneyra-Domínguez, E, Bernal-Morales, B, et al. (2019). Chrysin (5,7-dihydroxyflavone) Exerts Anxiolytic-like Effects through GABAA Receptors in a Surgical Menopause Model in Rats. *Biomed. Pharmacother.* 109, 2387–2395.
56. Said, UZ, Saada, HN, Abd-Alla, MS, Elsayed, ME, and Amin, AM (2012). Hesperidin Attenuates Brain Biochemical Changes of Irradiated Rats. *Int. J. Radiat. Biol.* 88, 613–618.
57. Sairam, K, Dorababu, M, Goel, RK, and Bhattacharya, SK (2002). Antidepressant Activity of Standardized Extract of Bacopa Monniera in Experimental Models of Depression in Rats. *Phytomedicine* 9 (3), 207–211.
58. Kayani S, Ahmad, M; Zafar, M; Sultana, S; Khan, MPZ; Ashraf, MA; Hussain, J; Yaseen, G: Ethnobotanical uses of medicinal plants for respiratory disorders among the inhabitants of Gallies-Abbottabad, Northern Pakistan. *J. Ethnopharmacol.* 156, 47–60 (2014).
59. Kadir, MF; Sayeed, MSB; Shams, T; Mia, MMK: Ethnobotanical survey of medicinal plants used by Bangladeshi traditional health practitioners in the management of diabetes mellitus. *J. Ethnopharmacol.* 144(3), 605–611 (2012).
60. Kadiri SK, Khobragade DS and Roy SP. *Preclinical Appraisal of the Aphrodisiac Effects of Emblica officinalis Seed Extract on Stress-induced Sexual Behavior in Albino Rats.* Current Drug Therapy (Bentham Science Publishers), 2024, 19, 594–603.
61. Amjad, MS; Qaeem, MF; Ahmad, I; Khan, SU; Chaudhari, SK; Zahid Malik, N; Shaheen, H; Khan, AM: Descriptive study of plant resources in the context of the ethno-medicinal relevance of indigenous flora: a case study from Toli Peer National Park, Azad Jammu and Kashmir, Pakistan. *PLoS One* 12(2), e0171896 (2017).
62. Majeed, M; Bhatti, KH; Amjad, MS; Abbasi, AM; Bussmann, RW; Nawaz, F; Rashid, A; Mehmood, A; Mahmood, M; Khan, WM: Ethno-veterinary uses of Poaceae in Punjab, Pakistan, *PLoS One* 15(11), e0241705 (2020)
63. Cheung, S; Tai, J: Anti-proliferative and antioxidant proper ties of rosemary Rosmarinus officinalis. *Oncol. Rep.* 17(6), 1525–1531 (2007)
64. Oluwatuyi M, Kaatz GW, Gibbons S, Antibacterial and resistance modifying activity of Rosmarinus officinalis. *Phyto Chem.* 65(24), 3249–3254 (2004)

65. Li W, Lee MK. Antiapoptotic property of human α-synuclein in neuronal cell lines is associated with the inhibition of caspase-3 but not caspase-9 activity. *J. Neurochem.* 2005;93(6):1542–1550.
66. Gunasekaran R, Narayani RS, Vijayalakshmi K, Alladi PA, Shobha K, Nalini A, et al. 'Exposure to cerebrospinal fluid of sporadic amyotrophic lateral sclerosis patients alters Nav1.6 and Kv1.6 channel expression in rat spinal motor neurons. *Brain Res.* 2009;1255:170–179.
67. Ge P, Luo Y, Wang H, Ling F. Anti-protein aggregation is a potential target for preventing delayed neuronal death after transient ischemia. *Med. Hypotheses* 2009;73(6):994–995.
68. Sinclair HL, Andrews PJ. Bench-to-bedside review: hypothermia in traumatic brain injury. *Crit. Care* 2010;14(1):204.
69. Bhattacharya SK, Chakrabarti A, Chatterjee SS. Activity profiles of two hyperforin-containing Hypericum extracts in behavioral models. *Pharmacopsychiatry* 1998;31;Suppl 1:22–29.
70. Watanabe CM, Wolffram S, Ader P, Rimbach G, Packer L, Maguire JJ, et al. The in vivo neuromodulatory effects of the herbal medicine ginkgo biloba. *Proc. Natl. Acad. Sci. USA* 2001;98(12):6577–6580.
71. Maitra I, Marcocci L, Droy-Lefaix MT, Packer L. Peroxyl radical scavenging activity of Ginkgo biloba extract EGb 761. *Biochem. Pharmacol.* 1995;49(11):1649–1655.
72. Galasso C, Orefice I, Pellone P, Cirino P, Miele R, Ianora A, Brunet C, Sansone C. On the neuroprotective role of astaxanthin: new perspectives?. *Mar. Drugs* 2018 Jul 24;16(8):247.
73. Agarwal A, Malini S, Bairy K, Rao MS. Effect of Tinospora cordifolia on learning and memory in normal and memory deficit rats. *Indian J. Pharm.* 2002;34:339–349.
74. Nadkarni K *Indian plants and drugs with medicinal properties and uses*. Madras: Norton and Co., India; 1908.
75. Damodaran M, Ramaswamy R. Isolation of l-3:4-dihydroxyphenylalanine from the seeds of Mucuna pruriens. *Biochem. J.* 1937;31(12):2149–2152.
76. Vaidya AB, Rajagopalan TG, Mankodi NA, Antarkar DS, Tathed PS, Purohit AV, et al. Treatment of Parkinson's disease with the cowhage plant-Mucuna pruriens Bak. *Neurol. India* 1978;26(4):171–176.
77. Venturi VM, Piccinin GL, Taddei I. Pharmacognostic study of self-sown Galanthus nivalis (var. gracilis) in Italy. *Boll. Soc. Ital. Biol. Sper.* 1965;41(11):593–597.
78. Wilcock GK, Lilienfeld S, Gaens E. Efficacy and safety of galantamine in patients with mild to moderate Alzheimer's disease: multicentre randomised controlled trial. Galantamine International-1 Study Group. *BMJ* 2000;321(7274):1445–1449.
79. Alghamdi BS. The neuroprotective role of melatonin in neurological disorders. *J. Neurosci. Res.* 2018 Jul;96(7):1136–1149.
80. Bhattacharya SK, Kumar A, Ghosal S. Effects of glycowithanolides from Withania somnifera on an animal model of Alzheimer's disease and perturbed central cholinergic markers of cognition in rats. *Phytother. Res.* 1995;9(2):110–113.
81. Bhattacharya A, Ghosal S, Bhattacharya SK. Anti-oxidant effect of Withania somnifera glycowithanolides in chronic footshock stress-induced perturbations of oxidative free radical scavenging enzymes and lipid peroxidation in rat frontal cortex and striatum. *J. Ethnopharmacol.* 2001;74(1):1–6.
82. Siddiqui AN, Siddiqui N, Khan RA, Kalam A, Jabir NR, Kamal MA, Firoz CK, Tabrez S. Neuroprotective role of steroidal sex hormones: an overview. *CNS Neurosci. Ther.* 2016 May;22(5):342–350.
83. Ambikar DB, Harle UN, Khandare RA, Bore VV, Vyawahare NS. Neuroprotective effect of hydroalcoholic extract of dried fruits of Trapa bispinosa Roxb on lipofuscinogenesis and fluorescence product in brain of D-galactose induced ageing accelerated mice. *Indian J. Exp. Biol.* 2010;48(4):378–382.

14 Safety Assessment of Nanoparticle-Based Herbal Formulations

Haris Khan, Sumiya Mustafa Alvi, Muhammad Ibrahim Khan, Hazrat Nabi, and Muhammad Imran Khan
Department of Biomedical Sciences, Pak Austria Fachhochschule: Institute of Applied Sciences and Technology, Haripur, Pakistan

14.1 INTRODUCTION

Nanotechnology has its applications in various technological and scientific fields, including pharmaceuticals and medicine. The formulations based on nanoparticles are among the promising applications of nanotechnology in biomedical research [1]. These formulations, referred to as nanomedicines, are designed to increase the efficacy and delivery of therapeutic agents. In recent years, there has been growing interest in nanoparticle-based herbal formulations, involving incorporation of nanoparticle technology into herbal medicine. This approach involves incorporating time-tested efficacy of herbal medicines with the advance delivery properties of nanoparticles, therefore resulting in formulations which offers improved targeted delivery, bioavailability, and controlled release of active compounds [2].

For thousands of years, herbal medicine has been an essential part of healthcare systems, used in various traditional practices. These natural compounds derived from plants are found to have various therapeutic properties [3]. Despite their significant potential, the applications of herbal medicines in clinical practice involves several challenges. Under physiological conditions, numerous active herbal compounds have rapid metabolism, low bioavailability, poor water solubility, and instability. Thus, these issues significantly limit the effectiveness and absorption of herbal compounds and often require higher doses to produce the desired effect [4]. However, there are numerous unwanted side effects associated with the usage of high doses of such herbal compounds.

Nanoparticles are very small particles with a size range of 10–100 Nm. Due to their large surface area and small size, nanoparticles possess unique chemical and physical properties. Nanotechnology has offered promising solutions to overcome challenges associated with using herbal medications [5]. Studies have shown that therapeutic effectiveness of herbal medicines can be increased by engineering nanoparticles as their carriers. The properties of nanoparticles make them suitable for delivering the drugs because these properties help nanoparticles interact at molecular and cellular levels in biological systems [6].

Nanoparticles can be integrated by encapsulation and conjugation to herbal compounds, which helps in the transport of herbal compounds across biological barriers, protecting them from degradation, and improving their solubility. Moreover, the controlled release of herbal compounds, and the targeted delivery of herbal compounds, can be ensured by using target specific nanoparticles [7].

14.1.1 Types of Nanoparticles Used in Herbal Formulations

The incorporation of nanoparticles in herbal medications has shown numerous therapeutic benefits, including in various neurological, cardiovascular, and infectious diseases [13]. For instance,

DOI: 10.1201/9781003544418-14

TABLE 14.1
Different Type of Nanoparticles with Their Composition and Advantages

Type	Composition	Advantages	References
1. Lipid nanoparticles	Solid lipid nanoparticles (SLNs), and nanostructured lipid carriers (NLCs)	For encapsulation by providing solid matrix to lipophilic herbal compounds	[8]
2. Liposomes	Lipid bilayer spherical	Encapsulating both hydrophobic and hydrophilic herbal compounds. Increasing the bioavailability of drugs that have poor solubility	[9]
3. Polymeric nanoparticles	Synthetic or natural polymers	Reduce dosing frequency and increase therapeutic efficacy	[10]
4. Dendrimers	Tree-like structures	Increasing stability, solubility and targeted delivery of compounds	[11]
5. Metal nanoparticles	Silver, gold, and other metal nanoparticles	Diagnostic and therapeutic applications	[12]

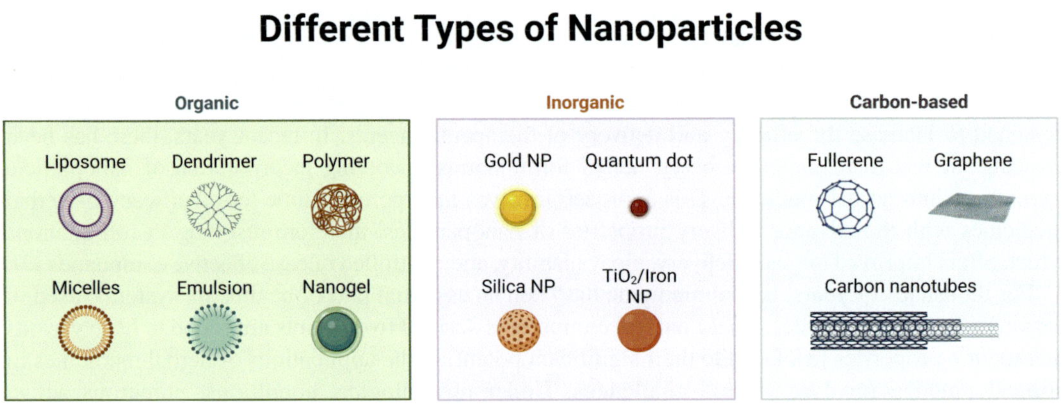

FIGURE 14.1 Types of nanoparticles used in herbal medications.

curcumin has lower bioavailability when taken orally; however, encapsulating it within nanoparticles significantly increases the stability, effectiveness, and bioavailability of curcumin [13]. While nanoparticle-based herbal formulations have numerous benefits, several challenges exist for their development and commercialization processes. These are scalability of nanoparticle production, regulatory hurdles, potential toxicity, and the necessity for thorough preclinical or clinical evaluations to be conducted. Moreover, there are some complexities that come with herbal medicines especially due to multiple active compounds they contain, thereby making standardization difficult [14] (Table 14.1 and Figure 14.1).

14.1.2 Importance of Safety Assessment in Nanotechnology-Enhanced Herbal Medicines

In the development of nanotechnology-incorporated herbal medicines, it is crucial to have safety assessment in addition to efficacy evaluation. On the other hand, their distinctive nature poses threats which should be well comprehended and managed. Nanoparticle-based products' safe profiles are determined by factors like nanoparticles' physicochemical properties, the nature of encapsulated herbal compounds and interaction between biological systems with the nanoparticles [15].

14.1.2.1 Physiochemical Characteristics and Biological Interactions

These medicines have unique physicochemical characteristics such as small size and large surface area, as well as the ability to penetrate biological membranes. These features may alter biodistribution and pharmacokinetic behaviors of entrapped herbal substances, hence causing unpredicted biological reactions. For example, nanoparticles can pass through physiological barriers such as the blood–brain barrier due to their reduced size which may enhance therapeutic efficiency but also increase the chances of inadvertent accumulation into susceptible tissues [16].

The cellular uptake or endocytosis, hydrophobicity, surface charge, and targeting of ligands all have impacts upon the nanoparticle's cellular internalization, distribution, and clearance. Therefore, it is crucial to understand these interactions to regulate and predict the biological behavior of these nanoparticles-based formulations [17].

14.1.2.2 Potential Toxicity Concerns

Herbal medicines are typically harmless. However, the incorporation of nanotechnology may raise new safety issues. Cytotoxicity, oxidative stress, and inflammation can be associated with using nanoparticles due to their high reactivity and capacity for generating ROS. Retaining these particles in the body leads to their accumulation, thereby enhancing their effects which could result in long-term poisoning [18].

Moreover, toxicity can be influenced by nanoparticle composition, such as the choice of coating materials and core. For example, silver or gold metal nanoparticles have been shown to cause cytotoxicity at higher concentrations or upon prolonged exposure. Moreover, some polymeric nanoparticles may degrade into toxic byproducts. Consequently, careful material selection and characterization is needed for nanoparticle-based formulations [19].

14.1.2.3 Regulatory and Ethical Considerations

The evaluation of safety on nanotechnology-enhanced herbal medicines is not only a scientific and technical challenge but also an ethical and regulatory issue [20]. Various regulatory agencies globally have recognized the necessity for specific guidelines for assessing the safety and efficacy of nanomedicines. These guidelines usually entail extensive preclinical and clinical studies inclusive of toxicological evaluations done to ensure that benefits from using nanoparticulate-based formulations outweigh any risks associated with it [21].

Ethically, it is crucial to ensure that these advanced formulations are developed and applied in a responsible manner. This includes open disclosure of all safety data, full consent from clinical trial participants, and equal access to the benefits of these technologies. Another crucial factor that must be considered is nanoparticles' potential environmental impact [22]. Incorporation of nanotechnology into herbal medicine has significant potential to improve therapeutic efficacy. However, a thorough safety assessment is required to ensure that these developed formulations are both human-safe and environmentally friendly. As the field evolves, development efforts and ongoing research must prioritize optimizing the safety profile of nanoparticle-based herbal products [23]. This chapter explores more thoroughly the approaches and methodologies used to assess the safety of nanotechnology-enhanced herbal medicines, emphasizing the importance of taking a balanced and cautious approach to their development and use.

14.2 PRECLINICAL SAFETY ASSESSMENT

Clinical trials are conducted after the important stage of preclinical safety assessment in the development of nanotechnology-based herbal medicines, which is aimed at identifying any potential risks associated with such formulations [24]. At this stage both in vitro and in vivo models are employed to thoroughly investigate the toxicity, pharmacokinetics, and biocompatibility of nanoparticles [25]. The components of preclinical safety assessment are explained in detail below:

14.2.1 In vitro Toxicity Testing

At this stage both in vitro and in vivo models are employed to thoroughly investigate toxicity, pharmacokinetics, and biocompatibility of nanoparticles These studies are performed on cultured cells and tissues using a variety of parameters, including morphological changes, cell population doubling time and viability. Common techniques include MTT assay and XTT assay, the LDH release assay, flow cytometry procedure, and comet assay.

Various specialized assays are commonly used in in vitro investigations to evaluate the production of reactive oxygen species (ROS), oxidative stress, and mitochondrial dysfunction. Because of their affordability, ease of use and capacity to regulate test settings, in vitro assays are an invaluable resource for preliminary toxicity screening [26].

14.2.2 In vivo Animal Studies

In vivo animal studies are vital for assessing the systemic toxicity, pharmacokinetics, and overall safety profile of herbal medicines based on nanoparticles, even while in vitro research offers important preliminary data. In these investigations, the formulations are given to mouse models of animals and their effects on different physiological systems are observed [27].

In vivo study involves acute and chronic toxicity studies, dose–response studies, and assessment of organ toxicity.

14.2.3 Evaluating the Pharmacokinetics and Biodistribution of Nanoparticles

In order to understand the absorption, distribution, metabolism, and excretion (ADME) of nanoparticles, pharmacokinetics and biodistribution studies must be considered. Overall, these investigations help us in understanding the body processes that determine what happens to nanoparticles (Figure 14.2).

Absorption: To determine the bioavailability, the uptake of nanoparticles from the site of administration, including oral, intravenous, or topical applications, is evaluated.

Distribution: Various imaging techniques, including magnetic resonance imaging (MRI), positron emission tomography (PET) and fluorescence imaging, are commonly used to determine the movement of nanoparticles in various tissues and organs. Therefore, the sites of accumulation may be recognized from this study.

Metabolism: Scientists are looking at the metabolic pathways through which nanoparticles pass through to better understand its biotransformation. This involves determining toxic metabolites.

Excretion: Elimination routes such as urine, fecal matter or other channels of nanoparticle removal are analyzed for clearance rate determination and likely deposition sites.

14.2.4 Immunogenicity and Biocompatibility Testing

The interaction of nanoparticle-based formulations with the immune system can result in immune responses that may jeopardize their safety and efficiency. Immunogenicity testing considers nanoparticles' ability to cause allergic, inflamed, or immunosuppressive reactions [28].

Cytokine release assays reveal pro-inflammatory cytokines produced by immune cells when exposed to particles, indicating inflammation. By contrast, complement activation assay is a test which determines complement system activation, leading to immune-mediated adverse effects. Moreover, in vivo immunogenicity studies involve animal models that are used to study the response of the immune system towards nanoparticles, such as potential anaphylaxis, activation of immune cells, and production of antibodies. Biocompatibility tests seek to assess how nanoparticles interact with biological tissues with a view to establishing if they have any harmful consequences [29].

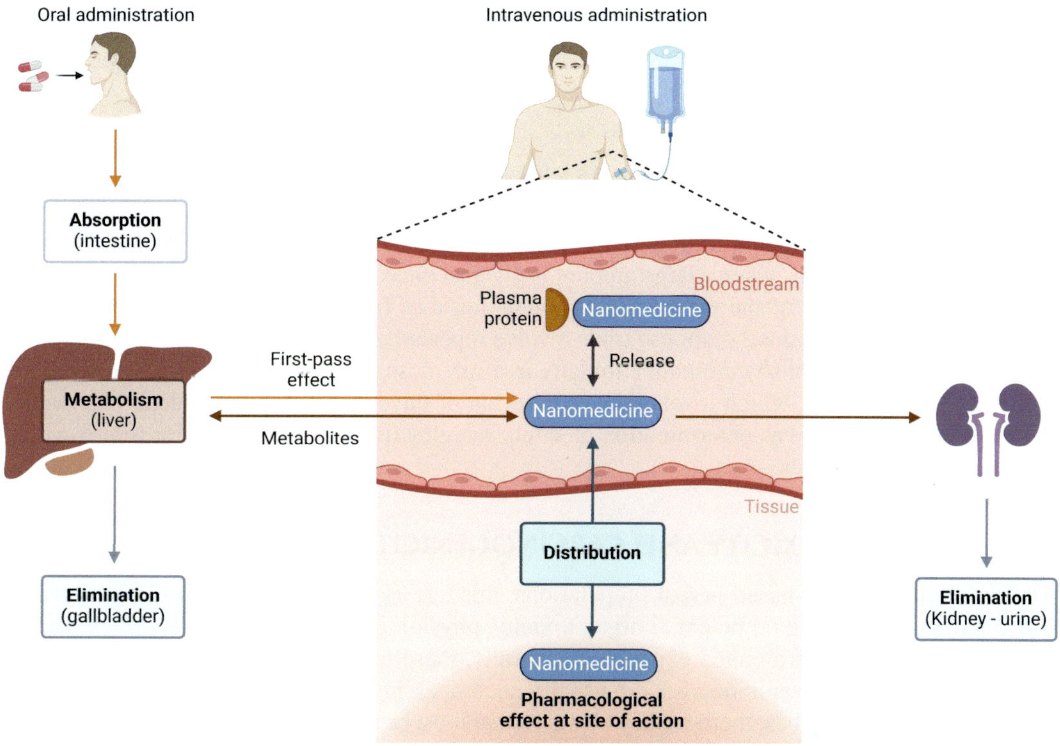

FIGURE 14.2 Pharmacokinetics and Biodistribution of Nanoparticles within the body.

14.3 TOXICOLOGICAL PROFILING

This is a complete evaluation of nanoparticle-based herbal medicines to identify, characterize, and quantify potential toxicities. This process ascertains the safety of these formulations by assessing their impacts on biological systems at cellular and organismal levels [30].

14.3.1 Identification and Characterization of Possible Toxins

Possible toxins in nanoparticle-based herbal formulations are associated with the following parameters:

14.3.1.1 Nanoparticle Components

Materials used for nanoparticle construction (core, coating, functionalization) can be toxic. For instance, metal nanoparticles may generate oxidative stress and cause cellular impairment [31].

14.3.1.2 Contaminants and Impurities

Toxicity can also result from residual solvents, reagents, or byproducts present during manufacturing [32].

14.3.1.3 Herbal Compounds

Some phytochemicals which are considered safe can show toxicity at high doses or prolonged exposure [33].

Characterization entails detailed analysis of nanoparticle physicochemical properties such as size, shape, surface charge, and composition by means of transmission electron microscopy (TEM), dynamic light scattering (DLS), and inductively coupled plasma mass spectrometry (ICP-MS).

Impurities and herbal compounds are identified and quantified using high-performance liquid chromatography (HPLC) and gas chromatography-mass spectrometry (GC-MS).

14.3.2 Dose–Response Relationships

Understanding the dose–response relationship is essential. This links the level of exposure with how degree of toxicity. No observed adverse effect level (NOAEL) and least observed adverse effect level (LOAEL) are established through dose–response studies [34]. Acute dose–response experiments assess what happens when a large amount is given at once, thus determining the median lethal dose; that is, where half of the experimental population dies from exposure to this amount [35]. Subacute and subchronic dose–response trials involve repeated administration at various doses over weeks or months to establish cumulative toxicity and safe dosage ranges [36]. Moreover, there are in vitro and in vivo tests, which involve the monitoring of physiological, biochemical, histopathological parameters as well as determination of safety margins from collected data for future dosing schedules on other animals.

14.4 CHRONIC TOXICITY AND CARCINOGENICITY STUDIES

Concerning nanoparticle-based herbal preparations, this test assesses the long-term effects, focussing in particular focusing on potential organ toxicity, physiological changes, and system reactions which may result from prolonged exposure. Generally, these tests involve repeated dosing of the formulation to animals over a long period, usually encompassing much of the animal's life [37].

In chronic toxicity studies, there are key elements that must be considered, including health parameters monitoring, specific organ toxicity assessment, biochemical, and hematological analysis.

Additional studies may be conducted to understand the mechanisms behind observed genotoxic effects. These studies may investigate oxidative stress, DNA repair mechanisms, and the activation of specific signaling pathways.

14.4.1 Genotoxicity and Mutagenicity Testing

Genotoxicity and mutagenicity tests assess a substance's ability to damage genetic material, resulting in mutations, chromosomal aberrations, or other genetic alterations. These tests are critical in determining the carcinogenicity and reproductive safety of nanoparticle-based formulations [38].

In vitro genotoxicity tests include the Ames test (bacterial mutagenicity), the comet assay (DNA strand breaks), and the micronucleus assay (chromosomal damage in cultured mammalian cells) [39]. In vivo genotoxicity tests include the bone marrow micronucleus test, which causes chromosomal damage in bone marrow cells, and the dominant lethal test, which detects mutagenic potential in germ-cells [40].

Genotoxicity and mutagenicity test results assess cancer and genetic disease risk, which helps with overall toxicological assessment and regulatory safety requirements.Genotoxicity and mutagenicity tests yield critical information for determining the risk of cancer and genetic diseases. They are an important part of the overall toxicological assessment and are mandated by regulatory agencies to ensure the safety of new formulations [40]. Overall, toxicological profiling is an important aspect for developing safe nanoparticle-based herbal medicines.

14.5 CLINICAL SAFETY ASSESSMENT

Evaluation of the clinical safety for a drug is a very important stage before it can be introduced in the market for human consumption. It includes phases of clinical trials, adverse effects monitoring, long-term safety evaluation, and drug surveillance. Different methods and safety protocols must be followed in clinical safety trials to ensure the maximum safety and efficacy of the drug [41]. In this section, we will discuss the clinical safety assessment procedures and protocols for nanoparticle-based herbal formulation.

14.5.1 Phases of Clinical Trials for Nanoparticle-Based Herbal Formulation

Clinical trials give us the information and data for the safety, efficacy, and effectiveness of treatments. They are categorized by their purpose, different phases, and design of trials. Clarifying the study question and population, identifying treatment and comparison groups, selecting methods for treatment group allocation, clarifying primary and secondary outcomes, power analyses, analytic plans, and reporting of results are all important parts of a clinical trial design. Nanoparticle-based herbal formulations involve five phases [42]. These phases are explained in Table 14.2.

14.5.2 Monitoring Adverse Effects and Long-Term Safety in Human Subjects

Monitoring the adverse effects and long-term safety of a drug in humans is very crucial and a step-by-step process [46]. Several methodologies and procedures are designed to collect and analyze data. Here, we will discuss this process in detail.

14.5.2.1 Initial Reporting Systems

Initially, any adverse effect is reported by physicians, pharmacists, or other healthcare professionals who play an important role in the identification and reporting of adverse effects associated with herbal drugs [46]. Patients using herbal medicines can also report any adverse effects they experience on government or healthcare authorities' given platforms.

14.5.2.2 Clinical Monitoring

After an herbal drug is distributed in the market, post-approval clinical trials (Phase 4) are conducted to gather more data on its safety profile. In these trials, we can identify less common adverse effects that may not have been seen in earlier phase trials due to smaller sample sizes [45]. In these studies, a group of patients are monitored who are taking the herbal drug and their health outcomes are compared to a control group not using the drug [47].

TABLE 14.2
Description for Clinical Trial Phases

Phases for Clinical Trials	Purpose	Number and Type of Patients
Phase 0	To gather primary data about the new compound and how it affects the subjects.	10–15 healthy volunteers [43]
Phase 1	To assess and evaluate the short-term safe dosage value, tolerability, clinical pharmacology, pharmacokinetics, and pharmacodynamics of the new compound.	50–100 healthy volunteers or patients [43]
Phase 2	To check the effects and efficacy of the drug on different subjects and determine the effective and safe dosage ranges.	100–300 patients with the targeted disease [43]
Phase 3	To check dosage impact, efficacy, and any adverse effects and ensure its safety on a larger scale [44].	Up to 1000 patients with the targeted disease [43]
Phase 4	To check the long-term positive and adverse effects and other indications are monitored with the passage of time [45].	Thousand or million patients with the targeted disease [43]

14.5.2.3 Pharmacovigilance Networks

Several healthcare organizations initiate programs for collecting data monitoring the reports of adverse events [48]. These include, for example, the WHO's International Drug Monitoring Program, which collects and analyzes data on adverse effects reported globally. These databases help us identify different patterns and signals that might indicate a safety issue with a particular herbal drug.

14.5.2.4 Regular Safety Updates

It is necessary for manufacturers and producers of different herbal drugs to regularly submit safety updates about products to the health regulatory authorities. These updates include all known data on adverse effects and new safety information gathered since the drug was last reviewed [49]. There are also risk management plans (RMP) that clearly state how the manufacturer intends to monitor the risks associated with the herbal drug, including strategies made for long-term safety monitoring [50].

14.5.2.5 Post-Marketing Studies

These studies are used to observe the effects of the herbal drug in real-world scenarios over a prolonged period. It includes large populations and is beneficial for the detection of rare or long-term adverse effects [51]. There are different patient registries designed for specific diseases or treatments. These registries can provide real-time data on the safety and effectiveness of herbal drugs used to treat those conditions [52].

14.5.2.6 Pharmacogenomics Studies

In these studies, scientists analyze how genetic differences among patients affect their response to herbal drugs. This study can help us in the identification of those patients who are at higher risk for adverse effects. In this way, more personalized and safer use of herbal medicines can be given to them [53].

14.5.3 Post-Market Surveillance and Pharmacovigilance

As discussed above in sections 5.2.5 and 5.2.6, these studies are essential for the long-term evaluation of drug's adverse effects and to analyze how different responses of herbal drugs are generated among different patients because of the individuals' genetic dissimilarities.

14.5.3.1 Real-World Evidence Collection

Data from electronic health records (EHR) is analyzed to check and monitor the effects of herbal drugs on a large scale. This includes the tracking of patient outcomes, adverse effects, and their interactions with other medications.

14.5.3.2 Active Surveillance Programs

Specific populations known to use certain herbal drugs is surveyed regularly to gather detailed information of their experiences and any adverse effects. Healthcare facilities can be set up as sites to monitor and report on the use and effects of herbal drugs [54].

14.5.3.3 Signal Detection

Advanced algorithms and data mining techniques should be used to analyze large datasets from various sources to detect signals indicating potential adverse effects.

14.5.3.4 Risk Communication

When a new risk is detected, regulatory authorities issue safety alerts to healthcare providers and the public, advising on the risks and recommending actions to mitigate them.

14.5.3.5 Regulatory Actions

Based on new safety information, regulatory authorities may require updates on the labeling of herbal drugs to include warnings about the potential adverse effects. In cases in which the risks of an herbal drug outweigh its benefits, regulatory authorities announce the withdrawal of drug from the market to protect public health [55].

14.6 ANALYTICAL TECHNIQUES FOR SAFETY ASSESSMENT

Analytical techniques give us detailed insights into the characterization, behavior, aggregation, distribution, adsorption, and degradation of nanoparticles in biological environments. Thus, analytical techniques are essential for the assessment of safe nanoparticle-based herbal formulations [56]. To assess the safety of nanoparticle-based herbal formulations, advanced imaging and spectroscopy techniques are used such as TEM or SEM. For nanoparticle tracking and quantification, methods such as DLS is used and for surface characterization and stability analysis techniques like X-ray Photoelectron Spectroscopy (XPS) were employed (Figure 14.3).

14.6.1 Advanced Imaging and Spectroscopy Methods

It is essential to characterize the specialized nanoparticles to access the nano-bio interactions and for this advanced imaging and spectroscopic techniques are used. These techniques offer comprehensive knowledge of the structural, functional and compositional characteristics of nanoparticles which assists us in better understanding of their behavior, interaction, and possible toxicities within the biological systems [57]. It is necessary to establish different analytical techniques for the identifying of nanomaterials in accordance with the recommendations of European Commission.

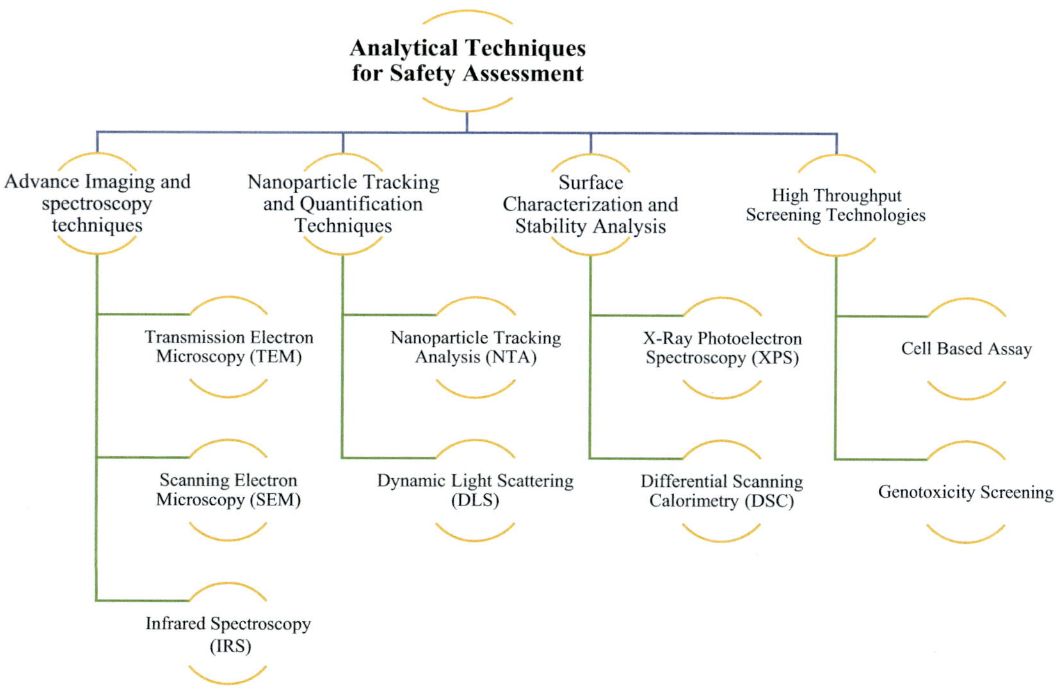

FIGURE 14.3 Various analytical techniques used for safety assessment of nanoparticles.

14.6.1.1 Transmission Electron Microscopy (TEM)

Transmission electron microscopy (TEM) allows for the visualization of nanoparticles at the atomic level in order to determine whether the nanoparticles are located in a cell or tissue, which is crucial for the determination of their possible toxicity and biodistribution due to which it can provide the most detailed knowledge about the in vitro nanoparticle uptake and localization [56]. However, when employing this method for the nanoparticle absorption studies, the analytical efficiency for the biological sample preparation and image processing is severely restricted by the length of time required [56].

14.6.1.2 Scanning Electron Microscopy (SEM)

Scanning electron microscopy (SEM) is a technique used for surface observations as well as to describe materials at the nanoscale. Lu and colleagues utilized both SEM and TEM to describe the DNA-mediated Ag growth on Au nanocrystals and the impact of the DNA capping ligand on the morphology of Au-Ag core shell nanoparticles [58]. Accordingly, we can state that SEM clarifies the interaction of biological membranes with nanoparticles and the cellular response that is generated by this specific interaction.

14.6.1.3 Infrared Spectroscopy (IRS)

For molecular and material characterization, the most widely utilized technique is infrared spectroscopy (IRS) [57]. Fourier transform infrared spectroscopy (FTIS) is suitable for noninvasive live-cell observation of physiological reactions that are happening with the potentially harmful nanosized particulates or other factors. By employing this unique approach, live-cell cultures can be monitored without causing any disruptions and toxicity testing can also be conducted swiftly with the help of this technique [59].

14.6.2 Nanoparticle Tracking and Quantification

The evaluation of dispersion, bioavailability, and potential cytotoxic consequences of nanoparticles in biological environments require precise tracking and quantification of these particles. The assessment of nanoparticle uptake quantification typically involves analytical techniques, including cell isolation. The current work uses a red-luminescent ruthenium transition metal complex coated with gold nanoparticles to measure and monitor particles uptake and localization [60].

14.6.2.1 Nanoparticle Tracking Analysis (NTA)

Nanoparticle tracking analysis (NTA) is a technique based on two important fundamental concepts, i.e. light scattering and Brownian motion. The camera records for the nanoparticles as a point scatter in solution of a plane at a 90° angle enabling visualization. After the identification of each particle, the hydrodynamic diameter is determined and tracked separately [61]. This method is particularly beneficial for examining the stability and aggregation behavior of nanoparticle compositions [62].

14.6.2.2 Dynamic Light Scattering (DLS)

The primary method that is used for the measurement of size distribution in a suspension of nanoparticles is dynamic light scattering (DLS) [63]. For the quantification of nanoparticles, light scattering provides an easy-to-understand, highly sensitive, and generally selective analytical approach without the need for expensive or specialized test equipment [64].

14.6.3 Surface Characterization and Stability Analysis

To ensure the efficacy and safety of nanoparticles in clinical applications, it is important to understand the surface properties and stability of nanoparticles. Surface characterization contains the information regarding interactions, functional groups, and chemical properties of nanoparticles while stability analysis analyzes how they behave in different physiological environments. Stability analysis ensures that nanoparticles maintain their integrity and efficacy under various conditions [65]. This analysis offers comprehensive data which allows us to develop a safe and effective nanoparticle herbal based formulation.

14.6.3.1 X-Ray Photoelectron Spectroscopy (XPS)

X-ray photoelectron spectroscopy (XPS) is used to discover the shell thickness of core-shell nanomaterials as well as to describe the ligands on the nanoparticles, which is essential for understanding surface modifications and functionalization that alters the interactions between nanoparticle and biological systems [66]. XPS has become one of the most popular methods for surface analysis.

14.6.3.2 Differential Scanning Calorimetry (DSC)

Heat stability and phase behavior of nanoparticle herbal formulation can be evaluated using an efficient analytical method known as differential scanning calorimetry (DSC). It gives the details on thermal stability, crystallization, and melting behavior of the nanoparticles by measuring the temperature fluctuations, i.e. heat flow related to phase transitions in the material [67].

14.6.4 High-Throughput Screening Technologies

A variety of automated approaches known as high-throughput screening (HTS) enable scientists to efficiently conduct numerous chemical, genetic, or pharmacological experiments. It can be used to screen the possible cytotoxicity, genotoxicity, and other harmful biological effects in conjunction with the safety assessment. The main objective of HTS is to discover members of a chemical library which react with the specified system in a particular manner [68]. These techniques are crucial for the evaluation of nanoparticle herbal formulation safety profiles.

14.6.4.1 Cell-Based Assay

For toxicological assessment, different cell-based assays are employed. These assays measure the cell viability, proliferation, and apoptosis to assess the cytotoxicity of nanoparticles' compositions [56]. The lactate dehydrogenase (LDH) assay is commonly used to measure the leaks of functioning enzymes into the cell medium or to track the uptake of supravital dyes such as propidium iodide (PI), Trypan Blue (TB), and Neutral Red (NR) [69]. In the proliferation assay, the proportion of metabolically active cells is determined and the production of formazan-based dyes are measured by the optical absorbance. The most widely used of these tetrazolium salts is to evaluate the in vitro toxicity of a wide range of nanostructures [56].

14.6.4.2 Genotoxicity Screening

Unrepaired single and double strands DNA breaks could be an outcome of being exposed to nanoparticles through apoptosis, alterations in an oxidative environment, or the physical contact between the DNA and nanoparticles. This physical interaction is known as genotoxicity (56). Genotoxic substances can harm DNA, which may result in genetic changes and raise the risk of tumor development. Among the numerous in vitro tests which have been used for the evaluation of DNA damage, which is caused by chemicals are the comet assay, the micronucleus test, the Ames test, and the mouse lymphoma test [70]. High content screening (HCS) assays offer greater understanding for the mode of action of the genotoxic substances due to their capacity to measure several parameters (such as micronucleus and γH2AX) (71).

14.7 REGULATORY FRAMEWORKS AND GUIDELINES

Regulatory frameworks for any medical product provide a backbone and a clear pathway regarding the use of the relevant product. These frameworks are basically guiding roadmaps which direct the user to select an optimal usage route which helps in mitigating the unnecessary risks. Due to nanomedicines being an emerging class, high legislative frameworks are required for them and undergo additional quality and safety assessments [72]. To meet up with the updated regulatory criteria, the International Pharmaceutical Regulators Forum (IPRF) has formed a Nanomedicines Working Group, which has been given the role of tackling the questions related to new and emerging regulatory needs. This working group is currently acting as a platform for sharing unclassified information related to the usage of nanomaterials in drug products. Moreover, the group works as a supporting body towards the achievement of regulatory harmonization and the European Medicines Agency (EMA) is the chairing body for this IPRF group [73].

14.7.1 International and National Regulatory Frameworks

To address the concerns related to the utilization of nanoparticles in different agriculture-related products, various regulatory bodies have designed some guidelines, so that the nanotechnology could be administered safely. Some of these important regulatory necessities are described below:

14.7.1.1 Regulation Management

Many countries have formed regulatory sectors for implementing nanotechnology. For example, the Environmental Protection Agency (EPA) is the functional regulatory body for overseeing regulatory parameters in the United States, while the European Food Safety Authority (EFSA) oversees these regulations within the European Union [74].

14.7.1.2 Risk Analysis

The analysis of the toxicity of nanoformulations and their unseen negative impacts on the environment and on human health are among the tasks for which regulatory bodies are responsible [74].

14.7.1.3 Labelling and Informed Consent

In order to help the public in making relevant decisions regarding the purchase and usage of nanoproducts, proper identification and tagging is required, which is a crucial responsibility of regulatory bodies [74].

14.7.1.4 International Standards

Various international standards have been developed by the International Organization of Standardization (ISO) related to the safety of nanotech utilization in agro-sciences. Some standards, such as ISO/TS 800041, contain a large amount of information regarding nanomaterials [74].

14.7.1.5 Regulation in Research and Development

Particularly in the research sector, regulatory bodies need combined efforts from researchers, manufacturers, and consumers so that risk monitoring could become effective, thereby ensuring the proper safety and effectiveness of products containing nanoparticles [75].

14.7.2 Risk Assessment Models and Safety Thresholds

A wide range of assays are used to evaluate nanomaterial-related cytotoxic effects and receive help from different chemical reagents so that the cellular metabolic conditions could be evaluated. But these assays also face some hurdles, as, for example, when these nanoparticles interact with the cell culture media. The outcome is the generation of false positive toxic effects [76]. Some of the novel techniques/assays for the purpose of nanomaterials' toxicity analysis are descripted in this section.

14.7.2.1 Invitro Toxicity Assay

This is a novel, non-invasive, and in-vitro assay that aids in visualizing cell growth events. Parameters like cell-proliferation kinetics, cell growth, real-time tissue cells, and morphological effects can be studied through this assay, and it also utilizes label-free techniques, which is why chemicals, and dyes are avoided [77]. This assay has the benefit of excluding false-positive and false-negative outcomes as these outcomes present a major problem in other cytotoxic assays. It also utilizes an electrical-impedance tool which analyses morphological and proliferation changes [78].

14.7.2.2 Green Algorithms

Complex and time-consuming tasks can be performed with ease because of the involvement of various machine learning and artificial intelligence models. One particular algorithm, which goes by the name of 'Hartung', which was recently discovered, has greatly aided in the field of toxicology. It supported 3R principles, which is why it was believed that the software could replace in vivo analytical techniques. The algorithm designs a chemical map which harbours the data of hundreds of chemicals from a variety of databases. The toxicity is predicted by the comparison and substitution of different moieties extracted from thousands of nano chemistry databases [79].

14.7.2.3 Nanoprobes for Measuring ROS

Nanoprobes are manufactured by utilizing a dye that is enveloped in a nanoparticle delivery system. Because of this, the flaws of traditional fluorescent dyes can be minimized [80]. As the envelopment of probes is done with a chemically neutral material matrix, such as PVC, gold colloid or polyacrylamide, which produces a shielding effect against non-specific interactions, this property results in no cytotoxic effects. Moreover, because of their miniature size, conventional methods can be utilized to inject them into the cells. These methods include lipofection, microinjection and TAT-protein delivery [80]. The first nanoprobe for biomedical use was termed as PEBBLE abbreviated as probe encapsulated by biologically localized embedding and it came with the diameter of 20–600 nm [81].

14.8 RISK MITIGATION STRATEGIES

The pace of nanotechnology integration in medicine was very rapid and it quickly transitioned from basic-level research and experimentation to advanced clinical trials. Many nanomedicine formulations have found their way onto the market. A recent study mentions 247 nanomedicine products; some of them have been approved while some are almost ready for in-human use. The most important of these are the 'first in human trials' in nanotechnology medical applications because they have the highest degree of uncertainty in them during clinical trials [82].

The important thing in 'first-in-human' (FIH) nanomedicine trials is the mention of the purpose of study being conducted in the consent form. Primarily, the main objective of these FIH trials is safety; whenever a new drug is developed, the classical trial methods usually do not serve the particular purpose. This is why the information added in the consent form should focus on the parameters related to risk identification and safety testing [83]. Because of the high degree of unfamiliarity regarding the true advantages and risks of nanoparticles, many obstacles were seen through the transitional pathway of their development. This is the reason why risk assessment and management, as well as communication, became highly challenging problems in nanomedicine clinical research [84].

An analytical research infrastructure by the name of QualityNano was formed for NM safety assessment, characterization, and initiation of reliable or reproducible approaches to nanometrology. This was the first European initiative, finishing in 2015. It included the SOPs' development procedure to address the possible risks of nanomaterials when they were being administered to the living systems. Assay reproducibility, the usage of correct positive and negative controls, and controls overdose delivery were the areas of focus for this research infrastructure [85].

14.8.1 Designing Safer Nanoparticle-Based Formulations

Nanomedicine development is a difficult and complex mechanism involving the careful consideration of several parameters like chemistry details, manufacturing procedures, economic, and

TABLE 14.3
The Three Common Techniques Utilized for Designing Herbal Nanoformulations

Common Techniques Used in the Formulation of Nanomedicines		
High-Pressure Homogenization	**Solvent Emulsification–Evaporation**	**Solvent Emulsification–Diffusion**
This technique utilizes the pressure of approx. 100–200 bars so that it could push the coarse emulsion harbouring the drug through a narrow path of few microns with the help of a microfluidizer. [91]	This technique utilizes an organic solvent that is not mixable with water, so that it could dissolve hydrophobic drug components. After this emulsification is done in an aqueous phase through the utilization of a homogenizer, the organic solvent is evaporated through stirring and lowered pressure resulting in the formation of solid lipid nanoparticles. [92]	Mutual saturation of solvent and water is done first in this process; the drug and lipid are then added in the aqueous phase, which is saturated with the solvent of interest. This is then followed by the emulsification of saturated solvent with water by utilizing a stirrer. [93]

regulatory aspects. These parameters are considered as greatest challenges in the production and scaling up of nanomedicine-based formulations [86]. To obtain optimal characterization of nanomedicines, different techniques, such as high-performance liquid chromatography, nucleic magnetic resonance, and mass spectrometry, are also crucial. It was also further reported that nanomedicine characterization should include the particle size, zeta potential, percentage purity, viscosity, and pH of different components [87].

There are a wide range of techniques which are being utilized to make formulations of different nano-phytomedicines. These different techniques include the salting-out method, the co-precipitation method, nanoprecipitation, the supercritical fluid method and complex coacervation. But the most utilized techniques are the high-pressure homogenization (HPH) method, the solvent emulsification–evaporation technique, and the solvent emulsification–diffusion technique (88–90). These three techniques are briefly explained in tabulated form in Table 14.3, which focuses mainly on their involved mechanisms.

14.8.2 Controlled Release Systems and Targeted Delivery

There have been many developments in recent decades regarding the manufacturing of drugs and advances are also being made in nanoscience to include nanoscale materials so that the problematic areas of formulation development could be tackled, because until recently the nanoscale materials incorporation is only being done in the field of cosmetics [94]. The process of the application of nanotechnology on the plant extracts has been recorded in various papers because there are many benefits of nanostructured systems to boost the actions of plant extracts, like enhancing the release of active components and decreasing the side effects by reducing the needed dose [95, 96].

To elevate the absorption parameter of the active constituents in some formulations, Bhattacharya and Ghosh utilized lipid-based systems in which they integrated ginseng and green tea extracts [97]. The manufacturing of liposomes with *Artemisia arborescens* L. (Asteraceae) was also reported by Sinico et al, and it was noted that these systems help the active components extracted from the mentioned plant to effectively cross the cytoplasmic viral barrier [98]. Rajendran et al. manufactured nanoparticles by utilizing the methanolic extract from the plant *Ocimum sanctum* L. (Lamiaceae). After studying the antimicrobial activity, the researchers reported that the extract that was encapsulated gave better results than the free-form preparation. They studied this antimicrobial activity on *Escherichia coli*, *Bacillus subtilis*, *Staphylococcus aureus*, and *Pseudomonas aeruginosa* (96).

Additionally, the integration of nanoscale particles can reduce the problems associated with the use of medicinal plants [99]. Through the use of different nanotechnology-based drug delivery systems, one can obtain the formulation's required properties. These drug delivery systems include polymeric nanoparticles, liquid crystal (LC) systems, liposomes, microemulsions and solid lipid nanoparticles (SLNs) [100]. As these, two systems, LC system and SLNs, are described above.

14.8.3 Reducing Off-Target Effects and Enhancing Selectivity

Some nanoparticles are dedicated towards genome editing, but delivering these genome editing systems is difficult because of the sensitive cargo they harbour and the intracellular and extracellular biological barriers they must cross to reach the target cells genome, which is why these systems become multicomponent. Both lipid-based and polymer-based nanoparticles were recorded as being successful in delivering the nucleic acids in vivo and are currently under clinical development [101, 102]. One such example of a LNP siRNA drug, which has been given by the name of Onpattro (patisiran), has recently approved by the FDA for the treatment of amyloidosis [103]. In the case of genome editing, NPs show less toxic and immunogenic effects than viral vectors [104].

14.8.3.1 Nanoparticle-Based Systems for Intracellular Targeting

Electrostatic complexation of nucleic acids in combination with cationic substances is the basis for the formulation of most NP-based systems designed for genome editing, as these systems are delivered intracellularly by utilizing mechanisms such as phagocytosis and receptor-mediated endocytosis [105]. Cationic substances help to give responsive properties to NPs, which are helpful for endosomal escape like for delivering nucleic acids, lipofectamine (a common transfection reagent) is utilized which comes under ionizable lipid-like materials [106, 107]. These systems are very beneficial towards the intracellular environment and can be stabilized to ensure endocytic uptake [108]. As DNA is the main target of interest for gene editing, two different approaches are employed to target the nucleus: the first approach is to reduce the size of the particles so that they can easily enter the nuclear pore; the second approach is to make the particles functional so that they can be utilized after endosomal escape [109]. These particles can also be utilized to target some specific intracellular environments or organelles such as mitochondria [110]. There are some nanoparticles designed specifically for this scenario which have enhanced mucus-penetrating properties, and which can be administered through oral routes. Improved penetration was demonstrated by those NPs that were smaller than the mucus mesh pores while some systems which employ hydrophilic coatings such as polyethylene oxide or PEG have also demonstrated positive penetration [111]. It was also demonstrated that PEGylation improved penetration the capability of the particles through cystic fibrosis mucus [112].

14.8.4 Engineering Biodegradable and Biocompatible Nanoparticles

Polymeric nanoparticles are one of the most dependable nanocarriers to be under production and they fall into the category of nanostructured systems. They are designed with a diameter range of between 10 and 1000 nm. Synthetic biodegradable polymers are utilized to construct these nanoparticles, of which the most often cited is poly caprolactone (PCL). Because of some of its positive attributes, and because it has been approved, PCL is reported to be the polymer of choice. These attributes include biocompatibility and low commercial costs [113, 114].

In a study, the root extract of a plant, *Clerodendrum infortunatum* L, was used to manufacture herbal nanoparticles, which had been formulated to treat a notorious metabolic disorder, hypercholesterolemia, which is a major cause of hypertension and cardiovascular diseases. The surface morphology, surface charge, entrapment efficiency, and drug-loading capacity of these biodegradable nanoparticles were analyzed and the values recorded: 608 nm was the recorded particle size, -30.0 mV was the zeta-potential, while the entrapment efficiency and the drug-loading capacity were 98.40% and 32.8%, respectively [115]. Table 14.4 shows the categorization of some biodegradable nanoparticles along with their advantages and disadvantages.

TABLE 14.4
The Categorization of Some Biodegradable Nanoparticles Along with Their Advantages and Disadvantages [116]

Biodegradable Nanoparticles		Positive Attributes	Negative Attributes
Polymer-based particles	Chitosan nanoparticles	Higher absorbability, easy degradation, less toxicity and optimal moisture retention	Degradation rate effected by alteration in environmental pH and long-term stability is minimal
	PLA micelles	Optimal pharmacokinetic and good hydrophobicity	Drug-loading capacity and encapsulation capacity are both minimal
	PLGA micelles	Modifiable particles and changeable degradation rates	Acidic nature of PLGA is not optimal for certain drugs and biodistribution gets changed easily
	PCL nanoparticles	No production of acidic byproducts and slow degradation rate	Limiting factor is hydrophobicity
Lipid-based particles	Liposomes	Can be made into different forms, surface modification is also easy and greatly defends the encapsulated drugs from early inactivation.	Manufacturing method is very complex, and stability is minimal.

14.9 CASE STUDIES OF SAFETY ASSESSMENT

To guarantee that nanoparticle-based herbal formulations are both safe and effective for human consumption, it is essential to conduct a safety assessment [117]. Through the analysis of several case studies, we can develop a deeper understanding of the approaches utilized, the obstacles faced, and the accomplishments made in guaranteeing the security of these formulations [118]. This section explores a number of successful cases in which herbal formulations which were based on nanoparticles underwent comprehensive safety evaluations which indicated that they could serve as secure treatment choices.

14.9.1 Successful Examples of Safe Nanoparticle-based Herbal Formulations

The incorporation of nanotechnology into herbal medicine is the creation of herbal formulations based on nanoparticles that maintain safety while simultaneously enhancing therapeutic efficacy. The unique characteristics of nanoparticles are that they increase the bioavailability and delivery of herbal active compounds when utilized in these formulations [119]. This section discusses several case studies which show effective and safe herbal formulations based on nanoparticles.

14.9.1.1 Curcumin-Loaded Nanoparticles

The use of curcumin-loaded nanoparticles is one prominent example of a successful herbal formulation based on nanoparticles [120]. Curcumin, which is derived from turmeric, has been shown to have anti-inflammatory and anti-cancer properties, but it is difficult to get it absorbed into the body [121]. Researchers have created curcumin nanoparticles to improve the drug's effectiveness and absorption. The formulation was well tolerated in Phase 1 clinical trials, and no serious side effects were reported. Curcumin nanoparticles could provide therapeutic benefits at lower doses than conventional formulations, thereby reducing potential side effects, as demonstrated by further Phase 2 and 3 trials [122].

14.9.1.2 Green Tea Polyphenol (EGCG) Nanoparticles

Another successful example is the creation of the green tea polyphenol (EGCG) nanoparticles. Like curcumin, EGCG is well-known for its limited bioavailability, while it does possess strong anticancer and antioxidant qualities. In vitro cytotoxicity tests and in vivo animal studies, as well as preclinical safety assessments, have demonstrated that EGCG nanoparticles are safe and well-tolerated. In addition, the nanoparticle encapsulation reduces the gastrointestinal discomfort which is frequently associated with the high doses of EGCG [123]. These trials not only demonstrate the formulation's efficacy but also highlight its superior safety profile in comparison to conventional green tea extracts [124].

14.9.2 Lessons Learned from Safety Failures and Recalls

Some formulations have experienced safety concerns despite their thorough testing, which has resulted in recalls and failures. The examination of these cases emphasizes the significance of comprehensive safety evaluations, and it also provides valuable insights into potential dangers. In order to enhance the future formulations as well as the regulatory methods, this section examines the lessons learned from notable safety failures and recalls.

One well-known case is that of an arthritis-treating herbal extract formulation based on nanoparticles. In this instance silver nanoparticles were added into the formulation to strengthen the extract's anti-inflammatory qualities. Positive outcomes which emerged from the initial preclinical studies related to there being significant anti-inflammatory effects with no apparent toxicity in animal models. However, a wide number of participants in Phase 2 clinical studies reported adverse effects, such as skin discoloration, digestive issues and signs for systemic toxicity. Upon investigation, the scientists discovered that the body gradually became overloaded by the silver nanoparticles, resulting in argyria (a condition caused by an accumulation of silver in the body) as well as other harmful consequences. The significance of this case demonstrates that an understanding of the biodistribution and excretion pathways of nanoparticles in the human body is essential. It also underlines the necessity for long-term toxicity studies.

14.9.3 Comparative Safety Profiles of Conventional vs. Nanoparticle-Enhanced Formulations

Nanotechnology has been incorporated into herbal medicine to create formulations with improved nanoparticles which provide better bioavailability and targeted delivery of active ingredients [125]. In contrast to the conventional herbal formulations, these advancements bring their own set of unique safety considerations. To understand the risks as well as the benefits of both conventional and nanoparticle-enhanced formulations, a comparison of their safety profiles is necessary [124] (Table 14.5).

Despite the fact that nanoparticle-enhanced herbal formulations provide significant therapeutic advantages through enhanced bioavailability and targeted delivery, these formulations also bring

TABLE 14.5
Difference between Conventional and Nanoparticle-Based Formulations

Parameters	Conventional Formulations	Nanoparticle-Enhanced Formulations
Bioavailability	Moderate or low solubility [121]	High solubility because of enhanced absorption and targeted delivery[120]
Dosage	High dose to achieve therapeutic effect [125]	Low doses are effective [120]
Side effects	Related to high doses and systemic distribution [121]	Related to nanoparticle specific effect to the targeted side [127]
Interactions	Well documented and understood [126]	Less known interactions [128]
Regulatory Considerations	Established guidelines and pathways [128]	Emerging guidelines, require more thorough safety assessment [129]

their own unique safety issues that require careful management [124]. For the safe and effective use of the nanoparticle-enhanced herbal medicines, it is essential to comprehend these differences to ensure that the benefits might outweigh the risks over time [124].

14.10 ETHICAL CONSIDERATIONS

Research ethics is a pivotal part of any process and should not be ignored. Before the utilization of nanomedicine products in diagnosis, prevention or disease treatment, all these nanomaterials first undergo substantial preclinical and clinical testing. Different toxicological, pharmacological, and immunological properties are already under examination with regard to these nanoparticles. Additionally, risk assessment programmes are being initiated by institutes such as the US Environmental Protection Agency, the National Institute of Environmental Health Sciences, the National Science Foundation, and the National Institute of Occupational Safety and Health [130]. Ethical guidelines are necessary because nanoparticles can translocate from the site of exposure to other parts of the body; they can easily cross the cell membranes, including the extremely tight junctions of the blood–brain barrier [131].

Ethics guidelines for natural nanoparticles is one thing. However, ther is also a pressing need to establish guidelines related to manufactured nanomaterials such as plant-derived fullerenes and C60 carbon shells as these manufactured entities pose significant risk to human health. This is because while the human body may possess modified and evolved biological mechanisms to deal with natural nanoparticles, they lack any such advanced biological defense mechanism against synthetic ones [84, 132].

14.10.1 Ethical Issues in Nanotoxicology Research

Before considering the ethical concerns related to nanotoxicological research, let's go through a structural layout to better understand, 'How does the toxicological profiling of a particular nanoparticle begin?

Ethically, there are a range of opinions among different regulatory bodies regarding how to deal with nanomaterials. On one side, there are some regulatory bodies, such as the UK's Royal Society and Royal Academy of Engineering, who proposed in 2004 that ''if products harbour nanomaterials as a form of ingredients, then they must undergo through comprehensive safety trials and should only be marketed for use when approved through the scientific advisory body'. Additionally, they also recommended that these nanomaterials should be regarded as toxic and hazardous and hence their neutralization should be carried out in waste streams generated by different laboratories and factories [133].

On the other side, ther are regulatory bodies such as the Pacific Research Institute, who claim that nanotechnology needs no new regulations as the many advantages and innovations this field will bring in the areas of food sciences, medicine, and energy technologies would be either slowed down or halted completely. Such bodies argue that the only one thing needed is self-regulatory measures. Many industry-funded institutes claim that the risks of stopping nanotechnology development are far greater than those which are potentially posed by the technology. Several scientists also argue that at present knowledge about nanotechnology is limited, meaning that nano-related regulatory decisions might be made prematurely. They argue instead that any regulation of the area should focus on actual scientific evidence related to any toxicity and exposure risks [134], which are the main objectives of nanotoxicology.

14.10.2 Informed Consent and Patient Safety in Clinical Trials

Several ethical conditions must be met in order to initiate a defensive approach towards nanotech regulation. These conditions are crucial for devising ethical decisions covering the risks accompanied by nanotechnology and nanotoxicology [135]. Disclosure, competence, understanding, and voluntariness are the four conditions described by traditional ethical theories and are pivotal to ensuring free and informed consent. As designed in the system of biomedical ethics, (i) the associated risk disclosure should be made clear from the regulatory bodies, (ii) the potentials patients must understand the risk they are in, and (iii) after accepting it voluntarily, the victims should be able enough to withhold their consent [136].

These three rights are also termed as, "The Rights to Know" as these enable the people to know about the things which can harm them, and highlighting these rights is important for achieving the goal of informed consent. If a candidate knows about some risks which were not disclosed to them then they can't give consent to them. According to a US survey conducted in 2006, 42% of the Americans weren't even familiar with the term nanotechnology, and only 20% of them had some awareness. These findings are matters of huge concern. The older public masses who were using nanoparticles within consumer products, such as cosmetics and skin care products, were also ignorant with regard to this technology, which greatly violates the condition of disclosure. Once these conditions are fully covered, both the nanotechnology and the ethical understandings related to it will progress healthily [137].

14.11 CONCLUSION

In this chapter we discussed the detailed safety assessment of nanoparticle-based herbal formulations, which highlights the potential to transform herbal therapy into nanoparticle-enhanced formulations by improved targeted delivery, bioavailability, and efficacy. However, these advantages come with unique safety issues that call for careful evaluation and regulatory oversight [125]. Firstly, the significance of safety assessment is discussed in these formulations which highlights the potential benefits and risks of these advanced formulations [124]. Preclinical safety assessment involves in vitro toxicity testing, in vivo animal studies, biodistribution, immunogenicity, biocompatibility, and pharmacokinetics. Before commencing the clinical trials, these studies provide a comprehensive understanding of how nanoparticles interact with biological systems [139]. The goal of toxicological profiling is to gain a deeper understanding of the dose-response relationships, to identify and characterize potentially harmful compounds, and to carry out tests for mutagenicity, carcinogenicity, genotoxicity, and long-term toxicity. This analysis is important in assessing the long-term safety and possible harmful effects [140]. Clinical safety assessment is crucial to evaluate the side effects as well as to guarantee the long-term safety in human subjects [41]. Pharmacovigilance and post-market surveillance are considered as important components of ongoing safety monitoring which addresses any emerging risk over the course of a product's lifecycle. The regulatory framework and guidelines provide a backbone necessary for risk assessment models, international and national standards and good manufacturing practices (GMP) [129]. Different case studies provide lessons from safety failures and highlight successful examples. These real-world incidents demonstrate the significance of comprehensive safety reviews and impart knowledge for important lessons. Moreover, ensuring ethical procedures and maintaining the public confidence is essential for the effective implementation of these technologies.

The key recommendations are standardizing safety assessment protocols in order to ensuring consistency as well as reliability amongst investigations [141]. Moreover, there is a requirement for strong regulatory frameworks that are only customized for nanoparticle-based formulations [117]. Interdisciplinary cooperation between the researchers, clinicians, administrative offices, and industry partners must be encouraged to bring about a better understanding of nanoparticle-based herbal formulations. In addition, to conduct thorough safety assessment, we must invest our resources in advanced analytical techniques. Ensuring patient safety should be a top priority. This can be achieved by prioritizing patient-centric techniques in both clinical trials and post-market surveillance [142].

REFERENCES

1. Haleem A, Javaid M, Singh RP, Rab S, Suman R. Applications of nanotechnology in medical field: a brief review. *Global Health Journal*. 2023;7(2):70–7.
2. Srinivasan N. Recent advances in herbal-nano formulation: A systematic review. *Asian J Biol Life Sci*. 2023;12(1):23.
3. Jamal A. Embracing nature's therapeutic potential: Herbal medicine. *International Journal of Multidisciplinary Sciences and Arts*. 2023;2(1):117–26.
4. Yuan D, Guo Y, Pu F, Yang C, Xiao X, Du H, et al. Opportunities and challenges in enhancing the bioavailability and bioactivity of dietary flavonoids: A novel delivery system perspective. *Food Chem*. 2023;430:137115.
5. Ahmed HM, Nabavi S, Behzad S. Herbal drugs and natural products in the light of nanotechnology and nanomedicine for developing drug formulations. *Mini Rev Med Chem*. 2021;21(3):302–13.
6. Yusuf A, Almotairy ARZ, Henidi H, Alshehri OY, Aldughaim MS. Nanoparticles as drug delivery systems: a review of the implication of nanoparticles' physicochemical properties on responses in biological systems. *Polymers (Basel)*. 2023;15(7):1596.
7. Javed MN, Dahiya ES, Ibrahim AM, Alam MS, Khan FA, Pottoo FH. Recent advancement in clinical application of nanotechnological approached targeted delivery of herbal drugs. In *Nanophytomedicine: Concept to Clinic*. 2020;151–72.
8. Xu L, Wang X, Liu Y, Yang G, Falconer RJ, Zhao CX. Lipid nanoparticles for drug delivery. *Adv Nanobiomed Res*. 2022;2(2):2100109.
9. Hameed H, Khan MA, Paiva-Santos AC, Faheem S, Khalid A, Majid MS, et al. Liposomes like advanced drug carriers: from fundamentals to pharmaceutical applications. *J Microencapsul*. 2024;30: 1–23.
10. Begines B, Ortiz T, Pérez-Aranda M, Martínez G, Merinero M, Argüelles-Arias F, et al. Polymeric nanoparticles for drug delivery: Recent developments and future prospects. *Nanomaterials*. 2020;10(7):1403.
11. Huang D, Wu D. Biodegradable dendrimers for drug delivery. *Materials Science and Engineering: C*. 2018;90:713–27.
12. Chandrakala V, Aruna V, Angajala G. Review on metal nanoparticles as nanocarriers: Current challenges and perspectives in drug delivery systems. *Emergent Mater*. 2022;5(6):1593–615.
13. Yetisgin AA, Cetinel S, Zuvin M, Kosar A, Kutlu O. Therapeutic nanoparticles and their targeted delivery applications. *Molecules*. 2020;25(9):2193.
14. Farjadian F, Ghasemi A, Gohari O, Roointan A, Karimi M, Hamblin MR. Nanopharmaceuticals and nanomedicines currently on the market: challenges and opportunities. *Nanomedicine*. 2019;14(1):93–126.
15. Oberdörster G. Safety assessment for nanotechnology and nanomedicine: concepts of nanotoxicology. *J Intern Med*. 2010;267(1):89–105.
16. Zhang TT, Li W, Meng G, Wang P, Liao W. Strategies for transporting nanoparticles across the blood–brain barrier. *Biomater Sci*. 2016;4(2):219–29.
17. Panariti A, Miserocchi G, Rivolta I. The effect of nanoparticle uptake on cellular behavior: disrupting or enabling functions? *Nanotechnol Sci Appl*. 2012;5:87–100.
18. Jain K, Kumar Mehra N, Jain K Nanotechnology in drug delivery: safety and toxicity issues. *Curr Pharm Des*. 2015;21(29):4252–61.
19. Soenen SJ, Rivera-Gil P, Montenegro JM, Parak WJ, De Smedt SC, Braeckmans K. Cellular toxicity of inorganic nanoparticles: common aspects and guidelines for improved nanotoxicity evaluation. *Nano Today*. 2011;6(5):446–65.
20. Morris J, Pena C, Bronaugh R, Corley E, Frankel MS, Geraci C, et al. Risk Management Methods, & Ethical, Legal, and Societal Implications of Nanotechnology. In 2009–2010 *NNI Series of EHS Workshops and Reports*; 2010.
21. Sainz V, Conniot J, Matos AI, Peres C, Zupanǒiǒ E, Moura L, et al. Regulatory aspects on nanomedicines. *Biochem Biophys Res Commun*. 2015;468(3):504–10.
22. Halwani AA. Development of pharmaceutical nanomedicines: from the bench to the market. *Pharmaceutics*. 2022;14(1):106.
23. Parasuraman S, Balamurugan S, Vanishya R. Overview of safety assessment and toxicological screening of dermal formulations. *J Basic Clin Appl Health Sci*. 2020;3:96–103.
24. Dobrovolskaia MA. Pre-clinical immunotoxicity studies of nanotechnology-formulated drugs: Challenges, considerations and strategy. *Journal of Controlled Release*. 2015;220:571–83.

25. Khlebtsov N, Dykman L. Biodistribution and toxicity of engineered gold nanoparticles: a review of in vitro and in vivo studies. *Chem Soc Rev*. 2011;40(3):1647–71.
26. Horie M, Kato H, Fujita K, Endoh S, Iwahashi H. In vitro evaluation of cellular response induced by manufactured nanoparticles. *Chem Res Toxicol*. 2012;25(3):605–19.
27. Lama S, Merlin-Zhang O, Yang C. In vitro and in vivo models for evaluating the oral toxicity of nanomedicines. *Nanomaterials*. 2020;10(11):2177.
28. Ray P, Haideri N, Haque I, Mohammed O, Chakraborty S, Banerjee S, et al. The impact of nanoparticles on the immune system: a gray zone of nanomedicine. *J Immunol Sci*. 2021;5(1).
29. Kus-Liśkiewicz M, Fickers P, Ben Tahar I. Biocompatibility and cytotoxicity of gold nanoparticles: recent advances in methodologies and regulations. *Int J Mol Sci*. 2021;22(20):10952.
30. Nikzamir M, Akbarzadeh A, Panahi Y. An overview on nanoparticles used in biomedicine and their cytotoxicity. *J Drug Deliv Sci Technol*. 2021;61:102316.
31. Sengul AB, Asmatulu E. Toxicity of metal and metal oxide nanoparticles: a review. *Environ Chem Lett*. 2020;18(5):1659–83.
32. Reid GL. Residual solvents. In: *Specification of Drug Substances and Products*. Elsevier; 2020. pp. 345–65.
33. Thakur M, Singh K, Khedkar R. Phytochemicals: Extraction process, safety assessment, toxicological evaluations, and regulatory issues. In: *Functional and preservative properties of phytochemicals*. Elsevier; 2020. pp. 341–61.
34. Moffett DB, Mumtaz MM, Sullivan Jr DW, Whittaker MH. General considerations of dose-effect and dose-response relationships. In: *Handbook on the Toxicology of Metals*. Elsevier; 2022. pp. 299–317.
35. Macko P, Palosaari T, Whelan M. Extrapolating from acute to chronic toxicity in vitro. *Toxicology in Vitro*. 2021;76:105206.
36. Haridas H, Das P, Kadeejath Azmeena SA, Khairunnisa T, Farsana K. A Review On: Toxicity Studies in Animals. *World Journal of Pharmaceutical Research*. 2023;12(4):176–94.
37. Garcés M, Cáceres L, Chiappetta D, Magnani N, Evelson P. Current understanding of nanoparticle toxicity mechanisms and interactions with biological systems. *New Journal of Chemistry*. 2021;45(32):14328–44.
38. Tirumala MG, Anchi P, Raja S, Rachamalla M, Godugu C. Novel methods and approaches for safety evaluation of nanoparticle formulations: A focus towards in vitro models and adverse outcome pathways. *Front Pharmacol*. 2021;12:612659.
39. Landsiedel R, Honarvar N, Seiffert SB, Oesch B, Oesch F. Genotoxicity testing of nanomaterials. *Wiley Interdiscip Rev Nanomed Nanobiotechnol*. 2022;14(6):e1833.
40. Luan Y, Honma M. Genotoxicity testing and recent advances. *Genome Instab Dis*. 2022;3(1):1–21.
41. Hudson PT, Guchelaar HJ. Risk assessment in clinical pharmacy. *Pharm World Sci [Internet]*. 2003;25(3):98–103. Available from: http://www.ncbi.nlm.nih.gov/pubmed/12840962
42. Parveen A, Parveen B, Parveen R, Ahmad S. Challenges and guidelines for clinical trial of herbal drugs. *J Pharm Bioallied Sci [Internet]*. 2015;7(4):329–33. Available from: https://www.ncbi.nlm.nih.gov/pmc/articles/PMC4678978/
43. Siddiqui MR, AlOthman ZA, Rahman N. Analytical techniques in pharmaceutical analysis: A review. *Arabian Journal of Chemistry [Internet]*. 2017;10:S1409–21. Available from: https://linkinghub.elsevier.com/retrieve/pii/S1878535213001056
44. Buyse M. Phase III design: principles. *Chin Clin Oncol [Internet]*. 2016;5(1):10. Available from: https://cco.amegroups.org/article/view/4672
45. Suvarna V. Phase IV of Drug Development. *Perspect Clin Res [Internet]*. 2010;1(2):57–60. Available from: http://www.ncbi.nlm.nih.gov/pubmed/21829783
46. Coleman JJ, Ferner RE, Evans SJW. Monitoring for adverse drug reactions. *Br J Clin Pharmacol [Internet]*. 2006;61(4):371–8. Available from: https://europepmc.org/articles/pmc1885044?pdf=render
47. Andrade C. Research Design: Cohort Studies. *Indian J Psychol Med [Internet]*. 2022;44(2):189–91. Available from: https://www.ncbi.nlm.nih.gov/pmc/articles/PMC9120971/pdf/10.1177_02537176211073764.pdf
48. Jose J, Al Rubaie MH, Al Ramimmy H, Varughese SS. Pharmacovigilance: Basic concepts and an overview of the system in Oman. *Sultan Qaboos Univ Med J [Internet]*. 2021;21(2):e161–3. Available from: https://journals.squ.edu.om/index.php/squmj/article/download/4398/3203
49. Ebbers HC, Mantel-Teeuwisse AK, Sayed-Tabatabaei FA, Moors EHM, Schellekens H, Leufkens HGM. The role of Periodic Safety Update Reports in the safety management of biopharmaceuticals. *Eur J Clin Pharmacol [Internet]*. 2013;69(2):217–26. Available from: https://www.ncbi.nlm.nih.gov/pmc/articles/PMC3548093/

50. Esslinger S, Quinn L, Sampat S, Otero-Lobato M, Noël W, Geldhof A, et al. Risk Management Plans: reassessment of safety concerns based on Good Pharmacovigilance Practices Module V (Revision 2)—a company experience. *J Pharm Health Care Sci [Internet]*. 2022;8:14. Available from: https://www.ncbi.nlm.nih.gov/pmc/articles/PMC9069847/
51. Chen Y, Katayose T, Nagaoka S, Piao Y, Yamaguchi K, Asou H. A post-marketing observational study of ramucirumab in patients with gastric cancer in Japan. *Gastric Cancer [Internet]*. 2021;24(6):1320–9. Available from: https://link.springer.com/content/pdf/10.1007/s10120-021-01199-0.pdf
52. Gliklich RE, Dreyer NA, Leavy MB. Registries for Evaluating Patient Outcomes: A User's Guide [Internet]. 3rd ed. *AHRQ Methods for Effective Health Care*. Rockville (MD): Agency for Healthcare Research and Quality (US); 2014. Available from: http://www.ncbi.nlm.nih.gov/books/NBK208616/
53. Nogueiras-Álvarez R. Pharmacogenomics in clinical trials: an overview. *Front Pharmacol [Internet]*. 2023;14. Available from: https://www.frontiersin.org/journals/pharmacology/articles/10.3389/fphar.2023.1247088/full
54. Lu M, Yin J, Zhu Q, Lin G, Mou M, Liu F, et al. Artificial Intelligence in Pharmaceutical Sciences. *Engineering [Internet]*. 2023;27:37–69. Available from: https://www.sciencedirect.com/science/article/pii/S2095809923001649
55. Ekor M. The growing use of herbal medicines: issues relating to adverse reactions and challenges in monitoring safety. *Front Pharmacol [Internet]*. 2014;4:177. Available from: https://www.ncbi.nlm.nih.gov/pmc/articles/PMC3887317/
56. Marquis BJ, Love SA, Braun KL, Haynes CL. Analytical methods to assess nanoparticle toxicity. *Analyst [Internet]*. 2009;134(3):425. Available from: https://xlink.rsc.org/?DOI=b818082b
57. Jayawardena HSN, Liyanage SH, Rathnayake K, Patel U, Yan M. Analytical Methods for Characterization of Nanomaterial Surfaces. *Anal Chem [Internet]*. 2021;93(4):1889–911. Available from: https://pubs.acs.org/doi/10.1021/acs.analchem.0c05208
58. Wang Y, Counihan MJ, Lin JW, Rodríguez-López J, Yang H, Lu Y. Quantitative Analysis of DNA-Mediated Formation of Metal Nanocrystals. *J Am Chem Soc [Internet]*. 2020;142(48):20368–79. Available from: https://pubs.acs.org/doi/10.1021/jacs.0c08604
59. Mudunkotuwa IA, Minshid A Al, Grassian VH. ATR-FTIR spectroscopy as a tool to probe surface adsorption on nanoparticles at the liquid–solid interface in environmentally and biologically relevant media. *Analyst [Internet]*. 2014;139(5):870–81. Available from: https://xlink.rsc.org/?DOI=C3AN01684F
60. Dosumu AN, Claire S, Watson LS, Girio PM, Osborne SAM, Pikramenou Z, et al. Quantification by Luminescence Tracking of Red Emissive Gold Nanoparticles in Cells. *JACS Au [Internet]*. 2021;1(2):174–86. Available from: https://pubs.acs.org/doi/10.1021/jacsau.0c00033
61. Gast M, Sobek H, Mizaikoff B. Nanoparticle Tracking of Adenovirus by Light Scattering and Fluorescence Detection. *Hum Gene Ther Methods [Internet]*. 2019;30(6):235–44. Available from: https://www.liebertpub.com/doi/10.1089/hgtb.2019.172
62. Filipe V, Hawe A, Jiskoot W. Critical Evaluation of Nanoparticle Tracking Analysis (NTA) by NanoSight for the Measurement of Nanoparticles and Protein Aggregates. *Pharm Res [Internet]*. 2010;27(5):796–810. Available from: http://link.springer.com/10.1007/s11095-010-0073-2
63. Jagadeesh P, Rangappa SM, Siengchin S. Advanced characterization techniques for nanostructured materials in biomedical applications. *Advanced Industrial and Engineering Polymer Research [Internet]*. 2023;S2542504823000180. Available from: https://linkinghub.elsevier.com/retrieve/pii/S2542504823000180
64. Brar SK, Verma M. Measurement of nanoparticles by light-scattering techniques. *TrAC Trends in Analytical Chemistry [Internet]*. 2011;30(1):4–17. Available from: https://linkinghub.elsevier.com/retrieve/pii/S0165993610002700
65. Khan NU, Ali A, Khan H, Khan ZU, Ahmed Z. Stability Studies and Characterization of Glutathione-Loaded Nanoemulsion. *J Cosmet Sci [Internet]*. 2018;69(4):257–67. Available from: http://www.ncbi.nlm.nih.gov/pubmed/30311901
66. Baer DR, Artyushkova K, Cohen H, Easton CD, Engelhard M, Gengenbach TR, et al. XPS guide: Charge neutralization and binding energy referencing for insulating samples. *Journal of Vacuum Science & Technology A [Internet]*. 2020;38(3):031204. Available from: https://pubs.aip.org/jva/article/38/3/031204/1063946/XPS-guide-Charge-neutralization-and-binding-energy
67. Gill P, Moghadam TT, Ranjbar B. Differential scanning calorimetry techniques: applications in biology and nanoscience. *J Biomol Tech [Internet]*. 2010;21(4):167–93. Available from: http://www.ncbi.nlm.nih.gov/pubmed/21119929

68. Entzeroth M, Flotow H, Condron P. Overview of High-Throughput Screening. *Curr Protoc Pharmacol [Internet]*. 2009;44(1). Available from: https://currentprotocols.onlinelibrary.wiley.com/doi/10.1002/0471141755.ph0904s44
69. Korzeniewski C, Callewaert DM. An enzyme-release assay for natural cytotoxicity. *J Immunol Methods [Internet]*. 1983;64(3):313–20. Available from: https://linkinghub.elsevier.com/retrieve/pii/0022175983904386
70. Hendriks G, Van De Water B, Schoonen W, Vrieling H. Cellular-signaling pathways unveil the carcinogenic potential of chemicals. *Journal of Applied Toxicology [Internet]*. 2013;33(6):399–409. Available from: https://analyticalsciencejournals.onlinelibrary.wiley.com/doi/10.1002/jat.2845
71. Li S, Xia M. Review of high-content screening applications in toxicology. *Arch Toxicol [Internet]*. 2019;93(12):3387–96. Available from: http://link.springer.com/10.1007/s00204-019-02593-5
72. Bartlett JA, Brewster M, Brown P, Cabral-Lilly D, Cruz CN, David R, et al. Summary Report of PQRI Workshop on Nanomaterial in Drug Products: Current Experience and Management of Potential Risks. *AAPS Journal*. 2015 Jan 6;17(1):44–64.
73. Bremer-Hoffmann S, Halamoda-Kenzaoui B, Borgos SE. Identification of regulatory needs for nanomedicines. *J Interdiscip Nanomed*. 2018 Apr;3(1):4–15.
74. Kumari R, Suman K, Karmakar S, Mishra V, Lakra SG, Saurav GK, et al. Regulation and safety measures for nanotechnology-based agri-products. Vol. 5, *Frontiers in Genome Editing*. Frontiers Media SA; 2023.
75. Arya SS, Tanwar N, Lenka SK. Prospects of nano- and peptide-carriers to deliver CRISPR cargos in plants to edit across and beyond central dogma. *Nanotechnology for Environmental Engineering*. 2021;
76. Tirumala MG, Anchi P, Raja S, Rachamalla M, Godugu C. Novel Methods and Approaches for Safety Evaluation of Nanoparticle Formulations: A Focus Towards In Vitro Models and Adverse Outcome Pathways. *Front Pharmacol*. 2021 Sep 9;12.
77. Özdemir A., Ark M. xCELLigence Real Time Cell Analysis System A New Method for Cell Proliferation and Cytotoxicity. *Niche Journal*. 2014;2(2):15–17.
78. Kustermann S, Boess F, Buness A, Schmitz M, Watzele M, Weiser T, et al. A label-free, impedance-based real time assay to identify drug-induced toxicities and differentiate cytostatic from cytotoxic effects. *Toxicology in Vitro*. 2013 Aug;27(5):1589–95.
79. Hartung T. Food for thought … 1. *ALTEX*. 2010;27(2):87–95.
80. Woolley JF, Stanicka J, Cotter TG. Recent advances in reactive oxygen species measurement in biological systems. *Trends Biochem Sci*. 2013 Nov;38(11):556–65.
81. Koo YEL, Fan W, Hah H, Xu H, Orringer D, Ross B, et al. Photonic explorers based on multifunctional nanoplatforms for biosensing and photodynamic therapy. *Appl Opt*. 2007 Apr;46(10):1924–30.
82. Kimmelman J, London AJ. Predicting harms and benefits in translational trials: Ethics, evidence, and uncertainty. Vol. 8, *PLoS Medicine*. 2011.
83. King NMP. *Nanomedicine First-in-Human Research: Challenges for Informed Consent*. US; 2012.
84. Resnik DB, Tinkle SS. Ethics in nanomedicine. *Nanomedicine*. 2007 Jun;2(3):345–50.
85. Senjen. R. Nanomedicine new solutions or new problems? *[Internet]*. 2013. Available from: http://marianarei.pt.vu
86. Desai N. Challenges in development of nanoparticle-based therapeutics. Vol. 14, *AAPS Journal*. 2012. pp. 282–95.
87. Lin PC, Lin S, Wang PC, Sridhar R. Techniques for physicochemical characterization of nanomaterials. Vol. 32, *Biotechnology Advances*. Elsevier Inc.; 2014. pp. 711–26.
88. Kalani M, Yunus R. Application of supercritical antisolvent method in drug encapsulation: a review. Vol. 6, *International journal of nanomedicine*. 2011. pp. 1429–42.
89. Diab R, Jaafar-Maalej C, Fessi H, Maincent P. Engineered nanoparticulate drug delivery systems: the next frontier for oral administration. Vol. 14, *AAPS Journal*. 2012. pp. 688–702.
90. Mahapatro A, Singh DK. Biodegradable nanoparticles are excellent vehicle for site directed in-vivo delivery of drugs and vaccines. Vol. 9, *Journal of nanobiotechnology*. 2011. pp. 55.
91. Dahiya S. Drug nanonization: An overview of industrially feasible top-down technologies for nanocrystal production. *Bulletin of Pharmaceutical Research*. 2017 Jul 18;7(2):144.
92. Babar Iqbal M Pharm, Javed Ali M Pharm, Sanjula Baboota M. Recent advances and development in epidermal and dermal drug deposition enhancement technology. *Int J Dermatol*. 2018 Feb 12;57(6):646–60.
93. Kaur S, Nautyal U, Singh R, Singh S, Devi A. 14. Nanostructure Lipid Carrier _NLC_ the new generation of lipid nanoparticles. ASIAN PACIFIC JOURNAL OF HEALTH *SCIENCES [Internet]*. 2015;2(2):76–93. Available from: www.apjhs.com

94. Mark M Bailey, Cory J Berkland. Nanoparticle formulations in pulmonary drug delivery. *Med Res Rev* . 2009 Jan;291(1):196–212.
95. Vijayalakshmi Ghosh, S Saranya, Amitava Mukherjee, Natarajan Chandrasekaran. Antibacterial microemulsion prevents sepsis and triggers healing of wound in wistar rats. *Colloids Surf B Biointerfaces*. 2013 May 1;105:152–7.
96. R Rajendran, R Radhai, T M Kotresh, Emilia Csiszar. Development of antimicrobial cotton fabrics using herb loaded nanoparticles. *Carbohydr Polym*. 2013 Jan 16;91(2):613–7.
97. S Bhattacharya, Ghosh A. Phytosomes: The New Technology for Enhancement of Bioavailability of Botanicals and Nutraceuticals. *International Journal of Health Research*. 2009;2(3):87–91.
98. Chiara Sinico, Alessandro De Logu, Francesco Lai, Donatella Valenti, Maria Manconi, Giuseppe Loy, et al. Liposomal incorporation of Artemisia arborescens L. essential oil and in vitro antiviral activity. *European Journal of Pharmaceutics and Biopharmaceutics* . 2005 Jan;59(1):161–8.
99. Ajazuddin, S Saraf. Applications of novel drug delivery system for herbal formulations. *Fitoterapia*. 2010 Oct;81(7):680–9.
100. Santos FK, Oyafuso MH, Kiill CP, Gremião MPD, Chorilli M. Nanotechnology-based drug delivery systems for treatment of hyperproliferative skin diseases – a review. *Curr Nanosci*. 2013;9(1):159–67.
101. Livia Palmerston Mendes, Jiayi Pan, Vladimir P Torchilin. Dendrimers as Nanocarriers for Nucleic Acid and Drug Delivery in Cancer Therapy. *Molecules* . 2017 Aug 23;22(9).
102. Dong Y, Love KT, Dorkin JR, Sirirungruang S, Zhang Y, Chen D, et al. Lipopeptide nanoparticles for potent and selective siRNA delivery in rodents and nonhuman primates. *Proc Natl Acad Sci U S A*. 2014 Mar 18;111(11):3955–60.
103. Riley RS, June CH, Langer R, Mitchell MJ Delivery technologies for cancer immunotherapy. Vol. 18, *Nature Reviews Drug Discovery*. Nature Publishing Group; 2019. pp. 175–96.
104. Cavazzana-Calvo M, Hacein-Bey S, de Saint Basile G, Gross F, Yvon E, Nusbaum P, et al. Gene Therapy of Human Severe Combined Immunodeficiency (SCID)-X1 Disease [Internet]. Vol. 39, *Proc. Natl. Acad. Sci. USA* 1972. Available from: www.sciencemag.org
105. Whitehead KA, Langer R, Anderson DG Knocking down barriers: Advances in siRNA delivery. Vol. 8, *Nature Reviews Drug Discovery*. Nature Publishing Group; 2009. pp. 129–38.
106. Kulkarni JA, Witzigmann D, Leung J, Tam YYC, Cullis PR. On the role of helper lipids in lipid nanoparticle formulations of siRNA. *Nanoscale*. 2019 Nov 21;11(45):21733–9.
107. Cheng X, Lee RJ. The role of helper lipids in lipid nanoparticles (LNPs) designed for oligonucleotide delivery. Vol. 99, *Advanced Drug Delivery Reviews*. Elsevier B.V.; 2016. pp. 129–37.
108. Chen WH, Luo GF, Zhang XZ. Recent Advances in Subcellular Targeted Cancer Therapy Based on Functional Materials. *Advanced Materials*. 2018 Sep 10;31(3):1–39.
109. Xu CF, Chen GJ, Luo YL, Zhang Y, Zhao G, Lu ZD, et al. Rational designs of in vivo CRISPR-Cas delivery systems. *Adv Drug Deliv Rev*. 2019 Nov 21. https://doi.org/10.1016/j.addr.2019.11.005
110. Zhang CX, Cheng Y, Liu DZ, Liu M, Cui H, Zhang B Le, et al. Mitochondria-targeted cyclosporin A delivery system to treat myocardial ischemia reperfusion injury of rats. *J Nanobiotechnology*. 2019 Jan 25;17(1).
111. Witten J, Samad T, Ribbeck K. Selective permeability of mucus barriers. *Curr Opin Biotechnol*. 2018 Aug;52:124–33.
112. Suk JS, Kim AJ, Trehan K, Schneider CS, Cebotaru L, Woodward OM, et al. Lung gene therapy with highly compacted DNA nanoparticles that overcome the mucus barrier. *Journal of Controlled Release*. 2014 Mar 28;178:8–17.
113. Rao JP, Geckeler KE. Polymer nanoparticles: Preparation techniques and size-control parameters. *Prog Polym Sci*. 2011 Jul;36(7):887–913.
114. Woodruff MA, Hutmacher DW. The return of a forgotten polymer—Polycaprolactone in the 21st century. *Prog Polym Sci*. 2010 Oct;35(10):1217–56.
115. Gupta R, Suman, Tewari S. Development of Herbal Biodegradable Polymeric Nanoparticle from Clerodendrum infortunatum L. *Journal of Bionanoscience*. 2013 Aug;7(4):341–7.
116. Su S, Kang PM. Systemic review of biodegradable nanomaterials in nanomedicine. Vol. 10, *Nanomaterials*. MDPI AG; 2020.
117. Ramos TI, Villacis-Aguirre CA, López-Aguilar K V, Santiago Padilla L, Altamirano C, Toledo JR, et al. The Hitchhiker's Guide to Human Therapeutic Nanoparticle Development. *Pharmaceutics [Internet]*. 2022;14(2):247. Available from: https://www.mdpi.com/1999-4923/14/2/247

118. Ridder HG. The theory contribution of case study research designs. *Business Research [Internet]*. 2017;10(2):281–305. Available from: https://link.springer.com/10.1007/s40685-017-0045-z
119. Ansari S, Islam F, Sameem Mohd. Influence of nanotechnology on herbal drugs: A Review. *J Adv Pharm Technol Res [Internet]*. 2012;3(3):142. Available from: https://journals.lww.com/10.4103/2231-4040.101006
120. Chopra H, Dey PS, Das D, Bhattacharya T, Shah M, Mubin S, et al. Curcumin Nanoparticles as Promising Therapeutic Agents for Drug Targets. *Molecules [Internet]*. 2021;26(16):4998. Available from: https://www.mdpi.com/1420-3049/26/16/4998
121. Kunnumakkara AB, Hegde M, Parama D, Girisa S, Kumar A, Daimary UD, et al. Role of Turmeric and Curcumin in Prevention and Treatment of Chronic Diseases: Lessons Learned from Clinical Trials. Vol. 6, *ACS Pharmacology and Translational Science*. American Chemical Society; 2023. pp. 447–518.
122. Mukherjee D, Krishnan A. Therapeutic potential of curcumin and its nanoformulations for treating oral cancer. *World J Methodol [Internet]*. 2023;13(3):29–45. Available from: https://www.wjgnet.com/2222-0682/full/v13/i3/29.htm
123. Dube A, Nicolazzo JA, Larson I. Chitosan nanoparticles enhance the intestinal absorption of the green tea catechins (+)-catechin and (−)-epigallocatechin gallate. *European Journal of Pharmaceutical Sciences [Internet]*. 2010;41(2):219–25. Available from: https://linkinghub.elsevier.com/retrieve/pii/S0928098710002435
124. Wolfram J, Zhu M, Yang Y, Shen J, Gentile E, Paolino D, et al. Safety of nanoparticles in medicine. *Curr Drug Targets [Internet]*. 2015;16(14):1671–81. Available from: https://www.ncbi.nlm.nih.gov/pmc/articles/PMC4964712/
125. Bonifácio BV, da Silva PB, Aparecido dos Santos Ramos M, Maria Silveira Negri K, Maria Bauab T, Chorilli M. Nanotechnology-based drug delivery systems and herbal medicines: A review. Vol. 9, *International Journal of Nanomedicine*. 2013. pp. 1–15.
126. Wachtel-Galor S, Benzie IFF. Herbal Medicine: An Introduction to Its History, Usage, Regulation, Current Trends, and Research Needs. In: Benzie IFF, Wachtel-Galor S, editors. *Herbal Medicine: Biomolecular and Clinical Aspects [Internet]*. 2nd ed. Boca Raton (FL): CRC Press/Taylor & Francis; 2011. Available from: http://www.ncbi.nlm.nih.gov/books/NBK92773/
127. Augustine R, Hasan A, Primavera R, Wilson RJ, Thakor AS, Kevadiya BD. Cellular uptake and retention of nanoparticles: Insights on particle properties and interaction with cellular components. *Mater Today Commun [Internet]*. 2020;25:101692. Available from: https://www.sciencedirect.com/science/article/pii/S2352492820327033
128. Pandey RP, Vidic J, Mukherjee R, Chang CM. Experimental Methods for the Biological Evaluation of Nanoparticle-Based Drug Delivery Risks. *Pharmaceutics [Internet]*. 2023;15(2):612. Available from: https://www.ncbi.nlm.nih.gov/pmc/articles/PMC9959606/
129. Wang H, Chen Y, Wang L, Liu Q, Yang S, Wang C. Advancing herbal medicine: enhancing product quality and safety through robust quality control practices. *Front Pharmacol [Internet]*. 2023;14:1265178. Available from: https://www.ncbi.nlm.nih.gov/pmc/articles/PMC10561302/
130. Resnik DB, Tinkle SS. Ethical issues in clinical trials involving nanomedicine. *Contemp Clin Trials*. 2007 Jul;28(4):433–41.
131. Hoet PHM, Brüske-Hohlfeld I, Salata O V. Nanoparticles - Known and unknown health risks. Vol. 2, *Journal of Nanobiotechnology*. 2004.
132. Mieno T Production of fullerenes from plant materials and used carbon materials by means of a chip-injection-type J×B arc reactor. *Fullerene Science and Technology*. 2000;8(3):179–86.
133. The Royal Academy of Engineering RPD 35/06. Royal *Society*. 2004 Oct.
134. Nel A, Xia T, Mädler L, Li N. Toxic Potential of Materials at the Nanolevel. Vol. 311, *Source: Science, New Series*. 2006.
135. Gillett G A History and Theory of Informed Consent. *J Med Ethics*. 1987 Dec;13(4):219.2–220.
136. Horton J Principles of biomedical ethics. *Trans R Soc Trop Med Hyg*. 2002 Jan;96(1):107.
137. Shrader-Frechette K Nanotoxicology and ethical conditions for informed consent. *Nanoethics*. 2007 Mar;1(1):47–56.
138. Cobb MD, Macoubrie J. *Public perceptions about nanotechnology: Risks, benefits and trust*. USA; 2004 Jul.
139. Højelse F Preclinical safety assessment: in vitro -- in vivo testing. *Pharmacol Toxicol [Internet]*. 2000;86 Suppl 1:6–7. Available from: http://www.ncbi.nlm.nih.gov/pubmed/10905745

140. Pollutants NRC (US) C on RA of HA. Assessment of Toxicity. In: *Science and Judgment in Risk Assessment [Internet]*. National Academies Press (US); 1994. Available from: https://www.ncbi.nlm.nih.gov/books/NBK208246/
141. Safety I of M (US) C on DS for P, Aspden P, Corrigan JM, Wolcott J, Erickson SM. Standardized Reporting. In: *Patient Safety: Achieving a New Standard for Care [Internet]*. National Academies Press (US); 2004. Available from: https://www.ncbi.nlm.nih.gov/books/NBK216081/
142. Algorri M, Cauchon NS, Christian T, O'Connell C, Vaidya P. Patient-Centric Product Development: A Summary of Select Regulatory CMC and Device Considerations. *J Pharm Sci [Internet]*. 2023;112(4): 922–36. Available from: https://www.sciencedirect.com/science/article/pii/S0022354923000515

15 Novel Drug Delivery Methods for Herbal Medicine

Mukul Machhindra Barwant
Department of Botany, Sanjivani Rural Education Society's (SRES), Sanjivani Arts, Commerce and Science College, Kopargaon, India

Odangowei Inetiminebi Ogidi
Department of Biochemistry, Faculty of Basic Medical Sciences, Bayelsa Medical University, Yenagoa, Nigeria

Shreya Singh
Department of Agriculture, Ramlalit Singh Mahavidyalaya, Kailhat, Chunar, India

Balwant Singh
Department of Botany, Dr. Ram Manohar Lohia Avadh University, Ayodhya, India

15.1 INTRODUCTION

In addition, of course, Many contemporary pharmaceuticals owe their origins to compounds derived from plants, such as aspirin from willow bark and digitalis from foxglove (Chakraborty et al., 2016). Despite the long-standing history and proven benefits of herbal remedies, the field faces significant challenges in terms of consistency, bioavailability, and the precision of drug delivery (Bruna et al., 2014).

The increasing integration of traditional herbal medicine with modern pharmaceutical technologies has opened new frontiers in healthcare, particularly in the realm of drug delivery systems (Afroj et al., 2019). Herbal medicine, with its long-standing history and diverse pharmacological potential, offers a plethora of bioactive compounds that have been used for centuries to treat various ailments (Dewi et al., 2022). However, the full therapeutic potential of these natural compounds is often limited by challenges related to their bioavailability, stability, and targeted delivery. This necessitates the development of novel drug delivery systems that can enhance the efficacy and safety of herbal medicines (Harika et al., 2021).

Traditional methods of administering herbal medicines, such as decoctions, tinctures, and powders, often result in inconsistent therapeutic outcomes due to variations in absorption and metabolism. Additionally, many herbal compounds exhibit poor solubility and stability, further limiting their clinical effectiveness (Mishra et al., 2022). The advent of modern drug delivery technologies offers promising solutions to these challenges by improving the pharmacokinetic and pharmacodynamic profiles of herbal medicines. These advanced systems can modulate the release, distribution, and targeting of active compounds, thereby optimizing their therapeutic benefits (Afroj et al., 2019; Narwade et al., 2024).

Nanotechnology, in particular, has emerged as a pivotal tool in the enhancement of herbal medicine delivery (Ruturaj et al., 2023). Nanocarriers such as liposomes, solid lipid nanoparticles, and polymeric nanoparticles can encapsulate bioactive compounds, protecting them from degradation and enhancing their absorption in the body. (These nanocarriers can be engineered to release the

DOI: 10.1201/9781003544418-15

herbal compounds in a controlled manner, ensuring sustained therapeutic levels over extended periods (Mathpati et al., 2021). Furthermore, surface modifications of nanocarriers can facilitate targeted delivery to specific tissues or cells, minimizing systemic side effects and enhancing the efficacy of the treatment (Ruturaj et al., 2023).

Another innovative approach is the use of transdermal delivery systems, which offer a non-invasive route of administration for herbal compounds (Gopi et al., 2016). Transdermal patches and gels can provide a steady release of active ingredients through the skin, bypassing the gastrointestinal tract and avoiding first-pass metabolism (Murkute et al., 2021). This method not only improves bioavailability but also enhances patient compliance due to its ease of use (Arifin et al., 2019; Mishra et al., 2022).

Additionally, the development of advanced oral delivery systems, such as gastroretentive and mucoadhesive formulations, can improve the bioavailability of poorly soluble herbal compounds (Chakraborty et al., 2016). These systems can prolong the residence time of the herbal medicine in the gastrointestinal tract, enhancing its absorption and therapeutic efficacy. The use of bioenhancers, which are compounds that can increase the bioavailability of other drugs, is also gaining traction in the formulation of herbal medicines (Gopi et al., 2016). By co-administering bioenhancers with herbal compounds, the overall therapeutic effect can be significantly amplified (Murkute et al., 2021).

Moreover, the safety and toxicity profiles of these advanced delivery systems need to be thoroughly evaluated in order to ensure their suitability for clinical use. Regulatory frameworks also need to evolve to accommodate the unique aspects of herbal medicines and their advanced delivery systems (Ansari et al., 2012; Sharma, 2014).

The fusion of traditional herbal medicine with modern drug delivery technologies represents a significant stride towards maximizing the therapeutic potential of natural compounds. By addressing the limitations of conventional administration methods, these novel delivery systems can enhance the efficacy, safety, and patient compliance of herbal treatments (Mathpati et al., 2021). Continued research and development in this field holds great promise for the future of integrative medicine, offering new hope for the treatment of various diseases through the synergistic benefits of nature and technology.

15.2 NOVEL DRUG DELIVERY APPROACHES

Numerous drug targeting and delivery systems are currently in development with the objective of enhancing medication bioavailability and localized drug concentration, reducing drug degradation and loss, and reducing undesirable adverse effects. Microcapsules, cells, cell ghosts, lipoproteins, liposomes, micelles, insoluble or biodegradable synthetic and natural polymer microparticles, and soluble polymers are all examples of drug carriers. The carriers can be designed to degrade slowly, respond to environmental stimuli (such as changes in pH or temperature), and bind to specific components of the target area to selectively target them with antibodies. To 'target' means to be able to point the drug-loaded device in a certain direction. Passive targeting and active targeting are the two main ways to localize drug discharge. When chemotherapy drugs are deliberately accumulated in solid lesions, this process is called passive targeting (Ogidi, 2024). This happens because blood artery permeability is higher in malignant tissues than in healthy ones (Graham & Neil, 1990).

One potential approach to active targeting is to modify drug carriers with ligands that can bind to target cell receptors. The high selectivity of ligand-receptor interactions enables more precise targeting of the region of interest, which is one of their advantages. It is imperative to regulate the discharge of medications and their biodegradation in order to develop successful formulations. The following are some of the potential methods by which medications can be released: (i) the removal of drugs from surfaces; (ii) the movement of drugs through carrier matrices; (iii) the movement of drugs through nanocapsule walls; (iv) the erosion of carrier matrices; and (v) a process that comprises erosion and diffusion. The effectiveness of a drug can be significantly influenced by the manner in which it is administered, as the mode of delivery is frequently a determining factor in its selection (Graham & Neil, 1990). Sustained or continuous drug release is facilitated by the use of polymers that enable the medication to be released at a controlled rate over time, either through diffusion out of the polymer or by polymer breakdown. Due to its ability to replicate the body's natural process of manufacturing hormones such as insulin, pulsatile release is a widely used method of drug delivery. The breakthrough in this technology was enabled by the use of drug-carrying polymers that could respond to specific stimuli, such as variations in temperature, pH, or light (Sungthongjeen et al., 2004).

For more than two decades, researchers have investigated the potential advantages of nanotechnology in terms of enhancing the delivery and targeting of medications. Pharmaceutical and drug delivery companies are presented with new opportunities as we strive to enhance delivery systems that enhance efficacy and minimize toxicity. This presents a significant opportunity for patients. Alternative drug delivery systems try to overcome certain physical barriers, such as the blood–brain barrier, in order to enhance the targeting and efficacy of medication. Additionally, researchers are engaged in the process of determining how to administer protein medications to patients in a manner that prevents breakdown in the digestive system. The innovative drug delivery methods that are currently in use have limitations, with the majority of them being used for allopathic medications. Therefore, it is advisable to consider the traditional use, safety, and efficacy of herbal medicine compositions (Sahoo & Labhasetwar, 2003).

15.3 POTENTIAL OF NOVEL DRUG DELIVERY FOR HERBAL DRUGS

It is only recently that the extensive herbal knowledge held globally has begun to be appreciated in terms of its immense potential. Nevertheless, the efficacy of the treatment is diminished by the conventional and antiquated drug delivery method that is employed to administer medication to the patient. Herbal extracts are susceptible to significant degradation in the stomach due to that organ's exceptionally acidic pH. The liver may metabolize other components prior to their entry into the circulation. Consequently, the substance may not be able to reach the bloodstream in the required quantity. The medicine will not have any medicinal effect if it does not reach the bloodstream at a specific concentration, known as the 'minimum effective level.' Phytopharmaceuticals are pharmaceuticals that derive their active ingredients from plants, rather than synthetic ones (Ogidi & Emaikwu, 2024). The body can metabolize natural substances with greater comfort and efficiency. Consequently, they produce fewer side effects and facilitate greater absorption into the bloodstream, thereby enabling more comprehensive and effective therapies. Medications that are synthesized from compounds are susceptible to adverse effects (Norman, 2001).

The human body has a propensity to reject a number of synthetic compounds. Manifestations of these rejections include a broad range of adverse effects, including mild ones such as headaches and severe ones that could be fatal. It is important to remember that phytopharmaceuticals may have chemical interactions with other prescription medications, despite the fact that they frequently have fewer or no side effects. Furthermore, their individual and refined nature enables them to be seamlessly integrated into contemporary drug delivery systems, in contrast to botanicals, which makes it possible to standardize them. Thanks to an abundance of research on this subject, researchers have demonstrated that lipid-based drug delivery systems have the potential to deliver medications precisely to the desired location. Pharmacosomes are formed when medications bind to phospholipids and contain active hydrogen. This type of amphiphilic complex has the ability to bind to molecules that are both hydrophobic and hydrophilic. They enhance the bioavailability of the medication by enhancing its biopharmaceutical properties. Phytosomes are innovative compounds that are composed of phospholipids and plant-derived components, such as *silybum marianum*, *ginkgo biloba*, and ginseng (Semalty et al., 2009; Ogidi & Ajoko, 2024).

Another term for these is phytolipid delivery systems. They exhibit enhanced lipophilicity, absorption, and therapeutic properties. The pharmacological and pharmacokinetic properties of these herbal extracts have been enhanced (Izah et al., 2023; Sawyer et al., 2023). Their utilization can be advantageous for acute liver diseases, regardless of whether they are precipitated by infection or metabolism. The polar ends of herbal extract components, such as flavonolignans and terpenoids, are molecularly attached to phospholipids, such as phosphatidylcholine, using a patented process to construct phytosomes. The cosmetics industry is just one of the many industries that utilize phytosomes, which are also employed as a medicinal agent. Numerous phytosome characteristics may be illuminated by future pharmaceutical discoveries. Phytosomes serve as a bridge between the traditional and the contemporary in the realm of delivery systems. By developing innovative drug delivery methods that integrate phytosomes, unadulterated phytopharmaceuticals, or botanicals, we can achieve the best of both worlds. It is imperative to combine Indian Ayurvedic remedies with modern medication delivery technology in order to effectively treat severe illnesses (Amit et al., 2008).

15.4 TYPES OF NOVEL HERBAL DRUG DELIVERY SYSTEMS

This section encompasses all the methods by which new herbal pharmaceuticals can be administered. These methods, which include proniosomes, nanoparticles, microspheres, ethosomes, liposomes, phytosomes, pharmacosomes, museums, and transdermal drug delivery systems (TDDS), are discussed below.

15.4.1 Mouth-Dissolving Tablets

The first polyherbal mouth-dissolving tablet ever manufactured was Res-Q, introduced by Asoka Lifescience Limited. Dissolving rapidly in the oral cavity, this medication commences its action immediately. This introduces a novel method of administering medications that enhances their efficacy. This was the first attempt in traditional medicine to improve the efficacy of pharmaceuticals in the treatment of chronic health conditions. Res-Q, a potent polyherbal medication, is highly effective in the treatment of asthma and other respiratory conditions. This innovative sublingual medication delivery method guarantees rapid absorption into the bloodstream by circumventing the initial metabolic process. After oral administration, it dissolves in saliva and is subsequently assimilated. Upon utilizing the Res-Q product, you will experience alleviation from any respiratory discomfort within 15 minutes. Sorbitrate is a novel oral disintegrating medicine that is utilized to treat cardiac distress (Parakh & Gothoskar, 2003).

15.4.2 Controlled-Release Formulations

This patent delineates a method for the preservation of granulated botanicals for long-term use or the regulation of their release when consumed orally. The ingredients in this mixture are a carrier and granulated herb. It takes between 4 and 18 eighteen hours after taking it for the majority of its active components to be released. Echinacosides, hypericin, and hyperforin are the active ingredients selected from this group. A more convenient oral dose form of herbs is the goal of the innovation, which intends to give superior herbal medicines. The user will be more likely to comply with the treatment plan and the plasma concentration of physiologically active components will be optimal. Matrix formulations, like matrix tablets, and multiparticulate formulations, such as microcapsules implanted inside two-piece capsules, are the two oral administration alternatives for the powdered herb. The goal of these formulations is to provide a sustained-release method of medicine delivery that will guarantee the active components are released over an extended period of time (Blatt et al., 2002).

Furthermore, a new stable herbal medicine composition has been patented in the US. The microgranules in this formulation are intended for extended release and contain Ginkgo biloba extract and its production process. Compressibility and flowability are not the most desirable characteristics of plant extracts. These extracts are challenging to express in sustained-release tablets due to the necessity of maintaining consistent mixtures of pharmaceutical excipients and extracts during compression. There are numerous methods available for the removal of microgranules, such as the extrusion–spheronization process, the fluid-air bed approach, and the cutting-pan method. Extrusion–spheronization can be employed to make pellets with a high concentration of active substances; however, this process requires additional equipment. The cutting-pan method is the most effective method for producing the granules in this innovation. This method was chosen due to the ease of use of the requisite equipment and technique (Marechal et al., 2009).

15.4.3 Liposomes

The formation of liposomes, which are small carriers with a diameter typically ranging from 0.05 to 5.0 μm, is a result of the natural hydration of specific lipids in water-based solutions (Sterer et al., 2008). This liquid is enclosed by these spherical particles, which are capable of freely moving within and entering them. Their transport capacity for concentric membranes can be one, two, three, or an endless number. Plurophilic and hydrophilic groups are found in polar lipids, the building blocks of liposomes. The moment polar lipids come into contact with water, they self-organize into particles. It is conceivable that liposome-based drug delivery systems could be more effective in the delivery of anticancer medications. This can be achieved by either decreasing the drug's exposure to healthy tissues or increasing its concentration in cancer cells (Sharma & Sharma, 1997). This is achieved through a combination of targeting strategies and the enhanced permeability and retention impact. Liposomes are primarily advantageous due to their high biocompatibility, ease of manufacture, adaptability in substance loading, and chemical composition-based property modification (Sharma et al., 2006).

15.4.4 Phytosomes

Flavonoids, which make up most phytomedicines, are poorly absorbed when consumed orally. Polyphenols and other phytoconstituent molecules can be modified into lipid-compatible molecular complexes. Phytosomes characterize these complexes. Phytosomes have higher bioavailability than standard herbal extracts due to their improved capacity to traverse lipid-rich biomembranes and arrive at their intended location. Phospholipids, especially phosphatidylcholine derived from soy, are the primary lipid-phase compounds used to increase the lipocompatibility of phytoconstituents (Zhong et al., 2005). The cosmetics industry was the first to investigate phytosomal complexes. Nonetheless, there is significant evidence of their drug-delivery potential from recent studies. Cardiovascular health, inflammation reduction, liver protection, and cancer prevention are just a few of the areas where these complexes have demonstrated favourable effects. The pharmacokinetics and therapeutic effects of phytosome complexes are superior to those of uncomplexed herbal extracts. The absorption of several phytochemicals has been significantly enhanced via phytosome technology (Ogidi & Joshua, 2024, 2023; Zhong et al., 2005).

15.4.5 Nanoparticles

Nanoparticles are an exceptional method for transporting medications, regardless of their solubility in water. Nanoparticles are particles that are smaller than a micrometre, with a dimension ranging from tens of millimetres to one thousand millimetres (Mohanraj & Chen, 2006). The primary objective of employing nanoparticles as a drug delivery vehicle is to modulate the release of compounds with pharmacological activity, as well as their size and surface characteristics. This regulation enables the precise administration of medications at the most effective dose and rate. In the past few years, there has been significant interest in biodegradable polymeric nanoparticles as a potential medication delivery technology. The active ingredient is dispersed throughout the matrix in nanospheres, while it is enclosed in a polymeric membrane in nanocapsules. The advantages of nanonization over the corresponding unrefined medication formulations include increased compound solubility, reduced medicinal dosages, and improved absorption of herbal remedies (Brigger et al., 2002).

15.4.6 Niosomes

The niosome building blocks include cholesterol, nonionic surfactants from the alkyl or dialkylpolyglycerol ether family, and other components arranged in multilayered vesicles. Niosomes, like liposomes, possess properties that render them appealing candidates for medication delivery, as indicated by prior research conducted in collaboration with L'Oréal (Tangri & Khurana, 2011). Conversely, niosomes exhibit evident advantages over liposomes. The potential drawbacks of liposomes include their exorbitant cost, the susceptibility of their chemical components, particularly phospholipids, to oxidative degradation, the necessity for specialized storage and handling, and the variability in the quality of naturally occurring phospholipids. These issues are not present in niosomes (Gupta et al., 2011).

15.4.7 Proniosomes

The proniosome gel system is a refined version of the niosome that has a multitude of potential applications in the delivery of targeted drugs (Shukla & Tiwari, 2011). When proniosomal gels are exposed to water from the epidermis, they can undergo a transformation into niosomes. Promiosomes, which are carriers, are soluble in water and have a surfactant coating (Goyal et al., 2011). They can be moistened to produce a niosomal dispersion by rapidly agitating them in heated water just prior to use (Raja et al., 2011).

15.4.8 Transdermal Drug Delivery System

The topical delivery of pharmaceuticals through the skin has recently attracted a lot of attention from transdermal drug delivery system (TDDS) researchers, who are interested in this method for both localized therapeutic effects on sick skin and systemic drug distribution. Conversely, numerous medications have failed to produce the anticipated outcomes. However, there is significant potential for transdermal medication administration to serve as a cutting-edge smart drug delivery system in the future (Garala et al., 2009). Transdermal delivery systems offer a variety of benefits, including precise medication administration, improved drug absorption, fewer adverse effects, and ease of application. Films, which contained herbal drug constituents were developed in an early attempt to employ herbal medications using TDDS. TDDS is a technique that involves the continuous injection of medications into the bloodstream through the epidermis. Therefore, this delivery method bypasses the initial phase of the drug's disintegration in the body, as opposed to injecting it. Additionally, the technology enables a consistent rate of drug release, which enables less frequent administration, resulting in a protracted period of medicine release. Furthermore, the patient has the option to initiate the therapy at their convenience. The use of turmeric in TDDS is a potential contemporary approach to the age-old traditional method of administering turmeric through poultices or leaps. TDDS enable localized pharmaceutical action at the site of administration (Verma et al., 2007; Ogidi, 2023).

15.4.9 Microspheres

Isolated spherical particles, ranging in size from 1 to 50 μ, are known as microspheres (Verma et al., 2007). The goal of microparticulate medication delivery systems is to minimize negative effects while achieving the desired concentration in the target location. They have been extensively studied and are widely used because of how reliably they transport drugs to their specific locations. Microencapsulation is a beneficial approach that substantially extends the duration of medicinal effectiveness and increases patient adherence. The total dosage can be reduced and adverse reactions can be minimized by maintaining a constant plasma concentration (Meena et al., 2011). As a result, microspheres have been created that contain a diverse array of active plant compounds. Zedoary oil, rutin, camptothecin, quercetine, tetrandrine, and *Cynara scolymus* extract are among the compounds in question. The frequency of publications pertaining to magnetic microspheres and immunological microspheres has accelerated considerably throughout the past few years. The immunological microsphere's immune competence is derived from antibodies; the antigen is either adsorbed or deposited onto the polymer microspheres (Lakshmana et al., 2009).

15.4.10 Emulsions

An emulsion is a dispersion system that is formed when two incompatible substances combine. In this system droplets of one liquid dissolve in droplets of the other. Water, oil, surfactant, and subsurfactant are the usual ingredients of an emulsion. The material looks like a liquid and is transparent. There are several ways to categorize the particle size of emulsions, including ordinary emulsion, microemulsion, sub-microemulsion, and many more in between (Gavini et al., 2005). Nanoemulsion and microemulsion are synonyms; lipid emulsion and sub-microemulsion are also synonyms. Because of its strong attraction to lymphatic fluids, the emulsion can be targeted to particular areas of the body and used to deliver medication there. The medicine is packaged within the inner phase, which precludes direct contact with the body and tissue fluid, allowing for sustained release over an extended period of time. Making an oil/water or oil/water/oil emulsion from oily medications or lipophilic pharmaceuticals yields a large volume of dissolved medicine. The oil droplets are eaten by macrophages and end up in the kidneys, spleen, and liver in large quantities (Kun et al., 2015).

Emulsions of water and oil or oil and water work wonderfully to encapsulate water-soluble pharmaceuticals. For optimal lymphatic system absorption, these emulsions may be administered intramuscularly or subcutaneously. The particulate size of an emulsion influences its target distribution. There are numerous advantages to converting the herbal medication into an emulsion, such as targeted sustained release, enhanced hydrolysed material stability, increased drug penetration through skin and mucous membranes, and reduced drug irritation to tissues. Several herbal remedies are now available in emulsion form, such as camptothecin, oil of *Brucea javanica*, oil of coixenolide, and oil of zedoary (Ogidi & Emaikwu, 2023a, 2023b). One study looked at how the aluminium emulsion affected protein synthesis and the A549 human lung cancer cell line (Kun et al., 2015). The aluminium emulsion substantially reduced the growth and multiplication of A549 cells in a laboratory setting, as indicated by the results. The inhibitory effect was also discovered to be influenced by the concentration of the emulsion and the duration of exposure to the substance. Elemenum emulsion, a novel anticancer drug, holds significant potential for the future. Furthermore, it does not impede bone marrow activity and nor does it have any adverse effects on the liver or tenderness (Kun et al., 2015).

15.4.11 Ethosomes

Ethosomal patches, which are medications that contain ethosomes, are a recent development in patch technology. Ethosomal systems are composed of water, ethanol, and soybean phosphatidylcholine (Song et al., 2005). They possess an extraordinary ability to entrap particles with differing degrees of lipophilicity and can self-assemble into multilamellar vesicles. Elastic vesicles and transferosomes are capable of transporting vaccines, peptides, proteins, and minuscule compounds. Ethosomes possess exceptional deformability and entrapment efficacy, enabling them to penetrate the epidermis entirely, thereby enhancing transdermal medication delivery (Vicentini et al., 2008). Ethosomes are distinguished from liposomes by their physical and chemical properties, which enable the safe and effective transportation of medications through the stratum corneum, deeper layers of the epidermis, and into the bloodstream. This makes ethosomes a promising approach to the delivery of transdermal medication (Aggarwal et al., 2009). Additionally, the ethosomes transporter has the capacity to effectively transport hydrophilic and lipophilic medications into cells (Touitou et al., 2001). As a consequence, the percutaneous absorption of the herbal anti-inflammatory medicine matrine is improved. It also facilitates the entrance of the antimicrobial peptide into the fibrocyte (Ogidi et al., 2023; Dayan & Touitou, 2000).

The reviewed literature indicates that only three clinical trials have employed human volunteers and ethosomal systems. The ethosomal acyclovir preparation considerably enhanced all clinical measures, including pain indications, duration of crust formation and disappearance, and overall efficacy, in comparison to Zovirax cream (Zhaowu et al., 2009). Forty acne patients participated in a preliminary clinical study that assessed the efficacy of an ethosomal gel that contained salicylic acid and clindamycin phosphate. The substance was administered to the patients twice daily for a period of eight weeks. In contrast to the placebo group, participants who received ethosomal gel for their acne experienced a substantial improvement in their condition, as evidenced by a reduction in the number of comedones, pustules, and overall lesions (Ogidi & Enenebeaku, 2023). In a pilot clinical study, the efficacy of prostaglandin E1 in an ethosomal formulation for erectile dysfunction patients was examined. Twelve of the fifteen individuals who were evaluated demonstrated enhanced peak systolic velocity and penile rigidity. Ten to sixty minutes was the typical duration of an erection. None of the cited clinical trials identified any cutaneous side effects (Zhaowu et al., 2009).

Transferosomes Transferosomes, which are modified particles or vesicles, are capable of promptly and energy-efficiently altering their shape in response to environmental stress (Abdulbaqi et al., 2016). The development of innovative methods, such as transferosomes, has significantly mitigated issues associated with transdermal medication delivery, including the stratum corneum's sluggish penetration and the inability to transport larger molecules. The physicochemical properties of pharmaceuticals can also serve as a deterrent to their transdermal delivery; however, transferosomes are capable of overcoming this challenge (Walve et al., 2011). These pliable vesicles have the capacity to transport larger molecules through epidermis pores that are significantly smaller than their own. Transferosomes can traverse the lipid lamellar regions of the stratum corneum when applied in a nonoccluded manner due to epidermal moisture or osmotic force (Kulkarni et al., 2011). They are capable of transporting a diverse array of small molecules, peptides, proteins, and botanical substances. Transferosomes, a type of elastic or flexible vesicle, are capable of penetrating the stratum corneum, the outermost layer of epidermis. Their ability to deliver nutrients locally can maintain the skin's functionality (Benson, 2006). Transferosomes, which have been around since the early 1990s, are able to bend and twist because an edge activator is part of their lipid bilayer structure (Benson, 2006). Xiao-Ying et al. (2006) are among the researchers who have developed transferosomes that contain capsaicin, a chemical component of chilli chiles. These transferosomes exhibited superior epidermal absorption when tested against pure capsaicin (Xiao-Ying et al., 2006).

15.4.12 Other Novel Approaches

For their study, Ma et al. (2000) looked at how and why Shuanghua aerosol (SHA) affected upper respiratory tract infections in children between the ages of three and fourteen. *Radix Bupleurum*, menthene, *Herba Houttuynia, and Flos Chrysanthemum Indicum* make up SHA. *Radix Scutellaria, Fructus Forsythia*, and *Flos Lonicera* aerosols were used as a control in the experiment (Tanwar et al., 2006). According to the research, babies suffering from respiratory infections can benefit from SHA's antiviral and anti-inflammatory characteristics. Another standardized extract from *Commiphora wightii*'s oleo gum resin that has been shown to decrease harmful blood lipid levels is gumulipid. The microparticles' physicochemical properties were determined by an evaluation. Gugulipid was confirmed to be present in the synthesized microparticles by the HPLC profile, which clearly differentiated Guggulsterone-E and -Z (Borodina et al., 2008).

Two plant extracts, *Calendula officinalis* L. and plantain (*Plantago major*), are contained in the microcapsules. First, carrageenan and oligochitosan were progressively adsorbed to prepare calcium carbonate microparticles. The composite was then treated with ethylenediaminetetraacetic acid to dissolve it, resulting in polymeric composite emulsions (PCE) (Shen et al., 2008). Using coprecipitation and adsorption methods together, PCE was collected. In the carbonate matrix, PCE was more efficiently captured by coprecipitation than by adsorption. By monitoring the in vitro behaviour with the addition of artificial stomach fluid, the release kinetics were studied. Evidence from the rat acetate ulcer model suggests that microcapsule-released PCE speeds up gastrointestinal tissue repair. The absorption and distribution of traditional Chinese herbs (TCHs) can be improved by encasing them in nanoparticles (Shen et al., 2008). The rapid clearance of arterial embolism and the reduction in thrombi were outcomes of the remarkable thrombolytic capabilities displayed by the TCH nanoparticles. Compared to their non-nanoparticle equivalents, TCH nanoparticles have far more powerful thrombolytic actions. Research on oral prolonged-release formulations of traditional Chinese medicinal medicines has focused on their pharmacological efficacy, pharmacokinetics, and integrative assessment. A herb extract sustained-release implant based on chitosan has also shown promising results. In order to assist in the healing of tissues and muscles around the abdominal incision site, an implant made of CS-gelatin containing a medicinal herbal extract of danshen (*Radix Salvia miltiorrhiza*) was created (Zhao et al., 2002).

In order to determine the material's ability to continuously release the marker component tanshinone IIa, in vitro tests were carried out. The solvent was analysed using high-performance liquid chromatography (HPLC). The substance's biodegradation potential was evaluated in vitro (in a controlled environment) and in vivo (in an actual living organism). The film produced by this chemical demonstrated a consistent and enduring release effect. The Higuchi equation is succeeded by the release profile (Zhao et al., 2002). Within the initial 15 days, 20% of the medication contained in the CS-gelatin (1:2) matrix was released. It appeared that the release of the drug was effectively regulated by the quantity of drug applied to the matrix. Lysozymes can degrade the enhanced film (CS/gelatin ratio: 1:16) in as little as four days in a controlled laboratory environment. The abdominal incision wounds of the rats healed correctly, and a 0.5 cm^2 piece of this film degraded completely after 28 days of implantation. Arthri Blend-SR™, a commercially available formulation that contains nutrient-rich herbal extracts, to improve the health of your joints and connective tissue (Arthri, 2009).

Scientific trials have confirmed the effectiveness of this unique blend of natural active substances in joint care. The formula's sustained-release technology is an additional benefit that assists in the long-term management of arthritis symptoms. The combination consists of Glucosamine sulphate, Curcumin C3 Complex (curcuminoids derived from *Curcuma longa*), and Boswellin (an extract of *Boswellia serrata*). When combined, these factors enhance the management of inflammatory diseases, including arthritis. The product's active ingredient will be released in a controlled and gradual manner over the course of 8 hours, reaching 80–90% of its total concentration. One of the primary advantages of a sustained-release formulation is enhanced glucosamine bioavailability (Devi et al., 2010).

15.5 FUTURE OPPORTUNITIES AND CHALLENGES

While bio-enhancers in the administration of drugs have proven effective, some methods have proved equally effective. The development of new bio-enhancers presents obstacles that must be addressed. Nevertheless, a few of the difficulties have already been and continue to be addressed by altering the physicochemical properties of nanomaterials to enhance characteristics that include long circulation in the blood, raised functional surface area, drug degradation protection, biological barrier crossing, and site-specific targeting (Kumari et al., 2010). Large-scale manufacturing is another hurdle in herbal bio-enhancer advancement and research. Here continually exists a requirement for building up laboratory or pilot innovations for future commercialization (Mukerjee et al., 2016). Moving up presents obstacles such as low nanomaterial concentrations, agglomeration, and the chemical process; it is simpler to change nanomaterials at the laboratory size for better performance than it is on an industrial scale. Controlling the dimension and structure of nanoparticles that improve bioavailability on a wide scale is also a difficulty. Developments in herbal bio-enhancers present new issues for surveillance by regulators. Here there is an increasing demand for rules that take into consideration the physicochemical and pharmacokinetic features of nanodrug services, that differs from that which is appropriate for traditional drug products. The US Food and Drug Administration and the European Medicines Evaluation Agency have taken the initiative to identify potential technical and regulatory problems (Patra et al., 2018).

Although there are many unknowns surrounding the development of nanodrugs, the discovery of pharmacologically active substances in natural sources is not as prevalent as it was fifty years ago. Nevertheless, using nanotechnology to increase the effectiveness of previously identified natural bioactive compounds has grown commonplace. Examples include the medicinal use of nanotechnology for berberine, curcumin, ellagic acid, resveratrol, and quercetin. The application of nanocarriers has developed with gold, silver, cadmium sulphide, and titanium dioxide polymeric nanoparticles, as well as solid lipid nanoparticles, crystal nanoparticles, liposomes, micelles, superparamagnetic iron oxide nanoparticles, and dendrimers, has significantly increased the medicinal value of these organic substances.

Novel natural biomaterials remain to be in high demand due to their biodegradability, biocompatibility, ease of supply, renewable nature, and low toxicity. Besides recognizing including polysaccharides and proteins as natural biopolymers, research into ways to render them more stable in industrial processing environments and biological matrixes via methods such as crosslinking is one of the most advanced study areas at present (Sharma et al., 2016). Bioenhancers lower doses while minimizing the risk of medication resistance. A decreased dose reduces medication toxicity, which is especially true in the case of anticancer medicines such as taxol.

There are also environmental advantages. Taxol, which is utilized to treat ovarian and breast cancers, is made by the bark of the Pacific yew tree, one of the world's slowest-growing trees. Currently, six trees, ranging in age from 25 to 100 years old, must be cut down to treat a single patient. Bioenhancers will result in fewer destructions (Suresh & Jain, 2016). The medical field of nanomedicine is one of the most exciting fields of inquiry. Over the previous two decades, extensive research in this subject has resulted in the filing of 1500 patents and the completion of dozens of clinical trials. As described in the different sections above, cancer appears to be the finest example of a disease whereby neither diagnosis nor therapy have benefitted from nonmedical technology. Utilizing diverse forms of nanoparticles to precisely administer medication to impacted cells, like cancerous or tumorous cells, while avoiding any disruption to the normal cell's physiology, the field of nanomedicine and nano-drug delivery systems is poised to remain a prominent area of study and innovation for many years to come.

The use of such metals, particularly gold and silver, in diagnosis and therapy is an area of research that might lead to broader applications of nanomedicines in the years to come. Gold nanoparticles appear to be efficiently assimilated in soft cancer tissues, rendering the tumour vulnerable to radiation-based heat treatment for selective eradication (Graham & Neil, 1990).

Although this chapter presents a broad overview of nanomedicine's future prospects and nanodrug delivery systems, its actual influence in the healthcare system, including with respect to cancer therapy/diagnosis, remains limited. This reflects the field's status as a nascent area of science, with just two decades of meaningful study on the subject and many crucial fundamental qualities still unknown. The basic indicators of sick tissues, including important biological markers that allow absolute targeting without changing the normal cellular process, is one of the primary future study areas (Sahoo & Labhasetwar, 2003). Finally, the use of nanomedicine will develop when we gain a better understanding of illnesses at the molecular level or that reflects a nanomaterial-subcellular scale equivalent marker identification, opening up new pathways for diagnosis/therapy. As a result, future developments in nanomedicine applications will be driven by a knowledge of the molecular fingerprints of diseases. Beyond what we have indicated in this study utilizing known nanoprobes and nanotheragnostics products, further research is needed for the broader use of nanomedicine (Ogidi and Ajoko, 2024; Semalty et al., 2009).

The notion of the controlled release of particular medications at beleaguered places, technology for assessing these events, drug action in tissues/cellular level, and theoretical mathematical models of prediction have yet to be developed. Many investigations in nanomedicine are focused on biomaterials and formulation studies, which appear to be the early phases of biomedicine applications (Izah et al., 2023). Animal studies and transdisciplinary research, which involve a large amount of time and research resources, will provide valuable data for possible pharmacological therapeutic and diagnostic investigations. With the expanding worldwide trend toward more accurate medications and diagnosis, the future of nanomedicine and nano-drug delivery technology appears promising.

Recently, there has been considerable excitement about the simple concept of developing nanorobots (and nanodevices) that work in tissue diagnostics and healing with full external control. Such technology has not yet become a reality and remains a futuristic research goal that humanity may be able to achieve in the near future (Semalty et al., 2009; Ogidi & Ajoko, 2024). However, as with their advantages, the potential risks of nanomedicines to individuals and the ecosystem as a whole necessitate further research. As a result, a thorough examination of the potential acute or chronic toxicity effects of novel nanomaterials on humans and the environment is required. As nanomedicines become more common, their affordability will be another area that requires more investigation. Finally, as previously discussed, the regulation of nanomedicines will change in tandem with improvements in nanomedicine applications (Parakh & Gothoskar, 2003).

15.6 CONCLUSION

Integrating traditional herbal medicine with innovative drug delivery technologies improves natural medicines' therapeutic potential by increasing bioavailability, stability, and targeted administration. Nanotechnology, transdermal systems, and oral formulations have improved therapy efficacy while reducing toxicity. Nevertheless, issues such as standardization and safety evaluation continue. Continued research and development, as well as developing regulatory frameworks, are critical to realize the full potential of these advances, which provide bright future possibilities for integrative medicine and disease treatment.

REFERENCES

Abdulbaqi, I.M., Darwis, Y., Khan, N.A., Assi, R.A., & Khan, A.A.. (2016). Ethosomal nanocarriers: The impact of constituents and formulation techniques on ethosomal properties, in vivo studies, and clinical trials. *Int J Nanomedicine*, 11, 2279–304.

Aggarwal, G., Garg, A., & Dhawan, S.. (2009). Transdermal drug delivery: Evolving technologies and expanding opportunities. *Indian J Pharm Educ Res*, 43, 251–9.

Ahmed, Hiwa M., Nabavi, Seyed, & Behzad, Sahar. (2020). Herbal Drugs and Natural Products in the light of Nanotechnology and Nanomedicine for Developing Drug Formulations. *Mini-Rev Med Chem* 20, 1–12.

Amit, J., Sunil, C., Vimal, K., & Anupam, P. (2008). Phytosomes: A revolution in herbal drugs. *The Pharma Review*. New Delhi: Kongposh Publications Pvt. Ltd, p. 24–8.

Ansari, S.H., Islam, F., & Sameem, M.. (2012). Influence of nanotechnology on herbal drugs: A Review. *J Adv Pharm Technol Res* 3, 142–146.

Arifin S.F., AlShami A., Omar S.S.S., Jalil M.A.A., Khalid K.A., & Hadi H. (2019). Impact of Modern Technology on the Development of Natural-based Products. *J Ayu Herb Med* 5(4): 133–142.

Arthri, B.S. (2009). A formulation containing herbal extracts and nutrients to support healthy joints and connective tissues in the body. *Nutraceuticals World*. Available from: http://findarticles.com/particles/mi_hb223/is_6_7/ai_n29102045/. (Last accessed on 2023 November 5.)

Benson, H.A.. Transfersomes for transdermal drug delivery. (2006). *Expert Opin Drug Deliv*, 3, 727–37.

Blatt, Y., Kimmelman, E., Cohen, D., & Rotman, A. (2002). Microencapsulated and controlled-release herbal formulations. *United States Patent*.

Borodina, T.N., Rumsh, L.D., Kunizhev, S.M., Sukhorukov, G.B., Vorozhtsov, G.N., & Feldman, B.M., et al.. (2008). Entrapment of herbal extracts into biodegradable microcapsules. *Biochem Suppl Series B Biomed Chem*, 2, 176–82.

Brigger, I., Dubernet, C., & Couvreur, P.. (2002). Nanoparticles in cancer therapy and diagnosis. *Adv Drug Deliv Rev*, 54, 631–51.

Bruna, Vidal Bonifácio et al. (2014). Nanotechnology-based drug delivery systems and herbal medicines: a review. *Int J Nanomedicine*. 9, 1–15.

Chakraborty, K., Shivakumar, A., & Ramachandran, S. . (2016). Nano-technology in herbal medicines: A review. *Int J Herb Med*4(3), 21–27.

Chhalwadi, A. S. Babar, A.S. Dahiwade, L. K., & Payghan, S. A.. (2022). Recent Advances in Herbal Nanotechnology. *Int J Creat Res Thoughts*, 9(5, 312–321.

Dayan, N., & Touitou, E. (2000). Carriers for skin delivery of trihexyphenidyl HCl: Ethosomes vs. liposomes. *Biomaterials*, 21, 1879–85.

Devi, V.K., Jain, N., & Valli, K.S. (2010). Importance of novel drug delivery systems in herbal medicines. *Pharmacogn Rev*, 4, 27–31.

Dewi, M.K., Chaerunisaa, A.Y., Muhaimin, M., & Joni, I.M. (2022). Improved Activity of Herbal Medicines through Nanotechnology. *Nano*, 12, 4073.

Garala, K.C., Shinde, A.J., & Shah, P.H. (2009). Formulation and in vitro characterization of monolithic matrix transdermal systems using hpmc/eudragit s 100 polymer blends. *Int J Pharm Pharm Sci*, 1, 108–20.

Gavini, E., Alamanni, M.C., Cossu, M., & Giunchedi, P. (2005). Tabletted microspheres containing *Cynara scolymus* (var. *spinoso sardo*) extract for the preparation of controlled release nutraceutical matrices. *J Microencapsul*, 22, 487–99.

Gopi S., Amalraj A., & Haponiuk J.T., Thomas S, 2016. Introduction of Nanotechnology in Herbal Drugs and Nutraceutical: A Review. *J Nanomed Biother Discov* 6(2), 1000143.

Goyal, C., Ahuja, M., & Sharma, S.K. (2011). Preparation and evaluation of anti-inflammatory activity of gugulipid-loaded proniosomal gel. *Acta Pol Pharm Drug Res*, 68, 147–50.

Graham, S., & Neil, B. (1990). Controlled drug delivery systems. *Chem Ind*, 6, 25–38.

Gupta, S., Singh, R.P., Lokwani, P., Yadav, S., & Gupta, S.K. (2011). Vesicular system as targeted drug delivery system: An overview. *Int J Pharm Technol*, 3, 987–1021.

Harika P., Deepthi, B.V.P., Vinitha, B., Baherji, R., Ali, J., & Sharma, J.V.C. (2021). Herbal Nanoparticles. *World J Pharm Med Res* 7(3): 127–130.

Harma, A., Jadhav, S. S., & Tripathi, C. D. (2016). Herbal drug delivery systems: An emerging area in herbal drug research. *J Pharm Sci Res*, 8(10), 111–124.

Izah, S. Ogidi, O.I., Ogwu, M., Salimon, S., Yusuf, Z., Akram, M., Raimi, M., & Iyingiala, A. (2023). Historical Perspectives and overview of the value of herbal medicine. In: *Reference Series in Phytochemistry. Herbal Medicine Phytochemistry*. Springer Nature Singapore Pte Ltd.

Kulkarni, P.R., Yadav, J.D., Vaidya, K.A., & Gandhi, P.P. (2011). Transfersomes: An emerging tool for transdermal drug delivery. *Int J Pharm Sci Res*, 2, 735–41.

Kumari, A., Yadav, S. K., & Yadav, S. C. (2010). Biodegradable polymeric nanoparticles based drug delivery systems. *Colloids Surf B: Biointerfaces*, 75(1), 1–18.

Kun, Z., Caigang, L., Zhuo, Z., & Lijuan, Z. (2015). *The effect of elemene on lung adenocarcinoma A549 cell radiosensitivity and elucidation of its mechanism Clinics (Sao Paulo)*, 70, 556–62. doi: 10.6061/clinics/2015(08)05.

Lakshmana, P.S., Shirwaikar, A.A., Shirwaikar, A., & Kumar, A. (2009). Formulation and evaluation of sustained release microspheres of rosin containing aceclofenac. *Ars Pharm*, 50, 51–62.

Ma, B., Duan, X., & Wang, Z. (2000). Clinical and experimental study on Shuanghua aerosol in treating infantile upper respiratory tract infection. *Zhongguo Zhong Xi Yi Jie He Za Zhi*, 20, 653–5.

Marechal, D., Yang, W., & Yuzhang, H. (2009). Sustained-release microgranules containing Ginkgo biloba extract and the process for manufacturing these. *United States Patent*, p.7569236.

Mathpati, R. V., Harangule, Y. R., Bavage, S. B., Bavage, N. B. (2021). Nano-Technology in Herbal Medicines: Advancements in Herbal Treatment. *Int J Innov Res Technol* 8(3), 116–122.

Meena, K.P., Dangi, J.S., Samal, P.K., & Namdeo, K.P. (2011). Recent advances in microspheres manufacturing technology. *Int J Pharm Technol*, 3, 854–93.

Mishra, Y. et al., 2022. Application of nanotechnology to herbal antioxidants as improved phytomedicine: An expanding horizon. *Biomed Pharmacother* 153(2022), 113413.

Mohanraj, V.J., & Chen, Y. (2006). Nanoparticles: A review. *Trop J Pharm Res*, 5, 561–73.

Mukherjee, P. K., Harwansh, R. K., Bhattacharyya, S., & Efferth, T. (2016). Integration of herbal medicine in modern drug delivery system: Emerging trends and challenges. *Planta Med*, *82*(05), 333–345.

Murkute, P. S., Kathar, N. P., & Sanap, G. S. 2021. Nanotechnology; Its Application, Process Development and Nano-Technology Based Herbal Medicine. *Int J Pharm Res App* 6(5), 593–605.

Narwade, V. V., Game, M. D., & Kadam, S. G. (2024). A Review on Nanotechnology in Herbal Medicine. *Int J Pharm Sc*, 2(4): 411–416.

Norman, G.B. (2001). *Herbal drugs and phytopharmaceuticals. A Handbook for Practice on a Scientific Basis*. 2nd ed. New York: Medpharm Scientific Publishers, Stuttgart and CRC Press, p. 230–48.

Ogidi, O.I. (2023). Sustainable Utilization of Important Medicinal Plants in Africa. In: *Sustainable Utilization and Conservation of Africa's Biological Resources and Environment, Sustainable Development and Biodiversity*. Springer Nature Singapore Pte Ltd. https://doi.org/10.1007/978-981-19-6974-4_12

Ogidi, O.I. (2024). Recent advances in anticancer activity and Bioinformatics Approach from Potential Plants. In: *Computational Approaches in Biomaterials, Bioinformatics and Biomedical Engineering Applications*. CRC Press, Taylor and Francis Group, Florida, USA.

Ogidi, O.I., & Ajoko, I. (2024). Neuroprotective Potentials of Phytochemicals. In: M.S. Rajput, T. Sarachana, P.D. Sarkar, & M. Raval (eds) *NeuroPhytomedicine*. CRC Press, Routledge Taylor and Francis Group.

Ogidi, O.I., Carbom, H.E., & Tawariowei, M.A. (2023). Effects of Methanol and N-hexane Extracts of *Raphia vinifera* Fruit and *Elaeis guineensis* Seeds against *Staphylococcus aureus* and *Escherichia coli*. *Int J Appl Sci Dev* 2, 177–187. DOI: 10.37394/232029.2023.2.19

Ogidi, O.I. & Emaikwu, N.G. (2023a). Utilization methods and practices of herbal medicine in Africa. In: *Reference Series in Phytochemistry. Herbal Medicine Phytochemistry*. Springer Nature Singapore Pte Ltd.

Ogidi, O.I., & Emaikwu, N.G. (2023b). Adoption and Application of Biotechnology in Herbal Medicine Practices. In: *Reference Series in Phytochemistry. Herbal Medicine Phytochemistry*. Springer Nature Singapore Pte Ltd.

Ogidi, O.I., & Emaikwu, N.G. (2024). Plant Phenolic compound isolation and its Bioinformatics Approaches of Molecular Mechanisms in Antimicrobial activities and Resistance. In: *Computational Approaches in Biomaterials, Bioinformatics and Biomedical Engineering Applications*. CRC Press, Taylor and Francis Group, Florida, USA.

Ogidi, O.I., & Enenebeaku, U.E. (2023). Medicinal Potentials of Aloe Vera (*Aloe barbadensis* Miller): Technologies for the Production of Therapeutics. In: *Sustainable Utilization and Conservation of Africa's Biological Resources and Environment, Sustainable Development and Biodiversity*. Springer Nature Singapore Pte Ltd. https://doi.org/10.1007/978-981-19-6974-4_11

Ogidi, O.I., & Joshua, M.T. (2023). Investigation of Phytochemical Compounds of Selected Nigerian Poly-Herbal Formulations. *Biomed J Sci & Tech Res* 53(1) 44313–44317. https://doi.org/10.26717/BJSTR.2023.53.008345

Ogidi, O.I. & Joshua, M.T. (2024). Invitro antioxidant and anti-inflammatory activities of selected polyherbal formulations sold in Nigeria. *Future Nat Prod* 9(2), xxx–xxx. https://doi.org/10.34172/fnp.2307-1252

Pandey, A., & Pandey, G.. (2014). Nanotechnology for Herbal Drugs and Plant Research. *Res Rev J Pharma Nanotechnol* 2(1), 13–16.

Parakh, S.R., & Gothoskar, A.V. (2003). Review of mouth dissolving tablet technologies. *Pharmaceutical Technology*. Duluth, MN: Advanstar Communications, 47–52.

Patra, J. K., Das, G., Fraceto, L. F., Campos, E. V. R., Rodriguez-Torres, M. Del P., Acosta-Torres, L. S., Diaz-Torres, L. A., Grillo, R., Swamy, M. K., Sharma, S., Habtemariam, S., & Shin, H.-S. (2018). Nano based drug delivery systems: Recent developments and future prospects. *J Nanobiotechnol*, 16, 71.

Raja, K., Ukken, J.P., Athul, P.V., Tamizharasi, S., & Sivakumar, T. (2011). Formulation and evaluation of maltodextrin based proniosomal drug delivery system containing anti-diabetic (glipizide) drug. *Int J Pharm Technol Res*, 3, 471–7.

Rupali, Sanjay Patare. (2022). Nanotechnology in herbal drug: a review. *Int J Sci Dev Res*, 7(12), 127–135.

Ruturaj, D. Dhanokar, A., Jatale, P., Swati, P. D., & Mangesh, K. H. (2023). Nanotechnology for Enhancing the Bioavailability of Herbal Drug. *Int J Creat Res Thoughts*. 11(10), 790–801.

Sahoo, S.K., & Labhasetwar, V. (2003). Nanotech approaches to drug delivery and imaging. *Drug Discov Today*, 8, 1112–20.

Sawyer, W.E., Jacob, D.E., Ogidi, O.I., Izah, S.C., & Popoola, A.M. (2023). Threats and Conservation strategies of Common edible vegetables that possess pharmacological potentials in Nigeria In: *Reference Series in Phytochemistry. Herbal Medicine Phytochemistry*. Springer Nature Singapore Pte Ltd.

Semalty, A., Semalty, M., Rawat, B.S., Singh, D., & Rawat, M.S. (2009). Pharmacosomes: The lipid-based new drug delivery system. *Expert Opin Drug Deliv*, 6, 599–612.

Sharma, A., & Sharma, U.S. (1997). Liposomes in drug delivery: Progress and limitations. *Int J Pharm*, 154, 123–40.

Sharma, G., Anabousi, S., Ehrhardt, C., & Ravi-Kumar, M.N. (2006). Liposomes as targeted drug delivery systems in the treatment of breast cancer. *J Drug Target*, 14, 301–10.

Sharma, M. (2014). Applications of Nanotechnology Based Dosage Forms for Delivery of Herbal Drugs. *Res Rev: J Pharma Nanotechno* 2(1), 23–30.

Shen, Y.J., Zhang, Z.W., Luo, X.G., Wang, X.F., & Wang, H.L. (2008). Nanoparticles of traditional Chinese herbs inhibit thrombosis in vivo. *Haematologica*, 93, 1457.

Shukla, N.D., & Tiwari, M. (2011). Proniosomal drug delivery systems – Clinical applications. *Int J Res Pharmaceut Biomed Sci*, 2, 880–7.

Song, Y.M., Ping, Q.N., & Wu, Z.H. (2005). Preparation of silybin nano emulsion and its pharmacokinetics in rabbits. *J Chin Pharm Univ*, 5, 427–31.

Sterer, N., Nuas, S., Mizrahi, B., Goldenberg, C., Weiss, E.I., Domb, A, et al. (2008). Oral malodor reduction by a palatal mucoadhesive tablet containing herbal formulation. *J Dent*, 36, 535–9.

Sungthongjeen, S., Puttipipatkhachorn, S., Paeratakul, O., Dashevsky, A., & Bodmeier, R. (2004). Development of pulsatile release tablets with swelling and rupturable layers. *J Control Release*, 95, 147–59.

Suresh, S., & Jain, V. (2016). Recent advances in herbal drug delivery systems and the challenge of bioavailability. *J Pharm Sci*, 105(8), 2505–2512.

Tangri, P., & Khurana, S. (2011). Niosomes: Formulation and evaluation. *Int J Biopharm*, 2, 47–53.

Tanwar, Y.S., Gupta, G.D., & Ramawa, K.G. (2006). Development and evaluation of microparticles of Gugulipid. *The Pharma Review*. New Delhi: Kongposh Publications Pvt., Ltd., p. 124–32.

Touitou, E., Godin, B., Dayan, N., Weiss, C., Piliponsky, A., & Levi-Schaffer, F., et al. (2001). Intracellular delivery mediated by an ethosomal carrier. *Biomaterials*, 22, 3053–9.

Verma, M., Gupta, P.K., Varsha, B.P., & Purohit, A.P. (2007). Development of transdermal drug dosage formulation for the anti-rheumatic ayurvedic medicinal plants. *Anc Sci Life*, 11, 66–9.

Vicentini, F.T., Simi, T.R., Del Ciampo, J.O., Wolga, N.O., Pitol, D.L., & Iyomasa, M.M., et al. (2008). Quercetin in w/o microemulsion: In vitro and in vivo skin penetration and efficacy against UVB-induced skin damages evaluated in vivo. *Eur J Pharm Biopharm*, 69, 948–57.

Wagner, V., Dullaart, A., Bock, A.-K., & Zweck, A. (2006). The emerging nanomedicine landscape. *Nat Biotechnol*, 24(10), 1211–1217.

Walve, J.R., Bakliwal, S.R., Rane, B.R., & Pawar, S.P. (2011). Transfersomes: A surrogated carrier for transdermal drug delivery system. *Int J Appl Biol Pharm Technol*, 2, 204–13.

Xiao-Ying, L., Luo, J.B., Yan, Z.H., Rong, H.S., & Huang, W.M. (2006). Preparation and in vitro and in vivo evaluations of topically applied capsaicin transfersomes. *Zhongguo Zhong Yao Za Zhi*, 31, 981–4.

Zhao, H.R., Wang, K., Zhao, Y., & Pan, L.Q. (2002). Novel sustained-release implant of herb extract using chitosan. *Biomaterials*, 23, 4459–62.

Zhaowu, Z., Xiaoli, W., Yangde, Z., & Nianfeng, L. (2009). Preparation of matrine ethosome, its percutaneous permeation in vitro and anti-inflammatory activity in vivo in rats. *J Liposome Res*, 19, 155–62.

Zhong, H., Deng, Y., Wang, X., & Yang, B. (2005). Multivesicular liposome formulation for the sustained delivery of breviscapine. *Int J Pharm*, 301, 15–24.

16 Targeted Delivery Systems for Herbal Drugs

Rizwan Ullah Khan and Syeda Pakeeza Fatima Naqvi
Institute of Chemical Sciences, Gomal University, D.I. Khan, Pakistan

Yubao Chen
Deputy Secretary General, Chinese Society of Biotechnology, China

Sohail Ahmad
Gomal Center of Biochemistry and Biotechnology, Gomal University, Dera Ismail Khan, Pakistan

16.1 FUNDAMENTALS OF HERBAL DRUG DELIVERY SYSTEMS

In recent years, herbal medicines have gained the worldwide attention of peoples and researchers due to their esthetic value, increased patient compliance and prominent therapeutic effects. Novel drug delivery systems for the delivery of herbal drugs possesses several advantages over conventional formulations. These include, among others, the enhancement of solubility, bioavailability, and protection from toxicity (Prajakta N. Dongare et al., 2021). The herbal drugs can be used in a more upright course with enhanced efficacy by incorporating them into suitable dosage forms (Guo et al., 2023). This can be achieved by designing novel drug delivery systems for such drugs. Such systems include liposomes, polymeric nanoparticles, micelle, nanocapsules, phytosomes, carbon nanotubes, DNA nano carriers, and aptamers-based delivery.

There has been an increased advancement in the field of herbal drug delivery systems, especially with the incorporation of the nanotechnology. This integration is intended to improve the solubility, effectiveness, and safety of the herbal products in which the different active compounds often have low solubility and bioavailability levels because of their complex molecular structures. Many of the preventive and curative formulations in the traditional system demand a greater amount of drugs to produce the desired therapeutic efficacy which invariably leads to more side effects and poor patient compliance. Some challenges require to be overcome in the development of herbal drugs through nanotechnology, which may enhance the pharmacokinetic properties of herbal extracts and deliver them to precise sites within the body. The application of nanotechnology in the herbal drug delivery system (HDDS) can effectively address these challenges. Nanotechnology-based drug delivery systems (NDDS) have been proven to be a revolutionary strategy to improve the medicinal efficacy of herbal drugs. These systems employ carriers in the nano range that are created using liposomes, solid lipid nanoparticles (SLNs), and polymeric nanoparticles to encapsulate valuable active constituents present in herbs. Moreover, it helps to enhance the efficacy profile of the active compounds mainly due to improved stability, solubility and bioavailability. This is because it has been discovered that nanoparticles of comparatively insoluble herbal extracts could improve their biopharmaceutical properties; in other words, the bioavailability of the compound, thereby optimizing therapeutic efficiency for minimum dosing (Jalili et al., 2023; Dewi et al., 2022; Verma et al., 2018). Further, it offers flexibility in minimizing the frequency of administration which may ease the

DOI: 10.1201/9781003544418-16

compliance of the treatment regimens especially compounded by the complex nature of the traditional herbal preparation dosage (Ambwani et al., 2018; Onyeji, 2022). This has also been made possible by the development of nano-sized herbal drug systems because the physicochemical barriers hampering the effectiveness of herbal remedies are also well addressed by this invention. A majority of the herbal constituents have been shown to be of low aqueous solubility and high molecular weights and thus cannot easily diffuse across biological membranes. These compounds, when formulated using nanotechnology, can be delivered in the form of nanoparticles that could easily cross lipid membranes and therefore augment their bioavailability (Kesarwani & Gupta, 2013; Azzahra et al., 2020). For instance, ethosomes and transfersomes are suggested to be a better method of transdermal delivery of herbal drugs, leading to better absorptive ability through the skin and a better response from the drugs (Chen et al., 2022; Rahman et al., 2020). Moreover, the application of nanocarriers in the system of herbal drug delivery enable the controlled and prolonged releases of the drug. This slow release is beneficial for the constant release of therapeutic drugs and regular upkeep of their blood concentration, thereby increasing patient compliance, especially in the context of compound chronic diseases (Jalali, 2022; Razavi, 2024). It is also possible to control the physicochemical properties of these nanocarriers so that they only release the herbal compounds at the desired site and time in response to some physiological stimuli such as pH or temperature (Bonifácio et al., 2013; Ansari et al., 2012). Indeed, nanotechnology is used in the improvement of the bioavailability of the active ingredients, controlled release and reduction of toxicity levels as associated with herbal medicines. The problem with these compounds is that they have several adverse effects. However, when herbal extracts are enclosed in nanoparticles, which are biocompatible and biodegradable, several of these issues are mostly likely to be avoided. This is specifically the case with plant extracts as these too may possess powerful biological effects, but are equally likely to cause toxicity when given in large doses (Darji et al., 2022; Wani et al., 2015). One advantage that is coupled with NDDS is the potential to give smaller concentrations of the active pharmaceutical ingredients but to give cure dosages. Furthermore, advances in technology has seen nanotechnology incorporated into the herbal medicine, which has provided a way through which diseases can be treated. Because the interaction between nanoparticles and tissues or cells depends on surface chemistry, researchers can modify the surface chemistry of nanoparticles so as to have a higher affinity to certain tissues or cells, thus enabling pharmacists to deliver herbal drugs to the desired location, such as tumors or inflamed tissues (Marella & Prasad, 2018; Gunasekaran et al., 2014). It also increases the therapeutic effectiveness and therapeutic window of the herbal medicine while minimizing the occurrence of systemic side effects and related risks to patients, thereby increasing patients' safety (Metkari, 2023; Gopi & Amalraj, 2016). The following are some of the real-life examples of nanotechnology in herbal drug delivery: Route of administration here also plays a significant role, the common routes being oral, transdermal and intravenous. Every route has its advantages and problems in terms of the improvement of delivery of the herbal products. For instance, oral delivery systems can be designed to protect herbal compounds from degradation in the gastrointestinal tract, while transdermal systems can facilitate the direct absorption of active ingredients through the skin (Arsude, 2023). Nanogels, and especially nanoemulsions have demonstrated improvements in stability and bioavailability of herbal formulations for numerous therapeutic uses (Alharbi et al., 2021).

16.1.1 Advantages of Herbal Drugs

Herbal formulations are preparations composed of one or more herbs or processed herbs, designed to provide nutritional, cosmetic, or therapeutic benefits for humans or animals. These formulations are produced through processes such as the extraction, distillation, or fermentation of whole plants, plant parts, or their derivatives, resulting in products like powders, tinctures, extracts, oils, and juices (Elkordy et al., 2021).

Herbal drugs have gained significant attention in recent years because of the various advantages that come with it as a natural product over chemically synthesized drugs. These benefits includes factors like: safety, effectiveness, cost and the prospect for fewer side effects. It is evident that the usage of herbs has constantly improved time by time due to the fact that these treatments have been used for thousands of years by the Ayurveda and traditional Chinese medicine practices. The use of herbal drugs has a number of benefits. of which one is that they are safe to use. Some of the herbal medicines are obtained from natural products, and thus it has been established that they have been in use for generations. When compared with synthetic drugs, the latter more frequently have severe side effects and toxicities (Kumari et al., 2021; Murshed et al., 2023; Zhou et al., 2017). This safety is especially beneficial where chronic conditions require long-term use of the intervention in question. For example, it has been confirmed that the herbal medicine is quite useful in the treatment and control of rheumatoid arthritis and there are fewer side effects compared to the traditional medicine (Li, et al., 2022; Vyshnevska et al., 2022). Economy is another advantage of herbal drugs that cannot be over-emphasized. In general, the cost of producing herbal medicine and procuring them is cheaper than synthesizing chemical-based medicines hence can reach many people (Akintelu et al., 2021; Murshed et al., 2023). This affordability is particularly importance in the low-income zones because people may not have adequate healthcare. Also, people that take herbal medicine do so at their own expense, hence they are less likely to burden the healthcare system with numerous charges, especially in the case of chronic diseases as they seek medical attention using these natural products, which are often cheaper than chemically produced drugs (Murshed et al., 2023; Mohammadi et al., 2020). Another factor that supports the effectiveness of herbal drugs is versatility of the preparation since they are often composite by nature. Most herbal medicines may contain several active ingredients which may, in a way, complement each other in an effort to boost their efficacy. Such mechanism of action of herbal drugs helps them modulate several pathways in the body, a valuable characteristic in treating complicated diseases such as cancer and diabetes (Yin et al., 2013; Yang et al., 2018). For instance, studies have suggested that herbal medicines enhance the body's ability to fight infections, also enhance the general well-being of cancer patients even if they use conventional treatments (Yin et al., 2013). In addition, the formulation of herbal drugs can be improved using newer concepts like phytosomes and nanocarriers. These new and complex drug delivery systems can enhance the dissolution and bioavailability of herbal ingredients, thereby enhancing therapeutic efficacy as reported by a number of researchers (Nandhini and Ilango, 2021; Gaikwad et al., 2021; Dongare et al., 2021). When such technologies are incorporated in the application of these plants, the curative value of the medicinal herbs can be enhanced, and therefore function as potential substitutes or adjuncts to modern synthetic drugs. However, the two options come with certain qualities that are linked with the usage of the herbal drugs; they have therapeutic values; they are also said to be holistic in the treatment. Most natural remedies for diseases are compounded in such a way that in addition to curing a particular disease, they help maintain health and harmony in the body. This view is of the whole-person approach which characterizes traditional systems of medicine and more so Ayurveda, where the focus is not on treating disease but on managing the affected individual. In addition, awareness regarding herbal medicines has grown significantly, creating awareness into formulation to expose new active ingredients and therapeutic uses. This is because this ongoing research is useful for the confirmation of the effectiveness of these herbal drugs destined for modern medicine and to establish their safety for use (Mohammadi et al., 2020; Kuruüzüm-Uz et al., 2012; Razavi, 2024). For instance, researchers have established that a number of phytochemicals possess strong anti-inflammatory and antioxidant characteristics that are useful for medical use (Kuruüzüm-Uz et al., 2012; Abo-Zeid et al., 2021). While appreciating the benefits of herbal drugs, one should not close eyes to the problem attendant with their use. Among the challenges that need to be considered are standardization, quality assurance and control, as well as possible herb–drug interactions (Dragoș et al., 2017; Ogawa-Ochiai & Kawasaki, 2019).

16.1.2 Challenges of Herbal Drugs

Around the globe to date, an estimated 422,000 species of plants have so far been identified for medicinal purposes. Of this total number, some 52,000 (around 12.5%) plant species are currently being used for different medical purposes. 4160 (nearly 8%) of the medicinally important plants are kept in the 'threatened' category. Many of these plants have not yet been fully explored for their true biological activities. Only a few of the plants have been identified, well-characterized and listed in official pharmacopoeias (Pan, 2014).

Therefore, herbal medicines and drugs are not without disadvantages, and hence, they cannot be applicable to all disease conditions. Although herbal drugs offer a lot of advantages, there is a downside as well, with some serious risks being associated with the consumption of herbal drugs (Studdert, et al. 1998).

With developments in biomedical sciences, advancements in isolation and characterization techniques, in vivo and in vitro studies, toxicological studies, novel drug delivery mechanisms and deeper understanding of biological systems, along with the discovery of new therapeutic targets in bodies, drug discovery overall remains a lengthy, very expensive, laborious, and inefficient process. Hence, there is still a low rate of new therapeutic drug discovery (Tang, K., 2011).

There are several notable challenges associated with the herbal drugs, including:

a. The regulations of herbal drugs. The US Food and Drug Administration (FDA) has much more stringent rules for the manufacturers of pharmaceutical drugs than are applied to herbal manufacturers.
b. Herbal drugs interact and interfere with other drugs and foodstuffs, which may cause serious, harmful effects (Raynor D. K 2011).
c. The lack of dosage instructions leads to either low dosage or over dosage since people take herbal medication by self-dosing.
d. Herbal drugs may pose serious allergic reactions. Before using herbal medicines, one must ensure the allergic test for that herbal drug. Conventional medicines are taken with a prescription, so they pose lower risks of allergy.
e. One major issue associated with herbal formulations is their adulteration with foreign substances like steroids and heavy metals. This can lead to serious renal problems.

The shelf-life of herbal medicines is another big issue. The herbal medicines are a mixture of different chemical substances. The variety of phytoconstituents may differ in terms of properties such as shelf-life, pharmacokinetics, and pharmacodynamics. Changes in these properties alter the biological actions of a drug. Currently, available knowledge cannot address these disadvantages of herbal formulations (Izzo, A. A. 2004).

16.1.3 Rise of Targeted Delivery for Herbal Drugs

The targeted drug delivery system is a technique that enables us to deliver the drug to the designated site of action and has a significant effect on the medicine's efficacy. What comes into play here is the role and the bioavailability at the target site alongside specific herbal drug activity (Dongare et al., 2021).

Traditional drug delivery strategies (oral, nasal, or transdermal injections as different routes of administration) display systemic adverse side effects (Hassan and Zhang, 2019). These effects may include inconsistent drug release, or off-target and non-specific biodistribution, which may disrupt the overall drug-targeting mechanism. Hence, the development of new drug delivery system facilitates the controlled release of drug payloads at the target site. Novel drug delivery systems can reduce herbal drug dosing frequency while keeping the drug concentration in the targeted cells, organs or tissue for a prolonged period (Singh & Sharma, 2023).

In the preceding sections, we will dive deeper into different carrier systems and detailed mechanisms involving the drug delivery systems overall, with the specific context of herbal drugs.

16.2 CARRIERS SYSTEMS FOR TARGETED DRUGS

Targeted drug delivery (TDD) is a precise method of delivering drugs directly to specific areas of the body, thereby reducing side effects and improving treatment effectiveness. TDD uses various carriers, such as polymers, nanoparticles, liposomes, and micelles, to transport drugs to their intended targets. The drug carrier systems used for the purpose of targeted drug delivery are in fact tiny cargo-carrying vehicles that transport drugs/active compounds or herbal compounds to a specific location in the body. The drug cargo-carrying vehicles are designed in such a way that they support the safe and controlled release of the drug molecules at the target site. They also avoid unloading the drug off target (Dunuweera, 2019).

Nanotechnology and nanomedicine are making great contributions to revolutionizing the horizon of drug delivery. The purpose of these domains of research is to offer better drug targeting as well as safe and controlled release of the drug compounds. Traditional delivery systems have many shortcomings, which include rapid drug release from cell or target site and low specificity. Nanotechnology offers a lot of variety in the form of nanocarriers, which provide a very precise and efficient way to transport drugs to their intended target sites.

Nanocarriers offer a variety of chemistries and combinations. Different polymers (bio-based polymers and inorganic polymers or a mixture of both), nanoparticles (metallic, non-metallic, biological), and liposomes (phospholipids) may all act as tiny delivery vehicles that enhance drug delivery, targeting, and safe release & effectiveness. By reducing side effects and minimizing adverse effects, nanomedicine offers a promising approach for improving the effective treatment of a number of diseases. Overall, nanocarriers enhance the cellular update of drug and therapeutic outcomes.

In this section, we will discuss liposomes, polymeric nanoparticles, nanocapsules and nanospheres, polymeric micelles, dendrimers, carbon nanotubes, aptamers, and DNA nanocarriers. We will discuss the currently used drug carriers as well as explaining the exciting domains of DNA nanotechnology and functional nucleic acids as potential drug delivery vehicles. These relatively new fields of science and biomedical research have not yet been explored and herbal drugs are not making full use of the potential of these fields.

16.2.1 Liposome-Mediated Drug Delivery System

Liposomal drug delivery systems can be considered to be one of the most significant advancements in the pharmaceutical sciences, especially in relation to the optimization of the pharmacodynamics and pharmacokinetic characteristics of different drugs. These systems incorporate liposomes which are spherical vesicles consisting of a lipid bilayer and are capable of encapsulating both aqueous soluble and water-insoluble drugs, thereby enhancing their stability, solubility, and pharmacokinetic properties. The relative size and shape of liposomes can be used to release their contents gradually and with a low risk of undesirable side effects at loci tissues, organs, and cells where the encased agents need to act (Gumulec et al., 2014; Smits et al., 2018). An important advantage of liposomal formulations is that they can improve the solubility and stability of compounds which are poorly soluble in water. For example, doxorubicin encapsulation in liposomal formulations was demonstrated to increase its therapeutic window through a decreased cardiotoxic effect without compromising the cancer cell-killing potential (Gumulec et al., 2014; Smits et al., 2018). This entrainment not only shields the drug from being degraded but also assists the drug in crossing biological membranes, thereby boosting its bioavailability. This has been done to show that liposomal formulations can produce a higher plasma level of drugs than the non-encapsulated drugs, which in turn enhance the therapeutic effects (Yi et al., 2013; Nogueira et al., 2015b). In addition, liposomes can be designed to have the capabilities of passive or active targeting of tissues or cells. Passive targeting makes use of what is known as the enhanced permeability and retention (EPR) effect, whereby, liposomes are allowed to penetrate and adhere to tumor tissues due to the hyperpermeability of their blood vessels (Shigehiro et al., 2014; Sercombe et al., 2015). Active targeting, however, concerns the alteration of liposomal surfaces with ligands that can interact with receptors present over the targeted cells, such as folic acid receptors in cancer cells Nogueira et al., 2015a). It not only increases the accumulation of the drug at the targeted site but also decreases the side effects at the other sites, further improving the overall safety of the treatment (Niu et al., 2015). The formulation versatility of liposomal delivery systems is an added bonus to the versatility of the system. Liposomes can incorporate almost any pharmacologic agents in their core or incorporated in the bilayer, such as small molecules, peptides, proteins, and nucleic acids for delivery (Akbarzadeh et al., 2013; Schwendener & Schott, 2017). Such flexibility also enable a technology that can encapsulate several drugs simultaneously, which is beneficial when the drugs are administered together in the cases of chronic illnesses such as cancer. For instance, the ability to deliver chemotherapeutic agent and siRNA simultaneously using liposomes has been found to improve the effectiveness of overall treatment also lower the dosages of each agents (Niu et al., 2015; Grace, 2014). In addition to its uses in therapy, liposomes have also been considered as carriers for vaccines. Liposomes have the capability to successfully delivering antigens and improve the immunogenicity of vaccines. This is well informed by the development of liposomal adjuvants for the enhancement of immune response to co-administered antigens, which is a process that is discussed by Schwendener & Schott (2017). Liposomes possessing capacity to encapsulate both the hydrophilic and the hydrophobic component make them a suitable approach for developing new vaccine formulations. Although the liposomal drug delivery systems have been associated with many benefits, there are several challenges that need to be overcome for the system to have higher value in clinical applications. One major concern is the low percentage of encapsulation that is sometimes reported with certain drugs, hence confining the benefits of liposome formulations (Skrinda et al., 2021; Sreekanth et al., 2017). There are great disparities in lipid composition, preparation methods, and drug properties which influence the efficiency of encapsulation and release kinetics (Teong et al., 2014; Pentak et al., 2011). The continual study is directed towards modifying these parameters to optimize situations with liposomal systems. In addition, the tendency of liposomal formulations to degrade in the course of storage, and their compatibility during administration, is of the essence. Liposomes are liable to aggregation, fusion, and leakage out of the sheltered pharmaceutical, and that again will affect its usefulness (Guimarães et al., 2019; Pauli et al., 2019). Other approaches include the addition of stabilizers or the employment of the microfluidic technology for the improvement of liposomal formulations' shelf stability (Elsana et al., 2019; Zheng et al., 2022).

More recently, advancements in liposome-mediated drug delivery have centered on bioconjugation strategies. Liposomes, a widely used nanoparticle carrier, have been modified to improve drug loading, targeting, and overall efficacy. These improvements aim to enhance the delivery of therapeutic agents to various sites of action (Almeida et al., 2020).

Despite all the potential applications of liposomes, the development and application of liposomes remain a challenge due to their complex production processes. Researches are being carried out with the aim to overcome these hurdles (Farooque et al., 2021).

16.2.2 Polymeric Nanoparticles as Drug Carriers

Polymeric nanoparticles rang in size from 1 to 1000 nm. These nanoparticles can be loaded with either drugs or active compounds. The loaded compounds are either trapped inside or absorbed onto the surface of the polymeric core (Zielińska, A.,2020). The term 'nanoparticle' can be used in the cases of both nanospheres and nanocapsules. Nanocapsules are made up of an oily core. In this oily core the drug is dissolved. The core is encircled by a polymeric shell, which controls the release of the entrapped drug from the core. By contrast, s nanospheres contain a continuous polymeric network. The loaded compound drug may be retained inside the core or adsorbed onto the surface of

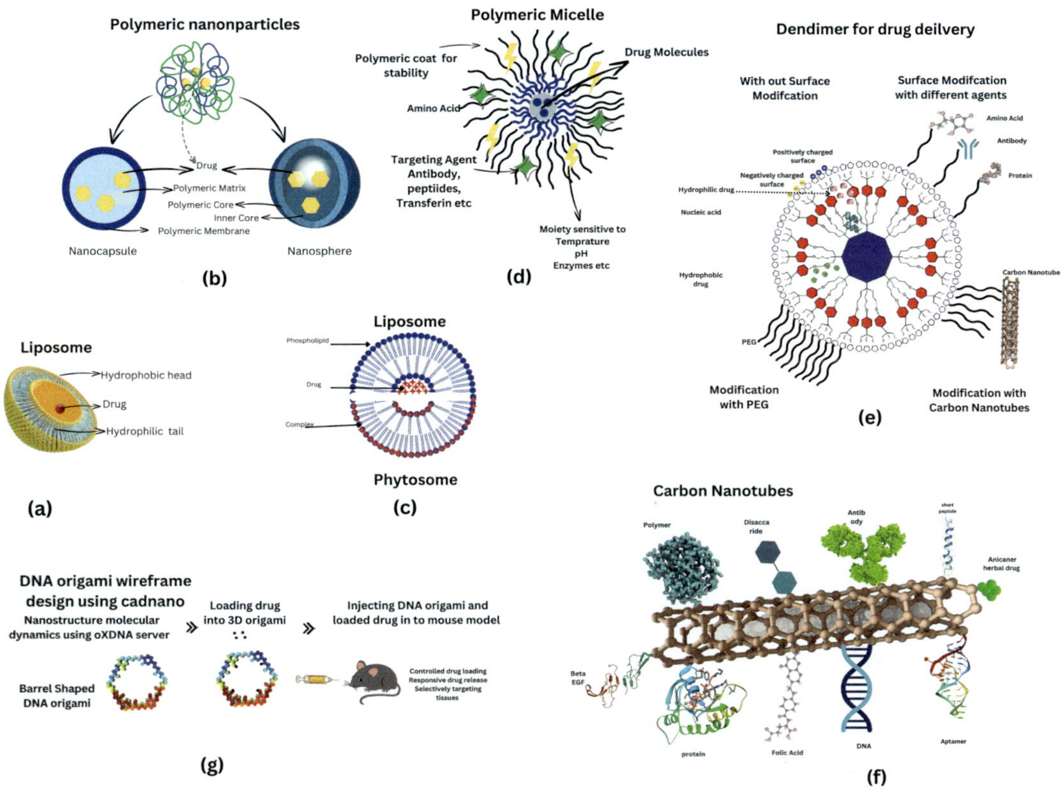

FIGURE 16.1 a) **Liposome** Hydrophobic heads towards the outside and hydrophilic tails towards the inner core, drug molecule in the center. b) **Polymeric Nanoparticle**, drug molecule in centre, Nanocapsule and Nanosphere shown containing the drug molecules. c) **Phytosome** with drug molecules . d) **Polymeric Micelle**, drug molecules, Targeting agent in green, moiety sensitive to stimuli, polymeric coat e) **Dendrimer**, upper left side without modification of functional groups, positively and negatively charged groups in, hydrophobic drug and nucleic acid shown. Upper right side, modification with different agents' amino acid, antibody and protein. Lower right side, modification with carbon nanotube. Lower left side, modification with PEGylation, f) **Carbon Nanotube** carrying different types of cargo, g) Barrel shaped **DNA origami**, structure made up of scaffold and staple strands, different cargos can be loaded on to origami, the cargo is injected in mice to deliver drugs to target site.

nanosphere (see Figure 16.1b). Essential oils have been efficiently delivered by polymeric nanoparticles (Lammari N., 2020).

The main characteristics of polymeric NPs, are that they are biocompatible and biodegradable and that they pose low-toxicity risks (Pinelli F., 2020).

According to a study conducted by Maurya et al. (2019), polymeric NPs can be safely used by humans. Similarly, polymeric NPs can significantly improve the bioavailability of the loaded drugs. They are stable and can encapsulate large amounts of guest molecules. Recent studies on biodegradable and non-biodegradable polymers suggest that both type of polymeric nanocarriers can be used for oral drug delivery (Maurya A., 2019).

It is worthwhile mentioning here the successful study in which pure curcumin was compared with curcumin-loaded polymeric NPs. The polymeric formulation provided 5.6-fold higher oral bioavailability (Chen, Y., 2020).

Polymeric nanoparticles are emerging as a groundbreaking solution to address the complexities of drug delivery. Key advantages of polymeric nanoparticles include enhanced bioavailability, targeted delivery, controlled release, and versatility. Applications include cancer therapy, ocular drug delivery, and nutraceutical delivery. The field of polymeric nanoparticles is rapidly evolving as scientists continue to explore the potential of these innovative materials. We can expect to see even more exciting advancements in drug delivery in the years to come (Begines et al., 2020).

Polymeric nanoparticles show promise in cancer treatment. They can deliver drugs directly to tumors, thereby reducing side effects and improving therapeutic outcomes. These nanoparticles have potential applications in various cancers: ovarian, colorectal and others, and brain tumors (Madej et al., 2022).

Nanoparticles can be modified to improve their effectiveness and they are being explored for various therapeutic applications (Dristant et al., 2023).

16.2.3 Micelles

Polymeric micelles are promising nanocarriers for the delivery of anticancer drugs, offering advantages such as targeted drug delivery, improved biocompatibility, and reduced side effects. (Elumalai et al., 2024).

Micelles involve self-assembly techniques. Micellar carriers are used in vast nanomedicine applications. They involve techniques, including:

- **Interaction with biological substances**
- **Kinetic stability**
- **Drug release**

These techniques are crucial for understanding drug delivery systems (Ghezzi et al., 2021).

Polymeric micelles are promising nanocarriers for delivering anticancer drugs. They can encapsulate both hydrophilic and hydrophobic drugs and release them in a controlled manner.

Micelles are useful for delivering different anticancer drugs. Considering factors such as methods and computational analysis, micelles are promising nanomaterials that can be used to target drug delivery (Guzmán Rodríguez et al., 2023. Physically, micelles are fluorescent materials that exhibit enhanced emission when aggregated. They have various applications in bioimaging, drug delivery, and therapy (Liu et al., 2023). Micelles are useful nanostructures that can be used in the targeting of cancer treatment. They can deliver both drugs and genes, enhancing cancer therapy. Various types of smart micelles, including pH-sensitive and multi-responsive micelles, can be used for targeted drug delivery. The combination of micelles and nanostructures can further enhance their potential in cancer treatment. Studies have shown that micelles are highly compatible and safe (Gao et al., 2024).

16.2.4 Dendrimers

Dendrimers are highly branched polymers. They were first developed in 1978 by Buhleier et al., when they were named cascade polymers. These cascade polymers laid the foundation of polypropylene imine dendrimers, which were developed in the 1990s. The name dendrimer is derived from the Greek words 'dendron,' which means tree or branch, and 'meros.' Which means part.

At molecular level, dendritic branching gives a semiglobular or globular structure to the dendrimer. The surface contains a small molecular volume with a high density of functionalities. The structure of a typical dendrimer comprises three different structural parts (see Figure 16.1e):

a. Core at the center (either a single atom or an atomic group), having at least two identical chemical functions.
b. Building blocks providing numerous interior layers made up of repeating units. It provides flexible space to different guest molecules.

c. Multiple functional groups at the peripheries of macromolecule exteriorly. It plays the major key role in the overall properties of the dendrimer. This part defines the macroscopic properties of the dendrimer.

This structural assembly and overall 3D structural complex of dendrimers confers unique properties, such as a globular shape at the nanoscale, functional groups at the peripheries, hydrophilic and hydrophobic cavities in the interior, and low polydispersity.

The globular structures, and the diameter of less than 10 nm, is very similar to the molecular size of biomolecules and proteins. Hence dendrimers act as biomimics. Overall, dendrimers provide structural complexities along with opportunities for accommodating and carrying drug molecules, targeting moieties like antibodies, and solubilizing groups like PEG on the surface (see Figure 16.1e).

Dendrimers have demonstrated their potential as nanocarriers for different types of drugs like antimicrobial, anticancer, and anti-inflammatory drugs.

16.2.5 Carbon Nanotubes and Fullerenes

Carbon nanotubes (which are abbreviated as CNTs) have a cylindrical shape. The sheets of carbons are, in fact, hexagonally ordered carbon atoms. This structural feature gives the nanotubes a diameter of few nanometers and length ranges of up to a few micrometers. Carbon nanotubes may be either single-walled (SWCNTs) or multi-walled (MWCNTs) (Zare, H., 2021).

Structurally, carbon nanotubes are hollow nanofibers, and this structural characteristic makes them attractive candidates for the delivery of drugs and diagnostic agents. Similarly, other properties of CNTs including high biocompatibility, enhanced conductivity, high surface-to-volume ratios, ease of functionalization, strength, and optical properties (Serpell, C. J., 2016) make them novel drug delivery carriers (see Figure 16.1f).

The biggest concerns associated with the use of CNTs in the field of biomedical sciences have centered around two main issues, biodegradability and toxicity.

In addition to these negative features of carbon nanotubes, a lot of research is still being carried out in this domain of science. Among the notable applications of CNTs are their functions in the fields of biosensors, drug delivery systems, bioimaging, vaccine delivery, gene delivery, gene therapy and diagnostic applications (Mahor, A., 2021).

Fullerene is an allotrope of carbon and usually described as a molecular form of carbon. In fullerene, carbon atoms ($n > 20$) are clustered on a spherical surface. The most studied of these forms is fullerene C60. The water solubility of fullerenes can be improved through the grafting and functionalization of different chemical groups. In this way new drugs can be obtained. Different studies have been reported in which functionalized fullerenes have been used for therapeutic and diagnostics applications. Functionalized derivatives of fullerenes are mainly focused in the areas of quenching reactive oxygen species (ROS), drug delivery and bioimaging (Debnath, S. K., 2021).

But keeping the real-life concern in biomedical sciences, the field of CNT still has a long way to go due to the inaccurate and relatively low number of toxicity studies (Nimushakavi, S., 2021). Nevertheless, it is hoped that more research in the coming times will make CNTs workable. In this respect, herbal medicines can benefit from this domain of nanotechnology for drug delivery purposes.

16.2.6 Phytosomes

Phytosomes are lipid-based complexes of herbal drug/plant extract with phospholipids. It's a relatively new technology, being invented by Indena in 1989. Another name for phytosomes is phytophospholipid complexes. The name itself is self-explanatory, since it describes the interaction between herbal drug/phyto compounds and the hydrophilic parts of phospholipids through hydrogen bonds (see Figure 16.1c).

Phytosomes make use of the power of liposomes and hence enhances the absorption & bioavailability of herbal drugs. Phytosome technology is applicable to help facilitate the delivery of different herbal drugs for various diseases. The pharmacokinetic and pharmacodynamic parameters can be improved via the use of phytosomes (Mahadev B. Khanzole, 2020).

Phytosomes enhances the bioavailability of phytoconstituents, especially hydrophilic groups containing herbal drugs, and hence increases drug efficacy. Phytosomes have been tested by different research groups for comparing the bioavailability of the herbal compounds alone as well as in phytosome form. Here are a few phytosome based herbal drugs available on the market. Ginkgo biloba extract, olive oil compounds, silymarin, ginsenosides, green tea catechins, anthocyanins and other flavonoids, curcumin, and visnadine (Prasad, S. B 2016).

16.2.7 DNA Nanocarriers for Targeted Drug Delivery

DNA nanostructures, also called DNA origami, make use of the basic building blocks, i.e. nucleotides. The nucleotides work as basic building material that self-assembles to form special kinds of nano-scale structures (see Figure 16.1g). The resulting DNA nanostructure may be 1D, 2D, or 3D molecular structures with controlled shapes and sizes. The Watson-Crick base pairing is the logic behind the design of the DNA origami structure. DNA origami is designed in software such as cadnano and the nano structure is made up of a large Scaffold (M13 phage genome) and short staple strands (Castro et al., 2011).

A number of webservers have been introduced over the past few years to facilitate the origami design. Desired structure and shape and size can be designed in the software, depending upon the requirements. The most notable servers which contributed to this field of science are cadnano and OxDNA. Once a desired structure is designed, it is large enough to carry drugs to the target site. Keeping in view that the basic structure is solely made up of nucleotides, its nonimmunogenic. DNA nanostructures has vast applications in the pharmaceutical and biomedical fields. DNA origami can be designed with spaces or voids in its structure that can excellently hold the drug cargo without any external disturbance or effect. Thus, DNA nanostructures serves as a good delivery vehicle for many types of drug molecules (Shishparenok et al., 2023).

DNA nanostructures serve as a versatile platform for a large number of application like drug delivery, biomaging, and theranostics. The main features of DNA origami include:

1. Improved drug effectiveness through effective solubility, circulation in body fluids, and cellular uptake.
2. Customizable size, shape, and chemical properties because DNA origami can be merged with different chemistries and surface modifications such as antibodies, proteins, nucleic acid aptamers, drugs, and other functional groups. This makes DNA nanostructures a versatile platform.
3. Controlled drug release mechanisms at target site with low off-targeting.
4. Size-dependent organ distribution of cargo and effective tumor targeting.

These features make DNA nanostructures/origami a promising tool for targeted drug delivery and other therapeutic applications (Kumar et al., 2023).

Cancer therapy, targeted gene therapy, vaccine delivery, and bioimaging are among the major areas where DNA nanotechnology is currently being applied at the research scale. Personalized medicines is another promising area where DNA nanocarriers offer the solution to the challenges posed by conventional drug delivery systems (Zheng et al., 2022).

Since DNA nanotechnology is still in its initial stages of development, it has a long way to go. Keeping in view the potential of DNA nanotechnology, the field of herbal drug delivery will benefit from it.

16.2.8 Aptamers for Drug Targeting

Aptamers are short stretches of single-stranded DNA or RNA. Previously, had been thought that nucleic acids only perform genetic roles. A few decades ago, however, scientist reconsidered this theory and stated that nucleic acids can also perform other functions. It was at this point that the term functional nucleic acids was introduced. The binding of nucleic acids to ligands laid the foundation of this field of science (Nimjee et al., 2017).

These are now termed aptamers and they can perform functions such as ligand binding. At present, the discovered aptamers can bind with a vast range of targets, starting from small organic molecules, inorganic groups to complex protein, and even entire cells.

Since aptamers can bind with the receptors on the cell membrane and facilitate themselves or conjugated nanostructures to enter the cells. Hence, aptamers can be used as excellent targeting ligands for novel drug delivery systems (He et al., 2018).

Aptamers can precisely deliver functional nucleic acids to target cells. They bind specifically to cell surface proteins, ensuring accurate delivery (Xie et al., 2023). Aptamers can improve the effectiveness of natural compounds in targeted drug delivery. They have many advantages like stability and simple synthesis and a wide range of targets. Combining natural products with aptamer-based delivery systems offers potential for broader applications in medicine (Joana Gamboa, 2024).

16.2.9 Microspheres and Micropellets

Microspheres are tiny spherical particles, with sizes ranging from 1 to 50 μm. It offers a promising approach to herbal drug delivery. These special microparticulates enhances drug concentration at the target site while minimizing the unwanted effects. Microspheres enhance the efficacy of the drug and hence patient compliance by prolonging the drug action as well as keeping plasma concentrations steady. It potentially reduces the overall herbal drug dosage and minimize side effects by lowering off-targeting (Elkordy et al., 2021).

To date, different herbal ingredients have been successfully formulated into microspheres. Rutin, camptothecin, and quercetin are the prominent examples of this application. Recent studies have made use of the variable microsphere, the immune and the magnetic microspheres. Immune microspheres incorporate the coating of antigen or antibody on the polymer surface (Sarangi and Padhi, 2018).

16.3 TARGETING STRATEGIES AND MECHANISMS

Nature has equipped plants with a huge array of secondary metabolites which serve two major roles: defense against predators and signaling within plant body. Generally speaking, these secondary metabolites demonstrate a huge number of biological and pharmacological properties. The early civilizations started using plant extracts and exploiting specific plant products and extracts for treating a number of ailments. Secondary metabolites exhibit two kind of bioactivities: a) specific and b) nonspecific.

Specific Bioactivities The secondary metabolites which interact with the specific targets within or on the surface of cells, such as membranes, proteins or nucleic acid molecules. For example, alkaloids on receptors of neurotransmitters.

Nonspecific Bioactivities Secondary metabolites like phenolics and terpenoids are less specific in their actions and attack large number of proteins by building different bonds like hydrogen, ionic bonds and hydrophobic bonds, which in turn change the 3D structures of proteins and, as a result, modulate their bioactivities (Wink, 2015).

Due to these variations and diversities, herbal drugs follow different mechanisms for targeting and hence their mode of actions differ.

16.3.1 Ligand-Receptor Mediated Targeting

There is a difficulty in getting drugs into the brain because of a protective layer called the blood–brain barrier. A special type of chemistry, called receptor-ligand chemistry, is used to help drugs cross this barrier and reach the brain. This could be useful for treating brain diseases like Alzheimer's, Parkinson's and Lou Gehrig's disease (Anthony et al., 2021).

The development of multi-target directed ligands is a promising approach for the treatment of complex diseases. It focuses on multi-protein-coupled receptors as key targets for multi-target directed ligands due to their involvement in various human diseases (Huang, *et al.*, 2021).

Ligand-decorated nanoparticles are used for targeted drug delivery in cancer therapy as they can selectively target cancer cells without harming healthy tissues (Bandyopadhyay *et al.* 2023). By incorporating specific ligands onto the polymeric nanomicelles (PNMs), they can bind to receptors on the surface of target cells, facilitating their uptake and improving therapeutic efficacy. This receptor-mediated targeting approach can help overcome the limitations of passive targeting and enhance the delivery of drugs to specific tissues and organs (Agwa *et al.*, 2023).

16.3.2 Antibody Drug Conjugates

Antibody drug conjugates have made significant advancements over the past decade. They are a type of chemotherapy that combines a cytotoxic agent with a monoclonal antibody through a spacer arm. While only one antibody drug conjugate had been approved by 2009, there are now nine approved ADCs and over 80 in clinical trials (Gogia et al., 2023).

Antibody-drug conjugates (ADCs) are a promising cancer treatment that combines a monoclonal antibody with a cytotoxic payload (Lu et al., 2023).

ADCs have evolved from broadly targeting cancer cells to precisely targeting them and bolstering the immune system to combat drug resistance. ADCs were initially developed in preclinical trials in 20th century & the second generation addressed the issue through innovative antibody modification techniques. At present, a third generation of ADCs is in development (Song et al. 2023a, 2023b).

ADCs have revolutionized targeted cancer therapy. Despite challenges, recent advancements have led to increased development and approvals. Novel conjugation and linker technologies are expanding the scope of antibody drug conjugates (ADCs) (Bulger et al., 2023).

16.3.3 Aptamers for a Targeted Delivery System for Herbal Drugs

Aptamers are single-stranded nucleic acids or short peptides, ranging in size from 15 to 80 nucleotides (ssDNA and ssRNA, XNA or peptide aptamers). Aptamers were discovered in the 1990s and selected or evolved through a process called Systemic Evolution of Ligands with Exponential Enrichment (SELEX). Aptamers can bind selectively and specifically to proteins, DNA, RNA, small molecules, carbohydrates, vitamins, metabolites, drugs, and any other ligand. Aptamer can even bind whole cells or target a specific organelle inside cells or surfaces of bacteria and other microorganisms (Gold et al., 2012). Aptamers are comparable to antibodies in their binding behavior. But antibodies are more costly to produce at industrial scale where aptamers are cost-effective. However, despite all the research and development in the promising domain of aptamers, they have still a long way to go to enter into real competition with the antibodies in the global market (Liu et al., 2022). Aptamers have a vast number of applications:

a. It can act as a binder against ligands (Gold et al., 2012).
b. It can work as direct biologic therapeutics (Wang et al., 2023a)
c. It can be used as an imaging molecule in diagnostics (which can be detected based on chromogenic material like fluorescence labelling attached to aptamer bound to specific target site in a sample) (Liu et al., 2022).

d. It plays a role in drug delivery mechanisms, as a cargo carrier molecule.
e. Aptamers can also work as binding to nanomaterials to make complex systems for drug delivery (Liu et al., 2022).

Based on these significant applications, aptamers open a window on research in drug designing, drug delivery, drug targeting, the diagnosis and prognosis of diseases and a lot more is yet to come (Wang et al., 2023b).

Keeping in view the challenges of herbal medicines, e.g. efficacy, targeting and retention, aptamers can be effectively utilized for addressing the aforementioned issues the potential of aptamers has the answer. Recently, many groups have conducted studies in this regard, by merging the power of aptamers to counter the challenges of plant-based therapeutics. One study, conducted by Gamboa et al. (2024), recently offered a thorough account of the use of aptamers in drug delivery for plant herbal medicines. The study analyzed published literature on the above said topic from January 1, 2018 to December 31, 2023. Their analysis concluded 43 such studies, which made use of aptamers for herbal drugs delivery. In the same study it was demonstrated that natural compounds extracted from plants, with known therapeutic potential but with low expression in clinical practice, were more significantly used for aptamer-related studies (Gamboa et al., 2024).

In these studies, herbal compounds were logically loaded into different kind of nanoparticles and the complexes were modified with different aptamers (see Figure 16.2). This approach was

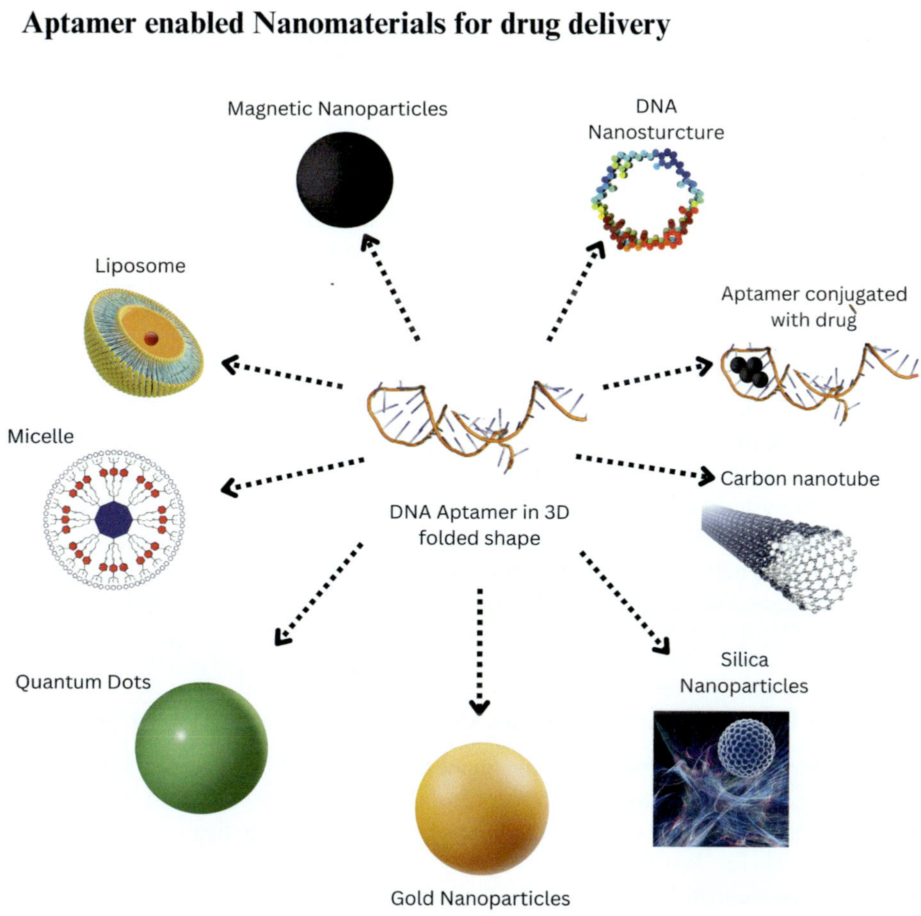

FIGURE 16.2 Aptamer-based therapeutics and the variety of nanomaterials utilized for drug delivery.

promising to overcome herbal drugs' generic issues and problems, like bioavailability, solubility, and stability and tissue permeation with extended retention time (Liu et al., 2022). This slightly unexplored area can be explored further with the addition of more innovative technologies like combining herbal drugs, aptamers, and nanobiotechnology for more effective and novel drug delivery systems.

16.3.4 Stimuli-Responsive Delivery Systems

Different targeting strategies and mechanisms are useful for drug delivery systems. The nanomaterials (e.g polymers, liposomes, micelles, etc.) encapsulate drugs and then external stimuli, such as temperature, pH and so on trigger drug release. One excellent method for targeting cancer cells is stimuli-responsive medication delivery, which makes use of particular characteristics and triggers. Choosing nanomaterials with efficient drug encapsulation and taking into account the stimuli that would cause drug release at the tumor location are essential. With this strategy, side effects could be reduced and therapeutic efficacy could be increased through the possibility of precise and targeted therapy.

Nanomaterials encapsulate drugs and then target specific sites. Biological ligands facilitate active targeting and uptake. Thus, it is also another effective mechanism. The strategies that can be adopted are the following: Optimize nanoparticle properties, use biological ligands & prioritize in vivo studies (Yijie Wu, 2019); Stimuli-responsive drug delivery systems to release drugs only in tumor environments, maximizing effectiveness in cancer treatment (Sun and Davis, 2021). Ligand-based active therapy is a promising approach for treating various cancers and it also improves targeting.

16.4 APPLICATIONS OF TARGETED DRUG DELIVERY SYSTEMS FOR HERBAL DRUGS

16.4.1 Herbal Drugs for Communicable Diseases

Communicable diseases, those transmissible from person to person, pose a persistent global health challenge. Viruses, in particular, have demonstrated an alarming capacity to mutate and evade immune responses, necessitating a continuous search for effective countermeasures.

Polyphenols, a vast class of naturally occurring compounds, are a promising strategy to combat these infectious diseases. Polyphenols, which number more than 10,000, are ubiquitous in the plant kingdom. Viral infections, such as COVID-19, SARS, and MERS, are all communicable diseases and spread from person to person. Thus, polyphenols which are derived from plants are used as herbal medicines to treat the communicable diseases.

Herbal drugs, with their rich phytochemical profile, offer a promising avenue for exploring the therapeutic potential of polyphenols. These plant-based remedies have been used for centuries in traditional medicine systems to cure communicable diseases. For example, thyme, oregano, mint, and basil have been employed for centuries to alleviate and combat contagious diseases (Garcia, 2020).

Many plants contain natural substances that work together to fight viruses and bacteria that cause communicable diseases. We can cure *E. coli* infections and skin infections through the use of herbal medicines.

16.4.2 Herbal Drugs for Communicable and Non-Communicable Diseases

Non-communicable diseases are those which cannot be transmitted from person to person.

Natural medicine has ancient roots, with records dating back to 2600 BCE in Mesopotamia. Key historical documents showcase the widespread use of plant-based remedies:

1. Mesopotamian records(2400 BCE): ~1000 plant-derived compounds
2. Egyptian Ebers Papyrus (1500 BCE): >700 natural drugs

3. Chinese Materia Medica (1100 BCE): 52 medicinal preparations
4. Indian Ayurvedic texts (1000 BCE): >800 natural extracts

Notably, Hippocrates incorporated herbal healing (phytotherapy) into his medical practice, demonstrating the enduring influence of natural remedies across diverse ancient cultures and medical traditions (Shaito et al., 2020).

The World Health Organization's Traditional Medicine Centers identified 122 key medicinal compounds derived from just 94 plant species. These compounds are used similarly across different countries for traditional treatments. For example:

a. Galegine, extracted from *Galega officinalis* L. (goat's rue), led to the development of metformin and other antidiabetic drugs.
b. Papaverine, from *Papaver somniferum* (the opium poppy), inspired the creation of verapamil, a medication for high blood pressure.

Despite a shift towards synthetic drug development in the late 1980s, natural products have regained prominence in drug discovery due to facts such as advances in scientific research demonstrating their efficacy in disease prevention and treatment. Economic accessibility among communities and perceived safety by general public, especially in developing countries. There is unparalleled biological activity and structural diversity compared to synthetic libraries, and huge potential for optimization with minor structural changes.

However, challenges remain such as the need for interdisciplinary approaches in natural product-based drug development for communicable diseases. There are concerns about herb–drug interactions and safety profiles in living systems. There is a need for more focus on structural modifications to improve potency and pharmacokinetics for herbal drugs. African herbs that interact with the cytochrome P450 (CYP) enzyme system & P-glycoprotein (P-gp), which are crucial for drug metabolism and transport, can also affect the efficacy of drugs used for the prevention of non-communicable diseases such as kidney and liver diseases, heart diseases, and inflammatory processes and neurological diseases (Ondieki et al., 2017).

Overall, natural products remain a rich, largely untapped source for drug leads, offering 'privileged scaffolds' for drug design, despite only a fraction of nature's biodiversity being tested for biological activity.

Antioxidants are also crucial in the prevention of non-communicable diseases, such as aging, kidney and liver disease, heart diseases, inflammatory processes and neurological diseases.

16.5 CONCLUSION AND FUTURE PERSPECTIVE

In the era of AI, there have been advancements in research in the domains of drug discovery, drug delivery, natural products, drug repurposing, the identification of new herbal compounds, the discovery of new drug targets within human bodies, nanobiotechnology, metabolomics and bioinformatics. Scientists from these multidisciplinary areas of research are joining hands together and finding new avenues and possibilities to bring more herbal drugs to the market. The limitations associated with herbal compounds and drugs such as poor pharmacokinetics, lower specificity and side effects like potential toxicity can be effectively encountered through the strategic use of precise combinations of nanomaterials, targeting molecules like antibodies and aptamers. In the years to come, we expect to see more studies on herbal drugs delivery in which more complex chemistries will be utilized for the delivery of herbal drugs to target sites with lower side effects and off-targeting.

Given the pace of developments and the way in which AI is being integrated into daily lives, AI will obviously play a future role as a catalyst in herbal drug discovery and drug delivery.

REFERENCES

Abo-Zeid, Y., Diab, R., Sanad, R., & Skran, W. (2021). Recent advances in herbal-based nanomedicine for anti-inflammatory purposes. *Journal of Advanced Pharmacy Research*, 5(4), 387–397. https://doi.org/10.21608/aprh.2021.79946.1135

Agwa, M. M., Elmotasem, H., Elsayed, H., Abdelsattar, A. S., Omer, A. M., Gebreel, D. T., Mohy-Eldin, M. S., & Fouda, M. M. G. (2023). Carbohydrate ligands-directed active tumor targeting of combinatorial chemotherapy/phototherapy-based nanomedicine: A review. *International Journal of Biological Macromolecules*, 239, 124294. https://doi.org/10.1016/j.ijbiomac.2023.124294

Ajazuddin, and Saraf, S. (2010). Applications of novel drug delivery system for herbal formulations. *Fitoterapia*, 81(7), 680-689. https://doi.org/10.1016/j.fitote.2010.05.001

Akbarzadeh, A., Rezaei-Sadabady, R., Davaran, S., Joo, S., Zarghami, N., Hanifehpour, Y., … & Nejati-Koshki, K. (2013). Liposome: classification, preparation, and applications. *Nanoscale Research Letters*, 8(1). https://doi.org/10.1186/1556-276x-8-102

Akbarzadeh, A., Rezaei-Sadabady, R., Roointan, A., Namn-Rahmat, A., & Jafari, S. (2023). Liposomes: Classification, preparation, and applications in drug delivery. https://www.sciencedirect.com/topics/agricultural-and-biological-sciences/liposome

Akintelu, S., Folorunso, A., & Folorunso, F. (2021). Speculative carcinogenicity assessment of polycyclic aromatic hydrocarbons found in some anti-diarrheal herbal drugs sold in Ondo state of Nigeria. *Iraqi Journal of Science*, *62*(4), 1046–1053. https://doi.org/10.24996/ijs.2021.62.4.1

Alharbi, W., Almughem, F., Almehmady, A., Jarallah, S., Alsharif, W., Alzahrani, N., … & Alshehri, A. (2021). Phytosomes as an emerging nanotechnology platform for the topical delivery of bioactive phytochemicals. *Pharmaceutics*, 13(9), 1475. https://doi.org/10.3390/pharmaceutics13091475

Almeida, B., Nag, O. K., Rogers, K. E., & Delehanty, J. B. (2020). Recent Progress in Bioconjugation Strategies for Liposome-Mediated Drug Delivery. Molecules, 25(23), 5672. https://doi.org/10.3390/molecules25235672

Alostad, A. H., Steinke, D. T., & Schafheutle, E. I. (2020). Herbal Medicine Classification: Policy Recommendations. *Frontiers in Medicine*, 7. https://doi.org/10.3389/fmed.2020.00031

Alpert, J. S. (2005). The Vioxx debacle. In *American Journal of Medicine* (Vol. 118, Issue 3, pp. 203–204). Elsevier Inc. https://doi.org/10.1016/j.amjmed.2005.01.020

Ambwani, S., Tandon, R., Ambwani, T., & Malik, Y. (2018). Current knowledge on nanodelivery systems and their beneficial applications in enhancing the efficacy of herbal drugs. *Journal of Experimental Biology and Agricultural Sciences*, 6(1), 87–107. https://doi.org/10.18006/2018.6(1).87.107

Ansari, S., Islam, F., & Sameem, M. (2012). Influence of nanotechnology on herbal drugs: a review. *Journal of Advanced Pharmaceutical Technology Amp Research*, 3(3), 142. https://doi.org/10.4103/2231-4040.101006

Anthony, D. P., Hegde, M., Shetty, S. S., Rafic, T., Mutalik, S., & Rao, B. S. (2021). Targeting receptor-ligand chemistry for drug delivery across blood-brain barrier in brain diseases. *Life Sciences*, 274, 119326. https://doi.org/10.1016/j.lfs.2021.119326

Arsude, J. (2023). Herbal nano formulations for topical drug delivery: prospective for multiple skin disorders. *International Journal of Drug Delivery Technology*, 13(04), 1568–1577. https://doi.org/10.25258/ijddt.13.4.69

Azzahra, L., Mita, S., & Sriwidodo, S. (2020). Formulation, characterization, and herbal drug delivery applications of ethosome, transfersome, and transethosome. *Indonesian Journal of Pharmaceutics*, 2(3). https://doi.org/10.24198/idjp.v2i3.29154

Bansal, D., Hota, D., & Chakrabarti, A. (2010). Open Access Journal of Clinical Trials ISSN: (Print) (Online) Journal homepage: www.tandfonline.com/journals/djct20 Research methodological issues in evaluating herbal interventions. https://doi.org/10.2147/OA

Bandyopadhyay, A., Das, T., Nandy, S. et al. Ligand-based active targeting strategies for cancer theranostics. *Naunyn-Schmiedeberg's Arch Pharmacol* 396, 3417–3441 (2023). https://doi.org/10.1007/s00210-023-02612-4

Begines, B., Ortiz, T., Pérez-Aranda, M., Martínez, G., Merinero, M., Argüelles-Arias, F., & Alcudia, A. (2020). Polymeric Nanoparticles for Drug Delivery: Recent Developments and Future Prospects. *Pharmaceutics*, 12(10), 1019. https://doi.org/10.3390/pharmaceutics12101019

Bonifácio, B., Silva, P., Ramos, M., Negri, K., Bauab, T., & Chorilli, M. (2013). Nanotechnology-based drug delivery systems and herbal medicines: a review. *International Journal of Nanomedicine*, 1. https://doi.org/10.2147/ijn.s52634

Bulger, P. G., Conlon, D. A., Cink, R. D., Fernandez-Cerezo, L., Zhang, Q., Thirumalairajan, S., Raglione, T., Liang, R., Zhou, J., & Chalgeri, A. (2023). Drug-linkers in antibody–drug conjugates: Perspective on current industry practices. Organic Process Research & Development, 27(7), 1248–1257.

Castro, C., Kilchherr, F., Kim, D. N. et al. A primer to scaffolded DNA origami. *Nat Methods* 8, 221–229 (2011). https://doi.org/10.1038/nmeth.1570

Chen, R., Chavda, V., Patel, A., & Chen, Z. (2022). Phytochemical delivery through transferosome (phytosome): an advanced transdermal drug delivery for complementary medicines. *Frontiers in Pharmacology*, 13. https://doi.org/10.3389/fphar.2022.850862

Chen, Y., Lu, Y., Lee, R. J., & Xiang, G. (2020). Nano Encapsulated Curcumin: And Its Potential for Biomedical Applications. *International journal of nanomedicine*, *15*, 3099–3120. https://doi.org/10.2147/IJN.S210320

Darji, S., Rani, K., & Patani, P. (2022). A review on herbal nanoparticles. *Journal of Pharmaceutical Negative Results*, 1990–2002. https://doi.org/10.47750/pnr.2022.13.s03.296

Debnath, S. K., & Srivastava, R. (2021). Drug delivery with carbon-based nanomaterials as versatile nanocarriers: Progress and prospects. Frontiers in *Nanotechnology*, 3. https://doi.org/10.3389/fnano.2021.644564

Dewi, M., Chaerunisaa, A., Muhaimin, M., & Joni, I. (2022). Improved activity of herbal medicines through nanotechnology. *Nanomaterials*, 12(22), 4073. https://doi.org/10.3390/nano12224073

Dongare, P., Motule, A., Dubey, M., More, M., Patinge, P., Bakal, R., … & Manwar, J. (2021). Recent development in novel drug delivery systems for delivery of herbal drugs: an updates. *GSC Advanced Research and Reviews*, 8(2), 8–18. https://doi.org/10.30574/gscarr.2021.8.2.0158

Dragoş, D., Gîlcă, M., Găman, L., Vlad, A., Iosif, L., Stoian, I., … & Lupescu, O. (2017).Phytomedicine in joint disorders. *Nutrients*, 9(1), 70. https://doi.org/10.3390/nu9010070

Dristant, U., Mukherjee, K., Saha, S., & Maity, D. (2023). An Overview of Polymeric Nanoparticles-Based Drug Delivery System in Cancer Treatment. *Technology in Cancer Research & Treatment*, 22, 15330338231152083. https://doi.org/10.1177/15330338231152083

Dunuweera, S. P., Rajapakse, R. M. S. I., Rajapakshe, R. B. S. D., Wijekoon, S. H. D. P., Thilakarathna, M. G. G. S. N., & Rajapakse, R. M. G. (2019). Review on Targeted Drug Delivery Carriers Used in Nanobiomedical Applications. *Current Nanoscience*, 15(4), 382–397.

Ekar, T., & Kreft, S. (2019). Common risks of adulterated and mislabeled herbal preparations. *Food and Chemical Toxicology*, 123, 288–297. https://doi.org/10.1016/J.FCT.2018.10.043

Elkordy, A. A., Haj-Ahmad, R. R., Awaad, A. S., & Zaki, R. M. (2021). An overview on natural product drug formulations from conventional medicines to nanomedicines: Past, present and future. In *Journal of Drug Delivery Science and Technology* (Vol. 63). Editions de Sante. https://doi.org/10.1016/j.jddst.2021.102459

Elsabahy, T., & Wooley, K. L. (2012). Design of polymeric nanoparticles for efficient drug delivery. *Chemical Reviews*, 112(11), 5407–5492. https://doi/10.1021/acs.chas.3c00008

Elsana, H., Olusanya, T., Carr-Wilkinson, J., Darby, S., Faheem, A., & Elkordy, A. (2019). Evaluation of novel cationic gene based liposomes with cyclodextrin prepared by thin film hydration and microfluidic systems. *Scientific Reports*, 9(1). https://doi.org/10.1038/s41598-019-51065-4

Elumalai, K., Srinivasan, S., & Shanmugam, A. (2024). Review of the efficacy of nanoparticle-based drug delivery systems for cancer treatment. Biomedical Technology, 5(1), 109–122. https://doi.org/10.1016/j.bmt.2023.09.001

Farooque, F., Wasi, M., & Mughees, M. M. (2021). Liposomes as Drug Delivery System: An Updated Review. *Journal of Drug Delivery and Therapeutics*, *11*(5-S), 149–158. https://doi.org/10.22270/jddt.v11i5-S.5063

Gaikwad, A., Ahire, K., Gosavi, A., Salunkhe, K., & Khalkar, A. (2021). Gaikwad abhijeet r. phytosome as a novel drug delivery system for bioavailability enhancement of phytoconstituents and its applications: a review. *Journal of Drug Delivery and Therapeutics*, 11(3), 138–152. https://doi.org/10.22270/jddt.v11i3.4847

Gamboa, J., Lourenço, P., Cruz, C., & Gallardo, E. (2024). Aptamers for the Delivery of Plant-Based Compounds: A Review. In *Pharmaceutics* (Vol. 16, Issue 4). Multidisciplinary Digital Publishing Institute (MDPI). https://doi.org/10.3390/pharmaceutics16040541

Gao, W., Bigham, A., Ghomi, M., Zarrabi, A., Rabiee, N., Saeb, M. R., Ertas, Y. N., … Jiao, T. (2024). Micelle-engineered nanoplatforms for precision oncology. *Chemical Engineering Journal*, 495, 153438. https://doi.org/10.1016/j.cej.2024.153438

Garcia, S. (2020). Pandemics and traditional plant-based remedies: A historical-botanical review in the era of COVID-19. Frontiers in *Plant Science*, 11, [571042]. https://doi.org/10.3389/fpls.2020.571042

Gatt, A. R., Vella Bonanno, P., & Zammit, R. (2024). Ethical considerations in the regulation and use of herbal medicines in the European Union. In *Frontiers in Medical Technology* (Vol. 6). Frontiers Media SA. https://doi.org/10.3389/fmedt.2024.1358956

Ghezzi, M., Pescina, S., Padula, C., Santi, P., Del Favero, E., Cantù, L., & Nicoli, S. (2021). Polymeric micelles in drug delivery: An insight of the techniques for their characterization and assessment in biorelevant conditions. *Journal of Controlled Release*, 332, 312–336. https://doi.org/10.1016/j.jconrel.2021.02.031

Gogia, P., Ashraf, H., Bhasin, S., & Xu, Y. (2023). Antibody–Drug Conjugates: A Review of Approved Drugs and Their Clinical Level of Evidence. *Cancers*, *15*(15), 3886. https://doi.org/10.3390/cancers15153886

Gold, L., Janjic, N., Jarvis, T., Schneider, D., Walker, J. J., Wilcox, S. K., & Zichi, D. (2012). Aptamers and the RNA world, past and present. *Cold Spring Harbor Perspectives in Biology*, 4(3). https://doi.org/10.1101/cshperspect.a003582

Gopi, S. and Amalraj, A. (2016). Introduction of nanotechnology in herbal drugs and nutraceutical: a review. *Journal of Nanomedicine & Biotherapeutic Discovery*, 6(2). https://doi.org/10.4172/2155-983x.1000143

Grace, V. (2014). Anti-metastatic study of liposome-encapsulated alltransretinoic acid (atra) in b16f10 melanoma cells-implanted c57bl/6 mice. *Cancer Investigation*, 32(10), 507–517. https://doi.org/10.3109/07357907.2014.964408

Guimarães, D., Noro, J., Silva, C., Cavaco-Paulo, A., & Nogueira, E. (2019). Protective effect of saccharides on freeze-dried liposomes encapsulating drugs. *Frontiers in Bioengineering and Biotechnology*, 7. https://doi.org/10.3389/fbioe.2019.00424

Gumulec, J., Fojtů, M., Raudenská, M., Sztalmachová, M., Skotakova, A., Vlachova, J., … & Masařík, M. (2014). Modulation of induced cytotoxicity of doxorubicin by using apoferritin and liposomal cages. *International Journal of Molecular Sciences*, 15(12), 22960–22977. https://doi.org/10.3390/ijms151222960

Gunasekaran, T., Haile, T., Nigusse, T., & Dhanaraju, M. (2014). Nanotechnology: an effective tool for enhancing bioavailability and bioactivity of phytomedicine. *Asian Pacific Journal of Tropical Biomedicine*, 4, S1-S7. https://doi.org/10.12980/apjtb.4.2014c980

Guo, M., Qin, S., Wang, S., Sun, M., Yang, H., Wang, X., Fan, P., & Jin, Z. (2023). Herbal Medicine Nanocrystals: A Potential Novel Therapeutic Strategy. In *Molecules* (Vol. 28, Issue 17). Multidisciplinary Digital Publishing Institute (MDPI). https://doi.org/10.3390/molecules28176370

Guzmán Rodríguez, A., Sablón Carrazana, M., Rodríguez Tanty, C., Malessy, M. J. A., Fuentes, G., & Cruz, L. J. (2023). Smart Polymeric Micelles for Anticancer Hydrophobic Drugs. *Cancers*, *15*(1), 4. https://doi.org/10.3390/cancers15010004

Hassan, S., & Zhang, Y. S. (2019). Microfluidic technologies for local drug delivery. In A. de Mello (Ed.), *Micro and Nano Technologies* (pp. 281–305). Academic Press. https://doi.org/10.1016/B978-0-12-812659-2.00010-7

He, F., Wen, N., Xiao, D., Yan, J., Xiong, H., Cai, S., Liu, Z., & Liu, Y. (2018). Aptamer-Based Targeted Drug Delivery Systems: Current Potential and Challenges. *Current Medicinal Chemistry*, 25(42), 5943–5962. https://doi.org/10.2174/0929867325666181008142831

Huang, B., St. Onge, C. M., Ma, H., & Zhang, Y. (2021). Design of bivalent ligands targeting putative GPCR dimers. Drug Discovery Today, 26(1), 189–199. https://doi.org/10.1016/j.drudis.2020.10.006

Izzo, A. A. (2004). Drug interactions with St. John's Wort (Hypericum perforatum): a review of the clinical evidence. *International Journal of Clinical Pharmacology and Therapeutics*, 42(3), 139–148. https://doi.org/10.5414/CPP42139

Jalali, R. (2022). Application of nanotechnology in biomedicine. Academic Journal of Environmental Biology, 3(3), 44–51. https://doi.org/10.38007/AJEB.2022.030306

Jalili, A., Bagherifar, R., Nokhodchi, A., Conway, B., & Javadzadeh, Y. (2023). Current advances in nanotechnology-mediated delivery of herbal and plant-derived medicines. *Advanced Pharmaceutical Bulletin*, 13(4), 712–722. https://doi.org/10.34172/apb.2023.087

Kallepalli, B., & Garcia-Fuentes, A. (2024). Stimuli-responsive drug delivery systems: Advances and applications. https://www.mdpi.com/2079-4983/10/3/34

Kesarwani, K. and Gupta, R. (2013). Bioavailability enhancers of herbal origin: an overview. *Asian Pacific Journal of Tropical Biomedicine*, 3(4), 253–266. https://doi.org/10.1016/s2221-1691(13)60060-x

Khanzole, M. B. (2020). Phytosomes: Novel drug delivery systems. https://www.researchgate.net/publication/283318972_Phytosome_A_Novel_Drug_Delivery_System_for_Herbal_Medicine

Kumar, A., Pandey, S., & Verma, S. (2023). Phytosomes: A promising approach for improved delivery of herbal bioactives. *Pharmaceutics*, 15(7), 1475. https://www.scitcentral.com/article.php?journal=14&article=2300&article_title=Phytosomes:%20A%20Promising%20Approach%20for%20the%20Delivery%20of%20Bioactive%20Compounds

Kumari, I., Kaurav, H., & Chaudhary, G. (2021). Myristica fragrans (jaiphal): a significant medicinal herbal plant. *International Journal for Research in Applied Sciences and Biotechnology*, 8(2), 213–224. https://doi.org/10.31033/ijrasb.8.2.27

Kuruüzüm-Uz, A., Süleyman, H., Çadırcı, E., Güvenalp, Z., & Demirezer, L. (2012). Investigation on anti-inflammatory and antiulcer activities of anchusa azurea extracts and their major constituent rosmarinic acid. *Zeitschrift Für Naturforschung C*, 67(7–8), 360–366. https://doi.org/10.1515/znc-2012-7-802

Lammari N., Louaer O., Meniai A.H., Elaissari A. Encapsulation of Essential Oils via Nanoprecipitation Process: Overview, Progress, Challenges and Prospects. *Pharmaceutics*. 2020;12:431. https://doi.org/10.3390/pharmaceutics12050431.

Li, R., Duan, H., Liang, Q., Huang, Y., Ly, W., Zhang, Q., … & Peng, W. (2022). Targeting matrix metalloproteases: a promising strategy for herbal medicines to treat rheumatoid arthritis. *Frontiers in Immunology*, 13. https://doi.org/10.3389/fimmu.2022.1046810

Liu, M., Wang, L., Lo, Y., Shiu, S. C. C., Kinghorn, A. B., & Tanner, J. A. (2022). Aptamer-Enabled Nanomaterials for Therapeutics, Drug Targeting and Imaging. In *Cells* (Vol. 11, Issue 1). MDPI. https://doi.org/10.3390/cells11010159

Liu, Y., Chen, X., Liu, X., Guan, W. J., & Lu, C. (2023). Aggregation-induced emission-active micelles: Synthesis, characterization, and applications. *Chemical Society Reviews* https://doi.org/10.1039/D2CS01021F

Lu, L., Niu, Z., Chao, Z. et al. Exploring the therapeutic potential of ADC combination for triple-negative breast cancer. *Cell. Mol. Life Sci*. 80, 350 (2023). https://doi.org/10.1007/s00018-023-04946-x

Madej, M., Kurowska, N., & Strzalka-Mrozik, B. (2022). Polymeric Nanoparticles—Tools in a Drug Delivery System in Selected Cancer Therapies. *Applied Sciences*, *12*(19), 9479. https://doi.org/10.3390/app12199479

Mahor, A., Singh, P. P., Bharadwaj, P., Sharma, N., Yadav, S., Rosenholm, J. M., & Bansal, K. K. (2021). Carbon-Based Nanomaterials for Delivery of Biologicals and Therapeutics: A Cutting-Edge Technology. *C*, 7(1), 19. https://doi.org/10.3390/c7010019

Marella, S. and Prasad, T. (2018). Nanotechnological approaches for the development of herbal drugs in treatment of diabetes mellitus – a critical review. *IET Nanobiotechnology*, 12(5), 549–556. https://doi.org/10.1049/iet-nbt.2017.0242

Maurya A., Singh A.K., Mishra G., Kumari K., Rai A., Sharma B., Kulkarni G.T., Awasthi R. Strategic use of nanotechnology in drug targeting and its consequences on human health: A focused review. *Interventional Medicine and Applied Science* 2019;11:38–54. https://doi.org/10.1556/1646.11.2019.04.

Metkari, V. (2023). Formulation and evaluation of herbal chewing gum. *American Journal of Biomedical Science & Research*, 18(6), 567–576. https://doi.org/10.34297/ajbsr.2023.18.002526

Mohammadi, S., Jafari, B., Asgharian, P., Martorell, M., & Sharifi-Rad, J. (2020). Medicinal plants used in the treatment of malaria: a key emphasis to artemisia, cinchona, cryptolepis, and tabebuia genera. *Phytotherapy Research*, 34(7), 1556–1569. https://doi.org/10.1002/ptr.6628

Mohan, A., Menon, A., Chacko, J., & Mohan, P. (2020). A new drug delivery strategy targeting mass population in combating epidemic crisis. *International Journal of Research in Pharmaceutical Sciences*, 11(4), 5994–5996. https://doi.org/10.26452/ijrps.v11i4.3261

Murshed, M., Al-Tamimi, J., Aljawdah, H., & Al-Quraishy, S. (2023). Pharmacological effects of grape leaf extract reduce eimeriosis-induced inflammation, oxidative status change, and goblet cell response in the jejunum of mice. *Pharmaceuticals*, 16(7), 928.https://doi.org/10.3390/ph16070928

Muyumba, N. W., Mutombo, S. C., Sheridan, H., Nachtergael, A., & Duez, P. (2021). Quality control of herbal drugs and preparations: The methods of analysis, their relevance and applications. *Talanta Open*, 4. https://doi.org/10.1016/j.talo.2021.100070

Nandhini, S. and Ilango, K. (2021). Development and characterization of a nano-drug delivery system containing vasaka phospholipid complex to improve bioavailability using quality by design approach. *Research in Pharmaceutical Sciences*, 16(1), 103. https://doi.org/10.4103/1735-5362.305193

Nimjee, S. M., White, R. R., Becker, R. C., & Sullenger, B. A. (2017). Aptamers as Therapeutics. *Annual Review of Pharmacology and Toxicology*, 57, 61–79. https://doi.org/10.1146/annurev-pharmtox-010716-104558

Nimushakavi, S., Haque, S., Kotcherlakota, R., & Patra, C. R. (2021). Biomedical Applications of Carbon Nanotubes: Recent Development and Future Challenges. In S. Jana & S. Jana (Eds.), *Nanoengineering Biomaterials*. (pp. 353–388). Wiley-VCH. https://doi.org/10.1002/9783527832095.ch29

Niu, N., Yin, J., Yuan, Y., Wang, Z., Zhou, Z., He, Z., … & Zhou, S. (2015). Novel targeting of pegylated liposomes for codelivery of siRNA and four antitubercular drugs to human macrophages for the treatment of mycobacterial infection: a quantitative proteomic study. *Drug Design Development and Therapy*, 4441. https://doi.org/10.2147/dddt.s79369

Nogueira, E., Lager, F., Roux, D., Nogueira, P., Freitas, J., Charvet, C., … & Cavaco-Paulo, A. (2015b). Enhancing methotrexate tolerance with folate tagged liposomes in arthritic mice. *Journal of Biomedical Nanotechnology*, 11(12), 2243–2252. https://doi.org/10.1166/jbn.2015.2170

Nogueira, E., Mangialavori, I., Loureiro, A., Azoia, N., Sárria, M., Nogueira, P., … & Cavaco-Paulo, A. (2015a). Peptide anchor for folate-targeted liposomal delivery. *Biomacromolecules*, 16(9), 2904–2910. https://doi.org/10.1021/acs.biomac.5b00823

Ogawa-Ochiai, K. and Kawasaki, K. (2019). Panax ginseng for frailty-related disorders: a review. *Frontiers in Nutrition*, 5. https://doi.org/10.3389/fnut.2018.00140

Ondieki, G., Nyagblordzro, M., Kikete, S., Liang, R., Wang, L., & He, X. (2017). Cytochrome P450 and P-Glycoprotein-Mediated Interactions Involving African Herbs Indicated for Common Noncommunicable Diseases. *Evidence-based complementary and alternative medicine : eCAM, 2017*, 2582463. https://doi.org/10.1155/2017/2582463

Onyeji, C. (2022). Prospects of integration of nanotechnology to antimalarial herbal remedies for improved therapeutic efficacy – a concise review. *African Journal of Traditional Complementary and Alternative Medicines*, 18(1), 27–35. https://doi.org/10.21010/ajtcamv18i1.3

Pan, S. Y., Litscher, G., Gao, S. H., et al. Historical perspective of traditional indigenous medical practices: the current renaissance and conservation of herbal resources. *Evid Based Complement Alter Med* 2014; 2014: 525340.

Patwardhan, B., Warude, D., Pushpangadan, P., & Bhatt, N. (2005). Ayurveda and traditional Chinese medicine: a comparative overview. *Evidence-Based Complementary and Alternative Medicine*, 2(4), 465–473. https://doi.org/10.1093/ecam/neh140

Pauli, G., Tang, W., & Li, S. (2019). Development and characterization of the solvent-assisted active loading technology (salt) for liposomal loading of poorly water-soluble compounds. *Pharmaceutics*, 11(9), 465. https://doi.org/10.3390/pharmaceutics11090465

Pentak, D., Sułkowski, W., & Sułkowska, A. (2011). Influence of some physical properties of 5-fluorouracil on encapsulation efficiency in liposomes. *Journal of Thermal Analysis and Calorimetry*, 108(1), 67–71. https://doi.org/10.1007/s10973-011-1822-0

Pinelli F., Perale G., Rossi F. Coating and functionalization strategies for nanogels and nanoparticles for selective drug delivery. *Gels*. 2020;6:6. https://doi.org/10.3390/gels6010006.

Prasad, S. B., Bhatia, S., & Singh, S. (2016). Phytosome: Phytoconstituent based lipid derived drug delivery system. *Journal of Chemical and Pharmaceutical Research*, 8(5), 664–670. Retrieved from www.jocpr.com

Rahman, H., Othman, H., Hammadi, N., Yeap, S., Amin, K., Samad, N., … & Alitheen, N. (2020). Novel drug delivery systems for loading of natural plant extracts and their biomedical applications;. *International Journal of Nanomedicine*, Volume 15, 2439–2483. https://doi.org/10.2147/ijn.s227805

Raynor D. K., Dickinson R., Knapp P., Long A. F., Nicolson D. J. (2011). Buyer beware? Does the information provided with herbal products available over the counter enable safe use? *BMC Medicine* 9:94 https://doi.org/10.1186/1741-7015-9-94

Razavi, M. (2024). harnessing nanotechnology for optimized herbal cancer treatment: a comprehensive review of nanoscale drug delivery systems. *Pharmaceutical Sciences*, 30(2), 153–186. https://doi.org/10.34172/ps.2023.27

Saggar, S., Mir, P. A., Kumar, N., Chawla, A., Uppal, J., Shilpa, S., & Kaur, A. (2022). Traditional and Herbal Medicines: Opportunities and Challenges. *Pharmacognosy Research*, 14(2), 107–114. https://doi.org/10.5530/pres.14.2.15

Salm, S., Rutz, J., van den Akker, M., Blaheta, R. A., & Bachmeier, B. E. (2023). Current state of research on the clinical benefits of herbal medicines for non-life-threatening ailments. In *Frontiers in Pharmacology* (Vol. 14). Frontiers Media SA. https://doi.org/10.3389/fphar.2023.1234701

Sarangi, Manoj Kumar, Padhi, Sasmita. Novel Herbal Drug Delivery System: An Overview. *Archives of Medicine and Health Sciences* 6(1): p 171–179, Jan–Jun 2018. https://doi.org/10.4103/amhs.amhs_88_17

Schwendener, R.A., Schott, H. (2017). Liposome Formulations of Hydrophobic Drugs. In: D'Souza, G. (eds) *Liposomes. Methods in Molecular Biology*, vol 1522. Humana Press, New York, NY. https://doi.org/10.1007/978-1-4939-6591-5_6

Sercombe, L., Veerati, T., Moheimani, F., Wu, S., Sood, A., & Hua, S. (2015). Advances and challenges of liposome assisted drug delivery. *Frontiers in Pharmacology*, 6. https://doi.org/10.3389/fphar.2015.00286

Serpell, C. J., Kostarelos, K., & Davis, B. G. (2016). Can Carbon Nanotubes Deliver on Their Promise in Biology? Harnessing Unique Properties for Unparalleled Applications. *ACS Central Science*, 2(4), 190–200. https://doi.org/10.1021/acscentsci.6b00005

Shaito, A., Thuan, D. T. B., Phu, H. T., Nguyen, T. H. D., Hasan, H., Halabi, S., Abdelhady, S., Nasrallah, G. K., Eid, A. H., & Pintus, G. (2020). Herbal Medicine for Cardiovascular Diseases: Efficacy, Mechanisms, and Safety. In *Frontiers in Pharmacology* (Vol. 11). Frontiers Media S.A. https://doi.org/10.3389/fphar.2020.00422

Shigehiro, T., Kasai, T., Murakami, M., Sekhar, S., Tominaga, Y., Okada, M., … & Seno, M. (2014). Efficient drug delivery of paclitaxel glycoside: a novel solubility gradient encapsulation into liposomes coupled with immunoliposomes preparation. *Plos One*, 9(9), e107976. https://doi.org/10.1371/journal.pone.0107976

Shishparenok, A. N., Furman, V. V., & Zhdanov, D. D. (2023). DNA-Based Nanomaterials as Drug Delivery Platforms for Increasing the Effect of Drugs in Tumors. *Cancers*, *15*(7), 2151. https://doi.org/10.3390/cancers15072151

Singh, H., & Sharma, G. (2023). Recent Development of Novel Drug Delivery of Herbal Drugs. *RPS Pharmacy and Pharmacology Reports*. https://doi.org/10.1093/rpsppr/rqad028

Singh, M. P., & Gohil, K. J. (2024). Therapeutic efficacy and cost effectiveness of herbal drugs – A reasonable approach. *Pharmacological Research – Natural Products*, 2, 100009. https://doi.org/10.1016/J.PRENAP.2023.100009

Singh, R. K. (2023). Carbon nanomaterials for drug delivery. *Materials Today Bio*, 18, 100322. https://www.sciencedirect.com/science/article/abs/pii/S1369703X23000232

Singh, S., & Shukla, V. K. (2021). Current regulations for Herbal Medicines in India. *International Journal of Drug Regulatory Affairs*, 9(2), 30–34. https://doi.org/10.22270/ijdra.v9i2.466

Skrinda, M., Dubņika, A., & Ločs, J. (2021). Liposomes as vancomycin hydrochloride delivery system. *Key Engineering Materials*, 903, 3–8.https://doi.org/10.4028/www.scientific.net/kem.903.3

Smits, E., Soetekouw, J., Pieters, E., Smits, C., Wijs-Rot, N., & Vromans, H. (2018). The availability of drug by liposomal drug delivery. *Investigational New Drugs*, 37(5), 890–901. https://doi.org/10.1007/s10637-018-0708-4

Song, C. H., Jeong, M., In, H., Kim, J. H., Lin, C. -W., & Han, K. H. (2023a). Trends in the Development of Antibody-Drug Conjugates for Cancer Therapy. *Antibodies*, *12*(4), 72. https://doi.org/10.3390/antib12040072

Song, C. H., Lee, J. H., & Kim, Y. S. (2023b). Aptamers for targeted cancer therapy. *Cancers*, 15(15), 3817. https://www.mdpi.com/journal/ijms/special_issues/1Y2FVAXQ9T

Sreekanth, V., Medatwal, N., Pal, S., Kumar, S., Sengupta, S., & Bajaj, A. (2017). Molecular self-assembly of bile acid-phospholipids controls the delivery of doxorubicin and mice survivability. *Molecular Pharmaceutics*, 14(8), 2649–2659. https://doi.org/10.1021/acs.molpharmaceut.7b00105

Studdert D., Eisenberg D., Miller F., Curto D., Kaptchuk T., Brennan T. (1998). Medical malpractice implications of alternative medicine. *Jama* 280 1610–1615. https://doi.org/10.1001/jama.280.18.1610

Studdert, D. M., Spittal, M. J., Mello, M. M., O'Malley, A. J., & Stevenson, D. G. (2011). Relationship between quality of care and negligence litigation in nursing homes. *The New England journal of medicine*, *364*(13), 1243–1250. https://doi.org/10.1056/NEJMsa1009336

Sun, Y., & Davis, E. (2021). Nanoplatforms for targeted stimuli-responsive drug delivery: A review of platform materials and stimuli-responsive release and targeting mechanisms. *Nanomaterials*, 11(3), 746. https://doi.org/10.3390/nano11030746

Tang, K., Zhu, R., Li, Y., & Cao, Z. (2011). Discrimination of approved drugs from experimental drugs by learning methods. *BMC Bioinformatics*, *12*, 1–7.

Teong, B., Kuo, S., Ch, C., Chen, Y., Cheng, Z., & Huang, H. (2014). Characterization and human osteoblastic proliferation- and differentiation-stimulatory effects of phosphatidylcholine liposomes-encapsulated propranolol hydrochloride. *Bio-Medical Materials and Engineering*, 24(5), 1875–1887. https://doi.org/10.3233/bme-140997

Teware, K. (2017). Benefits of nanoparticles in drug delivery of herbal drug components: a review. *World Journal of Pharmacy and Pharmaceutical Sciences*, 430–435. https://doi.org/10.20959/wjpps20175-9129

Torchilin, V. P. (2007). Micelles as pharmaceutical carriers. *Pharmaceutical Research*, 24(1), 145–160. https://www.ncbi.nlm.nih.gov/pmc/articles/PMC9964340/

Verma, R., Bhardwaj, S., & Gupta, J. (2018). Broad view of nano herbal medicine. *International Journal of Recent Advances in Science and Technology*, 5(02). https://doi.org/10.30750/ijarst.521

Vyshnevska, L., Severina, H., Prokopenko, Y., & Shmalko, A. (2022). molecular docking investigation of anti-inflammatory herbal compounds as potential lox-5 and cox-2 inhibitors. *Pharmacia*, 69(3), 733–744. https://doi.org/10.3897/pharmacia.69.e89400

Wang, D., Li, Y., Deng, X., Torre, M., Zhang, Z., Li, X., Zhang, W., Cullion, K., Kohane, D. S., & Weldon, C. B. (2023b). An aptamer-based depot system for sustained release of small molecule therapeutics. Nature *Communications*, 14(1). https://doi.org/10.1038/s41467-023-37002-0

Wang, H., Chen, Y., Wang, L., Liu, Q., Yang, S., & Wang, C. (2023a). Advancing herbal medicine: enhancing product quality and safety through robust quality control practices. In *Frontiers in Pharmacology* (Vol. 14). Frontiers Media SA. https://doi.org/10.3389/fphar.2023.1265178

Wani, K., Tarawadi, K., & Kaul-Ghanekar, R. (2015). Nanocarriers for delivery of herbal based drugs in breast cancer - an overview. *Journal of Nano Research*, 34, 29–40. https://doi.org/10.4028/www.scientific.net/jnanor.34.29

Wink, M. (2015). Modes of Action of Herbal Medicines and Plant Secondary Metabolites. *Medicines*, 2(3), 251–286. https://doi.org/10.3390/medicines2030251

Wu, Y., Wang, J., Li, S., Zhao, B., & Sun, J. (2019). Targeting strategies and mechanisms for drug delivery. https://www.ncbi.nlm.nih.gov/pmc/articles/PMC5783816/

Xie, S., Sun, W., Fu, T., Liu, X., Chen, P., Qiu, L., Qu, F., & Tan, W. (2023). Aptamer-Based Targeted Delivery of Functional Nucleic Acids. *Journal of the American Chemical Society*, *145*(14), 7677–7691. https://doi.org/10.1021/jacs.3c00841

Yang, K., Zeng, L., & Ge, J. (2018). Exploring the pharmacological mechanism of danzhi xiaoyao powder on er-positive breast cancer by a network pharmacology approach. *Evidence-Based Complementary and Alternative Medicine*, 2018(1). https://doi.org/10.1155/2018/5059743

Yang, Z., Xu, M., Liu, Y., & Fu, L. (2022). DNA nanocarriers for targeted drug delivery. *Small*, 18(33), 2202325. https://onlinelibrary.wiley.com/doi/abs/10.1002/adma.201300875

Yi, C., Fu, M., Cao, X., Tong, S., Zheng, Q., Firempong, C., … & Yu, J. (2013). Enhanced oral bioavailability and tissue distribution of a new potential anticancer agent, flammulina velutipes sterols, through liposomal encapsulation. *Journal of Agricultural and Food Chemistry*, 61(25), 5961–5971. https://doi.org/10.1021/jf3055278

Yin, S., Wei, W., Jian, F., & Yang, N. (2013). Therapeutic applications of herbal medicines for cancer patients. *Evidence-Based Complementary and Alternative Medicine*, 2013, 1–15. https://doi.org/10.1155/2013/302426

Zare, H., Ahmadi, S., Ghasemi, A., Ghanbari, M., Rabiee, N., Bagherzadeh, M., Karimi, M., Webster, T. J., Hamblin, M. R., & Mostafavi, E. (2021). Carbon Nanotubes: Smart Drug/Gene Delivery Carriers. *International Journal of Nanomedicine*, 16, 1681–1706. https://doi.org/10.2147/IJN.S299448

Zielińska, A., Carreiró, F., Oliveira, A. M., Neves, A., Pires, B., Venkatesh, D. N., Durazzo, A., Lucarini, M., Eder, P., Silva, A. M., Santini, A., & Souto, E. B. (2020). Polymeric Nanoparticles: Production, Characterization, Toxicology and Ecotoxicology. *Molecules (Basel, Switzerland)*, *25*(16), 3731. https://doi.org/10.3390/molecules25163731

Zheng, H., Tao, H., Wan, J., Lee, K., Zheng, Z., & Leung, S. (2022). Preparation of drug-loaded liposomes with multi-inlet vortex mixers. *Pharmaceutics*, 14(6), 1223. https://doi.org/10.3390/pharmaceutics14061223

Zhou, Y., Zhang, M., Yuan, B., Yuan, J., Zhao, X., Zhao, L., … & Ren, H. (2017). Herbal carrier-based floating microparticles of diltiazem hydrochloride for improved cardiac activity. *Tropical Journal of Pharmaceutical Research*, 16(6), 1239. https://doi.org/10.4314/tjpr.v16i6.5

17 Nanotechnology-Enhanced Delivery of Herbal Extract

Formulation, Preclinical, and Clinical Efficiency

Mukul Machhindra Barwant
Department of Botany, Sanjivani Rural Education Society's (SRES), Sanjivani Arts, Commerce and Science College, Kopargaon, India

Shreya Singh
Department of Agriculture, Ramlalit Singh Mahavidyalaya, Kailhat, Chunar, Mirzapur, India

Balwant Singh
Department of Botany, Dr. Ram Manohar Lohia Avadh University, Ayodhya, India

Shalagha Ab Sharma
School of Biological Engineering, SIET, Meerut, India

Odangowei Inetiminebi Ogidi
Department of Biochemistry, Faculty of Basic Medical Sciences, Bayelsa Medical University, Yenagoa, Nigeria

17.1 INTRODUCTION

Herbal extracts have been used for millennia in traditional medicine systems around the world, from Ayurveda in India to traditional Chinese medicine (TCM). These natural compounds, derived from plants, have shown significant potential in treating various ailments due to their diverse pharmacological properties [14]). In recent years, This can result from their low solubility in water, instability in the gastrointestinal tract, rapid metabolism, and subsequent elimination from the body [4]. Additionally, the complexity of plant matrices, which contain numerous active and inactive constituents, can lead to inconsistent therapeutic outcomes [20]. These challenges necessitate the development of innovative delivery systems to enhance the bioavailability and stability of herbal extracts, thereby maximizing their therapeutic potential [20].

Nanotechnology, the manipulation of matter on an atomic, molecular, and supramolecular scale, offers revolutionary solutions to these challenges. By engineering nanoparticles that can encapsulate herbal extracts, scientists can significantly improve the delivery and efficacy of these natural compounds [12, 25, 26]. Nanoparticles can protect the active ingredients from degradation, enhance their absorption in the gastrointestinal tract, and ensure controlled and targeted release at the desired

DOI: 10.1201/9781003544418-17

site of action [4]. These include methods like solvent evaporation, nanoprecipitation, and emulsion-based techniques, each offering distinct advantages and challenges [5]. The choice of materials for nanoparticle synthesis, such as polymers, lipids, or inorganic substances, is crucial in determining the properties and effectiveness of the final product. Additionally, factors such as particle size, surface charge, and drug-loading capacity play a significant role in optimizing the delivery system [12].

These studies, typically conducted in vitro and in vivo using animal models, provide critical insights into the biological interactions, biodistribution, and therapeutic potential of the formulations. By understanding these aspects, researchers can fine-tune the formulations to enhance their clinical applicability ([14]; The ultimate goal of nanotechnology-enhanced herbal extracts is to translate their preclinical promise into clinical success. The most reliable way to determine whether a new treatment is safe and effective for humans is to conduct a clinical trial [8]. Phase I trials are smaller in size and focus on safety, Phase II trials are larger and examine efficacy and side effects, and Phase III trials are comprehensive and confirm therapeutic benefits while monitoring adverse reactions [36]. The successful completion of these phases is crucial for the regulatory approval and commercialization of nanoparticle-based herbal formulations [34]. It offers a pathway to harness the full therapeutic power of natural compounds, providing safer, more effective, and targeted treatments for a variety of diseases [7]. Future research in this field is expected to focus on further improving the design and functionality of nanoparticles, exploring novel herbal extracts with untapped therapeutic potential, and conducting comprehensive clinical studies to establish their efficacy and safety [36].

The convergence of nanotechnology and herbal medicine represents a significant advancement in the field of drug delivery and therapeutics [11]. By addressing the limitations associated with traditional herbal formulations, nanotechnology offers a promising approach to enhance the bioavailability, stability, and targeted delivery of herbal extracts. The formulation techniques, preclinical assessments, and clinical trials of herbal extracts enhanced by nanotechnology will be covered in detail in this section, which will also offer a thorough summary of the present status of research and potential directions in this fascinating multidisciplinary topic.

17.2 AN OVERVIEW OF PHYTOMEDICINE

In the last century, some 121 therapeutic medicines derived from traditional knowledge from various sources have been created. Plant-based remedies are employed in both traditional medicine and Native American folklore to treat a wide range of ailments. Herbal medicine and phytomedicines are currently focused on the healing and strengthening of body systems, especially the immune system, which can then appropriately resist foreign invaders, and the destruction of illnesses without harmful side effects. Historical records indicate that Chinese medicine has been using herbal remedies since 2800 BC. Herbal medicine has developed into a separate industry today since many people prefer it to synthetic drugs. They were employed by the Chinese and Indians, the Egyptians and Greeks, the Romans and Syrians, and others about 5000 years ago.

While there are around 500 medicinal plants mentioned in ancient texts, about 800 are used in native medical systems. The Indian subcontinent is an ideal location for medical studies due to its abundance of traditional knowledge and medicinal plants. Many plant species are utilized to treat a variety of illnesses in indigenous medical systems, such as Ayurveda, Unani, Siddha, and Allopathy [30]. Over 80,000 plant species are used in various parts of the world [17]. Approximately 800 plant species are used by more than 500 traditional Indian groups to treat a variety of illnesses, even though over 20,000 known medicinal plant species exist [10]. Medication made from plants is the first line of primary healthcare for humans, and it has no negative side effects. Herbal supplements, over-the-counter treatments, prescription pharmaceuticals, and traditional medicinal practices are all part of the modern approach to herbal medicine regulation. It is necessary to harmonize and enhance the regulatory framework, taking into account both conventional wisdom and scientific findings. The availability of trustworthy and top-notch raw materials, crucial to the efficacy and security of herbal remedies, will be enhanced by the growing movement towards domestication, biotechnological studies, and the genetic development of medicinal plants (Table 17.1).

TABLE 17.1
A List of Herbs with their Phytomedicine Effects

Herbs	Habitat	Uses	References
Aloe Vera	Hot, dry areas	Skin wounds, burn	[43]
Simmondsia chinensis	Dry, perennial plant	Wound healing [13]	[31]
Soyabean	Warm, fertile, well-drained, sandy loam	Anti-inflammatory and antioxidant effects	[43]
Ocimum sanctum	Tropical regions	Antidepressant, antioxidant, anti-inflammatory	[38]
Rosmarinus officinalis	Mediterranean, Portugal, and northwestern Spain.	Dyspepsia, high blood pressure and rheumatism	[33]
Carica Papaya	tropical or semi-tropical climates	Fever, asthma, colic, beriberi, jaundice, dengue fever, and another virus.	[37]

17.3 APPLICATION OF NANOFORMULATION

Nanoformulations are basic in different fields, particularly in drug store, medication, horticulture, and material science, because of their one-of-a-kind properties and benefits over traditional definitions that permits further developed solvency, bioavailability, designated drug conveyance, controlled discharge, and decreased harmfulness. From the early stages, nanopharmaceuticals have advanced to become a crucial component of clinical practice, significantly advancing outcomes in the clinical domains of neurological disorders, cancer treatment, irreversible illnesses, tailored medication, and advanced diagnostics [28].

17.3.1 Nanosuspension Technology

By increasing the dissolvability and oral bioavailability of pharmaceuticals and plant extracts, nanosuspension innovation offers an enticing nanotechnological solution to address drug capability issues. The method of nanoprecipitation proved effective in creating a nanosuspension. Plant extract was dissolved in ethanol to create a natural stage, and the plant extract was then separated using the nanoprecipitation technique to determine the plant-determined nanosuspension's readiness [42]. The ensuing mixture was gradually introduced while being continuously mixed into a fluid phase that contained a stabilizer, which might be either a polymer or a surfactant. *Terminalia arjuna*, and they settled with sodium lauryl sulfate and polysorbate-80, separately [21]. The resulting nanosuspensions showed bioactivities and were judged to be genuinely stable and non-poisonous.

17.3.2 Nano-Encapsulation

Drugs, nutraceuticals, and other bioactive mixes have shown promise in the transportation of nanocapsules [18]. Their unique design takes precise control over drug discharge energy, security, and focus into account. It has a central shell structure with a width usually between 10 and 1000 nm. This audit looks into the many materials and definition processes used to make nanocapsules, highlighting their potential uses and implications for medicine delivery and beyond [40].

17.3.3 Three-Dimensional Printing in Nanopharmacy (Nano Printing)

A paradigm change in drug discovery and delivery is being heralded by the nanopharmaceutical industry's use of 3D printing technology. This comprehensive review examines the game-changing possibilities of 3D printing, looking at its applications in creating perplexing nanostructures and customized pharmaceutical data [16]. With its unparalleled precision and adaptability, 3D printing has emerged as a formidable force in the development of nanoscale drug transporters and therapeutic devices, and this influence has only grown since its inception. Through a definite assessment of the combination systems, materials, and applications, this survey clarifies the developing scene of 3D imprinting in nanopharmaceutical improvement, preparing for a future where accuracy medication meets progressed fabricating. We dive into the reasoning behind incorporating 3D printing into drug improvement, featuring reforming drug conveyance frameworks and customized medicine potential. According to Zahmanova et al. [46], this section paves the way for a more in-depth examination of the methods, materials, and applications of 3D imprinting in nanopharmaceuticals.

17.3.4 Applications in Drug Delivery Systems

The adaptability of 3D printing in drug delivery frameworks is becoming increasingly important as its applications are expanded to include the production of oral, transdermal, and injectable measurement structures[16]. By offering exact control over medication discharge profiles and site-explicit emphasis, 3D printing innovation revolutionizes drug distribution. The topic covers patient-explicit inserts, mucoadhesive medicine transporters, and controlled discharge plans. For the production of nanopharmaceuticals, experts have resorted to 3D printing, which is a versatile instrument. According to Yadav and Malviya [44], highlighting its ability to personalize medicine schedules, control drug release energy, and enhance treatment practicality, this section investigates the several ways in which 3D printing can be used in drug delivery systems. From specialized drug transporters and implanted devices to designs for oral and transdermal dosing, 3D printing offers unmatched versatility and accuracy in drug administration. Among the most promising uses of 3D printing in medication delivery systems are oral measuring structures that incorporate personalized drug discharge characteristics (Morrison-Lanceta and colleagues [22]).

17.3.5 Biomimetics and Bioinspiration in Nanopharmaceuticals/Nanomedicines

Because it presents unique opportunities to develop and create creative healing techniques, nanotechnology has revolutionized a number of industries, including medicine. Inspired by nature's intricate designs and cycles, biomimetics and bioinspiration have emerged as potent ideal models in the fields of nanopharmaceuticals and nanomedicines. Nanoscale treatments, diagnostics, and pharmaceutical distribution are being transformed by these supplementary methods. According to Rao et al. [32], Biomimetics seeks to address complex challenges across disciplines by developing scalable, effective, and adaptive innovations by mimicking natural processes and standards. The multidisciplinary character of biomimetics, which involves coordinated work by academics, engineers, physicists, scientists, and fashion designers, is one of its key components. By combining data from different domains, scientists are more likely to understand and replicate the notable changes and functionality of live organisms [9].

17.3.6 Green Design

Nanopharmaceuticals may have harmful effects on the environment, despite their importance to human health and wealth. Always keep the patient and their natural surroundings in mind. When organizing nanopharmaceuticals, it is best to adhere to the GREENER concept, an acronym which stands for the following: G: Great Practice for Patients; R: Diminutive Adverse Impacts and High Specificity; E: Greater Openness Decrease through Less Emissions; E: Ecological (Bio)degradability; N: Absence of PBT (Steady, Bioaccumulative, and Harmful) Properties; E: Impact Decrease (Preventing Unfortunate Moieties); and R: Risk and Hazard Mitigation. As an example, consider the 'harmless by configuration' approach [15].

Therefore, the "green" approach to recycling and reusing materials is highly desirable in modern times. Plants have garnered special attention due to their abundance in nature and the bioactive components they contain. For both the development of adaptable standards and the preparation of "green" nanoparticles, plants that form the basis of these "green biologics" play a critical role. The significance of nanoparticle-vegetal concentrate combinations has been emphasized for their potential in advancing sustainable and innovative applications., opening a door to the possibility of developing nontoxic ('green') approaches to the organization of 'green' nanomedicines [29].

The creation of more precise and personalized patient-specific regimens has been made possible by the continuous developments in medicine delivery technology. With an emphasis on explicitness and precise control over drug release energy, nanotechnology-based drug delivery systems, including liposomes, micelles, and nanoparticles, offer versatile platforms for proving and delivering medicines. Advancements in three-dimensional printing have made it possible to build customized measurement structures that account for each patient's unique demands by modifying medication stacking, discharge profiles, and computations. One significant example of personalized medicine is accuracy oncology, which uses sub-atomic profiling to guide the selection of immunotherapies and treatments for malignant growth patients based on their individual genetic mutations and biomarker profiles [45]. By reducing underlying toxicity, boosting the survivability of therapeutic interventions, and directly delivering anticancer medications to cancer cells while protecting healthy tissues, patient-explicit criteria play a significant role in precision oncology. Drug delivery systems based on nanoparticles enable the best chemotherapeutic experts, prescribed drugs, and immunotherapies to be precisely delivered to cancer sites, improving treatment outcomes and patient satisfaction. Even though personalized medicine has the potential to significantly impact healthcare, there are still certain challenges in developing and implementing long-term clear information. Administrative barriers, physical restrictions, financial considerations, and the ethical implications of data security and informed consent are a few of them. Nevertheless, new advances in biomarker identification, medication delivery systems, and computational demonstration hold great promise for overcoming these obstacles and bringing about a new era of precision medicine tailored to each patient's unique needs according to Vasdev et al. [39].

17.4 FUTURE STUDY

The chapter explores the current apprehensions regarding the application of nanotechnology in the pharmaceutical and biotechnology industries. In order to develop a functional nanosystem capable of securely transporting therapeutic medication targets, it is imperative to evaluate the dimensions, shapes, surface charges, and design rigidity of nanocarriers. The avoidance of harmful ingredients and the use of modern, clean technologies for the preparation of NPs with minimal energy consumption and low cost were also emphasized as 'green' perspectives. Examples of these include biomimetics, bioinspiration, and bioderivation. The most recent advancements in the intricate nanopharmaceutical strategy were prioritized. One can transmit desirable characteristics, particularly with bionanoformulations. Biomimetic and bio-roused systems may be implemented to ensure biocompatibility and covertness, for example, by coating nanoparticles with synthetic or natural cells. The introduction of 3D printing is one of the most recent advancements in the production of NP. Nanopharmaceutical development, which depends on 3D printing for its precise medication and superior assembly, will soon experience additional advancements. The intriguing potential is suggested by recent advancements, such as the integration of AI with human cognition. Further research and development are required to gain a comprehensive understanding of the potential impact of 3D printing on the future of nanopharmaceuticals. This survey concludes with a compilation of fundamental experiences that underscore this important point. In order to comprehend the profession's maximum potential and assist patients, it is imperative to contemplate past experiences and anticipate future directions as patient-explicit definitions and individualized medication continue to transform healthcare. This section addresses forthcoming advancements in patient-specific terminology and personalized treatment, including the implementation of new technologies.Scientists may be able to obtain a more comprehensive understanding of disease systems, identify novel biomarkers, and develop personalized targeted therapeutics by integrating data from multiple omics domains.

Future research initiatives should investigate multi-omics data and apply discoveries to therapeutically relevant experiences by utilizing state-of-the-art bioinformatics tools and AI calculations. 3D printing innovations have the potential to revolutionize medication research and herald in a new era of precision medicine by enabling the creation of personalized treatment regimens and the visibility of blueprints to patients. In accordance with Vasdev et al. [39], nanomedicine is committed to revolutionizing the delivery and treatment of medication in the context of personalized medicine. One potential future for nanomedicine is the development of theranostic specialists who can simultaneously perform analytical imaging and medication delivery, as well as multifunctional nanoparticles and targeted drug transporters. Scientists can develop patient-specific definitions that are more useful and cause less fundamental damage by utilizing the controlled drug release, surface functionalization, and size-subordinate pharmacokinetics of nanoparticles. Personalized medication may eventually be replaced by computerized health technologies, including peripheral devices, remote monitoring systems, and flexible health applications. These advancements facilitate the ongoing assessment of patient well-being limits, ongoing information gathering, and personalized input, collaborating with early disease identification, improved therapy, and patient engagement. Future bearings in advanced well-being incorporate the mix of computerized reasoning calculations, prescient examination, and customized choice help apparatuses to enable patients and medical care suppliers in going with informed therapy choices. As customized medication develops, administrative contemplations and strategy systems will be basic in guaranteeing patient well-being, viability, and admittance to imaginative treatments. Future bearings in guideline incorporate the advancement of versatile pathways, facilitated survey cycles, and worth-based repayment models custom-made to the exceptional attributes of customized treatments. Medical services partners, business associates, and patient support groups should collaborate with administrative offices to establish guidelines, streamline procedures, foster growth, and safeguard public health [28].

17.5 CONCLUSION

Nanotechnology, combined with herbal medicine-produced pharmaceuticals, represent the importance of advancement in drug delivery and treatments. Nanotechnology increases the useful potential of herbal medicine extracts by improving their bioavailability, strength, and targeted conveyance while reducing accidental effects. This strategy focuses on the efficacy of conventional treatments while also reflecting the growing interest in personalized medicine. As research progresses, the future application of nanomedicine in medical care appears promising, with possible uses expanding beyond existing possibilities, providing more secure, more attractive medicines tailored to particular patient needs.

REFERENCES

1. Ahmed, H. M., Nabavi, S., & Behzad, S. (2021). Herbal drugs and natural products in the light of nanotechnology and nanomedicine for developing drug formulations. *Mini Reviews in Medicinal Chemistry*, *21*(3), 302–313.
2. Ahmed, Hiwa M., Nabavi, Seyed, Behzad, Sahar. (2020). Herbal Drugs and Natural Products in the light of Nanotechnology and Nanomedicine for Developing Drug Formulations. *Mini-Reviews in Medicinal Chemistry*. 20, 1–12.
3. Ali, F., Neha, K., & Parveen, S.. (2023). Current regulatory landscape of nanomaterials and nanomedicines: A global perspective. *Journal of Drug Delivery Science and Technology*, *80*, 104118. https://doi.org/10.1016/j.jddst.2023.104118
4. Ansari, S.H., Islam, F., Sameem, M. (2012). Influence of nanotechnology on herbal drugs: A Review. *Journal of Advanced Pharmaceutical Technology & Research*. 3, 142–146.
5. Arifin, S.F., AlShami, A., Omar, S.S.S., Jalil, M.A.A., Khalid, K.A., & Hadi H. (2019). Impact of modern technology on the development of natural-based products. *Journal of Ayurvedic and Herbal Medicine* *5*(4), 133–142.

6. Bonifácio, B. V. et al. 2014. Nanotechnology-based drug delivery systems and herbal medicines: a review. *International Journal of Nanomedicine*. *9*, 1–15.
7. Borase, Vaibhav A., Kapadnis, Rahul S., Chaudhari, Ashok Y., Bachhav, Rishikesh S.. 2020. A Review on Novel Drug Delivery System. *Journal of Emerging Technologies and Innovative Research*. *7*(9), 19–27.
8. Chakraborty, K., Shivakumar, A., Ramachandran, S. (2016). Nano-technology in herbal medicines: A review. *International Journal of Herbal Medicine*. *4*(3), 21–27.
9. De Jong, W. H., Geertsma, R. E., & Borchard, G. (2022). Regulatory safety evaluation of nanomedical products: Key issues to refine. *Drug Delivery and Translational Research*, *12*(10), 2042–2047. https://doi.org/10.1007/s13346-021-01024-5
10. Dev, S. (1997). Ethnotherapeutics and modern drug development: the potential of Ayurveda. *Current Science*, *73*(11), 909–928.
11. Dewi, M.K.; Chaerunisaa, A.Y.; Muhaimin, M.; Joni, I.M. 2022. Improved Activity of Herbal Medicines through Nanotechnology. *Nanomaterials*. *12*, 4073.
12. Dhanokar, Ruturaj D., Jatale, Ankita P., Deshmukh, Swati P., Mangesh, K. Hire. (2023). Nanotechnology for Enhancing the Bioavailability of Herbal Drug. *International Journal of Creative Research Thoughts*. *11*(10), 790–801.
13. Gavhane, Snehal A., Gade, Aditi T.. (2021). The novel drug delivery system. *International Journal of Creative Research Thoughts*. *9*(9), 373–385.
14. Gopi S., Amalraj A., Haponiuk J.T., Thomas S, 2016. Introduction of Nanotechnology in Herbal Drugs and Nutraceutical: A Review. *J Nanomedine Biotherapeutic Discov*. *6*(2): 1000143.
15. Han, X., Lu, Y., Xu, Z., Chu, Y., Ma, X., Wu, H., Zou, B., & Zhou, G. (2023). Anionic liposomes prepared without organic solvents for effective siRNA delivery. *IET Nanobiotechnology*, *17*(4), 269–280. https://doi.org/10.1049/nbt2.12184
16. Husni, P., Shin, Y., Kim, J. C., Kang, K., Lee, E. S., Youn, Y. S., Rusdiana, T., & Oh, K. T. (2020). Photo-based nanomedicines using polymeric systems in the field of cancer imaging and therapy. *Biomedicines*, *8*(12), 618. https://doi.org/10.3390/biomedicines8120618
17. Izah, S.C., Ogidi, O.I., Ogwu, M., Salimon, S., Yusuf, Z., Akram, M. Raimi, M., & Iyingiala, A. (2023). Historical Perspectives and overview of the value of herbal medicine. In: *Reference Series in Phytochemistry. Herbal Medicine Phytochemistry*. Springer Nature Singapore Pte Ltd.
18. Kümmerer, K. (2019). From a problem to a business opportunity: Design of pharmaceuticals for environmental biodegradability. *Sustainable Chemistry and Pharmacy*, *12*, 100136. https://doi.org/10.1016/j.scp.2019.100136
19. Mathpati, Rushikesh V., Harangule, Yogesh R., Bavage, Shyamlila B., Bavage, Nandkishor B. (2021). Nano-Technology in Herbal Medicines: Advancements in Herbal Treatment. *International Journal of Innovative Research in Technology*. *8*(3), 116–122.
20. Mishra, Y. et al., 2022. Application of nanotechnology to herbal antioxidants as improved phytomedicine: An expanding horizon. *Biomedicine & Pharmacotherapy*. *153*(2022), 113413.
21. Moermond, C. T. A., Puhlmann, N., Brown, A. R., Owen, S. F., Ryan, J., Snape, J., Venhuis, B. J., & Kümmerer, K. (2022). GREENER pharmaceuticals for more sustainable healthcare. *Environmental Science & Technology Letters*, *9*(8), 699–705. https://doi.org/10.1021/acs.estlett.2c00537
22. Moreno-Lanceta, A., Medrano-Bosch, M., & Melgar-Lesmes, P. (2020). Single-walled carbon nanohorns as promising nanotube-derived delivery systems to treat cancer. *Pharmaceutics*, *12*(9), 850. https://doi.org/10.3390/pharmaceutics12090850
23. Murkute, Pooja S., Kathar, Nakul P., Sanap, Gajanan S.. 2021. Nanotechnology; Its Application, Process Development and Nano-Technology Based Herbal Medicine. *International Journal of Pharmaceutical Research and Applications*. *6*(5), 593–605.
24. Narwade, Vaibhav V., Game, M. D., Kadam, Sanket G. (2024). A Review on Nanotechnology in Herbal Medicine. *International Journal of Pharmaceutical Sciences*. *2*(4), 411–416.
25. Ogidi, O.I., & Emaikwu, N.G. (2023a).Adoption and Application of Biotechnology in Herbal Medicine Practices. In: *Reference Series in Phytochemistry. Herbal Medicine Phytochemistry*. Springer Nature Singapore Pte Ltd.
26. Ogidi, O.I., & Emaikwu, N.G. (2023b). Utilization methods and Practices of herbal Medicine in Africa. In: *Reference Series in Phytochemistry. Herbal Medicine Phytochemistry*. Springer Nature Singapore Pte Ltd.
27. Pandey, A., & Pandey, G. (2014). Nanotechnology for herbal drugs and plant research. *Research and reviews: Journal of Pharmaceutics and Nanotechnology*, *2*(1), 13–16.

28. Petrovic, S., Bita, B., & Barbinta-Patrascu, M.-E. (2024). Nanoformulations in pharmaceutical and biomedical applications: Green perspectives. *International Journal of Molecular Sciences*, *25*(11), 5842. https://doi.org/10.3390/ijms25115842
29. Petrovic, S. M., & Barbinta-Patrascu, M.-E. (2023). Organic and biogenic nanocarriers as bio-friendly systems for bioactive compounds' delivery: State-of-the-art and challenges. *Materials*, *16*(20), 7550. https://doi.org/10.3390/ma16207550
30. Rabe, T., & van Staden, J. (1997). Antibacterial activity of South African plants used for medicinal purposes. *Journal of Ethnopharmacology*, *56*(1), 81–87. https://doi.org/10.1016/s0378-8741(96)01515-2
31. Ranzato, E., Martinotti, S., & Burlando, B. (2011). Wound healing properties of jojoba liquid wax: an in vitro study. *Journal of Ethnopharmacology*, *134*(2), 443–449. https://doi.org/10.1016/j.jep.2010.12.042
32. Rao, S. V., Anusha, L., Rathan, B. H., Soniya, S., & Padmalatha, K. (2023). Nanopharmaceutics: A novel drug delivery technology. *Journal of Population Therapeutics and Clinical Pharmacology*, *30*(3), 888–903. https://doi.org/10.15586/jptcp.v30i3.1071
33. Rotblatt, M. (2000). Herbal medicine: Expanded commission E monographs. *Annals of Internal Medicine*, *133*(6), 487. https://doi.org/10.7326/0003-4819-133-6-200009190-00031
34. Rupali, Sanjay Patare. (2022). Nanotechnology in herbal drug: a review. *International Journal of Scientific Development and Research*. *7*(12), 127–135.
35. Sahu, A.N. (2013). Nanotechnology in herbal medicines and cosmetics. *International Journal of Research in Ayurveda and Pharmacy* *4*(3), 472–474.
36. Sharma, M.. (2014). Applications of Nanotechnology Based Dosage Forms for Delivery of Herbal Drugs. *Research and Reviews: Journal of Pharmaceutics and Nanotechnology*. *2*(1), 23–30.
37. Singh, S. P., Kumar, S., Mathan, S. V., Tomar, M. S., Singh, R. K., Verma, P. K., Kumar, A., Kumar, S., Singh, R. P., & Acharya, A. (2020). Therapeutic application of Carica papaya leaf extract in the management of human diseases. *Daru: Journal of Faculty of Pharmacy, Tehran University of Medical Sciences*, *28*(2), 735–744. https://doi.org/10.1007/s40199-020-00348-7
38. Udupa, S. L., Shetty, S., Udupa, A. L., & Somayaji, S. N. (2006). Effect of Ocimum sanctum Linn. on normal and dexamethasone suppressed wound healing. *Indian Journal of Experimental Biology*, *44*(1), 49–54.
39. Vasdev, N., Pawar, B., Gupta, T., Mhatre, M., & Tekade, R. K. (2023). A bird's eye view of various cell-based biomimetic nanomedicines for the treatment of arthritis. *Pharmaceutics*, *15*(5), 1150. https://doi.org/10.3390/pharmaceutics15051150
40. Verma, A., Gautam, S. P., Bansal, K. K., Prabhakar, N., & Rosenholm, J. M. (2019). Green nanotechnology: Advancement in phytoformulation research. *Medicines*, *6*(2), 39. https://doi.org/10.3390/medicines6020039
41. Wang, S., Chen, Y., Guo, J., & Huang, Q. (2023). Liposomes for tumor-targeted therapy: A review. *International Journal of Molecular Sciences*, *24*(5), 2643. https://doi.org/10.3390/ijms24052643
42. Wang, Z., Qiao, R., Tang, N., Lu, Z., Wang, H., Zhang, Z., Xue, X., Huang, Z., Zhang, S., Zhang, G., et al. (2017). Active targeting theranostic iron oxide nanoparticles for MRI and magnetic resonance-guided focused ultrasound ablation of lung cancer. *Biomaterials*, *127*, 25–35. https://doi.org/10.1016/j.biomaterials.2017.03.020
43. Xu, L., Choi, T. H., Kim, S., Kim, S.-H., Chang, H. W., Choe, M., Kwon, S. Y., Hur, J. A., Shin, S. C., Chung, J. I., Kang, D., & Zhang, D. (2013). Anthocyanins from black soybean seed coat enhance wound healing. *Annals of Plastic Surgery*, *71*(4), 415–420. https://doi.org/10.1097/SAP.0b013e31824ca62b
44. Yadav, D., & Malviya, R. (2023). Novel nanomaterials as photo-activated cancer diagnostics and therapy. *Medicare Advantage*, *1*(3), 190–209. https://doi.org/10.1016/j.medadv.2023.100023
45. Yan, W., Leung, S. S., & To, K. K. (2020). Updates on the use of liposomes for active tumor targeting in cancer therapy. *Nanomedicine*, *15*(3), 303–318. https://doi.org/10.2217/nnm-2019-0354
46. Zahmanova, G., Aljabali, A. A. A., Takova, K., Minkov, G., Tambuwala, M. M., Minkov, I., & Lomonossoff, G. P. (2023). Green biologics: Harnessing the power of plants to produce pharmaceuticals. *International Journal of Molecular Sciences*, *24*(17), 17575. https://doi.org/10.3390/ijms241717575

18 Comparative Analysis of Traditional and Nano-Formulated Approaches in Viability, Targeted Delivery, and Patient Outcomes

J. Nithya Shree and A. Ronaldo Anuf
Department of Biotechnology, Kamaraj College of Engineering and Technology, Virudhunagar, India

Sohail Ahmad
Gomal Center of Biochemistry and Biotechnology, Gomal University, D.I. Khan, Pakistan

18.1 INTRODUCTION

Herbal medicine has long been a cornerstone of traditional healthcare systems worldwide, offering natural remedies for health conditions. The use of plant-based compounds for therapeutic purposes dates back thousands of years, reflecting a deep-seated belief in the healing power of nature [1]. Despite their historical significance and ongoing relevance, traditional herbal formulations often face significant limitations that hinder their effectiveness. These challenges include high doses required for therapeutic efficacy, frequent administration, and issues with the delivery of active compounds to their intended sites of action. Additionally, traditional herbal medicines can have off-target effects, which may reduce their overall therapeutic benefit and increase the risk of adverse outcomes [2].

One of the foremost obstacles to the effectiveness of herbal medicines is their poor bioavailability. Many herbal compounds suffer from low solubility and slow absorption rates, which can result in suboptimal levels of the active ingredients reaching the target tissues [3]. This issue is compounded by the rapid systemic clearance of these compounds, necessitating frequent dosing to achieve therapeutic effects. In response to these challenges, recent advancements in nanotechnology have emerged as a transformative approach to enhancing the delivery and efficacy of herbal medicines. Nano-formulated systems, utilizing a variety of nanocarriers such as liposomes, polymeric nanoparticles, solid lipid nanoparticles (SLNs), and nanoemulsions, have shown promise in addressing the limitations of conventional herbal formulations [4].

Nanotechnology offers several advantages over traditional delivery methods. By encapsulating herbal bioactives in nanoscale carriers, these advanced systems can significantly improve the stability and permeability of the active compounds [5]. This encapsulation not only protects the herbal ingredients from degradation but also facilitates their sustained and controlled release [6]. As a result, nano-formulations enhance the bioavailability of herbal medicines, allowing for more precise

DOI: 10.1201/9781003544418-18

targeting of specific tissues or cells and reducing the need for frequent dosing [7]. This targeted approach not only improves the therapeutic value of herbal treatments but also minimizes potential side effects, leading to better patient outcomes [8].

The integration of nanotechnology into herbal medicine is not without its challenges. The development of nano-formulated herbal products requires rigorous research to ensure their safety and efficacy [9]. Regulatory hurdles also pose a significant barrier, as there is a need for standardized manufacturing protocols and comprehensive safety evaluations for these novel delivery systems. Despite these challenges, the potential benefits of nano-formulated herbal medicines are substantial. They offer the promise of enhanced therapeutic efficacy, reduced side effects, and improved patient compliance, making them a valuable addition to modern healthcare practices.

This chapter aims to provide a detailed comparative analysis of traditional and nano-formulated approaches in herbal medicine. It will explore how modern nano-formulations address the limitations of conventional methods, offering insights into their mechanisms of action, therapeutic advantages, and potential impact on patient outcomes. Additionally, the chapter will discuss ongoing research efforts, regulatory considerations, and safety issues associated with nano-formulations, emphasizing the need for continued investigation and validation to fully harness the potential of these advanced drug delivery systems. By bridging the gap between traditional herbal remedies and cutting-edge technology, this analysis seeks to highlight the promising future of herbal medicine in the context of contemporary therapeutic practices.

18.2 HERBAL PHYTOCONSTITUENTS FOR DISEASE MANAGEMENT

Herbal phytoconstituents, the bioactive compounds derived from plants, have gained significant attention for their therapeutic properties in managing a variety of diseases [10]. These compounds, including alkaloids, flavonoids, terpenoids, glycosides, and phenolic acids, have shown promise in treating chronic diseases like diabetes [11], cardiovascular disorders [12], cancer [13], infectious ailments [14], and neurodegenerative diseases [15].

For instance, flavonoids and phenolic acids are known for their antioxidant and anti-inflammatory effects [16, 17], while alkaloids and terpenoids exhibit antimicrobial and antiviral properties [18, 19]. Phytoconstituents such as curcumin and resveratrol are being studied for their potential anticancer activities, particularly in inducing apoptosis in cancer cells [20, 21].

Despite their promising potential, one major challenge in applying herbal phytoconstituents is their low bioavailability due to poor solubility and absorption. To overcome this issue, modern techniques like phytosomes have been developed. Phytosomes bind water-soluble phytoconstituents to phospholipids, forming lipid-compatible molecular complexes that improve absorption and bioavailability [22].

Historically, herbal remedies have been vital sources of medicine, and, despite advances in allopathy, plant-derived drugs remain a major part of modern and traditional medicine [23]. Over 70% of marketed medications are derived from natural or semi-synthetic plant sources, with research showing that phytoconstituents not only prevent but also treat various conditions, including fungal infections [24]. Plants and their compounds have been extensively studied for antifungal activity, especially against *Candida albicans* [25] (Figure 18.1).

Herbs have also been widely used in traditional remedies for improving cognitive impairment and age-related memory loss. Medicinal plants, with their diverse active components, are emerging as alternatives to synthetic drugs for cognitive disorders, including Alzheimer's disease (AD) [26]. Nanocarrier systems have been developed to enhance the delivery of herbal treatments for AD, addressing challenges like low solubility and metabolism associated with natural compounds.

Herbal phytoconstituents have also played a crucial role in managing infectious diseases, particularly viral infections, before the advent of antibiotics. Many plant species containing antiviral molecules have been identified, showing activity against viruses like rabies, HIV, and influenza [27].

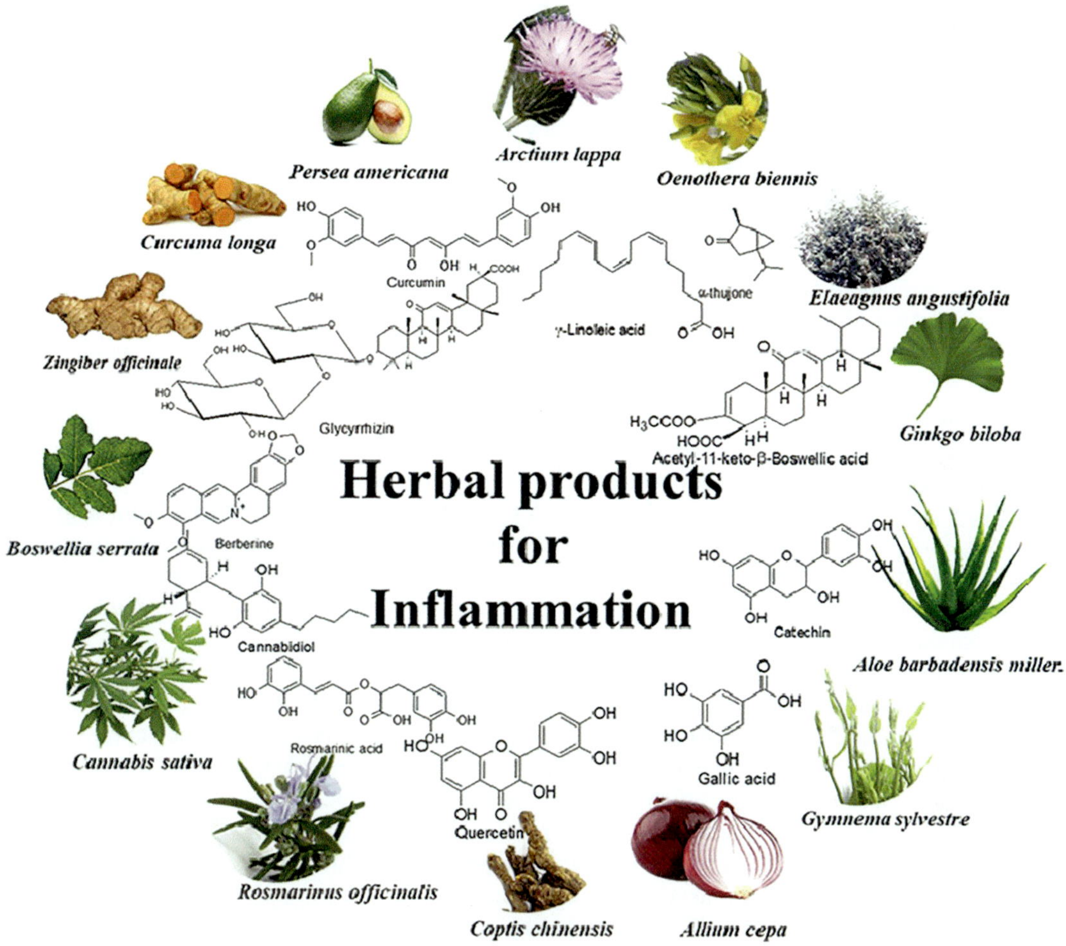

FIGURE 18.1 Herbal Products for Inflammation-Medication.

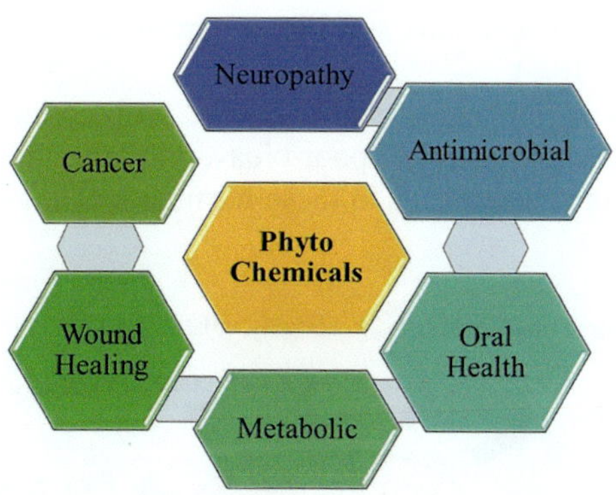

FIGURE 18.2 Diseases treated by phytochemicals.

With increasing drug resistance and the limited availability of suitable drug candidates, herbal sources provide significant potential for developing new treatments for viral diseases.

Ongoing research continues to explore the potential of herbal phytoconstituents to develop novel, effective, and safe therapeutic agents for a wide range of diseases, addressing challenges like bioavailability and standardization to improve their clinical efficacy (Figure 18.2).

18.3 BARRIERS TO HERBAL FORMULATIONS

Barriers to herbal formulations in modern healthcare systems arise from various challenges, including regulatory obstacles, standardization issues, and limited scientific validation [28]. Herbal products are often required to meet safety and efficacy standards similar to pharmaceutical drugs, yet the complex and variable nature of their compositions makes it difficult to achieve uniformity in quality and dosage. The lack of standardized manufacturing processes and variability in raw materials further complicate the consistent performance of these products [28]. Additionally, insufficient clinical research and randomized trials hinder broader acceptance by healthcare professionals, creating skepticism about their therapeutic efficacy [29]. Intellectual property rights, along with the commercialization of traditional knowledge, further complicate the integration of herbal formulations into mainstream medicine, while misconceptions and limited awareness influence public trust [28]. Another significant barrier is the poor bioavailability, stability, and inconsistent therapeutic outcomes of traditional herbal formulations [30]. Nanotechnology offers a promising solution to these barriers by enhancing the solubility, bioavailability, and stability of herbal bioactives [31]. Nanoformulations, such as polymeric nanoparticles, liposomes, SLNs, and nanoemulsions, improve the delivery and effectiveness of herbal medicines by protecting active compounds from degradation and ensuring sustained, controlled release [22]. These advancements help overcome traditional obstacles, allowing herbal treatments to be more potent and long-lasting. However, nanotechnology also introduces new challenges, including safety concerns and high production costs, which require further research and regulation to ensure its successful integration into the field.

Herbal formulations face numerous barriers in modern healthcare systems, despite the growing interest in their natural therapeutic potential. One of the primary challenges is the lack of standardization in the production and composition of herbal medicines [32]. Herbal products, composed of complex mixtures of phytochemicals, often suffer from variability in raw materials, making it difficult to ensure consistent quality, efficacy, and safety [33]. Unlike pharmaceutical drugs, which are subject to rigorous regulatory scrutiny, herbal medicines must meet similar safety and efficacy standards despite the difficulties posed by their complex and variable nature. This leads to significant regulatory hurdles and limits their acceptance in mainstream medicine. Additionally, the absence of extensive scientific validation, including clinical research and randomized trials [29], contributes to skepticism among healthcare professionals regarding the therapeutic effectiveness of herbal formulations. Intellectual property rights and commercialization issues further complicate the integration of traditional herbal knowledge into modern healthcare practices, while public perceptions are influenced by misconceptions and limited awareness about the benefits and risks of herbal medicines [34] (Figure 18.3).

Another major barrier is the inherent challenges in formulating effective herbal products. Traditional herbal formulations often exhibit poor bioavailability, stability, and inconsistent therapeutic outcomes. The active phytochemicals in these formulations may not reach the target tissues

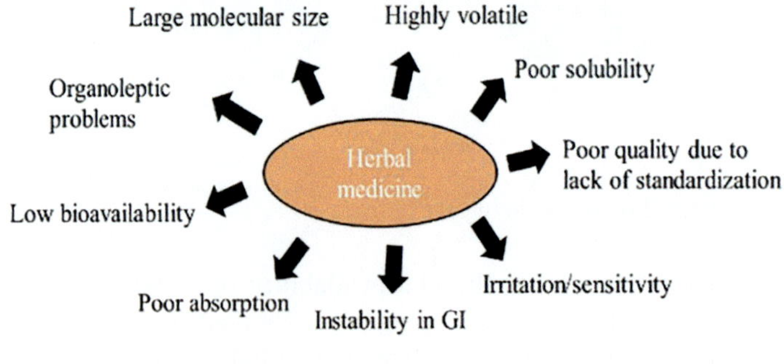

FIGURE 18.3 Barriers to herbal formulations.

in sufficient concentrations, reducing their overall therapeutic efficacy [35]. Moreover, issues such as bulk dosing, limited absorption, and rapid systemic clearance make it necessary to administer higher doses or frequent treatments, which reduces patient compliance and overall treatment effectiveness. Processing difficulties, such as the poor solubility and stability of plant-based compounds, further complicate the formulation of herbal medicines.

Nanotechnology has emerged as a promising solution to overcome many of these barriers. By integrating herbal bioactives into novel drug delivery systems such as nanoparticles, liposomes, and nanoemulsions, nanotechnology enhances the solubility, bioavailability, and stability of herbal medicines [31]. These nanoformulations protect active compounds from degradation, enable sustained and controlled release, and improve the distribution of herbal constituents to target tissues. This not only enhances the therapeutic value of herbal formulations but also reduces toxicity and the frequency of administration, increasing patient compliance. However, while nanotechnology presents exciting opportunities, it also introduces new challenges, including safety concerns regarding the potential toxicity of nanomaterials and the high production costs associated with these advanced delivery systems.

Furthermore, the regulatory landscape for herbal nanoformulations remains complex, with a need for standardized guidelines to ensure consistency, safety, and quality across products [36]. Therefore, while nanotechnology holds the potential to revolutionize herbal medicine by addressing key barriers, further research, collaboration, and regulation are required to fully integrate it into clinical practice and maximize its benefits.

18.4 STRATEGIES TO ENHANCE BIOAVAILABILITY

Enhancing the bioavailability of herbal formulations is crucial for maximizing their therapeutic efficacy and involves several strategic approaches. Nanotechnology-based delivery systems, such as nanoparticles, liposomes, and SLNs, are effective in improving the solubility, stability, and sustained release of herbal bioactives, thus enhancing their absorption and circulation time [37]. Micronization, or the reduction of particle size, increases the surface area of herbal constituents, promoting better dissolution and absorption [38].

Bioenhancers, such as piperine, curcumin, and quercetin, can further boost the bioavailability of herbal compounds by inhibiting metabolic enzymes and improving absorption [39]. Prodrug formation, where inactive compounds are converted into active forms after metabolism, also enhances solubility and stability. Cyclodextrin complexation and phospholipid complexation improve the solubility and permeability of herbal constituents, facilitating their absorption. Self-emulsifying drug delivery systems (SEDDS) enable the formation of emulsions in the gastrointestinal tract, enhancing dissolution and absorption.

Permeation enhancers, like fatty acids and surfactants, increase the permeability of herbal compounds across biological membranes [40, 41], while enzyme inhibitors prevent premature metabolism, resulting in higher systemic concentrations. Mucoadhesive formulations, which adhere to mucosal tissues, prolong residence time, thus enhancing the bioavailability of herbal medicines [42]. These strategies collectively address the challenges of poor bioavailability in herbal formulations, leading to more effective therapeutic outcomes and improved patient compliance [43] (Figure 18.4).

Moreover, novel drug delivery systems, such as the Phytolipid Delivery System (Phytosome), improve the therapeutic effects of plant extracts by forming lipid-compatible complexes that enhance absorption and bioavailability [44]. Phytosomes, which involve binding herbal ingredients to phospholipids, have demonstrated improved pharmacokinetic and pharmacological parameters, making them beneficial for treating a range of conditions.

Natural bio-enhancers, advanced formulation strategies like micronization and complexation, and the incorporation of nanotechnology in drug delivery are vital to optimizing the bioavailability of herbal medicinal products (HMPs) and integrating them into modern healthcare systems.

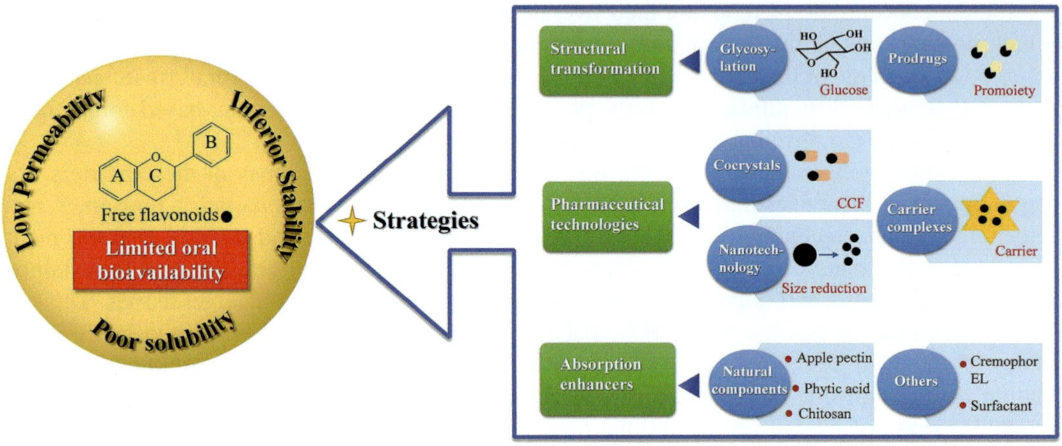

FIGURE 18.4 Strategies to enhance bioavailability.

18.5 HERBAL FORMULATIONS – CONVENTIONAL DOSAGE FORMS

Herbal formulations have traditionally been available in various conventional dosage forms, each designed to cater to different therapeutic needs and patient preferences [45]. Powders, which are simply ground herbal substances, can be consumed directly or mixed with liquids, providing versatility but often facing challenges related to bioavailability and stability [7]. Capsules offer a convenient and precise dosage by encapsulating herbal powders or extracts, effectively protecting the contents from degradation [46]. Tablets, compressed forms of herbal powders or extracts, provide stability and ease of administration but require specific excipients to ensure proper dissolution [47]. Herbal teas involve infusing herbs in hot water, a straightforward method that extracts water-soluble compounds, although it may not be suitable for all herbal constituents [48] (Figure 18.5).

Tinctures are highly concentrated alcoholic or hydroalcoholic extracts, offering a potent and long-lasting form of herbal medicine [49]. Extracts, concentrated preparations made with solvents

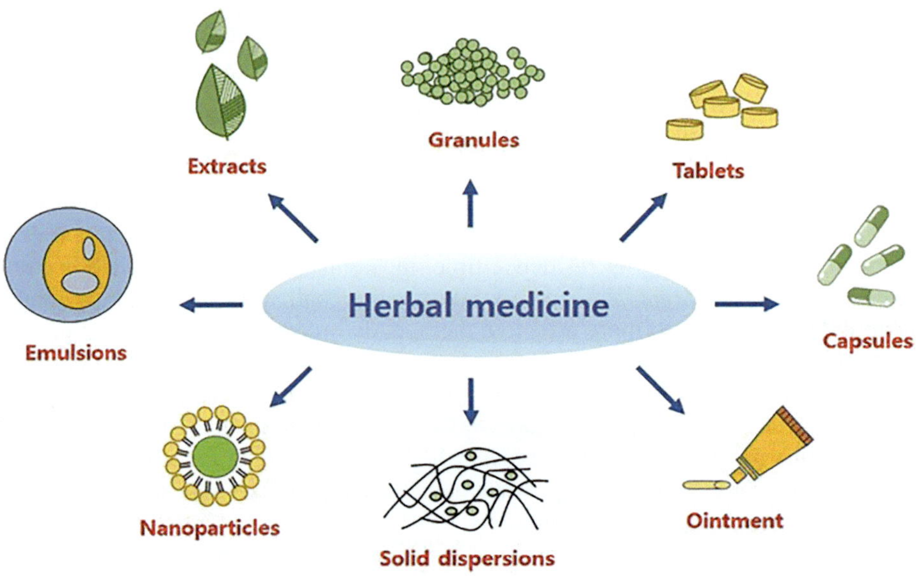

FIGURE 18.5 Conventional dosage forms.

like alcohol or water, can be standardized to contain specific active ingredients and are often used in capsules or tablets. Syrups blend herbal extracts with sweeteners, making them more palatable and suitable for children [50]. Oils, including essential oils, are used for topical applications or flavoring [51], while liniments and ointments incorporate herbal extracts into oily or waxy bases for localized treatment [52]. Poultices, fresh preparations applied directly to the skin, offer relief from pain and inflammation [53].

Lozenges dissolve slowly in the mouth, providing gradual relief for throat and cough issues, and suppositories are designed for insertion into body cavities, releasing herbal constituents where needed [54]. Each dosage form has distinct advantages and limitations that impact its effectiveness and suitability for various therapeutic purposes. As traditional medicine gains renewed importance in modern healthcare, there is a focus on improving solubility, bioavailability, pharmacological activity, and stability while reducing toxicity and enhancing patient compliance [55]. Contemporary oral dosage forms include granules (such as effervescent and rapid-release types), tablets (like fast-dissolving and rapid-disintegrating varieties), capsules, suspensions, and syrups. Topical dosage forms have evolved to include products like toothpaste, natural hair dyes, face packs, hand washes, liniment sprays, roll-ons, gels, creams (emulsions), and shampoos. Modern traditional pharmacies, including Ayurvedic and Unani establishments, have embraced these new trends, creating innovative dosage forms that go beyond conventional practices [56]. Integrating these advancements into Unani and other traditional Indian medicine systems is crucial for aligning with global standards. The shift towards improved dosage forms in traditional medicine offers numerous advantages, such as enhanced solubility and bioavailability, reduced toxicity, improved pharmacological activity, better stability, effective drug delivery systems, and increased patient compliance, leading to better therapeutic results. Herbal formulations, with their conventional dosage forms, have evolved to address contemporary therapeutic needs effectively [57].

Among these, tablets are favored for their convenience and customizable release profiles, including immediate, sustained, or controlled release, allowing for tailored therapeutic effects [58]. Capsules, another popular form, can be filled with powdered extracts or essential oils, providing a palatable alternative to tablets and masking the taste of herbal ingredients.

Granules, such as effervescent and rapid-release types, dissolve in water for easy consumption and improved absorption. Suspensions offer a liquid option for patients who struggle with solid forms, while syrups, with their sweetened taste, cater to both children and adults seeking a more pleasant intake method. Topical forms like ointments and creams deliver herbal remedies directly to the skin, addressing localized conditions with anti-inflammatory or analgesic effects [59] (Figure 18.6).

Liniments applied externally, offer relief for muscle and joint pain, and gels, with their non-greasy texture, are used for conditions like burns and rashes [60]. Additionally, herbal extracts are incorporated into everyday products such as shampoos and kinds of toothpaste, providing therapeutic benefits for hair and oral care [61]. These varied dosage forms enhance the accessibility, efficacy, and patient acceptability of herbal medicines, ensuring their continued relevance in modern healthcare. Recent advancements in drug delivery systems (NDDS) for plant actives and extracts, including polymeric nanoparticles, nanocapsules, liposomes, phytosomes, nanoemulsions, microspheres, transferases, and ethosomes, offer enhanced therapeutic outcomes compared to traditional formulations [62].

These novel systems provide notable advantages, such as improved solubility and bioavailability of herbal compounds, enhanced protection from toxicity, increased pharmacological activity, greater stability, and better distribution within tissues [63]. They also offer sustained release, protecting active ingredients from physical and chemical degradation. For example, liposomes, capable of encapsulating both hydrophilic and hydrophobic substances, are notable for their high biocompatibility and ability to enhance therapeutic efficacy, particularly in cancer treatments [64]. Phytosomes, formed by binding herbal extracts to phosphatidylcholine, significantly improve absorption and effectiveness, as evidenced by the increased bioavailability of silybin from silybin phytosome

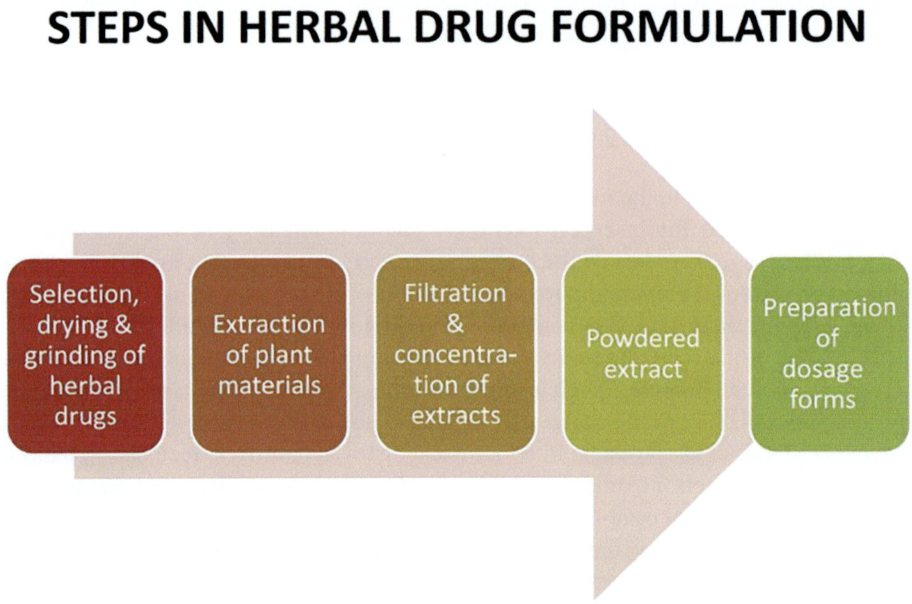

FIGURE 18.6 Steps involved in herbal drug formulation.

compared to traditional milk thistle extract. Nanoparticles, with their ability to enhance solubility and pharmacokinetics, are also crucial in overcoming limitations of traditional plant medicines [65].

The development and application of these NDDS technologies represent significant advancements in herbal medicine, addressing longstanding challenges and improving therapeutic outcomes [66]. Although naturally produced plants have been integral to herbal medicine for decades, major pharmaceutical companies largely overlooked these products, often focusing on synthetic compounds [67]. However, natural products and their derivatives have gained significant traction in healthcare, with approximately one-third of top-selling pharmaceuticals being derived from plants or microorganisms [68]. Their appeal lies in their therapeutic efficacy, lower side effects, and cost-effectiveness compared to synthetic drugs. Despite this, many natural products face challenges in clinical trials due to toxicity and biocompatibility issues [69].

18.6 NANOCARRIERS IN HERBAL DRUG DELIVERY

Herbal medicines have been used worldwide since ancient times, valued for their therapeutic benefits and often associated with fewer adverse effects compared to modern pharmaceuticals. Recent advancements in drug delivery systems, particularly through nanotechnology, have greatly enhanced the effectiveness of herbal therapeutics by reducing toxicity and side effects while increasing bioavailability. The application of nanotechnology in herbal medicine is growing rapidly, particularly in the realm of drug delivery. Nano herbal drug delivery systems offer immense potential to enhance the therapeutic activity of medicinal plants while addressing traditional limitations [70] (Figure 18.7).

Nanocarriers, such as micelles, liposomes, polymeric nanoparticles, and dendrimers, are engineered to carry both hydrophobic and hydrophilic drugs, optimizing delivery and minimizing side effects [71]. These systems have shown particular promise in treating severe diseases like cancer and diabetes, offering targeted delivery, sustained release, and improved stability of herbal compounds. For instance, in chemotherapy, nanocarriers can more effectively target tumors, reducing the toxicity of anticancer drugs and improving efficacy [71]. Nanotechnology's integration with herbal medicine

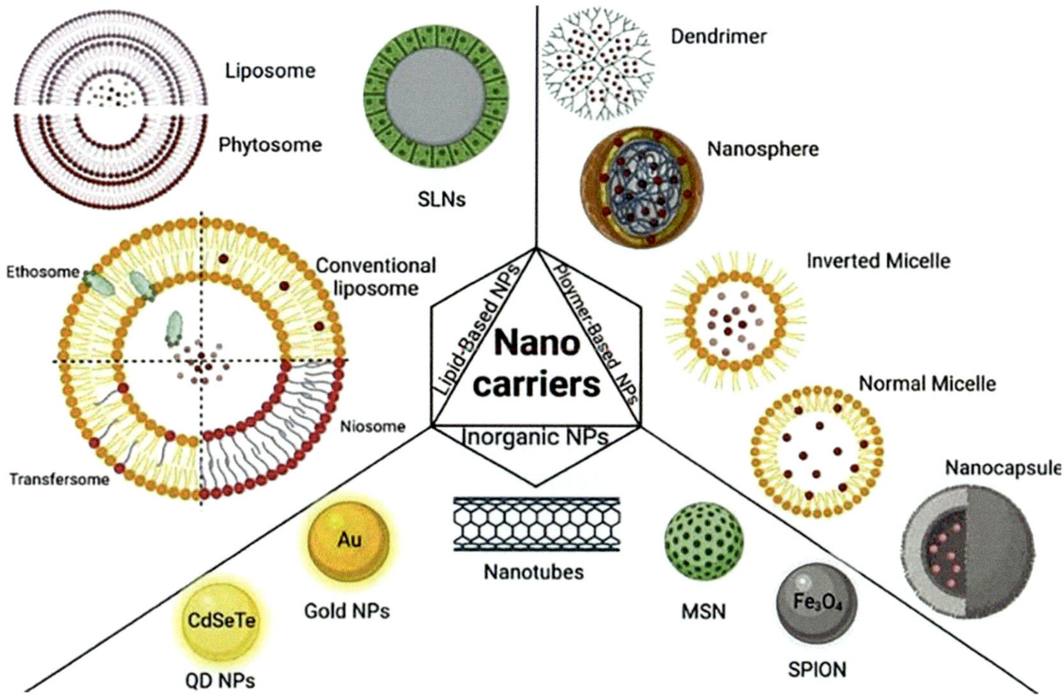

FIGURE 18.7 Types of nanocarriers.

is seen as a significant leap forward in the field of phytomedicine, enhancing treatments and opening new possibilities for safer, more effective therapies.

One of the key benefits of nano-drug delivery systems in herbal medicine is improved bioavailability. Nanocarriers enhance the solubility and absorption of poorly bioavailable herbal compounds, leading to more effective therapeutic outcomes [72]. Targeted delivery is another advantage, as nano-delivery systems can be designed to specifically target tissues, organs, or cells, improving the efficacy of herbal treatments while minimizing off-target effects. Additionally, encapsulating herbal compounds within nanocarriers protects them from degradation, ensuring long-term stability and efficacy [1] (Figure 18.8).

Controlled release is another major benefit, allowing for sustained delivery of herbal compounds, reducing the frequency of dosing, and improving patient convenience [58]. Nanocarriers also enable the combination of multiple herbal ingredients, which can lead to synergistic effects and more comprehensive treatments. Furthermore, nano drug delivery systems can be tailored to suit individual patient needs, making personalized medicine more achievable.

However, challenges such as regulatory considerations, long-term safety assessments, and the standardization of herbal formulations need to be addressed to ensure the widespread adoption and clinical translation of these innovations. Continued research and collaboration between scientists, herbalists, and clinicians will be essential in unlocking the full potential of nano drug delivery systems in herbal medicine [73].

Nanotechnology has emerged as a promising field in drug discovery and delivery, offering unique properties such as self-targeting due to the small size of nanoparticles, which can be directed to specific pathological areas without needing specific ligands [70]. Nanocarriers, used in drug delivery, demonstrate significant potential, particularly in chemotherapy, and are now being explored for herbal remedies. This combination is expected to enhance herbal medicines' effectiveness in treating chronic diseases and promoting health benefits. Nanocarriers have shown the ability to enhance the bioavailability and therapeutic efficacy of plant-based medicines, overcoming challenges such as

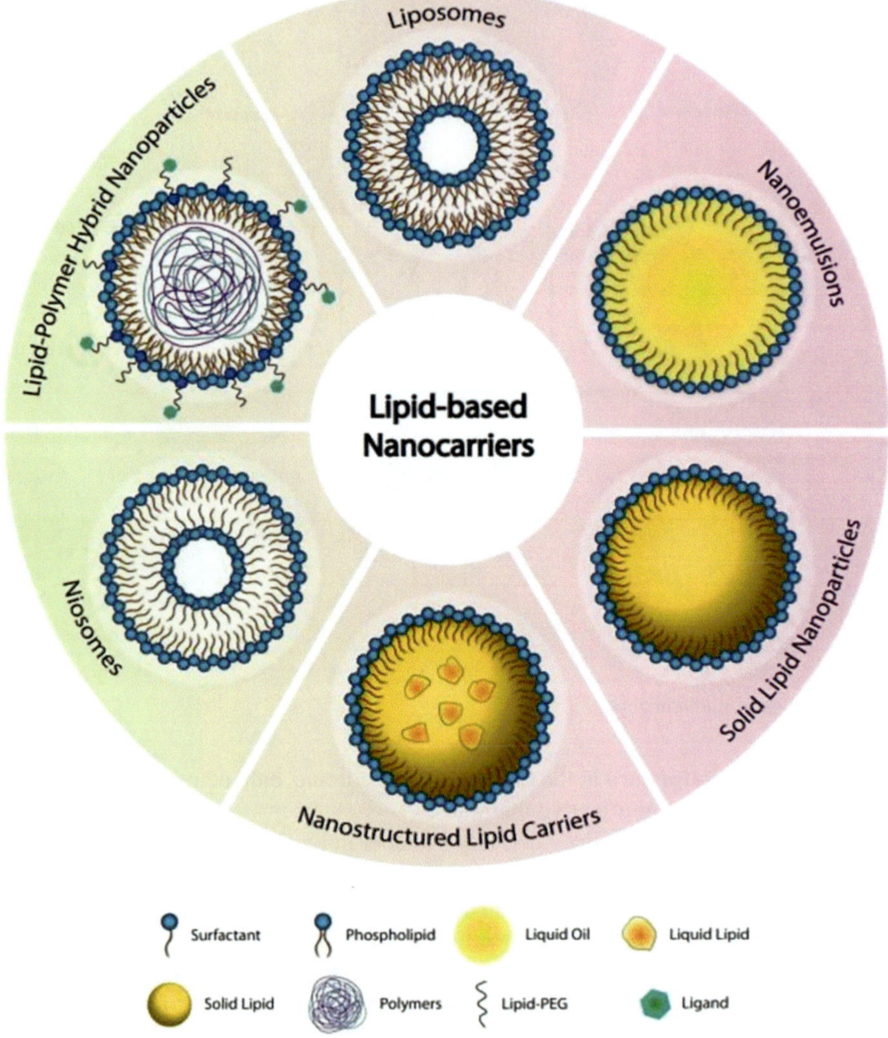

FIGURE 18.8 Types of lipid-based nanocarriers.

poor solubility, low stability, and rapid elimination from the body [74]. Liposomes, solid lipid nanoparticles, and metal nanoparticles are some of the common nanocarriers used in phytomedicine, offering unique therapeutic benefits and improving patient outcomes.

18.7 CLINICAL STATUS OF CURRENT DELIVERY STRATEGIES

The clinical status of current delivery strategies for herbal medicines, particularly those incorporating nanotechnology, reflects significant progress but also faces notable challenges. Nanocarrier systems, such as liposomes, micelles, and polymeric nanoparticles, have emerged as promising tools in enhancing the efficacy, stability, and bioavailability of herbal compounds [75]. These advanced delivery systems leverage their nanoscale size to improve targeted delivery, thereby reducing systemic toxicity compared to conventional administration routes. Extensive preclinical research has demonstrated their potential, especially in treating chronic conditions like cancer and diabetes [76]. However, translating these advancements into clinical practice remains complex.

While several nanocarrier formulations have shown promise in animal studies, only a few have advanced to early-phase clinical trials [77]. These trials primarily focus on evaluating safety, pharmacokinetics, and initial efficacy. One significant barrier to broader clinical adoption is the regulatory environment, which is hindered by a lack of standardized manufacturing protocols and concerns about the long-term safety of nanomaterials [78]. The pharmaceutical industry, with its stringent regulatory requirements, has seen slower progress in integrating nanotechnology compared to the nutraceutical and cosmeceutical sectors, where some nano-based herbal products have already been commercialized [79].

Emerging trends in nanomedicine, such as personalized medicine and hybrid nanocarriers, offer promising avenues for future research and development [72]. These approaches aim to further enhance the precision and effectiveness of drug delivery systems [57]. However, they remain in the early experimental stages. The successful integration of nanotechnology into mainstream herbal medicine will depend on overcoming these regulatory and safety challenges through ongoing research and collaboration among scientists, regulators, and industry leaders [80] (Figure 18.9).

Additionally, research efforts are concentrated on improving the scalability of nanoparticle production, understanding the interactions between nanoparticles and biological systems, and optimizing delivery systems for better therapeutic outcomes. These advancements are crucial for maximizing the potential of nanotechnology in medicine and ensuring its safe and effective use in clinical applications. Nanotechnology has revolutionized drug delivery strategies, particularly in enhancing the

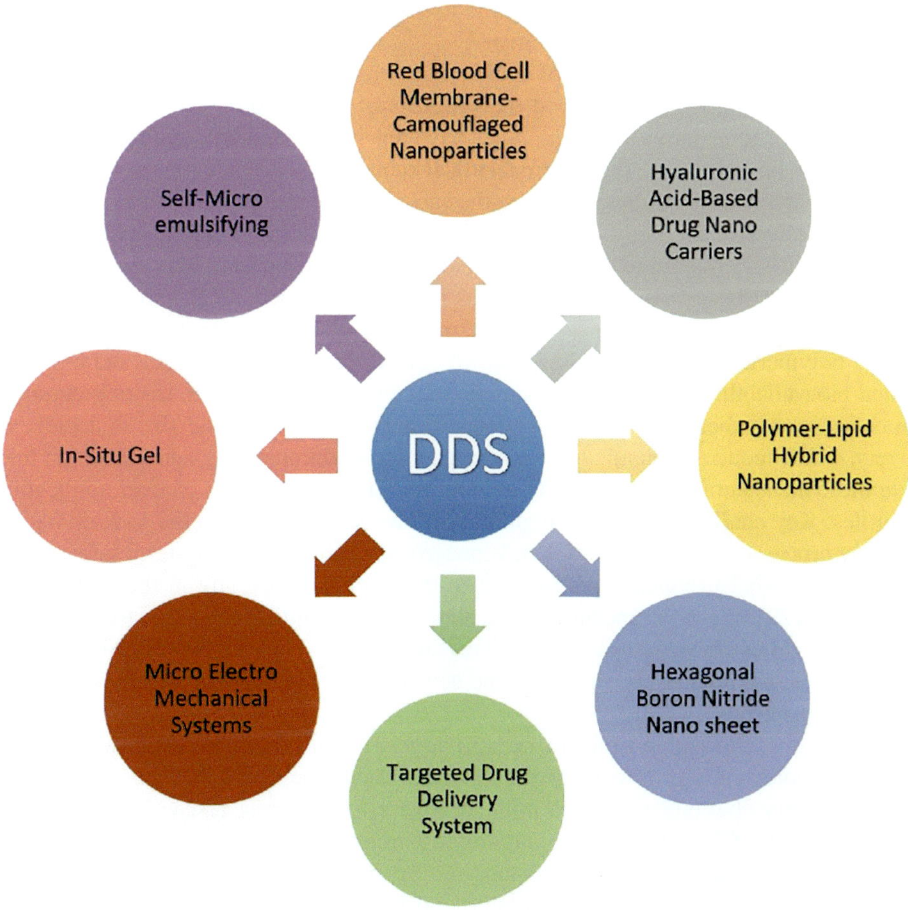

FIGURE 18.9 Current trends in DDS.

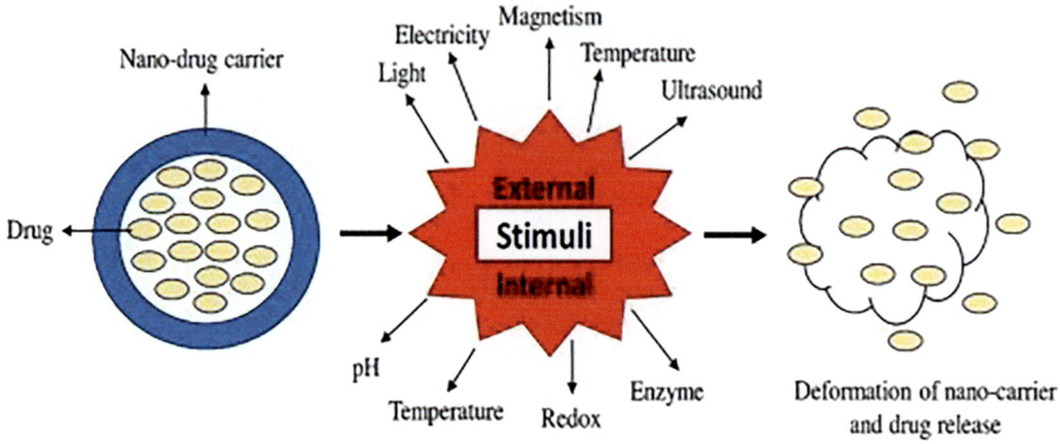

FIGURE 18.10 Recent advancements in drug delivery systems.

efficacy, stability, and bioavailability of herbal medicines. Preclinical studies have demonstrated the potential of nanocarrier systems like liposomes, micelles, and polymeric nanoparticles in treating chronic conditions such as cancer and diabetes. These nanomaterials offer advantages, such as improved targeted delivery and reduced systemic toxicity compared to conventional drug administration routes [81].

Regulatory hurdles, such as the lack of standardization in nanoparticle manufacturing and concerns about long-term safety, have slowed the progress of nanotechnology in the pharmaceutical sector [82]. Although some nano-herbal products have reached the market, mainly in the nutraceutical and cosmeceutical industries, the pharmaceutical sector faces stricter regulations and demands comprehensive toxicological studies to ensure the safety of nanomaterials for human use [83]. The demand for more effective and highly biocompatible nanomaterials continues to grow as researchers work to overcome issues related to degradation rates and biodistribution [84] (Figure 18.10).

The rapid advancement of nanotechnology has brought about significant improvements in drug delivery systems, particularly in the realm of herbal medicine. Nanocarrier systems, such as liposomes, micelles, and polymeric nanoparticles, have shown considerable promise in enhancing the efficacy, stability, and bioavailability of herbal compounds, making them effective in treating chronic conditions like cancer and diabetes [85]. These systems leverage their nanoscale size to achieve targeted drug delivery, which reduces systemic toxicity compared to traditional drug administration methods.

Despite these advancements, the transition from preclinical success to clinical application faces several hurdles. Key challenges include regulatory issues stemming from the lack of standardized manufacturing processes for nanoparticles and concerns regarding their long-term safety [86]. While some nano-based herbal products have made it to the market, especially in the nutraceutical and cosmetic sectors [79], the pharmaceutical industry remains cautious due to stringent regulatory requirements. To address these issues, extensive research is ongoing to improve the biocompatibility and effectiveness of nanomaterials, including the development of hybrid nanocarriers and personalized medicine approaches.

These innovations, although promising, are still largely in the experimental phase. The successful integration of nanotechnology into mainstream herbal medicine will hinge on overcoming these regulatory and safety challenges through collaborative efforts among researchers, regulators, and industry professionals [87]. Additionally, ongoing research focuses on optimizing nanoparticle production processes for scalability and understanding the in vivo interactions between nanoparticles and biological systems [88]. This research is crucial for maximizing the therapeutic potential of nanotechnology and ensuring its safe and effective use in clinical settings [89].

18.8 CONCLUSION

The integration of nanotechnology into herbal medicine represents a groundbreaking advancement in the field of therapeutic interventions. Traditional herbal remedies have long been celebrated for their natural origins and therapeutic potential, yet their effectiveness has often been constrained by issues such as poor bioavailability, high dosage requirements, and frequent administration. These limitations have underscored the need for innovative approaches to enhance the delivery and efficacy of herbal medicines. Nano-formulated systems, leveraging cutting-edge technologies such as liposomes, polymeric nanoparticles, and nanoemulsions, have emerged as transformative solutions to these longstanding challenges.

The application of nanotechnology in herbal medicine offers several key advantages. Enhanced stability and solubility of herbal bioactives, facilitated by nanocarriers, address the issues of poor absorption and rapid clearance that have historically impeded the therapeutic potential of traditional formulations. Through improved bioavailability and targeted delivery, nano-formulations not only increase the efficacy of herbal treatments but also reduce the frequency of dosing and minimize side effects. This refined approach to drug delivery exemplifies the promise of nanotechnology in bridging the gap between conventional herbal practices and modern therapeutic needs.

Despite the considerable benefits, the transition to nano-formulated herbal products is not without its challenges. Rigorous research and development are essential to ensure the safety, efficacy, and regulatory compliance of these advanced formulations. The complexity of manufacturing processes and the need for comprehensive safety evaluations present significant hurdles that must be addressed to facilitate the widespread adoption of nano-formulated herbal medicines. Moreover, regulatory frameworks need to evolve to accommodate these novel delivery systems, ensuring that they meet the necessary standards for quality and safety.

Ongoing research and development efforts are crucial for realizing the full potential of nano-formulated herbal medicines. Continued exploration into their mechanisms of action, therapeutic advantages, and long-term effects will provide deeper insights into how these advanced systems can be optimized for clinical use. As the field advances, collaborative efforts between researchers, regulatory bodies, and industry stakeholders will be essential in addressing the challenges and maximizing the benefits of these innovative therapies.

In conclusion, the intersection of nanotechnology and herbal medicine holds significant promise for the future of healthcare. By overcoming the limitations of traditional herbal formulations and enhancing therapeutic efficacy through advanced delivery systems, nano-formulated herbal medicines represent a valuable evolution in the field. As research progresses and regulatory pathways are established, these innovations have the potential to transform herbal medicine, offering more effective, targeted, and patient-friendly treatments. The continued exploration and development of nano-formulated herbal products will undoubtedly contribute to advancing therapeutic practices and improving patient outcomes in the years to come.

REFERENCES

1. Liu, H., Y. Bai, C. Huang, Y. Wang, Y. Ji, Y. Du, et al. "Recent Progress of Electrospun Herbal Medicine Nanofibers." *Biomolecules*, 2023. Available from: https://www.mdpi.com/2218-273X/13/1/184. Accessed September 12, 2024.
2. Doench, J. G., N. Fusi, M. Sullender, M. Hegde, E. W. Vaimberg, K. F. Donovan, I. Smith, and Z. Tothova. 2016. "Optimized sgRNA Design to Maximize Activity and Minimize Off-Target Effects of CRISPR-Cas9." *Nature Biotechnology*. Accessed September 12, 2024. Available from https://www.nature.com/articles/nbt.3437
3. Bhalani, D., B. Nutan, and A. Kumar. 2022. "Bioavailability Enhancement Techniques for Poorly Aqueous Soluble Drugs and Therapeutics." *Biomedicines*. Accessed September 12, 2024. https://www.mdpi.com/2227-9059/10/9/2055

4. Viegas, C., A. B. Patrício, J. M. Prata, A. Nadhman, P. K. Chintamaneni, and P. Fonte. 2023. "Solid Lipid Nanoparticles vs. Nanostructured Lipid Carriers: A Comparative Review." *Pharmaceutics*. Accessed September 12, 2024. Available From: https://www.mdpi.com/1999-4923/15/6/1593
5. Horcajada, P., T. Chalati, C. Serre, B. Gillet, C. Sebrie, T. Baati, J. F. Eubank, and D. Heurtaux. 2010. "Porous Metal–Organic-Framework Nanoscale Carriers as a Potential Platform for Drug Delivery and Imaging." *Nature Materials*. Accessed September 12, 2024. Available from: https://www.nature.com/articles/nmat2608
6. Haseli, S., M. Pourmadadi, A. Samadi, F. Yazdian, M. Abdouss, H. Rashedi, et al. 2022. "A Novel pH-Responsive Nanoniosomal Emulsion for Sustained Release of Curcumin from a Chitosan-Based Nanocarrier: Emphasis on the Concurrent Improvement of Loading, Sustained Release, and Apoptosis Induction." *Biotechnology Progress* 38 (5): e3280.
7. Zhang, M., X. Wang, C. Wei, M. Wang, W. Han, Z. Yuan, et al. 2023. "Effects of Exogenous Phosphates on Speciation and Bioavailability of Arsenic and Cadmium in Farmland Soils." *Journal of Soils and Sediments* 23 (4): 1832–43.
8. Lee, H. G., M. A. Wheeler, and F. J. Quintana. 2022. "Function and Therapeutic Value of Astrocytes in Neurological Diseases." *Nature Reviews Drug Discovery*. Accessed September 12, 2024. Available from: https://www.nature.com/articles/s41573-022-00390-x
9. Polack, F. P., S. J. Thomas, N. Kitchin, J. Absalon, A. Gurtman, S. Lockhart, J. L. Perez, and G. Pérez Marc. 2020. "Safety and Efficacy of the BNT162b2 mRNA Covid-19 Vaccine." *New England Journal of Medicine* 383 (27): 2603–15. Accessed September 12, 2024. Available from: https://www.nejm.org/doi/full/10.1056/nejmoa2034577
10. Kumar, R., P. Saha, P. Lokare, K. Datta, P. Selvakumar, and A. Chourasia. 2022. "A Systemic Review of Ocimum Sanctum (Tulsi): Morphological Characteristics, Phytoconstituents and Therapeutic Applications." *International Journal for Research in Applied Sciences and Biotechnology*. Accessed September 12, 2024. Available from https://ijrasb.com/index.php/ijrasb/article/view/251
11. Marso, S. P., Daniels, G. H., Brown-Frandsen, K., Kristensen, P., Mann, J. F., Nauck, M. A., ... & Buse, J. B. 2016. Liraglutide and Cardiovascular Outcomes in Type 2 Diabetes, *Drug and Therapeutics Bulletin* 54 (9): 101.
12. John, A., H. H. Yang, S. Muhammad, Z. I. Khan, H. Yu, M. Luqman, M. Tofail, and M. I. Hussain. 2022. "Cross Talk Between Synthetic Food Colors (Azo Dyes), Oral Flora, and Cardiovascular Disorders." *Applied Sciences*. Accessed September 12, 2024. https://www.mdpi.com/2076-3417/12/14/7084.
13. Bray, F., J. Ferlay, I. Soerjomataram, R. L. Siegel, L. A. Torre, and A. Jemal. 2018. "Global Cancer Statistics 2018: GLOBOCAN Estimates of Incidence and Mortality Worldwide for 36 Cancers in 185 Countries." *CA: A Cancer Journal for Clinicians* 68 (6): 394–424. Accessed September 12, 2024. https://acsjournals.onlinelibrary.wiley.com/doi/abs/10.3322/CAAC.21492
14. Bhimraj, A., R. L. Morgan, A. H. Shumaker, L. R. Baden, V. C. C. Cheng, K. M. Edwards, J. C. Gallagher, et al. 2024. "Infectious Diseases Society of America Guidelines on the Treatment and Management of Patients with COVID-19 (September 2022)." *Clinical Infectious Diseases*. Accessed September 12, 2024. https://academic.oup.com/cid/article-abstract/78/7/e250/6692369
15. Wilson, D. M., M. R. Cookson, L. Van Den Bosch, H. Zetterberg, D. M. Holtzman, and I. Dewachter. 2023. "Hallmarks of Neurodegenerative Diseases." *Cell*. Accessed September 12, 2024. https://www.cell.com/cell/fulltext/S0092-8674(22)01575-6.
16. Al-Khayri, J. M., G. R. Sahana, P. Nagella, B. V. Joseph, F. Alessa, and M. Al-Mssallem. 2022. "Flavonoids as Potential Anti-Inflammatory Molecules: A Review." *Molecules* 27. Accessed September 12, 2024. https://doi.org/10.3390/molecules27092901
17. Saini, R., A. Ranjit, K. Sharma, P. Prasad, X. Shang, K. G. M. Gowda, and Y. S. Keum. 2022. "Bioactive Compounds of Citrus Fruits: A Review of Composition and Health Benefits of Carotenoids, Flavonoids, Limonoids, and Terpenes." *Antioxidants* 11 (2): 239. Accessed September 12, 2024. https://www.mdpi.com/2076-3921/11/2/239
18. Bureau, J., M. E. Oliva, Y. Dong, and C. Ignea. 2023. "Engineering Yeast for the Production of Plant Terpenoids Using Synthetic Biology Approaches." *Natural Product Reports*. Accessed September 12, 2024. https://pubs.rsc.org/en/content/articlehtml/2023/np/d3np00005b
19. Dhyani, P., C. Quispe, E. Sharma, A. Bahukhandi, P. Sati, D. C. Attri, et al. 2022. "Anticancer Potential of Alkaloids: A Key Emphasis to Colchicine, Vinblastine, Vincristine, Vindesine, Vinorelbine and Vincamine." *Cancer Cell International* 22(1). Available from https://doi.org/10.1186/s12935-022-02153-4

20. Urošević, M., L. Nikolić, I. Gajić, V. Nikolić, A. Dinić, and V. Miljković. 2022. "Curcumin: Biological Activities and Modern Pharmaceutical Forms." *Antibiotics* 11(2): 135. Accessed September 12, 2024. https://www.mdpi.com/2079-6382/11/2/135
21. Meng, T., D. Xiao, A. Muhammed, J. Deng, L. Chen, J. He. 2021. "Anti-Inflammatory Action and Mechanisms of Resveratrol." *Molecules* 26(1): 229. Accessed September 12, 2024. https://www.mdpi.com/1420-3049/26/1/229
22. Dutt, Y., R. P. Pandey, M. Dutt, A. Gupta, A. Vibhuti, V. S. Raj, C. M. Chang, and A. Priyadarshini. 2023. "Liposomes and Phytosomes: Nanocarrier Systems and Their Applications for the Delivery of Phytoconstituents." *Coordination Chemistry Reviews*. Accessed September 12, 2024. https://www.sciencedirect.com/science/article/pii/S0010854523002400
23. Djahra, A. B., I. Lmhanat, M. Benkaddour, S. Benkherara, I. Laib, and C. Benine. 2023. "Traditional Herbal Remedies from Algeria for Treating Digestive Disorders." *Journal of Drug Delivery and Therapeutics* 13(1): 1–10. Accessed September 12, 2024. http://www.jddtonline.info/index.php/jddt/article/view/5906
24. Kumar, R., P. Saha, P. Lokare, K. Datta, P. Selvakumar, and A. Chourasia. 2022. "A Systemic Review of Ocimum sanctum (Tulsi): Morphological Characteristics, Phytoconstituents and Therapeutic Applications." *International Journal for Research in Applied Sciences and Biotechnology*. Accessed September 12, 2024. https://ijrasb.com/index.php/ijrasb/article/view/251
25. Hashem, A., A. M. Abdelaziz, A. A. Askar, H. M. Fouda, A. M. A. Khalil, K. A. Abd-Elsalam, and M. M. Khaleil. 2021. "Bacillus megaterium-Mediated Synthesis of Selenium Nanoparticles and Their Antifungal Activity Against Rhizoctonia solani in Faba Bean Plants." *Journal of Fungi* 7 (3): 195. Accessed September 12, 2024. https://www.mdpi.com/2309-608X/7/3/195
26. Burstein, O., A. Shamir, N. Abramovitz, and R. Doron. 2022. "Patients' Attitudes Toward Conventional and Herbal Treatments for Depression and Anxiety: A Cross-Sectional Israeli Survey." *International Journal of Social Psychiatry* 2022 (3): 589–99. Accessed September 12, 2024. https://journals.sagepub.com/doi/abs/10.1177/0020764021992385
27. Claus-Desbonnet, H., E. Nikly, V. Nalbantova, D. Karcheva-Bahchevanska, S. Ivanova, and G. Pierre. 2022. "Polysaccharides and Their Derivatives as Potential Antiviral Molecules." *Viruses* 14 (2): 426. Accessed September 12, 2024. https://www.mdpi.com/1999-4915/14/2/426.
28. Rezaei-Tazangi, F., H. Roghani-Shahraki, M. Khorsand Ghaffari, and F. Abolhasani Zadeh. 2021. "The Therapeutic Potential of Common Herbal and Nano-Based Herbal Formulations Against Ovarian Cancer: New Insight into the Current Evidence." *Pharmaceuticals* 14 (12): 1315. Accessed September 12, 2024. Available from: https://www.mdpi.com/1424-8247/14/12/1315.
29. Portney, L., and M. Watkins. 2009. Foundations of Clinical Research: Applications to Practice. Accessed September 12, 2024. Available from : https://babymariam.gm/sites/default/files/webform/pdf-foundations-of-clinical-research-applications-to-practice-3rd-e-leslie-g-portney-mary-p-watkins-pdf-download-free-book-d3094a3.pdf
30. Di Lorenzo, C., F. Colombo, S. Biella, C. Stockley, and P. Restani. 2021. "Polyphenols and Human Health: The Role of Bioavailability." *Nutrients* 13: 273. Available from: https://doi.org/10.3390/nu13010273
31. Iordache, T., N. Badea, M. Mihaila, S. Crisan, A. L. Pop, and I. Lacatusu. 2021. "Challenges in Coopted Hydrophilic and Lipophilic Herbal Bioactives in the Same Nanostructured Carriers for Effective Bioavailability and Anti-Inflammatory Action." *Nanomaterials* 11 (11): 3035. Accessed September 12, 2024. https://www.mdpi.com/2079-4991/11/11/3035.
32. Marsden, A., and A. Shahtout. 2024. "International Organization for Standardization." *Clinical Laboratory Management. Wiley Online Library*. Accessed September 12, 2024. Available from: https://onlinelibrary.wiley.com/doi/abs/10.1002/9781683673941.ch19.
33. Hesarzadeh, R., A. Bazrafshan, and S. Aqel. 2022. "Audit Committee and Regulatory Scrutiny." *Revista de Administração de Empresas*. SciELO Brasil. Accessed September 12, 2024. https://doi.org/10.1590/S0034-759020220603.
34. Westlake, S., C.H.D. John, and E. Cox. 2023. "Perception Spillover from Fracking onto Public Perceptions of Novel Energy Technologies." *Nature Energy*. Accessed September 12, 2024. https://www.nature.com/articles/s41560-022-01178-4
35. Kumar, A., P. Nirmal, M. Kumar, A. Jose, V. Tomer, E. Oz, C. Proestos, M. Zeng, T. Elobeid, K. Sneha, and F. Oz. 2023. "Major Phytochemicals: Recent Advances in Health Benefits and Extraction Method." *Molecules*. Accessed September 12, 2024. https://www.mdpi.com/1420-3049/28/2/887

36. Townsend, B.A., I. Sihlahla, M. Naidoo, S. Naidoo, D.L. Donnelly, and D.W. Thaldar. 2023. "Mapping the Regulatory Landscape of AI in Healthcare in Africa." *Frontiers in Pharmacology* 14: 1214422.
37. Li, W., J.Y. Chen, R.N. Terry, J. Tang, A. Romanyuk, S.P. Schwendeman, and M.R. Prausnitz. 2022. "Core-Shell Microneedle Patch for Six-Month Controlled-Release Contraceptive Delivery." *Journal of Controlled Release*. Elsevier. Accessed September 12, 2024. Available from https://www.sciencedirect.com/science/article/pii/S0168365922002656
38. Lai, S., Z. Chen, Y. Zhang, G. Li, Y. Wang, and Q. Cui. 2022. "Micronization Effects on Structural, Functional, and Antioxidant Properties of Wheat Bran." *Foods*. MDPI. Accessed September 12, 2024. https://www.mdpi.com/2304-8158/12/1/98
39. Narade, S. B., and Y. V. Pore. 2023. "Pharmacokinetic Assessment of Natural Anticancer Berberine Chloride in Presence and Absence of Some Herbal Bioenhancers in Rabbit Model." *Research Journal of Pharmacy and Technology* 16 (11): 5121–29. https://doi.org/10.52711/0974-360X.2023.00830
40. Smith, P. M., M. R. Howitt, N. Panikov, M. Michaud, C. A. Gallini, M. Bohlooly-Y, et al. 2013. "The Microbial Metabolites, Short-Chain Fatty Acids, Regulate Colonic T Reg Cell Homeostasis." *Science* 341 (6145): 569–73.
41. Badmus, S. O., H. K. Amusa, T. A. Oyehan, and T. A. Saleh. 2021. "Environmental Risks and Toxicity of Surfactants: Overview of Analysis, Assessment, and Remediation Techniques." *Environmental Science and Pollution Research* 28 (44): 62085–104.
42. Kumar, A., P. K. Naik, D. Pradhan, G. Ghosh, and G. Rath. 2020. "Mucoadhesive Formulations: Innovations, Merits, Drawbacks, and Future Outlook." *Pharmaceutical Development and Technology* 25 (7): 797–814.
43. Yu, Z., J. P. Huang, J. Yang, C. Liu, Y. Yan, L. Wang, J. Zhao, Y. Chen, W. Xiang, and S. X. Huang. 2023. "Discovery and Biosynthesis of Karnamicins as Angiotensin Converting Enzyme Inhibitors." *Nature Communications*. Available from: https://doi.org/10.1038/s41467-023-35829-1
44. Pérez-Sánchez, A., Cuyàs, E., Ruiz-Torres, V., Agulló-Chazarra, L., and Verdura, S. "Intestinal Permeability Study of Clinically Relevant Formulations of Silibinin in Caco-2 Cell Monolayers." *International Journal of Molecular Sciences*, 2019. Available from: https://doi.org/10.3390/ijms20071606
45. Kagathara, C., K. Odedra, and N. Vadia. 2022. "Development of HPTLC Method for the Simultaneous Estimation of Quercetin, Curcumin, and Ascorbic Acid in Herbal Formulations." *Journal of the Iranian Chemical Society* 19 (3): 4129–38. https://doi.org/10.1007/s13738-022-02586-9
46. Sabour, S., N. Frosst, and G. E. Hinton. 2017. "Dynamic Routing Between Capsules." *Advances in Neural Information Processing Systems*. Available from https://proceedings.neurips.cc/paper_files/paper/2017/hash/2cad8fa47bbef282badbb8de5374b894-Abstract.html
47. Baishya, H. (2017). Application of Mathematical Models in Drug Release Kinetics of Carbidopa and Levodopa ER Tablets. *Journal of Developing Drugs*. 06. 10.4172/2329-6631.1000171.
48. Yu, Y., G. Wang, X. Yin, C. Ge, and G. Liao. 2021. "Effects of Different Cooking Methods on Free Fatty Acid Profile, Water-Soluble Compounds and Flavor Compounds in Chinese Piao Chicken Meat." *Food Research International*. https://www.sciencedirect.com/science/article/pii/S0963996921005950
49. Vazquez-Quintal, Patricia, Jesús Barrón-Zambrano, Salvador Medina-Peralta, Yolanda Moguel-Ordoñez, Jenny Nelson, and David Muñoz-Rodríguez. 2023. "Elemental Analysis of Propolis Tinctures by Microwave Plasma–Atomic Emission Spectrometry (MP-AES)." *Analytical Letters* 56 (14): 2249–2261. https://doi.org/10.1080/00032719.2022.2163401
50. Martinello, M., R. Stella, A. Baggio, G. Biancotto, and F. Mutinelli. 2022. "LC-HRMS-Based Non-Targeted Metabolomics for the Assessment of Honey Adulteration with Sugar Syrups: A Preliminary Study." *Metabolites* 12 (10): 985. https://doi.org/10.3390/metabo12100985
51. Amorati, R., Foti, M. C., and Valgimigli, L. 2013. "Antioxidant Activity of Essential Oils." *Journal of Agricultural and Food Chemistry* 61 (46): 10835–47. https://doi.org/10.1021/jf402634r
52. Blauvelt, A., Kempers, S., Lain, E., Schlesinger, T., Tyring, S., Forman, S., et al. 2021. "Phase 3 Trials of Tirbanibulin Ointment for Actinic Keratosis." *New England Journal of Medicine 384* (6): 512–20. https://doi.org/10.1056/NEJMoa2028912
53. Mansour, M. M. A., and M. Z. M. Salem. 2023. "Poultices as Biofilms of Titanium Dioxide Nanoparticles/Carboxymethyl Cellulose/Phytagel for Cleaning of Infected Cotton Paper by Aspergillus sydowii and Nevskia terrae." *Environmental Science and Pollution Research* 30 (53): 114625–45. https://doi.org/10.1007/s11356-023-27814-x

54. Wang, S., X. Xiao, A. Li, and P. Li. 2021. "The Herbal Constituents in An-Gong-Niu-Huang Wan (AGNH) Protect Against Cinnabar-and Realgar-Induced Hepatorenal Toxicity and Accumulations of Mercury and Arsenic in Mice." *Evidence-Based Complementary and Alternative Medicine* 2021: 1–10. https://doi.org/10.1155/2021/5538765
55. Jiang, J., X. Ma, D. Ouyang, and R. O. Williams III. 2022. "Emerging Artificial Intelligence (AI) Technologies Used in the Development of Solid Dosage Forms." *Pharmaceutics* 14 (11): 2257. https://doi.org/10.3390/pharmaceutics14112257
56. Sharma, R., P. Bedarkar, D. Timalsina, A. Chaudhary, and P. K. Prajapati. 2022. "Bhavana, an Ayurvedic Pharmaceutical Method and a Versatile Drug Delivery Platform to Prepare Potentiated Micro-Nano-Sized Drugs: Core Concept and Its Current Relevance." *Bioinorganic Chemistry and Applications* 2022: Article ID 5580731. https://doi.org/10.1155/20225580731
57. Yusuf, A., A. R. Z. Almotairy, H. Henidi, O. Y. Alshehri, and M. S. Aldughaim. 2023. "Nanoparticles as Drug Delivery Systems: A Review of the Implication of Nanoparticles' Physicochemical Properties on Responses in Biological Systems." *Polymers* 15 (7): 1596. https://doi.org/10.3390/polym15071596
58. Li, R., Y. Pan, D. Chen, X. Xu, G. Yan, and T. Fan. 2022. "Design, Preparation and In Vitro Evaluation of Core-Shell Fused Deposition Modelling 3D-Printed Verapamil Hydrochloride Pulsatile Tablets." *Pharmaceutics* 14 (2): 437. Available from: https://doi.org/10.3390/pharmaceutics14020437
59. Chen, Y., D. Li, N. Li, P. Y. Loh, Y. Guo, X. Hu, et al. 2023. "Role of Nerve Signal Transduction and Neuroimmune Crosstalk in Mediating the Analgesic Effects of Acupuncture for Neuropathic Pain." *Frontiers in Neurology* 14 (January 23). https://doi.org/10.3389/fneur.2023.1002423
60. Jayaprakash, A.A., P.R. Aswin, and S.R. Rathinam. April 2022. "Joint Pain Liniments and Gum Lotion in the Eye—A Consecutive Case Series of Mistaken Identity." *Indian Journal of Ophthalmology* 70 (4): 1159–1162. https://doi.org/10.4103/ijo.IJO_2328_21
61. Dubey, S.K., Parab, S., Achalla, V.P.K., Narwaria, A., Sharma, S., Jaswanth Gowda, B.H., et al. 2022. "Microparticulate and Nanotechnology Mediated Drug Delivery System for the Delivery of Herbal Extracts." *Journal of Biomaterials Science, Polymer Edition* 33 (12): 1531–1554.
62. Paunovska, K., Loughrey, D. & Dahlman, J.E. 2022. Drug delivery systems for RNA therapeutics. *Nature Reviews Genetics* 23, 265–280.Available form: https://doi.org/10.1038/s41576-021-00439-4
63. Yang, Yang, Gang Xiao, Pengcheng Cheng, Jing Zeng, and Yun Liu. 2023. "Protective Application of Chinese Herbal Compounds and Formulae in Intestinal Inflammation in Humans and Animals." *Molecules* 28. Accessed September 12, 2024. https://www.mdpi.com/1420-3049/28/19/6811
64. Wang, Shu, Yuhua Chen, Jian Guo, and Qian Huang. 2023. "Liposomes for Tumor Targeted Therapy: A Review." *International Journal of Molecular Sciences* 24 (3). Accessed September 12, 2024. https://www.mdpi.com/1422-0067/24/3/2643
65. Khan, Imran, and Khalid Saeed. 2019. "Nanoparticles: Properties, Applications and Toxicities." *International Journal of Chemical Sciences* 24 (3). Accessed September 12, 2024. https://www.sciencedirect.com/science/article/pii/S1878535217300990
66. Giuili, Elsa, Renaud Grolaux, Carla Z.N.M. Macedo, Laurence Desmyter, Benjamin Pichon, Sophie Neuens, et al. 2023. "Comprehensive Evaluation of the Implementation of Episignatures for *Diagnosis of Neurodevelopmental Disorders (NDDs)." Human Genetics* 142 (12): 1721–35.
67. Booth, Amy, Alexandra Jager, Stuart D. Faulkner, Christopher C. Winchester, and Sara E. Shaw. 2023. "Pharmaceutical Company Targets and Strategies to Address Climate Change: Content Analysis of Public Reports from 20 Pharmaceutical Companies." *International Journal of Environmental Research and Public Health* 20 (4). Accessed September 12, 2024. https://doi.org/10.3390/ijerph20043206
68. Newman, David J., and Gordon M. Cragg. 2016. "Natural Products as Sources of New Drugs from 1981 to 2014." *Journal of Natural Products* 79, (3): 629–61. https://doi.org/10.1021/acs.jnatprod.5b01055
69. Abbasi, R., G. Shineh, M. Mobaraki, S. Doughty, and L. Tayebi. 2023. "Structural Parameters of Nanoparticles Affecting Their Toxicity for Biomedical Applications: A Review." *Journal of Nanoparticle Research* 25 (3). https://doi.org/10.1007/s11041-023-01592-2
70. Mitchell, M.J., M.M. Billingsley, R. M. Haley, M. E. Wechsler, N. A. Peppas, and R. Langer 2021. Engineering Precision Nanoparticles for Drug Delivery. *Nature Reviews Drug Discovery*. Accessed September 12, 2024. Available from: https://www.nature.com/articles/s41573-020-0090-8
71. Mura, S., J. Nicolas, and P. Couvreur 2013. Stimuli-responsive Nanocarriers for Drug Delivery. *Nature Materials*. Accessed September 12, 2024. Available from: https://www.nature.com/articles/nmat3776

72. Shi, J., P. W. Kantoff, R. Wooster, and O. C. Farokhzad 2017. Cancer Nanomedicine: Progress, Challenges, and Opportunities. *Nature Reviews Cancer*. Accessed September 12, 2024. Available from: https://www.nature.com/articles/nrc.2016.108
73. Pulipaka, S., M. R. Kumar, P. Sriram, and A. Suttee 2021. A Review On Nano Drug Delivery Systems of Herbal Medicine. *Journal of Emerging Technologies and Innovative Research* Accessed September 12, 2024. Available from: https://www.academia.edu/download/83411255/JETIR2102189.pdf
74. Lombardo, D., and M. A. Kiselev 2022. Methods of Liposomes Preparation: Formation and Control Factors of Versatile Nanocarriers for Biomedical and Nanomedicine Application. *Pharmaceutics*. Accessed September 12, 2024. Available from: https://www.mdpi.com/1999-4923/14/3/543
75. Chavda, V. P., L. V. Nalla, P. Balar, R. Bezbaruah, V. Apostolopoulos, R. K. Singla, A. Khadela 2023. Advanced Phytochemical-based Nanocarrier Systems for the Treatment of Breast Cancer. *Cancers*. Accessed September 12, 2024. Available from: https://www.mdpi.com/2072-6694/15/4/1023
76. Makarova, M. N., and V. G. Makarov. 2023. Dogs in Preclinical Research. *Laboratornye Zhivotnye dlya nauchnych issledovanii (Laboratory Animals for Science)* 6 (4): 26.
77. Negi, P., A. Singh, S. Pundir, A. Parashar, N. Upadhyay, S. Agarwal, et al. 2023. Essential Oil and Nanocarrier-based Formulations Approach for Vaginal Candidiasis. *Therapeutic Delivery* 14 (3): 207–25.
78. Bird, R., Z. J. Baum, X. Yu, and J. Ma 2022. The regulatory environment for lithium-ion battery recycling. *ACS Energy Letters Journal* 7 (2): 736–40.
79. Pouryousefi-Koodehi, T., M. S. Kahrizi, H. Roghani-Shahraki, F. Rezaei-Tazangi, and R. Arefnezhad 2024. Nano-based herbal medicine: A new candidate for prostate cancer treatment? *Current Traditional Medicine*. Accessed September 12, 2024. Available from: https://www.ingentaconnect.com/content/ben/ctm/2024/00000010/00000007/art00018
80. Leskovac, A., and S. Petrović 2023. Pesticide Use and Degradation Strategies: Food Safety, Challenges, and Perspectives. *Foods*. Accessed September 12, 2024. Available from: https://www.mdpi.com/2304-8158/12/14/2709
81. Ru, J., P. Li, J. Wang, W. Zhou, B. Li, C. Huang, et al. 2014. TCMSP: A Database of Systems Pharmacology for Drug Discovery From Herbal Medicines. *Journal of Cheminformatics* 6 (1). Accessed September 12, 2024. Available from: https://link.springer.com/article/10.1186/1758-2946-6-13
82. Shepherd, S.J., X. Han, A. J. Mukalel, R. El-Mayta, A.S. Thatte, J. Wu, et al. 2023 Throughput-scalable Manufacturing of SARS-CoV-2 mRNA Lipid Nanoparticle Vaccines. *Proceedings of the National Academy of Sciences of the United States of America* 33: e2303567120.
83. Augustyniak, J., A. Bertero, T. Coccini, D. Baderna, L. Buzanska, F. Caloni. 2019 Dec 1. Organoids are Promising Tools for Species-specific in Vitro Toxicological Studies. *Journal of Applied Toxicology*. 39 (12): 1610–22.
84. Handa, M., S. K. Maharana, K. Pal, and R. Shukla 2022. Biocompatible Nanomaterials for Burns. *Current Pharmaceutical Biotechnology*. Accessed September 12, 2024. Available from: https://www.ingentaconnect.com/content/ben/cpb/2022/00000023/00000012/art00010
85. Rakotondrabe, T., M. Fan, F. Muema, M. Ghosh 2023. Modulating inflammation-Mediated Diseases via Natural Phenolic Compounds Loaded in Nanocarrier Systems. *Pharmaceutics*. Accessed September 12, 2024. Available from: https://www.mdpi.com/1999-4923/15/2/699
86. Harris, A.R., M.J. Walker, and F. Gilbert 2022. Ethical and Regulatory Issues of Stem Cell-derived 3-Dimensional Organoid and Tissue Therapy for Personalized Regenerative Medicine. *BMC Med* 20(1).
87. Malik, S., K. Muhammad, and Y. Waheed (2023) Nanotechnology: A Revolution in Modern Industry. *Molecules*. Accessed September 12, 2024. Available from: https://www.mdpi.com/1420-3049/28/2/661
88. Palmieri, V., G. Perini, M. De Spirito, and M. Papi (2019) Graphene Oxide Touches Blood: In Vivo Interactions of Bio-coronated 2D Materials. *Nanoscale Horizons*. Accessed September 12, 2024. Available from: https://pubs.rsc.org/en/content/articlehtml/2018/nh/c8nh00318a
89. Slika, H., Mansour, H., Wehbe, N., Nasser, S.A., Iratni, R., Nasrallah, G., Shaito, A., and Ghaddar, T. 2022. "Therapeutic Potential of Flavonoids in Cancer: ROS-Mediated Mechanisms." *Biomedicine & Pharmacotherapy*. Elsevier. Available from: https://www.sciencedirect.com/science/article/pii/S0753332221012282

19 Implementing Herbal Nanomedicine in Clinical Settings with Patient Care Histories Treatment Protocols and Observed Benefits

Saleem Ur Rahman, Asma Zahoor, Mashaal Nageen, Afsheen Tajummal, Ehtisham Shafique, Nasir Ali, and Farakh Javed
Department of Biological and Health Sciences, Pak-Austria Fachhochschule: Institute of Applied Sciences and Technology, Haripur, Pakistan

19.1 INTRODUCTION

The year 1999 saw the emergence of nanomedicines, while herbal nanomedicine is a new branch that combines nano science principles with traditional herbal medicine. This process involves the manipulation of materials at the nanoscale (typically 1 to 100 nm) with unique physical and chemical properties, thereby enabling the fabrication of novel therapeutic agents [1]. In herbal nanomedicine, bioactive compounds from plants are designed as nanoparticles to improve their delivery, activity, and safety in the treatment of different medical conditions.

19.1.1 Definition and Scope

Herbal medicines have been increasingly used across the world for diseases due to their superior characteristics to other synthetic drugs, which are more toxic in nature. A considerable number of infections and life-threatening diseases can be treated by various useful and medicinal plants, providing vital resources for modern pharmaceutical industries. Over a long period, researchers have investigated the activities related to the uses of herbs as therapeutics and it has been found out that such drugs are environmentally friendly, in contrast to their counterparts, which are synthesized because they possess low stability, solubility, and problem with bioavailability [2].

Nanotechnology is a groundbreaking field that has managed to address many challenges related to herbal medicine through its widespread effectiveness in diverse situations with no side effects. For instance, over the past few decades Ayurvedic herbs have exhibited strong activity against various chronic diseases such as asthma [3], cancer [4], diabetes [5], heart diseases [6], and multiple sclerosis [7]. These are the most in-demand products on the market having therapeutic properties such as being anticancer, antioxidant, and gene protecting, all of which can challenge the upcoming problems for the pharma industries.

Nanotechnology is one of the main scientific branches, and includes physics, chemistry, biology, and computer science. In this field, structures and components are studied within the 1 to 100 nm

DOI: 10.1201/9781003544418-19

range. The incorporation of nanotechnology into various fields has permanently transformed mankind's future; it will bring positive changes into people's lives, the organization of healthcare, environmental challenges, and other matters [8].Nanotechnology focuses on prevention strategies and lifestyle changes for maintaining good health and preventing diseases.

19.1.2 History

Botanical medicine or phytomedicine, also known as herbal medicine, has a very rich history that encompasses the entire globe. Herbal medicine, which is the earliest form of medical practice, according to the written records of ancient civilization, was first used around 60,000 years ago [9] (Figure 19.1).

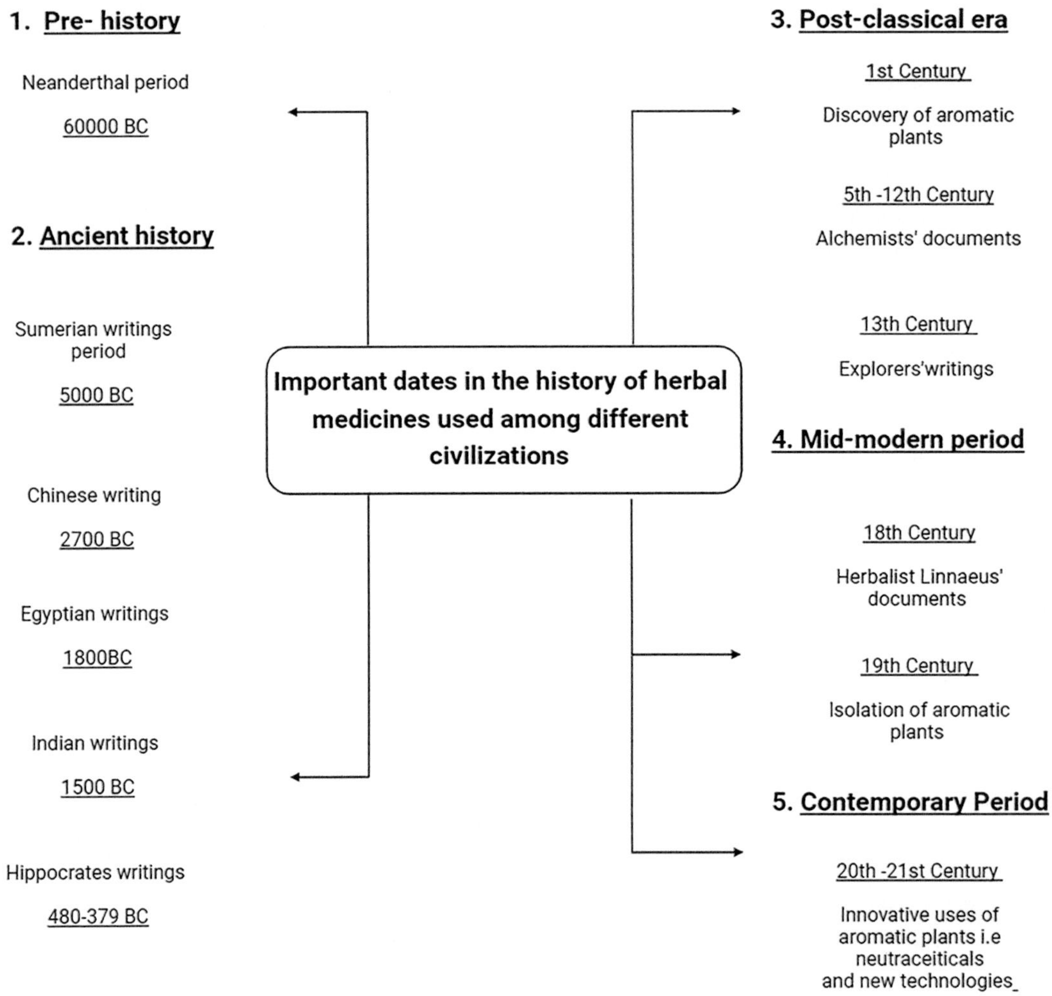

FIGURE 19.1 The history of herbs, medicinal and aromatic plants, and their extracts [30].

19.1.3 Importance and Relevance in Modern Medicine

One modern medical discovery which is proving to be very important is herbal nano-medicine, which combines ancient herbal wisdom with nanotechnology to optimize overall product. Some points can be mentioned here with regard to the significance of the use of medicinal plants in modern medicine.

- **Bioavailability**: Drugs are better absorbed and bioavailable when administrated as herbal nanomedicine. Furthermore, they are more effective and potent when administered at lower doses [10].
- **Drug Delivery**: The toxicity of nano herb compounds can be reduced while releasing their components to specific targeted sites. This makes it convenient for those patients who need repeated treatments because of chronic diseases such as cancer therapies, cardiovascular diseases, neurological disorders, and so on, with higher treatment efficacy [8, 11].
- **Personalized Medicine**: It can help in designing therapy against specific biomarkers by improving precision and effectiveness [12]. The use of herbal nanomedicine provides safer and more accurate treatment options than is the case with traditional ones.
- **Research and Novel Discovery**: Scientific research and clinical studies play a very crucial role in validating the safety and efficacy of traditional herbal medicine. This area needs vast research and studies to develop new and novel therapies. Additionally, the integration of nano technology with herbal extracts opens up the possibility to develop new and multifunctional therapeutic options [13].
- **Accessibility**: The use of herbal nanomedicine in medical practices could be an accessible and affordable option, which is much safer as well as cheaper for low-resource settings or as an alternative to expensive drugs.
- **Sustainability**: Utilizing plant-based medicines in a healthcare setting would promote the use of renewal natural resources among populations. This would also support an environment conservation agenda and the use of eco-friendly products.

19.2 BASICS OF NANOTECHNOLOGY AND HERBAL MEDICINES

19.2.1 Nanotechnology

As mentioned earlier, nanotechnology is a foremost scientific domain that unites the wisdom of physics, chemistry, biology, informatics, and engineering. A usual definition of this discipline is studying the structures and components within the 1–100 nanometers range. The inclusion of nanotechnology in different sectors has transformed the world with positive changes in people's lives, healthcare setup, and tackling environmental challenges, among others.

At nano scale level, a material can have physical, biological, and chemical properties that are quite different from their macroscopic counterparts, leading to significant changes in various branches of science. There are many things on which nanotechnology depends:

- Nanoscale dimension: In this field size and scale matters most; therefore, materials should be reduced to the nanoscale i.e., between 1 and 100nm in order to achieve desired properties. Nonetheless, at least one dimension less than 100nm means nanoscale dimensions having unique electrical, magnetic, or optical attributes of these substances. The number of dimensions that an object possesses at the nanoscale determines its name such as a nano rod, nano tube, nano fiber and nano film etc. Each mode gives rise to its own distinct behavior for each nanoparticle.
- Aspect ratio: Another name for surface-to-volume ratio is aspect ratio, in which nanoparticles often have higher aspect ratios than other particles. Nanomaterials with higher aspect ratios become more reactive and stronger. These gains added interaction with other molecules and also improved drug delivery.
- Quantum confinement: The electronic properties of a given material in nanotechnology are related to the concept of quantum confinement, which refers to the arrangement

TABLE 19.1
Top-Down and Bottom-Up Approaches

Top-Down Approach	Bottom-Up Approach
Scaling down bulk material into nanoscale by using different techniques, ball milling, laser ablation, lithography, and thermal evaporation	Assembling the material from atomic or molecular scale to the nanoscale. This could be done by using these techniques, Sol-gel, self-assembly, and hydrothermal.

of electrons at various dimensions. For its part, quantum confinement can be said to be study of how the electronic properties of materials change as they are scaled down to the nanoscale (1–100nm). In particular, this is when the sizes are comparable or smaller than the de Broglie wavelength of a particle. This holds truer in cases where one or more dimensions, such as those seen in quantum dots, nanowires and thin films, are restricted to the nanometer scale features.

The synthesis of nanomaterials could be achieved through the use of two different techniques, top-down and bottom-up approaches as outlined in Table 19.1.

- Applications: There are several benefits that nanotechnology can provide through its usage in different sectors. The most important at present is the field of medicine, in which nanoparticles can be engineered to enhance the therapeutic efficacy of drugs by targeting specific tissues and cells or, for diagnostics purposes, nano sensors may be employed for the early detection and precise monitoring of diseases. In this case, it may also be used in tissue engineering as a scaffold for tissue regeneration and repair. For instance, in electronics, nanoscale transistors and sensors in electronic devices could be created that are much more efficient and faster than those currently used. Finally, nanotechnology is also useful in environmental science because it can be employed in water filtration systems for purification purposes as well as to control air pollution emissions, among other emerging concerns using nanomaterials.

19.2.2 Basics of Herbal Medicines

Phytotherapy, also known as botanical medicine, herbal medicine, or herbalism, is one of the oldest practices in human history that has been used since antiquity [14]. This ancient practice has been practiced as a form of traditional culture by various cultures and tribes and hence some people now refer to it as phytotherapy. Phytotherapy is based on knowledge, skills and practices that have evolved through time among different ethnic communities from different cultures and racial backgrounds. For many years, even before technology took over the healthcare provision system, herbal medicine has been employed in healthcare for the prevention and treatment of diverse diseases. In fact, much effort is being made to explore this subject by encouraging studying herbal medicine in the present day due to its manifold advantages. Traditional Chinese medicine (TCM), which dates back approximately 3000 years, is one example of herbal medicine still being utilized within contemporary medical care systems [15]. Herbal medicine implies the use of plant-derived substances, including whole plants or parts such as roots, leaves, stem barks, flowers, and seeds in the preparation of different recipes for healing or treating disease conditions. The principal reason for using plants and herbs in alternative therapy is that they are less expensive than regular drugs. They are also more affordable, easily available, cost-effective, safe, and eco-friendly.

The use of herbal medicine was passed down from one generation to another depending on the type of civilization, their norms and traditions, as well as the practices in healthcare. The use of herbal medicine was interlaced with spiritual practices, combining both body and mind. It is well documented and described among different civilizations and based on ancient texts, as detailed below:

1. Ayurvedic herbal medicine, which originated in India.
2. Traditional Chinese medicine.
3. Graeco-Roman and Islamic medicine.
4. Middle European herbal medicine.
5. 19th-century North America herbal medicine.

19.3 IMPLEMENTING HERBAL NANOMEDICINES

19.3.1 Protocols for Implementation

In this discussion, we shall review the procedures related to the preprocessing and administration of herbal nanoparticles for efficient delivery as well as the effectiveness of plant-based compounds. The process includes various preparation methods that use no adverse reactions, such as solvent evaporation, green synthesis, and the homogenizer method; it also describes the dosing approaches due to patient Standard by Design (SBD) compliance with an herbal nanomedicine monitoring system.

19.3.1.1 Techniques for the Preparation of Herbal Nanoparticles

This is indeed a major advancement in the delivery and therapeutics of medication with herbal components integrated into nanomedicine. The formulation of herbal nanoparticles involves a complex process with multiple advanced methods of converting plant-derived chemicals to prepare herbal nanomaterials, which may be used for enhancing their bioavailability and efficacy. Table 19.2 details a wide range of methods for the preparation of herbal nanoparticles which are specific to transport, as well as the therapeutic potential of these natural ingredients. Furthermore, different nanoformulations are present, as shown in Figure 19.2.

Each of these processes has its own benefits, so one can select from a range depending on the characteristics that need to be achieved in the nanoparticles and compound. This table details how these techniques are used in the preparation of herbal nanomedicines which can revolutionize therapeutic interventions by providing more targeted and precise treatments with minimal side effects.

19.3.1.2 Dosage and Administration Strategies

The dose and pattern of administration for herbal-mediated nanomedicine should be customized to enhance therapeutic efficacy with minimal side effects. The characteristics make the nanoparticles likely to require specific dosage forms and administration routes which are established as targeted delivery, controlled release, and enhanced permeability and retention (EPR) effect. While smaller nanoparticles are often more readily absorbed, they might require less administration. However, recent research highlights how the surface chemistry of nanoparticles and their interactions with biological fluids may affect dosage prerequisites [16]. Herbal nanoparticles offer improved drug delivery technology such as liposomes, micelles, and polymeric nanoparticles into diagnostic formulations to enhance their output [17].

- Dosage Form Optimization: Capsules/tablets, injections, or suspensions can serve as dosage forms for herbal nanoparticles which are formulated and designed. The decision of the dosage form is based on nanoparticles' action site at the end-target, administration method employed, and physicochemical factors associated with them. In most cases, injectables are used for localized delivery to specific tissues or organs while oral dosage forms such as capsules and tablets are used for systemic administration [18, 19].

TABLE 19.2
Preparation Techniques for Herbal Nanoparticles [16, 29]

Technique	Description	Advantages	Considerations
Hot Homogenization Technique	High-temperature dissolving of lipids to integrate herbal medications. Herbal components are mixed with lipids dissolved in a heated aqueous surfactant solution. Maintains temperature above lipids' melting point to aid nanoparticle production.	Uniform dispersion, improved nanoparticle integrity, ideal for lipophilic herbal extracts.	Requires precise temperature control to avoid degradation of sensitive compounds.
Cold Homogenization Technique	Combines herbal remedy with melted lipids and cools quickly using cryogenic equipment. The mixture is homogenized after being milled into a fine powder. Ideal for temperature-sensitive bioactive substances.	Preserves herbal constituents' integrity by reducing heat exposure, maintaining medicinal efficacy.	Potential challenges in scaling up and ensuring uniform cooling.
High-Pressure Homogenization Technique	Uses mechanical forces to produce fine nanoparticles by forcing herbal-lipid mixtures through small apertures under high pressure. Parameters like pressure, flow speed, and number of passes are regulated.	Scalability, consistent size distribution of nanoparticles.	Requires careful control of process parameters to achieve desired nanoparticle properties.
Solvent Emulsification/ Evaporation Method	Dissolves a polymer and plant components in an organic solvent, then transfers to an aqueous phase with a stabilizer to create an oil-in-water emulsion. Nanoparticles form as the solvent evaporates.	Controlled release, increased bioavailability of hydrophobic herbal extracts.	Requires precise control of solvent evaporation and stabilization conditions.
Nanoprecipitation Technique	Dissolves plant components and a polymer or lipid in a solvent, then injects into a non-solvent causing nanoparticles to precipitate.	Minimal energy requirements, cost-effective, produces nanoparticles with limited size distribution.	Ensuring consistent particle size and encapsulation efficiency can be challenging.
Sonication	Uses high-frequency sound waves to produce nanoparticles from a solution of plant chemicals and a polymer. The sound waves split the solution into smaller droplets, forming nanoparticles as the solvent evaporates.	Consistent nanoparticle production, works well with both hydrophilic and hydrophobic extracts.	Potential issues with scalability and uniformity of nanoparticles.
Double Emulsion/Solvent Evaporation Technique	Creates nanoparticles with a core-shell structure using an oil-in-water-in-oil or water-in-oil-in-water emulsion.	Controlled release, protects hydrophilic herbal components from degradation.	Complexity in preparing and stabilizing core-shell structures.
Wet Chemical Approach	Uses chemical processes like reduction, sol-gel, and precipitation to create nanoparticles in liquid media. Involves dissolving precursors, triggering chemical reactions, and using stabilizers to prevent agglomeration.	Versatile, can produce nanoparticles with various compositions, sizes, and shapes.	Requires careful control of chemical reactions and potential toxicity of metal components.
Ball Milling	Finely crushes herbal material to reduce size. Useful for reducing particle size but may cause some breakdown of substances. Herbal extracts are used as capping and reducing agents.	Simple method for size reduction, can be integrated with other chemical processes.	Risk of substance degradation and limited control over particle size and distribution.
Supercritical Fluid Technology	Uses supercritical carbon dioxide ($scCO_2$) as a solvent for nanoparticle synthesis. Polymers and herbal compounds are dissolved in $scCO_2$ and precipitated by a sudden pressure drop.	Environmentally benign, precise control of particle size and shape, preserves bioactivity.	Requires specialized equipment and conditions for handling supercritical fluids.
Coacervation Method	Phase separation of two liquid phases in a colloidal system containing herbal chemicals. Oppositely charged polyelectrolytes form nanoparticle aggregates, which can be stabilized through cross-linking.	Versatile for encapsulating various herbal components, customizable release profiles.	Control over phase separation and cross-linking is crucial for consistent nanoparticle formation.

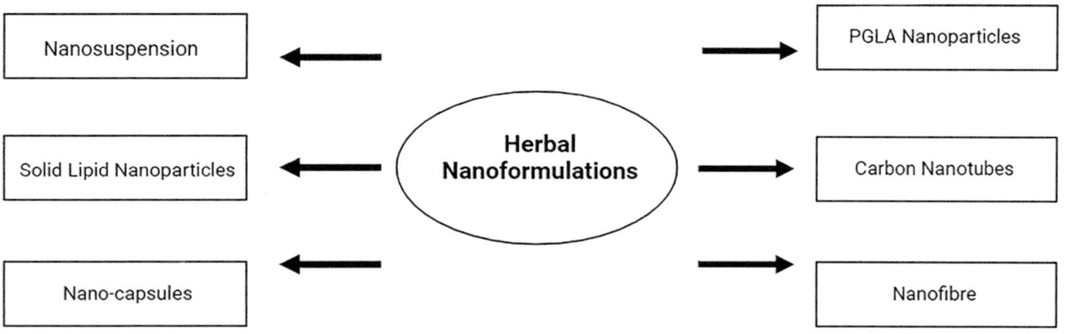

FIGURE 19.2 Different types of nanoformulations [31].

- Controlled Release Formulations: The most significant application has been in manipulating and controlling release characteristics. These formulations allow a slower release of the herbal constituents over time so that they can be given less frequently while still reaching therapeutic blood levels. One way of achieving this is by encapsulating the material within a relatively inert polymer that degrades over time. Nanoparticles like liposomes added to it also prolong the effect and prevent side effects caused by high doses [19, 20].

Optimization of Dose: The ideal concentration range for effective dosage has to lie between minimum toxic levels and the threshold, thereby providing an effective dose level. Preclinical studies are aimed at examining the pharmacokinetics and pharmacodynamics of nanoparticles, both in vitro and in vivo. They provide a basis for fixed dosing regimens in future clinical trials, MTD (Maximum Tolerated Dose), as well as OED (Optimal Effective Dose) planning [19, 20].

Table 19.3 shows the different administration techniques. The researchers can enhance the therapeutic potential of herbal nanomedicines and therefore develop unique therapies and improve patient results by modifying dosage forms, employing sustained drug release systems, and selecting appropriate administration routes. Research has shown that nanosizing of bioactive compounds leads to enhanced bioavailability, absorption compared to bulk materials as well as higher body levels [21].

19.3.1.3 Monitoring Patient Safety and Treatment Efficacy

Treatment using herbal nanoparticles focuses on patient safety and efficacy. This implies, for example, strict supervision like pre-treatment assessments consisting of: patient history, physical examination, laboratory tests for general health, among others. Early detection of safety hazards necessitates the use of adverse event (AE) reporting mechanisms with well-defined criteria as well as trained clinicians in healthcare [22]. In a bid to minimize possible negative consequences, information concerning adverse events (AEs) is always gathered while at the same time putting in place risk management strategies.

There must be clear outcome measures and relevant data collection if the effectiveness of treatments is to be assessed. In order for one to optimize treatment regimens as well as understand drug behavior, he/she needs to have a good understanding of pharmacokinetic and pharmacodynamic studies [16]. Measures that evaluate therapy's impact on quality of life would benefit from adding patient-reported outcomes. Trackings and maintaining such records becomes simple through the use of technology like electronic health records. At the same time, wearable tech and real-world evidence (RWE) may offer information about long-term efficacy and safety of drugs. However, there are still independent variables related to patient heterogeneity; complexity in herbal formulation; uniformity in assessment tools required. Further research is needed to develop new surveillance methods, as well as stringent safety and efficacy criteria.

TABLE 19.3
Administration Techniques [19–21, 23]

Administration Method	Description	Advantages	Considerations
Oral Administration	The most common method, where herbal nanomedicine is taken orally. Nanoparticles enhance the dissolution and absorption of herbal components in the gastrointestinal tract. Gastrointestinal coatings and mucoadhesive polymers protect nanoparticles from acidic conditions and aid their breakdown in the intestines.	Improves oral bioavailability of insoluble components.	Requires effective gastrointestinal coatings to protect nanoparticles from acidic conditions.
Parenteral Administration	Involves direct distribution of herbal nanoparticles into the bloodstream through intramuscular, subcutaneous, or intravenous injections. Useful for rapid action or targeting specific tissues.	Provides quick onset of action and precise targeting.	Requires attention to potential immunogenicity and sterility of nanoparticles.
Topical and Transdermal Delivery	Delivers herbal nanoparticles through the skin for systemic or targeted treatment of skin conditions. Enhances penetration and controlled release of herbal ingredients. Formulations include creams, gels, and patches.	Non-invasive, allows for sustained release and targeted treatment of skin conditions.	Effective formulation is needed to ensure adequate penetration and therapeutic effect.
Inhalation Therapy	Delivers herbal nanoparticles through the respiratory tract using nebulizers or inhalers. Suitable for treating respiratory conditions or achieving systemic effects.	High bioavailability and rapid absorption. Suitable for respiratory conditions like asthma and COPD.	Requires accurate delivery devices and formulation for effective lung deposition.
Targeted Delivery	Techniques like pH-sensitive release, magnetic targeting, and ligand–receptor interactions direct herbal nanoparticles to specific areas, such as tumors. Reduces systemic exposure and adverse effects.	Increases therapeutic efficacy and reduces toxicity by targeting specific cells or tissues.	Effective targeting strategies and delivery systems are essential for success.

- Pharmacokinetics and Pharmacodynamic Monitoring: Pharmacokinetics of plant nanoparticulate products have to be determined, especially their ADME (absorption, distribution, metabolism and excretion) properties for the purpose of making sure they possess an adequate pharmacokinetic profile. Biological samples can be analyzed using mass spectrometry (MS) and high-pressure liquid chromatography (HPLC) for the quantification of nanoparticles as well as metabolites. By these means one can define bioavailability, half-life and clearance rates of phytocompounds which are important parameters for optimal dosing regimens. Biochemical or physiological effects and their mode of actions must be investigated by focusing on pharmacodynamic profiling of herbal nanoparticles. Western blotting and enzyme-linked immunosorbent assay (ELISA) help in biomarker analysis for therapeutic evaluation of pharmacotherapy. This observation also shows how effective nanoparticles are in modifying biological pathways, thereby acting as therapeutic agents to specific diseases.
- Monitoring Safety: It is essential to perform toxicological research to monitor the safety profiles of herbal nanoparticles. Traditional toxicity studies in vitro and in vivo, such as

animal experiments, cytotoxicity testing, or haemolysis assays, are employed to screen for signs of harmful effects. Cell viability tests and histopathological examinations of organs (liver function markers) are assessed to detect any response associated with the exposure to nanoparticles. Nanoparticles may elicit an immunological response, termed immunogenicity. Monitoring protocols for immune activation and cytokine release involve in vitro assays using human immune cells. The immunogenicity of the nanoparticles and ensuring that they do not trigger harmful immune responses is evaluated through in vivo experimentation using animal models. Long-term safety assessment of the genotoxic and carcinogenic properties of herbal nanoparticles involves studying genetic damage and mutations through techniques like the Ames test, comet assay, or micronucleus test. Research on the clinical carcinogenicity model of animals helps determine if there is a risk of cancer in humans exposed long-term to nanoparticles.

- Clinical Trials: Human clinical trials are done to check whether herbal nanomedicine is safe or effective in humans. These trials are often multi-phased, where each phase corresponds to a different aspect of the treatment. While Phase II studies are undertaken to evaluate efficacy and side effects, followed by a larger population for long-term side effect tracking as well as the confirmation of effectiveness in the case of Phase III trials. Besides it also evaluates the safety and dosage in case of the Phase I process. It ensures the continued examination of the output for identifying and rectifying any potential issues that arise.
- Patient-Reported Outcomes: Collection of patient-reported outcomes (PROs) will be used to capture subjective experiences related to this herbal nanoparticle therapy strategy. Surveys and questionnaires rate everything from symptom reduction and quality of life to treatment satisfaction, among others. This information is also useful in identifying specific areas of deficit and how the treatment translates to real-world use.

One key aspect in monitoring the safety profile of herbal nanomedicine is of post-marketing surveillance on adverse events, for which a robust adverse event reporting mechanism will be required. Unexpected side effects or reactions should be reported by all patients and healthcare professionals. These data are then used to detect trends and safety signals in pharmacovigilance databases. In this way, it allows quick enforcement of regulatory interventions if required.

The positioning of a nanoparticle can be accomplished by high-tech imaging methods, particularly fluorescence, positron emission tomography (PET), and magnetic resonance imaging. Combined with the underlying non-invasive nature, these pairing technologies provide real-time information on nanoparticle biodistribution and target tissue accumulation that greatly advances therapeutic efficacy evaluation as well as potential toxicity.

This is the appropriate application to wear any device that monitors information continuously about physiological signals from patients. They facilitate real-time capture and transmission of measurements (like heart rate, blood pressure, glucose levels, or drug concentrations), allow strategy modifications for personalized patient management with maximal therapeutic gains.

19.3.2 Documenting Patient Case Histories and its Analysis

In this section a structured documentation of resolved patients' cases, the drug administered, and the drug reactions are discussed, focusing on the importance of proper medical record-keeping and uniformity in the case history. Comprehensive documentation of the preliminary evaluation, identification, management plans, and final outcomes can provide inputs to optimize disease handling allowing better understanding regarding herbal nanomedicine. Use of this method helps in the simplification, allowing focused care to patients with the best and most indicated therapy while obtaining data for potential future clinical applications or research efforts.

19.3.2.1 Documenting the Conditions Treated, Therapies Administered, and Patient Response

This emphasizes the importance of compiling case histories of patients in clinical practice, such as their experiences of herbal nanomedicine. Not recording medications accurately make it almost impossible to track whether or not a therapy is effective, monitor for any side effects, or see if there has been progress.

Comprehensive Baseline Assessment: This baseline consists of recording the medical history, symptoms and signs with which patients present (including complications) and is dedicated to measuring improvement during herbal nanomedicine therapy. Diagnosis testing validates the diagnosis and guides therapy options. An accurate record of imaging studies, blood tests, and other diagnostic procedures gives valuable insights that can then be customized when prescribing herbal nanomedicine treatments.

The exact prescription details, such as specific herb or formula information (dosing and/or route of administration), must be recorded for any treatment regimens applied to patients at during the study. Such detailed reports guarantee the dissemination of best practices and the continuity of service delivery. Detailed treatment application records are maintained to ensure that therapy is being administered as prescribed and to detect any errors in the timing or form of delivery.

Patient-reported outcomes (PROs) advance patient-centered insights. Under this system the evaluation of symptom relief, quality of life, and treatment satisfaction is performed via surveys and questionnaires. Measurements of the performance characteristics are useful in recording PROs, which can further evaluate and optimize herbal nanomedicine utility in practice. Biomarker analysis can be used to determine how these parameters affect inter-individual variability in clearance, particularly whether Antibody Drug Conjugates (ADCs) fulfil their theoretical advantages or real-life constraints by reliably predicting the response of a patient with additional pharmacokinetics biomarker information which has the potential value for establishing the predictive performance of cytostatic herbal nanoparticle-based therapy efficacy and further treatment modification as necessary.

As trends, correlations and outcomes in treatment can be assessed via statistical methods as well as through the use of specialized software. The patient responses are compared with the initial conditions and treatment objectives to evaluate whether or not herbal nanomedicine therapies have worked. This analysis could be employed for the verification of the results whether suitable or not. The adoption of a systematic process in the ongoing assessment and documentation leads to the standardization of treatment methods and the production of evidence-based therapy.

Detailed patient case histories can be turned into research articles or even presented as case studies, showcasing effective treatments and allowing lessons to be drawn. These studies are published in order to bring about the exchange of knowledge between researchers and also among other medical practitioners, thereby leading to increased scientific discovery Bringing together the medical system with modern engineering has paved the way for the advance of herbal nanomedicine. These nanoparticles utilize unique properties of the delivery system, such as high surface area and functionalization, to targeted delivery into specific tissues in order to innovate with traditional herbal therapies. In what follows we will discuss the medicinal applications of herbal nanoparticles, treatments, and patient responses.

19.3.2.1.1 Condition Treated

Their capacity to provide a positive influence on the bioavailability and controlled transport of therapeutic medications has prompted their status as practicable alternatives for therapy in many medical states. One of the most interesting ones is in their application in various areas of cancer treatment. For example, polymeric nanoparticle-loaded root extract of *Phytolacca decandra* [23] has been proven to produce beneficial chemo-preventive actions in the treatment of lung cancer. Similarly,, curcumin-loaded nanoparticles have shown better outcomes in the treatment of several types of cancer, such as pancreatic or prostate cancers in comparison with the existing ones. Because curcumin is poorly water-soluble, it has been encapsulated in p-hydroxybenzoic acid polymeric nanoparticles, which helps to enhance its efficacy in reaching and acting on cancer cells [23]. It has also been demonstrated that natural medicine can prove efficacious in the treatment of a huge range of oxidative strain-mediated diseases such as most cancers, neurological sicknesses (such as Alzheimer's disease), inflammatory illnesses (including arthritis and asthma), and cardiovascular ailments. Such disorders arise from the damage of essential biomolecules by free radicals [24].

In one study, nanoparticles were synthesized by employing the extracts of *Harungana madagascariensis* and *Ocimum sanctum*, which exhibited potent inhibitory action against bacterial and fungal pathogens. The resistance of a range of diseases, such as-resistant strains is reinforced by these formulas before the usage of any anti-microbial treatments [23].

Herbal nanoparticles and liver: To date, *Cuscuta chinensis*-coated nanoparticles have shown the most promising hepatoprotective properties. Thus, these nanoparticles might provide a potential therapeutic alternative for the treatment of liver illnesses with the attenuation of oxidative stress and protection against hepatic injury [23]. In another piece of research, Cardiovascular diseases (CVDs), such as atherosclerosis, hypertension, myocardial infarction and diabetic cardiomyopathy, were treated [25].

19.3.2.1.2 Treatment Provided

Medicinally, a herbal nanoparticle is a medicinal formulation which has been specifically prepared to carry out a specific treatment. It ensures the targeted/nano scale distribution of constituents. A variety of delivery methods are used, depending on the disease being targeted, and the selected properties of our nanoparticles.

In the same way as curcumin-containing herbal nanoparticles are usually administered orally [23], plant-mediated Ag, Au, Cu, ZnO and Fe core-shell NPs are also common in therapies. These plant-based nanoparticles are made highly antioxidative by being loaded with bioactives such as terpenoids, flavonoids and polyphenols. Various assays have been used to evaluate their antioxidant biological activity, including 1,1-diphenyl-2-picrylhydrazyl(DPPH), ferric reducing antioxidant power (FRAP), trolox equivalent antioxidant capacity (TEAC) and oxygen radical absorbance capacity (ORAC), but chemiluminescence has now emerged as one of the most popular approaches [24]. Metformin (MET) and curcumin (CUR), hTERT genes are targeted by dual drug-loaded PLGA/PEG nanoparticles in tumorous breast tissues. *Hybrid lipid-protein shell nanoparticles for lung cancer therapy with enhanced oral bioavailability*: To the best of our knowledge, in two reports hepatocellular carcinoma was investigated to determine the potential healing capabilities of quercetin (QRC) encapsulated with lactoferrin together and the term, "Protein core-shell nanoparticles co delivering quercetin (QRC)", is used for liver cancer treatment. [24]. Curcumin is delivered more efficiently when encased in nanoparticles, helping to ensure that the compound remains active as it moves through our gut and into target tissues. This delivery approach is particularly useful for systemic and chronic diseases where there is a need for sustained therapeutic activity. Various curcumin formulations have been developed including liposome, which enhances the stability and delivery of the drug with augmenting cardioprotective action; and polymeric nanoparticles, which are intended to boost solubility as well as bioactivity i.e. a nanoemulsion-based approach which helps to promote a wide range of benefits, including improved bioavailability and therapeutic efficacy in the management of myocardial infarction or diabetic-related complications. To improve the distribution of ginsenoside Rg3 and attenuate doxorubicin-induced cardiotoxicity, polymeric micelles were used in combination with rice endosperm-derived exosomal vesicles. The PEG-PE micelles were used in the purerarin treatments to enhance bioavailability and encapsulate cardiomyocytes from ischaemia-induced death. Therapy involving liposomal encapsulation of the berberine was aimed at increasing solubility and achieving no alteration in unfavorable cardiac remodeling after myocardial infarction. After this, lipid-polymer hybrid nanoparticles were proposed as potential delivery vectors for both salvianolic acid B and panax notoginsenoside, with the intention of ameliorating biochemical parameters and substantially lowering myocardial ischemia [25]. Accordingly, Aloe vera nanoparticles have been prepared for use as a topical treatment with the particles being directly administered onto the skin in the form of either lotions or gels or ointments to treat skin disorders and promote healing. They can also be used orally, for example in the form of powders or capsules as dietary supplements that enhance systemic immunity and general health [26]. Various nanomedicine delivery approaches, such as liposomes, nanoemulsions and microemulsions, were used to improve the bioavailability and therapeutic efficiency of these phytochemicals. These systems were developed with the intention of optimizing therapy outcomes for specific diseases and were intended to be administered via a variety of routes, including topical, oral, intranasal, and intravenous [23].

19.3.2.1.3 Patient Response

Herbal nanoparticles have received generally excellent feedback in both clinical and preclinical preferences, which is indicative of their improved therapeutic potential and sophisticated delivery mechanisms. Patients receiving therapy with these novel formulations have reported significant improvements in both their general health and their ability to manage their diseases.

For example, patients receiving *Phytolacca decandra*-encapsulated nanoparticles in cancer therapy have seen a notable decrease in tumor size and improved overall results when compared to patients getting conventional therapies [23]. Herbal nanomedicine therapies have demonstrated favorable outcomes for patients, as evidenced by increases in antioxidant defense and a decrease in symptoms associated with oxidative stress. The biocompatibility and stability of the nanoparticles contribute to their efficacy in reducing oxidative damage and battling microorganisms and cancer [24]. The promising results from these treatments include: large decreases in lung cancer metastatic foci dimensions and tumor biomarkers; strong antitumor activity in liver cancer models; and highly effective targeted drug release in breast cancer. As reported in the same study, curcumin-based micelles boosted pro-inflammatory cytokines and had synergistic antitumor effects in treating [24]. In a similar vein, patients have had improved treatment outcomes and fewer side effects with the use of nanocurcumin, proving its efficacy in treating a range of malignancies. Iin the field of antimicrobial treatment, patients utilizing *Ocimum sanctum* or *Harungana madagascariensis*-formulated nanoparticles have demonstrated successful control over infections, particularly those produced by resistant strains [23]. Treatments involving *Cuscuta chinensis*-loaded nanoparticles have enhanced liver health by lowering oxidative stress and improving liver function. Patients given this treatment have reported an improvement in liver health overall and improved treatment of liver diseases. Comparing ginsenoside Rg3 micelles to untreated controls revealed less myocardial damage and also retained cardiac function. Again, in treated animal models, the levels of oxidative stress indicators and inflammatory cytokines were reduced through the use of purrarin and curcumin treatments. Drug stability and distribution were enhanced using liposomal and polymeric nanoparticles, leading to better clinical results and fewer adverse effects than experienced under conventional therapies. Preclinical research revealed that nisin-loaded nanoparticles were well tolerated and non-toxic, improving ECG patterns and lowering indicators of myocardial injury [25]. In further evidence of the effects, aloe vera NP-based herbal nanomedicine therapies have been shown to improve skin health. Patients have reported better skin texture, less inflammation, and faster wound and skin condition healing. Most patients report excellent results from the therapeutic uses of herbal nanomedicines, and most have great tolerance and experience little in the way of side effects [26]. Following treatment with these nanomedicine therapies, patients and animal models experienced notable improvements. These included decreased tumor size, improved liver function, better protection against brain ischemia, increased antiviral activity, and an effective reduction of inflammation [23].

19.3.2.2 Analyzing Clinical Findings to Refine Treatment Approaches

The clinical results analysis is a crucial method of improving herbal nanomedicine treatment. Through the close examination of patient data, scientists can find out such things as relationships, patterns, and trends which in turn help to fine-tune the treatment plans. The major spheres of concern are to examine the patient response diversity, to consider the safety and tolerability, to measure the therapeutic effectiveness, and to relate the pharmacokinetics and the pharmacodynamics. The data is analyzed with the help of advanced methods such as artificial intelligence to obtain predictive models and apply personalized medicine. Constant observation and assessment are the two key factors in the improvement of long-term therapeutic efficacy and the generation of data that will prove the efficacy and safety of herbal nanomedicine.

Clinical symptoms are of great importance in enhancing the effectiveness of therapy and improve treatment practices. This can be achieved by examining the laboratory studies, patient data, and advanced diagnostic techniques. To illustrate, the microbiome of the gut has an important contribution in the pathogenesis of such diseases as colorectal cancer and inflammatory bowel disease. Probiotics and prebiotics are two personalized treatments that have been developed in clinical research which have found specific microbial signatures connected with various conditions. These strategies boost therapeutic efficacy by enabling accurate treatment adjustments and facilitating the restoration of health through continuous feedback from the patient.

Another key issue requiring in-depth clinical investigation is in the area of antibiotic resistance management. MRSA and MDR-TB are two deadly superbugs which pose serious health risks due to their resistance to antibiotics. Resistance pattern analysis helps to discover the best ways to fight against such types of bacteria through, for example, using combination medicines and new antibiotics, as well as alternative therapies like bacteriophage therapy. This insight also helps in the formulation of sustainable approaches which helps to resist the mutations of the germs and increase the effectiveness of the treatment.

Herbal nanomedicine is a novel framework of therapy that aims to achieve maximum effectiveness of medical treatment with minimum side effects and increased accessibility of drugs to the organism. Improvements in tumor markers or decreases in inflammatory markers are two particular patient responses to these drugs, which lead to drug delivery system optimization. To give just one example: In the case of patients with cancers, any data collected could be used to improve the effectiveness of the particular treatment, for example, curcumin nanoparticles; with this feedback, they will have more focus in lessening side effects. Therapy success evaluation takes into account all the data from kinase assays, biomarker analysis, clinical observations, and PROs. Through medical examination, we can gain immediate feedback on physical problems and disease markers; for example, curcumin nanoparticle treatment has been shown to reduce C-reactive protein (CRP). These PROs can be collected, for example, through surveys or questionnaires, which include topics such as symptom relief or quality of life. Biomarker analysis, like blood glucose monitoring in diabetes treatment, helps to point out possible complications and also provides definite proof of any therapeutic effect. The increasing integration of artificial intelligence and machine learning into clinical result analysis will surely improve the future accuracy and efficacy of treatment methods, leading the health of patients to be enhanced. Multivariate studies, which will be further developed with advanced statistical software and tools, recognize positive outcome predictors and create prediction models. These models permit the personalization of treatment plans, thereby ensuring that each patient receives the most suitable and targeted care. To bring herbal nanomedicine onto the same level as other medical fields, the methods of treatment should be continually improved through research and evidence-based modifications. This will result in an incremental method of clinical result investigation, which provides the basis for the improvement of the treatment process.

19.4 STANDARDIZED TREATMENT PROCEDURES

The developments discussed above translate into a new era in health as herbal nanomedicines are highly specific to a patient's condition and personal needs. In this approach, nanoparticle engineering is used for the precision targeting of disease processes or specific organs (e.g., cancer theragnostics) to enhance treatment efficacy and minimize side effects. This promise of more effective and personalized care, achieved by incorporating patient-specific data into their choices [30], will improve the efficacy of therapy outcomes as well as relieving patients' conditions, leading them towards compliance with prescribed treatments and transforming traditional herbal medicine into a new branch of medical sciences, known as nanomedicines.

19.4.1 Customization for Specific Ailments

One of the most significant capabilities of herbal nanomedicine is its tailored formulation for specific disease conditions. Treatment efficacy can be further enhanced while unwanted effects are minimized by fine-tuning the nanoparticles' composition for targeted disease protocols or organs.

Important features of personalization include:

- Disease-specific nanoparticles: The design of nanoparticles with physicochemical characteristics that target the root cause of a disease
- Combination therapy: The combination of herbal nanoparticles with other potential medicinal entities to treat the complex disorders and offer synergistic effects.
- Personalized medicine: Personalized treatments tailored for the patient in order to optimize results.

To give just one example: nanoparticles can be programmed to stick only on cancer cells and not in healthy tissues, or they could dock with the sore joints of rheumatoid arthritis sufferers so that doses of anti-inflammatory medications are delivered directly to the required sites. It has also been seen that in the case of rheumatoid arthritis, curcumin nanoparticles have shown a significant decrease in inflammatory markers including C-reactive protein (CRP) and erythrocyte sedimentation rate (ESR) [23]. This customization of the route increases tolerability and potency by avoiding systemic inflammation, while neutralizing a majority of gastrointestinal distresses experienced with higher doses seen in traditional formats.

Likewise, nanoparticles from herbal extracts found in garlic and hawthorn can be direct to the inner layers of blood vessels, thereby treating cardiovascular diseases [19]. It has also been shown that phytochemical quercetin-mediated herbal nanoaggregates show great promise for use as nanodelivery vehicles of cardioprotective compounds to combat cardiovascular disorders. Again, quercetin nanoparticles have been shown to reduce oxidative stress (OS) and improve endothelial function (EndoPF), which is an important method for controlling diseases such as hypertension and arterial sclerosis [24]. This targeted approach is more efficient at reducing blood pressure and the absorption of adjunct ingredients, thereby delivering increased diastolic blood pressure health when compared with other treatments. In cancer therapy, the customization of herbal nanomedicines is especially remarkable. This can be customized by the production of nanoparticles loaded with active anticancer herbal ingredients, which is especially effective in the case of conditions demanding targeted actions. In the management of diabetes, for example, the use of nanoparticles enhances the delivery of herbal extracts like kaempferol (KA) directly to pancreatic tissues, thereby bringing out the intended therapeutic effects with a minimum of adverse systemic effects. Likewise, as observed in several research works, targeted release applications for compounds from *C. arborea* extract portrayed higher efficacy toward the treatment of oxidative stress-related diabetes [27]. The plant alkaloid berberine, which is known for its anti-diabetic properties, often suffers from low bioavailability. This issue is addressed by encapsulating berberine in nanoparticle form, enhancing absorption and ensuring a sustained release of the active ingredient [24]. Clinical trials have demonstrated that berberine nanoparticle treatment can significantly lower fasting blood glucose levels, improve insulin sensitivity, and reduce HbA1c levels in patients.

Furthermore, herbal nanomedicine enables the production of nanoparticles that can penetrate the blood–brain barrier, delivering active substances directly to the brain. This is promising for the treatment of neurodegenerative diseases like Alzheimer's disease. For example, ginkgo biloba extracts encapsulated in nanoparticles have been shown to improve cognitive performance and preserve neural integrity more effectively than conventional formulations [23]. As is clear from the findings outlined here, herbal nanomedicine is a powerful tool in modern healthcare, offering significant advancements in personalized medicine by precisely targeting therapies to the affected areas.

19.4.2 Tailoring for Individual Patient Needs

The above discussion has shown the importance of individualizing the treated preparations for the patient's condition in order to achieve efficiency in the area of herbal nanomedicine. This approach is more advantageous and promising, as it takes into account the specifics of each patient. Tailoring is an important aim because the decisions in clinical practice should be patient-specific (i.e. specific to patients who enunciate their values, preferences, and health-related goals) and it is crucial to create methods in order to avoid (or at least reduce) the recognition of futility while increasing the ratio of success.

If genetic data is combined with predictors, then better treatment options can be employed. Precision medicine is a form of customized care regarding a particular patient, which is based on clinical information and the patient's genetic makeup and other factors. First, a comprehensive analysis of the patient's heredity background, current diseases, and history is carried out. This knowledge also defines the process of choosing the right herbal nanoparticles together with the right process of altering the pre-existing ones. During the course of the treatment, biomarkers, signs, and symptoms are being monitored constantly in order to check for the safety and effectiveness of the treatment. Adjustments are undertaken depending on the current conditions of the patient and/or the emergence of new signs and symptoms.

The patient's characteristics, including their age, weight or body surface area, genotype, metabolism, and renal/hepatic function, influence the doses of some drugs in order to avoid side effects while achieving the optimal therapeutic effect. In the case of gastric cancer, it may be appropriate to employ combination therapy, with nano-encapsulated herbs being prescribed alongside other forms of treatments in order to provide a multifaceted treatment plan to deal with such a chronic disease. Once again, genetic markers, the stage of the disease, and metabolic characteristics could all be taken into account in the process of nanoparticles creation.

Nanoparticles can be made to recognize specific tumour antigens or mutations that are diagnosed in a patient's cancer cells, especially in relation to the treatment of cancer as compared to routine drugs. For instance, nanoparticle encapsulants of medications can be changed to incorporate sustained release therapeutic levels for various patients, taking into account individual speeds of metabolism and the stage of the disease. This enables the delivery of the drug to the required part in small doses, which is a true strategy of adaptation.

The Individual Safety Assessments (ISAs) can play an important role in finding out the formulation and delivery system suitable for each patient, which will make the experience even safer for treatment and increase the likelihood of better clinical outcomes. For example, nanoparticles with herbal coatings can be designed to deliver drugs in response to blood glucose levels, increasing the efficacy and minimizing incidents of hypoglycemia. In this regard, meeting the patient's or client's needs and applying these complex technologies, it is possible to optimize the choice and usage of herbal extract-based replacement therapy (HERT) while enhancing the health status (HS) satisfaction with the applied herbal medicine and, simultaneously, guaranteeing the safety and effectiveness of the herbal nanomedicine.

19.5 ADVANTAGES OF HERBAL NANOMEDICINE IN CLINICAL SETTINGS

Herbal nanomedicines significantly increase patient adherence, minimize adverse effects, and enhance therapeutic efficacy by improving bioavailability and targeting capabilities. Because nanoparticles may mitigate the drawbacks of traditional herbal formulations, such as poor solubility and delayed action, more patient-friendly and successful therapies are now possible. This section looks at how these developments lead to improved clinical results and more effective application of herbal remedies.

19.5.1 Increased Patient Adherence

It has been shown that the patient compliance is far better in case of the herbal nanomedicine than is the case with synthetic drugs. Similarly,, herbal nanoparticles have the capability to act more rapidly and effectively than normal herbal extracts due to disadvantages associated with capsules and tablets such as low solubility and slow rate of dissolution. This can enhance patients' compliance with the recommended treatments and overall healthcare outcomes. Whereas a number of complications surface with traditional herbal therapy, the major handicap lies in achieving compliance from the patients due to their large size, the unpleasant bad taste, and their often frequent administration. These problems are solved by herbal nanomedicine as this increases the stability and bioavailability of the herbal constituents; it therefore reduces the frequency of administration per treatment. As a result, active herbal components encapsulated in nanoparticles can enhance patient compliance due to the elimination of some of the downsides associated with traditional herbal medicine. This patient-specific approach makes it easier to individualize the therapies to fit every individual, thereby enhancing the yields of herbal nanomedicines [20, 22, 28].

19.5.2 Reduced Side Effects

Traditional remedies are deemed to have fewer side effects than synthetic ones. Nanotechnology improves this advantage by allowing the practice of personalized medicine, decreasing side effects, and assessing and avoiding potentially fatal consequences. This can be beneficial to patients as it enhances comfort and therefore improves their quality of life. Herbal nanomedicine solves these problems due to the targeted release and delivery of therapeutic agent's desired site, minimum circulation in the system, and thus minimal doses. These nanoparticles can be made in such a way that they release their contents only in the intended area of the body, thus reducing harm to other organs. For instance, berberine nanoparticles contrasted positively with non-nano berberine in diabetic regulation with fewer gastrointestinal side effects [19, 20]. It removes the possibility of side reactions, hence making treatments more tolerable and safer and, thus, improving patient adherence. Nanotechnology in the drug delivery system is quite effective in the precise targeting of the location where the drug has to be delivered so that the incidence of side effects in the herbal medicinal drugs can be avoided as much as possible. For instance, silymarin-loaded liposomes for liver disorders exhibit a high level of hepatoprotective benefits with a minimal side effect level on the gastrointestinal tract (GIT) compared to other regular oral formulations [20, 28]. Nanotechnology has a capacity to reduce the chances of side effects since it regulates the rate of drug delivery, thereby reducing the chances of a sudden buildup of medication concentration in the patient's body. Herbal drugs and remedies have enhancing quality through the application of nanotechnology and thus the issues, such as solubility, stability, and bioavailability issues are sorted out. This allows for smaller amounts of the active ingredients to be packed into nanoparticles and nano-emulsions, thus allowing for the patient to take smaller doses at wider intervals and yet for the medicine to retain its effectiveness. For instance, curcumin-loaded nanoparticles can be given to the patients less frequently for chronic diseases such as diabetes and heart diseases, yet the patients will adhere to the treatment schedules. Among the advantages are that, occasionally, dosing regimens require the patients to take several doses in a day; however, with these preparative, dosing regimens are simplified resulting the reduced chances of patient noncompliance. The use of nanotechnology also enables the enhancement of the strength of therapeutic effects and lessening of undesirable effects.

19.5.3 Improved Efficacy

The encapsulation of herbals in nanoparticles improves the solubility and bioavailability of active ingredients, enhancing therapeutic efficiency. Nanoparticles can also deliver drugs directly to specific tissues or organs, increasing the concentration of active ingredients at the site of action, and resulting in more potent and efficacious therapies. Herbal nanomedicine is known for its increased effectiveness. The higher permeability of cellular units provided by nanoscale particles facilitates greater absorption and penetration at a cellular level, enhancing the overall therapeutic efficacy of herbal compounds. For example, ginseng extracts encapsulated in nanoparticles have shown improved bioavailability and therapeutic efficacy in reducing fatigue and enhancing cognitive abilities [27]. This precision is particularly important in treating diseases such as cancer or antibiotic-resistant infections, requiring high accuracy and powerful interventions. Many biologically active constituents in herbal extracts, such as terpenoids, tannins, and flavonoids, have large molecular sizes and poor membrane permeability, leading to low absorption across cell membranes. However, these components are also highly water-soluble. Nanoparticles eliminate these barriers, enhancing the delivery of therapeutic agents. In the case of ischemic cerebral injury, breviscapine encapsulated in multivesicular liposomes shows prolonged blood circulation and continued release, thereby improving therapeutic outcomes [23]. Nanotechnology provides better solubility, stability, and targeted delivery of conventional herbal treatments, enhancing their therapeutic efficacy with lower dose requirements. Nanomedicine offers a potential tool for delivering drugs more effectively, enabling better uptake and utilization of plant components, and broadening the applicability of disease management without deleterious side effects. These nanoparticles have shown greater anti-cancer efficacy in tumor growth inhibition and apoptosis induction when compared with conventional curcumin formulations [20]. Moreover, the controlled release properties of nanoparticles maintain therapeutic levels over the long term, ensuring sustained efficacy without drug resistance during chronic treatments.

19.6 THE FUTURE OF HERBAL NANOMEDICINE IN CLINICAL PRACTICE

Herbal nanomedicine is the new evolution in healthcare, and it has immense possibilities for revolutionizing clinical practice. The continuous improvement of nanotechnology yields new-generation herbal nanoparticle compositions that are even more advanced and well-suited. This includes the synthesis of nanoparticles with biocompatible properties, stimuli-responsive release profiles, and higher targeting capabilities. Herbal nanomedicine can be personalized for each individual by incorporating genomics, proteomics, and metabolomics, thereby making it more effective with fewer side effects.

Combining antipathogenic drugs and herbal nanoparticles will provide insights into synergistic drug interactions, offering novel opportunities to combat complex diseases. Mixing both could improve patient outcomes and lessen the reliance on synthetic drugs. There need to be strict regulatory guidelines for the development and therapeutic application of herbal nanomedicine. In a health system that strongly encourages innovation for the benefit of producers rather than all patients on grounds of safety and efficacy, stringent measures must regulate stem cell 'customization.'

Herbal nanoparticle guidelines should be developed through innovatively advanced effective, safe, and quality measures supported by standard regulatory frameworks with uniform testing protocols. The main thrust of future endeavors should be to set standard operating procedures for the clinical application of herbal nanomedicine, with a full compilation of preclinical studies, side effects, and follow-up. Meeting these requirements and developing a greater degree of consumer confidence in herbal nanomedicine demands the amalgamation of researchers, physicians, and regulatory authorities.

International collaboration between researchers, clinicians, and regulatory authorities is needed to catalyze the faster development and acceptance of herbal nanomedicine. The integration of research findings into clinical practice can also be assisted by sharing both knowledge and resources. To facilitate the adoption of herbal nanomedicine by health systems, public education is paramount. This will require educational initiatives and patient empowerment programmes to foster the adoption of these therapies. The transformation of these nanoparticles from bench to bedside, requiring additional research and clinical trials currently in progress, may move the management of chronic/co-morbidities in a completely different direction altogether, leading to better patient outcomes on Planet Earth.

REFERENCES

1. Bayda, S., et al., The History of Nanoscience and Nanotechnology: From Chemical-Physical Applications to Nanomedicine. *Molecules*, 2019. 25(1): p. 112.
2. Rehman, S.U., K. Choe, and H.H. Yoo, Review on a Traditional Herbal Medicine, Eurycoma longifolia Jack (Tongkat Ali): Its Traditional Uses, Chemistry, Evidence-Based Pharmacology and Toxicology. *Molecules*, 2016. 21(3): p. 331.
3. Wang, L., et al., Advances in nanotechnology and asthma. *Annals of Translational Medicine*, 2019. 7(8): p. 180.
4. Zhang, Y., et al., Nanotechnology in cancer diagnosis: progress, challenges and opportunities. *Journal of Hematology & Oncology*, 2019. 12(1): p. 137.
5. Li, D., Y. Liu, and N. Wu, Application progress of nanotechnology in regenerative medicine of diabetes mellitus. *Diabetes Res Clin Pract*, 2022. 190: p. 109966.
6. Guo, J., et al., Nanotechnology in coronary heart disease. *Acta Biomater*, 2023. 171: pp. 37–67.
7. Rahiman, N., et al., Recent advancements in nanoparticle-mediated approaches for restoration of multiple sclerosis. *J Control Release*, 2022. 343: pp. 620–644.
8. Farokhzad, O.C. and R. Langer, Impact of nanotechnology on drug delivery. *ACS Nano*, 2009. 3(1): pp. 16–20.
9. Barkat, M.A., et al., Herbal Medicine: Clinical Perspective and Regulatory Status. *Comb Chem High Throughput Screen*, 2021. 24(10): pp. 1573–1582.
10. Yin, C., et al., Nanotechnology improves delivery efficiency and bioavailability of tea polyphenols. *Journal of Food Biochemistry*, 2020. 44(9): p. e13380.
11. Mishra, M., et al., Nanotechnology: Revolutionizing the science of drug delivery. *Current Pharmaceutical Design*, 2018. 24(43): pp. 5086–5107.
12. De Matteis, L., R. Martín-Rapún, and J.M. de la Fuente, Nanotechnology in personalized medicine: A promising tool for Alzheimer's disease treatment. *Current Medicinal Chemistry*, 2018. 25(35): pp. 4602–4615.
13. Guo, M., et al., Herbal medicine nanocrystals: A potential novel therapeutic strategy. *Molecules*, 2023. 28(17): p. 6090.
14. Falzon, C.C. and A. Balabanova, Phytotherapy: An introduction to herbal medicine. *Prim Care*, 2017. 44(2): pp. 217–227.
15. Wu, Y., et al., SymMap: an integrative database of traditional Chinese medicine enhanced by symptom mapping. *Nucleic Acids Research*, 2019. 47(D1): p. D1110–d1117.
16. Hoshyar, N., et al., The effect of nanoparticle size on in vivo pharmacokinetics and cellular interaction. *Nanomedicine (Lond)*, 2016. 11(6): pp. 673–692.
17. Ezike, T.C., et al., Advances in drug delivery systems, challenges and future directions. *Heliyon*, 2023. 9(6): p. e17488.
18. Subramani, K., et al., Larvicidal, super hydrophobic and antibacterial properties of herbal nanoparticles from Acalypha indica for biomedical applications. *RSC Adv.*, 2017. 7: pp. 41763–41770.
19. Chatla, S. and A. Obilineni, *Nanao Herbal MedicinesChief Editor*. 2021.
20. Bojanić, A., R. Suručić, and M. Đermanović, Integration of nanotechnology and herbal medicine: Therapeutic potential for improvement of health care. *Contemporary Materials*, 2023. 14(2). https://doi.org/10.7251/COMEN2302149B
21. Abbasi, R., et al., Structural parameters of nanoparticles affecting their toxicity for biomedical applications: a review. *Journal of Nanoparticle Research*, 2023. 25(3): p. 43.
22. Wang, H., et al., Advancing herbal medicine: enhancing product quality and safety through robust quality control practices. *Front Pharmacol*, 2023. 14: p. 1265178.
23. Bonifácio, B.V., et al., Nanotechnology-based drug delivery systems and herbal medicines: a review. *International Journal of Nanomedicine*, 2014. 9: pp. 1–15.
24. Anwar, D.M., et al., Recent advances in herbal combination nanomedicine for cancer: delivery technology and therapeutic outcomes. *Expert Opin Drug Deliv*, 2021. 18(11): pp. 1609–1625.
25. Hesari, M., et al., Current Advances in the Use of Nanophytomedicine Therapies for Human Cardiovascular Diseases. *International Journal of Nanomedicine*, 2021. 16(null): pp. 3293–3315.
26. Subramani, K., et al., Screening the UV-blocking and antimicrobial properties of herbal nanoparticles prepared from Aloe vera leaves for textile applications. *IET Nanobiotechnol*, 2018. 12(4): pp. 459–465.

27. Mishra, Y., et al., Application of nanotechnology to herbal antioxidants as improved phytomedicine: An expanding horizon. *Biomedicine & Pharmacotherapy*, 2022. 153: p. 113413.
28. Kumar, R. and M. Sharma, Herbal nanomedicine interactions to enhance pharmacokinetics, pharmacodynamics, and therapeutic index for better bioavailability and biocompatibility of herbal formulations. *Journal of Materials NanoScience*, 2018. 5(1): pp. 35–60.
29. Sandhiya, V. and U. Ubaidulla, A review on herbal drug loaded into pharmaceutical carrier techniques and its evaluation process. *Future Journal of Pharmaceutical Sciences*, 2020. 6(1): p. 51.
30. Giannenas, I., et al., *Chapter 1 - The history of herbs, medicinal and aromatic plants, and their extracts: Past, current situation and future perspectives, in Feed Additives*, P. Florou-Paneri, E. Christaki, and I. Giannenas, Editors. 2020, Academic Press. pp. 1–18.
31. Teja, P.K., et al., Herbal nanomedicines: Recent advancements, challenges, opportunities and regulatory overview. *Phytomedicine*, 2022. 96: p. 153890.

20 Safety Profiles and Potential Toxicological Concerns of Herbal Nanomedicine

Sutapa Biswas Majee and Rachayeeta Bera
Department of Pharmaceutical Technology, NSHM Knowledge Campus, Kolkata – Group of Institutions, Kolkata, India

Dhruti Avlani
Department of Pharmaceutics, Dr. Prabhakar B Kore Basic Science Research Center, Off-Campus, KLE College of Pharmacy (A constituent unit of KAHER-Belagavi), Bengaluru, India

Ushasi Das
Department of Pharmaceutical Technology, Jadavpur University, Kolkata, India

Sudarshan Singh
Faculty of Pharmacy, Chiang Mai University, Chiang Mai, Thailand

20.1 HERBAL NANOMEDICINES: A BRIEF OVERVIEW

Herbal medicines are plant-derived valuable and renewable bioactives that offer potentially immense therapeutic benefits. They have a long well-documented history of traditional uses in Indian, Chinese, and other complementary and alternative systems of medicine. Folklore medicine and ethnic preparations constitute the basic foundation upon which modern-day herbal formulations have been developed. They have considerable potential in the management of several chronic diseases, cancers and the amelioration of critical life-threatening conditions. In many developing nations across the globe, a significant proportion of the population continues to depend on traditional herb-based medicinal systems for the relief and healing of disease symptoms. Herbal formulations are slowly beginning to gain popularity over the past few decades, even in the Western world, which has traditionally viewed such treatments as unscientific. Since the herbal preparations are composed of several characterized and uncharacterized active principles, they can attack a disease from different angles and thus prove to be highly beneficial over conventional allopathic systems of medicine (Ai ei al., 2024; Anand et al., 2024; Jalili et al., 2023; Mansingh et al., 2023; Subbiah et al., 2019).

Despite their popularity and health benefits, the realm of herbal formulation-based therapeutic strategies is frequently marred by several drawbacks during the initial stages of formulation development These include the presence of pharmacophores of complex chemical structures and stereochemistry, the presence of unidentified and un-characterized therapeutically active moieties, suboptimal physicochemical and pharmacokinetic attributes, the risk of inter-batch variation in the quality of extracts and content of active principles, and inadvertent and unavoidable exposure to living and non-living

DOI: 10.1201/9781003544418-20

contaminants and pollutants from soil, water, and air and the entry of impurities during cultivation and harvesting (Anand et al., 2024; Mansingh et al., 2023; Subbiah et al., 2019).

Nanotechnology is envisaged as a transformative platform to address the pitfalls of achieving desired therapeutic action from herbal formulations. The revolutionary approach has opened promising avenues and unlocked the doors to improving the bioavailability, efficacy, and efficiency of herbal formulations, culminating in the emergence of herbal nanomedicines. The integration of herbal medicine with nanotechnology is poised to expand the landscape of traditional practices with herbs and herb-based formulations, and has been instrumental in addressing the limitations and harnessing the potentials of these valuable drug resources. Herbal nanomedicine development is continuously expanding, advancing into domains where conventional allopathic therapies have failed and evolving as a stepping stone to unforeseen advances in future healthcare (Alemi et al., 2024; Jalili et al., 2023; Kad et al., 2022; Khan et al., 2021; Kumar et al., 2021).

Herbal nano-formulations have been developed using a multitude of phytochemicals and extracts, offering possibilities for the treatment of a wide spectrum of diseases. Among the phytochemicals that have been encapsulated in diverse types of nanocarriers are curcumin, berberine, quercetin, capsaicin, epicathechin, epigallocatechin-3-gallate, shogaol, glycyrrhizic acid, 10-hydroxycamptothecin, gallic acid, silymarin, silybin, genistein, rhein, lignin, flavonoids, babichi oil, lavender oil, thymoquinone, hesperitin, resveratrol, diosmin, mangiferin, withanolide A, andrographolide, piperine, boswellic acids, rutin, ursolic acid, and so on (Alemi et al., 2024; Anand et al., 2024; Pan, 2024; Razavi et al., 2024; Setia et al., 2024; Sharma et al., 2024; Jalili et al., 2023; Jia et al., 2023; Koklesova et al., 2023; Mahmood et al., 2023; Mansingh et al., 2023; Ray et al., 2021; de Oliveira Pacheco et al., 2022; Dewi et al., 2022; Kaur et al., 2022; Kumar et al., 2021; Kumar et al., 2022a; Li et al., 2022; Ohadi et al., 2022; Parusu et al., 2022; Zhou et al., 2022; Abo-zaid et al., 2021; Afzali et al., 2021; Ahmadi et al., 2021; Anwar et al., 2021; Barros et al., 2021; Dos Santos et al., 2021; Ebaid et al., 2021; Khan et al., 2021; Murkute et al., 2021; Paliwal & Paliwal, 2021; Rajasekar et al., 2022; Rehman et al., 2021; Zare et al., 2021; Ghafelehbashi et al., 2020; Ghaffarzadegan et al., 2020; Hashemian et al., 2019; Jain et al., 2020; Khan et al., 2020; Kim et al., 2020; Moradi et al., 2020; Onyeji, 2020; Sandhiya et al., 2020; Tiwari et al., 2020; Duse et al., 2019; Koupaei Malek et al., 2019; Pallela et al., 2019; Subbiah et al., 2019; Zheng et al., 2019; El-Far et al., 2018; Gupta et al., 2019; Shahgolian et al., 2018; Wu et al., 2018 ; Korth, 2014; Ansari et al., 2012). Table 20.1 offers a bird's-eye view of the applications and advantages of herbal medicines, obstacles to exploiting their full potential, the scope of nanotechnology in overcoming the hurdles and improving the therapeutic outcomes of herbal medicines. Nano-formulations have also been fabricated with *Aloe vera* extract and neem leaf extract (Alti et al., 2020; Subbiah et al., 2019).

It is interesting to mention that the nanovehicles or nanocarriers that have been developed by research scientists across the world for the delivery of herbal medicines or extracts or phytoconstituents can be broadly categorized as engineered or manufactured soft and hard nanoparticles(NPs). They have been fabricated artificially from biodegradable and non-biodegradable organic and inorganic constituents such as polymers, lipids, metals, metallic oxides, carbon in contrast to the ones that are formed naturally in living systems or are formed accidentally in the environment. The chemical component from which the nanomaterials are being constructed may be obtained from either natural or synthetic sources. Most of the nano-formulations developed to date are first-generation or second-generation NPs designed to improve the functionalities of the bulk counterparts and are involved actively in the process (Li et al., 2022; Anwar et al., 2021; Ray et al., 2021; Gupta et al., 2019; Qiu et al., 2018).

The chapter aims to offer a comprehensive analysis of the safety and toxicity concerns associated with nanomaterials used as carriers in the delivery of herbal medicines. It seeks to critically assess how nanomaterial attributes rather than their cargo contribute to the toxic and adverse outcomes, to investigate the impact of the engineered nanomaterials on the biotic and abiotic components of the ecosystem of which the end-user is a part. Individuals involved in manufacturing and

TABLE 20.1
Scope and Benefits of Nanotechnology in the Development of Herbal Nanomedicines

Herbal Formulations: Key Considerations	
Applications	**Advantages**
• Carcinoma of lung, breast, stomach, liver, pancreas, colon • Inflammation/neuro-inflammation • Neuro-degenerative disorders • Anti-cancer/Anti-oxidant • Wound healing • Dentistry	• ↓ Toxicity to normal tissues • ↓ Risk of tumor recurrence • ↓ Risk of development of drug resistance
CHALLENGES	
• Poor permeability & poor absorption • Rapid pre-systemic & systemic metabolism • Susceptibility to P-gp efflux transport • Risk of inconsistency in quality & quantity active principle	
NANOTECHNOLOGY: PROSPECTS AND PROMISES	
• ↑ Solubility • ↑ Biocompatibility • ↑ Stability • ↑ Site-specificity • ↑ Cellular uptake	• ↑ Retention in blood circulation • ↓ Toxicity to off-sites • ↑ Bioavailability • ↑ Duration of action • ↑ **THERAPEUTIC EFFICACY**
HERBAL NANOMEDICINES	
Applications	**Benefits**
• Nanoparticles • Nanospheres, Nanocapsules, Nanotubes • Nanogels & self-assembled nanogels • Nano lipid carriers • Solid lipid nanoparticles • Nanovesicles: liposomes, niosomes, pro-niosomes, ethosomes, transferosomes	• ↓ Frequency of administration • ↓ Cost of therapy • ↓ Incidences of adverse outcomes • ↑ Bioavailability • ↑ Duration of action • ↑ Patient compliances • ↑ **THERAPEUTIC OUTCOMES**

quality control of the nanomaterials face occupational hazards that are different from those observed in the production of conventional delivery platforms. The chapter delves into a detailed discussion of various approaches, techniques, and tools currently being employed in nanotoxicological assessment, highlights their benefits and drawbacks and identifies the existing lacunae in the regulatory guidelines, and frameworks for herbal nanomedicines specifically, the need for the development of standardized protocols for assays. It provides a brief overview of the classification of nanomaterials, based on their safety and toxicity profiles. The chapter focusses on the idea that future strides in the field of herbal nanomedicines should be oriented toward ensuring high standards of quality, and better accuracy and prediction of safety and toxicity profiles of nano-formulations with existing herbal constituents. By addressing these goals, the chapter seeks to provide a complete understanding of nanomaterial safety and toxicity, promote best practices and safe-by-design approaches in the manufacture of herbal nanomedicines, in an environment-friendly and sustainable manner, and provides the right impetus for future development in enhancing therapeutic outcomes from herbal nanomedicines.

20.2 SAFETY ISSUES AND TOXICOLOGICAL CONCERNS WITH HERBAL NANOMEDICINES

An objective understanding of safety issues and toxicological concerns of herbal nanomedicine should be based on neutral and fair approaches. It must be made clear that any form of medicine can lead to dire consequences due to misuse, abuse, or being used irrationally in target and non-target organisms and may also adversely affect the ecosystem. This happens because any therapeutic moiety is designed to interfere with various biochemical and signaling pathways in the pathological condition in the diseased cell, cancer cell, or microbes, ultimately causing their death. A classic example of an adverse effect in non-target species is the feminization of fish due to estrogen in the aquatic ecosystem. Other molecules that are frequently detected in water bodies are anticancer agents, anti-inflammatory drugs etc. (Mahapatra et al., 2018). Moreover, herbal-based medicines suffer from some inherent drawbacks which will be dealt with in subsequent sections. Nanosystems, or nano-formulations, themselves pose challenges due to their complexities in terms of both their fabrication and also their unique features. The withdrawal of ultra-small superparamagnetic iron oxide-based contrast agent for magnetic resonance imaging from the market after reports of adverse events should be taken into consideration during the risk assessment of nanoparticles. In several instances, herbal formulations are concomitantly administered along with conventional medicines such as antibiotics for bioburden containment. Graphene oxide NP has an affinity to adsorb these antibiotics and lower their efficacy in turn. Endogenous myeloperoxidase can reduce the cytotoxicity of graphene oxide NPs by accelerating their degradation (Martinez et al., 2021). On the other hand, co-administered NP can alter membrane integrity and may facilitate easy access to other NPs, leading to induced toxicity as a result of the interaction effect. The toxicity of copper oxide NPs was increased by zinc oxide NPs (Forest, 2022). Moreover, spray-based herbal nano-formulations or cosmetics or cosmeceuticals may release particles into the environment during application (Martinez et al., 2021; Foulkes et al., 2020). The effect of un-degraded but exhausted NPs on the patient system as well as on the outer environment need to be investigated (Mahapatra et al., 2018). Thus, this entire section will try to cover the various aspects associated with the safety, risk, and toxicity of herbal nanomedicines towards mankind and biotic and abiotic components of the environment from different angles, during manufacture, use, post-use storage, and disposal.

There is a lot of prejudice and misunderstanding with respect to the safety of herbal preparations. Herbal medicines may suffer from a lack of reproducibility due to variations in the content of active principles based on seasonal variation, geographical location, and biodiversity (Parusu et al., 2022). Simultaneously, inactive ingredients are often not quantified, which may adversely affect the composition of herbal extract-based dosage forms . One of the common problems associated with the use of herb or herb-derived phytochemicals is the risk of heavy metal contamination and pesticide residues in plant or plant products during the processes of cultivation and harvesting. Other hazardous contaminants include toxins secreted by fungi, pyrrolizidine alkaloids, and toxic chemicals like polycyclic aromatic hydrocarbons (Łuszczki et al., 2019). Decisions on dose and dosage regimen and duration of treatment with herbal finished products should be patient-specific and based on the age and sex, genetic constitution of patients, the existence of co-morbidities, the concomitant administration of drugs, dietary habits, smoking addiction, and alcohol consumption levels. Evaluation and maintenance of quality control standards for herbal formulations is a daunting task (Zhang et al., 2015; Korth, 2014). Hepatotoxicity, genotoxicity, and carcinogenicity have been reported as the most common adverse effects of herbal preparations (Qari et al., 2021; Korth, 2014). Herbal formulations are widely popular for the treatment of skin diseases; at the same time, however, they tend to induce photosensitization and phototoxicity (Tirumala et al., 2021). In a nutshell, herbal medicines may be said to suffer from intrinsic and extrinsic toxicity (Zhang et al., 2015). Therefore, the development of herbal preparations in a systematic manner should include investigations of their potential toxic effects on vital organs, and the elucidation of mechanisms of toxicity induction during preclinical stages (Guidelines by Ministry of Health, Malaysia, 2023; Jitareanu et al., 2023).

The fabrication of nanostructures with herbal ingredients in compliance with principles of Good Manufacturing Practices (GMP) is a complex process in which control of the manufacturing process and its associated parameters is pivotal to ensuring reproducibility in achieving critical quality attributes (CQA), performance, safety, and both in vitro and in vivo stability(Foulkes et al., 2020). Slight changes in the critical material attributes (CMA) of raw materials, variation in the critical process parameters (CPP), and deviation from the prescribed limits may change the CQAs with potentially serious consequences (Soares et al., 2018). The inter-relationship among CMA, CPP, and CQA is established by statistical methods of analysis and the design of experiment tools such as response surface methodology (Pan et al., 2019). From the manufacturing aspect, as well as from the viewpoints of therapeutic efficacy and toxicity, it must be mentioned that NPs are unique in the sense that the excipients and the active ingredient go together into the formation of a stable and efficacious particle of nano-dimension and non-active ingredients cannot be termed as excipient in the truest sense (Hemmrich & McNeil, 2023). The concept of the "nano-paradox" needs to be understood from the viewpoint of the negative impact of manufacturing and the handling of nanomaterials by the individuals involved in the manufacturing industry. The nano-dimension, which is instrumental in the beneficial biological effects, proves harmful for the industry personnel and the extent of occupational hazards is greatly governed by the route of exposure, and duration of exposure (Guidelines and best practices for safe handling of nanomaterials in research laboratories and industries compiled for nano mission. DST, Govt of India, Centre for Knowledge Management of Nanoscience & Technology (https://dst.gov.in/).

Safety and toxicity concerns regarding nanoparticles arise more out of the shell material design and composition, fabrication technique, surface modification strategies, presence of impurities (e.g. metal ion-based catalysts as in iron, molybdenum, nickel, unreacted monomer), organic contaminants, and resultant physicochemical properties and are influenced less by the core material of the nanosystem (Liu et al., 2022). The shell of NP thus assumes significance in the biocompatibility, safety, and toxicity assessment as it is the component that initially comes into contact with the living cell surface (Ahmed et al., 2019). In most of the cases, intracellular ROS levels and antioxidant defense mechanisms are affected by the above-mentioned factors, leading to a cascade of events (Shukla et al., 2005).

Engineered or manufactured nanomaterials are complex heterogeneous entities with respect to their physicochemical features and, hence, demonstrate variable pharmacokinetic (ADME – Absorption, Distribution, Metabolism and Excretion) and pharmacodynamic behaviors, efficacy, and downstream cytotoxicity. Biodegradation and clearance patterns also differ. These unique properties impart versatility to nanomaterials in their application and at the same time account for the risks and hazards associated with them. Nanoscale dimensions confer upon them features entirely different from the bulk counterparts from/with which they have been fabricated. They have been reported to possess superior optical properties, electrical conductivity, chemical reactivity, catalytic activity, and adsorption efficiency. Physicochemical properties that should be taken into consideration as influencing nanomaterial behavior in vitro and in vivo and governing their bio-interface interaction include size, size distribution, shape, surface area, composition, crystallinity, hydrophilicity/hydrophobicity, surface properties such as roughness, porosity, surface charge, surface chemistry, functionalization, surface coating/grafting, element doping, and the state of aggregation/agglomeration. Other factors which contribute to the double-edged sword of efficacy and toxicity are dose, route of administration/exposure, and duration of exposure (Abdelkader et al.;, 2023; Forest, 2022; Martinex et al., 2021; Shin et al., 2021; Akcan et al., 2020; Mahapatra et al., 2018; Soares et al., 2018; Mengying et al., 2015).

The size or hydrodynamic diameter of manufactured NP in vitro depends upon the approach adopted, whether it is top-down or bottom-up. Moreover, surfactants or stabilizers are usually added to prevent protection against aggregation which is otherwise an obvious phenomenon with natural NPs. The lower the size, the greater the risk of nanotoxicity due to an increase in the specific surface area. A tendency to aggregate/agglomerate in vitro or under physiological conditions can prove to be beneficial in this respect. The higher specific surface area promotes interaction and is potentially hazardous and responsible for causing damage to several vital organs such as lungs, liver, kidney, spleen, male and female reproductive organs, heart, and even brain. Nano-scale dimensions permit access across the blood–brain barrier and into the fetus by crossing the placental barrier and exhibiting teratogenicity. The trans-placental passage of smaller NPs may have a negative impact on fetal development (Tirumala et al., 2021). Owing to their small size, they are also removed rapidly by the macrophages (Tirumala et al., 2021; Akcan et al., 2020; Foulkes et al., 2020; Soares et al., 2018). Nano-size may also result in the photoreactivity of titanium dioxide NP (Martinez et al., 2021). It is possible to fabricate NP in various shapes as spheres, fibres, rods, cages, cluster, stars, and so on. Of these, conventional spherical NPs have demonstrated the least toxicity (Abbasi et al., 2023). Gold nanostars demonstrated maximum cytotoxicity in hFOB 1.19 (fetal osteoblastic cells), 143B, and MG-63 (osteosarcoma) cell lines (Steckiewicz et al., 2019).

The nanostructure plays a significant role in precipitating toxicity. Carbon nanotubes are toxic for mitochondria, for example, whereas silver nanoparticles interfere with cell membrane properties and morphology (Martinez et al., 2021). Metal nanoparticles are responsible for intracellular lipid peroxidation, leading to the damage of the plasma membrane, mitochondria, and endoplasmic reticulum. Metal oxide nanoparticles, such as titanium dioxide, iron oxide, and cupric oxide NP, cause an imbalance in intracellular ROS-oxidized glutathione machinery in the liver. Similarly, zinc oxide NP, administered at a dose of 50mg/kg body weight to rats, proved to be toxic for the intestine and generation of superoxide dismutase (Yang & Merlin, 2023; Catalano, 2021; Akcan et al., 2020). Another problem observed with metal/metallic oxide NP is their exceptionally low rate of clearance and, hence, the high degree of accumulation in the body. In addition to this, stable metallic NPs are more biocompatible than dissolvable metallic/metal oxide NPs. The type of NP component also affected genotoxicity. Gold and superparamagnetic iron oxide NP did not affect gene regulation or protein expression in human vein endothelial cells or adipose tissue-derived stem cells, respectively, and neither did it increase ROS production in RAW264.7 macrophages (Fröhlich, 2017). Liposomes are reported to be immunotoxic, triggering complement activation-related pseudo-allergy (CARPA) (Ray et al., 2021). Tailor-made NPs with biodegradable polymers pose a different challenge. Owing to biodegradation in vivo after predetermined and predefined time duration, there are changes in their surface properties and concentration, and that may lead to immunotoxicity. Moreover, the presence of anti-PEG antibodies affects the performance and may lead to the elicitation of toxic effects from PEGylated NPs. NPs of different compositions have been reported to be immunosuppressive and to act as immunomodulators, and capable of generating hypersensitivity reactions. Hydrophilic NPs as mesoporous silica NPs failed to gain access across the blood–brain barrier and are less toxic, owing to the presence of silanol groups on the surface (Yang & Merlin, 2023; Liu et al., 2022). Cerebral toxicity has been observed with charged gold and manganese dioxide nanoparticles, presumably due to an increase in endocytosis capacity (Tirumala et al., 2021). To secure successful intracellular drug delivery, positively charged NPs are preferred as they can escape the endosome–lysosome degradation pathway, induce less endoplasmic reticulum stress and are less deleterious for cellular membranes and organelles, (Yang & Merlin, 2023). For NPs exhibiting differences in crystallinity as with titanium dioxide NP, the two forms were found to have different toxicity concerns, when studied in Balb/3T3 mouse fibroblasts or in human bronchial epithelium cell line. The rutile lattice showed cytotoxicity as well as genotoxicity and the cellular uptake was higher with anatase form (Abbasi et al., 2023). Amorphous forms of nano-silica were more toxic to human cell lines as they generated ROS and caused DNA cleavage (Akcan et al., 2020). Nanoparticles may be coated to achieve certain specific functions, and the coating imparts charges on the NP surface which may alter biocompatibility, biodegradation, biological functions, and, ultimately, nanotoxicity phenotypes. Coated silver NPs were found to have higher biocompatibility even at higher concentrations and to produce fewer alterations in intracellular glutathione and superoxide dismutase levels in comparison to uncoated ones, although the effects were size-dependent. The coating material also affected toxicity as citrate-coated silver NP demonstrated lower toxicity than PVP-coated ones.

Similarly, titanium dioxide coating offered protection against zinc ion-induced damage from zinc oxide NPs (Abbasi et al., 2023). Surface modification of gold NPs with polyethylene glycol improved overall dispersion, and distribution in the circulation, prevented opsonization, retention in liver and spleen, and facilitated hepatobiliary and renal clearance. The results were just the opposite and detrimental when polyethyleneimine (PEI) was used for the surface modification of the same gold nanoparticles. Furthermore, PEI-modified gold nanoparticles demonstrated the formation of corona with evidence of agglomeration (Li et al., 2020; Wang et al., 2020). Doping reduces the dissolution rate of doped nanomaterials as observed with iron doped-zinc oxide NP, which led to higher in vitro and in vivo toxic manifestations such as mitochondrial damage, reactive oxygen/nitrogen species (ROS/RNS)-induced oxidative stress, interruption with development of zebrafish embryo, and, finally, eliciting inflammatory responses in rodent pulmonary systems (Liu et al., 2022; Yan et al., 2019). Cellular uptake studies with nanoparticle aggregates showed contradictory results. Usually, aggregation lowered the uptake by HeLa cells; in MDA-MB 435 cells, however, the uptake was increased. With zinc oxide NP, the effect of aggregation on uptake by RAW 264.7 cells was found to be concentration-dependent. A low concentration of zinc oxide NP aggregate induced apoptosis (Abbasi et al., 2023). Due to high surface energy, nanoparticles have a natural propensity to undergo aggregation and agglomeration to achieve thermodynamic stability and in the process are themselves modified substantially. However, due to a decrease in the specific surface area, the agglomerate may behave differently with reduced interaction potential and may prove to be less toxic than the native NP (Liu et al., 2022).

Thus, from the above study, it is evident that alterations in NP physicochemical attributes may have a serious impact on CQAs, cellular uptake, desired therapeutic outcomes, in vivo fate, excretion, and toxicity. In the light of these, the ISO/TR 13014:2012 guideline recommends estimation of the above parameters for safety and hazard assessment of nanostructures (Oberdörster, 2010).

Apart from the above-mentioned parameters, the dose of the nano-formulation, exposure time, and route of administration may affect cell viability and cytotoxicity. Monodisperse spherical mesoporous silica nanoparticles should have been ideally non-toxic, but their cytotoxic effects were found to be dose-dependent, as seen with titanium dioxide NP (Abbasi et al., 2023; Gandamalla et al., 2019). Dose-dependent toxic effects were also seen with platinum NPs on HepG2 (human hepatocellular cells) and titanium dioxide NPs in human lung epithelial cells. On the other hand, polylactic acid or magnetic mesoporous silica NPs were established as being non-toxic for both cancerous and non-cancerous cells, such as human melanocyte (NGM), fibroblast (FGH), and endothelial (HUVEC) line cells (Labrador-Rached et al. 2018; Gea et al., 2019; Helal Neto et al., 2019). An exposure time of short duration (in this case 5 minutes) of poly(N-isopropylacrylamide) (PNIPAM) and N-isopropylacrylamide/N-tert-butylacrylamide (NIPAM/BAM) polymeric NPs of varying ratios produced detectable toxicity in *Vibrio fischeri* (Naha et al., 2009). Nano-formulations with FDA-approved constituents and administered via the oral route may adversely affect the intestinal mucosal barrier and may alter colonic microbiota composition (Yang et al., 2023). Due to the difference in the composition of bacterial cell membranes, Gram-positive and Gram-negative bacteria are affected by NPs in different manners (Martinez et al., 2021). Intravenous administration of silica and titanium dioxide NPs affected the growth and development of fetus in mice (Tirumala et al., 2021). Dermal application or intravenous administration of nano-formulations of the same composition may affect the immune system in different ways (Keck & Müller, 2013).

Protein corona formation occurs following the intravenous administration of NP, when their actual size interacting with the biological surface is more than the original size of the manufactured NP. The adsorption of different types of proteins from biological fluid onto the nanomaterial surface compensates for the presence of excess energy on the particle surface (Foulkes et al., 2020; Lv et al., 2015). Alternatively, it can be said that the pristine, synthetic identity of NP acquires a new biological identity in the body which can exhibit potentially beneficial or deleterious effects. The phenomenon of adsorption is highly dynamic owing to the variable composition of the fluids coming into contact with solid nanoparticle surfaces and due to differences in the affinity of proteins. A change in composition of the corona is attributed to the health condition, physiological environment, internal localization of NP, co-morbidities, and the co-administration of different therapeutic moieties (Soares et al., 2018). Not only is the adsorption dynamic, but so also is the subsequent interaction with the cell membrane, and such dynamic events alter the NP surface properties in a cyclical fashion (Ahmad et al., 2022). The corona alters the movement of NP in the circulatory system, biodistribution, and metabolism, and it also accelerates NP clearance from the body as it is easily recognized by macrophages. Therapeutic outcomes have been reported to differ according to the presence or absence of protein corona on the NP surface (Forest, 2022; Bai et al., 2021; Cai et al., 2018; Ding et al., 2018).

Moreover, intentional or undesirable trafficking, bioretention, and biopersistence depend on the type of organism/cell; that is, whether it is a prokaryotic cell or an eukaryotic cell. In the case of simple cells, as is the case with microbes, entry occurs through the exposed cell surface; in complex organisms, by contrast, as in animals and humans, they enter the biological system via respiratory or gastrointestinal routes or through the skin. The nanoparticles are then taken up by cells and internalized by the processes of endocytosis or phagocytosis (Martinez et al., 2021).

The specific and non-specific biophysical interaction of nanomaterials with cellular organelles and subcellular components, serum proteins, blood cells, and intracellular compartments may alter their biological performance and functions and induce damage in DNA, leading to molecular initiating events (MIE), adverse outcomes (AO) via adverse outcome pathways (AOP), culminating in nanotoxicity. These interactions may be initiated by physical, chemical, and mechanical forces in vivo or may be mediated by receptors (Liu et al., 2022; Catalano, 2021; Tirumala et al., 2021). Adverse events normally observed with NP are loss in integrity of cell membrane and nuclear membrane (in extreme cases); the dysregulation of mitochondrial function; the permeabilization of the lysosomal membrane; the elevation of ROS/RNS-induced oxidative stress; interference with signal transduction; and the over-production of pro-inflammatory cytokines. These events are manifested as fibrosis, granuloma, organ damage, cardiac disturbances, genotoxicity, disruption in cell proliferation, apoptosis, necrosis, and other modes of cell death (Martinez et al., 2021; Korth, 2014). Among the various vital organs under potential risk from nanomedicines, liver, and kidneys arise first due to nanopores present in sinusoidal blood vessels and the basement membrane of liver and the glomerular membrane of kidneys. NP greater than 5 nm cannot be cleared from the body by kidneys. The presence of a primary tumor in an adjacent organ may restrict the advance of NP to the desired target organ and, if entry is facilitated, must be able to ward off endosome-mediated degradation (Mengying et al., 2015 , Lokugamage et al., 2018 , Koklesova et al., 2023).

One issue that has been frequently overlooked by nanomaterial manufacturers and regulatory bodies associated with the production of engineered NPs for pharmaceutical applications is the issue of contamination of raw materials, intermediates, or finished products by microbes and endotoxins (Siegrist et al., 2019).

20.3 APPROACHES TOWARDS SAFETY AND TOXICOLOGICAL ASSESSMENT: NANOTOXICOLOGY

In 2005 the International Life Sciences Institute Research Foundation/Risk Science Institute constituted a working group with experts in the domain of nanomaterials with the aim of evaluating the state-of-the-art concerning toxicological aspects and framing guidelines. They realized the need for the screening of nanomaterials based on their hazard levels with standardized protocols (Hussain et al., 2015). Nanotoxicology deals with the identification, determination, and establishment of unintentional adverse effects, interactions, and toxicokinetics of engineered or manufactured nanomaterials on target organisms(users), producers, and biotic and abiotic components of the environment at cellular and molecular levels throughout their lifecycle (Liu et al., 2022; Tirumala et al., 2021; Akcan et al., 2020).

Nanotoxicological studies become challenging as the same NPs can respond and behave differently and may exhibit different effects on different types of cells, as is reported in studies on the effect of silver NPs on stem, nervous, epithelial, phagocytic, endothelial cells, and alveolar macrophages. Moreover, site-specific nano-formulations and smart or intelligent nanoparticles are being envisaged to achieve targeted, controlled drug delivery at the site of action and release the cargo in response to various internal biological stimuli. The fabrication of such delivery platforms adds to the challenge in the assessment of toxicity and hazards (Abdelkader et al., 2023).

Development of ROS-induced oxidative stress has been elucidated as the primary toxic effect elicited by NPs in the biological environment and ecosystem, or as MIE, as defined previously by AOP. The excess of ROS alters protein conformation and alters the chemistry of lipids and nucleic acids, leading to apoptosis and necrosis (Martinez et al., 2021). Figure 20.1 summarizes the potential impact of physicochemical attributes, in vivo attributes, and other factors associated with nanoparticles on their safety and toxicity.

Prime factors that are considered in nanotoxicological assessment involve the physicochemical characterization of nanomedicines/nanomaterials, followed by in vitro and in vivo investigations (Kad et al., 2022).

Safety and toxicity evaluation of NP should include both in vitro and in vivo methods preferably at an estimated human dose to have a clear idea about the effect of NP concentration on target and non-target tissues and organs (AzoNano). Nanotoxicological assessment can be carried out by adopting empirical approaches involving experimental data and computational techniques, and in silico approaches (Forest, 2022).

In empirical methods, humans, animals, and excised tissues or cells or cell lines are exposed to nanomaterials under investigation and the effects produced are recorded and compared to control data. The responses that are observed include the extent of cellular uptake, biodistribution, cell

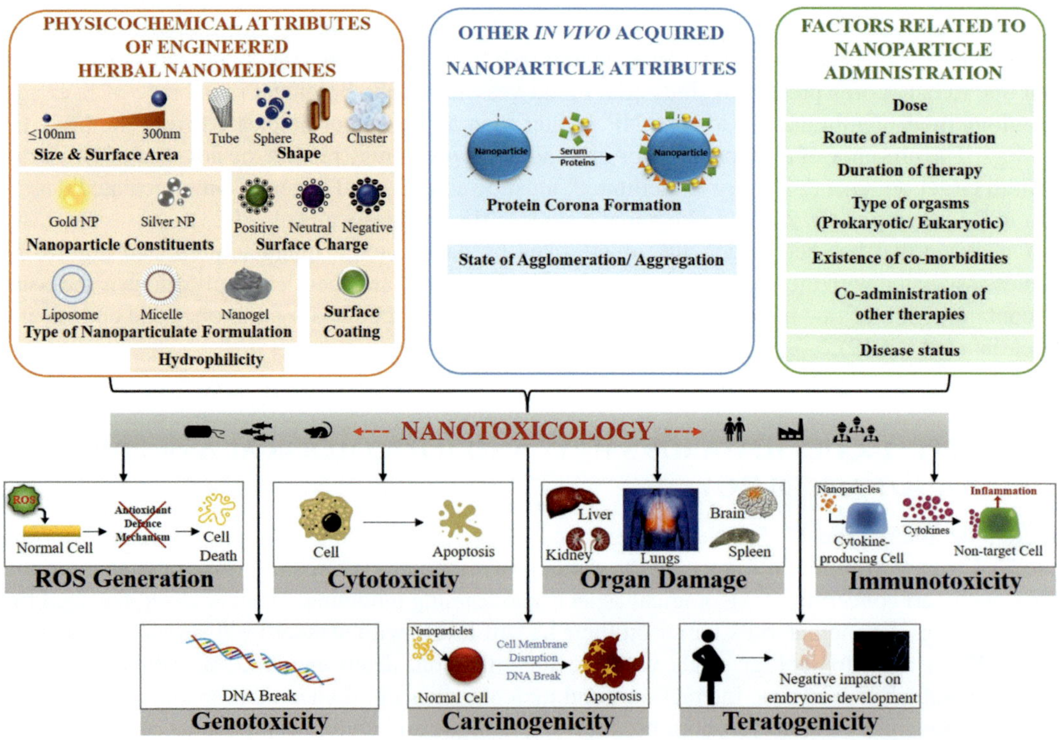

FIGURE 20.1 Potential impact of herbal nanomedicines on safety and toxicity to man and his environment.

viability, proliferation, effects on biochemical pathways, signs of cytotoxicity such as elevated ROS levels, the development of oxidative stress, the overproduction of pro-inflammatory cytokines, the induction of inflammatory responses, and the integrity of the genome (Abdelkader et al., 2023).

However, before subjecting NP to a battery of in vitro and in vivo tests, it is essential to ensure the chemical composition of NP and to characterize their size and size distribution as these properties affect their biological and toxic responses significantly. For example, the purity of NP polymeric components, and the presence of surface coating can be analyzed through thermogravimetric analysis (TGA), mass spectrometry, or inductively coupled plasma mass spectrometry (ICP-MS). Particle size and size distribution can be effectively monitored by transmission electron microscopy (TEM), dynamic light scattering, Brunauer–Emmett–Teller (BET) adsorption isotherm, and electrospray differential mobility analysis (ES-DMA) techniques (Halamoda-Kenzaoui et al., 2019b). In the later methods, however, processing is required before actual characterization and that may induce artificial changes in surface properties which will not occur in vivo (Foulkes et al., 2020).

20.3.1 In Vitro Methods

In vitro models are regarded as direct, convenient, reliable, accessible, and inexpensive surrogates of animal models, involving minimal ethical issues. During nano-formulation development, these should be carried out before animal studies. In in vitro studies, primary cell cultures (derived from cancer cells), cell lines artificially immobilized in monoculture systems, multicellular 3D models or organoids, spheroids, and co-cultures are all widely employed. Among the examples are cells obtained from lungs, gastrointestinal tract, neurons, blood, liver, stem cell-derived hepatocytes, pluripotent stem cells (iPSCs), human fetal hepatic progenitor cells (hFHPCs), and human skin-derived precursors (hSKPs), and those of co-cultures are alveolar epithelial type II cells, two types of immune cells (human monocyte-derived macrophages and dendritic cells), three-dimensional lung co-culture comprising of alveolar epithelial cells-fibroblasts-macrophages, Caco-2-HT29 co-culture, triple culture of Caco-2, HT29-MTX-E12, and THP-1 cells. Of these different systems, primary cell cultures and cell lines are useful in the preliminary stages of nanotoxicity assessment and produce reliable data. However, they fail to represent the heterogeneity and complexity of pharmacokinetics and toxicokinetics in the actual biological system resulting from crosstalk and interaction at multiple cellular levels and prolonged use may produce erroneous responses due to de-differentiation and phenotypic change. In such circumstances, co-cultures and 3D models have proven to be valuable and, in particular, the latter model can serve as a bridge between the 2D cell cultures and animal models (Forest, 2022; Tirumala et al., 2021).

During in vitro studies in different types of cell models, the endpoints are alterations in membrane integrity, cell viability, cytotoxicity, the estimation of oxidative stress, the lowering of antioxidant activities, nucleic acid, and chromosomal damage, alteration in genetic profile and expression, and apoptosis. The techniques widely employed are 3-(4,5-dimethylthiazol-2-yl)-2,5-diphenyltetrazolium bromide (MTT) assay, Trypan blue assay, and 2',7'-dichlorofluorescein diacetate assay (DCFDA). Instrumental methods of analyses adopted for detecting the endpoints are based on colorimetry, UV-Vis spectrophotometry, fluorescence spectroscopy, luminescence, scanning electron microscopy (SEM), transmission electron microscopy (TEM), scanning electron microscopy/energy dispersive X-ray spectroscopy (SEM-EDX), atomic force microscopy (AFM), video-enhanced differential interference contrast (VEDIC) microscopy, and so on (Tirumala et al., 2021; Akcan et al., 2020). An interesting in vitro study has been conducted with a modified lipopolysaccharide (LPS) membrane of Gram-negative bacteria to determine the effect of modification on binding of gold NP and subsequent penetration inside bacterial cells. Observations with modified bacteria and solid-supported LPS-containing lipid bilayers revealed LPS to be the essential component for NP-microbial cell surface interaction. To increase the sensitivity of toxicity screening assays, *E. coli* knockout mutants were employed which provided desirable information on the cellular response to NPs, in comparison to wild varieties (Qiu et al., 2018). Latest developments in in vitro methods for nanotoxicological studies involve the inclusion of cells, tissues or organs-on-a-chip producing cell-on-a-chip or organ-on-a-chip on a microfluidic platform. This system facilitates dynamic characterization. The toxicity of titanium dioxide or zinc oxide nanoparticles in lungs has been evaluated by lung-on-a-chip model (Forest, 2022; Akcan et al., 2020). 3D epidermal models, such as the EpiKutis model and the Ediderm skin model, have demonstrated a more accurate prediction of NP-induced skin toxicity. A 3D hepatocyte chip is yet to be developed for the accurate assessment of NP-induced liver injury and hepatotoxicity (Tirumala et al., 2021).

Genotoxicity assessment is a crucial part of in vitro assay portfolio in nanotoxicological studies, for which the tests commonly employed are *Salmonella typhimurium* reverse mutation assay (AMES), COMET, chromosomal aberration, micronucleus, and hypoxanthine phosphorybosyl transferase (HPRT) mutation assays (Verma, 2018). However, none of them is capable of providing a complete picture of genotoxic potential; there may even be false positive or negative results due to the interference and interaction of NPs with assay reagents such as that between cytochalasin B and nanomaterials in micronucleus assay, inability, or incomplete penetration of NP into bacterial cells in AMES test, risk of interaction with naked DNA in COMET assay (Tirumala et al., 2021). Interactions occur due to the unique characteristics of NPs such as adsorption efficiency, chemical reactivity, optical and magnetic properties, and other physicochemical characteristics (Soares et al., 2018).

Despite several merits of in vitro tools, they fail to mimic the effects in the case of repeated use in the management of chronic conditions. It is challenging to carry out such simulations for a prolonged duration. Growth promoters, nutrients, and proteins present in artificial culture media used for the growth and maintenance of cell lines and cell cultures may be absorbed by NPs. NPs may interact with dyes used in various colorimetric, spectrometric, and fluorescent methods. Assays involving estimation of optical parameters or measuring products of redox reactions may fail to give accurate results. In vitro test conditions do not emulate the tissue/tumor microenvironment and may be compromised leading to loss in reproducibility and precision of experimental data (Kad et al., 2022, Foulkes et al., 2020; Mengying et al., 2015). Neither of the in vitro cell-based models can mimic the actual pathophysiological response to potentially toxic NP, nor can they reproduce the compensation mechanism exhibited by humans or animals in response to challenge with NP in clinical settings (Soares et al., 2018).

20.3.2 In Vivo Assays

In vivo assays are performed in animal models such as drosophila, zebrafish, rats, mice, and other non-human primates, and attempts are usually made to use relevant doses at which they will be used for therapeutic effects. The objective of in vivo studies is to characterize biodistribution and uptake by organs such as the brain, with the help of microdialysis, sampling of cerebrospinal fluid, quantification of brain uptake index, quantitative radiography, and various imaging methods (Kad et al., 2022). A plethora of imaging techniques are available at hand such as laser confocal microscopy (LCM), laser ablation inductively coupled plasma mass spectrometry (LA-ICP-MS), transmission electron microscopy electron energy loss spectroscopy (TEM-EELS), transmission electron microscopy energy dispersive X-ray analysis (TEM-EDX), synchrotron radiation microbeam techniques like synchrotron radiation X-ray fluorescence (SRXRF), and synchrotron-based X-ray absorption spectroscopy (SRXAS), dark field microscopy. Techniques such as solution NMR, surface-enhanced Raman spectroscopy (SERS), attenuated total reflectance-Fourier transform infrared (ATR-FTIR) spectroscopy , synchrotron radiation-based circular dichroism (SR-CD) with high light flux in ultraviolet regions, surface plasmon resonance SPR, quartz crystal microbalance (QCM), and hyper-spectral imaging help in studying absorption and penetration kinetics, composition and conformational changes of adsorbed proteins, the identification of chemical composition of protein corona, intracellular agglomeration, and spatial and temporal biodistribution (Liu et al., 2022; Tirumala et al., 2021; Ramanathan, 2019; Hussain et al., 2015). In addition to these imaging techniques and instrumental methods of analysis, biochemical tests, hematological tests, and histopathological studies are performed. In vivo assays in whole organisms can simulate pathophysiological responses, and cellular and subcellular interactions at multiple levels, provide an idea of the various defense mechanisms, tissue-repairing phenomena, biopersistence, and the accumulation of nano-formulations. They can predict the pharmacokinetics and toxicokinetics of nanomaterials. Long-term efficacy and chronic toxicity studies can be done easily in animals. Although in the preceding discussion, it has been mentioned that genotoxicity can be predicted by in vitro techniques, advancements have been made in the development of genotoxicity assays in a single animal species, most similar to humans. The genotoxicity of titanium dioxide and silica nanoparticles has been investigated in rodents (Forest, 2022). Although different animal species have been tried for in vivo testing of NP for safety and toxicity, the accuracy of results may be compromised owing to inter-species biochemical and genetic differences. The same limitation is applicable in immunotoxicity assessment for anti-cancer nanomedicines in rodent models. The biodistribution of NPs can be studied in humanized mouse models to overcome the limitation of rodent model. Animal models are likely to help investigate T-cell-Dependent Antibody Response (TDAR) (Ray et al., 2021).

However, experimental nanotoxicological methods are not based on standardized protocols and the outcomes of the experiments may vary with the selection of specific cell lines, cell culture medium composition, and the dispersion technique of nano-formulations. Moreover, dose calculation, its basis, and allometric scaling of dose may be inappropriate in several circumstances. The availability of positive control is scarce (Forest, 2022).

One common problem associated with in vitro and in vivo techniques in experimental toxicological studies is the lack of certified reference materials for most nano-formulations and nanomedicines. Certain reference criteria are available in officially recommended guidelines for the size and surface area of titanium dioxide, cellulose nanocrystals, gold, silica, and silver NPs. In the case of commonly used liposomal preparations and NPs fabricated with biodegradable polymers, no such reference material is yet available (Halamoda-Kenzaoui et al., 2019b). Regarding dose administration in animals, it has been frequently observed that doses for in vitro and in vivo data acquisition are different, leading to complications in the interpretation of pharmacokinetics and toxicokinetics. Usually, acute high dose or concentration is used to elucidate mechanisms for inducing toxicity and predict dose-dependent effects during in vitro studies, which is irrelevant for in vivo studies. Moreover, effects and responses attributed to a single high dose will be entirely different from those with chronic multiple or repeated administration of low doses due to accumulation effect and retention in the systemic circulation. Although repeated dose administration can be adopted during in vivo tests, such a strategy escalates the cost of experimentation, and is also time-consuming (Forest, 2022; Mengying et al., 2015).

However, available techniques fail to detect the effects and endpoints of exposing in vitro cell-based models or in vivo animal models to chronic low doses, for which more sensitive, precise, reliable, and innovative methods need to be developed and an in-depth understanding of molecular biology is also essential (Tirumala et al., 2021; Mengying et al., 2015). Despite progress in in vitro and in vivo techniques for nanotoxicity assessment, it has been difficult to establish a definite cause-and-effect relationship due to the above-mentioned factors and lacunae associated with the techniques being currently practiced. The relationship may be established successfully if the problem of interference of NPs with assay reagents can be overcome and appropriate techniques can be developed for real-time in situ monitoring of NP-cell membrane interface interaction and changes in chemical composition and conformation of the adsorbed proteins in the corona (Qiu et al., 2018).

Developments in the design of engineered nanomedicines and nano-formulations with customized properties and multiple functions are increasing the complexities in assessing their toxicity and hazard potential. The preceding discussion reveals the lack of human relevance of in vitro and in vivo data obtained by conventional methods. Little correlation is observed between in vitro and in vivo data. These techniques are not able to give a clear picture of the in vivo fate of NPs. To address these problems, the European Union Reference Laboratory for Alternatives to Animal Testing recommended the use of donor-derived human cell lines, during in vitro tests , instead of artificial immortalized abnormal cell lines. However, the use of donor-derived human cell lines such as peripheral blood leukocytes resulted in inter-individual variations in immune responses during immunotoxicity assessment (Tirumala et al., 2021). Relevant and suitable tests should be developed and strategies adopted for profiling the immune function of the whole organism after the administration of NP by the desired route of administration (Halamoda-Kenzaoui et al., 2019b).

Some recent trends and advancements in nanotoxicology assessment are discussed below.

20.3.3 Utilization of Advanced Analytical Tools

xCELLigence utilizes a low electrolyte impedance interface for real-time monitoring of the various cellular processes. Since it does not use any dye or reagent, the risk of interference by NPs has been minimized. Results obtained with several inorganic NPs from this in vitro, non-invasive technique have been found to match with data obtained from conventional toxicity assays . The use of fluorescent probes provides sensitive and accurate quantification of intracellular ROS levels. However, the method involves irreversible oxidation of the cellular constituents, leading to permanent damage, and thus cannot be adopted during clinical trials. Interference with the conventional COMET assay has been overcome by incorporating NPs in a gel and the discovery of the COMET Chip. Development and improvements in the Conventional FADU (Fluorimetric Detection of Alkaline DNA Unwinding) method have been possible with the use of a robot for the handling of samples and the dispensing of reagents in a light- and temperature-protected controlled environment . The improvised technique utilizes fewer cells (Tirumala et al., 2021). Quartz crystal microbalance-dissipation monitoring (QCM-D) shows promise in assessing the extent of binding and deposition of NP onto the cell surface and subsequent losses in membrane integrity. The electrochemical technique is a high-throughput technique that measures impedance behavior and enables real-time and in situ monitoring of phenomena occurring at the NP–membrane interface in cells, tissues, organs, fetuses, and whole body of an organism (Qiu et al., 2018). The chemical reactivity of surface-coated NPs and biophysical interactions can be investigated and monitored by electrochemical impedance spectroscopy (ECIS) (Shinde et al., 2020). High-throughput screening (HTS) techniques facilitate the rapid detection of analytes in small volumes of samples in a single shot, saving time, money, manpower, and resources. Standard analytical methods can be transformed into HTS methods by automation and increasing the sensitivity and efficiency of existing methods, such as high-throughput flow cytometry, high-throughput micronucleus assay, high-throughput COMET assay, and H2AX assay. Large number of samples can be analyzed through the integration of HTS techniques with high-content analysis or high-content imaging. High-content image-based screening can be used for the training of predictive models and to enhance the accuracy of toxicity predictions (Singh & Gauri, 2023; Forest, 2022).

20.3.4 In Silico Approach: Nano-QSAR

Prediction of Quantitative Structure-Activity Relationship of NPs (or nano-QSAR) is a highly promising technique that saves time, money, and resources in comparison to the above-mentioned conventional experimental nanotoxicity assessment techniques. In this approach, suitable, well-conceptualized molecular descriptors, data-driven artificial intelligence, and machine learning-based tools are employed to establish the cause-effect relationship between in vivo pharmacokinetics and toxicity data with NP features, without performing any in vitro or in vivo experiments (Singh et al., 2023; Tirumala et al., 2021; Pikula et al., 2020). The concept of nano-QSAR is built on the similarity between biological responses from nanomaterials and identical molecular structures. Nano-QSAR has been successfully employed to explore the toxicity potential of several metallic oxide NPs towards *E. coli*. The interpretation of outcomes of the models revealed NP composition to contribute significantly to NP-induced toxicity with insignificant roles of size or shape, as discussed in the previous sections. The results from studies on bacteria were validated with experiments on mammalian cells, the HaCaT cell lines. Similar nano-QSAR models have also been utilized to explore and predict the uptake of metallic oxide NPs by pancreatic cancer cells (PaCa2). Nano-QSAR can generate 1D, 2D, and 3D models. Limitation with highly efficient predictive tools is the need for extensive reliable, and high-quality databases and datasets for the training and validation of models. Missing data produce conflicting results. Although sufficient data are available with nanomedicines, most of them are heterogeneous which are unsuitable for model development. Another disadvantage of the in silico approaches is that the data cannot be used for establishing in vitro–in vivo correlation. Molecular descriptors used during model development are based on pristine NPs rather than on the NPs in biological systems, where substantial alterations in their surface properties and characteristics occur (Verma et al., 2023; Forest, 2022). Data mining, which involves assembling data from different sources and various studies involving different media, cells, and chemicals, can generate a huge database that can be processed appropriately to extract useful information (Singh et al., 2023). Improvements in prediction capacity and, hence, fabrication and characterization strategies have been possible by the de-convolution of variables with a known tendency of interaction and interference, thereby facilitating the unraveling of the different types of interaction (Qiu et al., 2018). Another advancement employs periodic table-based molecular descriptors in nano-QSAR, which can bridge the current gap in nanotoxicity studies (Roy & Roy, 2023).

20.3.5 Grouping/Read-Across Technique

In this approach, the prediction of adverse effects is carried out by the acquisition of data from substances identical to test substances (in this case, nanomaterials) under investigation with respect to physicochemical and structural properties. For successful implementation, chemometric tools are employed, namely hierarchical clustering (HC), principal component analysis (PCA), and random forest variable selection (Forest, 2022).

20.3.6 Genetic Approaches

Since fluorescent probes(dyes) employed for ROS quantification induce permanent oxidation, they may be replaced by designing genetically encoded reporters or redox-sensitive fluorescent protein with the ability to target typical cellular compartments. This approach is likely to be a useful tool in the future in determining the genetic susceptibility of vulnerable individuals to certain phenotypes of nanotoxicity and will also form the foundation for personalized medicines. However, the strategy involves technical expertise and is not always economically possible at present. Conventional genotoxic assessment methods have been replaced by GreenScreen HC, BlueScreen HC, and ToxTracker reporter assays (Akcan et al., 2020; Tirumala et al., 2021).

20.3.7 Utilization of Validated Human Cell Lines in Immunotoxicity Assays

To overcome the issues of inter-individual variation in immune responses from donors, recently developed immunotoxicity assays utilize validated human cell lines such as human Jurkat T-cell, human lymphoid T-cell (MOLT-4) or B-cell (IM-9), human acute myeloid leukemia HL-60 cells, human-based skin explant, and so on. For accurate prediction of delayed hypersensitivity reactions from engineered nanomaterials, tests like human cell line activation test (hCLAT) and myeloid U937 skin (sensitization) test (MUSST) have been developed (Tirumala et al., 2021).

20.3.8 In Vitro Carcinogenicity Assessment with Transformed Cells

In this approach, the European Union Reference Laboratory for Alternatives to Animal Testing and other research groups have performed in vitro carcinogenicity assessment with transformed cell-based models such as in vitro Syrian hamster embryo cell transformation and the Balb/c3T3 A31-1-1 mouse model. The endpoints for toxicity identification in this technique have been recognized as changes in morphology, colony formation, and the growth and formation of criss-crossed cells or the piling of cell foci (Tirumala et al., 2021).

20.3.9 DNA Barcoding

DNA barcoding is an innovative technique employed for the simultaneous screening of whole organisms for the biodistribution of innumerable NPs (size > 100nm) to target sites and off-target accumulation, leading to organ toxicity. The data-intensive tool needs help from bioinformatics methods such as the unbiased Euclidean clustering strategy. It has been established from several studies that NPs interact with some specific categories of immune cells (Kantak et al., 2023). The technique possesses a high potential for assessing immune- and genotoxicity accurately in in vivo models (Lokugamage et al., 2018).

20.3.10 Systems Toxicology: 'Omics' Technology

Systems toxicology is a discipline that attempts to amalgamate conventional toxicity assessment techniques for the better and realistic interpretation of data and to ensure the availability of safe nanomedicinal formulation. 'omics' lies at the heart of systemic toxicology. The 'omics' technologies are versatile tools that enable the quantitative and qualitative description of complex, biological, adaptive changes, and behaviors. They can assess cellular stress at multiple levels and cross-talk or interactions among different levels in response to low doses of nanomedicines, applicable in clinical settings. The methods reflect and correlate genotypic and phenotypic experimental observations. They provide molecular information for in-depth analysis of underlying mechanisms and cellular pathways leading to toxic effects. 'omics' provides a functional readout of cellular state from all aspects. Nanoparticle-induced interference observed in conventional cytotoxicity assays is absent when 'omics' technologies are employed (Abdelkader et al., 2023). However, highly advanced infrastructure and expertise are essential prerequisites to the adoption of 'omics' in nanotoxicity assessment (Forest, 2022; Fröhlich, 2017).

Studies of 'omics' encompass proteomics, lipidomics, transcriptomics, genomics, metabolomics, and toxicogenomics. Of these, metabolomics is unique since the metabolome of individuals is generic, whereas protein, gene, or transcript varies with individuals. The integration of different 'omics' techniques across the multiple 'omics' layers, and the combination of integrated 'omics' with in vitro tests and advanced instrumentation techniques, such as mass spectrometry and artificial intelligence- and machine learning-enabled statistical tools, enhance sensitivity and facilitate the detection and profiling of several endogenous small molecules. The application of high-resolution magic angle spinning (HR-MAS) NMR-based metabolomics study revealed that silver nanoparticle-induced oxidative stress led to the metabolic conversion of lactate and taurine to pyruvate (Mengying et al., 2015). The intersection of metabolomics, mechanobiological tools, bioinformatics, and machine learning algorithms revealed ROS production, alteration in glucose metabolism, and lower ATP synthesis in HEK293 cells following treatment with a low dose of magnetic silica nanoparticles containing Rhodamine B isothiocyanate ([MNPs@SiO_2(RITC)) (Shin et al., 2021; Shim et al., 2012).

20.3.11 Nano-Informatics Database

European Union–United States Roadmap Nanoinformatics 2030 enlisted several databases related to nanomaterials of both the European Union and the United States such as, eNano-Mapper, nanoHub, DaNa, the Online Chemical Modeling Environment (OCHEM), the NanoExposure and Contextual Information Database (NECID), the NanoDatabank, the Nanomaterials–Biological Interactions Knowledgebase, Nanominer, NanoMILE, and ModNanoTox. These databases have organized, stored, shared, analyzed, and ensured the application of data obtained from various sources and laboratories on nanomaterials (Pikula et al., 2020).

20.3.12 Miscellaneous Advanced Approaches in Nanotoxicology Assessment

Approaches such as testing species sensitivity distribution can be employed to predict the maximum concentration of NPs in the surrounding environment and enable band gap analysis to investigate the effects of chronic exposure to metallic nanoparticles . Computational models have been designed to fulfill regulatory needs where potential risks from nano-formulations can be predicted before preclinical and clinical studies in animals and humans. In compliance with this specific requirement, projects that came into the forefront are NanoTEST, NANoREG, and NANoREG2. The outcomes of these projects will provide a new direction to the taxonomy of NPs based on toxicity (Kad et al., 2022). Several of the tools, techniques, and approaches employed in experimental toxicology, advanced instrumentation techniques, and systems toxicological approaches have been combined and incorporated into the framework of Adverse Outcome Pathways (AOPs). Some AOPs are relevant in the study and assessment of nanotoxicology, e.g., AOP 173, AOP 303, AOP 237, and AOP 302. These pathways helped in describing and elucidating hallmark features and endpoints of the disease and nano-formulations under investigation (Forest, 2022). To minimize toxicity, a new paradigm in the synthesis and fabrication of NPs is the eco-friendly transition to green nanotechnology or the green synthesis of metallic NPs with the help of microbes and secondary metabolites present in vegetable extracts (Anand et al., 2022; Hu et al., 2022; Martinez et al., 2021; Osman, 2019). Figure 20.2 highlights various techniques and tools used in the safety and risk assessment of nanomedicines.

20.4 THE NANOTOXICOLOGICAL CLASSIFICATION SYSTEM: A NEW HORIZON

A surge in the development of nano-formulations for herbal medicines has necessitated the systematic classification and categorization of nanomaterials based on their adverse effects, risks, hazards, and toxicity. The DF4nanoGrouping has devised a functionality-driven scheme for the categorization of NPs, starting from simple to complex events attributed to intrinsic characteristics, biological system-dependent properties, and toxicological manifestations. In this system, there are four classes of nanomaterials, namely, soluble, biopersistent –high-aspect ratio, passive with no biological effects and active formulations exhibiting adverse effects related to their surface

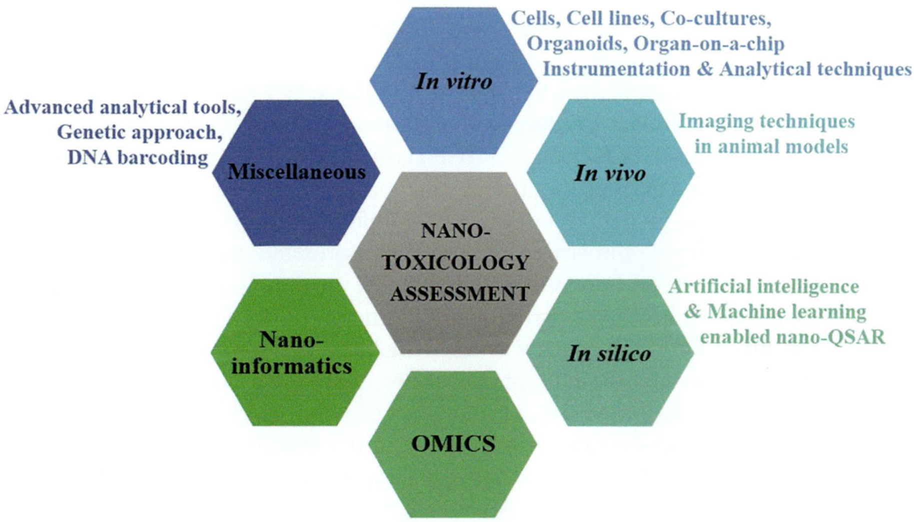

FIGURE 20.2 Techniques and tools for safety and risk assessment of nanomedicines.

properties. For this classification, both in vitro and in vivo data have been utilized. In response to the need to reduce the number of descriptors, DF4nanoGrouping has identified structural attributes of nanomaterials and calculated property with quantum mechanical significance. It has been previously reported by the National Nanotechnology Initiative that quantum effects are governed by size. However, DF4nanoGrouping is not ideal for nano-formulations for drug delivery (Gajewicz et al., 2018; Siegrist et al., 2019).

In the Nanotoxicological Classification System (NCS), nanomaterials have been classified into four types based on their size and biodegradability, governed by route of administration and biocompatibility of the surface. Class I and Class II are nanomaterials with size greater than 100 nm and are either biodegradable or non-bio-degradable, respectively. These NPs are not endocytosed. Class III and Class IV are meant for biodegradable and non-biodegradable NP with size less than 100 nm and hence are endocytosed and are more bioperistent, identical to Class II. These four classes can further be subclassified, leading to eight classes in which physicochemical surface properties are also taken into consideration. Other classification systems have integrated concepts of toxicology, risk assessment modeling, and the multi-criteria decision approach (Laloy, 2021; Keck & Müller, 2013).

A different system of classification of NPs can also be adopted based on the occupational hazards associated with the production of these nanomaterials. In this system, the NPs are categorized into three classes; namely, Class 1 for water-soluble NPs e.g. sucrose or siloxane NPs; Class 2 for synthetic, non-fibrous and persistent NPs as silver, gold, metal, and metallic oxide NPs; and, finally, Class 3 for fibrous, non-soluble carbon nanotubes (Osman, 2019).

Based on hazard assessment as per the Hazard Evaluation Strategy (HES), there are four categories of NPs: Category I, with low cellular uptake and low intracellular persistence (PEGylated superparamagnetic iron oxide nanoparticles or SPIONs); Category II, with high cellular uptake and low intracellular persistence (uncoated SPIONs); Category III, with low cellular uptake and high intracellular persistence (SPIONs coated with a layer of polyglucose sorbitol carboxymethylether); and Category IV, with high cellular uptake and high intracellular persistence (in situ-coated lauric acid ferrofluid) (Siegrist et al., 2019).

20.5 INTERNATIONAL GUIDELINES ON THE SAFETY AND TOXICITY ASSESSMENT OF HERBAL NANOMEDICINES

The domain of herbal nanomedicines is gradually expanding with more scope for development in the near future. Quality control requirements for the establishment of monographs for herbal raw materials, intermediates, and finished products and the subsequent validation of protocols and techniques can be found in different versions of guidelines from the European Medicines Agency, the Pharmaceutical Inspection Co-operation Scheme, and in the WHO guidelines (EMA 2022; WHO guidelines 2011). The International Conference on Harmonization of Technical Requirements for Registration of Pharmaceuticals for Human Use (ICH) issued guidelines for preclinical safety assessment, before conducting clinical trials for all medicinal products also including nano-formulations. However, lacunae exist in the guidelines as they were initially developed for conventional formulations, bulk ingredients of macro-dimensions, and not for formulations of nano-dimensions. Data obtained from such studies failed to represent the clinical situation in actual practice. In 2011, the Food and Drug Administration (FDA) first realized the need for developing guidelines for nanomedicines for regulatory control. The need was felt to make public aware about differences in properties between nanoparticles and the bulk chemical constituents of nanoparticles (Foulkes et al., 2020). Later, in 2017, FDA proposed the consideration of nano-sized medicinal formulation on a product-specific basis or a case-by-case basis, because of the unique features of each formulation, which was attributed to nano-dimensions. The regulatory authority has advised manufacturers to enter into consultation with the FDA in the development stage of nano-formulations, to avoid risk of rejection on the grounds of being toxic or being unable to provide sufficient data and documents in support of different phenotypes of nanotoxicity. Similarly, Medicines and Healthcare Products

Regulatory Agency (MHRA), the regulatory body for medicinal products in the UK also has no single guideline, designed specifically for nanomedicines, and proposes the manufacturers to consult with the MHRA Innovation Office. They also support case-by-case consideration of products with respect to toxicity assessment. The EU and Canada adopt a similar approach; indeed, international regulatory authorities are constantly refining the guidelines with new inputs from highly advanced techniques and novel strategies employed in nanotoxicological assessment to produce reliable experimental data and to minimize inter-batch variation in quality and toxicity profiles. The HES is the first-of-its-kind three-tier strategy, especially aimed at evaluating the toxicity of injectable, engineered nanomedicines, based on exclusion criteria. The HES enables application-oriented decision-making in an effective manner. HES is unique from several other toxicity control approaches in that it has focussed on monitoring microbial contamination and contamination with endotoxins in the case of nanotherapeutics. It has stressed that any attempt to minimize bioburden in the finished product should not affect or alter NP characteristics, essential for their clinical applications (Siegrist et al., 2019). The European Nanomedicine Characterization Laboratory (EUNCL) and the Regulatory Science Framework for Nano(bio)material-based Medical Products and Devices (REFINE) are two projects funded by the Horizon 2020 Research and Innovation programme. These projects are supported by the Decision Support System (Liu et al., 2022; Foulkes et al., 2020; Halamoda-Kenzaoui et al., 2019b). In one development, ISO/TC 229 (ISO/TC 229 N 673) developed proposals specifically for nanoparticles (Halamoda-Kenzaoui et al., 2019a). As mentioned previously, nano-formulations differ from conventional formulations with respect to excipients. International regulatory bodies should adopt consensus on the contribution of the excipient to the production of nanoparticles as a whole and the regulatory aspects that should be devised to fit the purpose and objective (Hemmrich & McNeil, 2023). During the assessment of nanotoxicity, little attention is paid to studying the eco-toxicological effects of the burgeoning development, use and the disposal of nanoparticles (Halamoda-Kenzaoui et al., 2019b). Moreover, the heterogeneity of nano-formulations, along with opportunities for administration by various routes for the amelioration of symptoms of different tissues, organs, and systems, create challenges in developing universal guidelines for safety and toxicological assessment for all nanomedicines under a single umbrella (Ramanathan, 2019). It has become obvious to show that the benefits of nano-drug delivery systems outweigh the risks. However, EMA has mandated the submission of data regarding environmental risk assessment (ERA) of drug products for human use, as an essential component of marketing authorization. Conventional methods for the estimation of the partition coefficient of materials in the octanol-water system will provide inaccurate data with nanomaterials as the method is based on the principle of equilibration. In the case of nanomaterials, interactions at the nanoparticle–cell membrane interface are highly dynamic manifesting non-equilibrium behavior. The Organization for Economic Co-operation and Development (OECD) test guidelines from 2006 are being revised to incorporate suitable guidelines for nanomaterials so that reliable data can be obtained to ensure safety and efficacy (OECD, 2017). For this, 91 reports on the safety of manufactured nanomaterials have been submitted up to 2019, revealing the increased awareness and scope of nanotoxicological aspects. In 2019, the OECD drafted two guidelines on the physical-chemical decision framework for informed decisions and principles for measurements and the reporting of physicochemical parameters for risk assessment of manufactured nanomaterials to assure relevance and reliability (Pikula et al., 2020). Under the aegis of the Government of India's Nano Mission, a Nanoregulatory Task Force has been constituted, which has guided the Center for Knowledge Management of Nanoscience and Technology to address the issues of occupational hazards associated with inadvertent exposure during the manufacture and quality control of nanomaterials in general, among the industry personnel. Extensive research has been undertaken to collect pieces of evidence of significant adverse effects of nanomaterials on human health. The draft guidelines aim to recommend strategies for the safe production, handling, use, and disposal of nanomaterials and implement them appropriately for societal benefits in a wider context (DST Guidelines, n.d.). In India, an important guideline for the evaluation of nanopharmaceuticals has been brought out by collaborative efforts from the Department of Biotechnology, the Ministry of Science and Technology, the Indian Council of Medical Research and Central Drugs Standard Control Organization, the Ministry of Health and Family Welfare (Gupta et al., 2019). Standardized genotoxicity assessment protocols have not yet been developed (Tirumala et al., 2021). The issuance of formal regulatory guidelines regarding safety and toxicity profiling of nanomedicines is essential to prevent disinvestment in this highly impactful technology (Mahapatra et al., 2018).

As discussed previously, personnel involved in manufacturing of nano-formulations may be inadvertently exposed to the same which may prove to be hazardous. Several authorities and organizations, such as the National Institute for Occupational Safety and Health (NIOSH), the Industrial Technology Development Organization (NEDO), and the American Conference of Governmental Industrial Hygienists (AGGIH), have recommended occupational exposure limits to nanomaterials (Tirumala et al., 2021; Osman, 2019).

The preceding discussion reveals that considerable efforts have been made internationally to formulate guidelines related to nanomaterials, and nanomedicines in general, but that very few have been formulated with herbal nanomedicines in mind. It is to be noted that as yet no uniform internationally and universally accepted and harmonized guideline exists for nanomedicines. The unifying concept of most of the guidelines is, however, to adopt a case-by-case approach for nanomedicines, taking into consideration the significant impact of nanomaterials on human health, industry personnel, non-target organisms in the surrounding environment, and, finally, on the abiotic components of the ecosystem. Moreover, none of these guidelines should be implemented for natural and biological nanomaterials or with conventional drug delivery systems with the insignificant and incidental presence of nanomaterials.

These guidelines will serve as useful documents for researchers, manufacturers, and stakeholders to engage in innovative and novel products, pave the way for the optimization of current research in the domain of development of nanopharmaceuticals in compliance with regulatory requirements, the translation of nano-formulations from bench to bedside with minimum risk of toxicity to man and environment and will promote ground-breaking advances for significant societal and economic impact.

One important aspect that needs to be strictly monitored by the regulatory bodies is the possible availability of nano-size pharmaceutical finished products as future over-the-counter (OTC) products. Herbal products are available as OTC products in the unregulated market of several of the world's developing nations. Yet the same marketing strategy may prove to be harmful to end-users. Nano-formulations are designed to provide controlled release over prolonged duration and if they cannot be validated to be safe for prolonged use, they may precipitate serious toxic consequences. Therefore, regulatory authorities should devise an appropriate framework to prevent the entry of nano-formulations as OTC products into the market (Foulkes et al., 2020) Further consideration of the durability of the complex formed between nanocarrier and the encapsulated payload in the environment is necessary as the drug, once released, may have undesirable effects on uptake by soil bacteria or other animals. The OECD is revising guidelines for the assessment of the hazard potential of NPs in the aquatic environment (Mahapatra et al., 2018). One area where work needs to be done is the framing of guidelines for the estimation of immunotoxic effects, especially for nano-formulations over and above ICH S8 guidelines (Halamoda-Kenzaoui et al., 2019b).

Recently, EU has adopted a "Green Deal", in which the ultimate objective is to adopt safe and sustainable practices in the design of nanomaterials with the ultimate goal of protecting the environment for future generations, giving birth to the concept of "Safe-and-sustainable-by-design" (SSbD), which is identical to the concept of Quality-by-Design adopted for the manufacture of any medicinal product and an improvement over earlier Safe-by-Design (SbD) frameworks. The GoNanoBioMat SbD approach is the SbD concept, applicable for polymeric nanomaterials for pharmaceutical applications, ensuring the use of safe materials, products, production techniques, safe use (especially with respect to suspensions, sunscreens, cosmetics, etc.), and the disposal of leftover products after use. Implementation of this production strategy from the onset ensures the affordability and availability of the product with minimal risk to humans, personnel in the manufacturing industry, non-target organisms, environment, in all stages of the lifecycle of the product and not only the final product (Bhat et al., 2023; Furxhi et al., 2023; Soeteman-Hernandez et al., 2019). ICH and OECD should incorporate the principles of SSbD in their revised guidelines for the complete nanotoxicological profiling of nanomedicines (Schmutz et al., 2020).

20.6 CRITICAL CHALLENGES IN THE DEVELOPMENT OF HERBAL NANOMEDICINES

No gold standard exists for the calculation and estimation of the dose, time, and duration of treatment with herbal nanomedicines. There is a lack of literature and scientific reports on the dosimetry of nanomedicines in general. Dosimetry should be clearly understood as biological response and toxic effects are directly related to dose. A consensus should be adopted in defining and using various terminologies associated with nanomedicine dosimetry. For conventional drug delivery platforms, the dose is usually expressed as amount or as concentration. However, with nano-formulations, this system may prove to be inefficient and incorrect. Since previous sections have repeatedly focussed on the role of size and specific surface area of nanoparticles with regard to safety and efficacy, the dose calculation should preferably be based on mass, surface area, or number of particles to avoid inconsistency in dose and interpretation of outcomes and effects (Hussain et al., 2015).

20.7 CONCLUSION

Continuous involvement and endeavors of the scientific fraternity in the area of nanotechnology have facilitated rapid and tremendous progress in enriching the knowledge base, resulted in innovative approaches in delivering herbal medicines, extracts, and phytoconstituents, enabled great strides in the development of characterization techniques, and have driven the regulatory authorities to formulate effective guidelines in controlling and monitoring the safety and toxicity of nanopharmaceuticals. The collaborative efforts and expertise of academia, industry, government agencies, and other stakeholders have contributed significantly to these advancements. Although, several in vitro, in vivo, and highly sophisticated instrumental techniques, in silico and computational predictive tools, and nano-informatics databases are available , no single assay or technique can provide a comprehensive quality-efficacy-safety profile of herbal nanomedicines and can establish a causal relationship between physicochemical attributes and pharmacokinetic behavior of nanoparticles. Intrinsic and extrinsic toxicity characteristics of herbal medicines, and existence of the "nano-paradox" are the barriers to translating the effective products from laboratory to the bedside of the patient. In an effort to overcome these formidable hurdles in the successful commercialization of nanomedicines, the regulatory authorities should work round the clock to bridge the gap between existing guidelines and future needs and formulate and implement guidelines, specifically for herbal nanomedicines, taking care not to overburden the manufacturers and investors with statutory prejudices and provision. Special attention should be paid to standardizing dosimetry protocols to avoid over- or under-estimating toxicity and efficacy. The dissemination of reports in the public domain should be ensured. The ultimate goal of the pharmaceutical industry remains amelioration of pathological conditions to enable mankind to lead a quality life, with minimal impact on Mother Nature. Safe and sustainable practices should be adopted proactively from concept initiation through all stages of the product lifecycle.

ACKNOWLEDGEMENT

The authors express their gratitude to NSHM Knowledge Campus, Kolkata – Group of Institutions, Kolkata for offering the necessary facilities to conduct the data search for this chapter.

REFERENCES

Abbasi, R., Shineh, G., Mobaraki, M., Doughty, S., & Tayebi, L. (2023). Structural parameters of nanoparticles affecting their toxicity for biomedical applications: A review. *J. Nanopart. Res.*, 25, Article 43. https://doi.org/10.1007/s11051-023-05690-w

Abdelkader, Y., Perez-Davalos, L., LeDuc, R., Zahedi, R. P., Labouta, H. I. (2023). Omics approaches for the assessment of biological responses to nanoparticles. *Adv. Drug Deliv. Rev.*, 200, Article 114992. https://doi.org/10.1016/j.addr.2023.114992

Abo-Zeid, Y., Essam, R., Sanad, R. A., & Sakran, W. (2021). Recent advances in Herbal-based Nanomedicine for Anti-inflammatory purposes. *J. Adv. Pharm. Res.*, 5 (4), 387–397. https://doi.org/10.21608/aprh.2021.79946.1135

Afzali, E., Eslaminejad, T., Yazdi Rouholamini, S. E., Shahrokhi-Farjah, M., & Ansari, M. (2021). Cytotoxicity effects of curcumin loaded on chitosan alginate nanospheres on the KMBC-10 spheroids cell line. *Int. J. Nanomedicine*, 16, 579–589. https://doi.org/10.2147/ijn.s251056

Ahmad, A., Imran, M., & Sharma, N. (2022). Precision nanotoxicology in drug development: Current trends and challenges in safety and toxicity implications of customized multifunctional nanocarriers for drug-delivery applications. *Pharmaceutics*, 14, Article 2463. https://doi.org/10.3390/pharmaceutics14112463

Ahmadi, M., Pourmadadi, M., Ghorbanian, S. A., Yazdian, F., & Rashedi, H. (2021). Ultra pH-sensitive nanocarrier based on Fe2O3/chitosan/ montmorillonite for quercetin delivery. *Int. J. Biol. Macromol.*, 191, 738–745. https://doi.org/10.1016/j.ijbiomac.2021.09.023

Ai, S., Li, Y., Zheng, H., Zhang, M., Tao, J., Liu, W., Peng, L., Wang, Z., & Wang, Y. (2024). Collision of herbal medicine and nanotechnology: a bibliometric analysis of herbal nanoparticles from 2004 to 2023. *J. Nanobiotechnol.*, 22(1), Article 140. https://doi.org/10.1186/s12951-024-02426-3

Akçan, R., Aydogan, H. C., Yildirim, M. Ş., Taştekin, B., & Sağlam, N. (2020). Nanotoxicity: A challenge for future medicine. *Turk. J. Med. Sci.*, 50(4), 1180–1196. https://doi.org/10.3906/sag-1912-209

Alemi, A., Karamallah, M. H., Sabaghan, M., Hosseini, S.A., Veisi, A., Karamallah, S.H., Farokhifar, M. (2024). Combination drug therapy by herbal nanomedicine prevent multidrug resistance protein 1: promote apoptosis in Lung Carcinoma. *J. Appl. Biomater. Funct. Mater.*, 22, 22808000241235442.https://doi.org/10.1177/22808000241235442

Alti, D., Veeramohan Rao, M., Rao, D. N., Maurya, R., & Kalangi, S. K. (2020). Gold-silver bimetallic nanoparticles reduced with herbal leaf extracts induce ROS-mediated death in both Promastigote and Amastigote stages of *Leishmania donovani*. *ACS Omega*, 5(26), 16238–16245. https://doi.org/10.1021/acsomega.0c02032

Anand, A., Gautam, P., & Ojha, S. (2024). Application of Nanotechnology for herbal medicine development: A review. *Lett. Drug Des. Discov.*, 21(8), 1325–1333. https://doi.org/10.2174/157018082066623030810 5723

Anand, U., Carpena, M., Kowalska-Góralska, M., Garcia-Perez, P., Sunita, K., Bontempi, E., Dey, A., Prieto, M. A., Proćków, J., & Simal-Gandara, J. (2022). Safer plant-based nanoparticles for combating antibiotic resistance in bacteria: A comprehensive review on its potential applications, recent advances, and future perspective. *Sci. Total Environ.*, 821, Article 153472. https://doi.org/10.1016/J.SCITOTENV.2022.153472

Ansari, S.H., Islam, F., & Sameem, M. (2012). Influence of nanotechnology on herbal drugs: A Review. *J. Adv. Pharm. Technol. Res.*, 3(3), 142–146. https://doi.org/10.4103/2231-4040.101006

Anwar, D. M., El-Sayed, M., Reda, A., Fang, J. Y., Khattab, S. N., & Elzoghby, A. O. (2021). Recent advances in herbal combination nanomedicine for cancer: delivery technology and therapeutic outcomes. *Expert Opin. Drug Deliv.*, 18(11), 1609–1625. https://doi.org/10.1080/17425247.2021.1955853

Arifin, S. F., AlShami, A. K., Omar, S. S. S., Jalil, M. A. A., Khalid, K. A., & Hadi, H. (2019). Impact of modern technology on the development of natural-based products. *J. Ayu. Herb. Med.*, 5(4), 133–142.

AzoNano (https://www.azonano.com/article.aspx?ArticleID=5837)

Bai, X., Wang, J., Mu, Q., & Su, G. (2021). In vivo protein corona formation: Characterizations, effects on engineered nanoparticles' biobehaviors, and applications. *Front. Bioeng. Biotechnol.*, 9, Article 646708.

Barros, C. H. N., Hiebner, D. W., Fulaz, S., Vitale, S., Quinn, L., & Casey, E. (2021). Synthesis and self-assembly of curcumin-modified amphiphilic polymeric micelles with antibacterial activity. *J. Nanobiotechnol.*, 19(1), Article 104. https://doi.org.10.1186/s12951-021-00851-2

Bhat, M. A., Gedik, K., & Gaga, E. O. (2023). Environmental impacts of nanoparticles: Pros, cons, and future prospects. In M. Ozturk, A. Roy, R. A. Bhatt, F. Vardar-Sukanf, F. M. P. Tonelli (Eds.), *Synthesis of Bionanomaterials for Biomedical Applications: Micro and nano technologies* (pp. 493–528). Elsevier. https://doi.org/10.1016/B978-0-323-91195-5.00002-7

Bonifácio, B.V., Silva, P.B., Ramos, M.A., Negri, K.M., Bauab, T.M., & Chorilli, M. (2014). Nanotechnology-based drug delivery systems and herbal medicines: A review. *Int. J. Nanomedicine*, 9, 1–15. https://doi.org/10.2147/IJN.S52634

Cai, R., & Chen, C. Y. (2018). The crown and the scepter: Roles of the protein corona in Nanomedicine. *Adv. Mater.*, 31, Article 1805740.

Catalano, E. Biophysical interaction, nanotoxicology evaluation, and biocompatibility and biosafety of metal nanoparticles. In *Proceedings of Materials Science, Medicine, Environmental Science, Biology*, 2021. https://doi.org/10.48550/arXiv.2108.05964

Chiu, C. F., Fu, R. H., Hsu, S. H., Yu, Y. A., Yang, S. F., Tsao, T. C., Chang, K. B., Yeh, C.A., Tang, C.M., Huang, S.C., & Hung, H.S. (2021). Delivery capacity and anticancer ability of the berberine-loaded gold nanoparticles to promote the apoptosis effect in breast cancer. *Cancers (Basel)*, 13(21), Article 5317. https://doi.org/10.3390/cancers13215317

de Oliveira Pacheco, C., de Gomes, M. G., da Silva Neto, M. R., Parisotto, A. J. M., Dos Santos, R. B., Maciel, T. R., Ribeiro, A. C. F., Giacomeli, R., & Haas, S.E. (2022). Surface-functionalized curcumin-loaded polymeric nanocapsules could block apomorphine-induced behavioral changes in rats. *Pharmacol. Rep.*,74(1), 135–147. https://doi.org/10.1007/s43440-021-00331-2

Dewi, M. K., Chaerunisaa, A. Y., Muhaimin, M., & Joni, I. M. (2022). Improved activity of herbal medicines through nanotechnology. *Nano*, 12, Article 4073. https://doi.org/10.3390/nano12224073

Ding, L., Yao, C., Yin, X., Li, C., Huang, Y., Wu, M., Wang, B., Guo, X., Wang, Y., & Wu, M. (2018). Size, shape, and protein corona determine cellular uptake and removal mechanisms of gold nanoparticles. *Small*, 14, e1801451.

Dos Santos, R. B., Nakama, K. A., Pacheco, C. O., de Gomes, M. G., de Souza, J. F., de Souza Pinto, A. C., de Oliveira, F. A., da Fonseca, A. L., Varotti, F., Fajardo, A. R., & Haas, S. E. (2021). Curcumin-loaded nanocapsules: Influence of surface characteristics on technological parameters and potential anti-malarial activity. *Mater. Sci. Eng. C Mater. Biol. Appl.*, 118, Article 111356. https://doi.org/10.1016/j.msec.2020.111356

DST Guidelines: Guidelines and best practices for safe handling of nanomaterials in research laboratories and industries compiled for nano mission. *DST, Govt of India BY Centre for Knowledge Management of Nanoscience & Technology*. (https://dst.gov.in/)

Duse, L., Agel, M. R., Pinnapireddy, S. R., Schäfer, J., Selo, M. A., Ehrhardt, C., & Bakowsky, U. (2019). Photodynamic therapy of ovarian carcinoma cells with curcumin-loaded biodegradable polymeric nanoparticles. *Pharmaceutics*, 11(6), Article 282. https://doi.org/10.3390/pharmaceutics11060282

Ebaid, H., Habila, M., Hassan, I., Al-Tamimi, J., Omar, M. S., Rady, A., Alhazza, I. M. (2021). Curcumin-containing silver nanoparticles prevent carbon tetrachloride-induced hepatotoxicity in mice. *Comb. Chem. High Throughput Screen.*, 24(10), 1609–1617. https://doi.org/10.2174/1386207323666201211100830

El-Far, A. H., Al Jaouni, S. K., Li, W., & Mousa, S. A. (2018). Protective roles of Thymoquinone nanoformulations: Potential nanonutraceuticals in human diseases. *Nutrients*, 10, Article 1369. https://doi.org/10.3390/nu10101369

EMA Guidelines. European Medicines Agency (EMA). (2022). Guideline on Quality of Herbal Medicinal Products/Traditional Herbal Medicinal Products (EMA/HMPC/CHMP/CVMP/201116/20051 Rev. 3).

Forest, V. (2022). Experimental and computational Nanotoxicology: Complementary approaches for Nanomaterial hazard assessment. *Nano*, 12, Article 1346. https://doi.org/10.3390/nano12081346

Foulkes, R., Man, E., Thind, J., Yeung, S., Joy, A., & Hoskins, C., (2020). The regulation of nanomaterials and nanomedicines for clinical application: current and future perspectives. *Biomater. Sci.*, 8, 4653–4664. https://doi.org/10.1039/D0BM00558D

Fröhlich, E. (2017). Role of omics techniques in the toxicity testing of nanoparticles. *J. Nanobiotechnol.*, 15(1), Article 84. https://doi.org/10.1186/s12951-017-0320-3.

Furxhi, I., Bengalli, R., Motta, G., Mantecca, P., Kose, O., Carriere, M., Haq, E. U., O'Mahony, C., Blosi, M., Gardini, D., & Costa, A. (2023). Data-driven quantitative intrinsic hazard criteria for Nanoproduct development in a safe-by-design paradigm: A case study of Silver Nanoforms. *ACS Appl. Nano Mater.*, 6(5), 3948–3962. https://doi.org/10.1021/acsanm.3c00173

Gajewicz, A., Puzyn, T., Odziomek, K., Urbaszek, P., Haase, A., Riebeling, C., Luch, A., Irfan, M. A., Landsiedel, R., van der Zande, M., & Bouwmeester, H. (2018). Decision tree models to classify nanomaterials according to the DF4nanoGrouping scheme. *Nanotoxicology*, 12, 1–17. https://doi.org/10.1080/17435390.2017.1415388

Gandamalla, D., Lingabathula, H., & Yellu, N. (2019). Nano titanium exposure induces dose-and size-dependent cytotoxicity on human epithelial lung and colon cells. *Drug Chem. Toxicol.*, 42(1), 24–34.

Gea, M., Bonetta, S., Iannarelli, L., Giovannozzi, A. M., Maurino, V., Bonetta, S., Hodoroaba, V. D., Armato, C., Rossi, A. M., Schilirò, T. (2019). Shape-engineered titanium dioxide nanoparticles (TiO(2)-NPs): Cytotoxicity and genotoxicity in bronchial epithelial cells. *Food Chem. Toxicol.*, 127, 89–100.

Ghafelehbashi, R., Tavakkoli Yaraki, M., Heidarpoor Saremi, L., Lajevardi, A., Haratian, M., Astinchap, B., Rashidi, A. M., & Moradian, R. (2020). A pH-responsive citric-acid/α-cyclodextrin-functionalized Fe3O4 nanoparticles as a nanocarrier for quercetin: An experimental and DFT study. *Mater. Sci. Eng. C Mater. Biol. Appl.*, 109, Article 11059. https://doi.org/10.1016/j.msec.2019.110597

Ghaffarzadegan, R., Khoee, S., & Rezazadeh, S. (2020). Fabrication, characterization and optimization of berberine-loaded PLA nanoparticles using coaxial electrospray for sustained drug release. *Daru*, 28(1), 237–252. https://doi.org/10.1007/s40199-020- 00335-y

Guideline by Ministry of Health Malaysia. Guideline for Herbal Medicine Research First Edition 2023 *Published by: National Committee for Research and Development of Herbal Medicine c/o: Herbal Medicine Research Centre, Institute for Medical Research (IMR)*, National Institutes of Health (NIH), Ministry of Health Malaysia

Gupta, Y. K., Dinda, A.K., Pradhan, A.K., Dhawan, A., & Ninawe, S. (2019). *Guidelines for Evaluation of Nanopharmaceuticals in India: Jointly prepared by Department of Biotechnology, GoI.* And Indian Society of Nanomedicine (ISNM). DBT, Govt. of India.

Halamoda-Kenzaoui, B., Baconnier, S., Bastogne, T., Bazile, D., Boisseau, P., Borchard, G., Borgos, S. E., Calzolai, L., Cederbrant, K., Di Felice, G., Di Francesco, T., Dobrovolskaia, M. A., Gaspar, R., Gracia, B., Hackley, V. A., Leyens, L., Liptrott, N., Park, M., Patri, A., Roebben, G., Roesslein, M., Thürmer, R., Urbán, P., Zuang, V., & Bremer-Hoffmann, S. (2019b). Bridging communities in the field of nanomedicine. *Regul. Toxicol. Pharmacol.*, 106, 187–196. https://doi.org/10.1016/j.yrtph.2019.04.011

Halamoda-Kenzaoui, B., Holzwarth, U., Roebben, G., Bogni, A., & Bremer-Hoffmann, S. (2019a). Mapping of the available standards against the regulatory needs for nanomedicines. *Nanomed. Nanobiotechnol.*, 11, Article e1531. https://doi.org/10.1002/wnan.1531

Hashemian, M., Ghasemi-Kasman, M., Ghasemi, S., Akbari, A., Moalem-Banhangi, M., Zare, L., & Ahmadian, S. R. (2019). Fabrication and evaluation of novel quercetin-conjugated Fe3O4-β-cyclodextrin nanoparticles for potential use in epilepsy disorder. *Int. J. Nanomedicine*, 14, 6481–6495. https://doi.org/10.2147/ijn.s218317

Helal Neto, E., Barros, A., Saldanha-Gama, R., Machado Brandão Costa, R., Rebelo, L., Alencar, R., Costa, C., Dos Santos, C. C., Martínez-Máñez, R., Ricci-Junior, E., Alexis, F., Morandi, V., Barja-Fidalgo, C., & Santos-Oliveira, R. (2019). Molecular and cellular risk assessment of healthy human cells and cancer human cells exposed to Nanoparticles. *Int. J. Mol. Sci.*, 21, Article 230.

Hemmrich, E., & McNeil, S. (2023). Active ingredient vs excipient debate for nanomedicines. *Nat. Nanotechnol.*, 18, 692–695. https://doi.org/10.1038/s41565-023-01371-w

Hu, F., Sun, D. S., Wang, K. L., & Shang, D. Y. (2022). Nanomedicine of plant origin for the treatment of metabolic disorders. *Front. Bioeng. Biotechnol.*, 9, Article 811917. https://www.frontiersin.org/articles/10.3389/fbioe.2021.811917/full

Hussain, S. M., Warheit, D. B., Ng, S. P., Comfort, K. K., Grabinski, C. M., & Braydich-Stolle, L. K. (2015). At the crossroads of nanotoxicology in vitro: Past achievements and current challenges. *Toxicol. Sci.*, 147(1), 5–16.https://doi.org/10.1093/toxsci/kfv106

Iriventi, P., Gupta, N. V., Osmani, R. A., & Balamuralidhara, V. (2020). Design & development of nanosponge loaded topical gel of curcumin and caffeine mixture for augmented treatment of psoriasis. *Daru*, 28, 1–18.

Jain, S., Saxena, N., Sharma, M. K., & Chatterjee, S. (2020). Metal nanoparticles and medicinal plants: Present status and future prospects in cancer therapy. *Mater. Today Proc.*, 31(4), 662–673. https://doi.org/10.1016/j.matpr.2020.06.602

Jalili, A., Bagherifar, R., Nokhodchi, A., Conway, B., & Javadzadeh, Y. (2023). Current advances in Nanotechnology-mediated delivery of herbal and plant-derived medicines. *Adv. Pharm. Bull.*, 13(4), 712–722. https://doi.org/10.34172/apb.2023.087

Jia, W., Zhou, L., Li, L., Zhou, P., & Shen, Z. (2023). Nano-based drug delivery of polyphenolic compounds for cancer treatment: Progress, opportunities, and challenges. *Pharmaceuticals*, 16, Article 101. https://doi.org/10.3390/ ph16010101

Jitareanu, A., Trifan, A., Vieriu, M., Caba, I.-C., Mârtu, I., & Agoroaei, L. (2023). Current trends in toxicity assessment of herbal medicines: A narrative review. *Processes*, 11, Article 83. https://doi.org/10.3390/pr11010083

Kad, A., Pundir, A., Arya, S. K., Bhardwaj, N., & Khatri, M. (2022). An elucidative review to analytically sieve the viability of Nanomedicine market. *J. Pharm. Innov.*, 17, 249–265. https://doi.org/10.1007/s12247-020-09495-5

Kantak, M., Batra, P., & Shende, P. (2023). Integration of DNA barcoding and nanotechnology in drug delivery. *Int. J. Biol. Macromol.*, 230, Article 123262. https://doi.org/10.1016/j.ijbiomac.2023.123262

Kaur, A., Kaur, L., Singh, G., Dhawan, R.K., & Mahajan, A. (2022). Nanotechnology-based herbal formulations: A survey of recent patents, advancements, and transformative headways. *Recent Pat. Nanotechnol.*, 16(4), 295–307. https://doi.org/10.2174/1872210515666210428135343

Kebebe, D., Liu, Y., Wu, Y., Vilakhamxay, M., Liu, Z., & Li, J. (2018). Tumor-targeting delivery of herb-based drugs with cell-penetrating/tumor-targeting peptide-modified nanocarriers. *Int. J. Nanomedicine*, 13, 1425–1442. https://doi.org/10.2147/IJN.S156616

Keck, C. M., & Müller, R. H. (2013). Nanotoxicological classification system (NCS): A guide for the risk-benefit assessment of nanoparticulate drug delivery systems. *Eur. J. Pharm. Biopharm.*, 84(3), 445–448. https://doi.org/10.1016/j.ejpb.2013.01.001

Khan, A., Qadir, A., Ali, F., & Aqil, M. S. (2021). Phytoconstituents loaded based nanomedicines for the management of psoriasis. *J. Drug Deliv. Sci. Technol.*, 64, Article 102663. https://doi.org/10.1016/j.jddst.2021.102663

Khan, F. A., Lammari, N., Muhammad Siar, A. S., Alkhater, K. M., Asiri, S., Akhtar, S., Almansour, I., Alamoudi, W., Haroun, W., Louaer, W., Meniai, A.H., & Elaissari, A. (2020). Quantum dots encapsulated with curcumin inhibit the growth of colon cancer, breast cancer and bacterial cells. *Nanomedicine (Lond.)*, 15(10), 969–980. https://doi.org/10.2217/nnm-2019-0429

Kim, D. W., Choi, C. H., Park, J. P., & Lee, S.J. (2020). Nanospheres loaded with curcumin improve the bioactivity of umbilical cord blood-mesenchymal stem cells via c-Src activation during the skin wound healing process. *Cells*, 9(6), Article 1467. https://doi.org/10.3390/cells9061467

Koklesova, L., Jakubikova, J., Cholujova, D., Samec, M., Mazurakova, A., Šudomová, M., Pec, M., Hassan, S. T. S., Biringer, K., Büsselberg, D., Hurtova, T., Golubnitschaja, O., & Kubatka, P. (2023). Phytochemical-based nanodrugs going beyond the state-of-the-art in cancer management: Targeting cancer stem cells in the framework of predictive, preventive, personalized medicine. *Front. Pharmacol.*,14, Article 1121950. https://doi.org/10.3389/fphar.2023.1121950

Korth, C. (2014). Drug-induced Hepatotoxicity of select herbal therapies. *J. Pharm. Pract.*, 27(6), 567–572. https://doi.org/10.1177/0897190014546117

Koupaei Malek, S., Gabris, M. A., Hadi Jume, B., Baradaran, R., Aziz, M., Karim, K. J. B. A., & Rashidi Nodeh, H. (2019). Adsorption and in vitro release study of curcumin form polyethyleneglycol functionalized multi walled carbon nanotube: kinetic and isotherm study. *Daru*, 27(1), 9–20. https://doi.org/10.1007/s40199-018-0232-2

Kumar, K., Singh, R. K., Tyagi, P. K., Gore, D. (2021). Assessment of toxicity and safety profiles of Nanoparticles. *Lett Appl NanoBioSci*, 10(1), 1877–1888. https://doi.org/10.33263/LIANBS101.18771888

Kumar, R., Mirza, M. A., Naseef, P. P., Kuruniyan, M. S., Zakir, F., & Aggarwal, G. (2022b). Exploring the potential of natural product-based nanomedicine for maintaining oral health. *Molecules*, 27, Article 1725. https://doi.org/10.3390/molecules27051725

Kumar, S., Singh, K. K., & Rao, R. (2019). Enhanced anti-psoriatic efficacy and regulation of oxidative stress of a novel topical babchi oil (*Psoralea corylifolia*) cyclodextrin-based nanogel in a mouse tail model. *J. Microencapsul.*, 2, 140–155.

Kumar, V., Garg, V., & Dureja, H. (2022a). Nanomedicine-based approaches for delivery of herbal compounds. *Tradit. Med. Res.*, 7(5), Article 48. https://doi.org/10.53388/TMR20220223001

Labrador-Rached, C. J., Browning, R. T., Braydich-Stolle, L. K., Comfort, K. K. (2018). Toxicological implications of platinum nanoparticle exposure: Stimulation of intracellular stress, inflammatory response, and Akt signaling In Vitro. *J. Toxicol.*., 2018, Article 1367801.

Laloy, J. (2021). 12 Ethics reflection and responsibility of nano drugs. In M. Van de Voorde, & G. Jeswani (Eds.), *Ethics in nanotechnology: Emerging technologies aspects* (pp. 259–274). De Gruyter. https://doi.org/10.1515/9783110701883-012

Li, M., Zhang, H., Hou, Y., Wang, X., Xue, C., Li, W., Cai, K., Zhao, Y. & Luo, Z. (2020). State-of-the-art iron-based nanozymes for biocatalytic tumor therapy. *Nanoscale Horiz.*, 5, 202–217.

Li, Z., Zhao, T., Li, J., Yu, Q., Feng, Y., Xie, Y., & Sun, P. (2022). Nanomedicine based on natural products: Improving clinical application potential. *J. Nanomater.*, 2022, Article 3066613. https://doi.org/10.1155/20223066613

Liu, Y., Zhu, S., Gu, Z., Chen, C., & Zhao, Y. (2022). Toxicity of manufactured nanomaterials. *Particuology*, 69, 31–48. https://doi.org/10.1016/j.partic.2021.11.007.

Lokugamage, M. P., Sago, C. D., & Dahlman, J. E. (2018). Testing thousands of nanoparticles in vivo using DNA barcodes. *Curr. Opin. Biomed. Eng.*, 7, 1–8. https://doi.org/10.1016/j.cobme.2018.08.001.

Łuszczki, J. J., Gustaw-Rothenberg, K., Chmielewski, J., & Florek-Łuszczki, M. (2019). Prospects for the use of herbal medicines in relation to progressing environmental pollution. *Med. Srod.*, 22(1–2), 5–8. https://doi.org/10.26444/ms/117884

Lv, M., Huang, W., Chen, Z., Jiang. H., Chen, J., Tian, Y, Zhang, Z., & Xu, F. (2015). Metabolomics techniques for nanotoxicity investigations. *Bioanalysis* 7, 1527–1544. doi: 10.4155/bio.15.83

Mahapatra, I., Clark, J. R. A., Dobson, P. J., Owen, R., Lynch, I., & Lead, J. R. (2018). Expert perspectives on potential environmental risks from nanomedicines and adequacy of the current guideline on environmental risk assessment. *Environ. Sci. Nano*, 5, 1873–1889. https://doi.org/10.1039/C8EN00053K

Mahmood, T., Shamim, A., Siddiqui, M. H., & Ahsan, F. (Eds.). (2023). *Nanotechnology in Phytopharmaceuticals*. Cambridge Scholars.

Mansingh, P. P., Adhikari, L., & Dhara, M. (2023). Herbal Nanoparticles: A commitment towards contemporary approach. *Ind. J. Pharm. Edu. Res.*, 57(Suppl. 3), S465–S480.

Martínez, G., Merinero, M., Pérez-Aranda, M., Pérez-Soriano, E. M., Ortiz, T., Begines, B., & Alcudia, A. (2021). Environmental impact of nanoparticles' application as an emerging technology: A review. *Materials (Basel)*, 14(7), Article 1710. https://doi.org/10.3390/ma14071710

Mengying, L., Huang, W., Chen, Z., Jiang, H., Chen, J., Tian, Y., Zhang, Z., & Xu, F. (2015). Metabolomics techniques for nanotoxicity investigations. *Bioanalysis*, 7(12), 1527–1544.

Moradi, S. Z., Momtaz, S., Bayrami, Z., Farzaei, M. H., & Abdollahi, M. (2020). Nanoformulations of herbal extracts in treatment of neurodegenerative disorders. *Front. Bioeng. Biotechnol.*, 8, Article 238. https://doi.org/10.3389/fbioe.2020.00238

Murkute, P. S., Kathar, N. P., & Sanap, G. S. (2021). Nanotechnology: Its application, process development and Nano-Technology based herbal medicine. *Int. J. Pharm. Res. Appl.*, 6(5), 593–605.

Naha, P., Casey, A., Tenuta, T., Lynch, I., Dawson, K., Byrne, H., Davoren, M. (2009). Preparation, characterization and ecotoxicological evaluation of four environmentally relevant species of N-isopropylacrylamide and N-isopropylacrylamide-co-N-tertbutylacrylamide copolymer nanoparticles. *Aquat. Toxicol.*, 92, 146–154.

Oberdörster, G. (2010). Safety assessment for nanotechnology and nanomedicine: Concepts of nanotoxicology. *J. Intern. Med.*, 267, 89–105. https://doi.org/10.1111/j.13652796.2009.02187.x

OECD, 2017: Environment Directorate Joint Meeting Of The Chemicals Committee And The Working Party On Chemicals, Pesticides And Biotechnology Alternative Testing Strategies In Risk Assessment Of Manufactured Nanomaterials: Current State Of Knowledge And Research Needs To Advance Their Use Series on the Safety of Manufactured Nanomaterials No. 80 -Unclassified ENV/JM/MONO(2016)63. Organisation de Coopération et de Développement Économiques. Organisation for Economic Co-operation and Development 30-Jan-2017.

Ohadi, M., Rezaei, P., Mehrabani, M., Behnam, B., & Ansari, M. (2022). Synthesis, characterization and toxicity assessment of the novel non covalent functionalized multi-walled carbon nanotubes with glycyrrhizin, curcumin and rutin. *J. Clust. Sci.*, 33(3), 975–984. https://doi.org/10.1007/s10876-021-02026-3

Onyeji, C. O. (2020). Prospects of integration of nanotechnology to antimalarial herbal remedies for improved therapeutic efficacy: A concise review. *Afr. J. Tradit. Complement. Altern. Med.*, 18(1), 27–35. https://doi.org/10.21010/Ajtcamv18i1.3

Osman, E. M. (2019). Environmental and health safety considerations of nanotechnology: Nano safety. *Biomed. J. Sci. Tech. Res.*, 19(4), 14501–14515.

Paliwal, R., & Paliwal, S. R. (2021). *Advances in Nanochemoprevention: Controlled Delivery of Phytochemical Bioactives*. Springer Singapore. https://doi.org/10.1007/978-981-15-9692-6

Pallela, P., Ummey, S., Ruddaraju, L. K., Gadi, S., Cherukuri, C. S., Barla, S., & Pammi, S. V. N. (2019). Antibacterial efficacy of green synthesized α-Fe2O3 nanoparticles using Sida cordifolia plant extract. *Heliyon*, 5(11), Article e02765. https://doi.org/10.1016/j.heliyon.2019.e02765

Pan, D. (2024). Nanomedicine and nanobiotechnology in India. *Wiley Interdiscip. Rev. Nanomed. Nanobiotechnol.*, 16(1), Article e1939. https://doi.org/10.1002/wnan.1939

Pan, J., He, S., Zheng, J., Shao, J., Li, N., Gong, Y., & Gong, X. (2019). The development of an herbal material quality control strategy considering the effects of manufacturing processes. *Chin. Med.*, 14, Article 38. https://doi.org/10.1186/s13020-019-0262-9

Parusu, K. T., Mithiya, J., Kate, A. S., Bairwa, K., & Chauthe, S. K. (2022). Herbal nanomedicines: Recent advancements, challenges, opportunities and regulatory overview. *Phytomedicine*, 96, Article 153890. https://doi.org/10.1016/j.phymed.2021.153890.

Patra, J. K., Das, G., Fraceto, L. F., Campos, E.V.R., Rodriguez-Torres, M.D.P., Acosta-Torres, L.S., Diaz-Torres, L.A., Grillo, R., Swamy, M.K., Sharma, S., Habtemariam, S., & Shin, H.S. (2018). Nano-based drug delivery systems: Recent developments and future prospects. *J. Nanobiotechnol.*, 16, Article 71. https://doi.org/10.1186/s12951-018-0392-8

Pikula, K., Zakharenko, A., Chaika, V., Kirichenko, K., Tsatsakis, A., & Golokhvast, K. (2020). Risk assessments in nanotoxicology: bioinformatics and computational approaches. *Curr. Opin. Toxicol.*, 19, 1–6. https://doi.org/10.1016/j.cotox.2019.08.006.

Pingale, P., Kendre, P., Pardeshi, K., & Rajput, A. (2023). An emerging era in manufacturing of drug delivery systems: Nanofabrication techniques. *Heliyon*, 9(3), Article e14247. https://doi.org/10.1016/j.heliyon.2023.e14247

Qari, S. H., Alrefaei, A. F., Ashoor, A. B., & Soliman, M. H. (2021). Genotoxicity and carcinogenicity of medicinal herbs and their nanoparticles. *Forum Nutr.*, 1, 31–41. https://doi.org/10.3390/nutraceuticals1010005

Qiu, T. A., Clementa, P. L., & Haynes, C. L. Linking nanomaterial properties to biological outcomes: analytical chemistry challenges in nanotoxicology for the next decade. *Chem. Commun.*, 2018,54, 12787–12803. https://doi.org/10.1039/C8CC06473C.

Rai, A., Kaur, S., Rawat, S. S., & Singh, I. (2023). Herbal bioactive encapsulated nano-formulations for the treatment of gastric cancer: A concise review. Nano *Futures*, 7(3), Article 5003. https://doi.org/10.1088/2399-1984/acf822

Rajasekar, A., Devasena, T., Suresh, S., Senthil, B., Sivaramakrishnan, R., & Pugazhendhi, A. (2022). Curcumin nanospheres and nanorods: Synthesis, characterization and anticancer activity. *Process Biochem.*, 112, 248–253. https://doi.org/10.1016/j. procbio.2021.12.007

Ramanathan, A. (2019). Toxicity of nanoparticles_ challenges and opportunities. *Appl. Microsc.*, 49, Article 2. https://doi.org/10.1007/s42649-019-0004-6

Ray, P., Haideri, N., Haque, I., Mohammed, O., Chakraborty, S., Banerjee, S., Quadir, M., Brinker, A. E., & Banerjee, S. K. (2021). The impact of Nanoparticles on the immune system: A gray zone of Nanomedicine. *J. Immunol. Sci.*, 5(1), 19–33.

Razavi, M. S., Ahmadi, F., Ebrahimnejad, P., Akbarzadeh, A., Farokhrou, M., Nokhodchi, A. (2024). Harnessing nanotechnology for optimized herbal cancer treatment: A comprehensive review of nanoscale drug delivery systems. *Pharm. Sci. Res.*, 30(2), 153–186. https://doi.org/10.34172/PS.2023.27

Rehman, K., Ali, I., El-Haj, B. M., Kanwal, T., Maharjan, R., Saifullah, S., Muhammad, I., Simjee, S. S. U., & Shah, R. (2021). Synthesis of novel biocompatible resorcinarene based nanosized dendrimer-vesicles for enhanced anti-bacterial potential of quercetin. *J. Mol. Liq.*, 341, Article 116921. https://doi.org/10.1016/j.molliq.2021.116921

Roy, J., & Roy, K. (2023). Evaluating metal oxide nanoparticle (MeOx NP) toxicity with different types of nano descriptors mainly focusing on simple periodic table-based descriptors: A mini-review. *Environ. Sci. Nano*, 10(11), 2989–3011. https://doi.org/10.1039/D3EN00598D

Rycroft, T., Trump, B., Poinsatte-Jones, K., & Linkov, I. (2018). Nanotoxicology and nanomedicine: Making development decisions in an evolving governance environment. *J. Nanopart. Res.*, 20, Article 52. https://doi.org/10.1007/s11051-018-4160-3

Sandhiya, V., & Ubaidulla, U. (2020). A review on herbal drug loaded into pharmaceutical carrier techniques and its evaluation process. *Futur. J. Pharm. Sci.*, 6, Article 51. https://doi.org/10.1186/s43094-020-00050-0

Schmutz, M., Borges, O., Jesus, S., Borchard, G., Perale, G., Zinn, M., Sips, Ä. A. J. A. M., Soeteman-Hernandez, L. G., Wick, P., & Som, C. (2020). A Methodological safe-by-design approach for the development of Nanomedicines. *Front. Bioeng. Biotechnol.*, 8, Article 258. https://doi.org/10.3389/fbioe.2020.00258

Setia, A., Vallamkonda, B., Challa, R. R., Mehata, A. K., Badgujar, P., & Muthu, M. S. (2024). Herbal theranostics: Controlled, targeted delivery and imaging of herbal molecules. *Nano*, 8(3), 344–379. https://doi.org/10.7150/ntno.94987

Shahgholian, N. (2018). Herbal medicines synchronized with the advancement of nanotechnology. *Ann. Med. Chem. Res.*, 4(1), Article 1023.

Sharma, U., Jain, S., & Khambete, H. (2024). A new arena of herbal nanotechnology. *Asian J. Pharm.*, 18 (1), Article 35.

Shim, W., Paik, M. J., Nguyen, D. T., Lee, J. K., Lee, Y., Kim, J. H., Shin, E. H., Kang, J. S., Jung, H. S., Choi, S., Park, S., Shim, J. S., & Lee, G. (2012). Analysis of changes in gene expression and metabolic profiles induced by silica-coated magnetic nanoparticles. *ACS Nano*, 6(9), 7665–7680. https://doi.org/10.1021/nn301113f.

Shin, T. H., Nithiyanandam, S., Lee, D. Y., Kwon, D. H., Hwang, J. S., Kim, S. G., Jang, Y. E., Basith, S., Park, S., & Mo, J. S. (2021). Analysis of Nanotoxicity with integrated omics and mechanobiology Nanomaterials (Basel), 11(9), Article 2385. https://doi.org/10.3390/nano11092385

Shinde, R. B., Veerapandian, M., Kaushik, A., & Manickam, P. (2020). State-of-art Bio-Assay systems and electrochemical approaches for nanotoxicity assessment. *Front. Bioeng. Biotechnol.*, 8, Article 325. https://doi.org/10.3389/fbioe.2020.00325

Shukla, R., Bansal, V., Chaudhary, M., Basu, A., Bhonde, R. R., & Sastry, M. (2005). Biocompatibility of gold nanoparticles and their endocytotic fate inside the cellular compartment: A microscopic overview. *Langmuir*, 21, 10644–10654.

Siegrist, S., Cörek, E., Detampel, P., Sandström, J., Wick, P., & Huwyler, J. (2019). Preclinical hazard evaluation strategy for nanomedicines. *Nanotoxicology*, 13(1), 73–99. https://doi.org/10.1080/17435390.2018.1505000

Singh, A. V., Laux, P., Luch, A., Sudrik, C., Wiehr, S., Wild, A. M., Santomauro, G., Bill, J., & Sitti, M. (2019). Review of emerging concepts in nanotoxicology: Opportunities and challenges for safer nanomaterial design. *Toxicol. Mech. Methods*, 29(5), 378–387. https://doi.org/10.1080/15376516.2019.1566425

Singh, A.V., Varma, M., Laux, P., Choudhary, S., Datusalia, A.K., Gupta, N., Luch, A., Gandhi, A., Kulkarni, P., & Nath, B. (2023). Artificial intelligence and machine learning disciplines with the potential to improve the nanotoxicology and nanomedicine fields: A comprehensive review. *Arch. Toxicol.*, 97, 963–979. https://doi.org/10.1007/s00204-023-03471-x

Singh, H., & Gauri. (2023). Recent development of novel drug delivery of herbal drugs. *RPS Pharm. Pharmacol. Rep.*, 2(4), Article rqad028. https://doi.org/10.1093/rpsppr/rqad028

Soares, S., Sousa, J., Pais, A., & Vitorino, C. (2018). Nanomedicine: Principles, properties, and regulatory issues. *Front. Chem.*, 6, Article 360. https://doi.org/10.3389/fchem.2018.00360

Soeteman-Hernandez, L. G., Apostolova, M. D., Bekker, C., Dekkers, S., Grafström, R. C., Groenewold, M., Handzhiyski, Y., Herbeck-Engel, P., Hoehener, K., Karagkiozaki, V., Kelly, S., Kraegeloh, A., Logothetidis, S., Micheletti, C., Nymark, P., Oomen, A., Oosterwijk, T., Rodríguez, I., Sabella, S. & Noorlander, C. (2019). Safe innovation approach: Towards an agile system for dealing with innovations. *Mater. Today Commun.*, 20, Article 100548. 10.1016/j.mtcomm.2019.100548.

Steckiewicz, K. P., Barcinska, E., Malankowska, A., Zauszkiewicz-Pawlak, A., Nowaczyk, G., Zaleska-Medynska, A., Inkielewicz-Stepniak, I. (2019). Impact of gold nanoparticles shape on their cytotoxicity against human osteoblast and osteosarcoma in in vitro model. Evaluation of the safety of use and anti-cancer potential. *J. Mater. Sci. Mater. Med.*, 30, Article 22.

Subbiah, U., Elango, S., & Jayesh, R. (2019). Herbals and green synthesized nanoparticles in dentistry. In K. Subramani & W. Ahmed (Eds.), *In micro and nano technologies, nanobiomaterials in clinical dentistry* (pp. 617–646). https://doi.org/10.1016/B978-0-12-815886-9.00025-5.

Tirumala, M. G., Anchi, P., Raja, S., Rachamalla, M., & Godugu, C. (2021). Novel methods and approaches for safety evaluation of nanoparticle formulations: A focus towards in-vitro models and adverse outcome pathways. *Front. Pharmacol.*, 12, Article 612659. https://doi.org/10.3389/fphar.2021.612659

Tiwari, J., Garg, A., & Jain, A. P. (2020). Synthesis and characterization of single walled carbon nanotubes (SWCNTs) anchored curcumin for breast cancer targeting. *Int. J. Adv. Sci. Technol.*, 29(5), 13708–13719.

Verma, M., (2018). *Critical Evaluation of Toxicity Tests in Context to Engineered Nanomaterials: An Introductory Overview*. In V. Kumar, N. Dasgupta & S. Ranjan. CRC Press.

Verma, S. K., Nandia, A., Simnania, F. Z., Singha, D., Sinha, A., Nasera, S. S., Sahooa, J., Lenka, S. S., Panda, P. K., Dutta, A., Kaushik, N. K., Singh, D., & Suar, M. (2023). In silico nanotoxicology: The computational biology state of art for nanomaterial safety assessments. *Mater. Des.*, 235, Article 112452.

Wang, L., Hartel, N., Ren, K., Graham, N. A., & Malmstadt, N. (2020). Effect of protein corona on nanoparticle-plasma membrane and nanoparticle-biomimetic membrane interactions. *Environ. Sci. Nano*, 7, 963–974.

WHO (World Health Organisation). (2011).

Wu, H., Yu, T., Tian, Y., Wang, Y., Zhao, R., & Mao, S. (2018). Enhanced liver-targeting via coadministration of 10-Hydroxycamptothecin polymeric micelles with vinegar baked Radix Bupleuri. *Phytomedicine*, 44, 1–8. https://doi.org/10.1016/j.phymed.2018.04.022

Yan, L., Zhao, F., Wang, J., Zu, Y., Gu, Z., & Zhao, Y. (2019). A safe-by-Design strategy towards safer nanomaterials in Nanomedicines. *Adv. Mater.*, 31, Article e1805391.

Yang, C., & Merlin, D. (2023). Challenges to safe Nanomedicine treatment. *Nano*, 13, Article 1171. https://doi.org/10.3390/nano13071171

Zare, M., & Sarkati, M. N. (2021). Chitosan-functionalized Fe3O4 nanoparticles as an excellent biocompatible nanocarrier for silymarin delivery. *Polym. Adv. Technol.*, 32(10), 4094–4100. https://doi.org/10.1002/pat.5416

Zhang, J., Hu, K., Di, L., Wang, P., Liu, Z., Zhang, J., Yue, P., Song, W., Zhang, J., Chen, T., Wang, Z., Zhang, Y., Wang, X., Zhan, C., Cheng, Y. C., Li, X., Li, Q., Fan, J. Y., Shen, Y., Han, J. Y., & Qiao, H. (2021). Traditional herbal medicine and nanomedicine: Converging disciplines to improve therapeutic efficacy and human health. *Adv. Drug Deliv. Rev.*, 178, Article 113964. https://doi.org/10.1016/j.addr.2021.113964.

Zhang, J., Onakpoya, I. J., Posadzki, P., & Eddouks, M. (2015). The safety of herbal medicine: From prejudice to evidence. *Evid. Based Complement. Alternat. Med.*, 2015, Article 316706. https://doi.org/10.1155/2015316706

Zheng, J., Fan, R., Wu, H., Yao, H., Yan, Y., Liu, J., Ran, L., Sun, Z., Yi, L., Dang, L., Gan, P., Zheng, P., Yang, T., Zhang, Y., Tang, T., & Wang, Y. (2019). Directed self-assembly of herbal small molecules into sustained release hydrogels for treating neural inflammation. *Nat. Commun.*., 10, Article 1604. https://doi.org/10.1038/s41467-019-09601-3

Zhou, Z., Li, D., Fan, X., Yuan, Y., Wang, H., Wang, D., & Mei, X. (2022). Gold nanoclusters conjugated berberine reduce inflammation and alleviate neuronal apoptosis by mediating M2 polarization for spinal cord injury repair. *Regen. Biomater.*, 9, Article rbab072. https://doi.org/10.1093/rb/rbab072

21 Regulatory Considerations for Herbal Products

Saleem Ur Rahman, Ehtisham Shafique, Afsheen Tajummal, Mashaal Nageen, Asma Zahoor, Nasir Ali, and Farakh Javed
Department of Biological and Health Sciences, Pak-Austria Fachhochschule: Institute of Applied Sciences and Technology, Haripur, Pakistan

21.1 INTRODUCTION

Preparations which contain plant materials exclusively are called herbal products [1]. Similarly, the knowledge, skills, beliefs and practices which are used to diagnose and treat diseases to maintain the health of individuals are defined as traditional medicines (World Health Organization) [1]. According to the literature, herbal medicines are described as the preparations made according to the therapeutic knowledge and experiences of different generations over the centuries and the systems of medicines practiced indigenously in societies. These firm believes that there is a high demand for such medicines in many developing countries not only because they are less expensive but also as a result of better cultural acceptability and minimal side effects on the human body, along with better compatibility [2]. Herbal products are utilized to treat various ailments such as depression and mental health disorders as well as to achieve weight loss, and to operate as analgesics and cold and allergy treatments [2]. Serious health conditions such as diabetes, asthma, heart diseases, and endometriosis are also treated through the use of herbal medicines. This suggests that whatever the disease condition, there is always a herbal remedy which can alleviate it. Food supplements are classified by Food and Drug Administration (FDA) as one aspect of herbal medicine [3]. The FDA does not usually govern the production of new herbal supplements through strict protocols of drug approval process, meaning that no pre-market approval is required [3]. One of the current FDA-approved drugs is Veregen® (sinecatechins ointment) [4]. Both benefits and risks of herbal supplements are discussed widely on various platforms and strong opinions both in favor and against their use are noted. Although this debate will focus on their use, healthcare professionals do need to acknowledge that people around the world use them commonly, considering them to be safe and appropriate sources for healthcare products [5].

In addition, to improve general health and well-being, various phytochemicals, along with other herbal remedies, are useful in treating ailments. In the local market, there is a concern that there is no scientific basis for believing the fact that herbal remedies are non-hazardous although they have a high success rate. In today's world of social and electronic media, due to the online marketing of such drugs by manufacturers, there has been an exponential increase in both their use and the related demand. Thus, there is a requirement for the introduction of regulations, which will developing legislation to regulate the manufacturers of herbal products and to regulate the sale of herbal medicine. These should be used to ensure that herbal remedies should be proven safe before entering the market and made freely available to the public. Some such regulations are already in place: In India, for example, the laws of 1945 and the Drug and Cosmetics Act of 1940 regulate herbal medicines [6]. For so long various controversies, suspicions and doubts led to increased use of herbal medicine [7, 8].

DOI: 10.1201/9781003544418-21

TABLE 21.1
Traditional and Complementary Medicine (TMC/CM) Products

Country	Terminology	Description
China	Classic Chinese medicines	Refers to Chinese medicines documented and established through historical texts and practices.
Taiwan	Standard formula of Chinese medicine	Refers to established formulas that are widely recognized and used within the framework of Chinese medicine.
Hong Kong	Established medicine	Refers to recognized and standardized herbal formulations used within traditional Chinese medicine.
Japan	Kampo medicine	Derived from ancient Chinese medicine and standardized in Japan for modern use.
Korea	Herbal drug	Refers to herbal medicines that are standardized and used within the practice of Traditional Korean medicine.

21.2 GLOBAL REGULATORY LANDSCAPE

Under the mentorship of the US Food and Drugs Administration (FDA), in 2013 the Global Coalition for Regulatory Science Research (GCRSR) was established. The GCRSR was established in 2013. At present, it has around 10 member countries or organizations, including the European Union (EU). The GCRSR has gone on to collaborate and start partnerships with international bodies and agencies to utilize bid data sciences and emerging technologies in order to improve the safety and efficacy of foods and drugs as well as their improved regulatory structure [9]. As noted earlier, the individuals using herbal products as medicine are continuing to increase in many countries. In addition, the regulation of herbal products varies from country to country; in some places, however, herbal products are only allowed for general health matters rather than being used for broader medical purposes [10]. Globally, the use of herbal medicines (HM) derived from natural sources as well as their usefulness in dietary or medicinal terms has increased substantially [11, 12]. Even though HMs have been present for thousands of years, their jurisdictions and regulation has varied across countries, and some of them have also begun to be regulated recently. Psychological binding is also involved in such practices as for example, if a drug is being used in families over centuries, then the consumers will have blind faith in it as a remedy. . Hence the knowledge of the previous practices can serve as a guide for the current use of dietary supplements.

Table 21.1 offers a summary of the terminology and description of traditional and complementary medicine (TMC/CM) products across five regions, highlighting their unique regulatory and cultural contexts.

21.3 REGULATORY AGENCIES AND THEIR ROLES

21.3.1 United States

Products marketed as dietary supplements are regulated by the US FDA under the umbrella of the Federal Food, Drug and Cosmetic Act (FD&C act), which was later amended by the Dietary Supplement Health and Education Act (DSHEA), 1994 [13–15]. Herbal products (HPs) make up a major part of US dietary supplements. Statistically, in the year 2000, it has been reported that about 591 billion USD have been spent by American consumers on herbal products [4]. Similarly, the practice of Complementary and Alternative Medicine (CAM) has risen dramatically in recent years. In order to explore unconventional medical practices, in 1992 the US Congress introduced the Office of Alternative Medicine (OAM). The term CAM comprises a vast range of healthcare practices, therapies and products that differ from conventional or allopathic medicine. With the increase in practice of CAM in the US, the confusion has been seen by the FDA, whether or not some of products enlisted in CAM are subjected to regulations by the FFDCA or the PHSA. Under the Federal Food, Drug and Cosmetic Act/Public Health Service Act, certain products being utilized in CAM are subjected to regulations. According to the FDA, several CAM products being imported into the US have had guidance issued indicating which types of products would be subjects under the Act. These guidelines made two fundamental points. Firstly, the CAM product would be subjected to regulations as a cosmetic, device, food along with additives, dietary supplements, and drug or biological product under the PHS Act. Secondly, the CAM products are neither exempted from PHS nor from the Act. This means, for example, that if a person wants to produce and sell juice extracted from raw vegetables as a therapy for optimal health, then the product would be subjected to the regulation by the Act as well as the FDA, along with the Hazards Analysis and Critical Control Point (HACCP) system for juices. In addition, if the seller looks to retail this as a treatment option for any disease, the juice will also be subject to the regulations of FDA as a drug under the Act. The FDA is supposed to regulate products and preparations containing botanical ingredients in a different method, as will the products of those botanical ingredients, which intend to diagnose, mitigate, or cure the disease. All of the products should follow the pre-market approval process that fall under this category [16].

21.3.1.1 Key Responsibilities of the FDA

The FDA is a government body that lies within health and human services (HHS) department.

It regulates the safety, security, and effectiveness of various products, including:

- Food (food additives, dietary supplements, infant formula and bottled water)
- Vaccines
- Microwaves and other equipment that radiate etc.
- Cosmetics and beauty items
- Animal and veterinary products
- Tobacco products.

21.3.2 Canada

In Canada, most of the natural products are classified as medicine. They are known as Natural Health Products (NHPs) and are regulated by Health Canada (HC), specifically under the Natural Health Products Regulations (NHPR), which were enacted in 2004 [17]. The government planned to balance the safety of the consumers with their freedom of choice in terms of accessing traditional medicines. The NHPs must be licensed by Health Canada in order to be sold in the market in order to ensure the safety and effectiveness of the products. The product is considered to be safe by the department if the benefits of the product outweigh its risks and it is intended to be used according to the directions. Simultaneously, it is considered to be effective if the department is presented with evidence supporting the described benefits. Since 2004, more than 91,000 licenses have been issued by Health Canada for NHPs. In general, herbal products are regulated either as foods or as drugs according to the Food and Drugs Act. Various factors such as an ingredient's pharmacological activity, product's purpose, representation regarding uses, and the medicinal claims define its designation. Responsibility for the NHPs has blurred boundaries between the Directorate of Drugs Protection and the Directorate of Food Safety within Health Canada. Various products such as khat and cannabis are controlled drugs pursuant to the Controlled Drugs and Substances Act [5]. Several public attempts have been made by Health Canada to address the issues related to the regulation of herbs. On two occasions in recent decades, 1986 and 1993, expert advisory committees have reviewed the issues of herbs which is provided for food. More recently, in 1997, an advisory panel on herbal medicines was established to advise the development of a framework for the regulation of herbal products. In addition, Health Canada has also issued guidelines, information letters, and policy statements on these issues.

21.3.3 Europe

21.3.3.1 European Medicine Agency

On behalf of the European Medicine Agency (EMA), the Herbal Medicinal Products Committee (HMPC) regulates and issues guidelines regarding the preparation, use, and safe conditions of herbal substances and medicinal product. It gives companies and national authorities a reference and standard point for the registration of herbal products and the assessment of their applications and intended usage in the European Union (EU) and its member states [6]. Such products fall into two categories under EU classification criteria, i.e. supplements or medicine. The products which are regulated as food are called supplements under directive 2002/46/EC of the European Commission [18]. Similarly, scientific guidelines on herbal medicinal products help in the development and preparation of market authorization for human and veterinary medicines under the European Medicine Agency. The EMA has categorized the guidelines into various sections, as discussed below.

21.3.3.2 Key Responsibilities of the European Medicine Agency

In the European Union, for the benefit of animal and public health, the EMA aims to foster the evaluation and supervision of medicines and herbal products in the light of scientific excellence. It is responsible for the following major responsibilities:

1. To facilitate access to medicines and their development.
2. The regulation and evaluation of applications for marketing.
3. The portrayal of information regarding humans and veterinary medicines in simple language.
4. Check and balance maintenance about the safety of medicines throughout their lifecycle [6].

21.3.3.3 Quality Guidelines of the European Medicine Agency

21.3.3.3.1 The Declaration of Herbal Preparations in Traditional Herbal Medicinal Products

Th proper description of different types of herbal substances, along with the preparation steps and qualitative and quantitative particulars, shall ensure clear differentiation as a common criterion for the declaration. As a result, precise, consistent, and clear information is to be provided to the community about all of the active compounds in the herbal products.

In order to confirm the identification and to help with the differentiation of herbal preparations, various factors which determine the complex composition of herbal products are to be stated. These factors include the production process, the genuine drug extract ratio (DER), extraction solvent and the type of herbal substances. However, in order to keep the declaration precise and short, it is not possible to include all characteristics. The purpose of the declaration is to ensure the quality and to identify the quantity of herbal preparations, the active compounds present in the preparations, and the useful substances that ensure the effectiveness and safety of the herbal products.

In order to achieve this, a system has been established that focuses on and reflects the characteristics and specifications of herbal products as described. Therefore, the general guidance, as outlined in the European Pharmacopeia, should be followed. This should draw, in particular, on the monograph Extracts/Herbal Drugs/Herbal Drug Preparations, as well as the guidelines listed under the references (EMEA/HMPC/CHMP/CVMP/287539/2005).

21.3.3.3.2 Practices for Materials Collection from Herbal Origin

Wild or cultivated land areas could be a source for herbal medicinal plants. Greenhouse and indoor cultivation are the principal sources for these plants. At present, there is an increase in indoor cultivation because this allows for the control of various environmental factors such as light, temperature, and humidity.

Active Pharmaceutical Ingredients (API) are directly influenced by the process of cultivation, production, and primary processing. For this purpose, an adequate quality assurance system is needed to cultivate, harvest, and for primary processing.

Similarly, the choice for the collection of plants is also important as they could affect both the quality and the efficacy of the substances. The final application of the herbal medicinal products is dependent on the cultivation method employed. For instance, wild collection methods are vulnerable to certain risks and confusion like similar plants, lack of control, environmental damage, a and lack of qualified personnel. Also, a variation in the composition of the products could be a challenge exhibited by these plants due to non-uniformity in natural growing conditions. However, in situations in which herbs cannot meet the cultivation practices and the agronomic conditions could not be achieved, manufacturers may opt for wild cultivation (EMEA/HMPC/246816/2005).

21.3.3.3.3 Specifications: Criteria for Testing Herbal Substances and Their Procedures for Preparation

A list of tests, analytical and biological procedures references, and defined acceptance criteria which could involve numbers, limits, or ranges are termed specifications. This defines the criteria that should be fulfilled by any herbal preparation or herbal medicinal product for its intended use. When a product has fulfilled the criteria as defined in the listed procedure, it will be considered to have been manufactured in terms of "Conformance to specification". In addition, specifications could be described as the quality standards proposed by the manufacturers that are legally binding and are approved by regulatory authorities.

Here, in the case of HMPs, specifications refer to herbal substances, or herbal medicinal product. This classification is quite different from characterization, as in a specification only the quality of the herbal substances is defined. There should be a focus on the characteristics that ensure the safety and efficacy of the herbal substances. Contrary to the medicinal products, which contain chemically defined active substances, herbal medicinal products that contain herbal preparations as active substances should also define the specification, even if the herbal substance serves solely as an active starting ingredient for the herbal preparation (EMA/HMPC/CHMP/CVMP/162241/2005) (Figure 21.1).

21.3.4 Non-Clinical Guidelines

21.3.4.1 Genotoxicity Assessment of Herbal Preparations

In these guidelines, practical approaches to test the expected genotoxicity of the herbal substances is described and the interpretation of the results is discussed.

HMPs are presented with several characteristics that can differentiate them from other MPs. Examples of important differences may include:

- HMPs are prepared by substances that could form part of the regular diet, and/or present in the environment, which means their contribution to overall health needs to be considered.
- They can also contain variable amounts of large numbers of constituents which are present in sometimes highly variable amounts.
- Similarly, their composition can also vary due to harvesting method, geographical location, and the time of harvesting etc.

The overall complete composition of the preparation is difficult to reveal. It is always stated that many unknown constituents are present in the preparation, as a result of which there may be hidden dangers. Besides, in many other aspects, HMPs are quite like medicinal products that contain synthetic active substances:

- Their legal position is determined by the same basic legislation [1].
- A portion of the population has been using many HMPs for a long time.

Although herbal natural substances are shown to have genotoxic aspects, the question arises: could they be exposed to humans via food, water and environmental transmission? What would be the

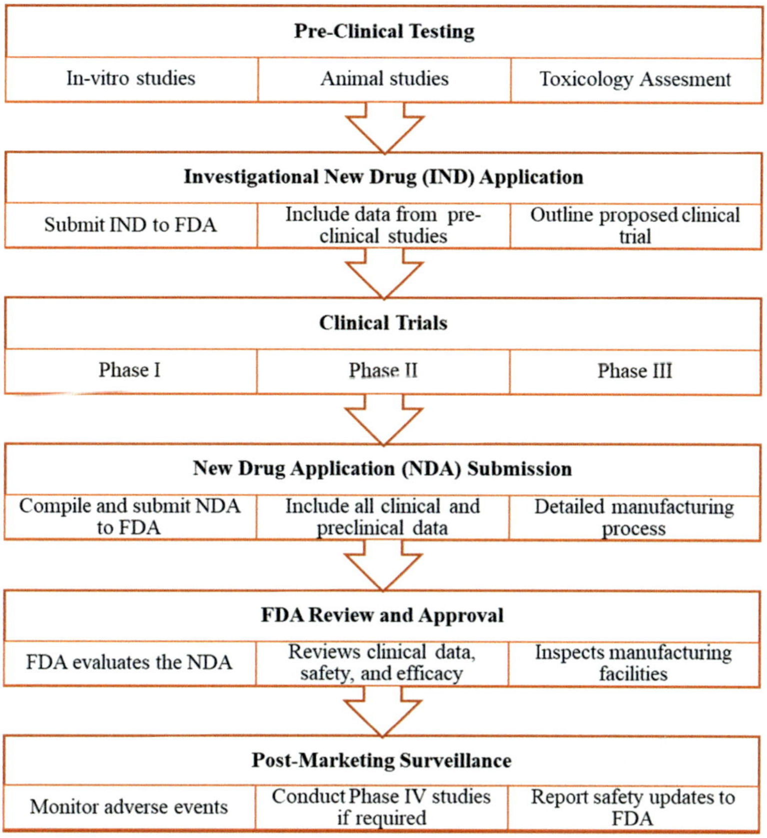

FIGURE 21.1 US and EU Regulatory aspects of herbal products.

burden to an individual of using HMPs on top of natural exposure? Is there a range that defines the acceptance of the level of exposure? Additionally, are there any significant standards to determine the acceptable range of exposures? All these questions have been addressed in the reference (EMEA/HMPC/107079/2007).

21.3.5 Asia

21.3.5.1 Traditional Chinese Medicines

One group of healthcare practices which may be integrated into the central healthcare system, up to an extent defined in various theories, beliefs and experiences, is referred to as Traditional Medicine (TM) and Complementary Medicine (CM) [19]. Traditional Chinese Medicine (TCM) has developed over thousands of years. TCM practitioners use a variety of physical and psychological methods, along with herbal products, to treat health issues. The historic status of cultural practices explains the lack of TCM regulation, which can be simply stated as the people's lives are unnecessary, even intrusive (http://www.chinadaily.com.cn/). The first recorded evidence of the widespread use of TCM on the mainland of China is provided by *The Yellow Emperor's Classic*. This old text was compiled before 85 BC for which details are mentioned in the traditional method of diagnosis and treatments. While China's rich medical history is further testified by written examples of old diagnosis and treatments [20].

From this perspective, having a proper system to regulate and ensure the quality, safety, and effectiveness of traditional medicines is crucial not only for public health but also for economic growth. The evaluation standards and the regulatory approaches are yet to be developed for TM/CM being used over a very long period of time [21].

21.3.5.2 Regulatory Approaches for TM/CM

There is a great difference in terms of the definitions, the scope, and the overall regulatory authorities and their approaches, which are associated with TM/CM products featured over the long history of use in selected regions or countries considering the risk-based approaches.

Four registration categories, innovative Chinese medicines, generic Chinese medicines, modified new Chinese medicines, and old Chinese medicines, are defined in China regarding TM/CM (22). A draft was issued in 2017, the Regulation of the Simplified Registration of the Simplified Registration for CCMF by NMPA, which was known as the China Food and Drug Administration at that time. Following the enactment of the provisions of drug registration in 1985, the CCMF listed all those medicines that were present in ancient Chinese medical books before 1911. All of these were strictly examined by a stringent examination process, approved standards, and regulation models. The approval and simplified registration process was also proposed accordingly. Compared with the modern biochemical drugs, the fact that CCMF has evident effects over centuries and their clinical applications are not applicable as of today's era of drugs.

21.3.6 Indian Ayurvedic Regulations

21.3.6.1 Food Safety and Standards Authority of India

Regulations regarding the safety and quality of food products have been formulated by the Ministry of Ayush, along with the Food Safety and Standards Authority of India under the Ministry of Health and Family Welfare (MoHFW). The motto of this initiative was to promote products made in India as well as to ensure the quality manufacturing of herbal products. The ministry was also confident enough that such initiatives will further strengthen India as a global custodian of the Ayush system.

This regulation made strict protocols that any Ayurveda Aahara shall be licensed under Food Safety and Standards Regulation of 2022. Following this, they will be ready for sale in the market. For this purpose, a special logo was designed to help the public identify examples of approved Ayurveda Aahara.

The ministry also focused on public messaging that food safety and quality is not solely the responsibility of the government; it is also a shared responsibility of each and every individual in India. The slogan was further strengthened during the COVID-19 pandemic, during which the focus on food, health, and nutrition was greater than in previous times.

According to the regulations, only food prepared in accordance with the instructions and recipes written in the authoritative book of Ayurveda will be considered to be genuine Ayurveda Aahara. In addition, foods that promote health, food specified for consumption, and food specific for physiological needs are referred to as Pathya in Ayurvedic and are covered under this regulation.

Different claims regarding Ayurveda Aahara, such as claims to reduce diseases, health claims and different categories of these Aahara shall fulfil the requirements specified in the regulations. It is to be noted that Ayurvedic medicines, cosmetics, narcotics, or ayurvedic drugs will not be included under this regulation. Furthermore, these are also prohibited for children under two years of age [7].

21.3.7 World Health Organization

In 1948, a specialized agency of the United Nations (UN) for improved public health conditions was established. This is better known as the World Health Organization (WHO). The WHO was assigned a broad mandate under its constitution to regulate and maintain the highest possible level of public health, although initially it was given the tasks of drug standardization, epidemic control, and quarantine measures. According to the definition by the WHO, health is defined as a state of physical and social well-being, or complete physical and mental well-being without disease or infirmity. Annually, April 7 is celebrated as the anniversary of the WHO's establishment and is named World Health Day.

The implementation of strategies and routine operations are operated by central headquarters or through regional WHO offices, as well as in its individual country offices around the world.

21.3.7.1 WHO Guidelines on the Safety Monitoring of Herbal Medicines

Safety is the critical component of quality control and is the basic principle in the provisions of herbal products for healthcare. Within the pharmacovigilance system, these guidelines provide basic technical guidance about the safety of herbal products and their regulations. There are various differences in the cultural and regulatory systems in pharmacovigilance, yet they are equally important [23].

The guidelines are developed with the view that the safety of medicines should be broadened and enhanced to allow the proper monitoring of herbal medicines (Table 21.2).

Four complementary actions are to be taken for the use of herbal medicines:

- Identification of the nature of the adverse events associated with the product.
- Risk management.
- Measurements to deal with the risks associated.
- Communication of benefits and risks of herbal medicines.

TABLE 21.2
Major Countries with Their Respective Regulatory Bodies and Their Responsibilities

Country	Regulatory Body	Roles and Responsibilities
United States	FDA (Food and Drug Administration)	- Evaluates and approves new herbal medicines for safety and efficacy. - Monitors post-marketing adverse events and compliance. - Sets standards for manufacturing practices (GMP).
European Union	EMA (European Medicines Agency)	- Reviews and authorizes herbal medicines based on quality, safety, and efficacy. - Coordinates pharmacovigilance activities across member states. - Provides scientific guidelines and assessments.
China	CFDA (China Food and Drug Administration)	- Approves herbal medicines based on traditional and modern safety and efficacy data. - Regulates manufacturing practices and quality control. - Conducts post-marketing surveillance and adverse event monitoring.
Japan	PMDA (Pharmaceuticals and Medical Devices Agency)	- Reviews and approves herbal medicines, integrating traditional usage with modern scientific data. - Conducts safety and efficacy evaluations. - Oversees post-marketing surveillance and regulatory compliance.
India	AYUSH (Ministry of Ayurveda, Yoga & Naturopathy, Unani, Siddha, and Homoeopathy)	- Approves and monitors herbal medicines based on traditional texts and modern research. - Ensures compliance with manufacturing standards and safety regulations. - Conducts post-marketing surveillance and adverse event reporting.
Australia	TGA (Therapeutic Goods Administration)	- Evaluates and registers herbal medicines for safety, quality, and efficacy. - Regulates manufacturing practices and standards. - Monitors adverse events and enforces compliance.
Canada	Health Canada	- Reviews and approves herbal medicines based on safety and efficacy data. - Sets guidelines for manufacturing practices (GMP). - Monitors post-marketing safety and adverse events.

21.4 CLASSIFICATION OF HERBAL PRODUCTS

In general, herbal products are classified into three major categories:

1. Dietary Supplements
2. Traditional Medicines
3. Over the Counter (OTC) Herbal Products

21.4.1 Dietary Supplements

The dietary ingredient is defined as a herb or botanical, mineral, dietary substance, vitamins, and/or amino acids used by a man to supplement the daily intake by the Federal Food, Drug and Cosmetics Act. These supplements could also be metabolites, constituents, or the combination of the previous substances, but not drugs to diagnose, prevent, or treat any disease. This means that these supplements should not be used for treating pains or claiming to treat any diseased condition. In addition, dietary supplements are marketed as soft gets, gelcaps, tablets, capsules, powders, and/or in liquid forms.

21.4.2 Herbal Supplements

Different jurisdictions interpret herbal medicine in a variety of ways. The WHO considers herbal medicines to be those that include herbal materials, herbal preparations, or herbs to bring harmonization. The herbal products could be comprised of ingredients that are either obtained from a single plant or multiple plants as active ingredients. However, some herbal products contain active ingredients from animals, rather than plants and plant sources. In USA, medicines classified as botanical would be termed herbal, out of which some are regarded as dietary supplements.

21.4.3 Functional Food

There is a concept that there are health benefits of some foods, and functional food is used as an alternative term for nutraceutical food. Food substances that could be extracted from normal food, such as peanuts, grapes, and contain resveratrol with antioxidant properties in processed or in powder form, are termed functional foods. Health Canada defines such substances as nutraceutical and as a product originated from food, while they are not recognized categories under EU legislation.

21.4.4 Traditional Medicine

In jurisdiction, these are the practices that have a wide history of practice. In addition to medicine, the practice might be combination of knowledge and beliefs. It could be employed as a diagnosis, and in the treatment of physical and mental health [23].

21.5 APPROVAL PROCESS

21.5.1 Pre-Market Approval

The process for the scientific and regulatory review for the evaluation of safety and effectiveness of Class III medical devices by the FDA is known as "premarket approval" (PMA). Class III devices are devices that are substantial in improving human health, sustaining human life irrespective of the risk of illness or injury. The FDA has determined that general and special controls are insufficient to ensure the safety and effectiveness of Class III devices, which carry a higher level of risk. Therefore, in order to obtain marketing approval, these devices require a "premarket approval" under section 515 of the FD&C Act.

The most stringent type of marketing device application by the FDA is a PMA. Prior to marketing the device, the applicant must receive FDA approval for its PMA application. The FDA approves a device by making sure it has enough valid scientific evidence to demonstrate that it is safe and effective for its intended use. The person who reserves the right and has authorized access to the data and other necessary information to be submitted to FDA is known as the PMA applicant. This could be any legal entity, organization, institution, or individual while the applicant is mostly the manufacturer or the developer of the device.

21.5.2 Post-Market Surveillance

In post-market surveillance, the developer or manufacturer of a certain device explores the market's opinion about the product so that they may take necessary actions for further improvements. This is crucial because it ensures that the device is working smoothly and helps in identifying common problems that occur. The evaluation of post-market surveillance can also help improve the medical device.

It is seen that in the past several developments have had a great impact on the post-market surveillance system. In 2017, the European Union published regulations on medical devices, and in the same year the WHO issued a global regulatory framework for medical devices and in vitro diagnostics (IVDs) [8]. Among these, EU regulations have set relatively strong and detailed requirements for post-market surveillance, including how post-market data is used, such as updating risk management files and clinical evaluations. Similarly, the United States FDA has emphasized on using data from past experiences gained using medical devices [9]. Real-world evidence for post-market surveillance is being considered by a number of regulators and other regulatory processes.

21.5.3 Clinical Trials

Herbal products have become a major and important part of the public healthcare around the world [24]. Their widespread use has been highlighted in various surveys on traditional and alternative medicines. However, clinical trials of such products must be encouraged in order to further widen their acceptance. It is advised that single and continued batches of formulations should be used to prove the efficacy of the clinical trials. This has become an integral part of its large-scale endorsement on national and international levels, although the herbal practitioners are not required to perform clinical trials [10].

21.6 DIVERSE REGULATORY STANDARDS

Difficulties in harmonizing herbal pharmacopoeia regulation has been greatly influenced by the diverse regulatory standards that are in place across various nations and regions [25]. This variation results from the different definitions and categorizations of herbal items, which, depending on the jurisdiction, might be classified as food, dietary supplements, or prescription medications [26]. Since various markets may have different regulations governing the same herbal product, these disparities may cause misunderstanding regarding safety and quality control standards (Figure 21.2).

21.6.1 Example of Divergence

There is a great deal of divergence in the ways of harmonizing the standards and approval process of herbal products to achieve global access among different agencies. For instance, the pre-market approval processes for herbal products are not essential according to the US FDA while the EMA strictly emphasizes the quality control and quality assurance of the herbal products before launching it into the market [27]. Rules in place for the herbal products by the EMA are relatively strict when compared with other regulation agencies.

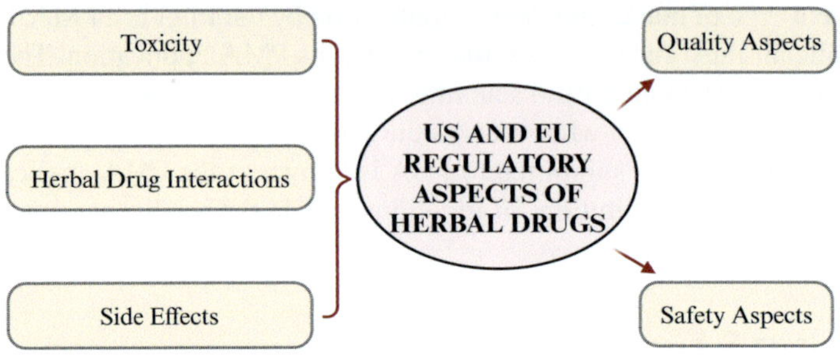

FIGURE 21.2 Regulatory approval process for herbal medicines.

There are huge differences in different regions regarding the consumption of herbal products. The lack of reliability and accessibility of the same herbal drug globally is a major obstacle in the approval process of herbal pharmaceuticals. Because of such differences in the laws of herbal pharmaceuticals, it is difficult to commercialize them in the market. To overcome these issues, the efforts of the international forum are required. There must be coherence in terms of the standards and quality control across the globe [10]. Although the WHO has developed a standardized set of regulations for quality control and quality assurance, the implementation of the guidelines still vary among the state members [27].

21.7 EFFORTS FOR INTERNATIONAL COLLABORATION

International collaboration in the area of herbal products is necessary because there is a complete difference between the consideration of herbal products when they are regarded as dietary supplements and when they are treated as pharmaceutical products. Due to the increase in globalization, there room for the introduction of one single herbal products classification across the world [28]. The role of the WHO in international collaboration is reflected through the development of an International Herbal Pharmacopeia (IHP). The main aim of the IHP is to standardize the quality control and safety measure of the herbal products across every individual country's pharmacopeia. Another prominent initiative in this regard is the formation of a Pharmacopeial Discussion Group (PDG), which will contain members from all the countries and states around the globe [29]. This group arranges the International Meeting of World Pharmacopeias for the harmonization of the regulations of herbal products.

21.8 IMPACT OF SCIENTIFIC ADVANCEMENTS

Scientific advancements are playing a crucial role in identifying and developing new herbal pharmaceutical products. Different technologies are used to check the chemical composition and potential effects of those components in order to identify their medicinal purposes. In this era of scientific advance, it is convenient to analyze and separate the active compounds from the herbal extracts. The analytical techniques of chromatography, characterization and spectroscopy, such as high-performance liquid chromatography (HPLC), X-ray diffraction (XRD), mass spectroscopy (MS), and nuclear magnetic resonance (NMR), are used for the detection of active and toxic compounds in herbal products [30]. The techniques also standardize the quality control measures and dosage ranges of the active compounds of these herbal products.

21.8.1 Combination of Modern Research and Traditional Knowledge

Herbal products have been used for medicinal purposes from the earliest times of human existence. Our forefathers used these products to cure the illnesses, and this traditional knowledge is enriched with invaluable insights for human health and the natural environment [31]. Through the integration of scientific technologies for experimenting on this traditional knowledge, researchers are able to overcome the gap between the effective usage of the herbal products and the modern research related to these medicines. The tradition of following customary rituals without any scientific evidence for the herbal products is a hindrance affecting herbal pharmaceutical development. This gap can only be overcome by integrating the scientific techniques in this traditional knowledge to evaluate the medicinal effect of herbal products.

21.8.2 Recognizing the Value of Traditional Knowledge

It is impossible to develop herbal products and their regulations without acknowledging traditional knowledge. The traditional knowledge transferred from generation to generation through the oral mode has played a significant role in the various fields of sciences such as ecology, medicines, resource management, and agriculture. In many instances, the early traditional methods are the basis for more scientific technologies [32]. Over the past few years, scientists have been emphasizing the fusion of traditional knowledge with scientific techniques. The knowledge gained from traditional practices is the basis for scientific development, and these two approaches must both be taken into consideration and valued.

21.9 APPROACHES TO INTEGRATION

Since both the scientific and traditional approaches each have their own significance, there are four different research approaches which can be used to combine these two major fields of information [33]:

- Collaborative Research
- Participatory Research
- Interdisciplinary Research
- Comparative Research

21.9.1 Collaborative Research

In this type of research, scientists collaborate with the communities and gain firsthand traditional knowledge from them. In turn, researchers provide maximum benefit to the subject of studies carried out within these communities.

21.9.2 Participatory Research

In this type of research, the indigenous communities voluntarily participate with the research groups and help them to collect and analyze the more relevant and meaningful information.

21.9.3 Interdisciplinary Research

In this approach, scientists from both the traditional and modern fields work together and analyze the evidence basis for the development of the herbal pharmaceutical.

21.9.4 Comparative Research

In such research, the scientists compare both the traditional and modern approaches and refine the available knowledge. In this way, they formulate new hypotheses and prove them through different experiments.

21.9.4.1 Benefits of Integration

The benefits of integrating both types of knowledge not only enhance the improvement in the development of herbal medicines, but also mitigate the risks associated with the side effects and contradictions of the herbal products. The synergy of both tradition and modern knowledge provides an improved and more effective formulation of the pharmaceutical products [34]. Modern technologies such as genetic engineering, characterization, and spectroscopy help in the conservation of medicinal compounds. The involvement of indigenous people in the research opens room for the development and deep understanding of the concept. Overall, the integration of traditional knowledge and the modern scientific approach expands the market for herbal products by advances in social, health, and environmental sustainability and innovation [35].

21.10 CHALLENGES & CONSIDERATIONS

Along with the benefits of this integration, there is an equal number of challenges and consideration for this collaboration. The foremost issue is the intellectual property right among the indigenous people and the scientists. There must be rules and regulations to protect the native people and scientists copyright claims and both parties must prioritize the protection of each other's rights for fair and equitable benefits. Cultural sensitivity is the other challenge in this integration [36]. Scientific researchers must consider the cultural context of the traditional knowledge and follow, as well as respect, the protocols of indigenous communities. Understanding cultural sensitivity can overcome challenges such as miscommunication, a lack of trust, and hindrance in collaboration. Investigating traditional knowledge within a scientific setting is itself a challenge. Not all the traditional techniques and easily replicated and optimized in a scientific setting.

In order to successfully integrate both of these approaches, it is necessary to communicate these issues within an international forum and set the regulations for overcoming the challenges associated with integration [37]. An ethical perspective for the research can also be achieved by considering and communicating these challenges.

21.11 ADVANCED TECHNOLOGIES IN QUALITY CONTROL

With the advancement of the technologies the quality, consistency, efficacy, and standard of the herbal pharmacopoeia landscape is improving. In this era, given the increase in global demand for natural rather than synthetic products, herbal products are in the limelight. It is now possible for the quality of these products to be secured through the enforcement of the strict quality assurance [38]. Cutting-edge technology can improve the proficiency of these products. These technologies include laboratory techniques and data resourcing through databases and the internet.

21.11.1 Analytical Techniques

Analytical techniques include all of the advancements of laboratory experiments so as to ensure the quality and efficacy of herbal products. These technologies include chromatography, characterization, and spectroscopy such as HPLC, XRD, MS, NMR. All these analytical techniques enable the identification and quantification of active compounds in herbal products [39, 40]. For example, XRD identifies the chemical compounds while HPLC segregates them from one another. Finally, the techniques of spectroscopy identify the molecular structures of these compounds.

21.11.2 Good Manufacturing Practices (GMP)

The quality of the herbal products can be affected significantly by the manufacturing processes and practices. Good Manufacturing Practices (GMP) include the stepwise quality assurance of the herbal products. Compliance with the guidelines and regulations of good manufacturing practices leads to the standard and quality products [41]. Each and every step, from obtaining raw material to the formation and packaging of the product, is inspected and validated during GMP practice. By complying with the requirements of GMP, the safety, quality, and effectiveness of products can be improved while there is a minimization of the danger of contamination and unpredictability.

21.11.3 Biological Assays

Biological assays are the essential quality control parameter for advances in quality control. The efficacy and potency of the herbal product is analyzed through performing clinically relevant bioassays and animal studies. In botanical and herbal products, it is mandatory to check the cytotoxicity and to study the synergistic effect in order to achieve the desired therapeutical effect in clinical trials [42]. Cell culture is one of the main advances in biological assay for the quality control and quality assurance of herbal products.

21.11.4 Standardization of Extraction Methods

In herbal pharmaceuticals, herbal extracts are used to obtain the active medicinal compounds. In order to gain pure and effective active compounds, it is essential to standardize extraction techniques. With advances in scientific knowledge, techniques such as supercritical fluid extraction (SFP) and ultrasound assisted extraction (UAE) are used to optimize the extractions of the compounds [43]. The techniques not only increase the yield of the active compound but also minimize the degradation of the sensitive phytochemicals.

21.11.5 Data Management & Traceability

With the advancement of technology, it is now possible to access and manage the large amount of data generated and to accurately assess each and every step of the manufacturing process [44]. The automation of machines allows the monitoring and recording of each stage, from the raw material acquisition up to the supply of product in the market. The degree of traceability ensures the degree of quality of the herbal product.

21.12 CHALLENGES & FUTURE DIRECTIONS

In spite of advances in technology, there are still many challenges before we achieve uniform quality standards. The chief issue in attaining these standards is the heterogenous nature of herbal products, and the fact that there is, at present, no comprehensive global framework for their standardization [45]. To overcome this issue, there is the need to establish greater collaboration among scientists, industrial stockholders, and regulatory bodies.

21.13 PERSONALIZED HERBAL MEDICINE

Personalized herbal therapy is the practice of customizing herbal remedies for each patient according to their specific genetic, environmental, and lifestyle characteristics [46, 47]. This approach recognizes that individuals may respond differently to herbal products, necessitating a more customized approach to treatment. The following steps are required in prototyping the personalized herbal medicines [48].

21.14 REGULATORY IMPLICATIONS

The development of guidelines for the formulation of personalized medicine is the key requirement in regulatory implications. These standards must ensure the safety and quality benchmark for the compounding formulation and practices. To ensure the effectiveness of safety measures, pre-market and post-market surveillance is required [49]. Accordingly, there must be rules for checking the cytotoxicity of personalized medicine through cell culture techniques. Regulatory bodies must require the manufacturer to conduct clinical trials for the effectiveness and side effects analysis of personalized medicines.

The regulatory implication includes introducing standards for the cultivation, harvesting, and consumption of the herbal products. Certification for the harvesting of the herbal products will help consumers to make informed choices and encourage manufacturers to adopt ethical practices [49]. Traceability and transparency in the supply chain of herbal products verify the sustainability of practices. Finally, regulations for the monitoring of conservation efforts with regard to endangered species help to achieve sustainability. Currently, the Convention on the International Trade in Endangered Species of Wild Flora and Fauna is the one of the main practical examples of sustainability [50].

21.15 SUSTAINABLE AND ETHICAL SOURCING

In the regulatory framework around herbal products, the sustainability and ethical sourcing of ingredients is a key factor. Both consumers and regulatory bodies are obliged to consider the sensitivity of prioritizing environmental and social responsibilities [51]. The unavailability and diversity loss of the herbal product due to overharvesting and unsustainable farming are among the major reasons for the decline in herbal products when this occurs. Accordingly, regulations must be put in place for sustainable and ethical sourcing.

21.16 CONCLUSION

In the area of herbal products and pharmaceuticals, a number of different regulatory bodies are playing major roles in the development of a comprehensive pharmacopeia. These bodies have divergent approaches to the development of rules and regulations for product development, which is acting as an obstacle to harmonization. The World Health Organization is playing a key role in trying to establish a global framework to achieve this. Any herbal pharmacopeia will have to consist of both traditional and modern knowledge and the integration of these two approaches will enhance the efficacy and safety of the herbal product. With the development of a suitable regulatory framework, the trend of personalized medicine, and ethical data resourcing and sustainability, will be effectively addressed and able to meet the needs of consumers.

21.17 FUTURE OUTLOOK FOR THE REGULATORY FRAMEWORK

In the future, the regulatory framework for herbal products will continue to evolve, incorporating advances in technology [52]. The focus of the regulatory bodies is for personalized medicine with a greater emphasis on environmental and health sustainability. The increased collaboration of the major regulatory bodies on international forums is one of the main goals for the future. Along with this, there should be support for the innovation and mutual benefits for the consumers and practitioners which are aligned with the objectives of the regulatory framework and agencies [53].

REFERENCES

1. Ernst E. Herbal medicines: balancing benefits and risks. *Novartis Foundation symposium*. 2007;282:154–67
2. Pal SK, Shukla Y. Herbal medicine: current status and the future. *Asian Pacific journal of cancer prevention: APJCP*. 2003;4(4):281–8.
3. Williams CT. Herbal Supplements: Precautions and Safe Use. *The Nursing clinics of North America*. 2021;56(1):1–21.

4. US Food and Drugs Administration. *Approval letter Veregen™ Ointment, 15%, NDA 021902*. 2006.
5. Heinrich M. Quality and safety of herbal medical products: regulation and the need for quality assurance along the value chains. *British Journal of Clinical Pharmacology*. 2015;80(1):62–6.
6. Wang G. Awards-Herbal, Traditional & Alternative Medicine 2020. *Asian Journal of Plant Science & Research*. 2020;9:2
7. Mawere M. Ethical quandaries in spiritual healing and herbal medicine: A critical analysis of the morality of traditional medicine advertising in southern African urban societies. *Pan African Medical Journal*. 2011;10.
8. Wah CL, Hock SC, Yun TK. Current scientific status and regulatory control of traditional/herbal medicinal products: Globalization challenges. *Pharmaceutical Engineering* 2012;32(6):10–20.
9. Healy MJ, Tong W, Ostroff S, Eichler H-G, Patak A, Neuspiel M, et al. Regulatory bioinformatics for food and drug safety. *Regulatory Toxicology and Pharmacology*. 2016;80:342–7.
10. Thakkar S, Anklam E, Xu A, Ulberth F, Li J, Li B, et al. Regulatory landscape of dietary supplements and herbal medicines from a global perspective. *Regulatory Toxicology and Pharmacology*. 2020;114:104647.
11. Mahady GB. Global Harmonization of Herbal Health Claims. *The Journal of Nutrition*. 2001;131(3): 1120S–3S.
12. Eisenberg DM, Kessler RC, Foster C, Norlock FE, Calkins DR, Delbanco TL. Unconventional medicine in the United States—prevalence, costs, and patterns of use. *New England journal of medicine*. 1993; 328(4):246–52.
13. United States Food and Drug Administration. *New dietary ingredients notification process*. Washington, DC: US Food and Drug Administration Available online at: http://www.fda.gov/Food/DietarySupplements/NewDietaryIngredientsNotificationProcess. 2016.
14. Denham BE. Dietary supplements—regulatory issues and implications for public health. *Jama*. 2011; 306(4):428–9.
15. Ross S. Functional foods: the Food and Drug Administration perspective. *The American journal of clinical nutrition*. 2000;71(6):1735S–1738S.
16. FDA. *Dietary Supplements: New Dietary Ingredient Notifications and Related Issues: Guidance for Industry*: FDA. 2016.
17. Walji R, Boon H. Natural health product regulations: perceptions and impact. *Trends in Food Science & Technology*. 2008;19(9):494–497.
18. European Parliament and Councill. Directive 2002/46/EC of the European Parliament and of the Council of 10 June 2002 on the approximation of the laws of the Member States relating to food supplements. *The Official Journal of the European Union* 2002;183:51.
19. World Health Organization. *WHO traditional medicine strategy: 2014-2023*: World Health Organization; 2013.
20. Schroeder T. Chinese regulation of traditional Chinese medicine in the modern world: can the Chinese effectively profit from one of their most valuable cultural resources? *Pacific Rim Law & Policy Journal* 2002;11:687.
21. Liang Z, Hu H, Li J, Yao D, Wang Y, Ung COL. Advancing the regulation of traditional and complementary medicine products: A comparison of five regulatory systems on traditional medicines with a long history of use. *Evidence-Based Complementary and Alternative Medicine*. 2021;2021(1):5833945.
22. China NMPAO. *Requirements for Registration, Categories and Application Documents of Traditional Chinese Medicine*. National Medical Products Administration of China Beijing, China; 2020.
23. Edwards IR, Biriell C. Harmonisation in pharmacovigilance. *Drug Safety* 1994, 10:93–102.
24. Edwards IR, Biriell C. Harmonisation in pharmacovigilance. *Drug Safety*. 1994;10(2):93–102.
25. Leung P-C, Xue C, Cheng Y. Clinical trials using Chinese medicine. *A Comprehensive Guide to Chinese Medicine*. 2003:275–300.
26. Lewith GT, Jonas WB, Walach H. *Clinical research in complementary therapies: Principles, problems and solutions*: Elsevier Health Sciences; 2010.
27. Verma N. Current regulatory challenges and approaches in the registration of herbal drugs in Europe. *Clinical Research and Regulatory Affairs*. 2016;33(1):9–24.
28. Drezner DW. Globalization, harmonization, and competition: the different pathways to policy convergence. *Journal of European Public Policy*. 2005;12(5):841–59.
29. Drichoutis AC, Lazaridis P, Nayga Jr RM. Consumers' use of nutritional labels: a review of research studies and issues. *The Academy of Marketing Science Review* 2006;2006:1.

30. Mukherjee PK, Bahadur S, Chaudhary SK, Kar A, Mukherjee K. Quality related safety issue-evidence-based validation of herbal medicine farm to pharma. *Evidence-based validation of herbal medicine*: Elsevier; 2015. p. 1–28.
31. Saggar S, Mir PA, Kumar N, Chawla A, Uppal J, Kaur A. Traditional and herbal medicines: opportunities and challenges. *Pharmacognosy Research*. 2022;14(2), 107–114.
32. Karunamoorthi K, Jegajeevanram K, Vijayalakshmi J, Mengistie E. Traditional medicinal plants: a source of phytotherapeutic modality in resource-constrained health care settings. *Journal of Evidence-Based Complementary & Alternative Medicine*. 2013;18(1):67–74.
33. Klein-Junior LC, de Souza MR, Viaene J, Bresolin TM, de Gasper AL, Henriques AT, et al. Quality control of herbal medicines: From traditional techniques to state-of-the-art approaches. *Planta medica*. 2021;87(12/13):964–88.
34. Seger C, Sturm S, Stuppner H. Mass spectrometry and NMR spectroscopy: modern high-end detectors for high resolution separation techniques–state of the art in natural product HPLC-MS, HPLC-NMR, and CE-MS hyphenations. *Natural Product Reports*. 2013;30(7):970–87.
35. Lemonnier N, Zhou G-B, Prasher B, Mukerji M, Chen Z, Brahmachari SK, et al. Traditional knowledge-based medicine: a review of history, principles, and relevance in the present context of P4 systems medicine. *Progress in Preventive Medicine*. 2017;2(7):e0011.
36. Fokunang C, Ndikum V, Tabi O, Jiofack R, Ngameni B, Guedje N, et al. Traditional medicine: past, present and future research and development prospects and integration in the National Health System of Cameroon. *African Journal of Traditional, Complementary and Alternative Medicines*. 2011;8(3). 10.4314/ajtcam.v8i3.65276
37. Brodnig G, Mayer-Schönberger V. Bridging the gap: the role of spatial information technologies in the integration of traditional environmental knowledge and western science. *The Electronic Journal of Information Systems in Developing Countries*. 2000;1(1):1–15.
38. Kennedy G, Wang Z, Maundu P, Hunter D. The role of traditional knowledge and food biodiversity to transform modern food systems. *Trends in Food Science & Technology*. 2022;130:32–41.
39. Casimirri G, editor Problems with integrating traditional ecological knowledge into contemporary resource management. *Submitted to the XII World Forestry Congress*; 2003.
40. Sen S, Chakraborty R. Revival, modernization and integration of Indian traditional herbal medicine in clinical practice: Importance, challenges and future. *Journal of Traditional and Complementary Medicine*. 2017;7(2):234–44.
41. Sarma N, Upton R, Rose U, Guo D-a, Marles R, Khan I, et al. Pharmacopeial standards for the quality control of botanical dietary supplements in the United States. *Journal of Dietary Supplements*. 2023;20(3):485–504.
42. Wang H, Chen Y, Wang L, Liu Q, Yang S, Wang C. Advancing herbal medicine: enhancing product quality and safety through robust quality control practices. *Frontiers in Pharmacology*. 2023;14:1265178.
43. Muyumba N, Mutombo S, Sheridan H, Nachtergael A, Duez P. Quality control of herbal drugs and preparations: The methods of analysis, their relevance and applications. *Talanta Open*. 2021;4:100070.
44. He T-T, Ung COL, Hu H, Wang Y-T. Good manufacturing practice (GMP) regulation of herbal medicine in comparative research: China GMP, cGMP, WHO-GMP, PIC/S and EU-GMP. *European Journal of integrative Medicine*. 2015;7(1):55–66.
45. Shinde VM, Dhalwal K, Potdar M, Mahadik KR. Application of quality control principles to herbal drugs. *International Journal of Phytomedicine*. 2009;1(1):4–8.
46. Prado JM, Veggi PC, Meireles MAA. Scale-up issues and cost of manufacturing bioactive compounds by supercritical fluid extraction and ultrasound assisted extraction. *Global Food Security and Wellness*. 2017:2017;377–433.
47. Choudhary N, Sekhon BS. An overview of advances in the standardization of herbal drugs. *Journal of Pharmaceutical Education and Research*. 2011;2(2):55.
48. Enioutina EY, Salis ER, Job KM, Gubarev MI, Krepkova LV, Sherwin CM. Herbal Medicines: challenges in the modern world. Part 5. status and current directions of complementary and alternative herbal medicine worldwide. *Expert Review of Clinical Pharmacology*. 2017;10(3):327–38.
49. Thomford NE, Dzobo K, Chimusa E, Andrae-Marobela K, Chirikure S, Wonkam A, et al. Personalized herbal medicine? A roadmap for convergence of herbal and precision medicine biomarker innovations. *OMICS: A Journal of Integrative Biology*. 2018;22(6):375–91.

50. Stargrove MB, Treasure J, McKee DL. *Herb, nutrient, and drug interactions: clinical implications and therapeutic strategies*: Elsevier Health Sciences; 2007.
51. Rehman SU, Shahzad M, Ding X, Razzaq A. Impact of corporate motives for sustainable sourcing: Key moderating role of regulatory pressure. *Environmental Science and Pollution Research*. 2023;30(27):71382–95.
52. Initiative CS, Golden JS, Vermeer D, Clemen B, Davie Nguyen M. An overview of ecolabels and sustainability certifications in the global marketplace. *Nicholas Institute for Environmental Policy Solutions Duke University Interim Report Document*. 2010:10–1.
53. Uhlenbrock L, Sixt M, Tegtmeier M, Schulz H, Hagels H, Ditz R, et al. Natural Products Extraction of the Future—Sustainable Manufacturing Solutions for Societal Needs. *Processes*. 2018;6(10):177.

22 Present Challenges and Future Perspective of the Herbal Drug Industry

Mehtab Yasmeen
Department of Botany, Telangana Mahila Viswavidyalayam, Hyderabad, India

Mukul Machhindra Barwant
Department of Botany, Sanjivani Rural Education Society's (SRES), Sanjivani Arts, Commerce and Science College, Kopargaon, India

Balwant Singh
Department of Botany, Dr. Ram Manohar Lohia Avadh University, Ayodhya, Uttar Pradesh, India

Odangowei Inetiminebi Ogidi
Department of Biochemistry, Faculty of Basic Medical Sciences, Bayelsa Medical University, Yenagoa, Nigeria

22.1 INTRODUCTION

Over the past few years, there has been an amazing growth in the herbal drug industry worldwide. This is mainly due to the increase in demands for plant products and a growing awareness about holistic therapy. Herbal drugs involve plants, or parts of plants, that provide an alternative method to modern medicines, treating the ailments but with less side effects. This new interest is mainly due to the practicing of traditional knowledge in many countries, increased awareness about dangerous outcomes of allopathy along with enormous body of scientific research data supporting the efficiency of many herbal drugs. Although the industry of herbal drugs has been growing positively, it has not fully realized its potential because it faces a number of hurdles. Among these challenges are regulatory barriers, quality control concerns, and standardizing herbal products. Various regions have different ways in which they regulate herbal medicines, which complicates approval processes and access to markets. Furthermore, the problem of maintaining uniformity in the quality and safety of herbal products is consistent given the fact that both plant sources and harvesting methods differ, and so do processing techniques. The sector's landscape is further complicated by unstandardized dosages and formulations which affect both consumer confidence and clinical efficiency. Moreover, the herbal drug market is plagued by issues of sustainability and ethical sourcing. The overharvesting of medicinal plants and the exploitation of biodiversity pose serious threats to the environment and local communities. Ethical sourcing practices and sustainable cultivation methods are imperative to safeguarding the future of the industry and the ecosystems it relies upon. The growth of the herbal drug industry therefore also depends on science and technology. The combination of modern biotechnological treatments with classical ones leads to the improvement in the standards and effectiveness of the herbal drugs. Many such effective methods of extraction and formulation of these products could have been used to arrive

DOI: 10.1201/9781003544418-22

at a more standard and acceptable form of 'herbal therapeutics'. Except for that, application of AI and digital technology in this area help to improve supply chain and traceability system for the herbal medicine and individualized prescription (Izah et al., 2023).

This chapter considers the current challenges facing the herbal drug industry and possible directions for its further growth. The use of science-based innovations addressing these concerns will ensure that the industry enjoys long-term sustainable development and makes a significant contribution to global health.

22.2 EMERGING TRENDS AND INNOVATIONS

22.2.1 Biotechnology and Genetic Engineering

As it has been ascertained that plants can produce some chemicals which are beneficial in the course of treatment, modern pharma research is gradually shifting towards these compounds. In this area, a lot of progress has been realized, and this has been as a result of the synthesis of these chemicals by employing biotechnology and synthetic biology methods (Ausländer et al., 2017). Since it turns out to be simple to produce enormous volumes of the pure constituent chemicals, this technique enhances the viability and safety of use of these medicine.

In fact, it is not wrong to say that a greater part of the modern medicines and therapies are derived from plant products and phytochemicals. Zajtchuk (1999) disclosed that biotechnology and synthetic biology have offered the new prospects for the drug development and approval. It will also enable the synthesis of new compounds in pharmaceutical industries that have even higher therapeutic values than the existing compounds in the market. This method has the propensity of altering the pharma industry through establishing productive and unique products which effectively address different diseases and diseases states (Gottweis, 1998). However, to that in cuts out the need to manually pick entire plants, an action that often leads to a destruction of the plant and space it occupies hence solving ecological problems. Bioreactors are containers that are cultivated under conditions that permit the growth of plant cells or microorganisms; these increase the synthesis rate of different compounds (Zhong, 2010). Probably, the genetic organization of these organisms may be redesigned to employ suboptimal metabolic engineering and the pathway engineering methods that will improve the efficiency of production of the encouraged compounds (Deparis et al., 2017). In addition to improving the sustainability of production, such innovations have a positive impact on the quality and trustworthiness of the plant-derived compounds which are synthesized. At the same time, bioinformatics as well as computer aided modeling and high-throughput screening are all mainly game changer in the discovery of new products. An advanced procedure called 'high-throughput screening' involves the screening of thousands to millions of isolated chemicals or extracts of plants for activity against specific biological targets using robotics and computers (Ismail et al., 2018). This makes it possible to find active chemicals rapidly. By contrast, virtual screening and molecular modeling methods are helpful in predicting the interaction among different chemical substances obtained from plants.

Virtual screening is a computer technique that assists in selecting possible medication that can be retrieved from large databases of chemical compounds. This process includes assessing the interactions between small molecule and proteins, together with other techniques in docking simulations and in molecular modeling (Reddy et al., 2007). It has been applied in optimizing currents drugs, identifying new lead molecules for drug development as a part of pharmaceutical realm (Singh and Bharadvaja, 2021). This enables one to grasp the hierarchical arrangements and the possible treatment regimes. As reported by Ambrosino et al. (2019), bioinformatics provides a plethora of information focuses on genes, metabolism, and pharmaceutics of these species. Such information is applied in the identification of new compounds and in the understanding of the biosynthesis processes. The technology integration also assists in identifying as well as enhancing plant-based compounds towards the enhancement of plant yielding, and quite successfully opens a new chapter of medication production. The molecular modeling method can be used to estimate the behavior and interaction of molecules and other molecular-based substances with the help of computers (Dhoundiyal and Alam, 2023). It is used to discover new drug molecules and therapeutic compounds against different diseases and also helps in studying the structure and functions of proteins, DNA, and other biological substances. These modern techniques provide a synergy of rather effective components to address environmental issues, as well as enhancing the methodical search and development of novelties on medications. Biotechnology and synthetic biology, which reduce environmental impact, are at the heart of the correct synthesis of compound parings originating from plants (Singh and Bharadvaja, 2021; Ogidi and Emaikwu, 2023a,b). As Sharma & Sarkar (2013) have stated, high-throughput screening and other computational methods are time-saving and systematic, on the one hand, and methodical, on the other hand, in recognizing the potential medicinal molecules. The knowledge of these molecules is enhanced by bioinformatics, thereby leading to the development of new medications (Wishart, 2005). In general, the application of biotechnology, which is influenced by powerful computing equipment, contributes to the enhancement of the herbal pharmaceutical sector with improved environmental outcomes and also the faster production of new medications.

22.2.2 Nanotechnology

The intrinsic components of the nanoparticles (NPs), which are surface-bound to the polymeric matrix, may include the active agents, which will have sizes ranging from 1 to 1000 nm. These can be used as drug carriers to increase the therapeutic window and bioavailability, provide regulated release, and shield different desired chemicals from the environment (Zielińska et al., 2020). Depending on the process used to make nanocapsules or nanospheres, one could coat nanoparticles. The medicine in nanocapsules is contained within a cell surrounded by a polymer membrane, whereas the drug in nanospheres is encased in a matrix that has particles physically scattered throughout it (Soppimath et al., 2001). Because of their exceptional stability and their capacity to transport both hydrophilic and hydrophobic molecules, nanoparticles are a promising option for the delivery of pharmaceuticals. It is also possible to administer them either orally or by means of inhalation. According to Gelperina (2005), these carriers might be made to enable the medication's gradual, controlled release. Furthermore, Mughees et al.'s 2021 research emphasizes the usefulness of nanoparticles in conjunction with phytochemicals and herbal extracts as a targeted anticancer therapeutic approach, albeit one with some drawbacks. Nanoparticles can be used as effective medicine carriers because they are highly stable, have a high carrier capacity, and can contain both hydrophobic and hydrophilic compounds. In addition, they can be given orally using a number of methods, including inhalation. According to Gelperina (2005), these carriers might also be made to make it easier for the drugs to release matrix-regulated over a longer time frame. Mughees et al. (2021) state that polymeric nanoparticles work against cancer by delivering medications in a safe and effective manner when paired with phytochemicals and herbal extracts. Similarly, liposomes containing lipid nanoparticles are among the few lipidic nanoparticles. Liposomes are tiny lipid-containing vesicles with a diameter of 0.1–10 nm that have a watery core surrounded by two layers of phospholipid. Agrahari et al. (2016) state that there are three different types of liposomes based on size and lipid bilayers: tiny unilamellar vesicles, big uniflagellar vesicles, and multilamellar vesicles. According to Eloy et al. (2017), it can be utilized to combine hydrophilic and lipophilic compounds. It is anticipated that using liposomes to target certain pharmaceuticals may improve the medications' therapeutic effects, reduce any potential toxicity, and eliminate localized irritation.

Liposomes help deliver drugs to particular places with controlled release, change biodistribution, and boost solubility and bioavailability. Another advantage of using liposomes for parenteral distribution is the avoidance of first-pass metabolism, limited gastrointestinal permeability, and deleterious effects on the gastrointestinal tract (Shah et al., 2020). In 2020, Shariare et al. (2020) developed a liposomal drug delivery technique using *Aphanamixis polystachya* leaf extract to treat neurological conditions such as Parkinson's disease and Alzheimer's disease. Solid lipid nanoparticles (SLNs) are the ideal alternative to colloidal approaches for targeted and controlled delivery. They contain hydrophilic and lipophilic medications and are made of biodegradable and biocompatible materials that range in size from 50 to 1000 nm (Manjunath et al., 2005). Since most lipids are biodegradable, SLNs offer remarkable biocompatibility. Both lipophilic and hydrophilic medications can be included in SLNs, which provide controlled and regulated drug release. Furthermore, SLNs are less expensive than carriers made of surfactants or polymers (Mukherjee et al., 2009).

Hibiscus rosasinensis extract-loaded solid lipid nanoparticles, or HSLNs, were initially made with glycerol monostearate or beeswax as the substrate by Vijayanand et al. (2018). The form, surface charge, and size of the HSLNs were assessed. Tests were conducted on male Swiss albino mice to assess the antidepressant efficacy of the enhanced HSLNs. It was found that HSLNs with a nearly spherical morphology and a negative charge may be manufactured at as small as 175 nm by using the optimized approach. The in vivo experiment data showed that the test animals' periods of immobility differed greatly. It was also found that HSLNs, at doses several times lower than those of the crude extract derived from native plants, could provide the same pharmacological effects. According to preliminary studies, solid lipid nanoparticle encapsulation of phytopharmaceuticals may enhance their in vivo efficacy (Vijayanand et al., 2018).

22.3 REGULATORY CHALLENGES AND OPPORTUNITIES

Each country has different legal standards for herbal medicines. In developing nations, the tradition of using plants for treating diseases is quite popular among the public. However, the drug laws in many of these countries do not regulate these widely used herbal remedies (Chavan et al., 2006). In most of the countries, herbal drug therapy relies on traditional knowledge of medicinal plants, provided do not harm for any regular ailments. However, in areas lacking traditional medical knowledge, many serious illnesses are being treated with alternative therapies. Despite this, the belief that herbal medications are free from allergies or negative effects is incorrect, as they contain various potentially harmful substances. Most herbal medicine purchases occur over the counter in a traditional setting. Therefore, legal regulations are essential to confirm the effectiveness, quality, and safety of herbal remedies. Conducting clinical trials and assessing the potential harm of herbal treatments are necessary steps for maintaining the regulatory status of herbal pharmaceuticals. Laws regarding herbal medications vary from country to country (Mukherjee et al., 2006).

22.4 INTELLECTUAL PROPERTY RIGHTS

Basically, whenever innovations are considered the findings of Western science are given importance. However, many inventions that are protected by intellectual property rights (IPR) are known which are known due to years of traditional knowledge build up, taking into consideration of experiences of other countries. Curing a sickness that modern medicine does not know how to treat is one of the most researched fields, and new issues about the IPR of the local population continue to arise. Diverse viewpoints on the matter are presented by the different domains and relation of intellectual rights and property (Dutfield, 2000).

The act of simply exploiting such plant-based knowledge without granting credit to the original author is known as "biopiracy," and it is a current reality that has various legal repercussions and poses a threat to inventors in the future (Gollin, 1999). IPR has progressed into a standard protocol for original ownership of all traditional information and has helped to pave the way for different possibilities for legal frameworks that guarantee benefits return to the original nations and cultures. The value of biodiversity for human health is emphasized by the IPR debates. Throughout history, plants are known to be basic suppliers for pharmacy. However, over time, the rate at which civilizations and species are going extinct, and non-curable diseases have worsened health conditions. Therefore, there has been global discussion about the IPR associated with the use of plants as medicines, and their importance has become increasingly acknowledged (Moran, 1998). The Agreement on Trade Related Intellectual Property Systems (TRIPS), which creates common standards for intellectual property protection, has been the subject of intense discussion since its beginnings. This is because there are concerns that the implementation of TRIPS may result in pharmaceutical price increases, which would decrease people's access to necessary medicines, particularly in developing nations. Much of the conversation thus far has surrounded the issue of contemporary medications. When considering the topic of IPR for such products, developing countries can see that the goals are well-defined and that established solutions exist to accomplish those goals; however, putting those strategies into practice may be challenging. With relation to conventional medicine, the same cannot be true. The preservation of traditional knowledge is not mentioned in TRIPS, and it does not recognize or differentiate between industry-based information and knowledge originating from indigenous communities. Access to biological and genetic resources, as well as local and indigenous groups' understanding of their medicinal qualities, are all entwined with traditional medicine. Therefore, a thorough evaluation of the goals and the potential ramifications of IPR protection for traditional medicine should be conducted at the national level (Timmerman, 2003).

22.4.1 Conventional Medicine and Rights to Intellectual Property

The traditional and Indigenous peoples of the world have made enormous contributions to the preservation of biodiversity and associated traditional medical knowledge. Regular ethnobotanical surveys and related research yield fresh insights into the treatment of a wide range of illnesses, therapies for skin and body enhancement, natural oils, and other health-related items. In an attempt to find additional "explorations" about modern diseases from traditional and indigenous groups, the globalized economy is facing hurdles that are giving rise to new and complicated issues. The property rights of traditional or indigenous communities, who are considered the "inventors," collide with the notion that sharing and access to the world's biodiversity resources is a fundamental human right. Trade is the utilization of all resources—including ideas, resources, and conventional medical procedures—for the goal of commercialization; it usually takes place with little benefit to the country or community of origin. Vinblastine and vincristine, which are made from the Madagascar rosy periwinkle (Vinca sp.), are a well-known example. The company makes about $100 million a year from these drugs, but neither the government of Madagascar nor the shamans who gave the researchers the information have received any payment for their services.

Commercializing what are collective resources—often of a secret or sacred nature—violates human rights and is demeaning to the local way of life. Thanks to the following, IPRs: -recognize individual rights rather than collective rights; -require a specific act of "invention"; -simplify ownership regimes; -stimulate commercialization; -recognize only market values; -are expensive, complicated, and time-consuming; -are subject to economic powers and manipulation (Posey, 2002). The Convention on Biological Diversity (CBD) and IPR Moreover, several medicinal plants are in danger of being extinct. The overuse of these plants to satisfy industrial and/or export demands may put people at risk (Newman, 1994). It is also a historical issue since it took place in the ancient Mediterranean region. For example, the plant recognized for its ability to prevent conception, Silphium, brought incredible wealth to the ancient residents of Cyrene, a coastal city in ancient Libya. But the overharvesting of the plant resulted in its demise and the loss of future economic benefits (Sumner, 2001).

Controlling the trade in medicinal plants is therefore essential, ideally as part of a larger regulatory framework addressing the sustainable use and protection of biological resources. Conservation is not only an essential ally in this struggle, but is also vital to the survival of traditional medicine in countries with high biodiversity. The Convention on Biological Diversity (CBD) is the only major international agreement that acknowledges indigenous communities as the owners of biodiversity and protects their right to keep it that way, although it is competing with the far more powerful advantages of TRIPS (Bodeker, 2000).

Indian literature dates back over two millennia, and makes early mention of the neem tree (*Azadiracthta indica*). It can be made into numerous goods, including fungicide, cosmetics, insect repellent, and pharmaceuticals for both humans and animals. Indian farmers used to soak neem seeds in water and then spray the resulting emulsion on their plants.

Neem was later popularized by Western merchants because it had fewer negative side effects than most chemical insecticides. At present, many patents covering neem products are pending in Europe, the US, and India. In 1993, for example, P.J. Margo Private Ltd. began to produce and distribute neem biopesticides in India. Two years later, in 1995, following public demonstrations against the initiative, a number of advocacy organizations united to contest the US and European patents on the grounds that the product or procedure was not an innovative one and that Indians had been utilizing such products in the same way for generations. However, the US patent is still enforceable, although the European Patent Office has revoked the patent in Europe (Finger, 2003).

22.5 GLOBAL MARKET TRENDS

Pharmaceutical firms have benefited greatly from products generated from traditional knowledge (TK), with indigenous plant knowledge playing a critical role in this. These businesses are especially drawn to TK because of to the costliness of screening new medications. Usually, only one in every 10,000 molecules is developed into a new medication; this process can take up to 15 years and may cost over $800 million. There are about 121 prescription plant-based medications in the world, with around half of them coming from tropical areas. Partly because of their climate, tropical countries are valuable because of their enormous biological and chemical variety. While tropical plants have evolved chemical defenses to withstand minimal seasonal respite from predators, temperate climates experience a reduction in plant predators throughout winter. Additionally, these plant-based compounds may have therapeutic advantages for human health (Moran, 1998). In the past, businesses from more developed northern nations have produced pharmaceutical goods using the traditional knowledge of therapeutic herbs from less developed, biologically diverse tropical nations without making any payment to the knowledge providers. There is an increasing trend in the use of herbal medicines globally, and the global market for plant-based medications is estimated to be worth $169.6 billion (Kanhere, 2024).

Traditional medicine has two distinct meanings in both modern Western countries and developing nations. Because it is accessible and affordable, traditional medicine plays a vital role in health systems in underdeveloped nations.

Peria (2001) claims that traditional medicine addresses concerns about technological transfer, cultural heritage preservation, and biodiversity conservation. On the other hand, traditional medicine is also increasingly seen as a choice or an alternative in contemporary nations. Popular and useful herbal species are in greater demand (8–15% yearly) in Europe, North America, and Asia (Grünwald and Büttel, 1996). Cultural variations impact the application of traditional medicine; according to a poll conducted in the United States, just 3% of respondents said they had used traditional medicine in the preceding year (Eisenberg et al., 1993). By contrast, 31% of respondents in Germany, a country with a high prevalence of traditional herbal therapy, stated they have used over-the-counter phytopharmaceutical preparations in a similar period (BAH, 2002).

22.6 CHALLENGES AND LIMITATIONS

The overextraction of plant species for medicinal components can place some plant species in danger and lead to the depletion of natural resources. Collection methods that are not sustainable can degrade habitats, upend ecosystems, and endanger biodiversity. In order to safeguard ecosystems and the indigenous populations that depend on them, these concerns emphasize the need for sustainable and ethical sourcing practices as well as the preservation of medicinal plant species. Plant genetics, environmental circumstances, and harvesting practices are among the variables that can affect the chemical composition of natural goods, making it difficult to guarantee the consistent quality and efficacy of compounds derived from plants (Yang et al., 2018).

Better techniques are needed to ensure the consistent replication of medicinal effects, which calls for the standardization of plant extracts. Ensuring the safety and effectiveness of natural goods requires the implementation of stringent quality control procedures, such as thorough testing and the establishment of exact quality standards. Variations in the concentration of contaminants or bioactive chemicals have an impact on the safety and efficacy of herbal treatments. Plant-derived compounds exhibit considerable heterogeneity in their pharmacokinetics, which encompasses their bioavailability, distribution, metabolism, and elimination within the human body (Lan and Jia, 2010). The efficacy of pharmacological drugs can be impacted by variables such as chemical stability, absorption rates, and interactions with other substances. For the purpose of creating drugs with dependable and constant therapeutic effects, it is essential to comprehend and enhance the pharmacokinetic characteristics of plant compounds (Kumar and Sharma, 2018). This may involve chemical modifications, formulation techniques, or drug delivery systems to enhance their clinical efficacy.

The laws governing drugs derived from plants are complicated and varies throughout nations and areas. Development takes longer and costs more because regulatory agencies frequently require evidence of safety and efficacy testing (Fan et al., 2012). Concerns concerning intellectual property are also brought up by the commercialization of traditional knowledge and plant resources, which sparks debates about just recompense for indigenous people and the preservation of their cultural legacy. Certain plant materials come from particular regions and might not be easily accessible or reasonably priced for everyone (Nicoletti, 2012). The restricted availability of certain medications may impede fair dissemination of their advantages, particularly for individuals residing in impoverished or financially challenged areas. A multidisciplinary strategy is needed to address these issues, including conservation initiatives, sustainable harvesting methods, strict quality control procedures, harmonized regulations, and moral responsibility of individuals.

22.7 FUTURE DIRECTIONS

Natural chemicals derived from plants are expected to continue to be a valuable resource for the creation of novel drugs and therapeutic approaches. According to Lautie et al. (2020), a broad range of molecules with potential medical applications are available due to the tremendous chemical variety present in plants. Research on novel bioactive compounds and plant modes of action continues to strengthen the medication development pipeline. This is especially important when researchers look for sustainable and alternative medicine sources. Developments in the study and cultivation of plant-based materials are intimately related to advances in genomes and metabolomics. A thorough knowledge of the genetic processes generating bioactive compounds is made possible by plant genome sequencing (Marchev et al., 2021). A comprehensive review of the tiny molecules found in plant systems is provided by metabolomics. These advanced methods allow scientists to accurately identify potential compounds, predict their functions, and modify plant genomes to enhance the production of valuable metabolites. This systematic approach streamlines the discovery process, reducing the need for exhaustive examination while facilitating the development of compounds tailored for specific therapeutic properties.

The application of chemicals generated from plants will continue to be greatly advanced by biotechnology and synthetic biology. It is anticipated that advances in genetic engineering, bioproduction, and route optimization will increase the variety of compounds that can be harvested from plants (Dörnenburg, 2009). Ecological concerns can be addressed by using bioreactors, which are controlled settings that provide a scalable and sustainable way to produce chemicals obtained from plants. Plant metabolites can be tailored to the unique requirements of drug development by genetic engineering and changes of pathways (Moon et al., 2019). In future, precision medicine and tailored therapies could benefit from the use of plant-derived compounds. Researchers will be able to customize medicines depending on a person's genetic composition and health features by using genomic and metabolomic data (Ottinger and Geiselman, 2023). By addressing each patient's distinct health demands, this strategy will pick or design specific plant-based chemicals to maximize therapeutic efficacy and minimize side effects.

Research on the synergistic effects of chemicals originating from plants and how they interact with traditional drugs is an area that is expanding quickly. In a number of medical disorders, combining synthetic medications with natural plant compounds may improve treatment results, lessen side effects, and overcome drug resistance (Yuan, 2016). Future studies on understudied plant species could be part of the process of developing plant-based pharmaceuticals (Shu, 1998). Various parts of the world are home to distinct and unexplored biological resources. Examining these creatures scientifically may reveal previously unidentified substances with important therapeutic promise.

22.8 CONCLUSIONS

Herbal remedies obtained from plants are known for ages to cure diseases and will continue to be important sources of natural medicines and help in the design and synthesis of different drugs for the treatment of diseases in humans and animals. With the increase in awareness about low-level side effects of herbal drugs, interest in exploring the medicinal properties of previously unknown natural substances is also increased. In the process of the development plant-based drugs, there is an optimization of phytochemicals done to produce potential analogs having drug effectiveness and safer to use. There is great surge of interest in herbal products and numerous innovative techniques and research advancements have been developed for selecting, identifying, isolating, characterizing, and biologically screening natural ingredients. These innovations can reduce the technical challenges associated with herbal drug development and throw light on the difficult process involved in discovering and producing new herbal remedies. It is assumed that plants will remain the main source to provide biomolecules which are not known, and will facilitate the discovery of new and better treatments for curing different diseases. However, their great demand in the global market challenges their survival. Therefore, it is important to ensure the conservation of vulnerable, threatened, and overexploited genetic resources to the greatest extent possible. This will allow future generations, equipped with advanced abilities, to conserve and utilize these species more efficiently and sustainably.

REFERENCES

Ambrosino, L.; Tangherlini, M.; Colantuono, C.; Esposito, A.; Sangiovanni, M.; Miralto, M.; Sansone, C.; Chiusano, M.L. Bioinformatics for marine products: An overview of resources, bottlenecks, and perspectives. *Mar. Drugs* 2019, *17*, 576.

Ausländer, S.; Ausländer, D.; Fussenegger, M. Synthetic Biology—The Synthesis of Biology. *Angew. Chem. Int. Ed.* 2017, *56*, 6396–6419.

BAH, Pflanzliche Arzneimittel heute. Wissenschaftliche Erkenntnisse und arzneirechtliche Rahmenbedingungen. *Bestandsaufnahme und Perspektiven*. 3rd edition. – Bonn, Bundesfachverband der Arzneimittelhersteller, 2002.

Bischoff, G.; Hoffmann, S. DNA-binding of drugs used in medicinal therapies. *Curr. Med. Chem.* 2002, *9*, 321–48.

Bodeker, G., *Indigenous Medical Knowledge: The Law and Politics of Protection* – This study was presented at the Oxford Intellectual Property Research Centre, 2000

Chavan, P.; Joshi, K.; Patwardhan, B. DNA microarrays in herbal drug research. *Evid. Based Complement. Alternat. Med.* 2006, *3*, 447–57.

Cravens, A.; Payne, J.; Smolke, C.D. Synthetic biology strategies for microbial biosynthesis of plant natural products. *Nat. Commun.* 2019, *10*, 2142.

Deparis, Q.; Claes, A.; Foulquié-Moreno, M.R.; Thevelein, J.M. Engineering tolerance to industrially relevant stress factors in yeast cell factories. *FEMS Yeast Res.* 2017, *17*, fox036.

Dhoundiyal, S.; Alam, M.A. Advances in Pharmacokinetic Modelling and Computational Approaches for Nanoparticles in Drug Delivery Systems. *Recent. Adv. Drug Deliv. Formul. Former. Recent. Pat. Drug Deliv. Formul.* 2023, *17*, 210–27.

Dipankar, G.; Laddha, K.S., Herbal safety and GMP: The most debatable argument in the present scenario. *Pharm. Rev.*, 2006, *2*, 25–29.

Dörnenburg, H. Progress in kalata peptide production via plant cell bioprocessing. *Biotechnol. J.* 2009, *4*, 632–45.

Dutfield, G., *Intellectual Property Rights, Trade and Biodiversity*. Earthscan Publications Ltd, 2000.

Eisenberg, D. M.; Kessler, R. C.; Foster, C.; Norlock, F. E.; Calkins, D. R.; Delbanco, T. L., "Unconventional medicine in the United States-prevalence, costs and patterns of use" *N. Engl. J. Med.*, 1993, *328*(4), 246–52.

Eloy, J.O.; Petrilli, R.; Trevizan, L.N. et al. Immunoliposomes: a review on functionalization strategies and targets for drug delivery. *Colloids Surf. B. Biointerfaces* 2017, *159*, 454–67.

Fan, T.-P.; Deal, G.; Koo, H.-L.; Rees, D.; Sun, H.; Chen, S.; Dou, J.-H.; Makarov, V.G.; Pozharitskaya, O.N.; Shikov, A.N. Future development of global regulations of Chinese herbal products. *J. Ethnopharmacol.* 2012, *140*, 568–86.

Gelperina, S.; Kisich, K.; Iseman, M. D. et al. The potential advantages of nanoparticle drug delivery systems in chemotherapy of tuberculosis. *Am. J. Respir. Crit. Care Med.*, 2005, *172*, 1487–90.

Gharib, R.; Greige-Gerges, H.; Fourmentin, S. et al. Liposomes incorporating cyclodextrin–drug inclusion complexes: Current state of knowledge. *Carbohydr. Polym.* 2015, *129*, 175–86.

Gollin, M. A., "Legal and practical consequences of biopiracy" *Diversity*, 1999, *15*, 7–9.

Gottweis, H. *Governing Molecules: The Discursive Politics of Genetic Engineering in Europe and the United States*; MIT Press: Cambridge, MA, USA, 1998.

Grünwald, J.; Büttel, K., "The European phytotherapeutics market." *Drugs Made Ger.* 1996, *39*, 6–11.

Ismail, F.M.; Nahar, L.; Sarker, S.D. High-Throughput Screening of Phytochemicals: Application of Computational Methods. In *Computational Phytochemistry*; Elsevier: Amsterdam, The Netherlands, 2018; pp. 165–192.

Izah, S.C.; Ogidi, O.I.; Ogwu, M.; Salimon, S.; Yusuf, Z.; Akram, M.; Raimi, M., & Iyingiala, A. Historical Perspectives and overview of the value of herbal medicine. In: *Reference Series in Phytochemistry. Herbal Medicine Phytochemistry*. Springer Nature Singapore Pte Ltd. (2023)

Kumar, R.; Sharma, M. Herbal nanomedicine interactions to enhance pharmacokinetics, pharmacodynamics, and therapeutic index for better bioavailability and biocompatibility of herbal formulations. *J. Mater. Nanosci*. 2018, *5*, 35–60.

Lan, K.; Jia, W. An integrated metabolomics and pharmacokinetics strategy for multi-component drugs evaluation. *Curr. Drug Metab.* 2010, *11*, 105–14.

Lautie, E.; Russo, O.; Ducrot, P.; Boutin, J.A. Unraveling plant natural chemical diversity for drug discovery purposes. *Front. Pharmacol.* 2020, *11*, 397.

Liu, B.; Li, S.; Hu, J. Technological Advances in High-Throughput Screening. *Am. J. Pharmacogenomics* 2004, *4*, 263–76.

Manjunath, K.; Reddy, J.S.; Venkateswarlu, V. Solid lipid nanoparticles Solid lipid nanoparticles as drug delivery systems. *Methods Find. Exp. Clin. Pharmacol.*, 2005 Mar, *27*(2), 127–44

Marchev, A.S.; Vasileva, L.V.; Amirova, K.M.; Savova, M.S.; Balcheva-Sivenova, Z.P.; Georgiev, M.I. Metabolomics and health: From nutritional crops and plant-based pharmaceuticals to profiling of human biofluids. *Cell. Mol. Life Sci.* 2021, *78*, 6487–503.

Moon, K.-B.; Park, J.-S.; Park, Y.-I.; Song, I.-J.; Lee, H.-J.; Cho, H.S.; Jeon, J.-H.; Kim, H.-S. Development of systems for the production of plant-derived biopharmaceuticals. *Plan. Theory* 2019, *9*, 30.

Moran, K., "*Health: Indigenous Knowledge, Equitable Benefits" World Bank, Indigenous Knowledge (IK) Notes*, No. 15, December, (1998).

Mughees, M; Wajid, S. Herbal Based Polymeric Nanoparticles as a Therapeutic Remedy for Breast Cancer. *Anti Cancer Agents Med. Chem.*, 2021, *21*(4), 433–44.

Mukherjee, P.K.; Maiti, K.; Mukherjee, K.; Houghton, P.J. Leads from Indian medicinal plants with hypoglycemic potentials, *J. Ethnopharmacol.*, 2006, *106*, 1–28.

Mukherjee, S.; Ray, S.; Thakur, R.S. Solid lipid nanoparticles: A modern formulation approach in drug delivery system. *Indian J. Pharm. Sci.*, 2009 Jul, *71*(4), 349–58.

Nicoletti, M. Nutraceuticals and botanicals: Overview and perspectives. *Int. J. Food Sci. Nutr*. 2012, *63*, 2–6.

Ogidi, O.I., & Emaikwu, N.G. (2023a). Adoption and Application of Biotechnology in Herbal Medicine Practices. In: *Reference Series in Phytochemistry. Herbal Medicine Phytochemistry*. Springer Nature Singapore Pte Ltd.

Ogidi, O.I., & Emaikwu, N.G. (2023b). Utilization methods and Practices of herbal Medicine in Africa. In: *Reference Series in Phytochemistry. Herbal Medicine Phytochemistry*. Springer Nature Singapore Pte Ltd.

Ottinger, M.A.; Geiselman, C. *One Health Meets the Exposome: Human, Wildlife, and Ecosystem Health*; Elsevier: Amsterdam, The Netherlands, 2023.

Pereira, G.C. Application of biotechnology in producing plant bio-active compounds. In *Natural Bio-Active Compounds*; Akhtar, M.S., Swamy, M.K., Eds.; Springer: Singapore, 2019; pp. 59–78.

Peria, E. V., The community protocol in the AEAN Framework Agreement on access to genetic resources. *Paper presented at the ASEAN workshop on the TRIPS Agreement and traditional medicine*, Jakarta, 13–15 February, (2001).

Poul, B.N.; Patil, S.S.; Kadam, C.S.; Mhaske, A.R.; Somnathe, N.D. Herbal drug with antidiabetic activity, *Pharm. Rev.*, 2007, *2*, 69–72.

Reddy, A.S.; Pati, S.P.; Kumar, P.P.; Pradeep, H.N.; Sastry, G.N. Virtual screening in drug discovery-a computational perspective. *Curr. Protein Pept. Sci.* 2007, *8*, 329–51.

Shah, S.; Dhawan, V.; Holm, R. et al. Advancements and innovation in the manufacturing process. *Adv. Drug Deliv. Rev.* 2020, *154*, 102–22.

Shariare, M.H.; Rahman, M.; Lubna, S.R. et al. Liposomal drug delivery. *Liposomal drug delivery of Aphanamixis polystachya leaf extracts and its neurobehavioral activity in mice model.* 2020

Sharma, V.; Sarkar, I.N. Bioinformatics opportunities for identification and study of medicinal plants. *Brief. Bioinform.* 2013, *14*, 238–50.

Shu, Y.-Z. Recent natural products based drug development: A pharmaceutical industry perspective. *J. Nat. Prod.* 1998, *61*, 1053–1071.

Singh, H.; Bharadvaja, N. Treasuring the computational approach in medicinal plant research. *Prog. Biophys. Mol. Biol.* 2021, *164*, 19–32.

Soppimath, K.S.; Aminabhavi, T.M.; Kulkarni, A.R. et al. Biodegradable polymeric nanoparticles as drug delivery devices. *J. Control. Release* 2001, *70*, 1–20.

Timmermans, K., "Intellectual property rights and traditional medicine: Policy dilemmas at the interface". *Soc. Sci. Med.*, 2003, *57*, 745–756.

Upadhyay, S.K.; Singh, S.P. *Plants as Bioreactors for Industrial Molecules*; John Wiley & Sons: Hoboken, NJ, USA, 2023.

Vijayanand, P.; Jyothi, V.; Aditya, N. et al. Development and characterization of solid lipid nanoparticles containing herbal extract: in vivo antidepressant activity. *J. Drug Deliv.* 2018, *2018*, 2908626.

WHO, *WHO Traditional Medicine Strategy* 2002–2005. Geneva: World Health Organization, (2002).

Wishart, D.S. Bioinformatics in Drug Development and Assessment. *Drug Metab. Rev.* 2005, *37*, 279–310.

Xia, X. Bioinformatics and drug discovery. *Curr. Top. Med. Chem.* 2017, *17*, 1709–26.

Yang, L.; Wen, K.-S.; Ruan, X.; Zhao, Y.-X.; Wei, F.; Wang, Q. Response of plant secondary metabolites to environmental factors. *Molecules* 2018, *23*, 762.

Yuan, H.; Ma, Q.; Ye, L.; Piao, G. The traditional medicine and modern medicine from natural products. *Molecules* 2016, *21*, 559.

Zajtchuk, R. New technologies in medicine: Biotechnology and nanotechnology. *Dis. Mon.* 1999, *45*, 453–95.

Zhong, J.-J. Recent advances in bioreactor engineering. *Korean J. Chem. Eng.* 2010, *27*, 1035–41.

Zielińska, A.; Carreiró, F.; Oliveira, A.M. et al. Polymeric nanoparticles: production, characterization, toxicology and ecotoxicology. *Molecules* 2020, *25*, 3731.

Index

Pages in *italics* refer to figures and pages in **bold** refer to tables.

G

H

O

P

Q

R